Artur Braun

X-Ray Studies on Electrochemical Systems

Also of Interest

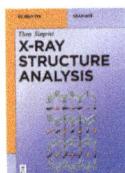

X-Ray Structure Analysis
Theo Siegrist, 2021
ISBN 978-3-11-061070-3, e-ISBN (PDF) 978-3-11-061083-3

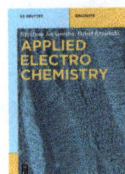

Applied Electrochemistry
Krystyna Jackowska and Paweł Krysiński, 2020
ISBN 978-3-11-060077-3, e-ISBN (PDF) 978-3-11-060083-4

Sodium-Ion Batteries
Advanced Technology and Applications
Man Xie, Feng Wu and Yongxin Huang, 2022
ISBN 978-3-11-074903-8, e-ISBN (PDF) 978-3-11-074906-9

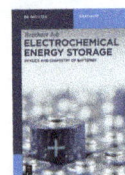

Electrochemical Energy Storage
Physics and Chemistry of Batteries
Reinhart Job, 2020
ISBN 978-3-11-048437-3, e-ISBN (PDF) 978-3-11-048442-7

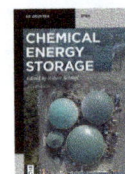

Chemical Energy Storage
Edited by: Robert Schlögl, 2022
ISBN 978-3-11-060843-4, e-ISBN (PDF) 978-3-11-060845-8

Artur Braun

X-Ray Studies on Electrochemical Systems

Synchrotron Methods for Energy Materials

2nd, Revised and Extended Edition

DE GRUYTER

Author
Dr. Artur Braun
Empa
Swiss Federal Institutes of Technology
Ueberlandstr. 129
8600 Duebendorf
Switzerland
artur.braun@alumni.ethz.ch

ISBN 978-3-11-079400-7
e-ISBN (PDF) 978-3-11-079403-8
e-ISBN (EPUB) 978-3-11-079426-7

Library of Congress Control Number: 2024930490

Bibliographic information published by the Deutsche Nationalbibliothek
The Deutsche Nationalbibliothek lists this publication in the Deutsche Nationalbibliografie;
detailed bibliographic data are available on the Internet at http://dnb.dnb.de.

© 2024 Walter de Gruyter GmbH, Berlin/Boston
Cover image: Artur Braun, with the kind permission of Weizmann Institute of Science, Rehovot Israel
Typesetting: VTeX UAB, Lithuania

www.degruyter.com

To Gye Weon, Agneta, and Lars

Preface for the second edition

The first edition of this book was received very well by researchers worldwide. Worldcat.org shows it is available in 774 libraries as of 2023. De Gruyter asked me whether I was willing to write a second edition. Meanwhile, researchers have made great progress with the use of synchrotron methods for electrochemical systems. There exists plenty of new material that should be included in a book like this. However, as the apparatus at the synchrotron beamlines and end stations becomes more developed and integrated, there comes a lack of transparency along with increased user friendliness. The new systems are not so open anymore. For researchers new to the topic, such high integration of the end station apparatus will pose a general difficulty. I think it is there-fore necessary to keep "the early stage stuff" in a scholarly book. Over 10 years ago, I visited my colleague, the Deputy Director of the Large Coherent Light Source (LCLS) in Menlo Park. Back then, I was interested in how the free electron laser there could be useful for my studies on photoelectrochemical cells. He gave me a tour through the dungeons of the LCLS. This second edition includes a new Section 5.3 on the free elec-tron lasers and electrochemical experiments performed there. Soon, more free electron lasers (FEL) were built and went in operation in several places in the world. The cover photo of this second edition shows the architecture around the Koffler accelerator at the Weizmann Institute of Science in Rehovot, Israel. It houses a Van de Graaf accelerator. In the 1980s, the building was equipped with a FEL, which came into operation in the 1990s. I felt I should present some more works in the field of tender X-rays, where the K-edges of elements like sulfur, phosphorous, and chlorine fall. Interesting is also the consideration of isosbestic points in X-ray spectra. I was not aware of this relatively old concept from optical spectroscopy. The concept is applicable to many other fields of spectroscopy, including XAS, and thus is covered in this second edition. Another recent development is the X-ray ptychography, where the real space is reconstructed by recovering the phase information from interference patterns. One of my current projects on batteries deals with metrology. Specifically, our international consortium develops experimental and computational methods for the *operando* diagnostics of lithium batteries by combining electrical impedance and X-ray spectroscopy.

Rehovot, Israel, October 7, 2023   Artur Braun

https://doi.org/10.1515/9783110794038-201

Preface

I wrote this book for scientists who are interested in moving their research on batteries, fuel cells, and solar cells to synchrotron radiation facilities, and neutron centers and free electron lasers. This is not one of the expert books about one particular experimental technique or one particular theory. It is a book for beginners who want to get an impression on how to combine experiments from physical chemistry, electrochemistry, chemical engineering, and biophysics with X-ray and neutron methods. It contains a collection of my original research work in electrochemical energy storage and conversion from 1997 onward, with the latest experiments performed in 2023. I also included important or pioneering works from other researchers.

Beginning in 2010, I organized several workshops with my colleagues on this synergetic topic at the Advanced Light Source Users Meeting, Symposia at the Materials Research Society Spring and Fall Meetings and two International Exploratory Workshops on X-rays and electrochemistry at Empa in Switzerland. In 2010, I was also approached the first time by a leading international publisher and asked to write a book about synchrotrons and electrochemistry. Since 2011, I have served on the synchrotron user representation boards of the Swiss Lightsource, the Advanced Light Source, and more recently as Swiss delegate for the European Synchrotron User Organization.

While the field of electrochemistry of synchrotrons used to be a niche for only a few pioneers, more and more researchers from prestigious schools have entered the field for their materials studies. Concomitantly, some synchrotron divisions hired staff with electrochemistry expertise for beamline work. This field has really grown in the last 15 years.

You may notice that I have tried to cover biophysics and X-ray and neutron studies on bioelectrochemical systems. This is a field where in the future more activities will emerge. There is more bioelectrochemistry in biophysics than we may be aware of. It may be a promising opportunity for curious researchers to invest time and interest in this field and eventually touch gold.

When you are reading this book and decide to move to electrochemistry with synchrotron methods or the other way around, I hope you get your inspiration, satisfaction, and joy with your own work at synchrotron centers, neutron facilities, and free electron lasers as much as I did. Then you will be well-off.

Ennenda, Switzerland, Spring 2022   Artur Braun

https://doi.org/10.1515/9783110794038-202

Acknowledgment

Many colleagues contributed to the success of the experiments shown in this book, and I apologize to those whom I forgot to mention. For the small angle scattering on glassy carbon, which I began during my PhD thesis at Paul Scherrer Institut, I am indebted to Alwin Frei, who always gave me access to his Phillips diffractometer, which I used for my first amateur and self-inspired SAXS work on glassy carbon plates and powders. I am grateful to Dr. H.-G. Haubold of Forschungszentrum Jülich, who gave me free access and beamtime at his JUSIFA Beamline in Hamburg, Germany. I owe him my debut in synchrotron radiation research. I met him on August 31, 1997, at the ISE/ECS Meeting in Paris, where he presented *operando* ASAXS on fuel cells: A gold mine of know-how for my own SAXS work. He and Dr. Günther Goerigk from JUSIFA provided me with practical tricks, which you learn only by doing, but not always by doing it all alone.

I also have to thank Dr. Rainer Saliger, Dr. Hartmut Pröbstle, Dr. Raino Petricevic (back then not yet finished with their doctoral research), and Dr. Andreas Emmerling from Universität Würzburg, with whom I traded electrochemistry knowledge against SAXS knowledge during our exchanges at Paul Scherrer Institut, HASYLAB Hamburg, and the University of Würzburg. It was because of their expertise in carbon aerogel analysis with SAXS that my supervisors Dr. Rüdiger Kötz, Dr. Otto Haas, and Professor Alexander Wokaun at PSI supported my SAXS studies at the synchrotron. The funding for my PhD thesis came from the Swiss Priority Program on Materials Research project "LOK2000" on glassy carbon supercapacitors.

Because of my experience with synchrotron radiation and my PhD expertise in applied electrochemistry, I was considered prepared and fit, and hired by Professor Elton J. Cairns to join his group in Berkeley as a postdoctoral fellow chemist on an exciting battery project funded by the Department of Energy. During his sabbatical at PSI, Elton had left a deep impression on me when he gave a talk on real-life electrocatalysts and their assessment with synchrotron spectroscopy. He gave me the great and invaluable opportunity to directly collaborate with Professor Stephen "Steve" P. Cramer from UC Davis, and his Berkeley group members, Dr. Hongxin Wang, Dr. Uwe Bergmann, Dr. Pieter Glatzel, Dr. Stephan Friedrich, Dr. Liliane Jacquamet, Dr. Tobias Funk, and Dr. Weiwei Gu. With them, I carried out the $K\beta$ and X-ray Raman spectroscopy and XAS (Uwe, Pieter) and EXAFS (Weiwei), and went into soft X-ray spectroscopy (mainly with Hongxin, Tobias, and Liliane).

My *in situ* and *operando* battery cell and the entire *operando* online concept are based on original design ideas of Dr. Craig Horne, who finished his PhD thesis just before my arrival at Berkeley in 1999; from Otto Haas on the occasion of his visit in Berkeley; and from Dr. Catherine "Kathy" Striebel. I learned a lot about battery synthesis with the guidance of Kathy and also UCB student Jeff Lowe, Dr. Seung-wan Song, and Dr. Joong-Pyo Shim. Joong-Pyo, an electrochemist, was actually the one who convinced me of the necessity of a reference electrode for an *operando* cell. From time to time, I sat with Elton over this *operando* cell design and project, and he recommended me to build it with

https://doi.org/10.1515/9783110794038-203

Eric Granlund from the UCB machine shop. Eventually, the cell worked, and it worked fine, and I tested it at the Advanced Photon Source with Dr. Sönke Seifert at his BESSRC-CAT ASAXS beamline. Thanks to Dr. Thiyaga P. Thiyagarajan and Dr. Randall "Randy" E. Winans from APS who supported me with the necessary beamtime. XANES, EXAFS, and anomalous XRD were tested *operando* at SSRL, where I have to appreciate the help of BSc student Shawn Shrout from San Francisco State University, whom I hired extra for beamline support. He did a great job with mounting and aligning the *in situ* cell at the diffractometer and doing the macro programming on the DEC VAX workstation.

ERULF student Alison Fowlks from Michigan State University and PST student Bopamo Osaisai assisted me during their DoE summer internship with the battery sample preparation and potentiostat programming, and onsite at BL 2-3 at SSRL for the *operando* EXAFS experiments. UC Berkeley MSc student Steven Lee was a great help in analyzing the lithium NEXAFS spectra from the ALS.

I should mention here the assistance of Dr. "Mike" Michael C. Tucker from LBNL MSD. For my later proton conductor experiments at the ALS, I always received help from him and his laboratory infrastructure. Mike, then PhD student, together with Professor Jeff Reimer at UC Berkeley was involved in my battery research with Elton Cairns' group "up-hill" at LBNL. It was a result of our many fruitful discussions of Mike's NMR work (a "bulk" method) on battery cathodes that I eventually became aware of the surface sensitivity and bulk sensitivity of some X-ray methods.

I thank Susan J. Lauer and Charlotte Standish, administrative assistants at LBNL's EETD who arranged my smooth travels to SSRL and APS. Susan was also there when my family and I needed her as domestic local parental support with the authorities and administrations (the admin foster mom), an invaluable and heartful personal service for international scientists and families with children in a foreign country like the United States.

Traumatized by the 9/11 tragedy, my family and I decided to leave the cosmopolitan Bay Area and move to rural Kentucky, where I joined the group of Professor Gerald "Jerry" P. Huffman, Professor Frank E. Huggins, and Professor Naresh Shah in Lexington at the University of Kentucky's Consortium for Fossil Fuel Liquefaction Sciences. Jerry and Steve Cramer collaborated at the PRT with Dow and Chevron at Beamline X19 at Brookhaven Lab. During my time in Kentucky, I did no *in situ* work and also no electrochemistry, but four years of carbon (the "organic" element) NEXAFS and some XAS on Fischer–Tropsch catalysts, and on both topics I did also SAXS.

In Lexington, it was Judith "Judy" Cromer, who made sure I would have smooth trips and travel to the four major synchrotrons in the United States—thank you, Judy.

Most of the spectroscopy work on iron perovskites on SOFC was done with assistance from Empa/ETHZ PhD student Selma Erat in my group. Selma also did the multiplet simulations. I have to thank to Dr. Cynthia Piamonteze from the Swiss Light Source, a former postdoc in Steve Cramer's group, who taught Selma the Cowan Code program package for the simulations.

My first PEC campaigns were done with assistance from Empa/Uni Basel PhD student Debajeet K. Bora. His lucky hand helped us in finding the right and spectacular transitions in the *ex situ* and *in situ* experiments, the latter at Jinghua Guo's end station at the ALS. Later, Empa/EPFL student Yelin Hu and Empa/Uni Basel student Florent Boudoire joined in and prepared the first *operando* PEC ambient pressure XPS experiment and also helped in the biofilm NAP-XPS and resonant battery studies at SSRL and ALS. Thanks to Zhi Liu from the ALS, who like Jinghua, gave the ALS community another world best *in situ* electrochemistry end station. The proton conductor work was done by Empa/ETHZ PhD student Qianli Chen, both at the neutron sources and at the ALS. Together with Empa/AGH Krakow PhD student Dorota Flak, she made the first ambient pressure XPS studies in my group at the ALS. I want to acknowledge the excellent and patient work of Jan Ilavsky and Pete R. Jemian at the Advanced Photon Source's UNICAT Beamline, who spent their family time on a major American holiday with me doing US-AXS for SOFC research, and Andrew J. Allen from NIST, who did a fantastic and unique job for the full quantitative analysis of thickness resolved anomalous USAXS. My most recent beamtime was at the synchrotron at Willy-Wien-Laboratorium of PTB in Berlin, thanks to the initiative of Dr. Burkhard Beckhoff with whom I have exploring industrial battery research at synchrotrons for several years now.

My travel and my group members' travel to synchrotrons and neutron sources was administered at Empa by Brigitte Schatzmann, Georgiana Schönberg, and Anita Städler. But most of the travel arrangements were made by Empa's former SwissAir travel agent Susanne "Susi" Kohler. I am grateful to my Abteilungsleiter Thomas Graule, who was always supportive of my synchrotron and neutron work, and my past group leaders Peter Holtappels and Ulrich Vogt, who hired me with Empa. Thank you to all of you.

It may sound surprising that an author of a book on X-ray and electron spectroscopy does not own an X-ray or electron spectrometer. I am therefore very grateful to Dr. Giuseppino "Pino" Fortunato († June 20, 2020) from Empa St. Gallen, who measured my samples right away without ever asking for any favor in return. Thanks also to Dr. Ulrich "Uli" Müller for XPS@Empa, who retired in late 2023.

I want to acknowledge also Dr. Joseph Sfeir from Hexis GmbH, the Swiss solid oxide fuel cell company. He provided me with the SOFC stack you will see in this book, and also with electrode assemblies for resonant USAXS. It is not always easy to get access to samples and specimen from industry. I am grateful that he walked the extra mile for me and my research.

A major part of my spectroscopy work was done at the ALS. David Malone did a lot of "underground" work for me to make sure that we worked efficiently and in a safe environment. David and I knew each other from my Berkeley time when I volunteered for him at the Community Relations Office at LBNL. Special thanks go to Professor Simon Bongjin Mun from the ALS and GIST Korea. Without Simon, I probably would have not gotten into XPS and VB XPS. He took care of several PhD students in my group worldwide, with funds from our joint Korean Swiss Science and Technology Programme project. Without Simon, I would not have discovered the work of Hiroki Wadati (University of

British Columbia and University of Tokyo), who like Simon, meanwhile has become a regular international visitor at my group at Empa. Hiroki's VB PES work opened the other side for me to approach the Fermi level. I am grateful to Dr. Paul Janssen from SCK·CEN in Mol, Belgium, who wrote the majority of Chapter 7.3.5. We now have a bilateral project together plus funding from ESA. I owe Paul a lot.

I was first approached in 2010 by a major international publishing house, which suggested to me to write this book. However, I got caught up with many things over the years and could not even start the book during my fabulous sabbatical in winter 2010/2011 at the University of Hawai'i at Manoa with my hosts Dr. Nicolas Gaillard and Dr. Eric L. Miller, mostly because of my research work there on PEC electrodes, but also the unintended and quite fortunate works on proton conductors with Professor Murli Manghnani and biophysics with Dr. Sam Wilson at C-MORE. This was, after all, for the benefit of the book because these Hawai'i works are part of this book. After my keynote speech at the 3rd Ertl Conference in the Harnack Haus in Berlin, Dr. Konrad Kieling from De Gruyter came and persuaded me to write this book. Thanks to Jaeyoung Lee and Simon Mun from GIST for inviting me to the Ertl conferences. This brought about the final call for the first edition of this book. Now, I am grateful to De Gryuter's, Jessika Kischke, who is in charge of managing the publication of the second edition of my book.

Finally, I want to express my utmost gratitude and apologies to my dear wife Gye Weon and our beloved children Agneta and Lars for their patience and great personal sacrifices. Much of my time spent on the actual research work presented in this book was supposed to be "family time."

My funding for this research, which extends now for almost 30 years, was from the funding agencies and projects listed in the Table 1 below. I am grateful for the funding, which I received from Empa in my career position. My thanks also go to the synchrotron centers HASYLAB, ALS, APS, SSRL, NSLS, SLS, ESRF, ELETTRA, BESSY, PTB, SPring8, and their funding sources.

Table 1: Projects which funded my research covered in this book.

Project name	Funding Agency	Project Number
OpMetBat	EURAMET, SBFI	21GRD01
PROTONIQUE3	Swiss National Science Foundation	https://data.snf.ch/grants/grant/188588
PHOGAM	Swiss National Science Foundation	https://data.snf.ch/grants/grant/189455
TD1102	COST	https://www.cost.eu/actions/TD1102/
Hawai'i	Swiss National Science Foundation	https://data.snf.ch/grants/grant/133944
Swiss Hungarian	SBFI	SCIEX 10.013, [Braun 2010d]
Swiss Lithuanian	SBFI	SCIEX 10.010
LIBEV	Swiss Polish PSRP	PSPB-080/2010
DeCaMa	EU Marie Curie	EMPAPOSTDOCS
SHINE	Swiss National Science Foundation	20NA21-145936
Indo-Swiss JRP	SBFI	138864
GIST	Swiss National Science Foundation	https://data.snf.ch/grants/grant/162232
SSAJRP	Swiss National Science Foundation	https://data.snf.ch/grants/grant/149031, [Braun 2013c]
PatternFormation	Swiss National Science Foundation	https://data.snf.ch/grants/grant/137868
PEC EPFL	Swiss National Science Foundation	https://data.snf.ch/grants/grant/132126
NANOPEC	European Commission	https://cordis.europa.eu/project/id/227179, [Augustynski 2008]
Protonique	Swiss National Science Foundation	https://data.snf.ch/grants/grant/124812
CEMTEC	Swiss CCEM	705
Bio-PEC	Velux Foundation	790, [Braun 2012h]
R'Equip	Swiss National Science Foundation	https://data.snf.ch/grants/grant/121306
BaselPEC	Swiss BfE	#152316-101883; #153613-102809
PhotoCAT	Empa Board of Directors	7. F&E
SOFC-LIFE	European Commission	https://cordis.europa.eu/project/id/256885
MetInsTra	Swiss National Science Foundation	https://data.snf.ch/grants/grant/116688
PROTONICS	Empa Board of Directors	6. F&E
HiTempEchem	European Commission	https://cordis.europa.eu/project/id/042095, [Braun 2006b]
REAL SOFC	European Commission	https://cordis.europa.eu/project/id/502612
CRAEMS	National Science Foundation	CHE-0089133
Energy	U.S. Department of Energy	DE-AC03-76SF00098
Lok2000	Swiss Confederation	SwissPPM

Contents

1 Introduction

The first Nobel Prize in physics was awarded to physicist Dr. Wilhelm Conrad Röntgen in 1901, for the discovery of the X-rays in 1895 [Röntgen 1901]. Röntgen's X-ray source was a Hittorf cathode tube, and he imaged the skeleton on a film by serendipity, or by coincidence.[1] This was the first X-ray imaging as it is done nowadays as routine diagnostics in medicine. Philipp Eduard Anton von Lenard received the Physics Nobel Prize in 1905 for his work on cathode rays and claimed a share of Röntgen's Nobel Prize, although Lenard did not provide a justification for his claim.

Asked by the German technology firm AEG, the former Deutsche Edison-Gesellschaft für angewandte Elektricität, whether Röntgen wanted to patent his invention, Röntgen responded that his discoveries belonged to the general public and he did not want to limit spreading of a new technology by patenting it—a remarkable insight that was reiterated recently by an electric vehicle entrepreneur (Tesla) Elon Musk, "All our patent are belong to you" [Musk 2014].

Many more Nobel prizes have a relation to X-rays: in 1903, the prize went to Marie Curie and Henry Becquerel for their work on radioactivity, where electromagnetic waves often include the X-ray wavelength range.

There are more Nobel Prizes where X-ray- and synchrotron-based methods have played a more or less significant role. From what is shown here, it becomes clear that X-ray methods were essential for the development of the technology and societies of the 20th century. I will use therefore the opportunity in this book to present the long list of Nobel laureates [Noble Prize 2014] whose work has a relation to X-rays. The list is provided in Appendix 1 and includes also some relevant Nobel Prizes in electrochemistry and physical chemistry.

Batteries, capacitors, electrolyzers, fuel cells, photo-electrochemical cells, and solar cells are energy converters and storage devices, some of which are omnipresent in our daily life, particularly batteries, capacitors, and solar cells, to a lesser but increasing extent, the fuel cells, and hopefully soon also the electrolyzers and photo-electrochemical cells. They virtually all operate on the basic principles of electrochemistry. The functionality and integrity of electrochemical energy storage and conversion devices depend on the comprehension and control of their transport properties. Transport includes the charge transport by electrons, electron holes, and ions, then the mass transport by fluids, i. e., gases and liquids, but also by ions, and finally, the heat and radiative transfer, which includes also optical or photonic processes.

---

[1] Sometimes an accident in the lab brings about a new scientific discovery or a patent. A doctoral student of one of my past supervisors discovered the magnetic skyrmion at the Technische Universität München by using neutron scattering. PhD student Sebastian Mühlbauer used a different magnetic measurement geometry than originally scheduled, and discovered spots in the scattering pattern, which later turned out to be projections of magnetic skyrmions, hitherto never experimentally observed [Muhlbauer 2009] suggested by Werner Heisenberg in the 1950s, and predicted theoretically by Tony Skyrm in the 1960s.

https://doi.org/10.1515/9783110794038-001

All this transport and transfer depends on the structure of the involved materials. Structure is not always a well-defined term. In this context, let us understand the word structure as something multiscale, something multidimensional, which starts at the atomic scale in the materials and extends to the design of components and architecture of devices.

This then leads to the inclusion of the crystallographic structure as it constitutes the structural unit cells, the resulting electronic structure, which defines the electric transport properties, and then the so-called microstructure, which actually means the topology of the material at the submicrometer scale down to the nanoscale.

With X-rays and with neutrons, we can probe and determine the structure of these relevant size scales to an extent that we can combine them with electroanalytical and electrochemical transport measurements, which I will show in the last chapter. This book presents a virtually complete set of the synchrotron X-ray-based methods, which can be used for the structural assessment (analysis, characterization, and diagnostics) of electrode and electrolyte materials in electrochemical systems.

Energy converters and energy storage devices must manage electronic and ionic and thermal transport, and thus need a sophisticated architecture of functional and structural components such as electrodes, electrolytes, current collectors, end plates, and sealing. X-rays and neutrons are great probes for studying the electronic structure and crystallographic structure and microstructure of materials, transport processes, kinetic processes, and dynamics. The inner life of a battery, fuel cell, or solar cell can be very "busy," and for physicists, chemists, materials scientists, and process engineers, looking into it is quite interesting.

What makes one material good for a battery, or better than another material for a battery? Why are the "good" materials often expensive; can we replace them by cheaper materials? Why is the lifetime of this particular type of fuel cell so short? Why does it require high-purity fuel to propel such fuel cell type? How can we mimic nature for solar energy conversion? How can we do that with low-cost materials?

At some point, scientists and engineers who address these questions must employ X-ray and neutron methods in order to understand how the above mentioned devices, their components and materials actually operate and work. Like any electrochemical energy converter and storage device, batteries are structurally and chemically very complex devices. Battery technology has made tremendous progress in the last two decades, partially because of advancement in understanding how materials work and interact at the molecular level.[2] X-ray and electron spectroscopy aided a lot to this purpose, as is

---

2 I am happy to share here the information that the European Partnership Program on Metrology has decided to fund the project "Operando metrology for energy storage materials" (OpMetBat). The project consortium is coordinated by the Physikalisch Technische Bundesanstalt in Berlin and has partners in Germany, Italy, Slovenia, France, Czech Republic, the United Kingdom, Switzerland, and Turkey. The original goal was to establish a metrology technology for the combined use of impedance analysis and X-ray methods for the determination of state of health of batteries at the assembly line during manufacturing.

evidenced by the vast number of literature that evolved parallel with the market growth of batteries. The chemical processes that take place in a battery and that make a battery work but also ultimately fail cannot entirely be monitored—and understood—without looking into the battery during operation.

Sometimes it is necessary to investigate these materials while they are actually working or under operation. Such investigations are possible with neutrons and X-rays, provided some technical preparations and adjustments are made.

While these devices, components, or materials were put under their true operation conditions, X-rays, neutrons, and photoelectrons were probing them and giving us structure information and dynamics information that would have not been possible few years ago. The experiments shown in this book needed many years of preparation and have consumed a considerable amount of public funding as well as enormous personal sacrifices from the involved technicians, engineers, and scientists worldwide to make it work. Some of these experiments are of the "we are the first" type and have not even been published yet elsewhere.

This book is organized in 10 parts with 39 chapters. The parts focus on particular experimental methods, whereas the chapters summarize particular materials properties. In doing so, I have sectioned some extensive experimental studies and allocated essential parts of them, in particular, chapters.

Part 2 begins with the X-ray and electron spectroscopy methods because these assess the molecular and electronic structure, such as occupied and unoccupied electronic states and valence band and conduction band of solids and highest occupied and lowest unoccupied orbitals in molecules. These can be resolved well with X-ray spectroscopy because this line of methods holds the element specificity by the core-level spectra of chemical elements, and thus the chemical and orbital sensitivity. One can exploit this property for the further refinement of structure analysis by chemical sensitivity, which is called contrast variation or resonant or anomalous spectroscopy.

Part 3 addresses the crystallographic structure and microstructure determination and includes diffraction, reflectometry, and scattering.

Part 4 deals with the imaging and other real space mapping methods, which are based on X-rays. With this, the basic set of methodologies is expressed.

Part 5 specializes in the utilization of the anomalous dispersion of chemical elements and its use in elemental or chemical or even orbital X-ray contrast variation.

In Part 6, I will review some of the aforementioned methods with respect to their surface sensitivity and bulk sensitivity. The overwhelming majority of X-ray and electrochemistry studies feature the inorganic materials.

A minor portion of the published works includes organic samples. Bio-organic systems are virtually absent, as far as their electrochemical properties are concerned. However, there exist many X-ray studies on biological systems such as the large field of protein crystallography and also protein spectroscopy. This will be covered in Part 7. For those readers who are interested in biological systems, I would also like to refer to my book "Quantum electrodynamics of photosynthesis," which was published by De Gruyter in 2020.

In Part 8, I will showcase several complex X-ray electrochemical case studies. This is actually a collection of experiments on batteries, fuel cells, and photo-electrochemical cells, which includes the *operando* and *in situ* studies.

In Part 9, I will show how X-ray-based data can very well match the conductivity data from some materials. This is the most important part of the book and shows the success of X-ray spectroscopy in helping understanding the transport properties of solids.

I have written this book for the nonexpert, who is interested in exploring synchrotron radiation methods for electrochemical systems. For most topics addressed in this book, there exists very good special literature, topical books that cover the sub-field in more detail with better theoretical rigor, books specializing in X-ray diffraction, small angle scattering, EXAFS, XPS, and certainly books about electrochemistry and impedance spectroscopy. On many occasions in this book, I share how experiments were designed and planned and what might have been the rational for doing it the way we did it. Sometimes it is only necessary to have some hands-on talent and a seemingly big problem can be solved with ease.

I have enjoyed almost 30 years of research as a user at synchrotron centers world-wide while working in quite different fields, and recently entering the field of biology and life sciences. It can be a challenge for a maturing researcher to enter new fields and start over again as a beginner, being lectured by experts who know only one field. I therefore want to finish this Introduction for those who follow my approach with a heads-up quote by 1977 Physics Nobel laureate Philip Warren Anderson [1972]:

> So it is not true, as a recent article would have it, that we each should "cultivate our own valley, and not attempt to build roads over the mountain ranges ... between the sciences." Rather, we should recognize that such roads, while often the quickest shortcut to another part of our own science, are not visible from the viewpoint of one science alone.

2 Molecular structure and electronic structure

The molecular and electronic structures of materials, in principle, reflect their physical and also chemical properties such as electronic transport properties, which determine the electronic and ionic conductivity; the optical, magnetic, and thermal properties and also to some extent their catalytic properties; and also the mechanical properties. X-ray methods are not the only ones to determine the electronic and molecular structure, but the fact that X-rays are very sensitive to the electron distribution in an atom and condensed matter makes it well suited for this pursuit. The electronic energy levels constitute a strong element specific signature, which can surpass that from optical methods or nuclear methods.

2.1 X-ray absorption spectroscopy

2.1.1 Hard X-rays: XAS and XANES

X-ray absorption spectroscopy (XAS) provides information on the unoccupied electron states of elements under observation. Imagine a specimen of some unknown chemical composition, which is irradiated with monochromatic X-rays, and the energy or wavelength of the X-rays is scanned from, say, 50 to 10,000 eV. At particular X-ray energies, the atoms of the corresponding elements in the specimen will absorb the X-rays while an electron from the corresponding orbital is ejected. The absorption of the X-ray radiation is measured around one of the absorption edges of the material under investigation. When one X-ray quantum has sufficient energy, it can kick an electron from the appropriate atomic orbital. At this X-ray energy $E = h\nu$, the X-ray absorption will increase considerably. This process will manifest in an X-ray absorption spectrum with absorption peak positions indicative of the particular elements.

The theory of X-ray spectroscopy boils down to what is frequently referred to as "Fermi's golden rule," which is a mathematical expression for the matrix elements, which constitute the resonances (electron transition rates from one energy eigenstate to the continuum upon some perturbation) that manifest in peaks in the X-ray spectra. Since the mathematical treatment is based on perturbation theory, we find a first-order approximation for the underlying physical process. For a review on the experimental determination of X-ray atomic energy levels, see [Bearden 1967] and [Fuggle 1980].

It is noteworthy that it was actually Paul Maurice Dirac [Dirac 1927a, Dirac 1927b] who presented the necessary mathematics and algebra for this problem. A fast derivation of "Dirac's golden rule" is presented in [Brillson 2010]. The transition rate Γ from an initial state $|i\rangle$ to a final state $\langle f|$ is in first-order approximation constant and given by

https://doi.org/10.1515/9783110794038-002

the product of the square of the transition matrix element $|\langle f|H'|i\rangle|^2$ and the density of final states ρ:

$$\Gamma_{i \to f} = \frac{2\pi}{h}|\langle f|H'|i\rangle|^2 \rho$$

Figure 2.1 shows the X-ray absorption spectrum of a powder sample of Pr-Sr-Mn-In oxide distributed over an adhesive tape [Richter 2008a, Richter 2008e]. The spectrum was recorded in the energy range from 5900 to 6800 eV and, therefore, shows the Pr L-edges LIII and LII and the Mn K-shell absorption edge. We have thus already some chemical element information about the sample, at least for Pr and Mn. Using the Bouguer–Lambert–Beer law of absorption, it is generally possible to determine concentration of an element in a complex sample, similar to X-ray fluorescence spectroscopy (XRF).

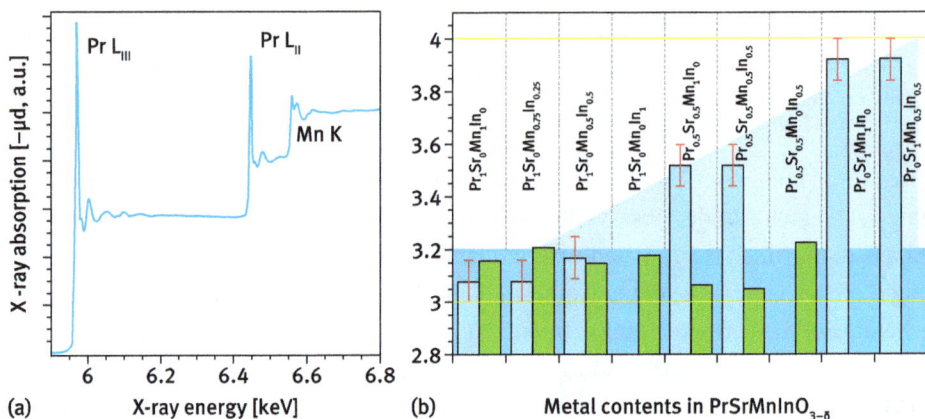

Figure 2.1: (a) X-ray absorption spectrum of a PrSrMnIn-oxide sample covering the Pr L-edges range and Mn K-edge range. (b) Oxidation state information on 9 specimens with different stoichiometries obtained from the quantitative analysis of the corresponding 9 absorption spectra. Blue bars show average oxidation state for Mn, and green bars show average oxidation state for Pr. Mn shows a wide variation from Mn^{3+} to Mn^{4+}, whereas Pr remains virtually around Pr^{3+}.

At this point, it should be mentioned that the attenuation of light in matter has been addressed reportedly by Pierre Bouguer in his 1729 essay, "Essai d'optique sur la gradation de la lumière" [Bouguer 1729]. The Bouguer–Lambert–Beer law describes the attenuation or extinction of a radiation intensity I_0 when it passes through a medium. The intensity of the radiation after it has passed the medium is

$$I_{(x,\mu_{(E)})} = I_0 \cdot \exp(-\mu_{(E)} \cdot x)$$

where $\mu(E)$ is the differential absorption cross-section. $\mu(E)$ is a chemical element specific "constant," which varies with or depends on the energy E of the radiation that passes through the material. x is the distance by which the radiation has passed through the body containing the particular element, e. g., the thickness of a copper sheet.

If the sample is an alloy, such as CuZn, then a second differential absorption cross-section comes into play, and the equation reads

$$I_{(x,\mu_{(E)})} = I_o \cdot \exp(-(\mu_{(E)}^{Cu} \cdot x^{Cu} + \mu_{(E)}^{Zn} \cdot x^{Zn})).$$

For multiple elements and more complex compounds, we can sum up all contributions. In medical radiography with many chemical components, a numerical integration is carried out:

$$I_{(x,\mu_{(E)})} = I_o \cdot \exp\left(- \sum_i^{\text{elements}} \mu_i \cdot x_i \right) \rightarrow I_o \cdot \exp\left(- \int \mu(x)dx \right)$$

Since the X-rays follow the exponential absorption law, the spectral intensity decays with an exponential with increasing X-ray energy. Convoluted with this global exponential decay is the characteristic spectroscopic signature of the absorption by the particular electrons in the atoms, which is typically set on with the absorption edge: the first upward deviation from the exponential intensity decay. There are also additional features, which cannot necessarily be assigned to particular electrons, but to collective behavior such as excitons, plasmons, charge transfers, or shape resonances. It is worthwhile to mention in this context the intensity oscillations that are often set 5–20 eV away from the absorption edge, which result from the constructive interference of electron waves in the proximity of the atoms whose electrons are excited by the X-rays. Distances between atoms can be determined by quantitative analysis of these so-called extended X-ray absorption fine structure (EXAFS) oscillations. To a wide extent, this is an elastic scattering process in the absorption experiment. Using Fourier transformations, one can obtain real space information on the geometric relations of atoms even when they are not showing any long-range order. EXAFS spectroscopy is a well-established X-ray method [Koningsberger 1988] and particularly suited for crystal structure analysis of noncrystalline materials, such as glasses and nanoparticles. We will shortly treat this in the next part of the book.

In the XAS spectrum in Figure 2.1, we see spectroscopically well-resolved L_3, L_2 multiplets of the Pr in the specimen, showing the EXAFS oscillations beyond the absorption edges but also fine structures between this edge range and the EXAFS range. This range is called the X-ray absorption near-edge structure (XANES) and contains information on the electronic structure of the material. I will address the multipletts in Section 2.1.

Figure 2.2 shows a magnified range of the Pr L_{III} spectrum for three different chemical compositions of the specimen. Specifically, the compositions of the specimen are $Pr_{0.5}Sr_{0.5}MnO_3$, $Pr_{0.5}Sr_{0.5}InO_3$, and $Pr_{0.5}Sr_{0.5}In_{0.5}Mn_{0.5}O_3$. This is an ABO_3-type perovskite structure material where we have substituted Mn and In on the B-site. The edge structure of the three spectra is virtually identical with onset energy of 5960 eV. The edge is

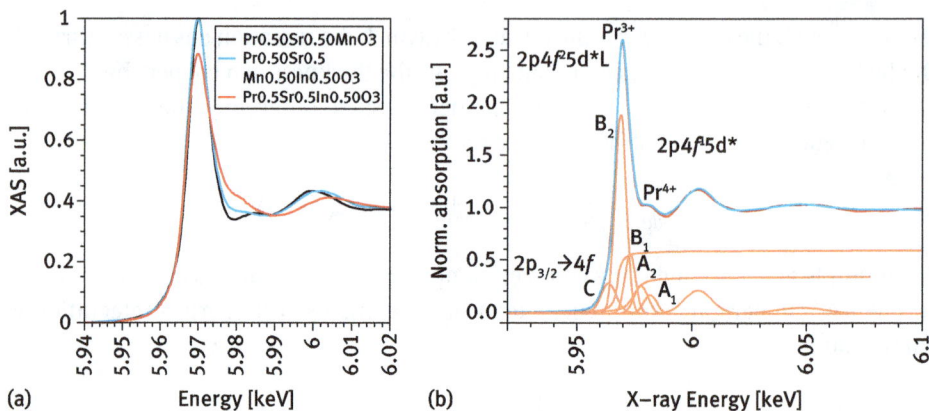

Figure 2.2: (a) Three XANES spectra of the LIII range of Pr for compositions $Pr_{0.5}Sr_{0.5}MnO_3$, $Pr_{0.5}Sr_{0.5}InO_3$, and $Pr_{0.5}Sr_{0.5}In_{0.5}Mn_{0.5}O_3$. The intensity of the white line for the Mn-containing specimen is normalized to unity (= 1). The intensity of the In-rich sample is arbitrarily matched at 5990 eV and 6020 eV. (b) The left panel shows the deconvolution of the spectrum into particular electronic transitions.

technically determined by determination of the maximum of the first derivative of the intensity versus the X-ray energy, which we can run in some specific XAS data analysis software, or one can do it with a generic data analysis program. We note that the peak intensity at 5970 eV is also very similar, except for the sample where the Mn is fully substituted by indium. The peak with the highest intensity is called the "white line" because historically, in X-ray experiments where camera films were used for recording the spectra, high intensity corresponded to a bright, light signature [Coster 1924]. Since the chemical element indium contains more electrons than Manganese (49 vs. 25), the self-absorption is higher in the indium-rich sample, and thus its white line has lower intensity. The intensity of the white line contains basically the information of the abundance of the elements of the irradiated specimen but care must be taken for proper normalization of the intensity and calibration of the spectrum.

We note, however, stark differences in the energy range after the white line at 5974 to 6010 eV. In this energy range, we are probing the Pr L_{III} structure. Note that the Pr and the Sr are the A-site atoms in the metal oxide, and thus central to the perovskite structure. They are intimate with the oxygen atoms, actually with the oxygen 2p ligands, which in turn encage the B-site atoms Mn and In, respectively. The stoichiometric change in these specimens cause changes in the electronic structure, which manifest in the observed spectroscopic changes. This is a very important insight in this matter, that X-ray spectroscopy is basically a key (method) to unlock electronic structure information. The complex arrangement of the electron shells makes that a whole range of transitions occurs during the experiment, which not always can be experimentally resolved for various reasons. Since the number of different transitions follows the theoretical algebraic conditions, one can "guess" the position of not well-resolved peaks, e. g., by their shoulders or unnaturally broad features, which could be double peaks, for example. Practically, one then uses software, which helps to deconvolute the spectra into peaks and other spectral features, with support from least square fitting procedures. This is demonstrated for one example in the right panel of Figure 2.2. For example, the peak at position 5956 eV, labeled A_2, has been identified with Pr^{4+}, whereas the white line is indicative of Pr^{3+}. It is a general principle in X-ray spectroscopy that a species with higher oxidation state has its spectral signature at an energy position higher than the reduced species. The reason for this is that with every electron removed from the electron shell, it becomes more difficult to remove a further electron from that shell, and thus a higher X-ray energy is necessary to remove this electron. The result that the corresponding peaks for this transition are shifted to higher X-ray energies is thus called "chemical shift" [Wong 1984c, Figueroa 2005e]. At later points in this book, we will come across samples where more detailed information on the quantitative analysis of electronic structure by XAS is demonstrated.

The relative height of the white lines of spectroscopic multiplets such as the PrL_{III}, L_{II} here is typically determined by the oxidation state, the valence state of a metal. However, we note from the wide spectrum in Figure 2.1 that the PrL_{II} spectrum is very close to the Mn K-edge spectrum, and thus could be affected with "spectral impurities" from the Mn spectrum. While this should not play much of a role for the white line of the PrL_{II} part of the spectrum, we yet abstain from further analysis of the branching ratio. Rather, we form the ratios of the white line and the peak at 5956 eV (A_2 in Figure 2.2), which is indicative of Pr^{4+}, whereas the white line peak B_2 is indicative of Pr^{3+}. We have determined these intensity ratios B_2/A_2 for all compounds in [Richter 2008a, Richter 2008e] and compared them with the white line ratios of spectra from compounds with well-known valence states of Pr [Ocana 1998, Dumschat 1995, Alleno 1999, Bianconi 1987]. With these data, we can make a simple calibration curve, and with interpolation and extrapolation steps, we can assign our specimen an approximate oxidation state of the Pr (Figure 2.3).

It is therefore obvious that the Pr in our samples is mostly in the Pr^{3+} valence state, slightly above 3+. We have listed this information in Table 2.1 and in the right panel of Figure 2.1, along with the oxidation states of the Mn.

In the table, the oxidation states are also included and obtained by so-called ELNES experiments, electron-loss near-edge spectroscopy, or EELS (electron energy-loss spectroscopy). Currently, EELS and ELNES are typically performed along with transmission electron microscopes, because these often come with a complementary EELS spectrometer setup. The EELS spectra reflect only, in principle, the same electronic structure information like XANES spectra, with respect to the electronic origin of the signal. The careful researcher will find that EELS spectra obtained with TEM microscopes usually will not bring about the same features, which XANES spectra will show [Braun 2005b]. One reason for this is that the electron beam in TEM results from 10 keV and higher acceleration voltages with according current densities, which normally dissipates a substantial power into the sample, which can create radiation damage in the worst case, but also sample heating, which renders a TEM EELS study very often a high-temperature study.

(a)

(b)

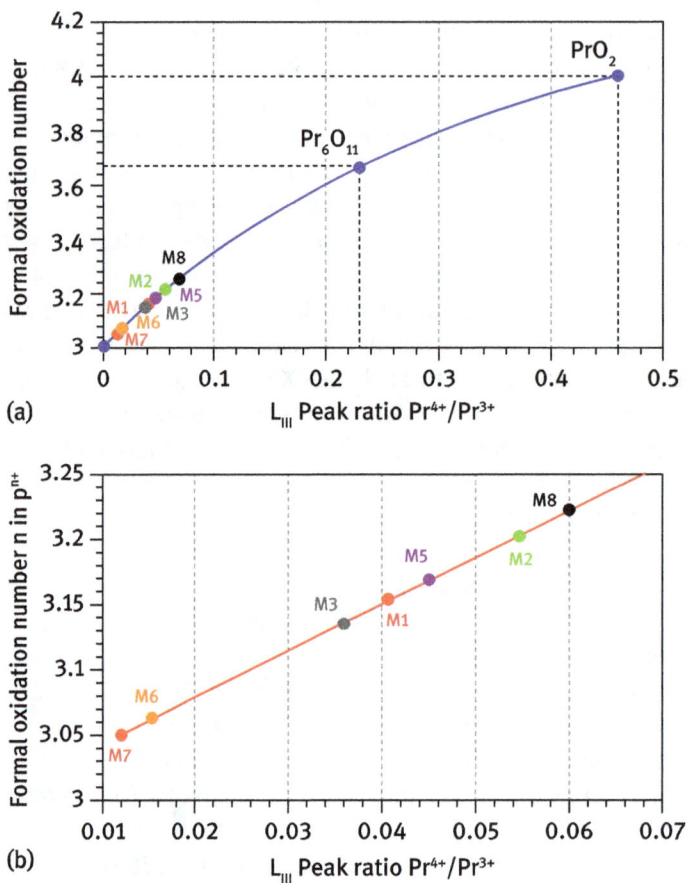

Figure 2.3: Formal oxidation states of PrO_2, Pr_6O_{11}, and Pr_2O_3 plotted versus the peak height ratio L_{III}/L_{II}. The magnified region in the bottom panel shows how our specimen fall in the oxidation number range between 3.05 and 3.25 for Pr.

Table 2.1: Peak height ratios B_2/A_2 and valence states n for praseodymium in specimen M1–M8.

		XANES		ELNES	
	Sample ID	$L_{III}(A_2)/L_{III}(B_2)$	n in (Pr^{n+})	M_V/M_{IV}	n (Pr^{n+})
$PrMnO_{3-\delta}$	M1	0.0408	3.16	1.81	3.06
$PrMn_{0.5}In_{0.5}O_{3-\delta}$	M3	0.0380	3.15	1.85	3.01
$PrInO_{3-\delta}$	M5	0.0460	3.18	1.83	3.04
$Pr_{0.5}Sr_{0.5}MnO_{3-\delta}$	M6	0.0160	3.06	1.90	2.98<3
$Pr_{0.5}Sr_{0.5}Mn_{0.5}In_{0.5}O_{3-\delta}$	M7	0.0130	3.05	1.87	3.00
$Pr_{0.5}Sr_{0.5}InO_{3-\delta}$	M8	0.0600	3.23	1.54	3.46

This situation is entirely different from the so-called high-resolution EELS spectrometer [Ibach 1982], where the primary electron energy is only 10 eV, which is used for surface vibration spectroscopy.

Let us now turn to the oxidation state of the manganese in the specimen. Have a look first at the Mn K-shell absorption edge in Figure 2.1, which shows also the two L-shell absorption edges of the praseodymium, which are at lower energies than the Mn K-edge. We have plotted in Figure 2.4 three different Mn K-edge spectra from the specimen with compositions $PrMnO_3$, $SrMnO_3$, and $Pr_{0.5}Sr_{0.5}MnO_3$. Their white line positions are shifted by some few electron volts. $PrMnO_3$ with 6558.5 eV has the lowest position thus revealing that the Mn in this specimen is in the lowest reduced oxidation state, whereas in $SrMnO_3$ (6559.6 eV), it is in the highest oxidation state. The mixed specimen $Pr_{0.5}Sr_{0.5}MnO_3$ has 6558.8 eV white line position. We also note that the energy range from 6540 to 6545 eV contains a noticeable so-called pre-edge peak, which is more pronounced for the two specimens, which contain Sr. This pre-edge peak may have various origins. We will come to this important feature at a later point.

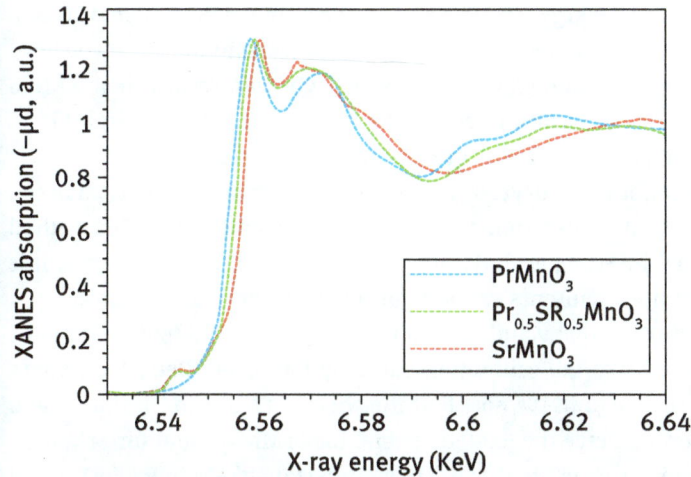

Figure 2.4: Mn K-edge XANES spectra of $PrMnO_3$, $SrMnO_3$, and $Pr_{0.5}Sr_{0.5}MnO_3$. The plot below shows the linear dependency of the Mn oxidation number from the energy shift of the white line in the XANES spectra for a wide range of samples.

The shift of the absorption spectrum, particularly its onset at the absorption edge, is directly related with the oxidation state of the elements. Higher oxidation state means generally a shift toward higher X-ray energy [Wong 1984a]. When a valence electron is kicked out of the orbital, the remaining electrons experience a stronger attraction by the nucleus, which is the reason why there is larger energy necessary to remove the next electron [Figueroa 2005a]. Using the Mn K-edge shift in XANES spectra, it is possible to obtain a quantitative estimation for the average oxidation state for all the manganese ions in a sample [Figueroa 2005a, Shiraishi 1997]. Coster has put it more general in his work from 1924 [Coster 1924], where he writes "New measurements are reported on the fine structure of the K-edge of elements Ti, V, Cr, and Mn and the L_{III}-edge of elements Sn, Sb, Te, and I, and it is being shown that the character of the fine structure depends on the chemical state of the corresponding element. An absorption line at the soft side of the principal edge is found only for these elements, when it is in a higher oxidation state. A qualitative interpretation of the occurring phenomena of the fine structure are attempted." (*"Es werden neue Messungen mitgeteilt über die Feinstruktur der K-Kante der Elemente Ti, V, Cr, und Mn und der L_{III}-Kante der Elemente Sn, Sb, Te, und J, und es wird gezeigt, daß der Charakter der Feinstruktur von dem chemischen Zustande des betreffenden Elementes abhängig ist. Eine Absorptionslinie an der weichen Seite der Hauptkante wird bei diesen Elementen nur angetroffen, wenn das Element sich in einer höheren Oxydationsstufe befindet. Es wird versucht, eine qualitative Deutung von den bei der Feinstruktur auftretenden Erscheinungen zu geben."*). While the white lines might be the visually most spectacular observation in an X-ray absorption spectrum, the absorption edges are typically determined by the position of the maximum of the first derivative of the spectrum versus the X-ray energy. We used an approach shown in [Figueroa 2005a], where an energy shift of 1 eV accounted for around 0.29 valence units for Mn. If you have a suite of samples with various known oxidation states, you can record their XANES spectra and determine their chemical shift versus the energy axis. The chemical shift of the corresponding reduced metal species is by default considered 0.

This constitutes the calibration curve. By interpolation and extrapolation, oxidation states from other samples can be determined or estimated to some extent. This method was originally introduced by Wong et al. [Wong 1984a] on vanadium compounds. It is worth noting that Wong and coauthors carried out their synchrotron-based research as employees of the Boeing Company and of General Electric. We will find in Section 2 that this corresponds exactly to the calibration curve by Dau et al. [Dau 2001], where 3.5 eV account for one valence unit. We note that the chemical shifts for $SrMnO_{3-\delta}$ and $PrMnO_{3-\delta}$ are very different. Since the oxidation state for $PrMnO_{3-\delta}$ and for $SrMnO_{3-\delta}$ are not accurately known, we estimate that the oxidation state of Mn in $PrMnO_{3-\delta}$ and $SrMnO_{3-\delta}$ is close to Mn^{3+} as the lowest and close to Mn^{4+} as the highest value. Their difference in peak shift was as high as 2.81 eV, which results in an oxidation number span of around 0.81 with an error of ±0.09. What do we mean by "shift of the white line"? In earlier times when photographic films were used for spectrometry, the highest peak in the absorption spectrum was the region on the film with highest intensity, and this was then historically termed "white line." With an intensity recorder, the intensity distribution on the film was then translated in what we know nowadays as spectrum. They were called "white lines" by Coster [Coster 1924] because of their characteristic trace left on photographic negatives in absorption measurements [Lye 1980], in lieu of the shift of the absorption edge determined by the maximum of the first derivative of the spectrum.

Figure 2.4 (bottom) shows the linear dependence of the chemical shift (white line position) on the Mn valence state according to Figueroa et al. [Figueroa 2005a], to which we have also added our data. The results indicate the average valence state of the manganese ions to be close to +3 for specimen M1 and M3. A distinct increase in manganese valence state could be observed when praseodymium is partly substituted by Sr^{2+}. The complete A-site substitution leads to a drastic increase in manganese valence almost reaching the expected tetravalent state. The XANES spectra of Pr and Mn have helped therefore to determine their oxidation states in the aforementioned compounds, which we are listing in Table 2.2. It is interesting to note that the Mn oxidation state for the specimen M1 to M9 is ranging from 3.09 to 3.91, virtually from Mn^{3+} to Mn^{4+}.

Table 2.2: Peak ratios and resulting valence states for manganese.

		XANES		ELNES	
	Sample ID	peak pos. (eV)	m (Mn^{m+})	L_{III}/L_{II}	m (Mn^{m+})
$PrMnO_{3-\delta}$	M1	6554.39	3.09	2.51	3.01
$PrMn_{0.5}In_{0.5}O_{3-\delta}$	M3	6554.68	3.18	2.66	2.95
$Pr_{0.5}Sr_{0.5}MnO_{3-\delta}$	M6	6555.85	3.52	2.34	3.26
$Pr_{0.5}Sr_{0.5}Mn_{0.5}In_{0.5}O_{3-\delta}$	M7	6555.85	3.52	2.26	3.51
$SrMnO_{3-\delta}$	M9	6557.19	3.91	–	–

A very primitive way of principal component analysis (PCA) can be done as follows. Imagine a Mn K-edge XANES of or Kβ spectrum (E, I_batt) from lithium manganite from a battery cathode. The Mn concentration in the sample may be to 40 % Mn^{3+} and to 60 % Mn^{4+}, just as a guess from electrochemical analysis. You also need to record or obtain from a spectrum library [Hitchcock 1994, Philip Ewels 2016], a XANES, or Kβ spectrum of a pure Mn^{3+} reference sample (100 % Mn^{3+}, E, I_Mn^{3+}) and also one from a Mn^{4+} reference sample (100 % Mn^{4+}, E, I_Mn^{4+}). They should not have any contamination from Mn of another oxidation state. You can then multiply the intensity column of the pure Mn^{3+} by an arbitrary weighing factor $0<a<1$ and the pure Mn^{4+} spectrum by the conjugated weighing factor $b = 1-a$. Then you add both multiplied intensity columns to generate a third, mixed-intensity column: E, I = aI_Mn^{3+} + bI_Mn^{4+}. We must ensure that both intensity columns refer correctly to the same energy column. If not, you have to spline the dissimilar energy columns and match them. Your generated new intensity column should be as close as possible to the spectrum recorded from the battery sample (E, I-batt). You can adjust the weighing factor a, b manually or program a least square-fitting routine, which compares the generated spectrum composed form the two pure "principal components" with the real spectrum from the battery sample. Manceau et al. [Manceau 2012] have explained this recently step-by-step. A more complex routine is the principal component analysis as shown for sulfur spectra by [Beauchemin 2002].

We thus can present a phase diagram where we place the oxidation states of the elements within their particular composition; see Figure 2.5 below.

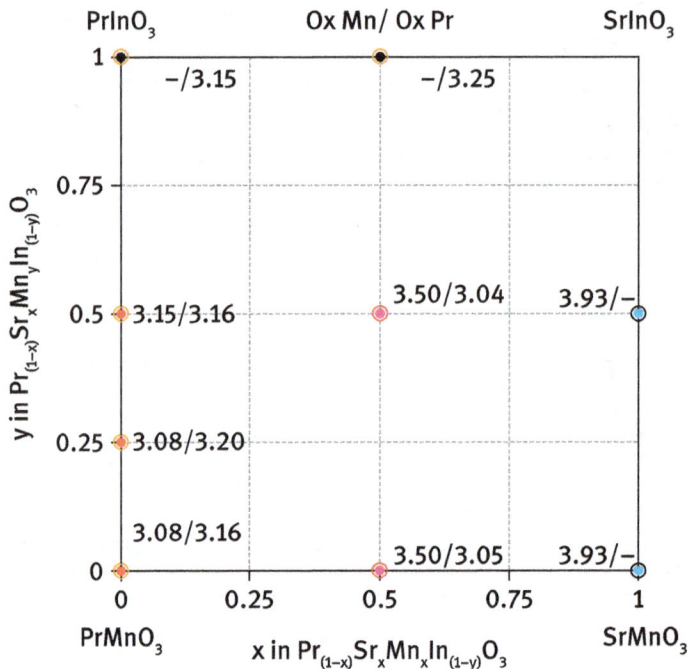

Figure 2.5: Phase diagram for the oxidation states of Pr and Mn as a function of relative concentration in the specimen.

What is the reason for doing all this? Why do we need to know the oxidation states of constituents of materials? It is known that many properties of materials depend on oxidation states of metals, for example. This holds particularly for nonmetal electrode materials, such as shown here. Cathode materials in solid oxide fuel cells are typically complex metal oxides with perovskite crystal structure, the most prominent being LaSrMnoxide [Mcintosh 2004]. The system we have begun this book with is one of the many model systems studied in this field [Richter 2008a, Richter 2008e]. The electric transport properties of metal oxides and virtually all compounds depend critically, e. g., on the oxidation states of metals. The charge transfer between say Mn^{3+} and Mn^{4+} is mediated by the oxygen ion, specifically by the O 2p orbitals with the Mn 3d orbitals, as illustrated by the superexchange unit in Figure 2.6. Depending on the constituents forming the superexchange unit, including their oxidation states and spin states, we may have a covalent exchange, double exchange or superexchange interaction [Goodenough 1998, Goodenough 2004, Goodenough 2008, Weihe 1997].

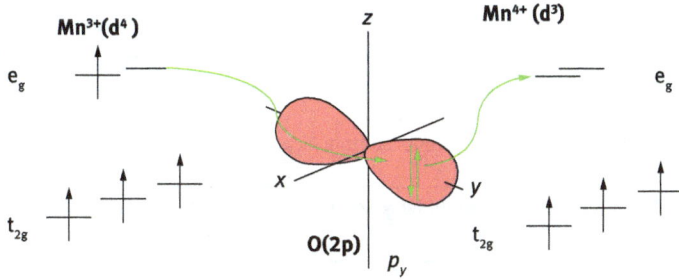

Figure 2.6: Schematic of electron transfer by double exchange from Mn^{3+} via the O 2p ligands to Mn^{4+} in a superexchange unit $Mn^{3+} - O\ 2p - Mn^{4+}$.

The oxidation state of the metal is certainly depending on its molecular, local environment, such as its proximity to oxygen, but also other anions with ligands such as nitrogen, sulfur carbon, and the like. We will see in the final part of this book how the oxygen ion is not just a mere filler atom in a material but an active agent for electronic charge transport, particular the exchange effects like superexchange, covalent exchange, double exchange, and the like.

Let us keep in mind that as of yet we looked into the Pr and Mn oxidation states only, but not into the In and Sr. We assume that strontium remains Sr^{2+} and indium remains In^{3+} to keep them as constants in our considerations. We can thus simply calculate the oxygen nonstoichiometry parameter "δ," which is listed in Table 2.3. The data show that A-site substitution results in a valence change of the B-site transition metal and also in formation of oxygen vacancies, which cause the $\delta > 0$.

Table 2.3: Estimated oxygen nonstoichiometry of samples measured with XANES and with ELNES.

		XANES	ELNES
	Sample ID	Oxygen stoichiometry	Oxygen stoichiometry
$PrMnO_{3-\delta}$	M1	3.13	3.04
$PrMn_{0.5}In_{0.5}O_{3-\delta}$	M3	3.12	2.99
$PrInO_{3-\delta}$	M5	3.09	3.02
$Pr_{0.5}Sr_{0.5}MnO_{3-\delta}$	M6	3.03	2.88
$Pr_{0.5}Sr_{0.5}Mn_{0.5}In_{0.5}O_{3-\delta}$	M7	2.89	2.88
$Pr_{0.5}Sr_{0.5}InO_{3-\delta}$	M8	2.81	2.87
$SrMnO_{3-\delta}$	M9	2.96	–

These situations are governed by the so-called Goodenough–Kanamori rules [Goodenough 2008, Weihe 1997]. Once the electronic structure is understood at the level of oxidation states and spin states, which is a still simplistic level of understanding, one can look at the transport properties, specifically the electric conductivity.

Table 2.4 shows the conductivities of samples M1–M10 as they were determined with conventional 4-point DC measurements from ambient temperature to 800 °C.

Table 2.4: Electric conductivities of ceramic samples (reproduced from Richter 2008b).

Stoichiometry	Sample ID	Conductivity S/cm			Activation energy eV
		25 °C	500 °C	800 °C	
$PrMnO_{3-\delta}$	M1	0.023	17.6	36.7	0.253
$PrMn_{0.75}In_{0.25}O_{3-\delta}$	M2	0.005	3.21	8.69	0.316
$PrMn_{0.5}In_{0.5}O_{3-\delta}$	M3	0.0006	0.35	1.16	0.366
$PrMn_{0.25}In_{0.75}O_{3-\delta}$	M4	≪5E-5*	71E-4	55E-3	0.578
$PrInO_{3-\delta}$	M5	–	–	–	–
$Pr_{0.5}Sr_{0.5}MnO_{3-\delta}$	M6	13	123.8	143.1	0.115
$Pr_{0.5}Sr_{0.5}Mn_{0.5}In_{0.5}O_{3-\delta}$	M7	0.06	4.62	8.50	0.224
$Pr_{0.5}Sr_{0.5}InO_{3-\delta}$	M8	≪5E-5*	50E-5	57E-4	0.600
$SrMnO_{3-\delta}$	M9	0.00016	0.58	3.7	0.346
$SrMn_{0.5}In_{0.5}O_{3-\delta}$	M10	≪5E-5*	24E-6	11E-4	1.006

*≪5E−5: Acquisition limit of the meter was 200 kΩ therefore no data values possible.

We note that the conductivities in Table 2.4 depend strongly on the stoichiometry. Since the Pr oxidation state is not significantly changing, we can assume that it has no significant influence on the conductivity. The Mn oxidation state, however, changes significantly.

In this particular study, the relationship between conductivity and electronic structure was rationalized along an empirical theory, which had been sketched in a graphical representation frequently referred to as Kamata–Nakamura map [Kamata 1973, Kamata 1974]. This map representation has originally been derived by Kamata during his PhD thesis [Nakamura 2006]. The information we need for drafting such map is the effective ionic charge and ionic radii of the A-site and B-site metal atoms. The ratio of the ionic charge Z_A over ionic radius r_A is plotted over the ratio of ionic charge Z_B over radius r_B. It turns out that the systems with localized electrons occupy the upper right quadrant of the map, whereas the lower left quadrant is occupied by the systems with itinerant electrons. Interestingly, the compounds which are at the borderline between both regions are also good SOFC cathodes.

Based on the values for Z_A, Z_B, r_A, and r_B, we are able to place the samples M1–M10 into the map as shown in Figure 2.7. The samples with the high conductivities range in the center of the map, which allows us to draw a diagonal separation line that indicates where the compounds with itinerant electrons are separated from those with localized electrons. It is gratifying to see that samples M5, M8, and M10, which have extremely low conductivites (Table 2.4), are placed in the lower portion with itinerant electrons, which shows metal-type conductivity behavior $d\rho/dT > 0$. Note that the conductivity in SOFC cathodes is not necessarily based on the metal-type electron transport processes.

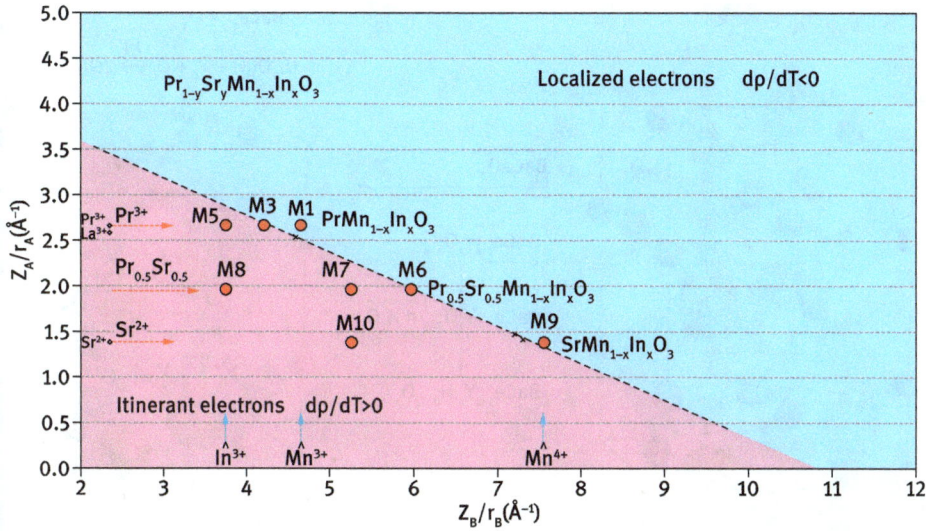

Figure 2.7: Kamata–Nakamura map for the system $Pr_{1-y}Sr_yMn_{1-x}In_xO_3$, reproduced from [Richter 2008a, Richter 2008e].

Let us shortly turn to a different part of the solid oxide fuel cell. By "solid oxide," the solid electrolyte [Goodenough 1995] is meant, as opposed to the conventional liquid electrolytes in electrochemistry. The electrolytes most be a solid in order to sustain the very high temperatures. Standard electrolytes for SOFC are (with yttrium-stabilized) zirconium oxide (YSZ) or cerium gadolinium oxide. Their ionic transport is based on diffusion of oxygen or oxygen vacancies, which becomes active at temperatures above 600 °C, up to 1000 °C. Such high temperature activates not only the necessary oxygen ion transport to warrant its electrolyte functionality, it can also activate unwanted processes such as electrode degradation. Ceramic proton conductors are alternative electrolytes, which work at around half the temperature where oxygen ion conductivity sets on; this is 400–500 °C. Ceramic proton conductors have protons as ionic charge carriers. Barium zirconate is an ABO_3-type perovskite which serves as parent compound. Substitution of the Zr^{4+} by Y^{3+} causes formation of oxygen vacancies, which will host oxygen ions from incoming water vapor molecules upon exposure to moisture. The two hydrogen atoms per water molecule will form bonds with the adjacent oxygen in the perovskite lattice. Upon thermal activation, the bonds between oxygen and hydrogen will break and the protons will be available as charge carriers. These steps from synthesis, processing and operation are illustrated in Figure 2.8. Parent compound $BaCeO_3$ is substituted with Yttrium to form $BaCe_{1-x}Y_xO_{3-\delta}$ with oxygen vacancies, accounted for by "δ." Exposure to water vapor forms a hydrate structure with "crystal water" $BaCe_{1-x}Y_xO_{3-\delta} \cdot n\,H_2O$. Eventually, some oxygen vacancies are filled and the protons become structural protons in the lattice to form $BaCe_{1-x}Y_xH_{0.02}O_{3-\delta}$.

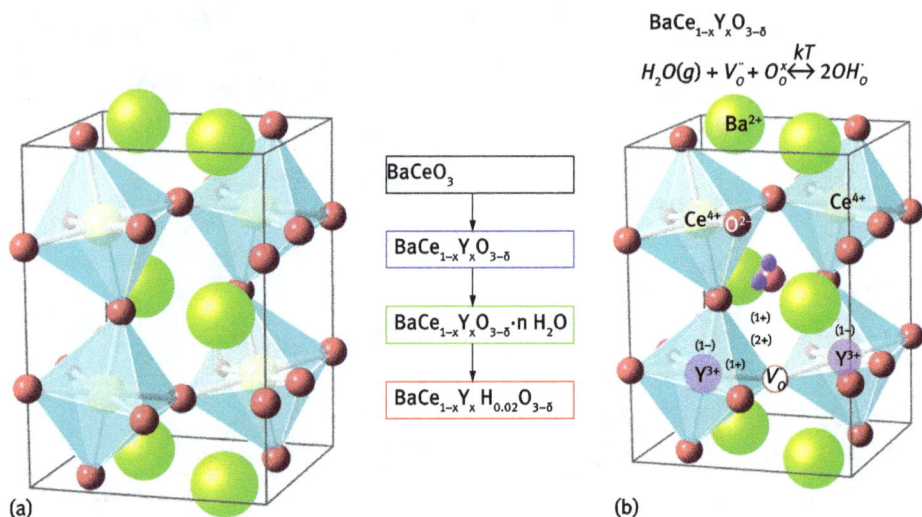

$$BaCe_{1-x}Y_xO_{3-\delta}$$
$$H_2O(g) + V_o^{..} + O_o^x \overset{kT}{\longleftrightarrow} 2OH_o^.$$

Figure 2.8: Representation of ABO_3-type perovskite crystal lattice with the large Ba ions on the A-sites (green), the "red" oxygen ions, and the Ce^{4+} in the centers of the oxygen octahedral. In the right panel, two Ce^{4+} ions are substituted by two Y^{3+} ions, forcing by charge balance an oxygen vacancy denoted by the red circle, above which a water molecule is approaching.

We have recorded the XANES spectra of dried ($BaZr_{1-x}Y_xO_{3-\delta}$) and of proton loaded ($BaZr_{1-x}Y_xO_{3-\delta} \cdot n\ H_2O$) BZY10 at the Swiss Norwegian Beamline (SNB) at the European Synchrotron Radiation Facility in Grenoble. The XANES pre-edge of the spectra in Figure 2.9 shows the Y K-edge spectra BZY 10. Only upon very close inspection of the entire XANES spectra, we notice a very minute change in the intensity distribution at around 17,037 and 17,038 eV. The proton-"loaded" samples have slightly higher intensity, some sort of step or kink, in comparison with the spectra from the "dry" samples. We measured and compared samples that were dried/protonated at Empa in Dübendorf and samples that were dried/hydrated on-site at ESRF Grenoble. This extra intensity at around 17,038 eV is probably from the Y dopant, which is bound to oxygen forming a hydroxyl bond with protons.

The spectra in Figure 2.9 were recorded from $BaZr_{0.9}Y_{0.1}O_{3-\delta}$, a well-known ceramic proton conductor model electrolyte material. The high energy "tail" of an X-ray absorption spectrum (XAS) shows often wiggles and oscillations, which have their origin in the near order structure and configuration of the atoms. This range is called the "extended" X-ray absorption fine structure (EXAFS) and provides information on the bond lengths and atomic coordination of the probed material. We see this in Figure 2.9 for energies larger than 17.1 keV, and we saw it in Figure 2.2 for energies larger than 6.03 keV. We will learn more about this in Chapter 3.4 of this book.

Substitution of the Zr^{4+} by the Y^{3+} drives the formation of oxygen vacancies in the crystal lattice, which are being accounted for by the inclusion of a "δ" in the ABO_3-type

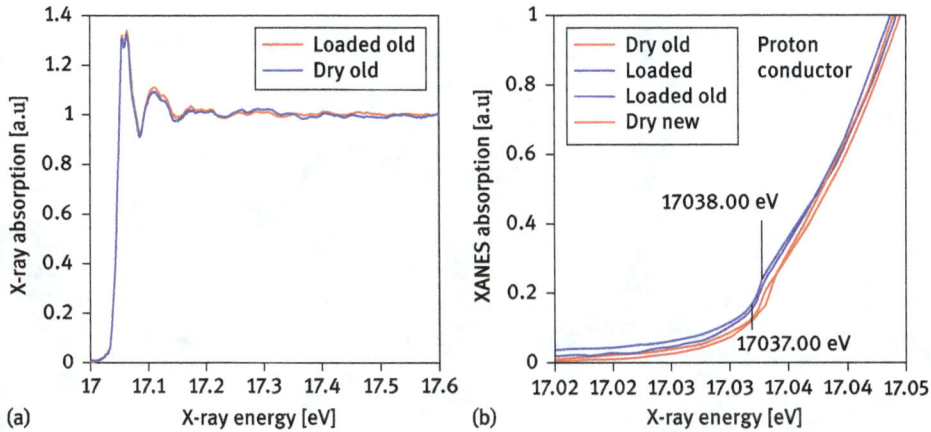

Figure 2.9: Pre-edge of the Y XANES spectrum of Y substituted barium zirconate in the "dry" and proton loaded condition. Suffix "old" indicates samples were prepared at Empa/Switzerland prior to synchrotron campaign in Grenoble/France. "Loaded" and "dry new" were prepared onsite at Grenoble. Spectra were recorded at the Swiss Norwegian Beamline (SNBL) ESRF, Grenoble.

perovskite material. Exposure to ambient conditions provides that the H_2O molecules from humidity in the ambient penetrate the material, where the oxygen ions from H_2O fill the oxygen vacancies and the protons H^+ from the H_2O form hydroxyl bonds inside the crystal lattice [Hempelmann 1995, Karmonik 1995, Matzke 1996]. Therefore, a proton conductor sample is readily filled with protons in a conventional laboratory atmosphere, and it is necessary to "dry the protons out" by a heat treatment procedure in a dry gas atmosphere. A dry sample can be easily mistaken for a loaded (protonated, hydrated) sample. It is possible to measure the weight difference of a dry and loaded proton conductor with a fine balance with milligram sensitivity.

The material undergoes through synthesis and processing different stages, including hydration and hydride, which we can summarize—and simplify—as follows:

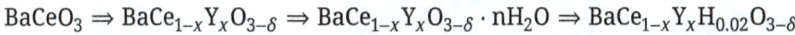

$$BaCeO_3 \Rightarrow BaCe_{1-x}Y_xO_{3-\delta} \Rightarrow BaCe_{1-x}Y_xO_{3-\delta} \cdot nH_2O \Rightarrow BaCe_{1-x}Y_xH_{0.02}O_{3-\delta}$$

Part of the talent of the researcher is therefore paying close attention to the experimental and instrumental details and also attention to details, which may vary from researcher to researcher when it comes to operating procedures. As already mentioned, the proton conductivity is a subtle mechanism, which easily may escape attention. In order to make sure that our observation and rational in Figure 2.9 (right) is correct, we carried out the experiment also *in situ* at the ESRF Swiss Norwegian Beamline (Figure 2.10). For this, we inserted the proton-conducting powder in boron nitride (BN) capillaries, which could be fed with dry or humidified carrier gas. The capillary was heated with a drying fan. The temperature was controlled with a thermocouple at the capillary.

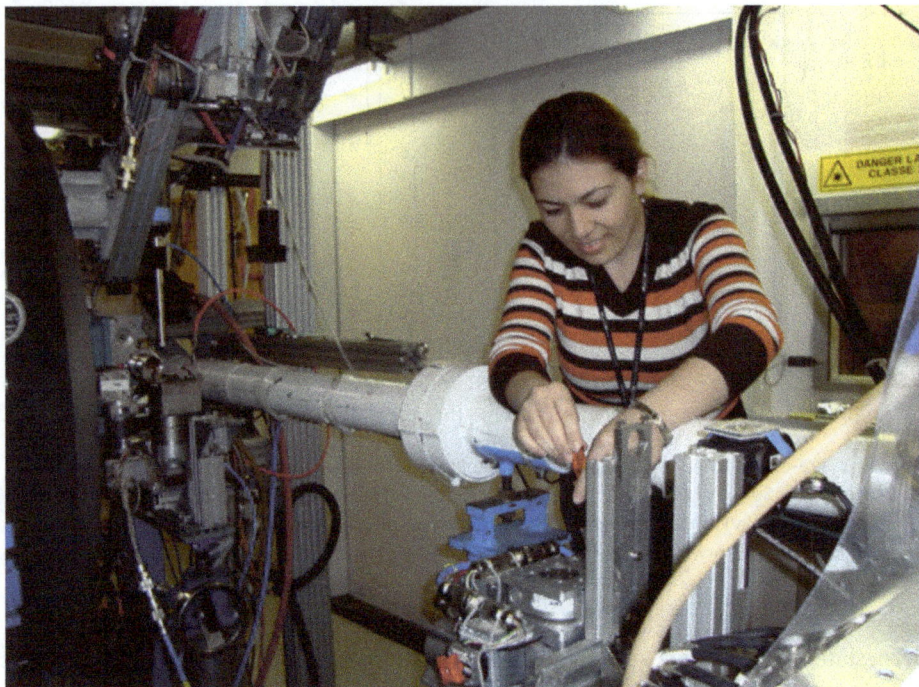

Figure 2.10: ETHZ/Empa PhD student Selma Erat (now Professor at Mersin University, Türkiye) preparing the *in situ* high-temperature vapor phase proton conductor experiment with hard X-ray absorption spectroscopy at the Swiss Norwegian beamline (SNBL) at ESRF in Grenoble, France (2009).

We now turn to solar energy conversion by CdSe solar cell materials. The overwhelming majority of the solar cells produced worldwide are still composed of costly monocrystalline silicon. CdSe thin films are a low-cost alternative, yet not a very environmentally benign alternative, which can be used as n-type window material of thin film solar cells. CdSe is a II–VI group semiconductor, the band gap of which ($E_g = 1.7$ eV) is in the visible range of the solar energy spectrum. CdSe has many beneficial properties for solar cell applications.

To make CdSe films for solar cells a low-cost alternative, chemical bath deposition is a favorable synthesis and processing route. The influence of the deposition temperature and annealing at different atmosphere on these properties has been investigated. The band gaps were determined from the optical absorption spectra as 1.83 and 1.76 eV at 60 and 70 °C. The electric resistivities range in the order of 106 Ω/cm at ambient temperature. Resistivities decrease with increasing annealing temperature, regardless whether annealed in oxidizing air or in reducing nitrogen atmosphere.

We have looked at CdSe films with Cd XANES at the Advanced Light Source in Berkeley. The films had been coated on glass by chemical bath deposition [Metin 2010]. Figure 2.11 shows the films, which have different yellowish color tones, mounted on a cop-

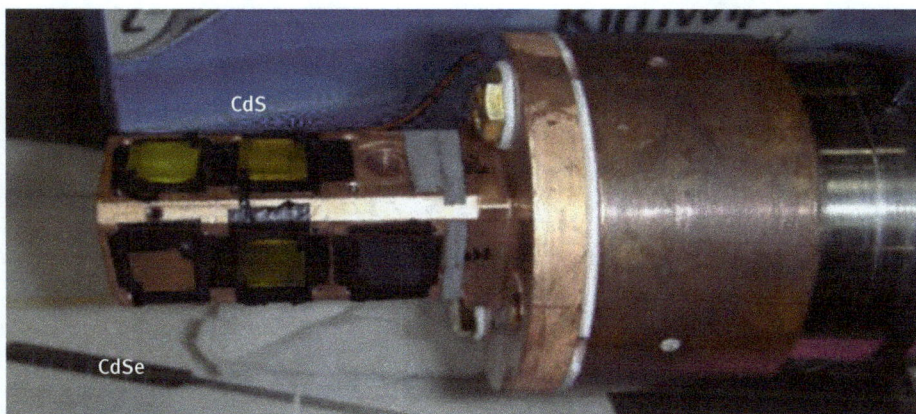

Figure 2.11: Sample holder at ALS Beamline 9.3.1 with yellowish CdSe and CdS thin films deposited on glass slides, attached with conducting carbon tape on top of the film for better electric conductivity between film and sample holder. The X-ray beam generates electric charges on the film, which have to be conducted away to the sample holder so as to prevent charging of the sample.

per metal sample holder. The samples are taped with sticky carbon tape to provide better electric contact between CdSe film surface and sample holder for better signal detection for target current spectroscopy (sample current detection mode). This warrants also a better electric grounding, and thus minimizes charging effects, which otherwise can cause noisy spectra.

Figure 2.12 shows the Cd L_3 and L_2 spectra of films, which had been heat treated at temperatures from room temperature (RT) to 773 K. With room temperature, we often mean the temperature of the ambient conditions under which we measure a sample. Often, this actual room temperature is not disclosed in experimental reports because it was not measured. Sometimes the RT or ambient temperature is not the same like the temperature of the sample, or the sample region on which the measurement (such as conductivity, spectroscopy, or other measurement) is actually performed. Sometimes it is just not possible to actually determine this temperature, and reference to RT is then just a hint that it could be around 300 K.

For many studies, it is also not necessary to disclose an accurate temperature, unless it is a specific aim to provide materials characteristics for one or several accurate temperatures. For phase transitions, the disclosure of an accurate temperature may be more relevant than for a temperature parameter study, which extends over a 1000 K for a handful of samples, for example. But in every case, this judgment depends on the leader of the study. Getting back to our CdSe study, we notice a strong variation of the XANES profiles upon the heat treatment. A well-pronounced double peak is at the L_3 edge for the sample heated to 773 K, which is not visible for the samples prepared at lower temperatures (lower temperature refers here to $T < 773$ K and not $T < 273$ K).

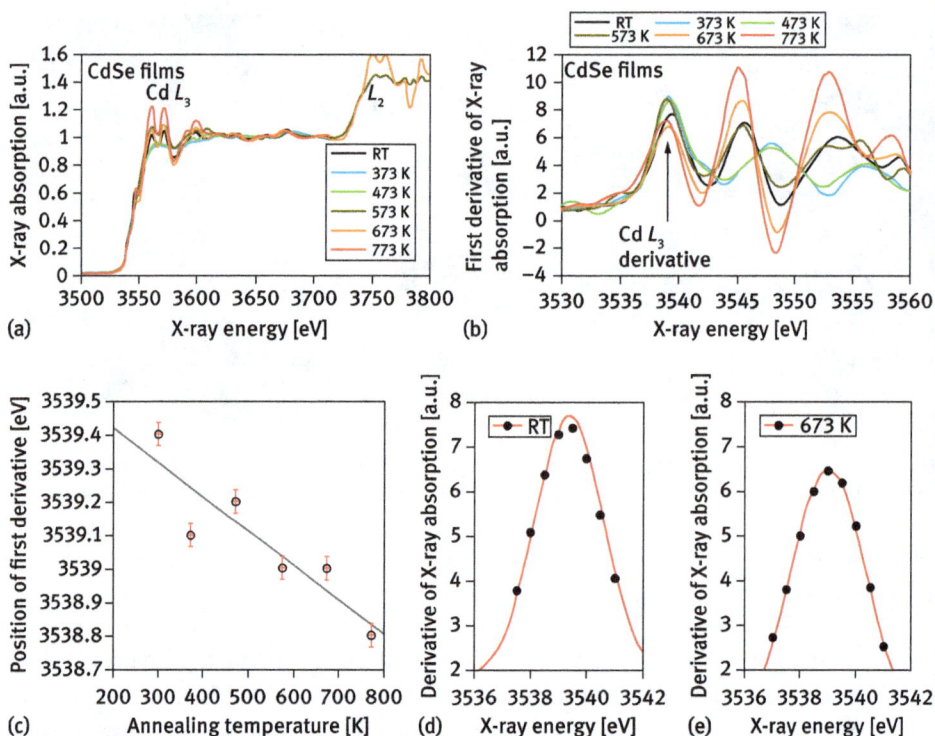

Figure 2.12: Cadmium XANES spectra of CdSe which was subject to heat treatment from 300 K to 773 K. Spectra were recorded at beamline 9.3.1 at Advanced Light Source.

For the determination of the suspected chemical shift of the spectra, I have formed the first derivative of the spectrum with respect to the X-ray energy, which is shown on the right top panel of Figure 2.12. The first maximum in the derivative is around 3539 eV, and thus constitutes the absorption edge. We notice that the positions of the maxima are shifted versus the energy axis. For the convenient determination of the position of the maximum of the first derivative, I have applied a Gaussian least square fit, which is demonstrated in the lower two right panel of Figure 2.12 for the samples synthesized at RT and heat treated at 673 K. The positions of the thus obtained absorption edges are summarized in the lower left panel of Figure 2.12. There we see that absorption edge (position of maximum of first derivative) is systematically shifting toward lower X-ray energy by around (3539.30 − 3538.75 eV) = 0.55 eV. For comparison, a shift of the Mn K-edge (~6500 eV) by around 14 eV accounts for 4 valence units, and a shift of the V K-edge (~5500 eV) by around 7 eV accounts for 3 valence units.

It is therefore obvious that the Cd becomes substantially reduced by annealing in air. This is an interesting finding because air as a processing gas is often considered an oxidizing atmosphere, which will oxidize a material. Air, in fact, is a gas mixture of around 20 % oxygen and 80 % nitrogen, and the latter is typically used as a reducing gas, similar to argon, carbon monoxide, or hydrogen, whereas carbon dioxide and water vapor are frequently used as mild oxidizing gases. However, it depends on the material under processing and the thermodynamic conditions such as composition of gas, gas concentrations, and temperature whether a material or a materials surface becomes oxidized or reduced. With X-ray spectroscopy, we have a relatively easy tool for the investigation of this process.

2.1.2 Soft X-ray spectroscopy: NEXAFS

The synchrotron methods were developed, at large, for an energy range, which allowed measuring the K-edges of the 3d metals and the L-edges for platinum metals. In the 1990s, this energy range was well developed, and 3rd-generation synchrotrons were designed, which featured also the lower energy range where one could measure the L-edges of the 3d metals and, more important, also the low Z elements like carbon and oxygen, or nitrogen, which had some relevance for the IT technology. This was the time when Joachim Stöhr summarized the readily existing X-ray and electron spectroscopy work on such elements and published it in his book NEXAFS spectroscopy [Stöhr 1992]. The energy range that I am talking about here covers approximately 100 to 1500 eV. By the end of the 1990s, it was realized that the tender X-ray range from 2000 to 3500 eV should deserve more attention, e. g., for the coverage of the elements sulfur and phosphorus.

I am not going to cover in this section specific soft X-ray studies. This book contains a whole range of various soft X-ray studies. One advantage of soft X-rays is that the radiation protection measures for experimenters is somewhat relaxed. The X-ray beams come through tubes directly in the UHV chamber and there is no exposure to X-rays to the experimenters. What we need to know is that this method typically requires an ultrahigh vacuum environment for the sample. This is the standard environment and a disadvantage because of more difficulties with sample handling. The pumping down of UHV sometimes cost much time. Some samples are also not compatible with UHV. There are, however, exceptions. The STXM instrument at Beamline X1A at NSLS in Brookhaven had a helium atmosphere around the sample position. Helium—because of its fewer number of electrons than oxygen and nitrogen in air—absorbs not so much soft X-rays, and from that perspective, it was easy to handle. You will later see in this book how we did *operando* photo-electrochemical experiments with soft X-rays in a liquid electrochemical cell at the Advanced Light Source at Beamline 7. This is a fantastic method from the scientific perspective, but the handling of the cell was difficult in the beginning, and frequently, the thin silicon nitride windows broke and the electrolyte splashed into the UHV chamber, which took away normally up to 24 hours of beamtime for cleaning and fixing. XPS is a typical soft X-ray method in UHV. Lately, more and more instruments allow for doing XPS at high pressures. The instruments are costly but the results are overwhelming.

I can quote here my colleague Hongxin Wang that the L-edges of the 3d metal spectra, which we measure with soft X-rays that have a number of advantages over the hard X-ray K-edge XANES spectra [Cramer 1992, Cramer 1997]. The multiplet in the L-edges is of some diagnostic ease and also easy and theoretically to be interpreted [De Groot 1990, Wang 1997]. The multiplet is caused by the interaction 2p holes with 3d electrons. The 2p holes can sometimes be detected with the oxygen spectra or other ligand spectra like C, N, S, and P, which are measured with soft X-rays. Of interest for analysis of spectra from 3d metals is the L-edge sum rule analysis, which Hongxin has worked on a lot; this tool can be used for directly observing the charge localization on particular elements [Wang 1998]. Therefore, the L-edges are sensitive for analyzing transition metal centers in inorganic compounds and organic matter, such as metalloproteins [George 2001, Ralston 2000, Wang 2000, Wang 2001c]. For the more elaborate handling of experiments because of the typical UHV environment, we have therefore an analytical positive tradeoff.

It is neither an advantage nor a disadvantage that, because of the smaller attenuation length of low energy X-rays, the soft X-ray methods are quite surface sensitive. We have to consider this as a technical fact, which we have to deal with. The low Z elements have their core levels in the soft and ultrasoft energy range. Hence, a spectrum of lithium or boron, e. g., will always be from the surface of the material. When the material is chemically reactive such as lithium, the surface is likely corroded and its spectrum will not represent lithium but some lithium compound. There has been early interest in obtaining bulk representative spectra of materials from low Z elements [Bergmann 2002]. There exists an X-ray analytical trick with which we can get around this problem. We will learn this in the chapter on X-ray Raman spectroscopy [Braun 2015j].

In the last 10 years, the field of resonant inelastic X-ray scattering (RIXS) has become popular [Braun 2013a, Glatzel 2005, Wang 2013b]. I would call this a method of obtaining spectral maps, which allow a deeper element specific and orbital specific insight in the electronic and molecular structure of materials. We can do this with soft X-rays [Braun 2013a] and ultrasoft X-rays. This is briefly covered in Section 5.6.5 in this book [Wang 2013b].

2.1.3 Tender X-ray absorption spectroscopy

2.1.3.1 Sulfur

The nomenclature in the X-ray spectroscopy makes that there is an energy range between the hard and soft X-rays, which is called the tender X-rays (cite Northrup 2016). The tender X-rays range between 2 and 5 keV. Prominent chemical elements, which have K-shell absorption edges in this range, are sulfur ($K\alpha_1$ 2307.84 eV), phosphor (2013.7 eV), chlorine ($K\alpha_1$ 2622.39 eV), potassium ($K\alpha_1$ 3313.8 eV), calcium ($K\alpha_1$ 3691.8 eV).

In Chapter 8.2, we will see a SOFC study on sulfur K-edge absorption spectra, where the sulfur poisoning of anodes is studied. Sulfur poisoning is an established topic in

SOFC research, because fossil fuels such as natural gas may contain sulfur naturally, and sometimes sulfur compounds may be added by intention so as to give the gas an alarming odor if there is a leak in a pipeline.

When an SOFC is operated with wood gas, there may be traces of chlorine or potassium, which over time react with the anode material and poison it. Therefore, the aforementioned chemical species may show up in some materials, which are used in the context of electrochemical energy storage and conversion. And there may be a need for chemical analysis to be done with X-rays. And the aforementioned chemical elements have their K-shell absorption edges in the tender X-ray range.

Lately, sulfur has become again a chemical element of interest for lithium sulfur (and sodium sulfur) ion batteries.[1] Lithium is a very light element and has the atomic mass number 7. Sulfur with atomic number 16 is also relatively light and has a mass number of 32.07. Compare the latter to the positive electrodes from the spinel structure metal oxides, such as $LiCoO_2$. Cobalt has a mass number of 58.93. The standard lithium ion battery has thus a larger weight than the lithium sulfur battery. Compare this against the lead acid battery, with Pb having mass number of 207. The lithium sulfur battery is therefore, at least theoretically, an interesting light weight alternative.

Figure 17 in [Song 2013] shows the specific capacity of the Li-S cell as a function of sulfur mass versus the sulfur content of the positive electrode mixture. The solid lines are the theoretical capacities. The data points are experimental values taken from literature, and values obtained at Lawrence Berkeley National Laboratory. The "equi-energy" lines are shifting toward the upper right corner with increasing sulfur content; from 200 Wh/kg to 800 Wh/kg in the graph. Note, however, that capacity or energy density is not the only important metric for a battery. Important is also, how fast the battery can store or release the energy. The metrics for this are the power density or the time constant. This is why many electrodes with "active" material contain also graphite or carbon, which shall facilitate electron transport in the electrode, and thus increase the power density.

We can figure the basic scheme of this battery in the sketch in Figure 2 of [Song 2013]. The lithium sulfur battery has a metal lithium anode and a sulfur cathode. It is a rechargeable battery, and thus the negative electrode is the lithium, and the positive electrode the sulfur. It is possible to run this battery with a liquid or a solid electrolyte. In the configuration of sketch (a) in Figure 2, the sulfur is mixed with carbon. When the electric circuit is closed, lithium ions move (along the difference, the gradient of the electrochemical potentials) from the lithium metal through the electrolyte to the sulfur positive electrode, producing the electromotoric force. You can imagine that the oxidation state of the sulfur is changing during this process, and this can be monitored with X-ray core level (or valence band) spectroscopy.

---

[1] There is a movie on YouTube and on www.exploratorium.edu website where Professor Elton J. Cairns makes a presentation on sulfur in lithium ion batteries, [Cairns 2016].

It is possible to use pre-lithiated sulfur, such as Li_2S, as positive electrode and then construct a cell with less harmful negative electrode materials, such as silicon or tin, which forms an alloy with Li. On the sulfur side, we can expect that the carbon might chemically bond with the sulfur, at least to some extent, depending on the synthesis and potential after-treatment of this electrode. Song et al. [Song 2013] refer to the C1s NEXAFS spectra recorded from such electrode. Peculiar changes in the C1s NEXAFS spectra between graphite oxide, and graphite oxide mixed with sulfur, can be made out (see Figure 5 in [Song 2013]).

With exposure of the active electrode materials to molecules from the ambient environment, such as water from humidity, oxygen, nitrogen, or even carbon dioxide, the electrode materials may be negatively affected, and this can be monitored by electrochemical methods such as cyclic voltammetry and impedance spectroscopy. Also, the charging and discharging of the battery (cycling) may affect the functionality of the material to the extent that degradation, passivation, and failure may happen in the end.

In this context, cycling stability is an important feature of any rechargeable battery. Figure 18 from the review paper of Song et al. [Song 2013] gives an overview of the specific capacity (mAh per gramm of active powder mix material in the positive sulfur electrode) versus the maximum number of cycles, based on literature data up to 2013 when their paper was published. The data range from 100 to 600 mAh/g with cycles not more than 100. The lithium ion equivalent as a benchmark standard is given to 500 cycles with 200 mAh/g. The authors express their hope with the green arrow "Future Direction" in the Figure 18. Indeed, in 2022, Pai et al. [Pai 2022] found that their Li-S battery could sustain 4000 cycles.

While the CV and EIS spectra may indicate such changes, they do not necessarily point to the structural origin of these effects. The latter, however, can be done with X-ray spectroscopy. A general problem for analysis of lithium battery components is the sensitivity of lithium to "air." It is therefore advised, to carry out experiments *in situ*, which sometimes requires a specifically designed spectro-electrochemical cell.

One experimental setup at a synchrotron beamline end station is illustrated in the sketch in Figure 2 in the paper [Müller 2014]. The cell is shown in the photo on the right. The cylindric flat cell has two concentrics through which the X-ray beam can pass in transmission geometry, probing the sample, which is inside the cell. Under a solid angle of 60° against incidence of X-rays, the fluorescence radiation can be detected with a silicon drift detector (SDD). Zech at al. [Zech 2021] have recorded sulfur XAS spectra from the polysulfides dissolved in the electrolyte by using altered coin cells, ready for *operando* XAS in the tender X-ray range.

The changes of the molecular structure of the sulfur during cycling can be anticipated already by looking at the cyclic voltammogram, as shown in Figure 6 of [Müller 2014], and further detailed in the review of [Huang 2019]. So-called polysulfides form during cycling of a lithium sulfur battery. This is a manifold of Li-S compounds with various stoichiometries, as listed below:

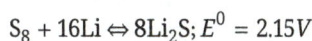

$$S_8 + 16Li \Leftrightarrow 8Li_2S; E^0 = 2.15V$$

$$S_8 + 2Li \Leftrightarrow Li_2S_8$$

$$Li_2S_8 + 2Li \Leftrightarrow 2Li_2S_4$$

$$Li_2S_8 \Leftrightarrow 2Li_2S_n + (8 - n)S$$

$$Li_2S_4 + 2Li \Leftrightarrow 2Li_2S_2$$

$$Li_2S_2 + 2Li \Leftrightarrow 2Li_2S$$

On a final note, the Materials Research Society. (MRS) Spring Meeting 2024 in Seattle has a Symposium (ES06) on "Sulfur and Sulfide Chemistry in High Performance Electrochemical Energy Storage."

2.1.3.2 Phosphorous XAS

Phosphorous is a rather elusive chemical element and hardly pops up in the daily life of a materials scientist and chemist, unlike oxygen, nitrogen, carbon, and sulfur. But there are fields where element P is used, e. g., as phosphate buffer solution (PBS), phosphoric acid H_3PO_4 as an antioxidant in soft drinks, as phosphate fertilizer in agriculture. P is an important element in biochemistry and contained in the energy carrier adenosintriphosphate (ATP). Important to know is that two atoms of P can bind five atoms of O to form the P_2O_5 phosophorpentoxide. Phosphorous has a high affinity to bind oxygen and can be used as flame retardant. Maybe this is more a matter of awareness, and not so much of presence. But there are not so many X-ray spectroscopy studies on phosphor.

One of the interesting materials for lithium ion cell positive electrodes is the lithium iron phosphate $LiFePO_4$. It has a high storage capacity. But the material is not well conducting. The former property gives it a high energy density. The latter property gives it a low power density. Notwithstanding the latter shortcoming, these batteries exist on the market and they compete well with the standard based on the lithium cobalt oxide based electrodes.

Schmidt et al. [Schmidt 2018] carried out *operando* XAS on a phosphor-based lithium ion cell at the PHOENIX beamline of the Swiss Lightsource and could prove that the phosphor participates in a second ligand redox process, the effect of which was recently attributed to parasitic side reactions such as solid electrolyte interface (SEI) or electrolyte reduction. The electrode material of choice was lithium iron methylenediphosphonate (FeMeDP).

There are not so many phosphor XANES spectra around in literature, which originate from electrochemical studies. It is therefore worthwhile to consult literature from the earth sciences—just for comparison—as an alternative, for example [Ingall 2011]. But let us continue with Schmidt et al. Figures 2 and 3 in [Schmidt 2018] show the XANES spectra recorded *operando* from 2146 to 2158 eV, which contains the absorption threshold of the phosphor K-edge. We see five spectra with prominent white lines at around 2151 eV. Shown are the spectra from five samples. This is the FeMeDP pristine sample (black spectrum), the electrode at 0.1 V potential after the 1st and 2nd cycle (green, red spectra), and the spectrum of the sample charged to 3 Volt (blue spectrum). Shown is also a spectrum from the reference material H4MeDP.

The spectra at the 1st and 2nd cycle recorded at 0.1 V look very much alike, underlining the reproducibility of the spectroscopy experiment. It appears both are shifted toward slightly lower X-ray energy compared to the pristine sample, suggesting a slight reduction of the phosphor. Their lowering of the intensity of the white line suggests a decreases in the ionicity of the bonding situation. This is accompanied by structural changes, as shown in the radial distribution from the EXAFS analysis in panel B of the Figure 2 in the paper from Schmidt et al. The change in potential moves apart the phosphor ions, by around 0.15 Å. Panel C in the figure shows the evolution of the spectra over the entire experiment time of 30 hours versus the potential in a contour plot.

Maybe interesting to mention is that the experimenters put their spectro-electrochemical cell into a 0.6 bar helium atmosphere. We may ask: why did they do that? During my *operando* XRD and XANES experiments at SSRL in the year 2000, I had put my battery cell into a zip-lock bag and filled it with argon. The purpose for this was the protection of the sample inside the cell from corrosion by potentially entering humidity or air, which contains oxygen and even more so nitrogen. We know that lithium likes to form a Li_3N with nitrogen. This is why we do not use nitrogen-filled glove boxes for lithium battery work. But argon as a protective gas is typically around in the chemistry lab and at the synchrotron.

Why were Schmidt et al. using the expensive and rare helium gas, and not argon? I guess they did so in order to replace the air between X-ray source (as far as not contained in vacuum pipes, but exit through X-ray windows) and sample (cell), and between sample (cell) and X-ray detector, with a less absorbing medium. This is helium at low pressure. We did so in 2012 for the X-ray Raman experiment at SSRL with our *operando* battery [Braun 2015j], which is described in more detail in Chapter 8.1.3, Figure 8.23. And if I recollect correctly, we did so 25 years ago at APS or SSRL when doing protein spectroscopy.

It is somewhat unfortunate that I cannot refer here to phosphor spectra from, e. g., phosphoric acid fuel cells, phosphor in photosynthesis, and P XANES spectra from solar cell materials such as InP and GaP (except maybe [Zerulla 2009]). On the other hand, the lack of such studies is your opportunity to do research on the topic.

2.1.4 Chlorine

Chlorine is another element, which has the K-edge in the tender X-ray range, i. e., around 2825 eV. Chlorine is the halogen ion in the rock salt (table salt, NaCl). Sea water contains

Cl^- ions. The mist from seawater transports the ions through the atmosphere.[2] When you operate vessels in or on the ocean, chances are that chlorine contamination may become an issue.

Baturina et al. studied the influence of chlorine exposure on the polymer electrolyte membrane fuel cells (PEMFC) and conducted in this context electrochemical studies and subjected some of the materials to Cl K-edge XAS spectroscopy [Baturina 2011]. For the foiling discussion, you will have to find this paper in the library and spread it in front of you on the desk, if you want to follow my outline here.

Chlorine can adsorb on the platinum catalyst and inhibit the oxygen reduction reaction (ORR). In addition to recording Pt XANES spectra (typically the L-edges because those are at the lower and easier accessible X-ray ranges), it is worthwhile to record the Cl K-edge spectra. The XAS study shown here was an *ex situ* study.

Baturina et al. used beamline X15B at the National Synchrotron Lightsource (NSLS) at Brookhaven National Laboratory, Upton, New York.[3] X-ray spectra were taken from chlorine containing reference samples and from the catalyst coated polymer membranes (CCM).

Let us first look at the cyclic voltammograms of the fuel cell cathode, as shown in Figure 3 in the paper by Baturina et al. The upper panel "a" shows with the black CV the begin of test (BOT). This would be the pristine cathode. The overall shape with multiple current waves in oxidizing and reducing direction is typical to the CV of a PEMFC. Upon mixing the gas with a very minute amount of chlorine by HCl, the area enclosed by the CV becomes slightly decreased (the authors mean the electrochemically active surface area (ECSA) as determined from the double layer capacity measured in the CV, from $71\,m^2$ to $56\,m^2$), possibly because of dissolution of the platinum catalyst (see the red CV). Upon reactivation of the catalyst by exposure to reducing H_2 gas, the ECSA increases slightly from $56\,m^2$ to $57\,m^2$ (see the green CV).

Moreover, one notices a positive shift of 10 mV at around the potential of 0.8 V for the adsorbed oxygen species in cathodic and anodic direction. And a negative shift of 20 mV near the hydrogen reduction potentials covering 0.08 V to 0.4 V. The shifts are indicated

---

2 A long time ago I worked at the University of Kentucky on a combustion aerosol project, where I developed the carbon C 1s NEXAFS spectroscopy for molecular speciation. Out of personal, even private interest, I collected all kind of samples from where ever I just could get them. The ceiling fans in my Kentucky home had dust on the top of the blades, which I measured at the Advanced Lightsource in Berkeley. The dust contained carbonate CO_3^{2-}. I asked my colleague Professor Frank E. Huggins, who had his office next to mine, if he had any idea why there would or could be carbonate on the ceiling fan of my bed room. He asked me, if there was a shower room nearby my sleeping room. Because the water in the Bluegrass region, where we lived, contains a lot of calcium carbonate. Yes, the bath room with shower was next to my bed room. And very likely the mist from the showring was carried through the first floor into my bed room and settled and accumulated the calcium carbonate on the ceiling fan.

3 This is on Long Island, a favorite place for vacation for New Yorkers, within the Hamptons, many villas and mansions owned by people working in the entertainment movie business.

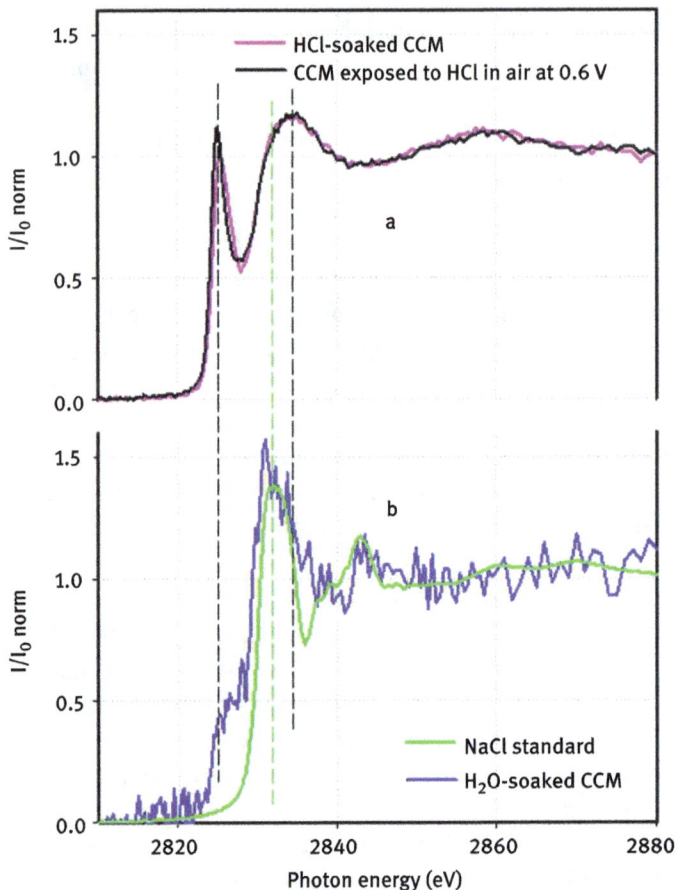

Figure 2.13: Pre-edge-background-corrected and normalized to a step size of unity XANES spectra of CCMs soaked in 0.01M HCl solution (a, red), CCM exposed to HCl in air for 20 h at 0.6 V (a, black), NaCl solid standard (b, green), and H2O (b, blue). Reprinted from Baturina et al. 2011 with permission by the source of the Work, including author, Insights into PEMFC Performance Degradation from HCl in Air, Journal of The Electrochemical Society, 158, 10, B1198, 31 August 2011; 10.1149/1.3621318; © The Electrochemical Society. Reproduced by permission of IOP Publishing Ltd. All rights reserved.

by colored vertical lines. The changes on the CVs are caused by the adsorbed Cl ions on the platinum surface.

Figure 2.13 shows the normalized[4] Cl 1s XAS spectra of the catalyst coated membrane (CCM) after the fuel cell was exposed to dilute HCl at a potential of 0.6 V (top panel (a), black spectrum), and after treatment with H_2 (top panel (a), blue spectrum).

---

4 The normalization intensity of the X-ray absorption spectra, which means that the absorption intensity at the pre-edge is set to 0, and the high energy tail of the spectrum is set to 1. This is a routine included in many software packages for EXAFS analysis. On this occasion, I want to point the reader to the UNIFIT software developed by Dr. Ronald Hesse; see www.unifit-software.de

The sharp pre-edge peak at around 2825 eV originates from a ligand to metal transition between chlorine and platinum. This peak becomes smaller, even a shoulder upon treatment with H_2, a reduction agent. This peak does not exist in the Cl 1s spectrum of NaCl, as is shown in panel (b), green spectrum below. Whereas the (red) spectrum of H_2PtCl_6, hexachloroplatinic acid reference shows a distinct such sharp peak.

2.2 X-ray emission spectroscopy

2.2.1 Hard X-ray emission

We know that the X-ray absorption process creates a core hole. This core hole can be filled when a valence electron "decays" into the lower lying core hole. This process goes along with the emission of an X-ray photon, the energy of which matches the energy of the decay process. X-ray emission (XES) hence probes the occupied valence states specific to the projection of an atom. It is therefore possible to record XES spectra, which show the element specific core levels of the compound under investigation. Moreover, the spectra contain similar fine structures like the XAS spectra, which allow therefore detailed electronic and molecular structure studies which go beyond the mere element specific analysis like EDAX and the like.

A brief review on the X-ray emission spectroscopy method using hard X-rays in photosynthesis research is given by Bergmann and Glatzel [Bergmann 2009]. Looking at the geometry of the experimental setup of the spectrometer, which is shown in the lower right panel of Figure 2.14, it becomes clear that the sample position constitutes a radiative "point source," but with a finite extension. For the sake of sufficient X-ray counts, it is therefore advised to compact the samples as much as possible. Monochromatic X-rays arrive at the sample, which emits an X-ray spectrum. Analyser crystals reflect the X-ray spectrum on the X-ray detector so that an X-ray emission spectrum is recorded.

One application of XES is Kβ spectroscopy, which is particularly sensitive to the spin states of ions. This method has been used for the study of Mn compounds such as the oxygen-evolving complex (OEC) in photosystem II (PS-II). The OEC in PS II is built from a pair of Mn_2O_2 networks, which are linked by an oxygen ion at two adjacent Mn ions. This cluster performs the water oxidation $2H_2O \Rightarrow O_2 + 4H^+ + 4e^-$. This process takes place in green biological matter, including the plankton in the oceans, and accounts basically for all produced oxygen on our globe. The interesting electronic structure of the 3d metal manganese atom is the origin for its rich redox chemistry, primarily because of its manifold of oxidation states, which varies between Mn(II), Mn(III) and Mn(IV) in the case of PS II water oxidation. This redox cycle is sketched in Figure 2.15 and shall illustrate that we are dealing here with a dynamic system. Capturing the oxidation state of Mn would be difficult in the course of this process. It is therefore important to preserve the sample of PS II in one particular phase of the redox cycle. A review paper on

(a)

(b)

(c)

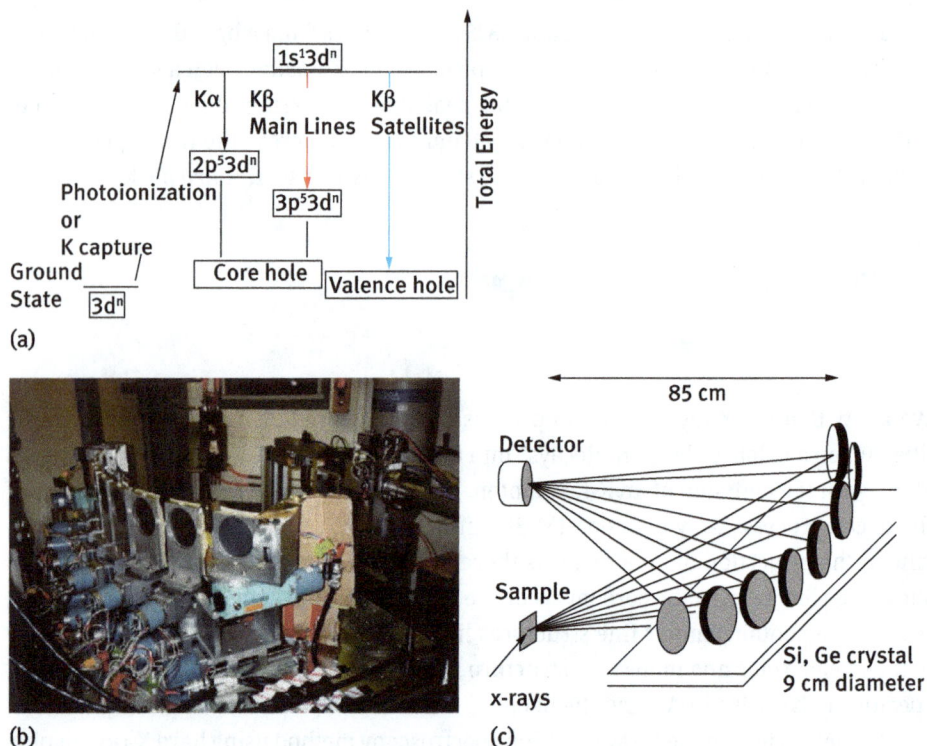

Figure 2.14: Schematic of energy levels and excitation processes in X-ray emission spectroscopy (XES, courtesy Pieter Glatzel). Spectrometer, photo courtesy Pieter Glatzel [Bergmann 2002, Glatzel 2005].

this issue was prepared by Robblee and Cinco and Yachandra [Robblee 2001]. An example of how these oxidation states can be very nicely read out from a calibration curve (compare Figure 2.3 in this book with L-edge calibration for Pr) of XAS edge positions is exercised in Figure 1 of the review paper by Dau Luzzolino, and Dittmer [Dau 2001], where a 3.5-eV chemical shift accounts for 1 oxidation state unit.

For the preparation of such relatively delicate experiments, it is common to carry out experiments on model compounds where the theory of spectroscopy can be verified. This is exemplified for the case of MnO (Figure 2.16, [Glatzel 7 May 2002]), where the K shell fluorescence $K\alpha$ and $K\beta$ principal lines and the $K\beta$ satellite lines are shown. It is obvious therefrom that these emission spectra reveal fine structures. Two structurally homologous Mn compounds with three different oxidation states were investigated by Visser et al. [Visser 2001]. They were considered as relevant for PSII.

The energy split between the $K\beta'$ and $K\beta1,3$ peaks originates from the exchange interaction of the unpaired 3d electrons with the 3p hole. This hole is final state following the filling of the core hole by a 3p->1s fluorescence transition.

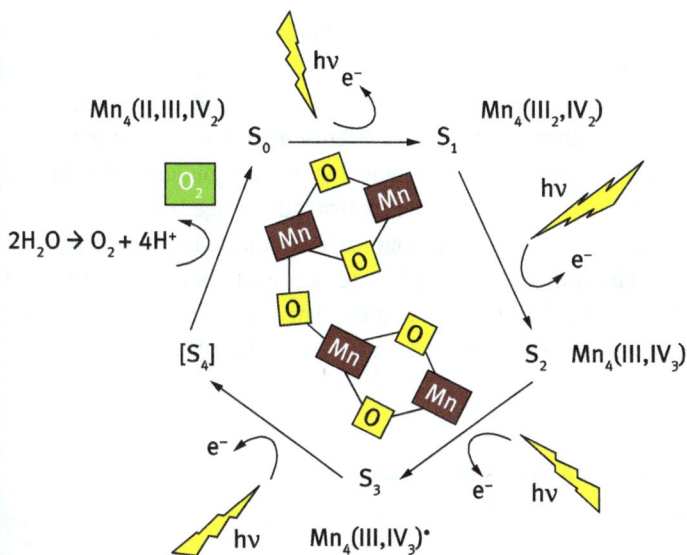

Figure 2.15: Schematic of the Mn redox cycle in the OEC in PS II, which extends over 5 states S0, ..., S4 and includes Mn_{II}, Mn_{III}, Mn_{IV}.

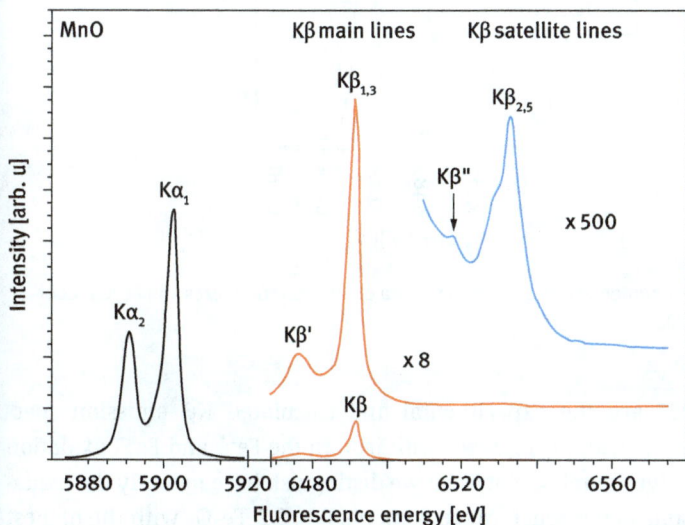

Figure 2.16: The K fluorescence lines for manganese of the model compound MnO. Graphics by courtesy of Pieter Glatzel.

With respect to this hole in the 3p level, the spin of the unpaired 3d valence electrons may be parallel or antiparallel. When fewer 3d electrons interact with this 3p hole, the energy split of $K\beta'$ and $K\beta1,3$ becomes smaller. The $K\beta1,3$ peaks shift therefore to lower energies when the ion has a higher valence. This is in contrast to the behavior of the XANES absorption edges [Robblee 2001].

Grush et al. have introduced a concept of obtaining chemical site selective XANES and EXAFS spectra from using high-resolution fluorescence detection [Grush 1995].

A general observation in the XES spectra is that they have broader, more diffuse features than XAS spectra. At times, this can be a practical advantage because an overwhelmingly large number of peaks in XAS, although corresponding to particular spectroscopic and structure information of the probed material, can sometimes carry away from simple aspects such as the oxidation state of an ion. The XES spectra show a chemical shift like the XAS spectra. Figure 2.17 shows a set of $K\beta$ spectra of Fe_2O_3 and of iron containing complexes.

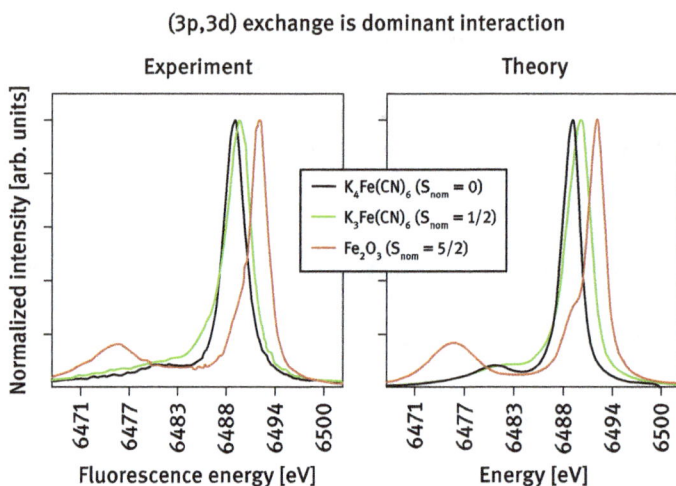

(3p,3d) exchange is dominant interaction

Figure 2.17: Comparison of experimental and simulated spectra of two iron complexes and Fe_2O_3. Courtesy P. Glatzel [Glatzel 7 May 2002].

Compared in Figure 2.17 are the experimental and calculated $K\beta$ emission spectra of Fe_2O_3, and potassium cyano complexes with iron in the Fe^{2+} and Fe^{3+} oxidation states, the Prussian blue dye complex. Note that we distinguish here not only the oxidation state, but also the spin states which range from net 0 to 5/2. Fe_2O_3 with the highest oxidation state for Fe has the largest chemical shift to higher energies, as anticipated. $K_3Fe(CN)_6$ has lower energy for the centroid (center of gravity) for the $K\beta$ peak position than for Fe_2O_3.

Yamaoka et al. [Yamaoka 2004] have investigated iron and iron oxides with $K\beta$ *resonant* emission spectroscopy. Specifically, they combined $K\beta$ resonant X-ray emission and partial fluorescence yield spectroscopy near the Fe K absorption edge. They looked at the spin states of these materials by considering the origin of the $K\beta'$ line and the pre-edge peak, along with computation of theoretical spectra, which produced for the case of Fe_2O_3 good agreement between experiment and theory. The authors observed and simulated the resonant Raman scattering at the absorption edge and in the pre-edge. A more recent study where Kβ spectroscopy was used is by Ceppi et al. [Ceppi 2014]. They investigated the "double-spinel" $Mn_{(2-x)}V_{(1+x)}O_4$ in order to determine coordination and oxidation state of Mn and V. Their corresponding XAS analyses did not fully agree with the Kβ-data with respect to the coordination, whereas the oxidation states were found to be similar for both methods.

The chemical shift of the Kβ spectra is obviously worthwhile to exploit for chemical analysis. We have therefore employed Kβ spectroscopy for the investigation of lithium battery cathode materials. Figure 2.18 shows two spectra from a cycled $LiMn_2O_4$ lithium battery cathode and the noncycled spinel powder [Braun 2002d]. At first glance, the two spectra of the two different samples look identical. When we zoom in the Kβ 1,3 peak, we notice a slight shift in both spectra. The spectrum of the cycled electrode is slightly shifted to a somewhat lower energy compared to the spectrum of the powder [Braun 2002d]. While this small shift may appear superficial, quantitative analysis shows that the "chemical difference" of the samples is indeed reflected by their different Kβ spectra.

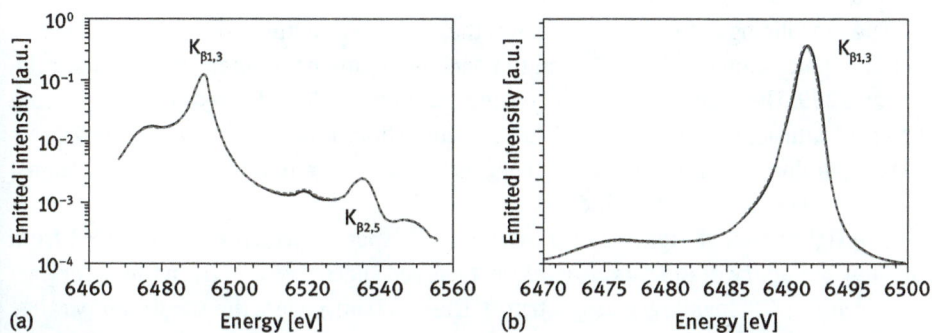

Figure 2.18: (left) Manganese Kβ emission spectra of a cycled and uncycled electrode. (right) Magnified manganese Kβ 1;3 emission spectra of a cycled (dashed line) and uncycled electrode from [Braun 2002d].

So far, we have looked into the hard X-ray method of XES, which turns out to be completely bulk sensitive within the limits of X-ray attenuation. We are irradiating, and thus exciting the sample/system with one particular monochromatic X-ray energy, which is has to be above the core level energies under investigation. With increasing excitation energy, the probing depth is increasing. However, the energy of the emitted photons is also subject to particular attenuation, which means the information depth is also within the limits of X-ray attenuation.

2.2.2 Energy-dispersive X-ray spectroscopy

We make now a short excursion to an often used method. This is energy-dispersive X-ray spectroscopy (EDS, EDX, or XEDS), frequently called energy dispersive X-ray analysis (EDXA) or energy dispersive X-ray microanalysis (EDXMA). EDX is used for the elemental analysis of a material. For the stimulation of the emission of the characteristic X-rays from a specimen, a high-energy beam of particles such as electrons or protons, or X-rays, is used. The energy resolution of the spectrometers are usually designed that it allows for only the elemental composition of the specimen to be identified. Fine structures, which would permit the assessment of the molecular or electronic structures are usually not resolved. Nowadays, electron microscopes are often equipped with an EDX setup so that a micrograph is accompanied by an elemental analysis with spatial resolution. This method is called "microprobe" [Goldstein 2003].

Below is shown a study on the chlorine poisoning of SOFC anodes [Nurk 2011]. Later in this book, we will learn more details on sulfur poisoning of SOFC anodes. Here, in this section, we begin with an electron micrograph of an anode sample which was also subject to EDX analysis in the electron microscope. The inspected sample had been operated for some long time under chlorine evaporation in order to learn about chlorine poisoning of SOFC anodes. Biomass fuel from wood may contain chlorine salts which exist as traces in the fuel. Accumulation of chlorine over long operation periods may be deleterious for the SOFC components.

The morphology of the anodes shows the after-service life with 3500-ppm concentration of potassium chloride KCl and the peeling-off and delamination of the material in Figure 2.19. The light gray material is the zirconium oxide stabilized with 3 % yttrium (3YSZ). The dark gray area is the nickel cerium gadolinium oxide anode material (Ni-CGO). Delamination of anode pieces from the anode covered with salt (white regions) and some cracking are noticeable.

The right panel of Figure 2.19 shows two EDX spectra, which were recorded from the outer boundary of the sample (EDX surface, red spectrum) and from the inner part of the sample (EDX inside, green spectrum). The spectrum covers the X-ray energy range from about 0 to 8500 eV and shows the emission lines Ni, Zr, Ce, O, to a lesser extent the Gd-emission but also the K-emission line of potassium K and chlorine. The two latter ones are more abundant on the EDX surface spectrum and give account of the exposure of the KCl evaporation.

Where the outer layers are peeled off, the KCl traces also fell off and, therefore, are not detected in large amounts. This is the morphological observation of degradation, which causes the performance degradation of the SOFC. The latter is typically observed during operation of the SOFC, in particular, by its electrical characteristics. The impedance spectra in Figure 2.20 give account of this electrical degradation.

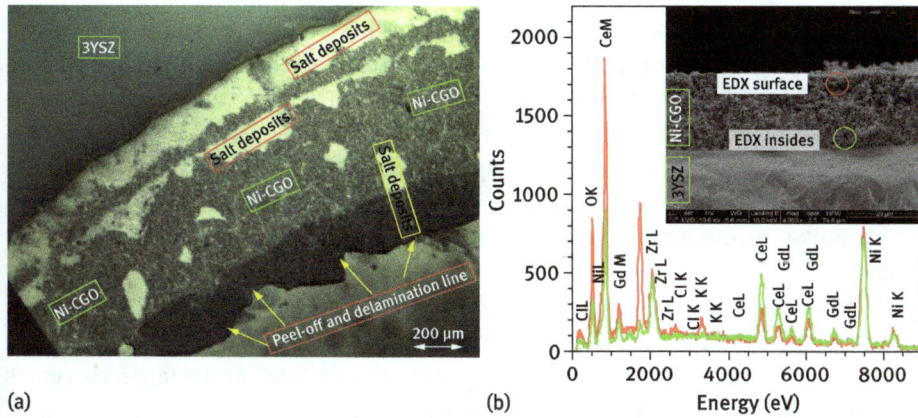

(a) (b)

Figure 2.19: (a) Micrograph of SOFC electrode assembly after exposure to salt vapor. Scale bar on top left is 200 micrometer [Nurk 2011]. (b) EDX spectra collected from the fracture cross-section of the cell assembly exposed to low salt concentration. The red spectrum was collected from the near anode surface region, as indicated by the SEM micrograph inset, showing a clear peak at the energy for the potassium K emission line around 3400 eV and noticeable chlorine K and L emission lines at 2620 and 200 eV, respectively. Reprinted from J. Power Sources, 196/6, Nurk, G., Holtappels, P., Figi, R., Wochele, J., Wellinger, M., Braun, A., Graule, T., A versatile salt evaporation reactor system for SOFC operando studies on anode contamination and degradation with impedance spectroscopy, 3134–3140. Copyright (2011), with permission from Elsevier.

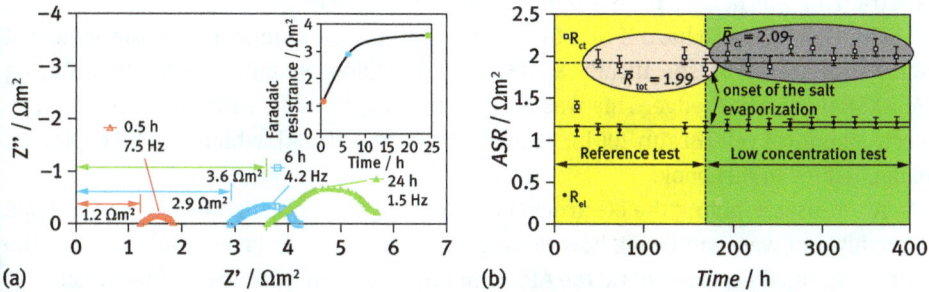

(a) (b)

Figure 2.20: (a) Nyquist plots for Ni–CGO | 3YSZ |LSMO single cell after 0.5, 6 and 24 h exposure in the fuel stream with high approximately 3500 ppm KCl concentration in 900 °C and 0.7 V. The inset shows the Faradaic resistance as a function of time. (b) Single cell total charge transfer resistance (Rtot) and high frequency ohmic resistance (Rel) dependence on time for Ni–CGO | 3YSZ |LSMO system, without (first 190 h) and with low concentration KCl vapor–aerosol impurity (6 ppm) in the fuel gas mixture in 900 °C and 0.7 V. Reprinted from J. Power Sources, 196/6, Nurk, G., Holtappels, P., Figi, R., Wochele, J., Wellinger, M., Braun, A., Graule, T., A versatile salt evaporation reactor system for SOFC operando studies on anode contamination and degradation with impedance spectroscopy, 3134–3140. Copyright (2011), with permission from Elsevier.

2.2.3 Soft X-ray emission spectroscopy

One advantage of hard X-rays is that they can be handled with ease in as far as they need no high vacuum environment. Because of the higher penetration of materials, hard

X-rays need more radiation shielding for a safe working environment. This is one trade-off. Soft X-rays have a short penetration depth and, therefore, need typically vacuum, high vacuum, or even ultrahigh vacuum environment. The intermediate range where, e. g., we have recorded the Cd and S spectra, is called "tender X-ray range." Having a short penetration depth means therefore that spectra from soft X-rays are surface sensitive. This can be an advantage or a disadvantage. We will learn later that it is good when one has bulk-sensitive and surface-sensitive methods available, particularly for battery studies.

Another aspect is the signal detection mode of the X-ray method. Soft X-ray methods come typically with a photodetector, which measures the fluorescence signal, or with an electron channeltron, which is the surface-sensitive method, or the target current is measured, which is a bulk-sensitive detection method.

One problem with soft X-rays is that insulating samples or samples, which are supported on electric insulators do not allow for target current of electron yield detection of spectra. The XANES study on CdSe films in Chapter 1 had a similar problem, which needed to be overcome. The CdSe films are not necessarily well conducting. We therefore contacted the surface of the CdSe films with sticky carbon tape, which is electrically well conducting.

In the current study, a different approach had to be taken to record spectra of a special suite of samples. A photodiode detector can still allow for recording of meaningful spectra. This was helpful for a project which I explain here.

I had performed a hard X-ray study on an operating lithium battery at the advanced photon source (APS), specifically XANES and anomalous small angle X-ray scattering (ASAXS) at the Mn K-edge. This was an *operando* experiment with an *in situ* spectro-electrochemical cell [Braun 2001a, Braun 2001f, Braun 2003a], which will be explained in detail later in this book.

After disassembling the electrochemically cycled cell, I had recovered the cathode assembly and was wondering how homogeneous the cathode layer would be after the aforementioned experiments at the APS. One way to determine this would be by "slicing" the electrode into thin layers. This seems to be not so trivial, however, since the cathode layer is around 50 micrometers thin. Moreover, it is not a compact monolith, but a porous composite material.

The procedure for slicing is as follows. The cathode is deposited on a thin aluminum metal disk-shaped foil of 25 micron with the 50 micron thick cathode layer on it. The adhesive tape (I used a Kapton® tape, which is normally abundant at X-ray laboratories and synchrotrons) with readily known mass (in milligrams, by weighing it on a microbalance) is firmly pressed onto the cathode layer and then removed from the aluminum foil. This adhesive tape with cathode layer is termed "sample 0," whereas the Al foil with residual cathode material on top is termed "sample 10." These are the two "ends" of the sequence of samples obtained by slicing.

Then another Kapton® tape with known mass (you have to cut the Kapton® tape in pieces beforehand and weigh them individually on the balance) is pressed onto the cathode layer sticking on the adhesive tape "sample 0," and gently removed again. This is then "sample 1," which contains a layer of cathode material, the mass of which is determined by weighing on a balance and comparing with the mass from the adhesive tape before pressing. Another adhesive tape with known mass is pressed on "sample 0," and thus "sample 2" is obtained. This is repeated until we have "sample 9."

The set of samples can be arranged as shown in Figure 2.21, which is a photocopy of my laboratory record book page on which, after all experiments were done, the Kapton® tapes, with the sample powder underneath, were stuck. Before that, the tapes with samples were subject to soft X-ray photoemission spectroscopy with a photodetector. Measuring in electron yield mode or sample current mode is not easily applicable because the sample material sticks on the nonconducting Kapton® tape. Back then, the Kapton® was the kind of tape, which could be handled very easily and which was also very light in weight. Since the powder sticking on the tape has a very lightweight, the tape should not significantly weigh more than the powder.

The spectra were recorded at the Advanced Light Source in Berkeley at Beamline 8 in ultrahigh vacuum conditions. Hence, the data are *ex situ* spectra. Figure 2.22 shows five representative spectra out of the 11 samples shown in Figure 2.21. We observe the Mn $2p_{3/2}$ and $2p_{1/2}$ doublet in the X-ray energy (binding energy) range from 635 to 665 eV. We observe a slight shift of spectral weight between sample 2 and sample 11, the two most extreme sample positions along the cathode thickness.

The emission spectra in Figure 2.22 originate from the occupied Mn 3d states and constitute transitions from Mn 3d to Mn 2p states. The fine structures have their origin in orbital and spin details, but the resolution is not sufficient for quantitative analysis beyond determination of chemical shift and relative spectral weight. The spectra for positions 2 and 5 are at lower energies than the other spectra. In photoemission spectroscopy, a shift to lower X-ray energy indicates a chemical oxidation, and a shift toward higher energy indicates a chemical reduction [Kanamura 1992]. For the quantification of the chemical shift of emission spectra, and to relate them with the specific properties, we determine the first statistical moment $\langle E \rangle$ of the spectral intensity by a weighed integration [Visser 2001, Braun 2002d, Solarska 2010d] in the energy range from 635 to 665 eV.

The manganese ions in $LiMn_2O_4$ are prevalent in two oxidation states, i. e., Mn^{3+} and Mn^{4+}. When we charge the cell, major part of the Li is electrochemically extracted from the spinel lattice. Therefore, the Mn^{3+} becomes oxidized to Mn^{4+}. At some arbitrary state of charge, the cathode material is written stoichiometrically as $Li_{(1-x)}[Mn^{3+}]_{(1-x)}[Mn^{4+}]_{(1+x)}O_4$. Concentration of Li^+ and Mn^{3+} should therefore evolve in the same way. Compare this with the spectrum of $Li_2Mn_2O_4$ in Figures 4 and 5 in [Song 2010]. In this highly lithiated sample, Mn exists as Mn^{3+}. The sharp peak at the L_α line is enhanced compared to our sample. The two spectra of the samples from positions 10 and 11 have a particularly low intensity feature at this energy position. This confirms that Mn in the sample is predominantly Mn^{4+}. This reaction should be more pronounced at the current collector of the electrode, because the resistance of the

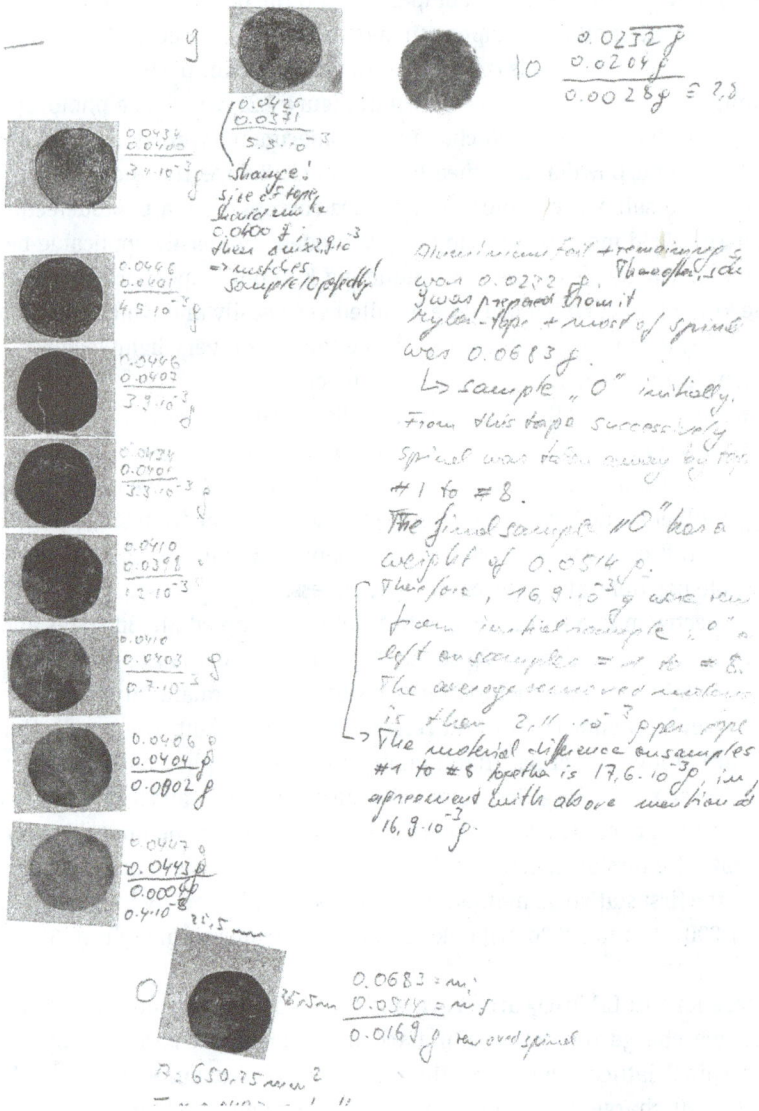

Figure 2.21: Photocopy of a sheet of paper in the laboratory record book with an arrangement of clipped translucent Kapton® tape samples with LiMn$_2$O$_4$ powder sticking on them after finishing of a battery operando experiment. This is from one battery sample sliced into 11 samples from consecutive thickness ranges. Numerals next to the tapes denote the weight of the tape before sticking power on it and after sticking powder on it and the weight difference.

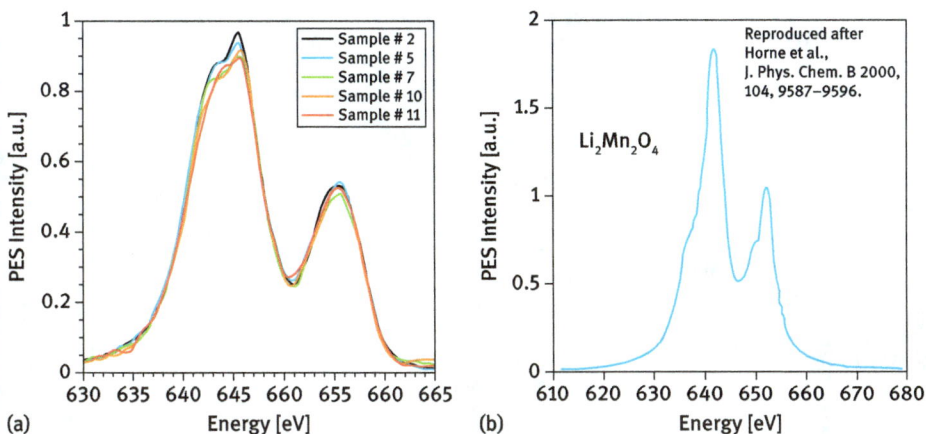

Figure 2.22: (a) Soft X-ray Mn2p emission spectra of cathode recovered from a cycled lithium battery cell, powder samples of which sliced on sticky Kapton® tape according to Figure 2.20. (b) Comparable spectrum from $Li_2Mn_2O_4$ sample obtained at the same beamline end station, reproduced from [Horne 2000]. Reprinted (adapted) with permission from (Horne, C. R., Bergmann, U., Grush, M. M., Perera, R. C. C., Ederer, D. L., Callcott, T. A., Cairns, E. J., Cramer, S. P., Electronic structure of chemically-prepared LixMn2O4 determined by Mn X-ray absorption and emission spectroscopies. Journal of Physical Chemistry B 2000, 104 (41), 9587–9596.). Copyright (2000) American Chemical Society.

pathway to the current collector is lower there. At the separator, this reaction should take place slowly. We therefore expect that after charging the current collector should face a higher concentration of Mn^{4+} than the separator. This scenario is confirmed by experiment, as is schematically displayed in Figure 2.23. The sketch shows as figure background the lateral cross-section of one-half cell, specifically the Al current collector on the left side, which meets the spinel electrode at the position $x = 0$. In the center, the spinel electrodes ranges from $x = 0$ to $x = 50$ μm, where is the sample top and where it faces the separator. The first moment $\langle E \rangle$ of the spectra is plotted on the abscissa, which is directly converted to relative Li^+ concentration versus sampling distance z. This has been performed for a number of samples. We are thus able to sketch the Li^+ concentration versus cathode depth position, as shown in Figure 2.23. This experiment then leads us to the conclusion that the highest Li^+ concentration should be at the separator side. With the exception of the minimum Li concentration 7 micrometers before the current collector, the general variation corresponds to the one obtained by multiscale simulation and modeling [Hellwig 2011, Hellwig 2013, Kupper 2015]. Inscribed in Figure 2.23 is also the mathematically modeled lithium concentration of a $LiFePO_4$ battery, which follows very well the data points which were experimentally obtained on $LiMn_2O_4$ 10 years prior [Braun 2002d] to the simulation [Hellwig 2011, Hellwig 2013]. This modeled $LiFePO_4$ battery had a total thickness of 150 micrometers, with 80-micrometer cathode thickness, 20-micrometer separator thickness, and 50-micrometer anode thickness.

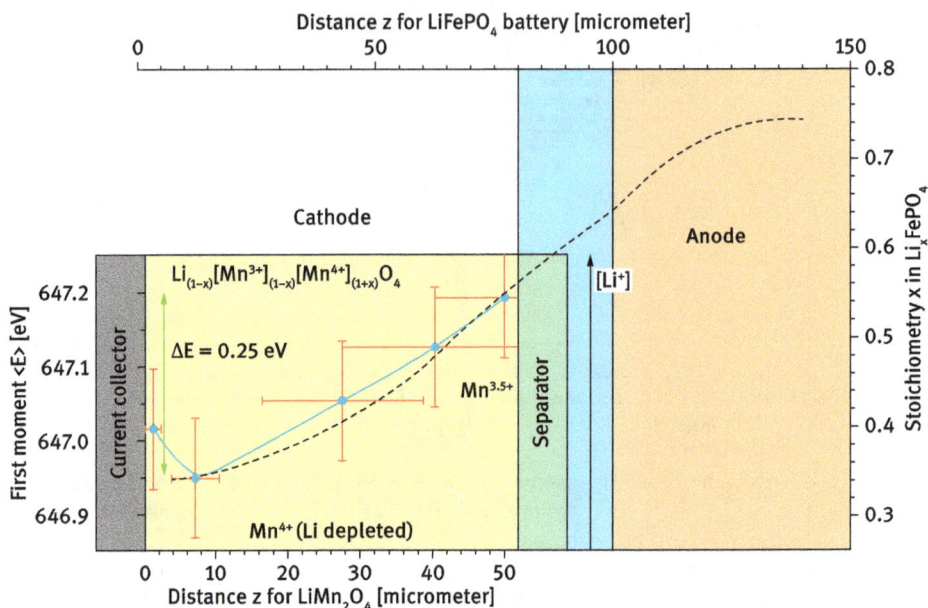

Figure 2.23: Schematic of a battery half cell with nominal 52 micrometer thickness active cathode, with inscribed first moment $\langle E \rangle$ of the Mn 2p emission peak vs. the sample position, blue data points with error bars. The blue solid line is a guide to the eyes [Braun 2010a]. The data were obtained from the spectra in Figure 2.22 from the samples shown in Figure 2.21. Highest lithium concentration is found at the separator side. The black dashed line is obtained from digital tracing the multiscale modeling and simulation of a LiFePO$_4$-based lithium-ion battery [Hellwig 2011, Hellwig 2013], where the distance z of the battery assembly was renormalized from an 80-micrometer cathode to a 52-micrometer cathode thickness.

There are numerous soft X-ray photoemission studies on battery materials. One of the early works of soft X-ray emission spectroscopy can be found in the paper by Grush et al., which deals with biologic and inorganic Mn samples [Grush 1996]. Following the work by Grush et al., Horne et al. [Horne 2000] made such investigations with XAS and XES for battery materials. They looked at LiMn$_2$O$_4$ and its chemically delithiated and lithiated derivatives, specifically λ-MnO$_2$ and Li$_2$Mn$_2$O$_4$ and also Mn^{3+} spinel model compounds. The purpose of their study was to comprehend associated changes in the molecular and electronic structure when lithium manganate is used in a lithium battery. The 2p emission spectra confirmed changes in covalency upon lithium concentration changes, and they were able to derive conclusions on the Li$^+$ diffusion coefficients and electronic conductivity mechanisms.

Hollmark et al. [Hollmark 2010] investigated the charge state behavior of LiNi$_{0.65}$Co$_{0.25}$Mn$_{0.1}$O$_2$ with hard X-ray XAS and soft X-ray emission (RSXE). They find dramatic changes in the hybridization, which is evident in XAS and in particular in the resonant soft X-ray emission at the O K edge. They have attributed this to strong screening of Li ions between the oxide layers. We will learn later how resonant X-ray studies are used for contrast variation and enhancement.

Jimenez-Mier et al. [Jimenez-Mier 2004] studied Mn metal, MnF_2, MnO, and $LaMnO_3$ with respect to their Mn 3s-2p X-ray emission after resonant and non-resonant photon excitation of a 2p electron and found differences in terms of ionic and covalent properties of the samples. They compare absorption and emission spectra and calculated spectra. We learn from such studies like this one that "oxidation state" and "valence" are actually a rather primitive concept of the electronic structure or molecular structure of an element. The fine structures of the spectra reflect delicate situations which the ions experience in intimate neighborhood to their counter ions. The discussion of such model compounds provides then the basis for discussing, e. g., battery materials which undergo cycling.

When the X-ray excitation energy is chosen close to an absorption threshold of an element, the absorption and emission spectrum of a compound contain a resonant enhanced signature from the specifically excited elements. This helps for the generation of an element specific contrast. Agui et al. [Agui 2005] applied Mn $L_{\alpha,\beta}$ resonant X-ray emission on $La_{1.2}Sr_{1.8}Mn_2O_7$. This method helped to investigate the Mn^{3d} d-d and charge transfer excitations. They carried out multiplet simulations and found that Mn^{3+} reproduced the experimental spectra best. This was consistent with the scenario that the $3d_{x^2-y^2}$ orbital was dominantly occupied, which would probably be the origin for a phase separation.

Kurmaev et al. [Kurmaev 1999] did XAS, XPS, and XES studies on $Pr_{0.5}Sr_{0.5}MnO_3$. We remember this compound from Chapter 2 in the context of SOFC cathodes. They carried out XES resonant to the Mn 2p and O K_α energies. They complemented their experimental studies with first principles calculations, i. e., local spin density approximation (LSDA) band structure calculations, based on which they found that the Mn3d states are localized near the top of the valence band. They found also differences in the XES and XPS spectra. These differences were not resulting from the investigated material, but because of the algebraic spin selection rules, which are different for XES and XPS. This is one example for the power and validity of X-ray and electron spectroscopy theory.

So far, we have looked into metal oxides for battery and fuel cell cathodes. An early carbon K_α emission spectroscopy study was done by Kurmaev et al. [Kurmaev 1986] in 1986. They showed that their technique can be used for the assessment of the bonding of carbon in materials and that the spectral characteristics are sensitive to differences in the material.

Fedorovskaya et al. [Fedorovskaya 2014] built supercapacitor electrodes from carbon nanotubes, which were catalytically deposited on silicon substrates by a chemical vapor deposition method. The catalyst is iron based and remains partially in the carbon electrode structure. They ran their supercap electrode in sulfuric acid and found with XPS that the iron forms sulfate compounds at the surface with Fe^{2+} or Fe^{3+} during electrode charging and discharging. Galakhov et al. [Galakhov 2013] dealt with a similar material, where iron was encapsulated by carbon shells. Their resonant X-ray emission spectra were compared with spectra simulated by density functional calculations, with an assumed graphene layer structure as input parameter for their calculations. Furthermore, they investigated the dispersion of energy bands of their "Fe@C" structure with carbon K_α resonant XES. The soft X-ray emission spectra of diamond-like carbon were presented by [Saikubo 2005].

A study on potassium intercalated graphite by Simunek et al. [Simunek 1988] reminds of its utility for the understanding of lithium ion intercatalation in battery anodes. They carried out angle resolved carbon K_α XES on stage-1 intercalation potassium graphite C8K, and on pure graphite. They accompanied their experimental study by ab initio calculations. They found that the intercalation creates a new pz-like valence state near the Fermi energy. Interestingly, they conclude that the conventional terms of ionicity, charge transfer, degree of hybridization, and electron occupancy cannot adequately describe the chemical bonding between carbon and the alkaline metal potassium.

Ilkiv et al. [Ilkiv 2013] have used ultrasoft X-ray emission spectroscopy (USXES) for the determination of the eletronic structure of graphene nanosheets synthesized by the reduction of oxidized graphene. These carbon sheets had a corrugation, which posed a signature on the fine structure of the C $K\alpha$ emission bands, as revealed by USXES. This corrugation was caused by overlapping of the π-orbitals and formation of mixed $(\sigma + \pi)$-states.

The method is not limited to specific elements. Amano et al. [Amano 2012] investigated specifically the nitrogen entrapped in carbon nanostructures.

Augustsson [Augustsson 2004] conducted his doctoral thesis on the soft X-ray emission of liquids and lithium battery materials. In one of his papers, Augustsson et al. [Augustsson 2005] used soft X-ray XAS (NEXAFS) and XES to view the changes in the electronic structure of Li_xFePO_4 during delithiation. The extraction of the lithium causes stronger hybridization between the Fe 3d and the O 2p states.

The work of geochemists and environmental scientists can be very inspiring for electrochemists. This is because the former often study minerals or materials under environmental influence, such as like exposure of minerals to gases, liquids, temperature, and biological species. Solid–liquid interface studies are relevant for electrochemists and for geochemists. Bluhm et al. [Bluhm 2006] wrote a review article on the various available synchrotron soft X-ray based microscopy and spectroscopy methods for the investigation of materials in contact with liquids and gases and bioorganic systems. X-ray emission is also a matter of interest in astronomy [see, e. g., [Maurellis 2000]].

2.3 X-ray photoelectron spectroscopy

2.3.1 Core level spectroscopy

Unlike X-ray absorption and X-ray emission spectroscopy, X-ray photoelectron spectroscopy (XPS) is a method, which has made it to commercially available instruments at relatively low cost for conventional laboratory use. Photoemission spectroscopy is frequently used synonymously for XPS ("PhotoEmission spectroscopy is an old experimental technique still in extensive use," wrote [Hüfner 1995]). They are particularly used for electron spectroscopy for chemical analysis (ESCA). Technically, a monochromatic light source or X-ray source is used, which excites the sample which is hosted in an ultrahigh vacuum (UHV) chamber. As a result, photoelectrons are generated in the sample, which escape from the sample through the UHV into the electron channeltron detector. These electrons have different kinetic energy E_{kin} and are collected and sorted in the channeltron accordingly. The excitation energy E_0 minus the kinetic energy E_{kin} yields the binding energy, along which the XPS intensities are recorded as XPS spectrum. The elements in the sample under investigation cause the core level spectra, provided their transitions are at lower energy then the excitation energy. A standard XPS textbook was authored by Stephan Hüfner [Hüfner 2003]. A textbook on photoemission was edited by Cardona and Ley [Cardona 1978].

The X-ray sources are typically light bulbs containing Mg (Mg K_α line, $h\nu$ = 1253,6 eV), Al (Al K_α line, $h\nu$ = 1486,6 eV), HeI or HeII (21.21 eV and 40.82 eV), the two latter of which are used for UPS, together with Ne and Ar. It makes a quantitative difference when a monochromator is used. This method will help getting sharper features in the core level spectra and also in the valence band spectra [Sherwood 1997]. Certainly, the availability of synchrotron radiation allows now to tune the excitation energy from the eV-range (UV-vis) to the hard X-ray energy range (HAX-PES).

The probe which is detected in XPS is photoelectrons. The mean free path of electrons is very low in solids and in fluids, very few nanometers only [Duke 2003, Seah 1979], as shown in Figure 2.24. This has three practical implications for the experiment. First, the generated photoelectrons which make it to the sample surface will originate only from a subsurface layer of few Ångstrom only. Therefore, XPS is a distinct surface-sensitive method. Second, the ambient environment will absorb the photoelectrons, which have made it from the subsurface to the surface of the material. Only in an ultrahigh vacuum (UHV) will it therefore be possible to guide the photoelectrons from the sample surface to the electron detector. The electron detectors are typically channeltrons (reference to Specs®, Scientia®, and others). The third implication is that XPS measurements cannot be done on a sample, which is in contact with a gas or with a liquid. *Operando* and *in situ* experiments on catalysis and electrochemistry are therefore by default not possible. Notwithstanding, researcher have tried from the early beginning of XPS technology (ambient pressure and near-ambient pressure XPS, AP-XPS, NAP-XPS) to improve pumping and detection. Since about 15 years from now, there has been considerable progress in this matter, which I will present in Part 8 of this book.

Figure 2.25 shows an XPS survey spectrum of a porous tungsten photoelectrode oxide film, doped with 5 at. % tin Sn [Solarska 2010a]. The intensity peaks in the spectrum signify the abundance of chemical elements detected on the photoelectrode surface and to some extent also underneath. We note strong spectral signatures from car-

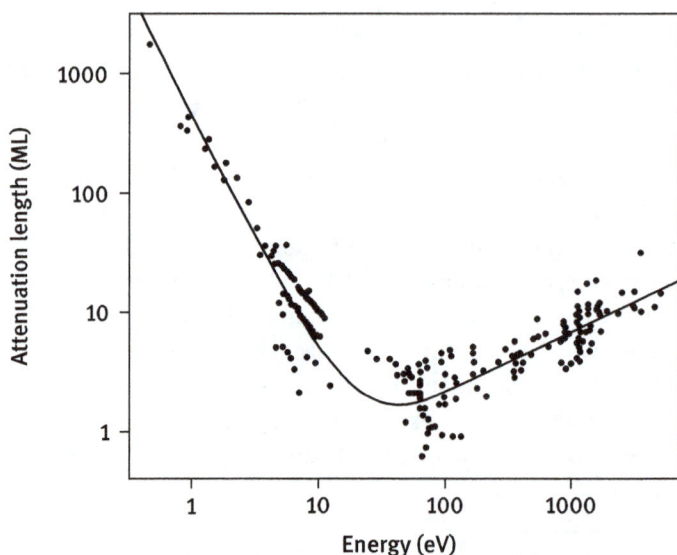

Figure 2.24: Attenuation lengths of electrons in solids as a function of their energy. An early compila-
tion of a variety of experimental data are given by the dots. An interpolation formula is shown by the
solid line. Reprinted from Surface and Interface Analysis, 1,1, 1979, Seah and Dench, Quantitative electron spectroscopy of surfaces: A standard data base for electron inelastic mean free paths in solids, 2–11,
Copyright (1979), with permission from Wiley [Adapted from Seah and Dench [Seah 1979] with permission;
[Duke 2003]].

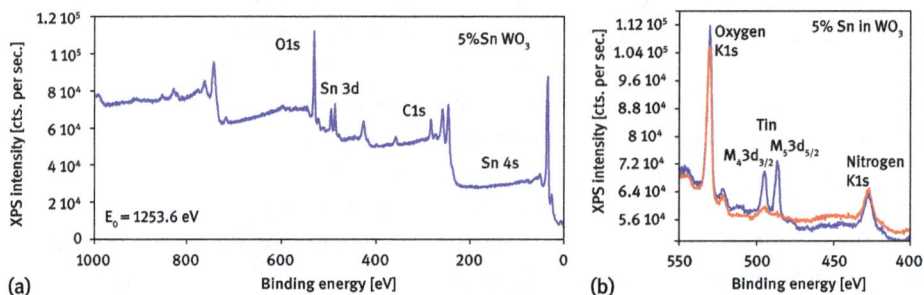

Figure 2.25: (a) XPS survey spectrum of WO_3 doped with 5% Sn, covering the detection range from 0 to
1000 eV. The excitation energy was 1253.6 eV. (b) Magnification of the survey spectrum in the range 550 eV
to 400 eV, highlighting the oxygen, tin, and nitrogen core level peaks. The undoped WO_3 has no relevant tin
peak intensity at around 500 eV, as indicated by the pink spectrum [Solarska 2010a].

bon at around 285 eV binding energy. This is most likely carbon from CO_2 and potentially
from CO adsorbed from the ambient environment on the film surface. Virtually every
sample exposed for some time to ambient conditions will exhibit the spectral features
from these adsorbents. The same holds for adsorbed oxygen and water and nitrogen.
The right panel of Figure 2.25 is a magnification of the spectrum in the range from 550

to 400 eV, which shows the well-known oxygen core level peak and also the Sn M4 and Sn M5 doublet in the spectrum of the Sn-doped sample. The nondoped sample shows no such strong Sn peaks, but a slight hump at the same energy, which could be an indication of Sn impurities from the FTO coated glass substrate (FTO = fluorine doped tin oxide). When you compare the positions of the peaks in this spectrum with tabulated values for the core level energies ([Thompson 2009] X-ray Data Booklet), you will note that they differ by around 6 eV. The oxygen peak noticeable at 528.8 eV is actually supposed to be at 543.1 eV according to the tabulated data from Henke et al. [Thompson 2009]. Therefore, the measured spectrum has to be shifted by the difference of this 14.3 eV shift. It is a common routine to carry out such correction. Typical spectral anchor points are, e.g., the carbon peak at 285 eV. Frequently, a spectrum of a gold metal sample is recorded in order to determine the position of the Fermi level, where the binding energy by default is $E = 0$.

The calibration of the energy axis as such is not always trivial. Noble metals with little reactivity or metal oxides, or other compounds which, contain metal ions with stable oxidation state are suited for calibration of the energy axis, as already mentioned in Chapter 1. We will learn in this book also that some seemingly "stable" compounds are not suited for such energy calibration in XPS measurements. This is because the surface of such compounds likes to interact with oxygen and nitrogen in humidity, and thus the oxidation state of the related metal deviates from the nominal stoichiometry.

Figure 2.26 shows magnified the core-level spectrum of tungsten. It shows the doublet comprising the N_4 $4d_{3/2}$ N_5 $4d_{5/2}$ states. These W4d peaks experience some chemical shift to higher binding energy for the 5 % doped samples. This is an indication that tungsten gets slightly reduced when doped with a meliovalent species. For the quantification of this chemical shift, we determine the first statistical moment $\langle E \rangle$ of the spectral weight [Braun 2002d] for the N_5 $4d_{5/2}$ peak from 240.97 eV–253.19 eV, and for the N_4 $4d_{3/2}$ peak from 253.19 eV–264.58 eV.

$$\langle E \rangle = \frac{\int I(E) \cdot E dE}{\int I(E) dE} \quad \langle E \rangle = \frac{\int_{88 \, eV}^{97 \, eV} I(E) \cdot E \cdot dE}{\int_{88 \, eV}^{97 \, eV} I(E) \cdot dE} \tag{2.1}$$

The shift for the WO_3 sample is set 0 by definition. The largest chemical shift is found for Sn-doped WO_3 (0.027 and 0.015 eV). While a complete quantitative analysis is challenging because of the small changes and the XPS instrumental resolution limit, it is interesting to note that the chemical shift observed is, in general, systematic with the valence of dopant atoms, i.e., Mo^{5+} has the smallest chemical shift because of its relatively high oxidation number close to W^{6+}, whereas Si^{4+} and Sn^{4+} with the lower oxidation numbers have a larger chemical shift (Figure 2.26, right).

The spectra shown in Figures 2.25 and 2.26 were recorded at the high performance XPS Beamline 9.3.2 at the Advanced Light Source in Berkeley. You see in Figure 2.27 Selma Erat checking for the sample position in the UHV chamber before running the actual XPS scans on tungsten oxide films deposited on glass substrates.

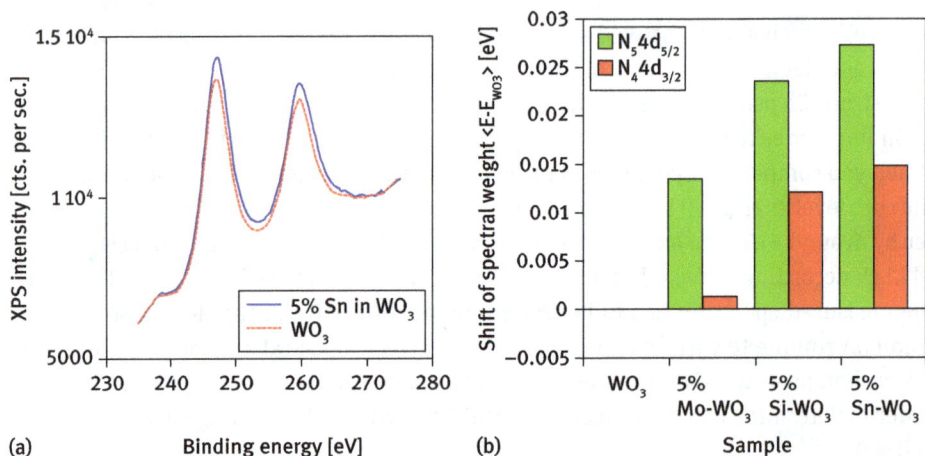

(a)   Binding energy [eV]   (b)   Sample

Figure 2.26: (a) Tungsten N_4 $4d_{3/2}$ N_5 $4d_{5/2}$ core level spectra of pure WO_3 (thin red line) and 5% Sn-doped WO_3 (thick blue line). (b) Shift of the first statistical moment $\langle E \rangle$ position of the N_4 and N_5 core level peaks of 5% Mo, Si, and Sn-doped WO_3 with respect to position of WO_3 (by default 0). The first moment was determined after equation (2.1).

Figure 2.27: ETHZ/Empa PhD student Selma Erat (now Professor at Mersin University, Türkiye) inspecting sample position for XPS measurements at end station 9.3.2 at the Advanced Light Source, Berkeley (2009) [Erat 2010a].

We have seen that absence and presence of oxygen in the tungsten oxide crystal lattice has influence on the band gap energy, and thus on many other properties. The manipulation of band gaps is one field of activity where material engineers engage. Note, e. g., that it can make a significant difference in functionality of a photoelectrode or photocatalyst when you add the Mo to the bulk, as was done here [Solarska 2010a], or to the surface, as was done by Bär et al. and Gaillard et al. [Baer 2010, Gaillard 2010].

Titanium oxide is still considered the archetype of photocatalyst. Because of its large band gap of 3.2 eV, TiO_2 will not absorb visible light, and thus is not a material for solar applications. Partial substitution of oxygen by nitrogen in TiO_2 makes TiNO and yields materials with smaller band gaps, more suitable for solar photocatalysis. We have looked into the titanium, oxygen, and also nitrogen NEXAFS spectra of TiNO samples, which had been used for photocatalyst measurements as well [Braun 2010b]. Figure 2.28 shows the N 1s XPS spectra of a "good" and a "bad" photocatalyst from the series of TiNO samples with different N concentration. We had found a defect state in the oxygen NEXAFS spectra, which was favorable for photocatalysis properties, and this defect state coincides with a well-pronounced N 1s peak at 400 eV in the XPS spectra.

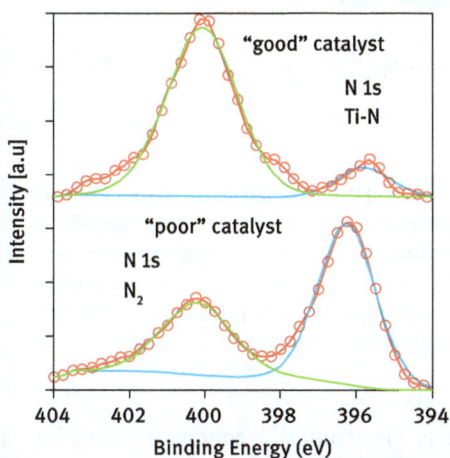

Figure 2.28: Nitrogen N1s core level spectra of TiON-based photocatalyst samples with a very good and bad performance on decomposing methylene blue dyes under UV irradiation [Braun 2010b]. The green peak is indicative of N_2, and the blue peak indicative of N in a Ti-N configuration.

Let us stay a little while more with the tungsten oxide-based materials for photocatalysts and photoelectrodes. We had coated TiO_2 single crystal substrates with tungsten oxide films by pulsed laser deposition (PLD). The stoichiometry of metal oxide films deposited by PLD depends on the deposition and processing conditions, particularly the substrate temperature and gas partial pressure. The samples we received after PLD had a blue color. Figure 2.29 shows photographs of nominally 10 nm thin and 100 nm thick films. The bluish appearance is obvious and stronger for the thick film. Stoichiometric WO_3 has a yellow color. The blue color is an indication that the tungsten oxide is substoichiometric and must be written as $WO_{3-\delta}$.

(a)

(b)

(c)

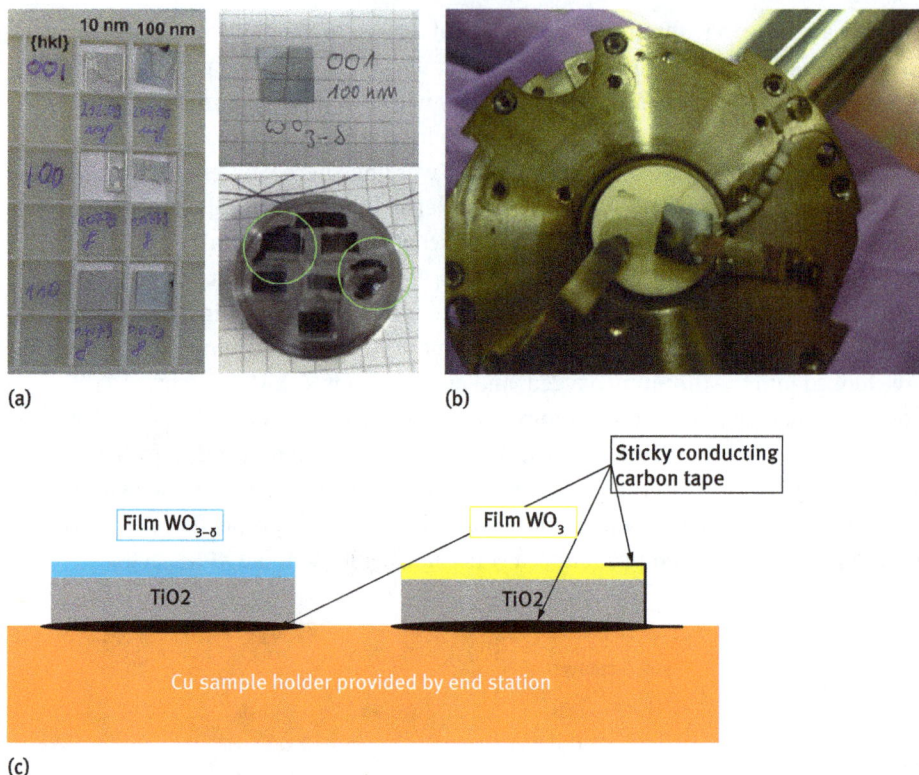

Figure 2.29: (a) Photograph of TiO$_2$ single crystals with (100) and (110) orientation coated with 10 nm (left column) and 100 nm (right column) nominal thickness tungsten oxide PLD films. Upper right panel in a) shows one 100 nm thick blue WO$_{3-\delta}$ film on (001) TiO$_2$ substrate cut into four aliquot pieces for further analyses. Lower right panel shows single crystals fixed on generic sample holder with black sticky carbon tape. (b) Two thin film samples (one transparent, one dark) clamped on heating element of the sample holder described in [Whaley 2010] as "Fixture". (c) Schematic for arrangement of single crystals on sample holder.

The substoichiometric blue tungsten oxide is electronically conducting and has a band gap energy too low to be able to perform photocatalysis or water splitting. Knowledge on the band gap energy, but also on the valence band and conduction band positions, is essential for determining whether a material or materials system is able to oxidize water (water molecule splitting). Weinhardt et al. have very accurately exercised the experimental determination of band gap and energy band positions with XPS and inverse photoemission spectroscopy for WO$_3$ [Weinhardt 2008, Weinhardt 2013].

We have investigated the electronic structure of as-deposited blue and post-treated yellow tungsten oxide films. We have therefore cut the samples in four aliquots as shown in the right of panel a) of Figure 2.29 and subjected them to heat treatment at 400 °C for 48 hours in 100 % oxygen atmosphere in a tube furnace. After this treatment, the bluish color had changed to a pale, barely visible yellow [Braun 2011e].

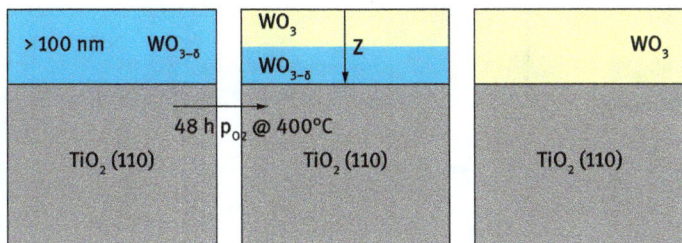

Figure 2.30: (left) Schematic stages for the thermal gas phase conversion of sub-stoichiometric 100 nm thick blue WO$_{3-\delta}$ film on (001) TiO$_2$ substrate. Exposure to oxygen at high temperature will cause oxidation of top surface towards stoichimetric yellow tungsten oxide WO$_3$, reaction front progressing in z-direction towards deeper film regions, enventually completing oxidation and turning entire film yellow.

This is illustrated in Figure 2.30, which shows what one might expect for the conversion of the nominally 100 nm thick blue WO$_{3-\delta}$ film into a yellowish WO$_3$ film when exposed at 400 °C to oxygen for 48 hours. The film should turn into stoichiometric yellow WO$_3$. We have carried out core level photoemission spectroscopy on the blue as-deposited and yellowish post-treated films at the Advanced Light Source in Berkeley [Braun 2011e]. The substrates with films are mounted on the copper sample holder with doubly sticky carbon tape, as shown in the lower right panel of Figure 2.29. The two green circles show how the films are also on their top covered with stripes of conducting carbon tape in order to prevent electric charging of the films during XPS measurements.

Figure 2.31 (upper left panel) shows the tungsten W 4f spectra of the as-deposited and post-treated tungsten oxide films. The X-ray energy for excitation was 700 eV. The range for the W 4f spectra is 32 to 42 eV. The spectrum from the as-deposited film (blue spectrum) appears broader than the spectrum from the post-treated film (yellow spectrum). The peak height of both spectra was neither calibrated nor normalized, but we take our observation as an indication of increased covalency of the blue film which is electrically more conducting. Spectra from covalent samples have often a more diffuse appearance than spectra from more ionic samples. The W 4f$_{7/2}$ and W 4f$_{5/2}$ doublet is well resolved and shows for both spectra intensity shoulders toward the high and low binding energy flanks [Braun 2011e].

One might expect that the oxidation during post-annealing has filled oxygen vacancies. This process would go along with partial reoxidation of the tungsten ions, as is indeed evidenced by the W4f core level XPS spectra. The spectrum of the post-annealed film has been deconvoluted and labeled the same way like exercised in [Bullett 1983, Bittencourt 2005]. The doublet 2 and 2* are indicative of W^{6+}, and the doublet 1, 1* is indicative of W^{5+}. Components 3, 3* supposedly arise from surface defects, which we will investigate in the section about valence band photoemission spectroscopy. The spectral weight of 1, 1*, and 3, 3* shows that the post-annealed film contains tungsten in the W^{5+} oxidation state plus surface defects, notwithstanding that W^{6+} is the dominant species. The core level spectrum of the as-deposited film has higher spectral weight at the lower

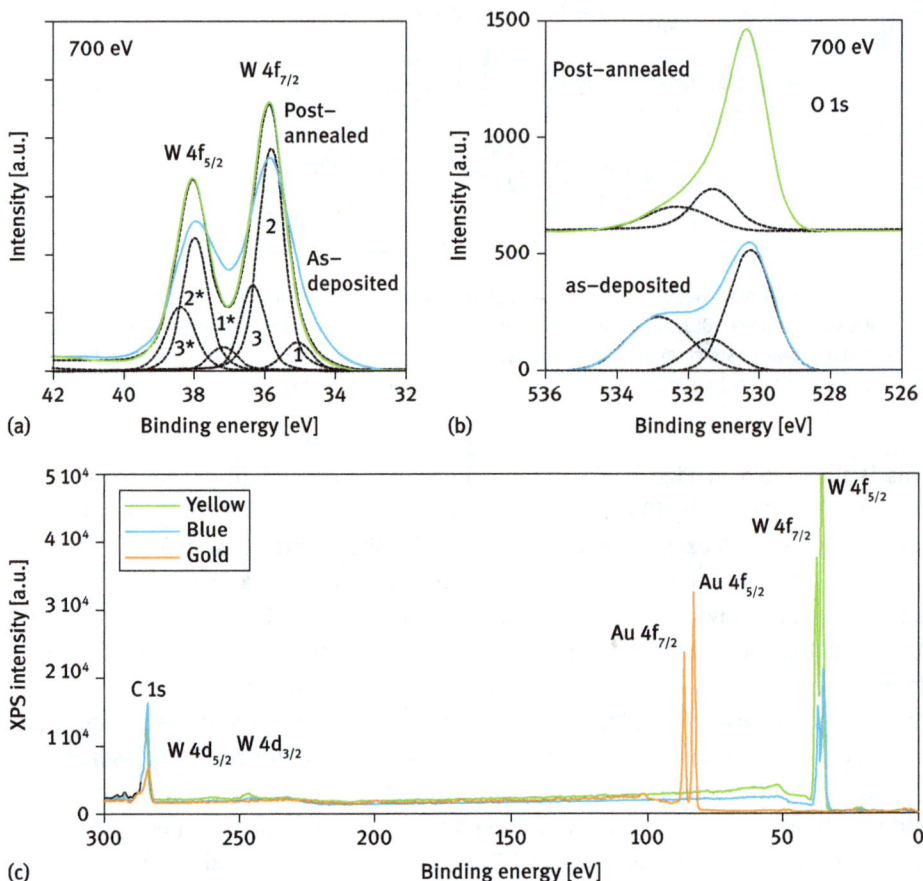

Figure 2.31: (a) W 4f spectra of as-deposited (blue) and post-annealed (yellow) 100 nm thick films. Peaks labeled 1,1* denote W^{5+}, and 2, 2* denote W^{6+}. Peaks 3, 3* originate from surface defects. (b) Oxygen 1s spectra of the as-deposited and post-annealed films (thick solid lines bottom and top), their deconvoluted Gaussians (gray dotted lines). (c) XPS survey spectrum for both films plus spectrum from gold foil for reference. Primary energy was 700 eV. "Reprinted (adapted) with permission from (A. Braun, S. Erat, X. Zhang, Q. Chen, F. Aksoy, R. Löhnert, Z. Liu, S. S. Mao, T. Graule, Surface and bulk oxygen vacancy defect states near the Fermi level in 100 nm $WO_{3-\delta}$/TiO_2 (110) films: A resonant valence band photoemission spectroscopy study, J. Phys. Chem. C, 2011, 115 (33), 16411–16417). Copyright (2011) American Chemical Society."

energies flanks of the W 4f doublet, revealing a relatively larger portion of W^{5+} in the film. There is a broadening of the spectrum toward the higher energy flanks, suggesting that the relative portion of surface defects in the as-deposited film is larger than in the post-annealed film. This suggestion is supported by the X-ray reflectometry data, which we will see in Section 3.5.1 in this book. There we will see that the surface roughness of the post-annealed film is lower than in the as-deposited film. Such surface roughness is a manifestation of disorder, and disorder in substoichiometric tungsten oxide films has been observed and assigned to reduced photocurrents [Marsen 2007].

The corresponding oxygen 1s core level spectra in Figure 2.31 (upper right panel) are virtually reminiscent to h-WO$_3$ and h-WO$_{2.8}$ in Figure 7 of [Solonin 2001]. For the as-deposited film, we identify two clear structures at around 533 eV and 530.5 eV, the latter one being dominant. The post-annealed spectrum shows predominantly only the feature at 530.5 eV, and a yet visible shoulder at 533 eV. While the two structures are clearly visible, deconvolution required actually a third structure at around 531.5 eV. Such deconvolution with three components has been made for the O 1s spectrum of a 200 nm thin WO$_3$ film obtained by electron beam evaporation [Ahsan 2009]. That study found that W exists as 4d on the surface, but at a depth of 10 nm, 4d and also 4f states were observed.

It is striking how well the core level XPS spectra of the tungsten and the oxygen reflect the changes of the molecular and electronic structure of the tungsten oxide film upon thermal gas phase treatment. We have to be aware in the case of the oxygen spectra that the oxygen, which is detected by the 700 eV X-rays, could also excite the oxygen from the TiO$_2$ substrate underneath the film. A quick check at the webpage of the Center for X-ray Optics at LBNL shows that the attenuation length of our tungsten oxide film with 7.1 g/cm^3 density is around 180 nm [Gullikson 1995, Henke 1993b]. We can only hope that the electrons excited in this depth, which already includes the TiO$_2$ substrate, are not escaping to the channeltron detector of the instrument. Figure 2.24 shows the attenuation length of below 10 monolayers (ML) for electrons with a kinetic energy of around 200 eV.

When we now look at a much thinner film of 10 nm, e. g., we have to count on the signature from the substrate to show up in the oxygen spectra. Before we look at the XPS spectra of 10 nm thin WO$_3$ samples, which are shown in Figure 2.32, let us inspect again the sketch in Figure 2.30. When the film is very thin, it should be possible that that diffusion processes in the substrate cause, e. g., Ti cations to diffuse into the tungsten oxide film. As a matter of fact, we have observed this by looking at the survey spectrum.

For thinner film, we should sooner observe a diffusing species from the substrate. We have provoked this process *in situ* during the XPS experiment by heating the sample on the sample holder at beamline 9.3.2 at the ALS in the vacuum chamber. The left panel of Figure 2.32 shows the Ti 2p XPS spectra of the 10 nm thin film coated on the TiO$_2$ substrate with (110). The spectrum recorded with the film at 623 K is the highest measured temperature where the Ti signature is not yet visible. Further heating to 673 K provides that the diffusion is more activated so that Ti cations are detected at the film surface or underneath. The X-rays will excite the Ti in the substrate but the generated electrons do not make it to the surface region where they can be detected. The 10-nm WO$_3$ film poses a barrier until the Ti cations diffuse further up. Therefore, we observe in Figure 2.32 the Ti 2p doublet. As we have described in the published paper [Braun 2012a], the effect is not observed for the 10-nm film deposited on the (001) substrate. So, apparently the orientation of the substrate has an influence on the high-temperature diffusion of Ti. We have carried out the XPS measurements with a pioneering ambient pressure system, which allows to record XPS spectra with the sample under a gas atmosphere. We have exposed the 10 nm thin tungsten oxide film to 100 mTorr oxygen pressure while heating the sample from ambient temperature to 673 K, as shown in the right panel of Figure 2.32.

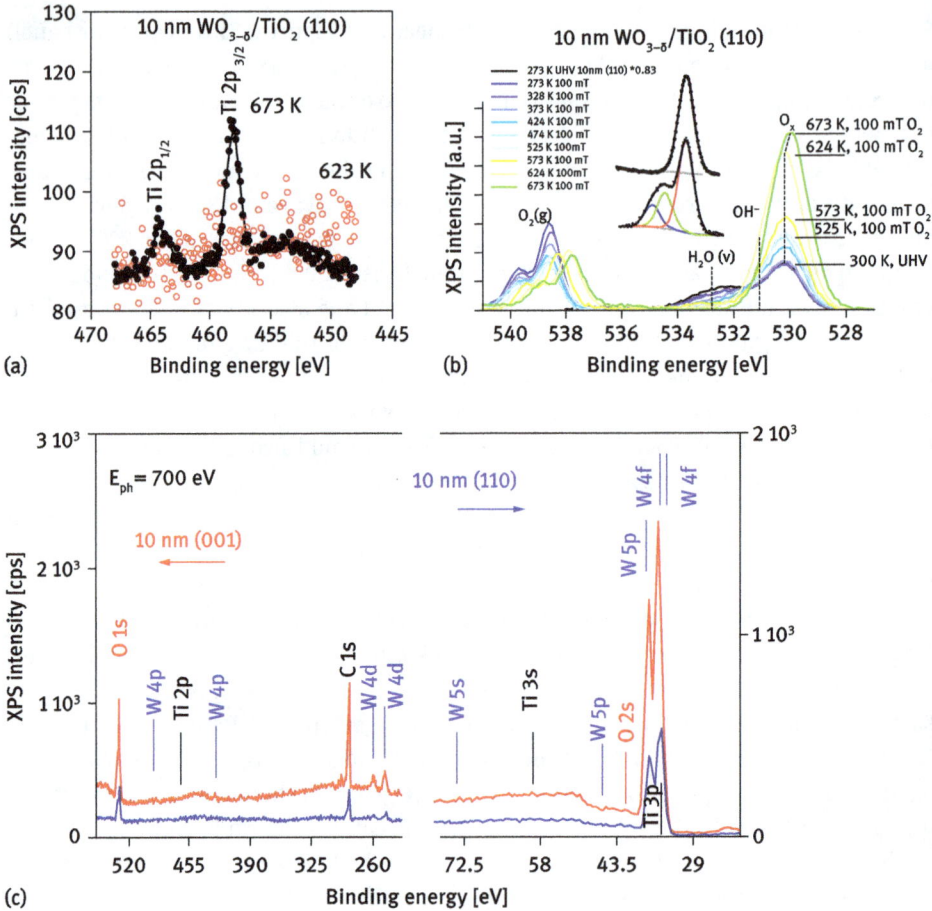

Figure 2.32: (a) Ti 2p spectra of as-deposited (blue) and post-annealed (yellow) tungsten oxide films of 10 nm on TiO$_2$ (110). (b) Oxygen 1s spectra of the as-deposited film during heating from 300 K–673 K under 100 mTorr oxygen pressure. (c) XPS survey spectrum for both films plus spectrum from gold foil for reference. Primary energy was 700 eV. Reprinted (adapted) with permission from (A. Braun, F. Aksoy Akgul, Q. Chen, S. Erat, T.-W. Huang, N. Jabeen, Z. Liu, B. S. Mun, S. S. Mao, X. Zhang, Observation of substrate orientation dependent oxygen defect filling in thin WO$_{3-\delta}$/TiO$_2$ pulsed laser deposited films with in situ XPS at high oxygen pressure and temperature, Chem. Mater., 2012, 24 (17), 3473–3480.). Copyright (2012) American Chemical Society.

At 300 K, we note a dominant peak at 530 eV from structural oxygen in the film lattice but also a strong contribution from surface water at 533 eV. At 531 eV, we have a strong shoulder from OH groups. At 538–540 eV, adsorber oxygen is measured. Upon annealing under oxygen, the water signature diminished and the peaks for structure oxygen and OH are increasing. We interpret this as the partial incorporation of water molecules into the oxygen vacancies of WO$_3$, with subsequent formation of hydroxyl groups. We know this effect from the ceramic proton conductors, as we will learn later in this book.

At this point, and at further points in this book, the reader may ask why are we seeing such study on PLD films grown on single crystal substrates and measured in UHV. This is not an electrochemical system, right? Well, this is not entirely true. Surfaces of objects in contact with a fluid, be it a gas or a liquid, will form double layers at this surface. In the extreme case, the ions of the fluid will form a chemical bond with the object, e. g., oxide or hydroxide or nitride scale formation on a metal under ambient atmospheric conditions. This is actually the normal case. Filling of oxygen vacancies is a relevant process for metal oxides when they are used as electrodes or electrolytes. It is therefore interesting to investigate the materials, such as tungsten oxide, which is a proton conductor, lithium conductor, and photoelectrode material, as ultrathin films with minimum porosity under UHV and low gas concentration conditions—and this in particular when no electrochemical bias in an electrolyte is applied. Such further conditions can be employed at a later stage of a project when the apparatus for this is available.

We have done similar studies on distinct ceramic proton conductors, specifically yttrium-substituted barium cerate and zirconate. Figure 2.33 shows the oxygen spectra of 20 % Y substituted barium cerate [Chen 2013d] while heated during XPS scanning from ambient temperature to 593 K and above. We have recorded around 500 XPS scans during this *in situ* experiment. During heating in oxygen, the peak from the crystal oxygen is decreasing, whereas the hydroxyl peak is decreasing. The relative gain of the latter

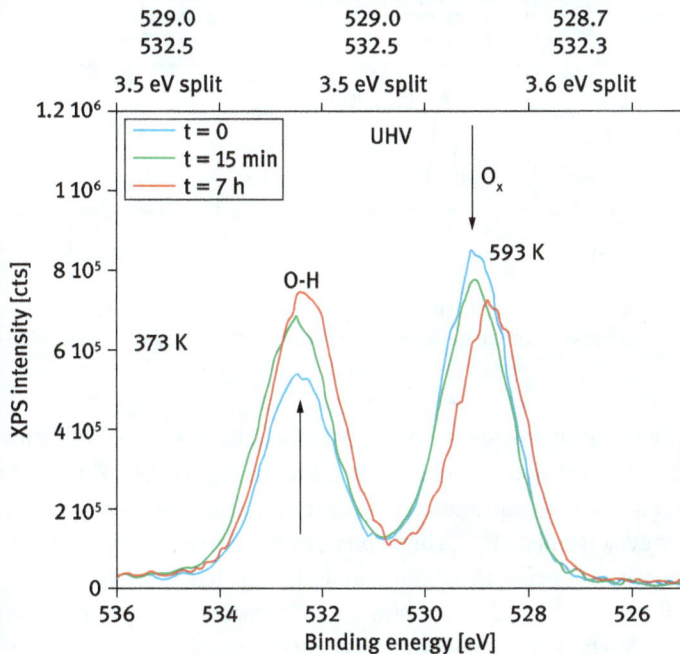

Figure 2.33: Three oxygen XPS spectra of BaCe0.8Y0.2O3 recorded during heating in UHV from 373 K to 593 K. From an overall 500 spectra, numbers 1, 15, and 485 are shown. Observe the redistribution of spectral weight from decreasing crystal oxygen toward increasing hydroxyl [Chen 2013d].

structure shows that water molecules from the ambient are being built in into the crystal lattice and the protons from water form hydroxyl bonds.

The filling of oxygen vacancies in a metal oxide constitutes chemically an oxidation of the metal central atom, in general. Oxidation and reduction can be very well monitored with X-ray spectroscopy, as we know by now, by either chemical shift or by the presence or absence of additional peaks in the spectra. Figure 2.34 exemplifies this for TiO_2, ZnO, and SnO_2 pellets, which were obtained from flame synthesis formed nanopowders. Note the analogy. The oxidized forms show an additional small peak labeled "c" in the oxygen spectra at higher binding energies [Flak 2013]. This observation is paralleled by the corresponding metal ion XPS spectra in Figure 2.35, where we have the 2p doublets for Ti and Zn, and the 3d doublet for Sn, respectively. For the observation of differences in oxidized and reduced specimens, titanium oxide is a gratifying system, whereas zinc oxide shows hardly any differences in nominally oxidized and reduced specimens.

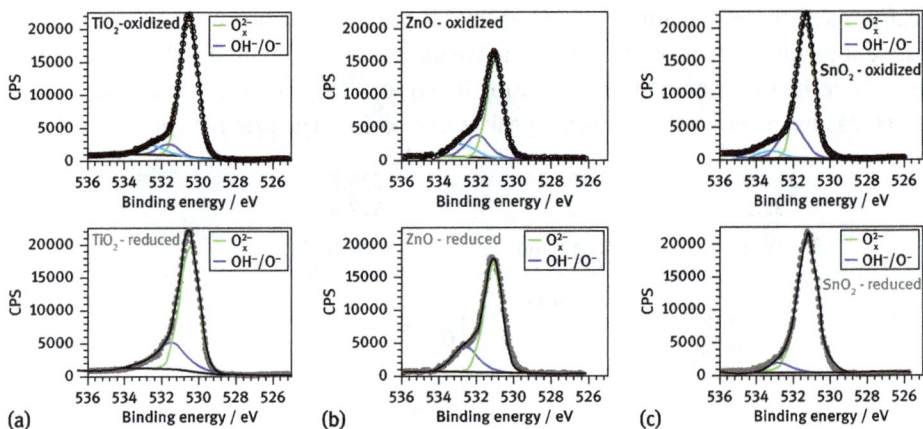

Figure 2.34: O 1s core level XPS spectra of TiO_2 (a), ZnO (b) and SnO_2 (c) of oxidized (bottom level) and reduced (top level) fitted with the Gaussian–Lorentzian function. Data courtesy of Dr. Dorota Flak, Empa/AGH Krakow.

This trend for the sensitivity becomes somewhat more obvious when we look into the corresponding metal core level spectra, which are shown in Figure 2.35. For the Ti 2p spectra, we see upon reduction a wide shoulder at lower binding energy from Ti^{3+}, which in some literature is even termed $Ti^{3.5+}$. The richness in structure can make it laborious for spectral deconvolution of peaks in Ti compounds, but the peaks are often very well resolved, and thus distinguishable. The website X-ray Photoelectron Spectroscopy (XPS), reference pages from Surface Science Western, provides many good examples for fitting of XPS spectra [Biesinger 2015]. The shoulders extend over the peaks from the oxidized sample both for the Ti $2p_{1/2}$ and Ti $2p_{3/2}$. The Zn 2p spectra look virtually identical within the limits of accuracy. It is difficult to make out spectral differences in the

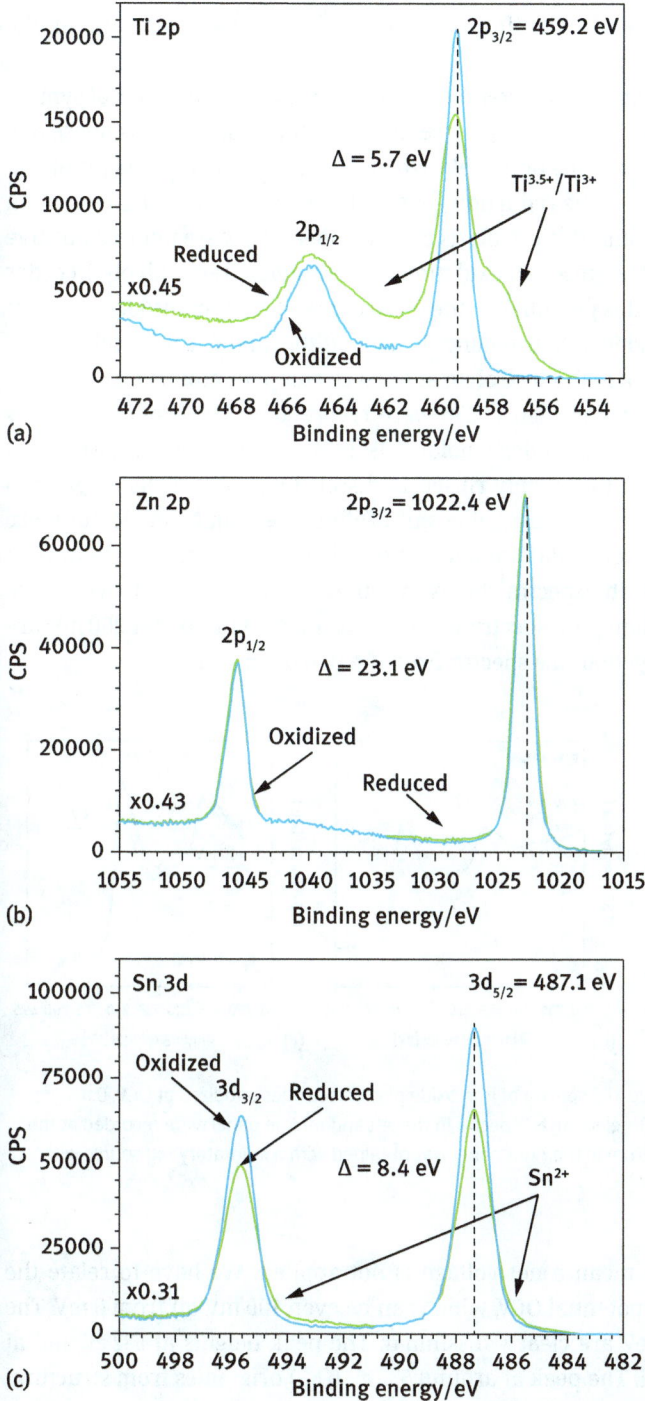

(a)

(b)

(c)

Figure 2.35: (a) Ti 2p core level spectra of TiO_2 of; (b) Zn 2p core level spectra of ZnO; (c) Sn 3d core level spectra of SnO_2. Reproduced from [Flak 2013] with permission from the PCCP Owner Societies.

oxidized and reduced zinc oxide. The tin oxide spectra show a slight difference in the widths of the peaks.

The aforementioned changes in materials were either obtained by ways of synthesis, such as doping and substitution, or by subsequent treatment such as gas phase oxidation. For electrochemists, anodization in electrochemical cells can be used for oxidation of materials, for example. The reduction goes in the reverse potential direction. In photo-electrochemical cells and in electrolysis cells, the electrodes are under a positive or under a negative bias. This causes an oxidation or a reduction, respectively. In order to investigate the impact of this potential on the electronic structure of the electrode, we used iron oxide as photoanode at bias potentials from 0, 200, and 600 mV, and an electrode which had been kept in KOH electrolyte at open circuit potential [Braun 2012c].

The oxygen 1s XPS spectra in Figure 2.36 were recorded at 635 eV photon energy from iron oxide photoanodes kept under 1 molar KOH in 600 mV bias, 200 mV, 0 mV, and open circuit (OCV), and then thoroughly rinsed in distilled water and rinsed prior to measurement. The spectra show a noticeable shift. Differences can be made out more easily here because of the better data quality or signal to noise ratio for the oxygen spectra. Close inspection of the spectra shows that they are not identical, even if we would shift and overlap them. The spectra of the films anodized at 0 and 200 mV are shifted toward lower energy than the spectra from the two other films.

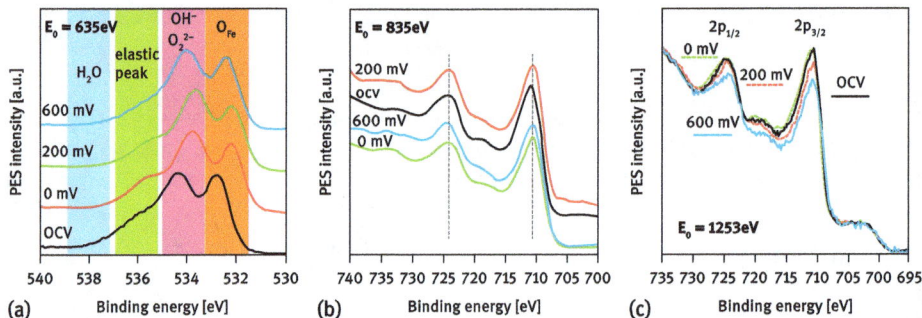

Figure 2.36: Oxygen 1s and iron 2p XPS spectra of iron oxide photoelectrodes anodized at OCV, 0 mV, 200 mV, and 600 mV in 1 molar KOH electrolyte. Spectra in the left and middle panel were recorded at the Advanced Light Source. The spectrum in the right panel was obtained with a laboratory Mg source with 1253 eV X-ray energy.

Note that "0 mV" does not mean a net voltage is not applied. We have to relate the 0 mV with the open circuit potential OCV, which can be even 100 mV off from 0 mV. The intensity shoulders at 536 eV are clearly dissimilar. The peak heights at 532 eV and at 534 eV are changing as well. The peak at around 532 eV (O_{Fe}) originates from structural oxygen in Fe_2O_3 or from FeOOH. The peak at around 534 eV typically originates from hydroxyl groups in FeOOH or from surface functional hydroxyl groups. The shoulder at 536 eV is the elastic peak mentioned. If there was any adsorbed water in our films, it would show up as a small sharp peak at around 538 eV. The spectral weight of the OH-peak at 534 eV of our films is large and must be attributed to the long exposure of the films to KOH, facilitating formation of some hydroxylated surfaces on the iron oxide. Such species can be interpreted as FeOOH in XPS or NEXAFS spectroscopy.

The right panel of Figure 2.36 shows the Fe 2p photoemission spectra of the four films recorded at the off-resonance energy of 835 eV, recorded at Beamline 9.3.2 at the Advanced Light Source. This excitation energy is lower than the one from the Mg source with 1253.6 eV. The information provided by the spectra in the middle panel of Figure 2.36 would seem to be less bulk-specific than the spectrum on the right in Figure 2.36 because photons of 1253 eV would have deeper penetration than photons at 835 eV energy. Differences are minute and can hardly be quantified. The spectra are similar to the spectra of FeOOH insofar supporting the suggestion from the relatively surface-sensitive TEY that the topmost layer of the anodized hematite films may contain FeOOH. A previous combined XAS and XES study showed that FeOOH is the major surface phase on Fe_2O_3 under alkaline conditions.

We finish this section on core level photoemission with one more example from solid oxide fuel cells (SOFC). We investigated a single crystal disk of $La_{0.9}Sr_{0.1}FeO_{3-\delta}$ as a model for fundamental investigations of the electronic structure and transport properties of iron perovskite SOFC cathode candidates [Braun 2010f].

Quick visual inspection of the survey spectra in the top panel of Figure 2.37, which were recorded *in situ* during annealing in UHV, reveals no spectacular changes. At this point, we can have a look into the X-ray Data Booklet [Thompson 2009] and see which peaks we can assign to particular elements. An excerpt of this booklet for our compound LaSrFe-oxide is shown in Table 2.5. The XPS survey scan has the purpose to survey for all chemical elements, which happen to be on the inspected sample. This survey scan certainly should show the chemical elements, which we have put in our material by synthesis. The sample may include oxygen, water, nitrogen, and carbon from carbon dioxide just from the exposure of the sample to ambient atmospheric conditions. An accidental fingerprint on the sample may cause sodium, potassium, and chlorine in the spectrum. Ball milling synthesis procedures may cause signatures from zirconia balls, and paper wipes may cause silicon signatures. During processing in a furnace, some elements may evaporate and then not show up in the survey spectrum. Note that the survey spectrum gives an account only of the surface and the near subsurface of the probed material. Diffusion processes can cause enrichment of an element at the surface with depletion of the same in the interior.

We have looked particularly into the Fe 3s spectra and plotted those recorded at 323 K, the lowest recording temperature, and at 723 K, the highest recording temperature, in the lower left panel of Figure 2.40. We note a shift of the spectrum toward higher binding energy upon annealing. We have quantified this shift by calculating the first statistical moment for the peak position, $\langle E \rangle$. We have done so for the spectra measured at

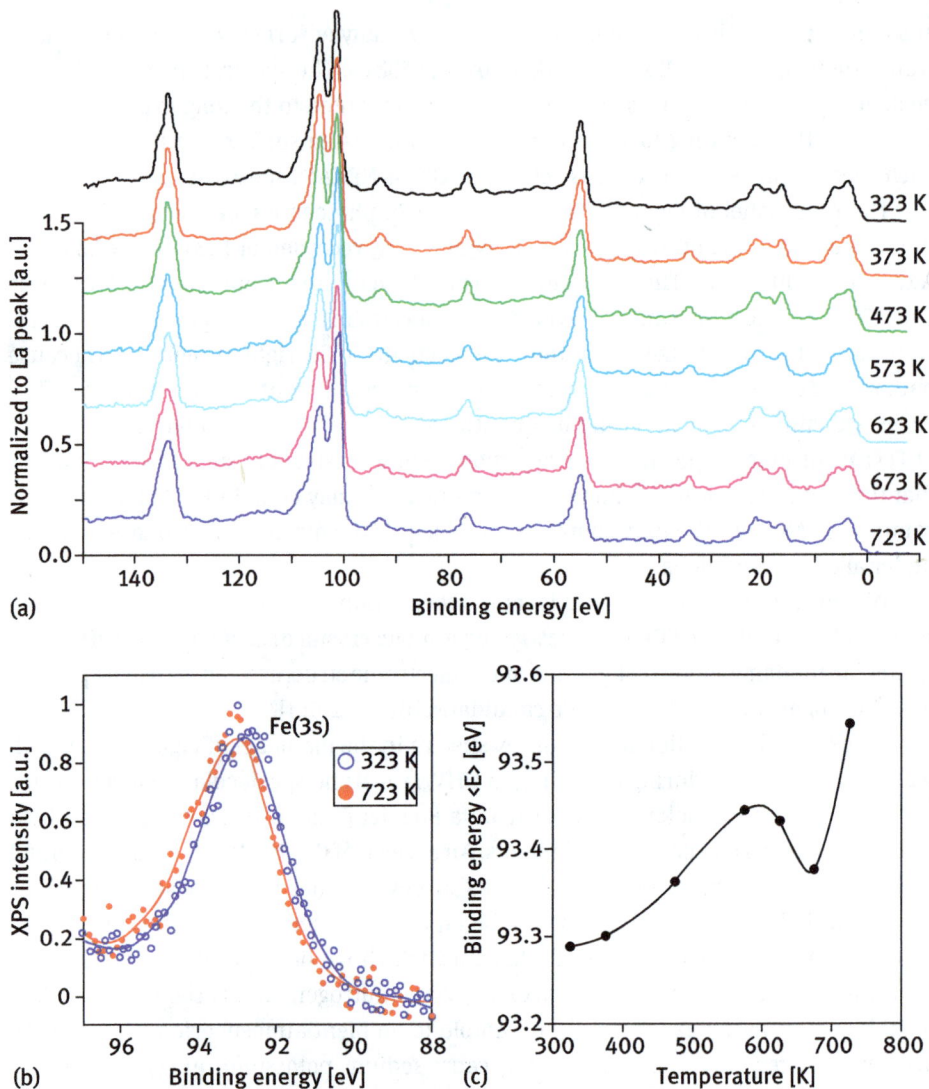

Figure 2.37: (a) Series of XPS survey scans, which were recorded when the sample was heated in UHV from ambient temperature up to 723 K. The relevant transitions are listed in Table 2.5. (b) Fe 3s XPS core level spectra of $La_{0.9}Sr_{0.1}FeO_3$ single crystal at 323 K and 723 K, showing a slight "chemical" shift of 0.3 eV, suggesting a reduction of the Fe with increasing T and revealing the temperature dependence of XPS spectra. (c) The statistical first moment $\langle E \rangle$ for the binding energy is plotted for all measured T in the right panel.

all seven different temperatures from 323 K to 723 K:

$$\langle E \rangle = \frac{\int_{88\,eV}^{97\,eV} I(E) \cdot E \cdot dE}{\int_{88\,eV}^{97\,eV} I(E) \cdot dE}$$

Table 2.5: Binding energies for La, Sr, Fe, and O. Date retrieved from the X-ray Data Booklet [Thompson 2009].

	L_1 2s	M_1 3s	M_2 3p$_{1/2}$	M_3 3p$_{3/2}$	M_4 3d$_{3/2}$	M_5 3d$_{5/2}$	N_1 4s	N_2 4p$_{1/2}$	N_3 4p$_{3/2}$	N_4 4d$_{3/2}$	N_5 4d$_{5/2}$	O_1 5s	O_2 5p$_{1/2}$	O_3 5p$_{3/2}$
													Binding energy [eV]	
La										105.3*	102.5*	34.3*	19.3*	16.8*
Sr					136.0†	134.2†	38.9†	21.3	20.1†					
Fe		91.3†	52.7†	52.7†										
O	41.6*													

The variation of the peak position with measurement temperature T is plotted in the right lower panel of Figure 2.37. The center of gravity of the peak, i. e., the X-ray emission intensity distribution, is shifting toward higher energies, with the exception of a turning point at around 600 K. One interpretation is that the sample becomes oxidized during annealing because the spectrum shifts to higher binding energies. This is surprising and counterintuitive. We expect that a metal oxide becomes reduced upon annealing in vacuum, which is reducing conditions.

2.3.2 Valence band spectroscopy

One advantage of XPS is that it permits to record the valence band (VB) spectrum of a solid [Hüfner 2003]. The VB contains essential information for the physical properties of solids, such as electric, magnetic, and optical properties. Particularly, the function of electrodes and electrolytes should be therefore addressable with VB XPS. In the survey scans, which we have seen in this chapter, the valence band was shown already near the Fermi energy E_F. This Fermi edge is typically observed for systems with very good electronic conductivity, particularly metals like gold [Ebel 1975, Kowalczy 1972, Hochst 1976]. Figure 2.38 shows the valence band spectrum of gold, which was measured by the staff of the GALAXIES hard X-ray XPS beamline at SOLEIL in France [Rueff 2012] with 9840 eV photon energy. The valence band spectrum is in general reminiscent to a Fermi distribution $f(E)$, which in theory at $T = 0$ K is a step function in undergraduate classes frequently referred to as "Fermi ice block," which "melts" with increasing temperature, resulting to the broadening of the distribution of states:

$$f(E) = \frac{1}{\exp\left(\frac{E-E_f}{k_B \cdot T}\right) + 1} \tag{2.2}$$

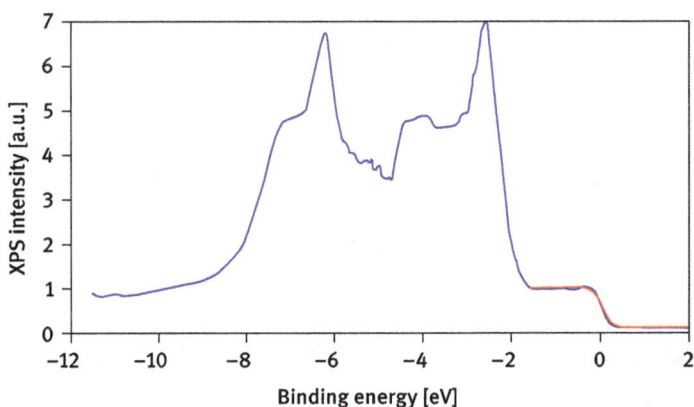

Figure 2.38: VB XPS spectrum of gold Au obtained with 9840 eV photon energy at the GALAXIES hard X-ray XPS beamline at SOLEIL, France [Rueff 2012]. The red line is the least square fit of a Fermi distribution $f(E)$ to the Fermi edge near binding energy $E_B = 0$.

Compounds which lack the metallic character do not have such a pronounced Fermi edge in the VB spectrum. The VB spectra reflect predominantly hybridized states and are therefore not beforehand element specific, which is an analytical disadvantage. We will see later in this book in the part on resonant methods and contrast variation how we can take advantage of the tunable X-ray wavelengths at synchrotrons to incur a contrast, which allows discrimination of spectral details with element specificity.

The spectra in Figure 2.39 were recorded from three different prepared iron oxide photoelectrode materials with a laboratory XPS instrument with an Mg X-ray source of 1400 eV. The green spectrum is from pristine α-Fe_2O_3, and the blue and purple spectra from such iron oxide, which had received a hydrothermal after-treatment in an autoclave container for 2 hours and 24 hours, respectively [Bora 2011f]. Below these three spectra plotted the computed XPS valence band spectrum of α-Fe_2O_3. The red vertical lines denote the electronic transitions. These are then typically broadened with Gaussians, which result in a computed spectrum which is comparable with the experimental spectrum. Figure 2.39 (right) shows the experimental and computed spectrum of the VB region of α-Fe_2O_3 by Fujimori et al. [Fujimori 1986] (compare also [Fujimori 1987]).

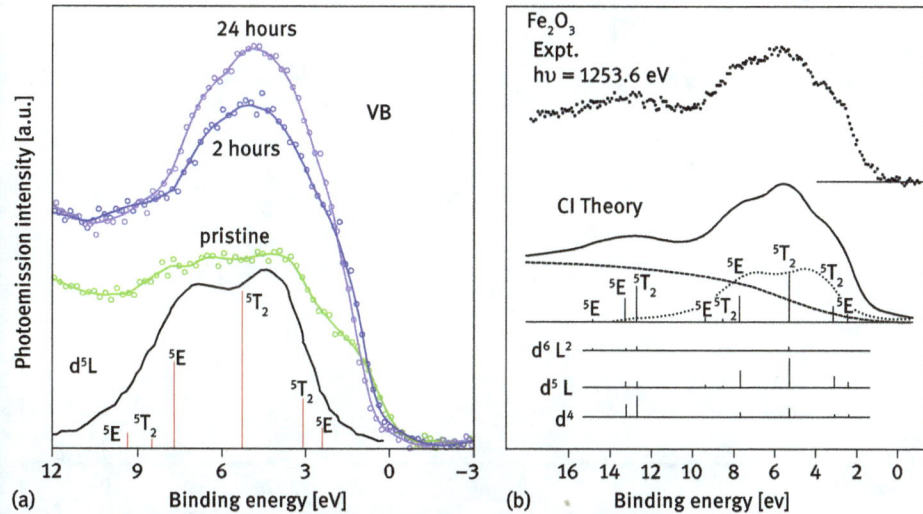

Figure 2.39: (a) VB XPS spectra of pristine α-Fe_2O_3 and hydrothermal treated iron oxide for 2 and 24 hours [Bora 2011f]. Inscribed at the bottom is a theoretical computed spectrum. (b) Computed valence band XPS spectra of Fe_2O_3 using the FeO_6^{9-} cluster configuration-interaction (CI) calculation compared with experimental spectrum. The dotted and dashed curves represent O 2p emission and integral background, respectively. Lifetime broadening has been assumed proportional to the binding energy E_B' relative to the valence band maximum ($2\Gamma = 0.44E_B'$) and a Gaussian full width of 1.5 eV has been employed to represent the combined effects of instrumental resolution and the d-d hopping (Reference 16). Bottom panel shows decomposition into configuration components for each final state line. Reprinted figure with permission from Fujimori, A., Saeki, M., Kimizuka, N., Taniguchi, M., Suga, S., PHOTOEMISSION SATELLITES AND ELECTRONIC-STRUCTURE OF FE_2O_3. Physical Review B 1986, 34 (10), 7318–7333. Copyright (1986) by the American Physical Society.

We will later turn to more VB XPS studies on iron oxide photoelectrodes. Here, I will continue first with iron perovskites, which serve as model compounds for SOFC cathodes. As we know already, SOFC operate at rather high temperatures. For a functional assessment of the material, it is therefore necessary to look at its electronic structure and transport properties under operational conditions. We have therefore carried out XPS measurements at beamline 9.3.2 at the ALS in Berkeley, which has a sample holder with heating and cooling system.

Figure 2.40 shows the VB XPS spectra of a $La_{0.9}Sr_{0.1}FeO_{3-\delta}$ single crystal sample [Braun 2008d]. The single crystal precursor material had been prepared by solid state synthesis, and the phase purity had been confirmed by X-ray diffractiometry (XRD). The gravimetrically obtained nonstoichiometry δ was 0.01. The single crystal (SC) had been grown in an optical floating zone furnace. A 1 mm thin disk in [111] orientation was cut and then the high-quality surface was finished. We also had determined the temperature-dependent DC conductivity with the 4-point DC method.

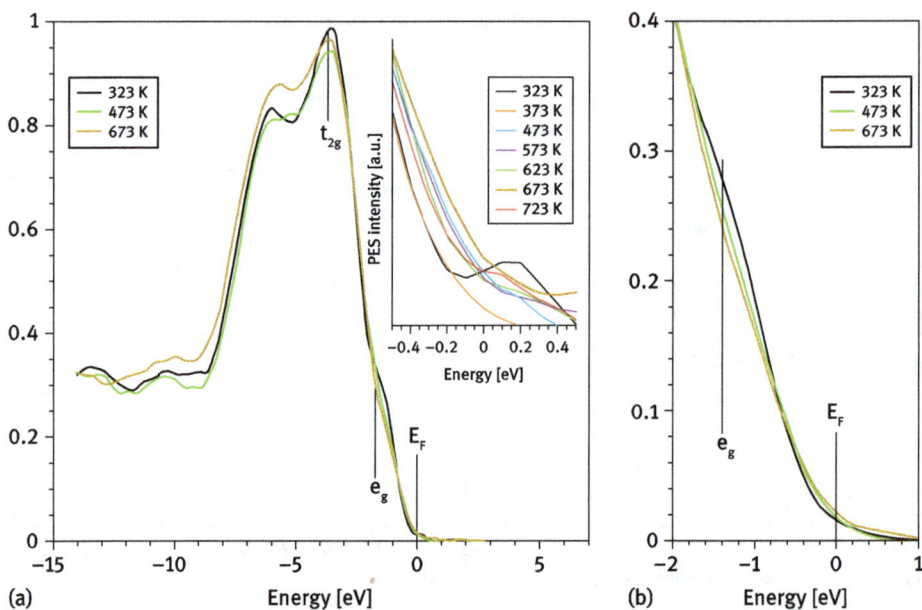

Figure 2.40: Valence band XPS spectra of $La_{0.9}Sr_{0.1}FeO_{3-\delta}$ recorded at sample temperatures for 323, 473, and 673 K. The inset in the left figure shows a magnification of the intensity near E_F for all measured spectra 323 K < T < 723 K. The right figure shows a magnification of the energy range from +0.5 to −3 eV for the spectra measured at 323 K and 623 K. Note the crossover in e_g and t_{2g} intensities with an isospastic point at −0.75 eV.

This single crystal was then subject to XPS at 450 eV excitation energy (Beamline 9.3.1; Advanced Light Source, Berkeley, CA, USA). Prior to the measurement, the single crystal received an Ar^+ bombardment at 500 eV for 30 minutes, a heating cycle to 500 K, and an XPS survey scan for residual carbon. This procedure was repeated two times, and then no carbon signal was detected in the XPS survey spectrum anymore.

The strong peak at –2.5 eV is the O 2p bonding peak. Upon temperature increase, the intensity of the t_{2g} band (–3.5 eV) decreases and the e_g band increases, which is developed as a shoulder at about –1.5 eV below E_F. The e_g peak is indicative of Fe_{2+}, whereas the t_{2g} peak is the spectral signature of Fe_{3+}. We therefore learn from this study that the sputtered and heated single crystal has Fe^{2+} at its surface. Upon annealing in UHV, the relative amount of Fe^{2+} is increasing while the spectral weight of the Fe^{3+} is decreasing. This is not surprising because heating in UHV constitutes a reducing atmosphere with the consequent changes of the crystallographic structure on their single crystal surface, i. e., formation of oxygen vacancies and the resulting reduction of the Fe^{3+} at and near the surface toward Fe^{2+}. We can consider the e_g state in the VB spectrum as a defect state, because it is associated with oxygen vacancies at the iron oxide surface.

Such defect states are easily noticeable in tungsten oxide. We remember the thin tungsten oxide films deposited by pulsed laser ablation on TiO_2 single crystal substrates from the previous section. We looked at these films with valence band XPS using different photon energies, ranging from 270 to 700 eV, (see Figure 2.41). The intensities of the spectra recorded with the high excitation energies from 400 to 600 eV needed to be multiplied by a factor 3000 in order to make them comparable with the two other spectra. Near the Fermi level, we observe a broad hump at a binding energy of 2 eV. The intensity is particularly pronounced for the spectrum with 525 eV excitation energy, which is the absorption threshold of oxygen in $WO_{3-\delta}$. Whereas the O 2p-W 6s hybridized peak is particularly pronounced at the excitation energy of 270 eV. We will learn in Part 5 how this is used as resonant photoemission spectroscopy.

The right panel of Figure 2.41 shows the VB region for the peaks with e_g orbital symmetry, and their deconvolution into a series of Gaussians. The deconvolution of spectra is not only used in order to make quantitative analyses. It is also useful for the mere illustration how states or transitions add up to complex spectra, like in the case with substoichiometric $WO_{3-\delta}$, which shows a rich spectrum in defect peaks, as we learn here. One aspect we have to consider when running XPS spectra with different primary energy is that the penetration depth of the X-rays increases with increasing photon energy. Therefore, the range of excitation, the depth of excitation becomes larger when high photon energies are applied. This does not, however, mean that the electrons which are generated in this range will actually find their way to the probed sample surface and then travel to the electron detector channeltron. The detected electrons originate typically only from the narrow range, which we have shown in Figure 2.30.

Table 2.6 shows the fit parameters from the deconvolution procedure, particularly the energy position, the height (spectral weight) and the position of the peak. It is interesting to learn that the e_g features originate from at least four different transitions.

We remember the core level spectra study from this 100 nm thick film in the previous chapter, which was measured in the pristine (blue) and in the oxidized (yellow) condition. In Figure 2.42, we compare the VB XPS spectra of the as-deposited and the

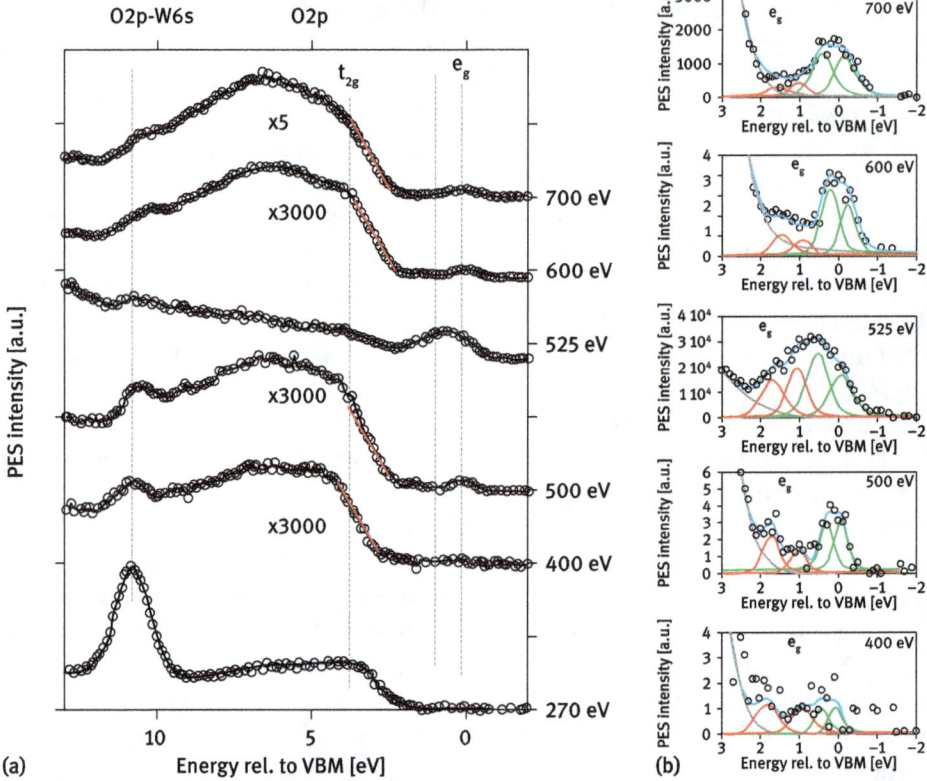

Figure 2.41: (a) VB PES spectra of the as-deposited 100 nm thick film for excitation energies from 270 to 700 eV. (b) Magnification of the spectral range near the Fermi energy with deconvolution of the defect states.

Table 2.6: Fit parameters for deconvolution of e_g orbital symmetry peaks near E_F.

E_0	H_1	E_1	σ_1	H_2	E_2	σ_2	H_3	E_3	σ_3	H_4	E_4	σ_4
400	1.1	−0.06	0.4	1.0	0.41	0.4	1.1	0.9	0.65	1.2	1.8	0.65
500	2.8	−0.08	0.5	2.6	0.3	0.5	1	1.1	.6	2.5	1.76	0.6
525	17000	−0.07	0.7	26000	0.52	0.66	20000	1.05	0.58	15000	1.67	0.7
600	2.02	−0.25	0.51	2.64	0.21	0.53	0.6	0.90	0.6	0.8	1.43	0.6
700	1200	−0.18	0.65	1300	0.39	0.65	424	1	0.55	315	1.57	0.65

post-annealed, oxidized film. The blue spectrum shows the pronounced e_g defect state W 5d, and the O 2p-W 6s peak at about 11 eV. Upon the heat treatment in 100 % oxygen at 400 °C for 48 hours, the intensity of W 5d becomes substantially decreased, but does not entirely vanish. The O 2p-W 6s peak, however, disappears almost completely after thermal oxidation.

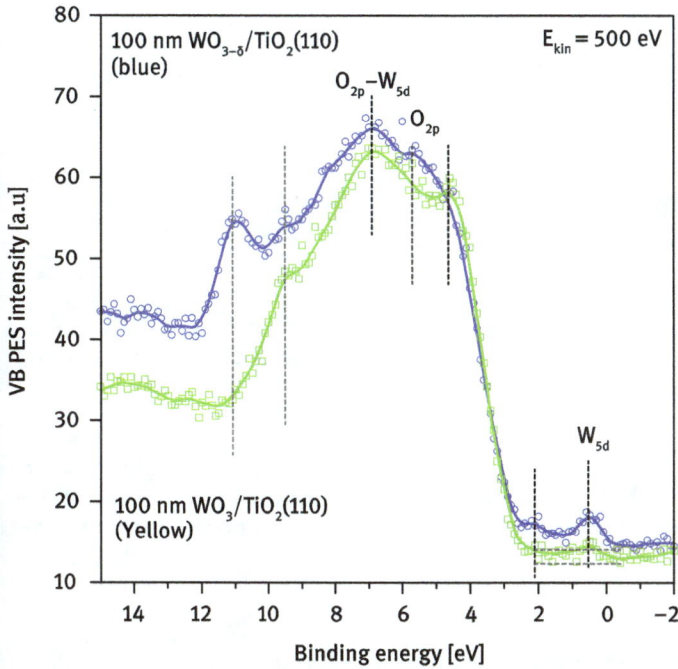

Figure 2.42: Comparison of VB PES spectra of as-deposited (blue) and post-annealed (yellow) films of 100 nm thickness, deposited on TiO_2 with (110) orientation for excitation energy 500 eV.

The spectra of the as-deposited and post-annealed films with nominal 100-nm thickness largely overlap and show the same global variation of intensity across resonance thresholds. We also notice peculiar differences. The peak at +10.5 eV is in the as-deposited films better developed than in the post-annealed films, particularly for photon energies of 400 and 500 eV. The post-annealed films have at this energy just a shoulder. Such feature is clearly visible in the VB-PES spectrum of an *in vacuo* fractured WO_3 single crystal, as is shown in spectrum L in Figure 4 in [Himpsel 1984]. This transition belongs to the p-like valence band. The defect states near the Fermi energy up to +2 eV in the post-annealed films have lesser spectral weight than in the as-deposited film. This is supporting the suggestion that oxidation during post-annealing has filled oxygen vacancies, and thus removed defect states associated with oxygen vacancies. This makes it difficult to carry out a meaningful deconvolution of the spectra of the post-annealed film near the Fermi energy.

We now turn to the 10-nm thin films of WO_3/TiO_2 (110). The as-deposited blue films were subject to ambient pressure XPS measurements during exposure to 100 mTorr oxygen, while at the same time these films were heated from ambient temperature to 678 K. Figure 2.43 shows magnified the evolution of the VB region of the film deposited on the (110) surface while being heated during exposure to 100 mTorr oxygen (left panel), along with the temperature ramp profile (right panel).

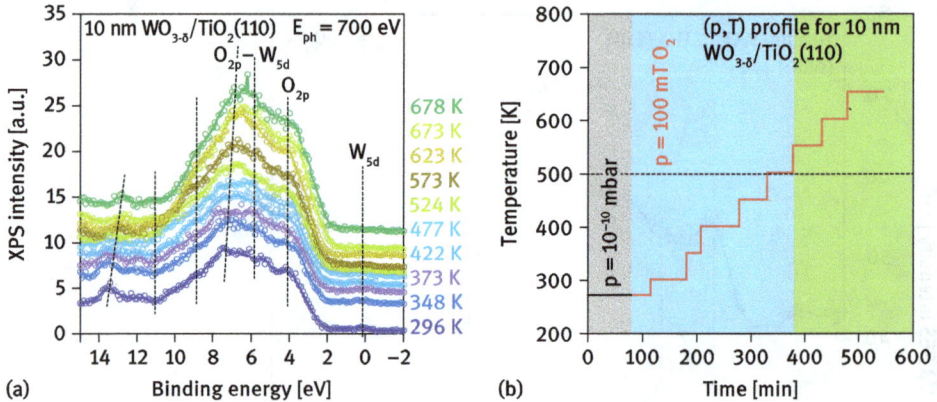

Figure 2.43: (a) Valence band XPS spectra of a 10 nm $WO_{3-\delta}/TiO_2$ (110) film recorded with E_{ph} = 700 eV *in situ* with near AP-XPS while the film was annealed from 296 K (lowest spectrum) to 678 K (highest spectrum) at 100 m Torr oxygen partial pressure in the XPS chamber. Spectra were shifted on the intensity axis for better comparison. (b) Temperature and gas concentration profile on the sample in the XPS chamber during the experiment.

The spectrum of the as-deposited film has near the Fermi energy E_F the shallow W 5d defect state from oxygen vacancies up to 524 K, beyond which it becomes barely noticeable. The O 2p bonding peak at 4 eV is followed by hybridized O 2p-W 5d peaks around 7 to 8 eV. We have in the previously shown *ex situ* study on the 100 nm thick WO_3 films deconvoluted the defect region near E_F into four peaks. We subtracted a linear background so as to be able to define an absolute height for the spectral weight of the defect state region. We attempted no elaborate deconvolution. Rather, we looked at the relative height of this intensity range. The data points have an appreciable scattering. We therefore averaged the data points in a data plot program. This is exercised for the spectra recorded at 296, 477, 524, and 678 K in Figure 2.44. The spectra at 296 and 477 K show a distinct defect peak around 0 eV, while this energy range is flat for the spectra above a critical temperature of 523 K, i. e., for the two spectra recorded at 524 and 678 K.

The relative peak height of this cumulated W 5d defect peak is plotted versus the temperature in Figure 2.44. At ambient temperature and above, this peak height is 0.2 units or slightly above. At 523 K, the peak height decreases abruptly to below 0.1 units and remains at about this value for 673 K, the highest temperature to which the film was exposed. The gray solid line in Figure 2.44 (right panel) is a least square fit of an arctan function for the determination of the critical temperature T_{crit}. 500 K where the defect state intensity critically decreases. This temperature is indicated in the temperature profile in Figure 2.45 (right panel) with a vertical dotted line. The height of the defect peak does not completely vanish above T_{crit}, suggesting the film contains still some remaining defect state concentration despite the oxidative treatment at high temperature.

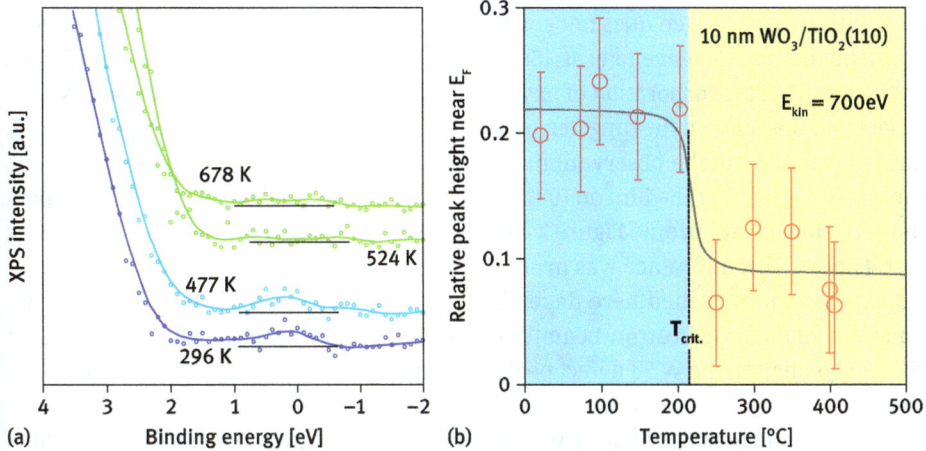

Figure 2.44: (a) Range of the valence band XPS spectra of 10 nm $WO_{3-\delta}/TiO_2(110)$ showing W 5d defect peaks near Fermi level, recorded *in situ* while the film was annealed from 296 K to 678 K at 100 mTorr oxygen pressure with $E_{ph} = 700$ eV. The spectra are shifted on the intensity axis for easier comparison. The horizontal black lines mark the level of zero intensity. (b) Relative peak height of the W 5d defect states range near E_F in the valence band XPS spectra of 10 nm $WO_{3-\delta}/TiO_2(110)$ recorded *in situ* with $E_{ph} = 700$ eV while the film was heated from ambient temperature to 678 K while maintaining 100 mTorr oxygen partial pressure.

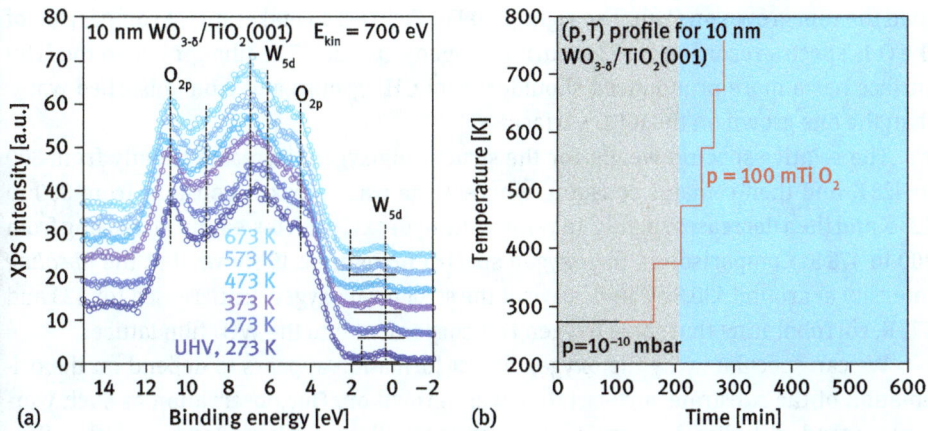

Figure 2.45: (a) Structural evolution of the film grown on the (001) oriented substrate during heating in 100 mT oxygen. (b) The figure on the right shows the temperature profile over time.

The situation is somewhat different for the film grown on the (001) substrate ($WO_3/TiO_2(001)$). The height of the defect signature in $WO_{3-\delta}/TiO_2(001)$, as shown in Figure 2.45, is larger than that of the film on the (110) substrate, revealing that tungsten oxide grows largely oxygen deficient on (001), more so than on (110). Even annealing to

673 K with 100 mT oxygen does not change the defect structure in the spectra, which is surprising because Schiavello et al. [Schiavello 1977] found that a temperature of 673 K was necessary for incorporation of oxygen into reduced tungsten oxide. We find not any signal from titanium on the tungsten oxide film surface after heating in oxygen to 673 K. This is in contrast to the observations with the film on the (110) substrate. We have to remark, however, that the film on (001) was subject to thermal oxidation only for about 2 hours (temperature profile Figure 2.45), whereas the film on (110) showed Ti diffusion after 4 hours. This difference was primarily caused by the condition that the kind of *in situ* experiment presented here depends on instrumental complexity, combined with limited available synchrotron beamtime. We thus cannot entirely rule out that after extended oxidation the W 5d defect peak in the film grown on (001) would decrease and eventually vanish.

In analogy to Figure 2.44 (left), we have made a quantitative analysis of the height of the defect peaks shown in the spectra in the left panel of Figure 2.45; see the bottom/left in Figure 2.46. The spectral weight of the defect peak remains constant up to at least 673 K. Above this constant trend, we have plotted the relative spectral weight of the O 1s spectra from structural oxygen O_x, hydroxyl OH^-, and water $H_2O(v)$ of the film grown on (001). Unlike with the film grown on the (110) substrate, there is no depletion of W 5d defect states during heating in oxygen up to 673 K. The four O 1s spectra of the film on (001) are shown on the right in Figure 2.46. The as-deposited film contains in UHV at 300 K a large spectral signature from H_2O and from OH groups, but to a smaller extent than the film grown on (110). The right panel of Figure 2.46 makes direct comparison of the O 1s spectra recorded at 373 K during oxygen exposure. The film grown on the (110) surface has a more pronounced shoulder from OH^- groups and from adsorbed water than the one grown on the (001) surface.

The relative spectral weight for the structural oxygen decreases slightly from 300 to 423 K and then remains constant, whereas the water peak is constant from 300 to 423 K and then decreases slightly. In return, the hydroxyl peak is steadily increasing from 300 to 473 K. Comparison of the oxygen spectra in Figure 2.46 shows that the *absolute* intensity at around 530.5 eV, indicative of the structural oxygen, is increased for 423 and 473 K, corroborating that some oxygen is incorporated into the WO_3 film lattice.

We can speculate why the oxygen defect formation appears to depend on the orientation of the substrate surfaces they were grown on. Our observation as such warrants extended studies, beginning with monitoring the growth modes, e. g., with reflexion high or medium energy electron diffraction in order to see to what extent layer by layer growth is possible. This should be complemented by a crystallographic analysis with glancing incidence XRD or low energy electron diffraction (LEED) in order to verify potential epitaxial growth.

Our observation that tungsten oxide grows as a 2-layer system on (110) and then relaxes toward a 1-layer film suggests that a strained phase may have been grown on (110), in contrast to (001). Probably the first 6 or 6.5 nm tungsten oxide is growing as a strained phase on (110), and the following 6 nm grow relaxed. The substrate thus

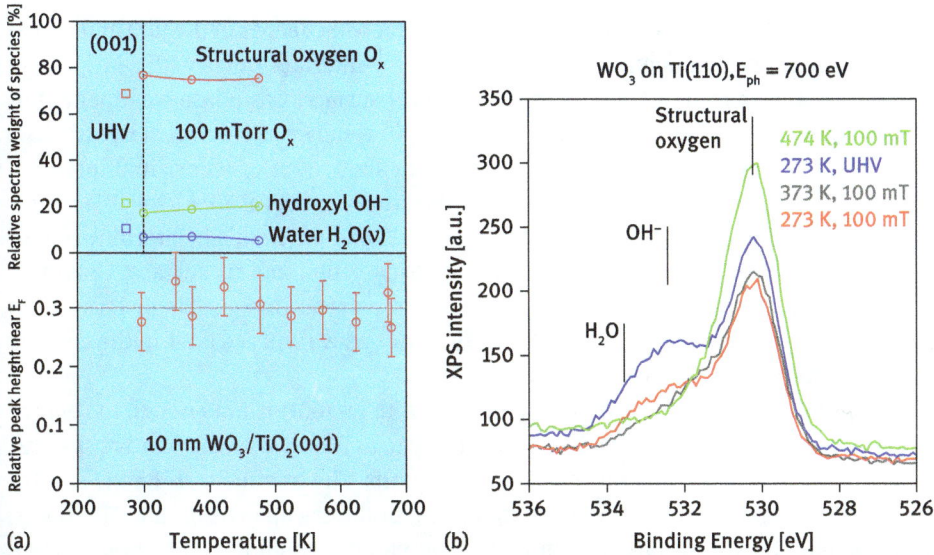

Figure 2.46: (a) Comparison of spectral intensity of the oxygen defect structure near the Fermi level of the (001) deposited film during heating in 100 mT oxygen (bottom), and variation of relative spectral weight of O_x, OH^-, and $H_2O(v)$. (b) The corresponding oxygen spectra have been aligned on the energy axis for facile comparison of the spectral shape.

preconditions the growth and possibly also the oxygen vacancies, even though the interaction of the surrounding gas with the film takes place at the film surface—and then possibly can propagate into the film interior. The review by Ganduglia-Pirovano et al. [Ganduglia-Pirovano 2007] deals with the oxygen vacancies at surfaces of TiO_2 and ZrO_2, CeO_2 and V_2O_5 and calculates the energies necessary to form oxygen defects. For hematite, the energies necessary to form such vacancies at surface and subsurface has been calculated in [Warschkow 2002]. The defect formation energy shows oscillations around 24 eV within the five oxygen layers underneath the (0001) hematite surface. Computational techniques are thus beneficial for deeper understanding of defect formation in metal oxides. Experimental progress has been made with isotope exchange and secondary ion mass spectroscopy on $SrTiO_3$, where an oxygen defect concentration profile of several tens of nanometers was mapped. This was attributed to an equilibrium space-charge layer with no oxygen vacancies, followed by a gradient reaching some micrometer into the solid, which is attributed to diffusion in a homogeneous bulk phase [De Souza 2012]. De Souza et al. found that a surface terminated with oxygen has an effect on the surface exchange coefficients. This approach is in principle applicable to tungsten oxide films as well and warrants further studies.

With the summary information now from the XPS core level and VB studies, we can draw some conclusions on our material. The $WO_{3-\delta}$ films on TiO_2 single crystals with surfaces with (110) and (001) orientation with a blue color are oxygen deficient. The spectra with E_{ph} = 700 eV are collecting information somewhat deeper in the film and reveal the films are substoichiometric in the bulk and in the interface range with a large W 5d defect state concentration for the film on the (001) surface. Gas phase treatment with oxygen under annealing of the film grown on (110) removes the defect states at 523 K to an extent that the spectral weight decreases to about 50 % of its original intensity. With increasing temperature, the defect concentration remains at 50 % level. We notice, however, a second process is activated between 623 and 673 K; this is interdiffusion of titanium ions from the TiO_2 substrate into the $WO_{3-\delta}$ film and its surface. Upon this treatment, the aforementioned bilayer structure transforms into one single layer without significant improvement of the X-ray reflectivity, as we will see later in the chapter on X-ray reflectivity.

This thermal treatment with oxygen on the film on (001) is not enough to remove its defect state signature in the VB XPS spectra up to at least 673 K. The relative spectral weight instead remains virtually constant at 0.3 units. The treatment shifts the spectra toward 0.5 eV higher binding energy, revealing n-type doping upon oxidation. This holds, however, only for the surface. In contrast to the film grown on the (110) substrate, we notice no interdiffusion of Ti into the bulk of the film grown on the (001) substrate or its surface. The film thickness grows slightly with oxidation and the X-ray reflectivity improves.

An extensive XPS study like this one shows how the electronic structure of tungsten oxide films can be altered depending on the surface orientation of the substrate. Further processing steps such as oxidative thermal after-treatment, e. g., can improve the electronic structure for a particular application, such as PEC photoelectrodes. But additional processes adverse to the function may be showing up as well. We have demonstrated how *in situ* XPS analyses can be helpful for monitoring the evolution of the electronic structure from synthesis to processing to assembly of components and devices, as we will show in this book also for batteries. It is desirable to complement such structural studies by *operando* studies of the electric transport properties, in the ideal case under realistic device operation, such as shown *operando* for iron oxide in a fully illuminated PEC using near-edge X-ray absorption fine structure spectroscopy [Braun 2012l].

3 Crystal structure and microstructure

3.1 X-ray diffraction

Our perception of a crystal lattice as a theoretical concept is more naïve than our perception of a real crystal that we find in nature. In our mind, we have the atomistic model of absolute order, of perfect and symmetric arrangement of atoms in some geometrical relation to each other. This naïve model is necessary so that we can perceive the concept of disorder—the deviations from order. Order and disorder are the two extremes of a property, which is of fundamental importance for transport properties. Figure 3.1 shows the crystallographic representation of the ABO_3-type perovskite structure of $LaFeO_3$. This is an electric insulator [Arima 1993]. The Bragg peak positions and their intensities were determined by Köhler and Wollan with neutron diffraction [Koehler 1957] and are listed in Table 3.1. The nice order that we see at first glance is impaired by some disorder. Let us first inspect which chemical elements are present. The large white ions are the rare earth La^{3+} cations on the A-site positions. The red ions are the oxygen anions, and behind the opaque triangles are the Fe^{3+} cations. These are "coordinated" by six oxygens in an octahedron.

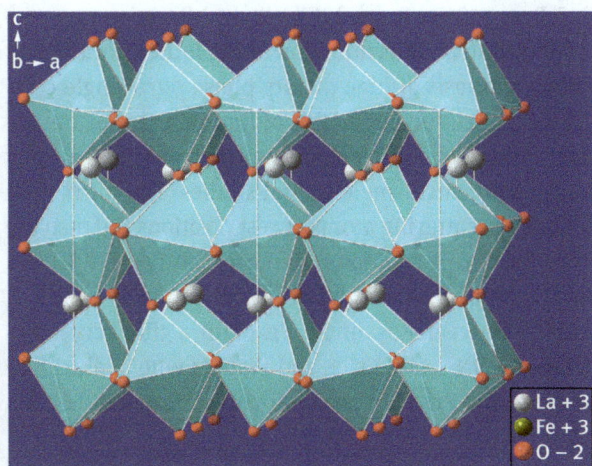

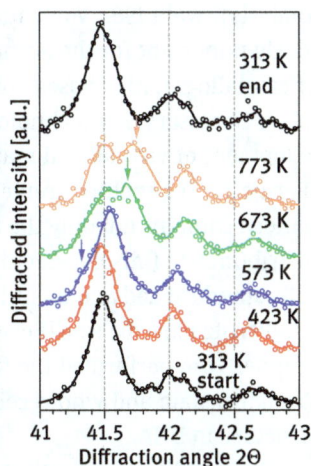

Figure 3.1: (a) Crystal structure representation of $LaFeO_3$. Red spheres indicate oxygen, white large spheres indicate lanthanum, and yellow spheres in the octahedron indicate iron. (b) X-ray diffractogtrams from LFNO for $41° < 2\theta < 43°$ heated from 313 K to 773 K and cooled down to 313 K. Arrows indicate redistribution of diffracted intensity [Braun 2008d].

These octahedra are tilted against the neighboring octahedra. This slightly distorted structure is not uncommon for materials and for energetic reasons in thermodynamic equilibrium. All atoms are where they are supposed to be according to the dictate of

https://doi.org/10.1515/9783110794038-003

Table 3.1: Experimentally observed neutron diffraction intensities for LaFeO$_3$ [Koehler 1957].

hkl	int.
011, 101	543
110, 002	161
111	10
200, 020, 112	553
210, 120	60
013, 103, 211, 121	450
202, 022	287
113	30
212, 122	90
004, 220, 203, 023, 221	935
031, 301, 213, 123, 310, 130, 222, 114	379

thermodynamic equilibrium. Neutron diffraction shows that LaFeO$_3$ has orthorhombic symmetry with lattice constants 5.556, 5.565, and 7.862 Å [Koehler 1957].

The most important X-ray method in materials science is X-ray diffraction (XRD). It permits the determination of the crystal lattice parameters of the unit cell. We discriminate Laue diffraction for single crystal analysis and Bragg diffraction for powder analysis [Ewald 1962, Von Laue 1915, Bragg 1915a, Bragg 1915c]. Laue diffraction is particularly important for the accurate quantitative determination of the lattice parameters of crystallographic phases, for which synthesis of high-quality phase pure single crystals is necessary. For technological purposes, this method is of limited use, but the development of virtually all materials science including modern protein crystallography [Perutz 1962] benefited tremendously from such single crystal studies. The diffraction pattern contains not only the positions of scattering atoms at rest positions but also the distributions of displacements from these rest positions [Hendrickson 1980]. XRD is a fully standardized and quantitative technique [Swanson 1951, Swanson 1966].

Perutz played a role [Lewens 2007] between Rosalind E. Franklin, a gifted crystallographer who performed the first XRD on DNA, and Crick and Watson who peeked into Franklin's data and who received the Nobel Prize for the discovery of the double helix structure in DNA.

Powder diffraction is the working horse for the synthesis and processing studies and is typically used for phase analysis and for systematic doping or substitution studies. The powder to be investigated is, e. g., dispersed over a silicon single crystal disk, the sample holder so to speak, and the powder diffractogram contains the XRD of the powder plus the single crystal peaks from the silicon disk underneath. These latter Si peaks are no experimental nuisance. Rather, they assist you in calibrating your Bragg angle scale because the Bragg reflections of Si are precisely known, for example. The powder should be finely ground and flat dispersed. You can make a slurry by immersing the powder in ethanol, then grinding it in a mortar, and then pouring some drops of the slurry on the Si disk. This makes a fine layer of powder for a decent diffractogram. Some researchers mix the powder with grease, which has no disturbing Bragg peaks of its own, and the grease assures that the powder remains well on the sample holder. Sometimes researchers use microscopy glass slides as sample holders. The glass has its own very broad glass peaks that should be measured separately for better distinction and discrimination.

The intercalation of ions into crystals lattices—and the corresponding deintercalation—causes changes in the crystal lattice, which can be checked with XRD. A simple example is the extraction and insertion of Li ions in $LiMn_2O_4$, a well-known standard cathode material in lithium battery research. The two diffractograms shown in Figure 3.2 were recorded before and after electrochemical cycling with a conventional laboratory diffractometer with an X-ray tube and copper target. Figure 3.2 shows the well-known Bragg reflections of $LiMn_2O_4$, i. e., the (111), (311), (104), (331), and (511) peaks according to JCPDS 89-0118. The (004) double peak with the highest intensity at around

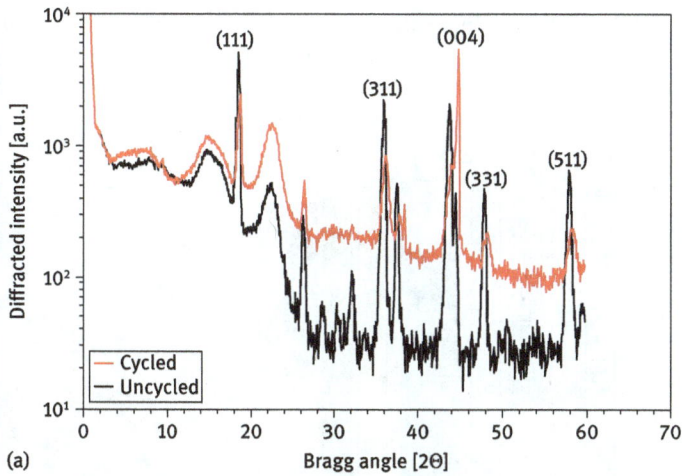

(a)

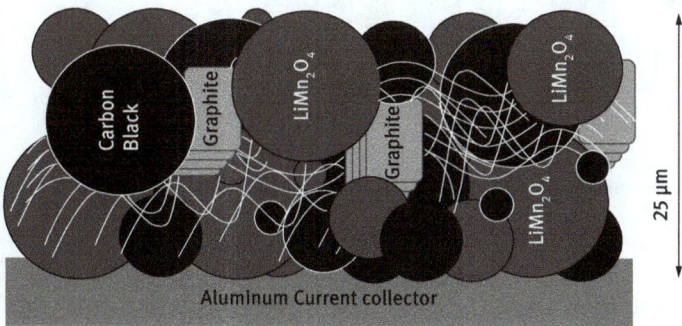

(b)

Figure 3.2: (a) Powder X-ray diffractograms of a $LiMn_2O_4$-based battery electrode before and after charging. Note the logarithmic intensity scale. (b) Schematic of a lithium manganite battery cathode with $LiMn_2O_4$ spinel particles, carbon black, and graphite, blended together with a polymer binder matrix and coated on aluminum foil.

45° undergoes an intensity redistribution to slightly below 44°, e. g., after cycling. We also observe that the incoherent background increases strongly after cycling in the range from 25° to 60°. This is because of structural changes at the atomic scale due to the repeated intercalation and deintercalation of lithium.

Note that this diffractogram contains the diffracted intensity not only from the spinel active material. The bottom panel of Figure 3.2 shows a simplified schematic sketch of the electrode assembly, which is a blend from spinel particles, the carbon black and graphite, potentially residual dried electrolyte with $LiPF_6$ crystals, the very weak scattering polymer matrix, and depending on the thickness of this electrode material, the aluminum current collector underneath. The X-ray beam probes the entire electrode sample, and hence the diffractogram contains the diffracted intensity from all these materials. It is therefore necessary to take account of all these potential contributions.

The situation becomes more complex when the experiment is conducted in an X-ray *in situ* spectroelectrochemical cell. This is illustrated in Figure 3.3, where the X-ray beam

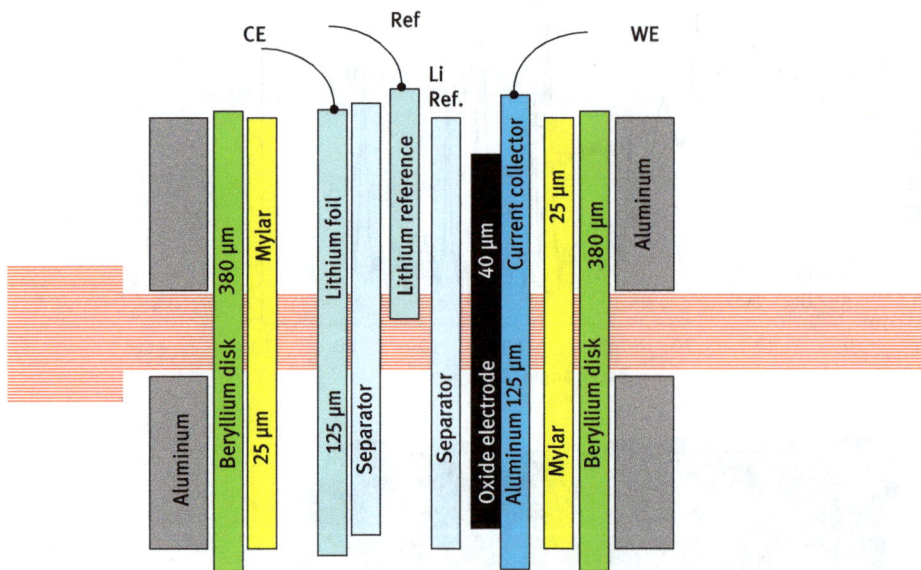

Figure 3.3: Illustrated here is the complexity of *operando* and *in situ* spectroelectrochemical cells for transmission X-ray experiments. The X-ray beam is passing from the left through an aperture in an aluminum plate (or Teflon®), which is pressing on a beryllium X-ray window. The bottom of the beryllium is protected by Kapton® foil or Mylar® foil against corrosive chemical reaction with the metal lithium negative electrode and $LiPF_6$ (and solvent) electrolyte in the cell. The lithium metal is separated by a polymer separator membrane against the positive electrode, which is cast on an aluminum foil. The right side of the aluminum foil is another beryllium X-ray window, which is also protected with a foil against corrosion from the electrolyte. Another aluminum disk or stainless steel disk closes the cell. The exiting X-ray beam on the right carries the structural information (molecular structure, crystallographic, and microstructure) of all the materials through which the beam has passed, including any potential path through air, and also materials in the beamline optics such as monochromator grids.

passes sample and cell components. Separate acquisition of diffractograms from either component thus helps to tell apart which peak in the diffractogram originates from which phase. This means one should record a diffractogram from the aluminium foil, carbon black powder, graphite, polymer binder, dried electrolyte, and maybe dry $LiPF_6$ electrolyte precursor, and so on. It is interesting to analyze materials, components, and devices during the complete stage of synthesis, processing, assembly, function, and degradation. Diffraction peaks (Bragg reflections) may overlap as proximate peaks from one single phase and form a double peak. A second crystallographic phase in the material under investigation may have Bragg peaks that overlap with peaks from the first phase. Alternatively, we have overlap with peaks from other materials that are in the sample under investigation. This may sound very laborious, but this procedure warrants a thorough and comprehensive analysis. Otherwise, unexpected issues may be overlooked.

The peak width of the Bragg reflections depends on the coherency length, the size of the crystal domains, and the instrument resolution of the diffractometer.

The metal oxide electrode material is the "center" of the experiment and is cast on an aluminum metal current collector, which is connected as working electrode to the potentiostat or power supply. A polymer separator is on the top of this working electrode. A stripe of lithium metal is on the other side of the separator and serves as reference electrode. To prevent short circuit with the lithium metal counter electrode, another separator is on top of the lithium strip for the reference electrode. On the second separator rests a lithium metal disk as counter electrode, which is connected with the potentiostat.

It is possible to design a complete battery cell with X-ray windows, thin windows that have a high transmission for X-rays, and thus permit to record XRD data from the battery while it is operated by a potentiostat (*operando* or *in situ*).

Materials for X-ray windows must have a low X-ray absorption, and thus must be elements with low Z number (few electrons in the shell). Aluminum and beryllium plates, graphite, and polyimide foils (Kapton®) are frequently used.

In the last 30 years, a whole range of electrochemistry and battery *in situ* studies has emerged. Nazri and Müller [1985] at Berkeley Lab did early XRD studies on surface layers on lithium. Vong et al. [Vong 1988] made a cell for gas-phase reactions (see also Puxley et al. [Puxley 1994]), and Samant et al. [Samant 1988] made an electrochemical cell for surface diffraction studies on Pb layers on single crystals. Toney et al. [Toney 1991] made a similar cell for synchrotron studies on electrodeposited films. XRD on thin films is not to be mistaken for surface XRD, where truncation rods originate from two-dimensional surfaces [Robinson 1992a, Robinson 1992b]. In the mid-90s, more and more *in situ* XRD studies were published. Fey et al. [Fey 1994] and Kotschau and Dahn [1998] made many contributions to lithium battery research with their *in situ* cells (see also Rodriguez et al. 1998, Rodriguez et al. 2000, Whitehead et al. 1996, Thackeray et al. 1998).

Today it is possible to "X-ray" commercial battery cells because with the advancement of packaging technology, cell housing has become ever thinner. For solid oxide fuel cells, this is not yet possible. Figure 3.4 shows the exploded view of a Solid Oxide Fuel Cell

Figure 3.4: Schematic of a solid oxide fuel cell stack with exploded view of a cell assembly. Structured metal interconnects are stacked, between which anode (green), electrolyte (light gray), and cathode (black) are stacked (courtesy HEXIS GmbH). The structures on the interconnectors provide paths for fuel gas, oxygen, and combustion gases.

(SOFC) cell assembly, where the top is the metal interconnect of around 1-mm thickness. They have to be so thick because of the necessary mechanical structural integrity during long-term operation at around 800 °C. The green layer underneath is actually three layers, where the green on top is the anode layer of some 10-micrometer thickness. The bright gray layer is the solid electrolyte of another 10- to 100-micrometer thickness. The bottom layer is the cathode with the similar thickness. At the bottom is the next metallic interconnect. It is basically impossible to look into such cell assembly with X-rays and then obtain a diffractogram or to obtain an X-ray Absorption Spectrum (XAS). XAS spectrum—unless maybe a small "peek" hole, an aperture, is built through the metal interconnects so that one can at least probe the cathode or anode layer. It needs therefore extra preparations to make *in situ* XRD or XAS studies on SOFC.

3.2 X-ray wide-angle scattering

In Chapter 3.1, I mentioned the presence of carbon in battery electrodes and that their signature would show up in diffractograms. Carbon is an element with an extreme molecular diversity and structural allotropy. Two crystallographically distinct forms are diamond and graphite, which have very different crystal structures and properties. Graphite is used in battery electrodes because of its very good electronic conductivity, the presence of which is necessary to provide conducting paths between the poorly conducting active materials such as the spinel particles or olivine particles. The carbon black contains large aggregates that support the binder in binding the electrode and binds also the electrolyte with its high porosity.

In XRD studies, it turns out that only graphite and diamond have distinct Bragg reflections, whereas other carbon materials, including soot and coal and pitch, have rather broad peaks, which interestingly are typically found at around below 26° when a Cu K$_\alpha$ wavelength is used. This intensity corresponds to the (002) Bragg reflex of graphite, and the large full width at half maximum (FWHM) of this peak is a result of the very short graphite-like coherence domains in carbons. The foundations for this field were actually laid by Rosalind E. Franklin [Franklin 1951a, Franklin 1951c, Franklin 1951e] in her early years of XRD studies [Harris 2001]. The study of the XRD of carbon materials (when we ignore the minerals diamond and graphite and the highly-ordered technological materials carbon nanotubes, fullerenes) shows that their structure is more defined by their disorder, by their deviations from sharp peaks.

In this case, it becomes common to use the term wide-angle X-ray scattering (WAXS) rather than XRD. Stevens and Dahn have made a number of *in situ* XRD studies featuring the structural changes in the carbon in battery electrodes [Stevens 2001].

Figure 3.5 (left) shows the diffractograms from glassy carbon (GC) powder type G (high pyrolysis temperature) and type K (lower pyrolysis temperature). G has a very broad intensity peak at around 25°. It is hard to call this a Bragg reflection, but it originates from the diffraction of small graphite-like crystallites that are built from graphene sheets in this turbostratic form of carbon. According to the reciprocal space relationship, their small size manifests in broad peaks. On the right panel, we have GC powder that was ground by corundum balls in ball mills. The sharp peaks around 28° are from corundum. The peak around 26° in the diffractogram for G could originate from highly crystalline graphite particles. On the left panel, we have XRD from 1-mm thick GC plates [Braun 2002a].

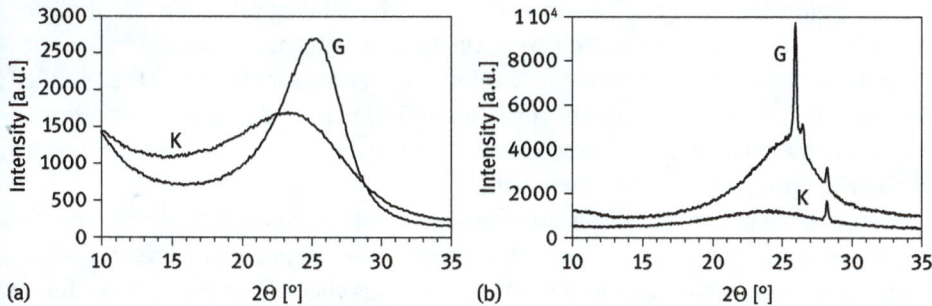

Figure 3.5: (a) X-ray diffractogram of glassy carbon (GC) plates and powder (b) type K and type G. X-ray diffractograms of GC powders K (lower curve) and per unit volume remain unchanged during pyrolysis [Braun 2002a].

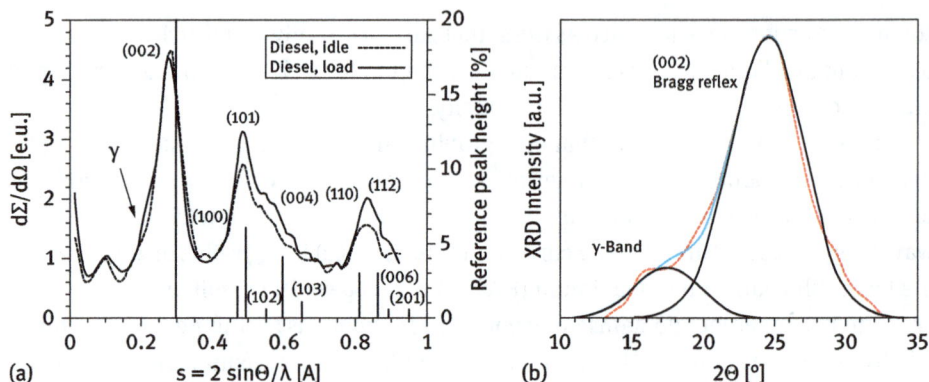

Figure 3.6: (a) WAXS curves of diesel soot obtained under idle and load engine operation conditions, recorded with Cu Kα radiation at the University of Kentucky [Braun 2005h]. (b) Deconvolution of the major peak at 25° into graphite Bragg reflection and γ-aliphatic side band.

Figure 3.6 shows the diffractograms from carbonaceous soot collected from the tailpipe of a diesel engine that was run under idle and under load condition. I have transformed the Bragg angle 2θ to the scattering vector (momentum transfer) $s = 2 \sin θ/λ\, 1/Å$ in order to compare it with literature data, which were given in the same quantity. I will explain in Chapter 5.2 (Part 5, Resonant Methods) why the Bragg angle 2θ is only a technical parameter and less useful than the momentum transfer s (depending on the literature, the momentum transfer is also termed k, q, Q).

The data reduction and analysis was done according to the protocols given for carbon in [Franklin 1951a, Franklin 1951c, Franklin 1951e, Lu 2001, Drits 1991]. On the right axis, I have plotted the calculated peak height of the reference sample graphite and the corresponding peak positions of graphite as vertical lines versus s. It is obvious that the three major intensity densifications at $s = 0.3, 0.5$, and $0.85\,1/Å$ correspond to the major Bragg reflections from graphite. The diffractograms are similar at large, but there are small differences due to the idle/load engine operation condition.

The right panel of Figure 3.6 shows the very broad "Bragg reflection" for the (002) graphite peak from load engine diesel soot. On the lower left flank is a considerable extra intensity that originates not from the graphitic core of the soot but from the so-called aliphatic side band (γ-band), which often observed in soot and originates from carbon atoms that are not arranged in an aromatic setting. The peak is deconvoluted therefore into a (002) "graphite" peak and the γ-band.

Suppression of soot formation is an important issue in combustion technology. Our colleagues in Utah operated a diesel engine with diesel mixed with ferrocene, which can be considered a molecular iron catalyst. There was virtually no soot formed during that experiment, and the little amount of collected material was very difficult to handle for an X-ray experiment. I therefore collected the little material available in a thin boron nitride capillary. See Figure 3.7 which shows a set of such capillaries filled with

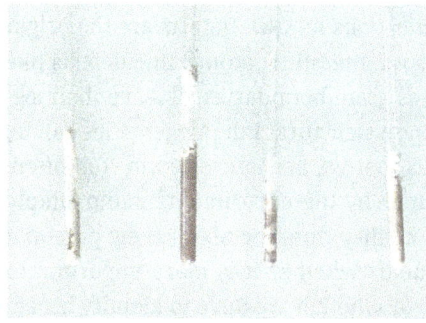

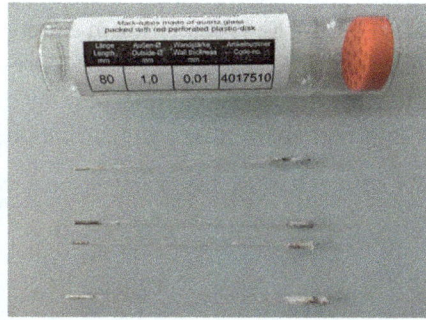

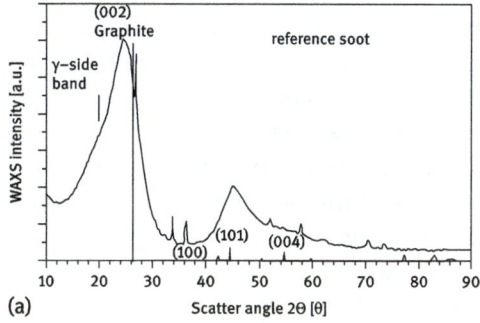

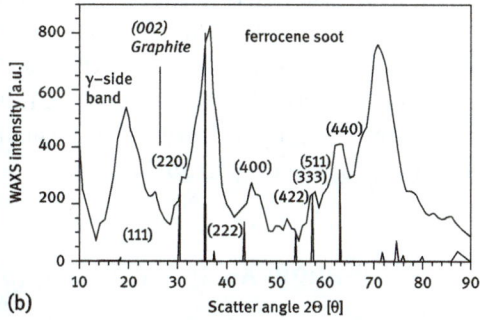

Figure 3.7: (Left) Boron nitride capillaries used for WAXS and SAXS measurements. The photos show capillaries filled with a silver oxide powder. They can also be filled with liquid samples or dispersions. (Right) WAXS patterns of reference diesel soot (a) with indicated graphite reference Bragg reflections. Bottom panel (b) shows WAXS pattern of soot obtained from diesel combustion with ferrocene, with absence of graphite peak and presence of maghemite Bragg reflections [Braun 2006c].

silver nanoparticles and glass fibers for closure of capillaries. WAXS was conducted by Steven N. Ehrlich at beamline X-18A at the National Synchrotron Light Source (NSLS), Brookhaven National Laboratory, Upton, NY, USA. The soot powder was stuffed in the boron capillary of 10-mm wall thickness and 25-mm outer diameter, held in an aluminum sample holder specifically designed for these experiments in transmission geometry. WAXS spectra were recorded from 5° to 45° in steps of 0.1°, with the X-ray energy being 10,000 eV. For the convenience of the reader, the 2θ scale for 10,000 eV was converted to the more familiar Cu Kα radiation. See WAXS diffractograms in Figure 3.7.

There is frequently confusion about the different terminology used for related X-ray methods: diffraction, scattering, wide-angle scattering, wide-angle diffraction, small-angle scattering, small-angle diffraction, and low-angle diffraction. I do not want to clarify these terms. It makes more sense when we try to understand what we see in the data and how they can be interpreted. The inverse relationship between real space and reciprocal space shows that structures with a large lattice constant or similar repeating unit distance are mapped at small momentum transfers. Objects that extend over a large coherency scale are characterized by sharp reflections in the scattering pattern. Disorder leaves a signature in the scattering patterns, which can be termed as "diffuse," as opposed to the distinct sharp reflections in XRD. Defects are the origin of such diffusivity. Point defects like vacancies and interstitial atoms, line defects like dislocations, and two-dimensional defects such as grain boundaries all leave their signature in the scattering patterns [Gray 1998]. These structural inhomogeneities can be very important for the function of the materials that we are interested in. Too often, these signatures are just considered "background" by the experimenters and simply removed from the scattering patterns without further thinking about their potential significance. We have to decide in our experiments whether it is more important to resolve Bragg reflections with high k-resolution, or whether we have to identify broad, diffuse structures where we have to use an instrument where we can accumulate as many counts as possible. Haubold has elaborated on diffuse X-ray scattering in a number of publications [Haubold 1975, Haubold 1976]. The WAXS scattering curve represents information of the reciprocal space. This constitutes a correlation function, which can be translated by a Fourier transformation into real space distance distribution [Rodriguez 2013, Rodriguez 2013a].

3.3 Small-angle X-ray scattering (SAXS)

In the past chapter, we heard about the diffuse scattering pattern, which less ordered materials exhibit in XRD experiments. When we look back in Figure 3.2, we notice that at very small angles the intensity is dramatically increasing. This small-angle scattering must originate from relative large objects or structures because of the reciprocal space relationship. On the basis of the Bragg condition $n\lambda = 2 \cdot d \cdot \sin\theta$, we can read that large lattice distances or superstructures have their signature at the small diffraction angles. When we approach the (000) reflection in a diffractogram [Hosemann 1962], we often notice increasing scattering intensity, which is not necessarily identical with the specular beam intensity. The shape of the specular beam originates from particle size, particle shape, fractal dimension, internal surface area, and porosity; in short, it originates from the microstructure of the probed sample. The theory of small-angle X-ray scattering (SAXS) is well explained in the book of Glatter and Kratky [Glatter 1982], which is now available online for free as a public domain. For SAXS data reduction and data analysis, I recommend the software package developed by Jan Ilavsky at Advanced Photon Source, Argonne National Laboratory [Ilavsky 2009, Ilavsky 2012], which runs on Wave Metrics' Igor Pro. However, a researcher can apply almost all of the SAXS theory with simple data manipulation and plotting software (MicroKal Origin, Sigma Plot, KaleidaGraph, Matlab, Maple). For the modeling of X-ray scattering patterns, it is necessary to use the atomic form factors for the particular elements that are scattering by the X-ray beam. The form factors are the Fourier transforms of the electron density distribution in an atom [Nelms 1955, Cromer 1965, Cromer 1967, Cromer 1970, Cromer 1981, Ilavsky 2006].

Historically, SAXS data were recorded with a pinhole camera [Guinier 1955]. SAXS cameras with prismatic slits cause smearing in the SAXS pattern, which needs to be accounted for with numerical routines after experiment such as "desmearing" and multiple scattering [Perret 1971, Ruland 1971, Ruland 1974]. The characteristic of SAXS instruments is that the approach of small scattering angles requires very long distance between sample and X-ray detector. This can extend over several meters, unless a Bonse-Hart camera is used. Small-angle scattering can be done with neutrons as well [Feigin 1987].

3.3.1 SAXS on GC supercapacitor electrodes

Figure 3.8 shows conventional X-ray diffractograms of four different types of GC, where we notice from the different sharpness of the Bragg reflections that the samples have a different degree of crystallinity because of different heat treatment temperature. This different crystallinity also manifests in different microstructure, which has a signature at the very low scattering angles of below roughly 10°. Small-angle scattering data are typically recorded from scattering angle to scattering vector or momentum transfer from 2θ to $k = 2 \cdot \pi \cdot \sin\theta/\lambda$.

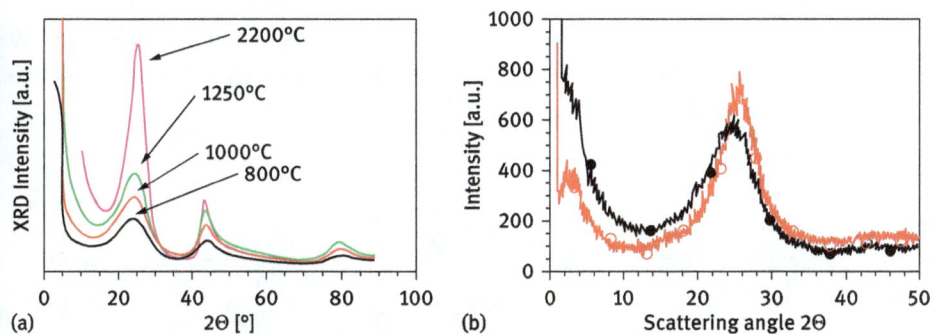

Figure 3.8: (a) Diffractograms (Cu Kα) of GC sheets pyrolyzed at four different temperatures 800 °C and 1000 °C (both HTW Sigradur K), 1250 °C (= K800 repyrolyzed to 1250 °C), and 2200 °C (HTW G2200). The (002) Bragg reflection of the K-type GC (the three lower temperatures) is around 24.2°, with a lattice spacing around 3.68 ± 0.1 Å. For the 2200 °C G-type GC, the peak is at 24.2° with a lattice spacing of 3.54 ± 0.1 Å. Note the decreasing FWHM with increasing T and the overshooting intensity at around 5°. (b) X-ray diffractograms of K-type GC sheet before (red) and after (black) thermal gas phase oxidation.

The intensity is transformed on the logarithmic scale, which makes the so-called log-log plot for the first qualitative assessment of the scattering curve. The first quantitative assessment goes via the Guinier plot, where the log intensity is plotted versus k^2. This produces the so-called Guinier shoulder, the position of which is directly numerated as the particle size. This is demonstrated again by the GC sample, which produces a Guinier radius of 5.9 Å for low crystallinity GC and 10.1 Å for high crystallinity GC. Various plotting routines exist from which one can determine several microstructure parameters.

The Porod plot is used to determine the internal surface area and the scattering background. The Kratky plot is used to determine the "length" of long structures in the sample.

The right panel of Figure 3.8 shows how the diffractogram of GC is changed when the sample is oxidized at high temperature in a muffle furnace (gas phase oxidation). The intensity at 2°–4° increases dramatically upon oxidation. Observe also the shift of the peak to lower angle and the redistribution of diffracted intensity from the graphite peak to the small-angle scattering range after this thermal oxidation.

GC is frequently used as electrode material in electrochemistry [Jenkins 1976]. GC is made from pyrolysis of phenolic resin. GC is a highly porous material, but the pores are not open and not connected. Oxidation at the surface opens and connects these pores and creates a high accessible porosity for porous electrodes. The heat treatment temperature (HTT) determines the microstructure of GC, including its pore structure. Figure 3.8 is a manifestation of the structural changes during pyrolysis and during oxidation.

The open porosity grows like a film into the GC material. This is shown in the electron micrographs in Figure 3.9, where two differently colored layers, films are growing on a differently colored bulk GC core layer. With the high internal surface area available in this activated GC, it is possible to assemble them as electrodes for electrochemical double-layer capacitors, so-called supercaps or ultracaps [Miklos 1980]. Figure 3.10 shows three such supercaps that were built during my PhD thesis at PSI by Dr. Martin Bärtsch within the LOK2000 [Schönborn 1998] Swiss PPM project.

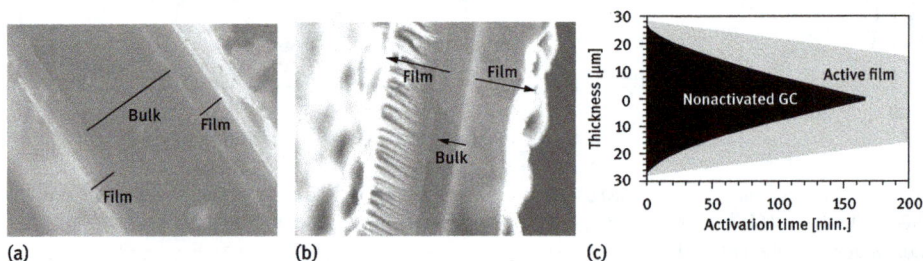

(a)   (b)   (c)

Figure 3.9: (Left) Scanning electron micrographs of GC sheets with nominal 100- (a) and 60-micrometer (b) thickness, showing films with open porosity on top and bottom sides. Pores are in nanometer size and not visible in SEM. (c) Schematic of the GC sheet with increasing film thickness and decreasing bulk thickness with increasing oxidation time. After 180 min, the nonactivated core vanishes and all material left is film with open porosity.

Thermochemical gas phase activation of GC is a relatively simple process suitable for use at the industrial scale. The kinetics of the growth of the active film with thickness D on a plate have been determined by multiple measurements of sample thickness and film thickness of number of samples grown under a wide range of conditions. This

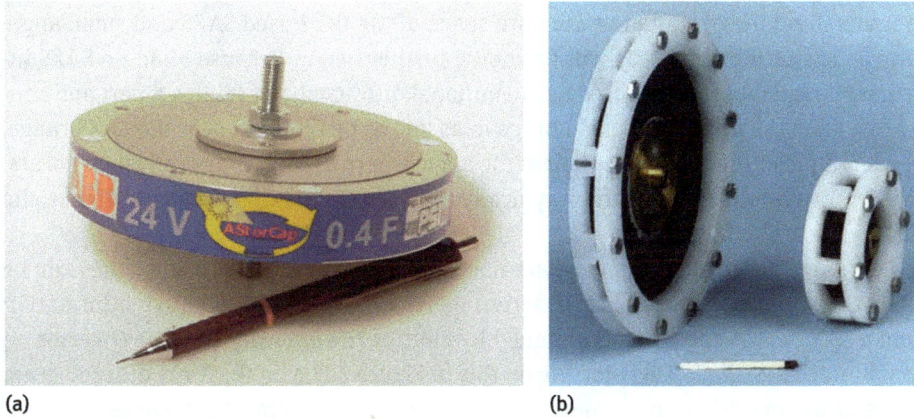

(b)

Figure 3.10: Operational supercapacitors built from oxidized GC plates, polymer separators, and three molar sulfuric acid as an electrolyte (photo credit Paul Scherrer Institut).

kinetics follows the generalized Lambert W function:

$$D(t) = \frac{b}{a}\left(1 + \text{Lambert } W\left(-\exp\left(-1 - \frac{a^2 t}{b}\right)\right)\right) \qquad (3.1)$$

Constants a and b are the reaction rate and diffusion constant, respectively, and are temperature dependent [Braun 2000b, Braun 2003j]. With a balance between reaction rate controlling the burn-off of the film and the diffusion constant, which is controlling the film growth, the film thickness remains after some specific activation time constant. This steady-state film thickness is given by the ratio b/a, which is approached with a time constant $t = b/a^2$. Activation time essentially determines the active film thickness $D(t)$ at a given temperature and oxygen concentration. Increasing the oxidant concentration accelerates the film growth.

   This situation is realized by activating the sample for a time t^*, which can be determined by solving equation (3.1) for the activation time t. A GC plate with initial thickness L will be totally activated after the time t^*,

$$t^* = -\frac{\ln(-\frac{2 \cdot L \cdot a - b}{b}) \cdot b + 2 \cdot L \cdot a}{a^2} \qquad (3.2)$$

The active film growth with thickness $L(t)$ in a shrinking sphere with outer radius $r(t)$ and unreacted inner core, burn-off rate α, and diffusion coefficient β, is governed by the following equation:

$$\left(\frac{r - L}{r}\right) \cdot L \cdot (a + \dot{L}) - \beta = 0$$

To understand the evolution of the pore space of the GC, I used SAXS and small-angle neutron scattering (SANS), as will be shown further below. Because I had no SAXS instrument available at first, I used a conventional diffractometer Philips X'pert and simply extended the diffraction angle range to as low as possible to cover the SAXS range, which is typically defined by the half of the angle where the first Bragg reflection occurs. For carbon materials like GC, this is typically the graphite peak at around 24°. Hence, the SAXS range is $Q < 12°$.

Although X-ray diffractometers are not designed for SAXS experiments, they can be useful for a quick comparison of different materials. The first qualitative information from a SAXS scattering curve is obtained by plotting the intensity over the Q-vector on logarithmic axes, the log-log plot. We do this in Figure 3.11 with the X-ray diffractogram of one of the previously presented carbon soot samples. On the right, we see the conventional XRD with linear intensity over linear 2θ scale. We see how the plot is changed by simple transformation from linear to logarithmic scale for both axes. The intensity plateau around $2\theta = 0.2°$ can be considered the Brag (000) reflection. The intensity minimum around $2\theta = 10°$ separates roughly the SAXS from the WAXS region.

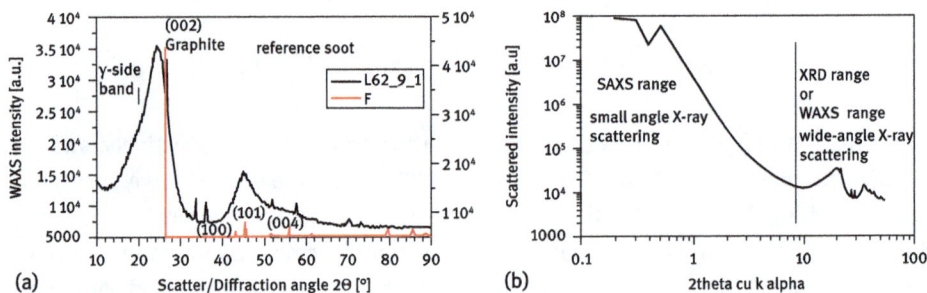

Figure 3.11: (a) X-ray diffractogram of diesel soot, along with vertical lines for Bragg reflections from reference graphite (red). (b) The same diffractogram after transformation to log-log scale.

We see in Figure 3.12 how the intensity at low Q and the slope of the high Q tail of differently oxidized GC is changing. SAXS is therefore a very good method for analyzing materials where the microstructure changes during synthesis and processing. The exponent of decay for typical carbon materials is around −2.5. When we have a porous material with an extended compact pore space, the exponent is exactly −3. Surfaces with a smooth electron density distribution have an exponent of −4 [Glatter 1982]. This latter case amounts to the so-called Porod background subtraction, where a power law background with an exponent of −4 accounts for a smooth electron density scattering background [Porod 1951, Porod 1952a, Porod 1952b]. The right panel of Figure 3.12 shows the variation of this power law exponent depending on how long the GC sample was oxidized. Obviously, the microscopic roughness of GC is heavily affected by the thermal gas phase oxidation, as the exponent of decay decreases from initially −2.4 to −2.2, and

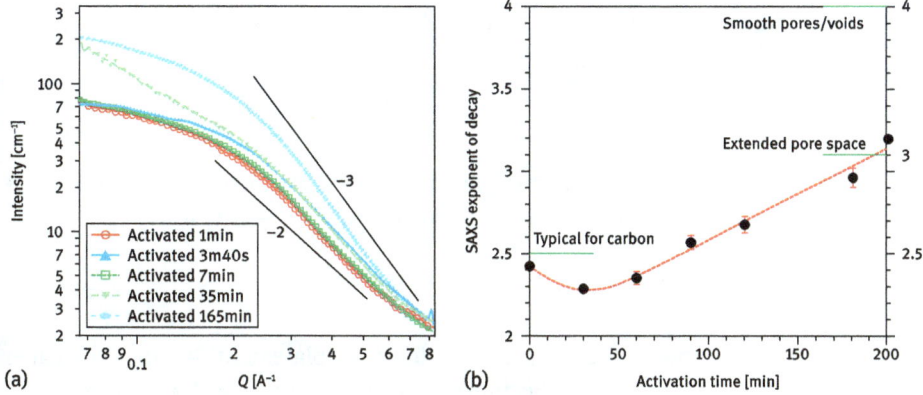

Figure 3.12: (a) SAXS curves of GC type K after thermal gas phase oxidation for times from 1 to 165 min. (b) Variation of the SAXS exponent of decay of GC type K after thermal oxidation for up to 200 min.

then upon opening and connecting of the micropores increases to −3. This is a special case in pore topology because it means the pore space extends connected over the entire probed sample volume. The pore space is compact.

We have studied diesel soot samples under different conditions, such as immersed in acetone, or compacted in pellets with different compaction pressure. The fractal aggregates of the soot particles then adjust to the outer environment, which is visible in the ultrasmall-angle X-ray scattering (USAXS) spectra [Braun 2005g].

The next step of assessment of the SAXS range is plotting the natural logarithm ln of the intensity versus Q^2 on ln scale. The justification for this is found by Guinier [Guinier 1939, Guinier 1964], where the radius of gyration R_g of an object, regardless whether particle or void can be determined by a plot of the logarithm of the scattered intensity $d\sigma/d\Omega$ versus the square of the scattering vector Q:

$$\frac{d\sigma}{d\Omega}(Q) = \Delta n_f^2 N V^2 \cdot \exp\left(-\frac{(QR_g)^2}{3}\right)$$

where N is the number of pores per volume, V is the volume of a single pore, and Q is the scattering vector. Q is a measure for the momentum transfer of the electrons with respect to the scattering atom and is related to the scattering angle θ and X-ray wavelength λ as

$$Q = \frac{4\pi}{\lambda} \sin \Theta.$$

Often, capital Q is reserved for the wave vector or momentum transfer in neutron scattering, whereas we often find k, s, q for the same meaning in X-ray scattering. The prefactor $2\pi, 4\pi$ may be sometimes missing, depending on which literature we are reading. The scattering contrast δn_f^2 is the difference of the electron densities in the vicinity of a particle (or void) in a matrix and depends on the chemical composition and mass density:

$$\Delta n_f^2 = \left(\sum_i f_i (n_{i,\text{particle}} - n_{i,\text{matrix}}) \right)^2$$

where f_i is the atomic structure factor [Ilavsky 2006] of atoms of sort i and n_i is the number density of atoms of sort i. For $Q = 0$, the measured intensity is

$$\frac{d\sigma}{d\Omega}(Q = 0) = \Delta n_f^2 N V^2$$

The Guinier plot in Figure 3.13 shows that the intensity plateau of GC type G is higher than that of type K. This is direct indication that the G-type has a larger surface area.

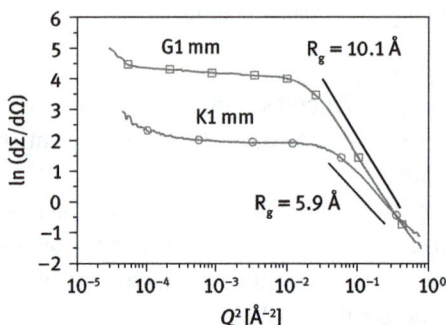

Figure 3.13: Guinier plot of nonactivated GC type K and G of 1 mm thickness. The high Q tail slope is the radius of gyration (Guinier radius) R_g [Braun 2006a].

This goes along with the steeper slope for G, revealing larger pore sizes. The size and shape, the topology, and the morphology of an inhomogeneous material puts its signature on the scattering pattern. For the SAXS range, we can resolve structural information from around 1 nm to 1 micrometer. The reciprocal space relation follows that if we want to resolve very large structures, we have to zoom in deeper into the scattering angle and approach $2\theta = 0$ as close as possible. Technically, this is warranted by increasing the distance between irradiated sample and X-ray detector (or neutron detector). SAXS instruments are therefore "long."

As there should be no material in the optical path between X-ray source, sample, and detector, so as to minimize parasitic scattering, e. g., by air, this optical path is usually evacuated within long metal tubes. Figure 3.14 shows the author of this book working at the SAXS instrument at the beamline sector ID-32 of the Basic Energy Science Synchrotron Radiation Center (BESSRC) at the Advanced Photon Source in Argonne National Laboratory.

The black unit on the left side of the metal beam guide vacuum tube is the detector unit. The X-ray beam is arriving from the right side of the tube. For increasing the Q-range to even smaller values, the detector unit can be moved to the left and another vacuum tube can be inserted to bridge the new, longer X-ray optical path.

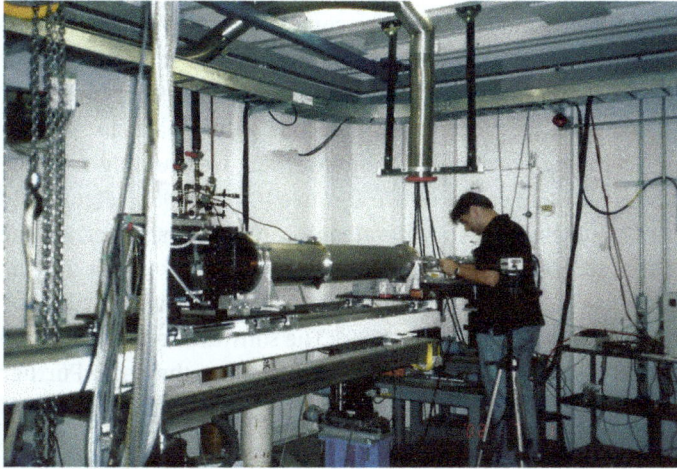

Figure 3.14: Artur Braun working in the experimental hutch at BESSRC-CAT beamline in Sector 32-ID, APS, Argonne National Laboratory. The horizontal tube is the X-ray waveguide evacuated to prevent parasitic scattering and absorption of X-rays. At the left end of the tube is the X-ray detector unit.

When a monolithic sample with some anisotropic structure is tilted in the X-ray beam, the anisotropy will impose its structure onto the scattering pattern. This can be easy seen when a two-dimensional detector is used. Figure 3.15 shows the SAXS pattern of a GC sheet with surface plane normal to the beam direction (left) and with surface normal 45° toward beam direction. We find a radial symmetric pattern in the first case but none in the second case. We can analyze such complex pattern by considering section-by-section and by reconstructing the anisotropy of the sample.

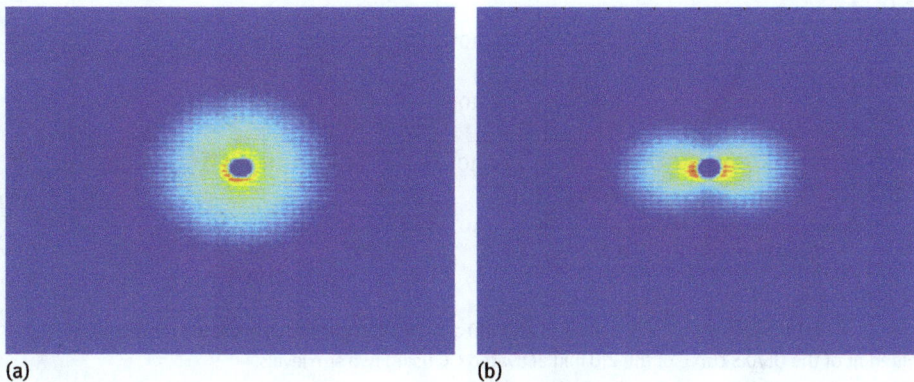

(a) (b)

Figure 3.15: Two-dimensional SAXS patterns of GC for short geometry ($0.03/°A \leq Q \leq 0.8/A°$). Pictures on left side were measured with the beam perpendicular to the sample surface normal. Pictures on right side were measured with a tilt angle of 45° between sample surface normal and beam direction. The color code determines the SAXS intensity. The (invisible) x- and y-axes denote the qx and qy scattering vector.

It is possible to mathematically model the entire complex SAXS curve by the various mathematical components. We can do this by considering a structure factor for the size and geometry of the objects in the sample, in the most primitive way we work with a Guinier function, and then we "fold," convolute this with a particle size distribution such as like a logarithmic normal distribution, which we multiply with the structure factor, and then integrate over the entire Q-range [Glatter 1982]. Figure 3.16 shows the USAXS curves of nonactivated and activated GC. In the right panel, we see how a Guinier fit and a Porod fit are simultaneously applied. We have used for this the IRENA and INDRA software packages from APS [Ilavsky 2009], which work with Igor Pro [Busbey 1999, Goroff 1997, Kline 2006, Marpet 1996]. The software returns, e. g., the pore size distribution, which is shown by the size and the width of the peak. Further quantitative analysis of the SAXS curves was performed using the Unified Fit method [Beaucage 1995], as shown in Figure 3.16.

$$I(q) = F_B + \sum_{i=1}^{n} S_i(Q)(G_i \cdot \exp(-Q^2 R_{g_i}^2/3) + B_i \cdot \exp(-Q^2 R^2/3) \times \{[\operatorname{erf}(QR_{g_i}/6^{1/2})]^3/Q\}^{p_i})$$

where F_B is a flat background, i represents the structural levels, and for each structural level i, G_i is the exponential prefactor, R_{g_i} is the Guinier radius of gyration, and B_i is a constant prefactor specific to the type of power-law scattering. We basically can include all structures and corrections in the so-called "Unified fit" [Beaucage 1995] in the macro that returns after good least squares fit all relevant structure parameters.

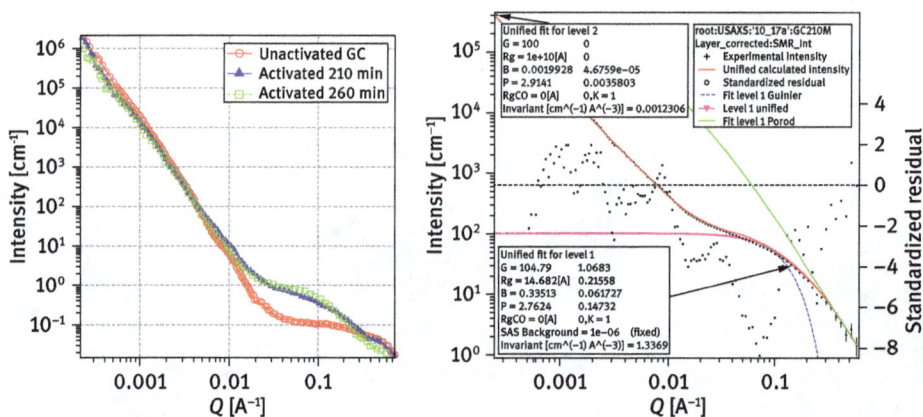

Figure 3.16: (a) USAXS curves for unactivated and 210 and 260 min gas phase activated (oxidized) GC. (b) Unified fit of the USAXS curve of the 210 min activated GC using two size levels.

Figure 3.16 shows ultrasmall-angle scattering curves of three types of K-type GC, two of which are thermal gas phase oxidized for 210 and 260 min. The "gain" in porosity and surface area is demonstrated by the increase of the plateau intensity in 0.02 and 0.2 1/Å.

The right panel of Figure 3.16 shows the USAXS curve of the 210 minutes oxidized sample along with a unified fit after [Beaucage 1995]. The blue dashed curve at the bottom is the Guinier range for size level 1 with Rg = 14.682 Å. The green solid line is the "enforced −4" Porod fit and fits the high Q range.

With the Unified fit applied, we can then plot, e. g., the pore volume of the GC as a function of the pore diameter. The 210-min activated GC has a size maximum at around 10 A, which upon further activation to 260 min enlarges to over 20 A, with a noticeable increase in the width of the pore size distribution (Figure 3.17).

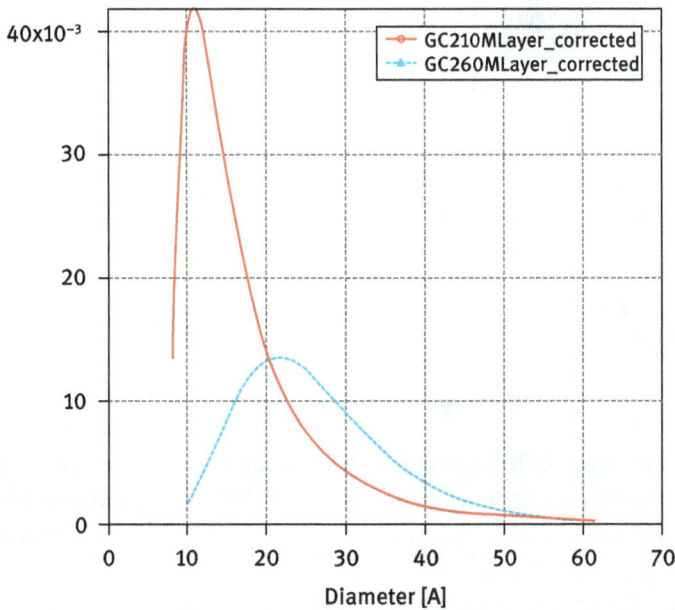

Figure 3.17: Pore size distribution of K-type GC, which was a gas phase oxidized in an air vented muffle furnace for 210 min (red curve) and 260 min (blue curve), respectively.

3.3.2 Excursion to ultrasmall-angle neutron scattering (USANS)

Small-angle scattering is not restricted to X-rays as probes. The physical principles apply to not only all photons but also electrons and neutrons and any other probes that satisfy the particle-wave dualism. SANS is often used for the investigation of polymers and soft condensed matter because neutrons are very good probes for hydrogen and protons and other low Z elements, notwithstanding that polymers, such as NAFION®, can also be measured with SAXS [Haubold 2001]. We have investigated GC with ultrasmall-angle neutron scattering at SINQ and prepared the samples in the following way, as illustrated in Figure 3.18. From GC disks of 100-micrometer thickness, pieces were cut of around 1 cm x 4 cm. The G-type was very soft and cut very easily with a doctor blade. This GC piece was then coated with Lacomite varnish (Agar Scientific) fully on the back and with

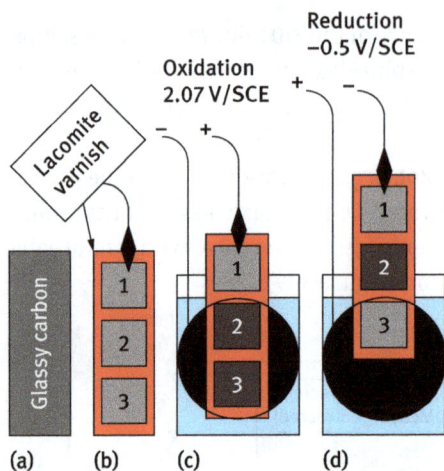

Figure 3.18: Schematic for the electrochemical preparation of GC sheet samples. We cut a highly graphitic G-type GC sheet with a doctor blade into a rectangle and arrive at stage (a). We then paint the pack of this slice fully and the front of it as shown with a frame pattern with Lacomite® varnish so as to have three sample compartments that can be separately exposed to electrolyte (b). We then clamp the electrode sample and dip the two lower compartments into the electrolyte and anodize (electrochemically oxidize them) and have thus situation (c). After oxidation, we pull out the sample so that only the bottom portion (3) is still in the electrolyte, and now we electrochemically reduce that portion of the electrode (d). The color change during oxidation and reduction was actually observed.

window frame compartments on the front so that only the open frames were exposed to the electrolyte. The GC piece was then connected as working electrode and dipped into the electrolyte with the two bottom windows and electrochemically oxidized (anodized or activated).

When the wanted oxidation was finished, the sample was partially pulled out from the electrolyte so that only the bottom compartment was exposed to electrolyte. Then the potential was reversed and the previously oxidized sample was electrochemically reduced. After this electrochemical processing, the sample was removed from the electrolyte, rinsed in distilled water, and dried in ambient atmosphere. Later, the sample was cut in three pieces and made ready for USANS experiments.

After the oxidation and after the reduction, a CV was recorded so as to get an impression on the accessible internal surface area of the GC. Before activation, we also took a CV of the pristine, untreated GC. Figure 3.19 shows the three CVs in one plot. The right abscissa shows the current density of the pristine GC in the microampere range. This very low current density is not surprising because the GC had a very smooth and flat surface with no apparent porosity. The CV after oxidation is plotted on the milliampere range on the left because of the current density increase by around three orders of magnitude. The CV has a slight linear trend versus the potential, which is an indication for an increased ohmic resistance of the electrode, likely because of the formation of a di-

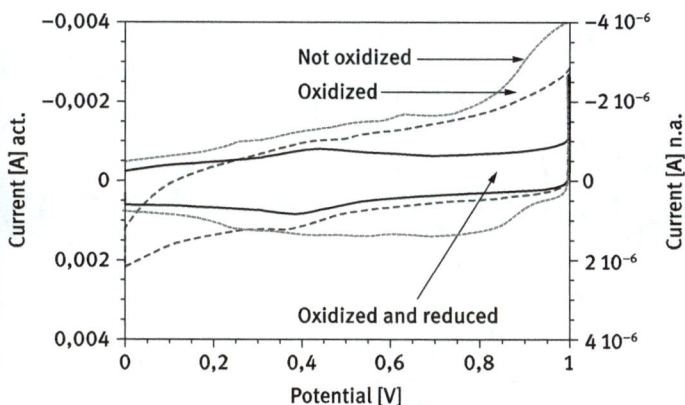

Figure 3.19: Cyclic voltamograms of nonactivated and electrochemically activated (5 min) and reduced (15 min) SIGRADUR G with 60 μm thickness. Note the two different abscissae on left side and right side.

electric layer during oxidation on the GC surface. After electrochemical reduction, this suspected layer is removed (reduced), but the current density is also noticeably reduced.

The corresponding USANS scattering curves of the three samples are displayed in Figure 3.20 as log-log plots. We notice a well-pronounced Guinier range for high Q values. The left panel shows how from the original USANS curve the Porod background was fitted and subtracted so as to arrive at the clear Guinier shoulder, which is indicated in the treated USANS curve with open symbols.

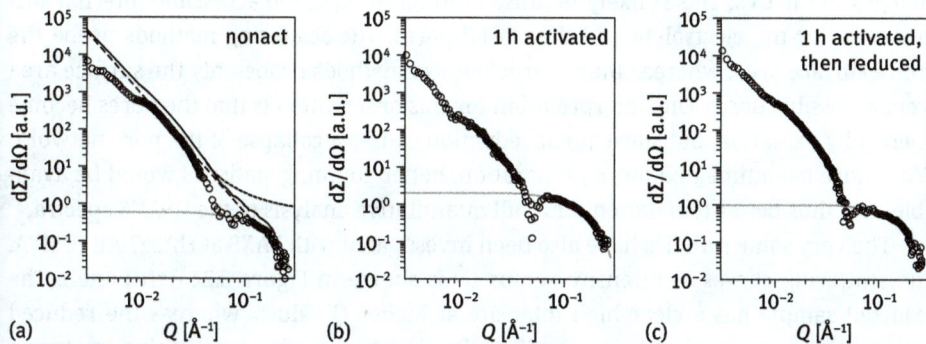

Figure 3.20: Log–log plots of nonactivated (a) and 1 h activated (b)/reduced (c) samples, after subtraction of a power law with Q^{-3} and constant background (open symbols). Left plot (a) shows the original SANS curve ($\cdots$), SANS curve after constant background subtraction (—), as well as the power law fit (- - -). Solid lines (—) in all three plots are least squares fits to Guinier function.

The differences in the scattering curves become more obvious when we plot them in a Porod plot as IQ^4 versus Q^4. This has been done in Figure 3.21. The "height" of the plateau in a Porod plot scales linear with the internal surface area of the probed vol-

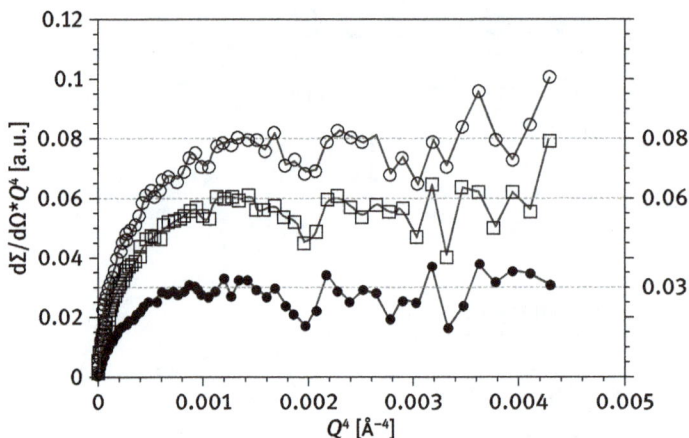

Figure 3.21: Porod plots (from SANS) of electrochemically treated SIGRADUR G with 60 µm thickness. Filled circles denote the nonactivated samples. Open circles denote the oxidized samples. Open squares denote the oxidized and subsequently reduced samples. The Porod constants are written in arbitrary units. Oxidation time was 1 h.

ume. It is obvious how the nonoxidized GC has the lowest intensity plateau with 0.03 a. u., whereas the both oxidized ones have higher intensity. Interestingly, oxidation increases this plateau to 0.08 but subsequent reduction decreases the surface area so that the plateau is decreased to 0.06 a. u.

The relative changes in the USANS intensity are not the same like the relative changes in the CVs. This is likely because of differences in the accessible internal surface area for the electrolyte and the closed pores. The scattering methods probe the entire surface area, whereas the electrochemical methods probe only the surface area from accessible pores. One interpretation for this observation is that the pores become opened by oxidation, but subsequent reduction causes a collapse of the pore network. With more beamtime and more preparation, better counting statistics would be available, and thus better foundation for a full quantitative analysis of the USANS spectra.

The very same samples have also been investigated with SAXS at HASYLAB JUSIFA. The scattering curves as intensity versus Q^2 is shown in Figure 3.22 (left panel). The oxidized sample has a clear high intensity at higher Q values, whereas the reduced and nonoxidized samples show actually quite similar, virtually overlapping scattering curves. Also, we did a quick Porod analysis for this set of curves, which is shown in Figure 3.22, right panel.

The trend for the change of the intensity plateau of the Porod plots is the same for the SAXS and for the USANS curves. However, the relative changes are different. What we observed from visual inspection of the scattering curves in Figure 3.22 (left panel) is reflected in the Porod plot in Figure 3.22 (right panel). The plateau height for the reduced and nonoxidized samples is quite similar. I believe the close proximity of these two latter samples is because the SAXS experiment was done later than the USANS experiment and,

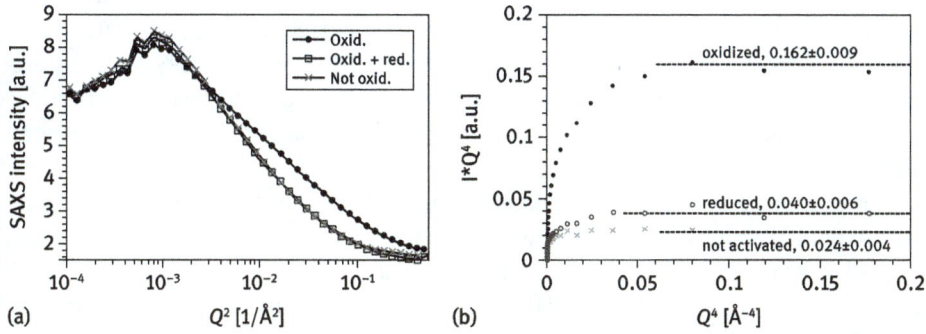

Figure 3.22: (a) SAXS Guinier plot of electrochemically treated thin SIGRADUR G sheets. State of sample is marked with symbols accordingly in the legend. Activation was 30 min at 2.07 V. Reduction was at −0.5 V. (b) Porod plot from SAXS data of electrochemically treated GC G with a thickness of 60 micrometers. Activation was 30 min at 2.07 V. Reduction was 15 min at −0.3 V.

therefore, the pore collapse had progressed because of the more thorough drying of the samples.

With the electrochemical impedance spectra, we are able to determine the specific double-layer capacitance $C_A(f)$ in Farad/cm^2 with the AC frequency f and the electrode area A, which is exposed to the electrolyte, and the imaginary part of the impedance Im(Z).

$$C_A(f) = \frac{-1}{2 \cdot \pi \cdot f \cdot \text{Im}(Z) \cdot A}$$

The double-layer capacitance is a direct measure for the internal surface area of an electrode to which the electrolyte has access. It depends on the specific adsorption of the electrolyte molecules on the electrode material, whether the specific double-layer capacitance in μF/cm^2 is large or small. It can be 10–20 μF/cm^2 for GC [Braun 1999c]. We have therefore to be prepared for a factor 2 in the difference between internal surface area as determined from this approach versus the surface area as determined from SAXS or from gas adsorption measurements.

3.3.3 USAXS on aerogel catalysts

SAXS and USAXS have been quite frequently applied to catalyst systems. We have to be clear here about the proper use of terminology. Catalysis is a very important topic in chemistry, physical chemistry, chemical engineering, and industrial engineering. Many surface scientists work on heterogeneous catalysis and use, e. g., single crystal substrates—the catalysts—such as Au, Pd, Cu, and so on to study catalytic reactions of these surfaces with particular molecular adsorbates. Such studies include also surface structure studies with Low Energy Electron Diffraction (LEED) LEED, STM, X-ray photoelectron diffraction (XPD), topography studies with STM, and also XPS, temperature-controlled desorption spectroscopy TDS, for example. Typically, such studies are called fundamental studies. These are model systems, which are only from a secondary point of view for industrial applications.

For example, in petrochemical industry, removal of sulfur from mineral oil is a traditional field of activity. This can be done by supplying hydrogen so as to form H_2S. This is called hydro-desulfurization and is frequently studied with deuterium tracer analysis [Katsapov 2010]. A realistic catalyst is typically a metal or metal oxide cluster deposited onto the surface of a highly porous catalyst support material, such as carbon, aluminum oxide, magnesium oxide, or more complex supports such as zeolithes. Here, I show cobalt catalysts deposited on silica aerogels. These aerogels were synthesized at the University of Utah in a glass tube with supercritical drying and doped with 2 % or 10 % cobalt and then reduced with drying, which led to monolithic porous bodies with cylindrical shape.

Aerogels with no cobalt showed three populations of scatterers with radius of gyration of 10, 40, and 65 Å, resembling the truly known high surface area material. Figure 3.23 shows the USAXS curve of the CO_2-dried aerogel in log-log plot (dotted line). The solid lines are a straight power law fit with exponent −3.6. The two other lines are Guinier functions. The structure factor $S(q, R)$ of a sphere is mathematically written as [Rayleigh 1910]

$$S(q, R) = \left\{ \frac{3 \cdot [\sin(q \cdot R) - q \cdot R \cdot \cos(q \cdot R)]}{(q \cdot R)^3} \right\}^2$$

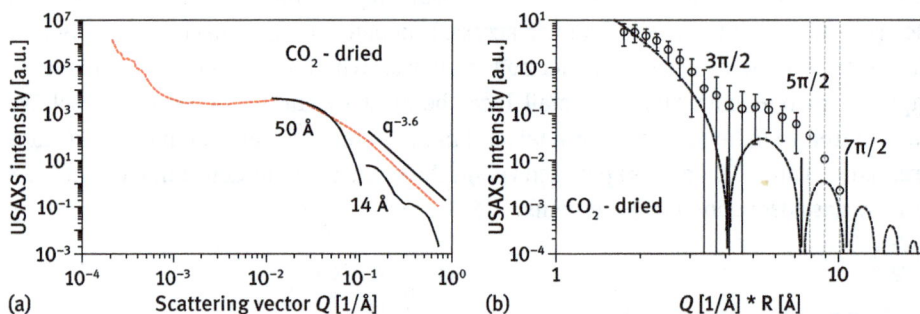

Figure 3.23: (a) Log-log plot of CO_2-dried aerogel. Two least squares fits of the Guinier function are marked with solid lines. The straight solid line to the right side is a power law function with −3.6 as the exponent of decay. The solid line below shows additional structures after subtraction of background scattering and the power law function. Numbers denote radii of gyration Rg. (b) Comparison of micropore scattering of aerogel and the structure function of spheres with 14 Å radii. Vertical lines denote the expected positions of the intensity minima after the equation by Rayleigh [Rayleigh 1910]. The ordinate axis is given in dimensionless units with = 14 Å. Reproduced with permission of the International Union of Crystallography from [Braun 2005d]. http://scripts.iucr.org/cgi-bin/paper?S0021889804029528.

This structure function shows intensity minima at the $(2n + 1)$ multiples of $\pi/2$. We indeed observe such minima in the SAXS curve of the aerogel, as shown in the right panel of Figure 3.23, where the SAXS intensity is plotted versus the product of qR, along with calculated function $S(q, R)$. When we integrate the structure function with a size distribution, such as a logarithmic normal distribution, the sharp minima will smear out, as can be seen in many SAXS curves.

When the "objects" in a sample is not spherical, this approach is not well applicable, and we have to try a different structure factor, such as for rods, disks, and other geometries [Glatter 1982].

When we dope the aerogel with cobalt, the scattering curves show new features that are only visible in the USAXS curves because the new structures formed by the cobalt are so large, around 1000 Å. Figure 3.24 shows and compares the USAXS curves of the nondoped and doped aerogel sample. The former USAXS curve has the readily shown plateau starting at 0.03 1/Å. The 2 % and 10 % Co-doped samples have intensity at this region, which is increased by two to three orders of magnitude. Interestingly, the low-concentration doped aerogel has structures with a larger Guinier radius (1747 Å) than the high-concentration doped aerogel (1044 Å).

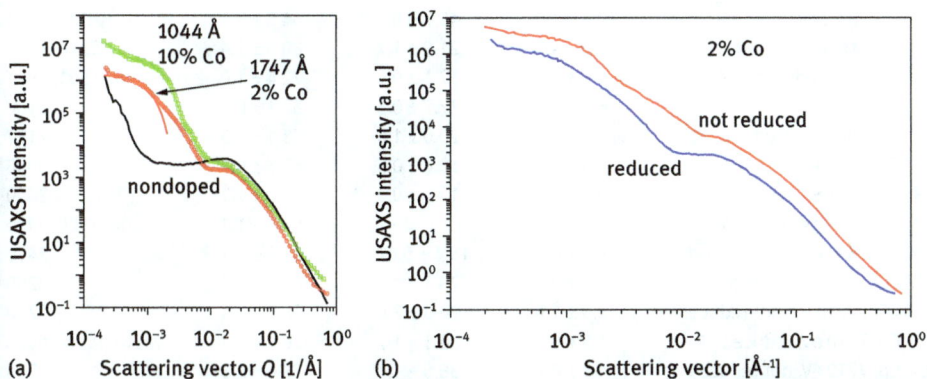

Figure 3.24: (a) Scattering curves of nondoped (thin scattering curve) and 10% doped (squares) and 2% doped (circles) samples. Drawn lines in the scattering curves from doped samples denote least squares fits of the Guinier function. (b) Comparison of the scattering curves of reduced and unreduced samples with 2% cobalt. Reproduced with permission of the International Union of Crystallography from [Braun 2005d]. http://scripts.iucr.org/cgi-bin/paper?S0021889804029528.

After catalyst synthesis, the metal nanoparticles are typically in an oxidized state. A subsequent reduction process is usually necessary to prepare the catalyst for further experiments. The right panel of Figure 3.24 shows how the USAXS curves change when the aerogel filled with Co catalyst is reduced.

Our silica aerogel samples were monolithic. They were grown in cylindrical glass tubes and had a color pattern with radial symmetry, reminiscent of Liesegang rings.

They had been synthesized by supercritical drying, doped to 2 % and 10 % with cobalt and were then reduced with hydrogen. The nondoped aerogels have three populations of scatterers with radii of gyration of about 1, 4, and 6–7 nm. The doped aerogels show an additional structure with a radius of gyration ranging from 105 to 300 nm. This structure causes intensity oscillations in the SAXS patterns, thus revealing a relatively narrow size distribution [Lifshitz 1961, Wagner 1961]. Scattering curves of the 10 % doped aerogels fitted well with a Lifshitz–Slyozov–Wagner (LSW) particle size distribution, thus revealing that Ostwald ripening might have occurred during aerogel preparation. Table 3.2 summarizes the structure parameters, which we obtained by the least square fits.

Table 3.2: Radii of gyration of the first three size levels, as obtained from the Unified fit [Beaucage 1995].

The largest size level was fitted to the structure factor of spheres, including a Gaussian size distribution function.			
Sample	Finest level Rg (A°)	Level 2 Rg (A°)	Level 3 Rg (A°)	Largest level Mean diameter; FWHM (A°)
CO$_2$ dried, nondoped	4.7 ± 3.5	34.5 ± 1.6	71.5 ± 2.5	–
EtOH dried, nondoped	11.0 ± 2.8	46.6	55.0 ± 5	–
10 % Co, no skin	10.6 ± 1.0	46.7 ± 0.3	64.5 ± 3.5	1880; 611
10 % Co, skin	7.5 ± 30.3	21.9 ± 1.7	80.6 ± 1.4	2248; 1189
10 % Co, not reduced	6.1 ± 1.1	17.1 ± 0.5	72.4 ± 1.0	1501; 1627
10 % Co, reduced	6.4 ± 0.7	17.7 ± 0.5	68.1 ± 0.7	2100
10 % Co, not reduced	7.2 ± 0.8	20.0 ± 1.6	68.1 ± 1.3	2127
2 % Co, reduced	7.3 ± 1.3	18.0 ± 0.6	62.8 ± 0.7	3000
2 % Co, not reduced	6.4 ± 0.6	17.6 ± 0.4	63.0 ± 1.2	2400
2 % Co, 7695 eV, reduced	8.9	25.3	61.1	1000, 2000, 4400
2 % Co, 7695 eV, not reduced	8.2 ± 0.8	18.1 ± 0.61	68.2 ± 1.1	2400
10 % Co, 7710 eV, reduced	7.4 ± 0.6	29.1 ± 4.5	67.0 ± 3.9	2200
10 % Co, 7710 eV, not reduced	8.3 ± 1.5	20.6 ± 2.0	68.2 ± 1.4	2210
2 % Co, 7710 eV, reduced	4.5 ± 0.9	36.1 ± 1.7	66.4 ± 3.7	More populations
2 % Co, 7710 eV, not reduced	4.8 ± 0.7	36.6 ± 1.7	73.6 ± 2.3	~2253

Following the quantitative analysis of the various USAXS curves from the samples subject to various synthesis and processing conditions, we list the numerical structure parameters in a table for easier overview and comparison. This is done exemplary in Table 3.2. Here, we list only the radii of gyration, Rg, for the four size levels that we have immediately observed by looking at the USAXS curves. Certainly, more structure parameters that are part of the scattering models and produced by the least squares fit routines can be listed, if wanted and necessary. Soon, the load of data can become overwhelming, and it needs very good bookkeeping during analysis. I will present such a case for SOFC where we applied anomalous USAXS at the Ni and Zr K-edge.

The same range shows also differences depending on whether the samples were reduced, or in their as-prepared condition. Scattering curves obtained from the cylinder axis region were different from the scattering curves obtained from the sample boundary, indicating a process-dependent skin effect.

Let me add some technical details for this experiment. The USAXS was conducted at synchrotron radiation beamline 33-ID at UNICAT, Advanced Photon Source, Argonne National Laboratory. At this beamline, a Bonse–Hart setup allows us to record USAXS scattering curves (SC) using a photo diode detector with an angular resolution of 0.0001 1/Å at a q-range from 0.0002 1/Å to 1.0 1/Å. The upper resolution range is 5000 Å. As we know already, the scattering vector q is the typically used quantity in small-angle scattering and relates to the diffraction angle 2θ, as known from XRD, via the relation $q = 4pl/\sin q$ with X-ray wavelength l. The data acquisition time for a SC with 150 data points was typically 15 min per sample. SC were recorded at X-ray energies of 7695 and 7710 eV, with the ultimate aim to make anomalous scattering (shown in Figure 3.25).

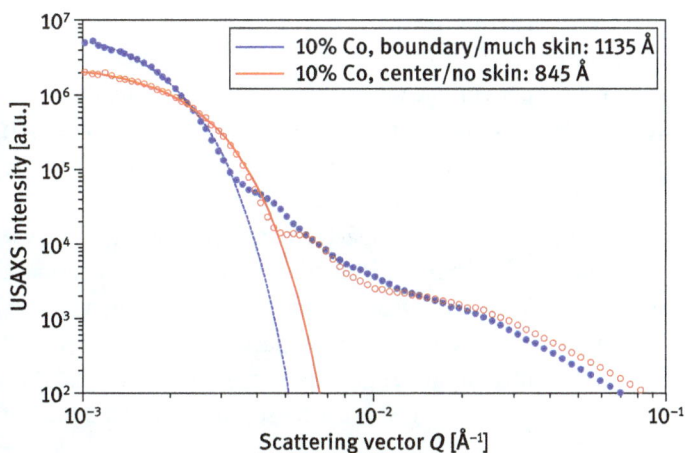

Figure 3.25: Scattering curves of 10% Co-doped samples from the center (no skin) and from the boundary (much skin). Reproduced with permission of the International Union of Crystallography from [Braun 2005d]. http://scripts.iucr.org/cgi-bin/paper?S0021889804029528.

All USAXS data were fully corrected for all instrumental effects, including background subtraction and desmearing. The samples came in cylinder-like shape and were cut in cylinder fragments with a thickness of approximately 1 cm. Accurate sample thickness data were not known, and scattered intensities were thus given in arbitrary units. For the skin studies, either the cylinder center (= no skin) or the cylinder boundary (= skin) was positioned in the X-ray beam, with the cylinder axis parallel to the X-ray beam. For all other cases, the cylinder axis of the sample was perpendicular to the beam, and the sample thickness at this position was at maximum.

It is possible that the cobalt particles could have undergone Ostwald ripening at high temperatures. The experimenter should be at least prepared for such possibility. An ensemble of particles that underwent Ostwald ripening would have a particular size distribution, which was calculated by LSW. We have applied the LSW distribution for the least squares fitting of the USAXS data. The agreement of fit and USAXS data suggests that the cobalt particles have indeed undergone Ostwald ripening. Such agreement, however, should not be taken as an ultimate proof for the proposition. The LSW distribution reads

$$P(R) = \frac{3^4 \cdot R^2 \cdot \exp[-1/(1 - 2 \cdot R/3)]}{2^{5/3}(R - 3)^{7/3}(3/2 - R)^{11/3}}$$

3.4 Extended X-ray absorption fine structure (EXAFS) spectroscopy

We have seen in Figure 2.9 in Part 2 that the X-ray absorption spectra may show ripples, wiggles, and oscillation far away high energy from the absorption edge. These oscillations are caused by the near-range order of the molecular environment of the central atom, which is energywise probed in XAS. Consider, e. g., a conventional metal atom such as Cu, being the central atom in a copper complex molecule. When you excite this atom with X-rays with energies around 9000 eV, photoelectron waves are generated by the Cu scattered by the surrounding atoms. The interference of these waves manifests in the oscillations observed as extended X-ray absorption fine structures (EXAFS).

When we transform the energy axis of the XAS spectrum into the k-space ($E = hv = p^2/2m = \hbar^2 k^2/2m$) and make a back-transformation from the k-space into the real space, we arrive at a distribution of the weighted X-ray scattering intensity, which in principle represents a radial distribution function, and thus the real space distances of first shell and second shell atoms around a central atom. This is illustrated in Figure 3.26. The X-ray intensity as a function of the k-vector is given by the so-called EXAFS equation [Koningsberger 1988]:

$$\chi(k) = \sum_j \frac{N_j f_j(k) e^{-2k^2 \sigma_j^2}}{k R_j^2} \sin[2k R_j + \delta_j(k)]$$

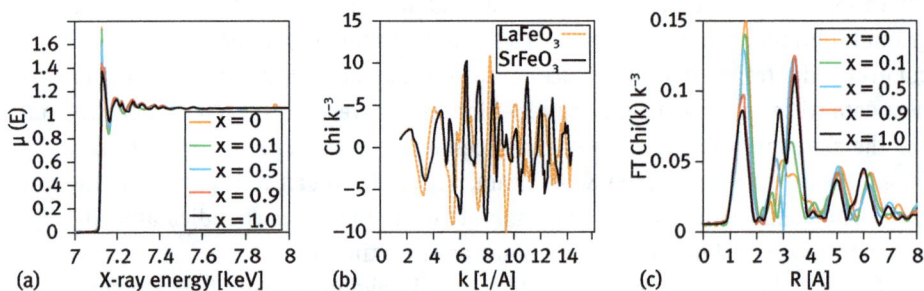

Figure 3.26: (a) Fe K-edge XAS with EXAFS oscillation of La$_{1-x}$Sr$_x$FeO$_{3-d}$ powder samples with $x = 0.0$–1.0. (b) Fourier transformed EXAFS oscillations for the LaFeO$_3$ and SrFeO$_3$. (c) Radial distribution of bond lengths obtained from EXAFS data analysis (reproduced from [Haas 2009]).

where $f_j(k)$ is the X-ray scattering amplitude of the N atoms j surrounding the central atom, which is exited by the X-rays, and $\delta_j(k)$ is the spectroscopic phase shift of these atoms. They depend on the atom number Z of the elements, which makes EXAFS an element specific analytical tool. The central atom has the distance R_j from the surrounding atom j. The mean deviation of the actual distance from the mean distance R is expressed by the disorder σ^2. Further disorder may come, e. g., by thermal vibrations, which are always in the sample at finite temperatures. This kind of disorder is typically accounted for by inclusion of a Debye–Waller factor [Debye 1913a, Waller 1923a] in the mathematical modeling of the spectra.

Because of the large energy distance from these wiggles/oscillations to the absorption edge, this spectrum part is called EXAFS. For an extensive treatment on the EXAFS method, the reader is referred to the book by Koningsberger and Prinz [Koningsberger 1988].

In materials with a crystalline long-range order, the distances between a central atom and surrounding atoms are determined by X-ray or neutron diffraction. The provided sample shows long-range order. Glasses or extremely small particulates like nanoparticles do not show such long-range order by default. At best, they have a near-range order, which can be identified by the aforementioned oscillations in the XAS and be analyzed with EXAFS spectroscopy.

EXAFS is likely the most important X-ray method for heterogeneous catalysis research. This is because the crystallographic structure of catalyst nanoparticles is hardly measurable with XRD. It is no coincidence that the EXAFS book was therefore written by researchers who are predominantly known for their catalysis research [Koningsberger 1988].

Similar to heterogeneous catalysis, EXAFS can be used in electrocatalysis. A very good example is the studies by H.-G. Haubold from KFA Jülich and JUSIFA at Hasylab Hamburg, Germany, who investigated the electrocatalytic properties of Pt nanoparticles in the context of fuel cell technology [Schlenter 1996]. Haubold is among the first who performed combined ASAXS and XANES *operando* at the synchrotron.

Figure 3.26 shows the Fe K-edge XAS spectra of La1-xSrxFeO3-δ for x = 0, 0.1, 0.5, and 0.9 in the X-ray energy range from 7000 to 8000 eV. EXAFS spectroscopy has one disadvantage. This is, you need to scan over a relatively large energy range, which will consume synchrotron beamtime. The advantage of EXAFS is that it is a distinct element specific method. This advantage does not hold per se for XRD, e. g., unless XRD is performed in the anomalous mode, which will be shown later in this book. Of interest are the EXAFS oscillations for E > 7200 eV. The minute differences in the EXAFS spectra reflect the structural differences in the samples, which arise from the substitution of La by Sr. Theoretically, it is necessary to extrapolate the spectrum to 0 eV and infinite energy so as to perform an accurate Fourier transform in these extreme ranges. Practically, this is not possible and not necessary. EXAFS software takes care of these extrapolations. With the subtraction of the exponential background and transformation to Chi(k)3 ($\chi(k)$3), we arrive at a shape of the EXAFS oscillations as shown in the middle panel of Figure 3.26

for the extreme cases $x = 0$ and $x = 1$. When we apply the Fourier transformation (FT) on the third power of the scattering amplitude, $\chi(k)3$, we obtain a probability distribution of spatial distances (radii) of the central atom versus the next and next-next neighbor atoms. The right panel in the Figure 3.26 below shows this radial distribution for the five compounds. The peak positions indicate statistically significant radial distances between the atoms. The given positions for R are not very accurate because phase relations have to be corrected for by a phase shift increment.

The EXAFS spectra can be mathematically modeled and then be compared and even least squares fitted to the experimental spectra. A typical data output that is obtained by applying the FEFF-fit code [Beckwith 2015, Cross 1998, Jorissen 2013, Rehr 1991, Rehr 1998, Rehr 2001, Rehr 2003] is provided in Table 3.3. The data allow us to sketch the variation of the distance between Fe and O depending on the composition ratio, as shown in Figure 3.27. Such information is generally interesting for assessment of the superexchange interaction and has importance for the transport properties of LaSrFe-oxide. One can get such information certainly also from XRD with Rietveld analysis [Rietveld 1966, Rietveld 1967, Rietveld 1969, Rietveld 2010, Rietveld 2014].

Table 3.3: FEFF-fit results of the Fe K-edge EXAFS Data for stoichiometry coefficients 0–1.SG = crystallographic space group [Haas 2009].

x	0	0	0.1	0.5	0.5	0.9	1
SG	Pm-3m	Pbnm	Pm-3m	Pm-3m	R-3c	Pm-3m	Pm-3m
So^2	0.9	0.9	0.9	0.9	0.9	0.9	0.9
r_{Fe-O}	2.006 +	2.007 +	1.990 +	1.952 +	1.950 +	1.919 +	1.913 +
	−0.003	−0.004	−0.010	−0.004	−0.004	−0.004	−0.007
NC	6	2,2,2	6	6	6	6	6
σ^2	0.0045	0.00465	0.00555	0.0066	0.0067	0.01	0.012
e_0	0.716	0.816	0.103	0.575	0.5156	−0.1281	−0.45
r_{Fe-La}		3.36 + −0.02			3.35 + −0.03		
r_{Fe-Sr}					3.28 + −0.07		3.290 + −0.05
r_{Fe-La}		3.42 + −0.02			3.40 + −0.03		
r_{Fe-La}		3.49 + −0.05					
r_{Fe-Sr}					3.40 + −0.03		
r_{Fe-Fe}		3.57 + −0.35			3.90 + −0.04		3.950−0.10
r_{Fe-Fe}		4.01 + −0.08					
$r_{Fe-Fe-O}$					3.93 + −0.02		3.950−0.10

Thus, why are researchers doing EXAFS then? In this particular case, here we have looked at LaSrFe-oxide, which can be considered a potential electrode material for SOFC cathodes. For this application, it is used in compact form and can be easy studies with XRD. Note, however, that a complete EXAFS study includes also the recording of the XANES, the near-edge spectral information includes much information on the electronic

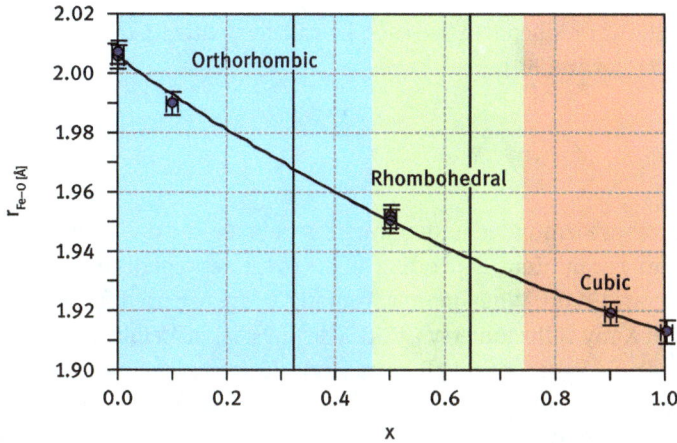

Figure 3.27: Variation of the bonding length between Fe and O in $La_{1-x}Sr_xFeO_{3-d}$ as a function of the strontium content x. The data were obtained by EXAFS. The low-level substitution yields orthorhombic phase, the mid-level substitution a rhombohedral phase, and high-level substitution a cubic phase [Haas 2009].

structure of the material, such as the oxidation state of the ion, potential charge transfer processes, and hybridization effects, for example. This cannot be obtained with XRD. Important for the preparation of EXAFS samples is that the method assumes a dilute dispersion of the material in a matrix with low X-ray absorption. For this purpose, the powder sample to be measured with EXAFS is typically diluted in a material like boron nitride or cellulose $(C_6H_{10}O_5)_n$. One purpose for this is to obtain a sample distribution, which has in the optimum case the X-ray optical thickness of one scattering length. This will avoid the multiple scattering, which can cause artifacts in the EXAFS and XANES data and yield inaccurate results. If a material in a real device does not satisfy this condition, then an in situ or *operando* experiment on this device with EXAFS yields insofar inaccurate results. This holds, e. g., in a battery, where the active metal oxide material is usually densely packed. If the researcher can live with this kind of accuracy, then there is no problem with making such EXAFS experiment, for example. No experiments are absolutely accurate because we are always dealing with inaccuracies, which are typically accounted for by error bars and other uncertainties—even if these are very small.

More important certainly is the use of EXAFS for noncrystalline materials, such as glasses, nanoparticles, and amorphous materials, where XRD is of little use, notwithstanding that a diffractogram of such materials could certainly allow for a qualitative assessment of the structure.

EXAFS is typically known in the context of hard X-rays. It is the K-edges of the 3d metals or the L-edges of platinum metals, which are applied for EXAFS. There are frequently exotic applications of the EXAFS formalism, such as for the carbon K-edge, which is at the rather low X-ray energy of 285 eV. It is indeed possible to apply the formalism and arrive at basic structure information, as was, e. g., shown by Diaz et al. for EXAFS on carbon [Diaz 2007] and with ELNES on silicon. In this case, we call it extended X-ray electron loss fine structure, EXELFS [Martin 1991].

3.5 X-ray reflectometry

When films with flat surfaces are irradiated under grazing incidence, the radiation is reflected with intensity maxima or minima, depending on the roughness conditions of the surface [Parratt 1954a, Parratt 1954b]. When monochromatic X-rays are used for irradiation, this method is called X-ray reflectometry (XRR). The physical principle of reflectometry also works with visible light (optical ellipsometry), electrons (medium energy electron diffraction MEED), neutrons (neutron reflectometry), or other particles, such as helium nuclei (helium scattering).

The reflectivity of flat films and multilayers may exhibit intensity oscillations, which can be mathematically modeling with respect to periodicity, layer thickness, and roughness at the nanoscale and stoichiometry changes across the films and interfaces [Kojima 1999]. Necas et al. have investigated iron oxide films with optical ellipsometry methods and with XRR to determine their optical constants and their electron density depth profile [Necas 2009].

$$\frac{R(Q)}{R_F(Q)} = \left| \frac{1}{\rho_\infty} \int_{-\infty}^{\infty} e^{iQz} \left(\frac{d\rho_e}{dz} \right) dz \right|^2$$

Here, $R(Q)$ is the reflected X-ray intensity, the reflectivity, $Q = 4\pi \sin(\theta)/\lambda$, λ is the X-ray wavelength, ρ_∞ is the density deep within the material, and θ is the angle of incidence. XRR is typically done on samples with surfaces that have an "optical quality."

3.5.1 XRR on photoelectrode materials

The shiny samples in Figure 3.28 are 10 nm thin WO_3 films deposited on TiO_2 single crystals with pulsed laser ablation of a commercial WO_3 ceramic target (American Materials Inc.). Thin compact films of 100 nm thickness were deposited on single crystal TiO_2 substrates with (110) orientation (CRYSTEC, Berlin) by pulsed laser ablation from a WO_3 target (99.9 % purity, American Elements) with 580 °C substrate temperature and 10 mTorr oxygen partial pressure. The films had a bluish color after deposition. An aliquot of the film was heated in a tube furnace for 48 h at 400 °C in oxygen, after which the color of the film had turned to a light yellow. These are the same films that were subject to XRD, XRR, and also contact angle measurements.

Figure 3.29 shows the X-ray reflectograms of a nominal 100-nm thick tungsten oxide film after deposition (blue bottom curve) and after subsequent thermal after treatment at 450 °C for 24 h at 100 % oxygen atmosphere in a sealed tube furnace.

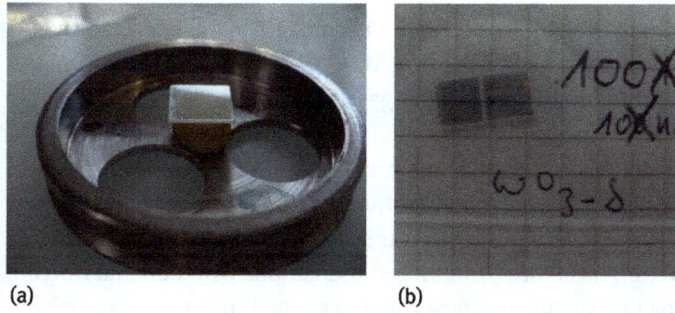

(a) (b)

Figure 3.28: (a) Metal sample holder for XRD with a 1-cm²-sized TiO_2 single crystal with bluish 100 nm PLD $WO_{3-\delta}$ film on top. (b) Two 5 x 5 mm-sized TiO_2 (100) single crystal substrates with 10 nm thin $WO_{3-\delta}$. Substrates were cut into several aliquots to allow for systematic heat treatment and *in situ* studies on comparable samples.

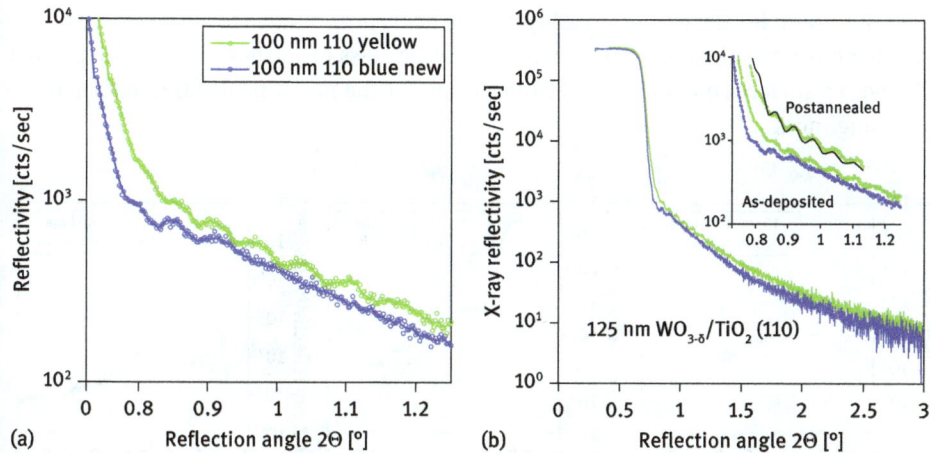

Figure 3.29: (a) X-ray reflectograms of two nominal 100-nm-thick tungsten oxide films deposited on TiO_2 (110) with pulsed laser ablation from a tungsten oxide target. The blue line is from the as-deposited stoichiometric film. The yellow line is from the film post-annealed in oxygen atmosphere. The reflection angle range is chosen where intensity oscillations are pronounced. (b) The same reflectograms extended over the range from 0° to 3°. The black solid line in the inset shows a simulated reflectogram.

The film after deposition had a bluish color, which is typical fur substoichiometric $WO_{3-\delta}$. Its XRR curve shows a flat high-intensity plateau until around 0.65° and then decreases steeply with the typical sixth power of the reflection angle. Close inspection of the range from 0.8 to 0.95 suggests oscillatory behavior that can be interpreted as an interference effect from a relatively flat film. The annealed film exhibits more oscillations extending from 0.8° to 1.25°. Obviously, the annealing in oxygen makes that the film becomes a better defined thickness and roughness. The reason for this is that the as-deposited film contains oxygen vacancies that lend the film its typical blue color.

Upon thermal treatment in oxygen, these vacancies become filled, which constitutes a structural ordering and the typical yellow color of WO_3. We learned already from the photoemission studies in Part 2 on the related changes of the electronic structure of this material.

The XRR curves in Figure 3.30 show intensity oscillations in the range $0.79° < 2\theta <$ $1.2°$ for the as-deposited (lower curve, blue) and for the post-annealed, oxidized film (upper curve, yellow). The as-deposited film shows four intensity maxima in the range $0.8°$ to $1.0°$. The post-annealed film has more pronounced oscillations with maxima ranging from $0.8°$ to $1.2°$, indicating that the film surface roughness and interface roughness decrease upon thermal oxidative after treatment, leading to an overall improvement of the film quality. By contrast, thermal oxidative treatment up to 350 °C of lithiated WO_3 films 50 nm thin films led to rougher surfaces [Maaza 1995]. The oscillations of the oxidized film move to somewhat larger angles, suggesting a slight overall decrease of film thickness upon oxidation. It was not possible to completely model the reflectivity curves, suggesting simple bilayer or diffuse layer were not representative to our films. However, results of our attempts fitted best with a film thickness of 100 nm for the as-deposited film with a surface roughness of 5 Å, and 105 nm for the postannealed film with 3 Å surface roughness.

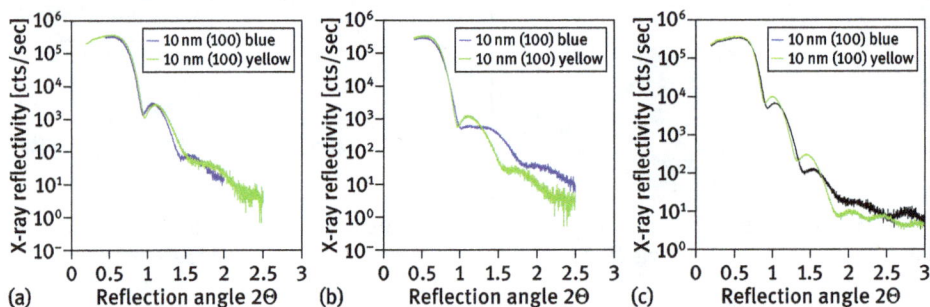

Figure 3.30: X-ray reflectograms of as-deposited (blue color) and thermally after-treated (yellow color) tungsten oxide grown by PLD on TiO_2 single crystals with different orientations (100, 110, and 001).

It is difficult to exactly model the X-ray reflectivity curves of the (001) and (110) grown films, likely because the films have a systematic yet unknown oxygen stoichiometry gradient. However, we observe that the (110) grown films has a two-layer structure that reconstructs upon thermal treatment to a one-layer film.

In any case, we should have a model of that in mind what we are going to deposit on a substrate. Figure 1 in [Necas 2009] illustrates with a simple sketch what every engineer or researcher should begin with when analyzing a multilayer electrode assembly.

The sketch in Figure 3.31 illustrates how a nominally stoichiometric tungsten oxide film with apparent yellow color is still oxygen deficient at the top surface and at the bottom interfacing the substrate. The red curve is the mass density as derived from the least

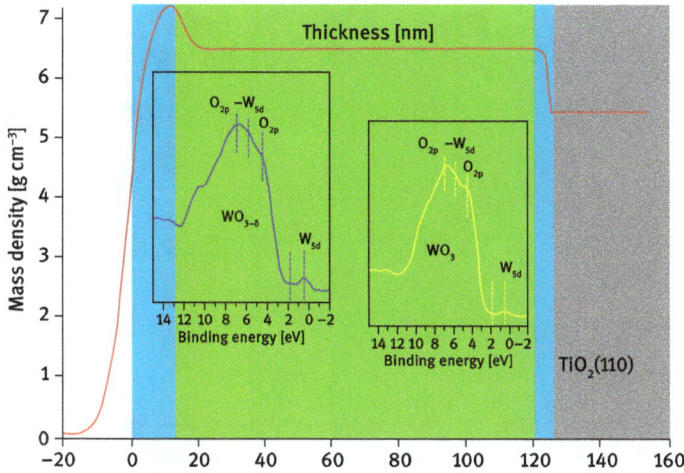

Figure 3.31: Schematic of single crystal substrate (on the right boundary) and the deposited tungsten oxide film. The blue shaded region denotes the oxygen deficient surface and interface. The red line is the mass density obtained by least squares fit to the XRR model. The two insets are photoemission spectra obtained from blue and yellow tungsten oxide samples.

squares fit of the reflectometry model to the actual XRR curve. The data are presented in arbitrary units but reflect the variation of the mass of the film, i. e., the stoichiometry, arriving from air or vacuum at the film surface and then increasing to a maximum value at the top 17 nm of the film (7 a. u.), where it decays to around 6.5 a. u. through the entire bulk of the film until it reaches the film-substrate interface with 5.5 a. u., where the substrate continues.

Figure 3.32 below shows two reflectograms from a pristine porous iron oxide photoanode before (red) and after (blue) electrochemical treatment for 2 h at 600 mV potential versus $Ag^+/AgCl$ reference electrode. We notice that the overall reflected intensity increases in the range 0.7° to 2°. The absence of intensity oscillations is evidence for the lack of film ordering. This sample was obtained from dip coating in supernatant liquid of iron nitrate derived iron oxide films [Bora 2011a]. The films have a very rough structure and are not suited for good XRR. We know that iron oxide as synthesized is oxygen deficient at the surface. Anodization causes oxidation of Fe^{2+} to Fe^{3+}. The anodization is an electrochemical oxidation process, which likely fills the oxygen vacancies, and thus causes oxidation of the Fe^{2+} at the iron oxide electrode surface toward Fe^{3+}. We have empiric evidence for this from the resonant photoemission spectra [Gajda-Schrantz 2013]. Probably, this causes a restructuring of the sample surface that amounts in a higher XRR intensity. This was an *ex situ* experiment, but it should in general be possible to make such studies *in situ* or *operando*.

What we learn here is that XRR requires samples with surfaces with optical quality. For a "good" porous PEC electrode, this is not always the case. Therefore, one has to switch to model systems and—for the time being—forget about high efficiencies or other

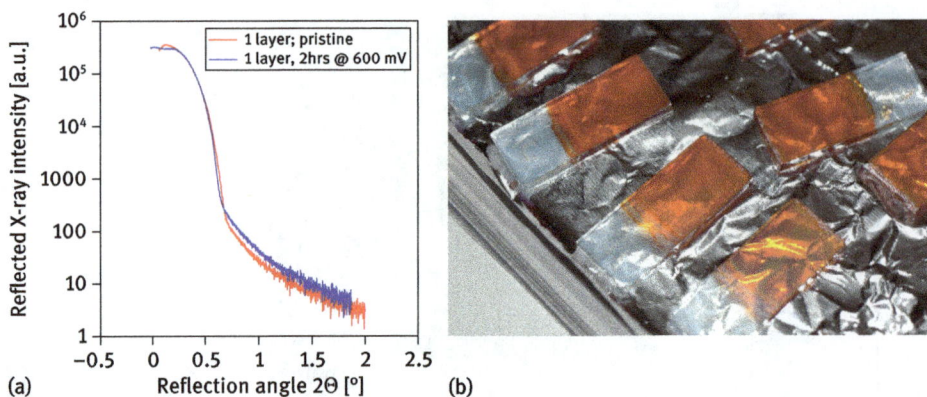

Figure 3.32: (a) XRR of 120-nm thick (b) Fe_2O_3 films dip coated and annealed on FTO glass. The red reflectogram is from the pristine film. The blue reflectogram is from the same sample after 2 h PEC operation at 600 mV versus Ag/AgCl reference.

utilitaristic properties of the samples but focus on the scientific scope of the study. We want to learn how the surface roughness and the surface topology may be affected by materials processing conditions or by device operation conditions. In this case, a well-polished single crystal or a shiny spin-coated film may do a better service for the study than a porous electrode, which has a very high photocurrent just because of this high porosity.

The deeper rational for doing XRR was the following. We will learn later in this book about photoemission that the operation of iron oxide photoelectrodes causes hole doping, which will shift the PES VB spectra toward lower binding energy upon continuing anodization. One question that arises from this observation is, which species causes the hole doping? Is it maybe protons from the electrolyte, from the oxidized water that diffuse into the interior of the electrode, which would shift the VB spectra toward lower binding energy? How can we probe the presence of protons in the iron oxide crystal lattice? The perfect probe would be by neutron diffraction, which is a very good probe for hydrogen, provided that deuterated samples are used. We would then have to use KOH in deuterated water and perform the anodization for a sufficient time so as to warrant that the deuterons would diffuse into the iron oxide crystal lattice and possibly occupy particular crystallographic positions that we could detect with neutron diffraction. However, neutron methods typically require a large amount of sample material. The material from a 150-nm thick electrode with $1\,cm^2$ area will not suffice. We probably would have to make the equivalent of a 1-m^2 large electrode and scrape the iron oxide off after anodization and then hope that we can see what was hypothesized.

A better suggestion is to use XRR, or better, neutron reflection for this purpose. We have seen now that the sensitivity of reflectometry is very high for changes on the surface and subsurface, provided the surface is flat enough. It would be interesting to learn whether protons or other positive charge carriers, such as potassium from KOH, would intercalate the electrode, and thus cause changes in the electronic structure of the photoelectrode and potentially affect its properties.

3.5.2 *In situ* and *operando* X-ray and neutron reflectometry on electrochemical systems

It is possible to place the sample for XRR in a sample holder, which allows for an electrochemical environment. Figure 3.33 below shows such *in situ* electrochemical cell made from polypropylene at ESRF Grenoble (Beamline Jörg Zegenhagen). It is basically an electrolyte container with a column in the center, which can host a flat sample, such as a disk or a single crystal plate. The center of the column has a hole through which an external applied low pressure vacuum pipe the sample can be "sucked" firmly on this column to provide a mechanically stable positioned sample during XRR. The small black O-ring in the center warrants that the vacuum is proper sealed and no electrolyte leaks through. The metal wire in the container is the counter electrode. There is also a metal wire pointing from the back of the column in the center, constituting the current collector for the working electrode.

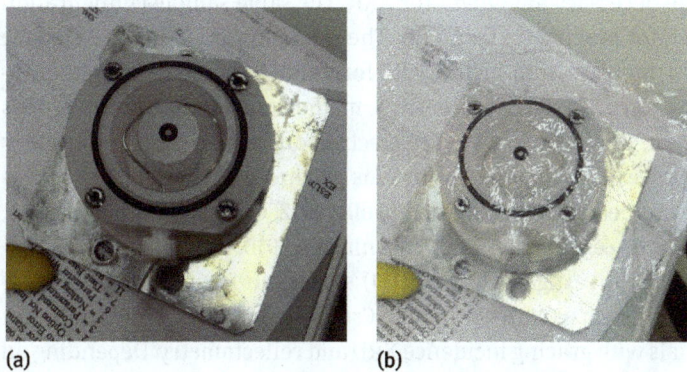

(a)   (b)

Figure 3.33: (a) Electrochemical cell for XRR and surface XRD at ID32 at ESRF, Grenoble. (b) The cell with a plastic foil on top. The corresponding diffractometer/reflectometer is shown in Figure 3.34.

Now we can place the sample on the smaller O-ring and apply the soft vacuum. Then a plastic foil is covered over the entire cell, and a solid ring is placed over the cell and pressed on the larger O-ring. Then the liquid electrolyte is filled with a hose into the cell and filled up until the liquid level is so high that the plastic foil is slightly bulging out. This means a thin liquid layer is over the sample surface in the cell center. This has to be done on a tripod stage at the beamline end station, which is shown in Figure 3.34.

Sample alignment is very important for XRR, more so than for a conventional XRD. Figure 3.34 shows a sample stage for *xyz* manipulation and on top a hexapod for the fine

Figure 3.34: Tripod table for diffractometer and reflectometry beamline ID32 at ESRF Grenoble. The cell is not yet mounted on the table. Numerous motors allow to lift, move, and tilt the table with the sample in the appropriate position for accurate X-ray surface diffraction and reflectometry experiments under *in situ* and *operando* electrochemical conditions.

adjustment to satisfy the necessary alignment for XRR. The same sample configuration and cell can also be used for making surface XRD. The symmetry breaking at a surface makes that the diffracted intensity from surface XRD forms not Bragg spots but rather so-called truncation rods. Surface XRD is a very sensitive method that to some extent probes also the subsurface region, similar to low-energy electron diffraction (LEED). The early stages of intercalation processes, where atoms or ions penetrate lattices, e. g., in catalysis experiments, can be studied with this method [Golks 2012, Golks 2015, Hussain 2016, Materlik 1984, Materlik 1987, Zegenhagen 1991, Zegenhagen 2010, Zegenhagen 2013].

Wang et al. [Wang 1992] investigated the Au(111) surface in various alkaline halogenide electrolyte solutions such as NaF, NaC1, LiCl, CsCl, KCl, and NaBr under different electrochemical potentials with gracing incidence XRD and reflectometry. Depending on the potential, they found different (p x $\sqrt{3}$) superstructures. They found confirmation for a unifying model for the surface, which depends on the induced surface charge.

3.5.3 Neutron reflectometry on electrochemical systems

Reflectometry experiments can be done also with neutrons [Colella 1996, Daillant 1999], helium, and visible light. Organic and bioorganic samples are very good candidates for neutron reflectometry. Figure 3.35 shows an electrochemical cell for neutron reflectometry at ANSTO in Australia, which was constructed for electrochemical XRR studies on ionic liquids [Lauw 2010]. It is a three-electrode electrochemical cell for the investigation of the electrochemical double-layer in ionic liquids. The counterelectrode was a

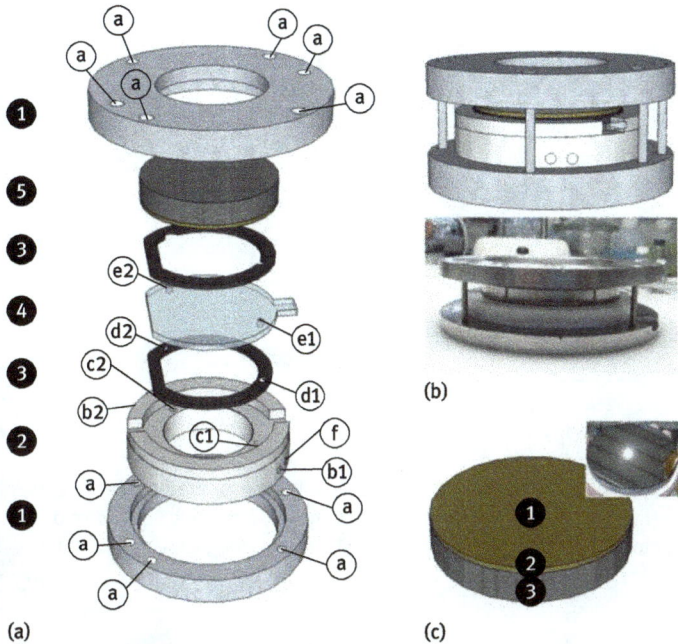

Figure 3.35: (Color) (a) Schematic of the cell: ❶ 10-mm thick stainless steel clamp-rings [155 mm outer diameter (OD), 125 mm inner diameter (ID)]; ❷ 15 mm Teflon® base (125 mm OD, 100 mm ID); ❸ 0.76-mm thick Kalrez® gasket (100 mm OD, 80 mm ID) with fluid ports; ❹ 4-mm-thick TEC 7™ glass counterelectrode with fluid ports (100 mm diameter), coated with a conducting film of fluorine-doped tin oxide; ❺ 10-mm thick Si–Cr–Au working electrode (100 mm diameter); (a) clamping screw holes; (b1) Luer-Lok® syringe port to inject liquid; (c1), (d1), (e1) holes for ionic liquid inlet; (b2) Luer-Lok® syringe port to eject liquid; (c2), (d2), (e2) holes for ionic liquid outlet; (f) hole for reference electrode. (b) Top: schematic of a fully assembled cell; bottom: the actual cell. (c) Schematic of the gold electrode: ❶ 250 nm Au layer, ❷ 25 nm adhesive Cr layer, ❸ 10 mm Si wafer. Inset: the actual gold electrode used in the experiment. Reprinted from [Lauw 2010] with the permission of AIP Publishing.

TEC 7 TCO glass, which allowed to visually monitoring gas bubble formation, for example. The working electrode was a 10-cm diameter silicon wafer, coated with an Au layer and Cr layer for adhesion. It appears that a Cr layer for adhesion and an Au layer for the electric contact is a suitable match of materials for providing contacts on electrochemical systems. We used this approach for contacting Si_3N_4 X-ray windows for photoelectrochemical experiments with NEXAFS spectroscopy [Braun 2012l].

Figure 3.36 shows the neutron reflectrogram of the Si–Cr–Au surface plus the least squares fit to the corresponding model of the multilayer system. The fine oscillations suggest that this is a well-defined system valid for quantitative analyses. The inset on the upper right panel of Figure 3.36 shows the scattering length distribution (SLD) obtained from the least squares fit (compare Figure 3.31 from our tungsten oxide film on titanium oxide single crystal). We read from the SLD that there is a 250-Å thin gold layer deposited on a Cr layer thinner than 30 Å.

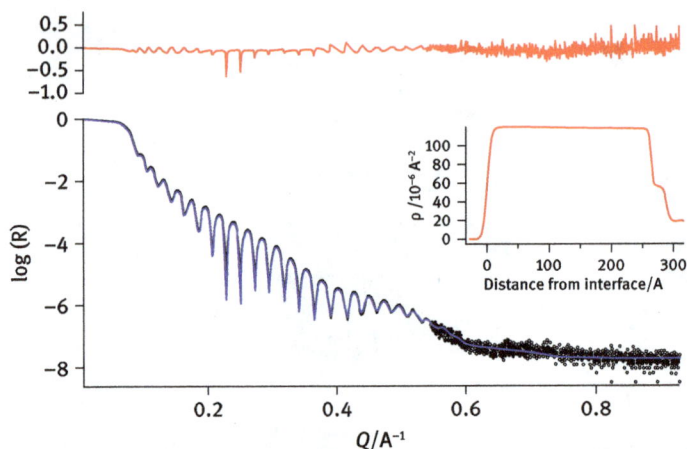

Figure 3.36: XRR curve for the Si–Cr–Au surface. The measured reflectivity data are presented as circles, with the fit as a continuous. The fit residuals are plotted above. Inset: the corresponding SLD profile of the surface based on the fitted model. Reprinted with permission from [Lauw 2010]. Copyright [2010], American Vacuum Society.

Jerliu et al. [Jerliu 2013] investigated the intercalation of lithium into amorphous silicon electrodes with neutron reflectometry. They deposited a 40-nm thin silicon working electrode on a 1-cm thick quarry substrate with a palladium current collector in between. Lithium metal was used as reference electrode and as counter electrode. Li was intercalated in propylene carbonate with 1 molar LiClO4 at 7.8 $\mu A/cm^2$. With the reflectometry, they could monitor *in situ* the volume expansion in the silicon upon lithium intercalation, and they monitored the solid electrolyte interphase layer.

The design of the electrochemical cell is shown in Figure 3.37 (left) and an image of the cell is displayed in Figure 3.37 (right). A thin film arrangement is used to produce a large area working electrode necessary for neutron reflectometry experiments. The cell has three electrodes. The working electrode was a quartz block as substrate coated with a thin palladium layer as back contact and a current collector. The actual electrode material, the study object, was made of amorphous silicon (about 40 nm) and was deposited on top of the palladium. Here, a circular design with an electrode diameter of 40.5 mm (contact area to electrolyte) is used. The dimensions of cells for neutron measurements are often large and massive, particularly because the neutron beams are relatively large. It is not always necessary to do millimeter work or even microfabrication. For the counterelectrode and the reference electrode, metallic lithium is used. The working and counterelectrodes are separated by Kalrezs gaskets forming a cell with a total volume of 1.4 ml. In addition, a separator with a thickness of 20 mm was introduced between the two electrodes. The lithium electrodes are sealed against high-density polyethylene housing by Kalrez® gaskets. An electrolyte can be supplied to the cell through two ports drilled into the polyethylene. Propylene carbonate (Sigma

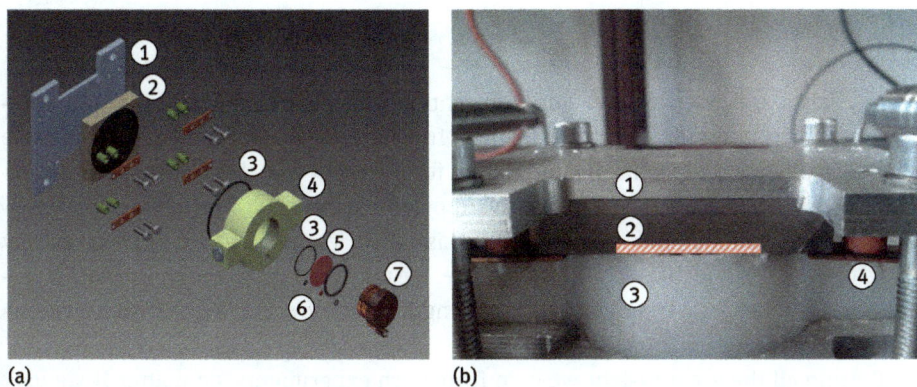

(a) (b)

Figure 3.37: (a) Schematic of the electrochemical cell: (1) 5-mm thick aluminum ground plate; (2) 10-mm thick Pd-coated quartz substrate with an amorphous silicon electrode on it (circle); (3) Kalrezs gaskets; (4) 20-mm thick high-density polyethylene base; (5) Li counterelectrode; (6) Li reference electrode; and (7) copper base. (b) Image of the cell: (1) aluminum ground plate; (2) quartz substrate and electrode; (3) polyethylene housing; and (4) contacts to potentiostat. The shaded area depicts the position and the cross section of the neutron beam, which is incident perpendicular to the paper plane (not to scale). Reproduced from [Jerliu 2013] Jerliu B., Doerrer L., Hueger E., Borchardt G., Steitz R., Geckle U., Oberst V., Bruns M., Schneider O., Schmidt H.: Neutron reflectometry studies on the lithiation of amorphous silicon electrodes in lithium-ion batteries. *Physical Chemistry Chemical Physics* 2013, 15:7777–7784 with permission from the PCCP Owner Societies.

Aldrich, anhydrous, 99.7 %) with 1 M LiClO4 (Sigma Aldrich, lithium perchlorate, battery grade, dry, 99.99 % trace metal basis) was the electrolyte. The cell was assembled in an argon-filled glove box. Later in this book (Section 7.3.2), we will see an example of neutron reflectometry on proteins and living cells [Junghans 2015].

4 Real space imaging and tomography

We physicists and chemists get our information typically from scattering and spectroscopy experiments. Scattering is the most fundamental process and even technique in modern physics represents the foundation for the many Nobel Prizes that were mentioned in the Introduction and listed in the Appendix to this book. The name of the country where scattering is spoken as a language is called "reciprocal space." Spectroscopy can be considered one dialect of this language. We may remember the coordinate transformation from energy, frequency, or wavelength $E = h \cdot v = h \cdot c/\lambda$ into k-space via the relation $2 \cdot m \cdot E = p^2 = (h \cdot k)^2$.

Despite all the great insight we gain from such experiments, no matter if we look as astronomers into the cosmos or as materials scientists in the microcosmos, the best way that we are able to communicate with the nonexperts in real life is via a real space picture.

A colleague of mine told me that he had discussed at a project meeting in Washington D.C. the performance of our iron-based Fischer–Tropsch catalysts based on our Mössbauer spectra: a deep and insightful analysis, which aimed at clarifying the molecular origin of activity. However, it was only until he had shown an electron micrograph of the catalyst in the presentation that one panel expert in D.C. spoke up and said, "Now I understand what you are talking about."

"Use a picture. It's worth a thousand words" is a well-known quote that appeared in a 1911 newspaper article referring to newspaper editor Tess Flanders discussing journalism and publicity [Flanders 1911]. This quote (*Ein Bild sagt mehr als tausend Worte*) is used today as an excuse rather than an actual technical rational for why a picture would express more contents than a thousand words. Since the invention of the spyglass lenses in ancient times [Sines 1987], there is a never-ending desire for higher and higher magnification microscopes.

This part of the book is about what we typically refer to as microscopy, i. e., the real space imaging of an object with substantial magnification. For more than 10 years, synchrotrons have been termed "microscopes," which in my opinion is not a very fortunate choice of wording. Electron microscopy is unsurpassed with respect to magnification. Like optical microscopy, electron microscopy is limited to surface morphology studies unless extremely small sample volumes are inspected. With EDX, electron microscopy can provide high-resolution images with elemental contrast in the Ångstrom size range. Moreover, with some limitations, it is possible to conduct chemical *operando* studies with electron microscopy. It is not possible today to look into devices with electron-based methods, mainly because of the limited penetration depth (mean free path) of electrons in materials. However, looking into electrochemical energy converters such as batteries or fuel cells while under operation is desirable for complete understanding of the manifold of processes that happen in these devices [Gelb 2012].

The penetration depth of neutrons and X-rays is considerably longer and allows also for the probing of devices. Solid oxide fuel cells (SOFC) from the HEXIS type can

https://doi.org/10.1515/9783110794038-004

be investigated with neutron tomography and neutron radiography [Braun 2016a]. It is not necessary to disassemble the SOFC electrode stack for such studies, as we will see in Chapter 8. The penetration depth of X-rays increases with the X-ray energy according to Lambert–Beer's law.

Figure 4.1 shows the penetration depths of neutrons, X-rays, and electrons in meters as a function of the atomic number Z of elements. The penetration depth is basically one scattering length or absorption length after which the beam intensity $I_{(d)}$ is attenuated to $1/e$ of its original value I_0.

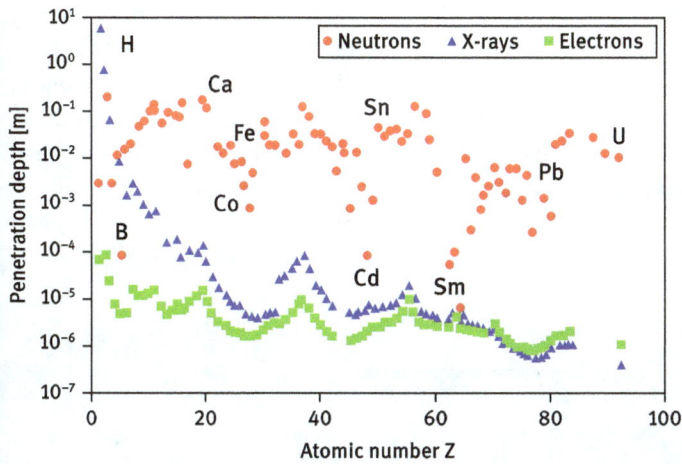

Figure 4.1: The plot shows how deeply a beam of electrons, X-rays, or thermal neutrons penetrates a particular element in its solid or liquid form before the beam's intensity has been reduced by a factor of i, i. e., to about 37% of its original Intensity. Neutron data were chosen for neutrons with wavelength of 1.4 Å. Reproduced from [Pynn 1990].

One shortcoming of neutron imaging is its relatively low spatial resolution. Neutron fluxes are also relatively poor in comparison with optical and X-ray methods, where photon fluxes are considerable higher. This makes it difficult for time-critical *operando* methods [Martin 2010]. A review on recent activities on improvement of X-ray imaging for a wide range of applications can be found in the study of Sakdinawat and Attwood [Sakdinawat 2010], whereas a historic overview on imaging with X-rays can be found in the study of Dommann [Dommann 2003].

4.1 X-ray imaging and X-ray microscopy

4.1.1 Polymer electrolyte membrane fuel cells

Polymer electrolyte membrane fuel cells (PEM-FC) require minimum humidity for the operation of the NAFION® proton conductor electrolyte. Water management is therefore an important issue for PEM-FC technology [Barbir 2012]. Lee et al. [Lee 2008] used a relative simple setup with an X-ray tube and X-ray camera (comparable with X-ray imaging at a physicians practice) in order to obtain images of the water distribution in a micro-PEM-FC.

The hydrophilicity of the graphite current collectors can be manipulated in order to influence the water distribution in PEM-FC electrode assemblies [Wang 2006b]. Zhu et al. [Zhu 2008], and [Wei 2008] conducted such manipulation and impregnated the graphite plate of a PEM-FC with oxygen plasma to make it hydrophilic. They used X-ray micro-computer tomography in order to look through the graphite flow channels of the fuel cell when it was under operation (Figure 4.2). They compared the difference of the water distribution in hydrophobic and hydrophilic graphite electrodes. They found in both cases that water drops in the flow channel distort their shape before they started to move. However, drops in the hydrophobic channel did not detach from the gas diffusion layer (GDL) before moving, whereas the water in the hydrophilic channel formed a thin water layer. It is obvious from Figure 4.2 that the dark-shaded water layer seems to be repelled from the hydrophobic surface of the graphite plate (top Figure 4.2) and attracted from the hydrophilic surface of the graphite plate (bottom Figure 4.2).

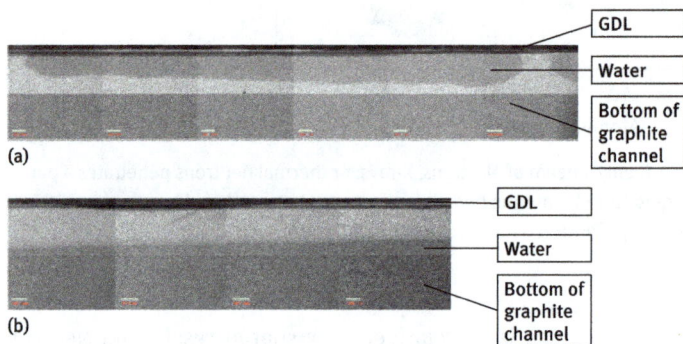

Figure 4.2: X-ray computer tomography (microCT) images of an assembly of graphite electrode with GDL for a PEM-FC. The top figure shows the water distribution in 1.5 mm depth hydrophobic flow channel after 2 h of fuel cell operation. The bottom figure shows the water layer formed on the hydrophilic graphite channel after 2 h of PEM operation (after [Zhu 2008]). Reproduced with permission from ECS Transactions, 16 (2) 995–1000 (2008). Copyright 2008, The Electrochemical Society.

4.1.2 Batteries

Nakata et al. [Nakata 2015] have built a Ni-Zn battery cell and investigated the dendritic growth of recrystallizing Zn near the electrode by *operando* X-ray fluorescence imaging with large time resolution. Their *in situ* cell is shown in Figure 4.3. Every 5 s, they recorded one XRF image from the operating cell. The right panel of Figure 4.3 shows the

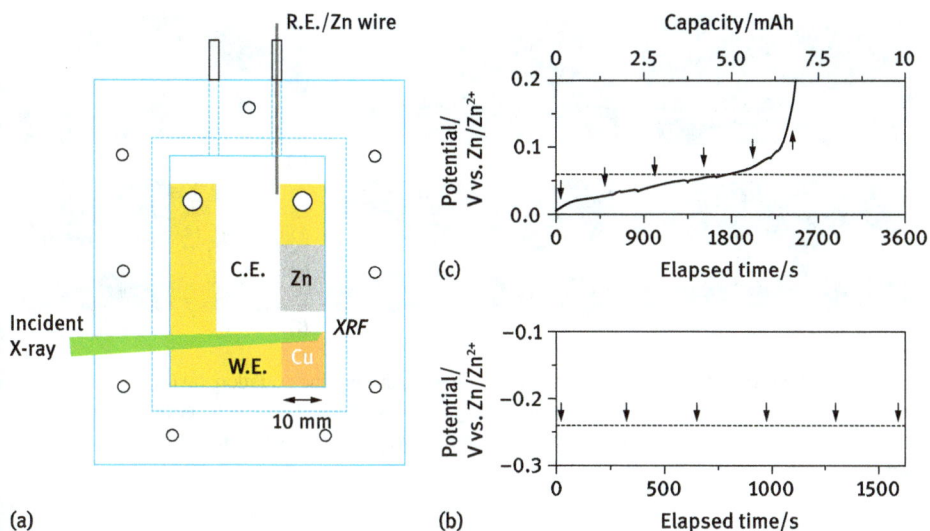

Figure 4.3: (a) Sketch of the electrochemical cell (polymethyl methacrylate [PMMA] Plexiglas®) for *operando* XRF imaging. Cu working electrode 10 × 1 mm with a 10-mm distance to Zn counterelectrode plate and Zn wire reference electrode. Electrolyte was an aqueous KOH ZnO solution. (b) Variation of electrochemical potential for reduction at 0.24 V versus Zn/Zn^{2+} for 10 mAh (broken line) and galvanostatic oxidation at 10 mA (solid line). Arrows indicate the snapshot points of *operando* XRF imaging shown in the figure below. Reproduced with permission from [Nakata 2015], Electrochemistry, 83 (10), 849–851 (2015). Copyright 2015, The Electrochemical Society of Japan.

evolution of the potential versus reduction time and oxidation time. The arrows indicate the times (= electrochemical conditions) where XRF images were recorded.

Within 1 h and less, they recorded six snapshots during a charging cycle of the battery, as demonstrated in Figures 4.4 and 4.5. One image took less than 5 min. The experiment was performed at BL28XU at SPring-8 (Hyogo, Japan).

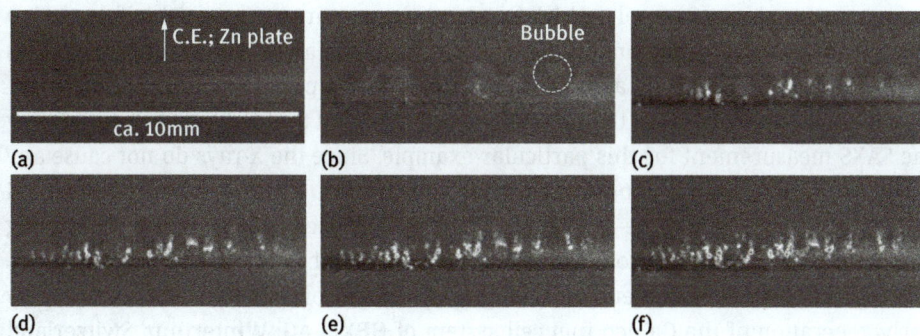

Figure 4.4: XRF *operando* snapshots during zinc reduction. (a) 0 s, (b) 325 s, (c) 650 s, (d) 975 s, (e) 1300 s, and (f) 1600 s after starting reduction. Reproduced with permission from [Nakata 2015], Electrochemistry, 83 (10), 849–851 (2015). Copyright 2015, The Electrochemical Society of Japan.

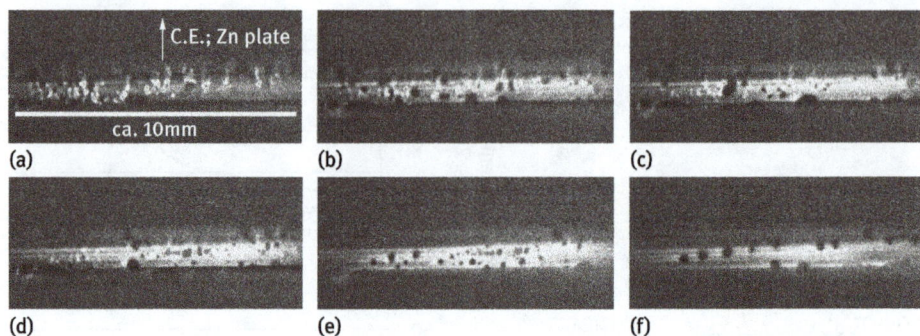

Figure 4.5: (Color online) Snapshots of *operando* XRF imaging during zinc oxidation. (a) 0 s, (b) 500 s, (c) 1000 s, (d) 1500 s, (e) 2000 s, and (f) 2450 s after starting oxidation. Reproduced with permission from [Nakata 2015], Electrochemistry, 83 (10), 849–851 (2015). Copyright 2015, The Electrochemical Society of Japan.

The incident X-ray energy was chosen to 9800 eV at the Zn K-edge so as to be able to have scattering contrast for Zn. The fluorescent X-rays were detected in reflection geometry with a charge-coupled device (CCD) camera. Acquisition of a full 2D image took 5 s exposure time and was recorded in the continuous mode so as to have a continuous imaging. The dendrite formation was provoked under potentiostatic condition of 0.24 V versus the Zn/Zn^{2+} reference until 10 mAh capacity was reached. The oxidation reaction was conducted with a 10-mA current.

4.2 Neutron imaging and radiography

4.2.1 Solid oxide fuel cells

Many analytical techniques are called "nondestructive," but this is not always correct. Doing tomography on a sample, which first has to be prepared as a very thin focused ion beam (FIB) lamella, is not nondestructive. Also, the small angle X-ray scattering (SAXS) measurements on electrode assemblies presented in Chapter 3.3 are not "nondestructive." This is simply because the sample had to be cut and thinned to make it ready for the SAXS measurement for this particular example. Since the X-rays do not cause any further destruction can be considered irrelevant. The issue with a destructive method is that the sample cannot be used again after measurement. This is not withstanding that X-ray and neutron methods are in principle useful for nondestructive methods.

Figure 4.6 shows the neutron radiography images of a SOFC stack from a very early generation of the Galileo fuel cell system of HEXIS AG, Winterthur, Switzerland [Mai 2011]. Sixty electrode assemblies are stacked together to a height of 30 cm and operate at 850 °C to 900 °C. The size of the NEUTRA beamline would not allow neutron radiography for the entire HEXIS system, which is as large (55 cm × 55 cm × 160 cm,

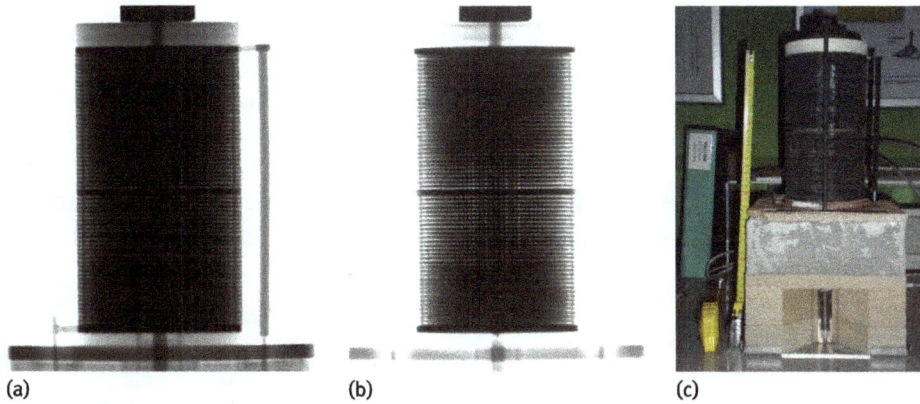

Figure 4.6: Neutron radiography images of a 30-cm tall SOFC stack from HEXIS AG. The images were recorded at NEUTRA neutron beamline at SINQ, Paul Scherrer Institut, Switzerland.

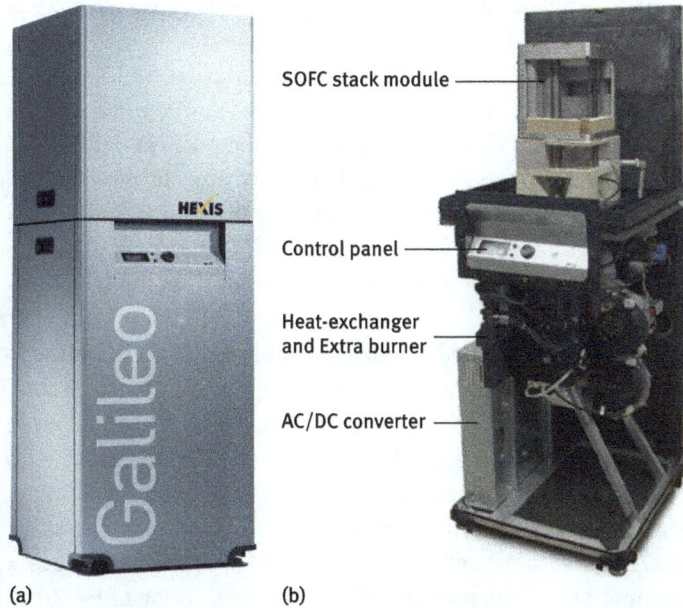

Figure 4.7: Galileo 1000 N SOFC system from HEXIS AG, photo credit HEXIS.

170 kg) as, e. g., a full size kitchen refrigerator, as shown in Figure 4.7 [Mai 2011]. However, the core of the system, the SOFC "stack," was fitted on the sample turntable at NEUTRA beamline and subject to neutron radiography and, as we see in Section 4.4.1, to neutron tomography [Braun 2016a].

Neutron imaging can be considered a truly nondestructive technique [Kardjilov 2011] because the large penetration depth of the neutrons warrants that many samples can be investigated without any significant disassembly or destruction of the sample or component or device. However, it is possible that the sample becomes radioactive after neutron irradiation (neutron activation). This applies, e. g., when the sample contains cobalt. In this case, the sample has to remain in quarantine at the neutron source until the radiation activity of the sample is below some specific threshold. This means that the sample cannot be used, although it has not been destructed for the measurement. Kardjilov et al. [Kardjilov 2011] provided an overview of the recent advancement in neutron imaging.

4.3 Complementarity of X-ray and neutron methods

The complementary of X-rays and neutrons for probing materials has been realized [Furrer 1998]. It has therefore been of interest to combine both methods in dedicated neutron/X-ray beamlines. One recent short review on neutron and synchrotron-based characterization methods can be found by Manke et al. [Manke 2011], who showcased examples for fuel cells, batteries, solar cells, catalysts, and gas separation membranes, which are studied with radiography, tomography, diffraction, scattering, and XANES.

In one particular study, Manke et al. [Manke 2007] have looked *in situ* into $Zn–MnO_2$ batteries at different charge stages with X-ray tomography with monochromatic X-rays. Specifically, they investigated the morphology changes that were caused by the chemical reduction of the MnO_2 and the dissolution of Zn with subsequent nucleation and growth of ZnO. The spatial resolution was around several micrometers. The authors used also neutron tomography in order to monitor the spatial distribution of H_2 in the MnO_2 matrix.

La Manna et al. [Lamanna 2015] presented a neutron beamline at NIST in Gaithersburg MD, which is also equipped with a laboratory X-ray source (Figure 4.8). Their combined neutron and X-ray imaging allow for the *in situ* visualization of reactants inside a battery, for example. They argue that whereas the X-ray attenuation increases with increasing atomic number, the neutron attenuation has strong attenuation by lithium. This provides complementary data by providing critical microstructure information. Their beamline points a 90-keV microfocused X-ray beam by 90° to the neutron beam and, therefore, allows for simultaneous tomography with spatial resolution of 15–50 micrometers, as illustrated in Figure 4.8. Matching image resolution of the two beams allows the overlaying of the two tomograms with different chemical sensitivity. Simultaneously, imaging of batteries will permit that reacting species, such as lithium ions, and the electrode matrix to be resolved at identical conditions that cannot be guaranteed for serial imaging, which would be typically at different facilities, i. e., a synchrotron source and a neutron source.

A similar combined instrument (Figure 4.9) has been built at Paul Scherrer Institut Swiss Spallation Source in the late 1990s. The NEUTRA beamline at SINQ was at its early stage already equipped with a complementary "laboratory" X-ray source

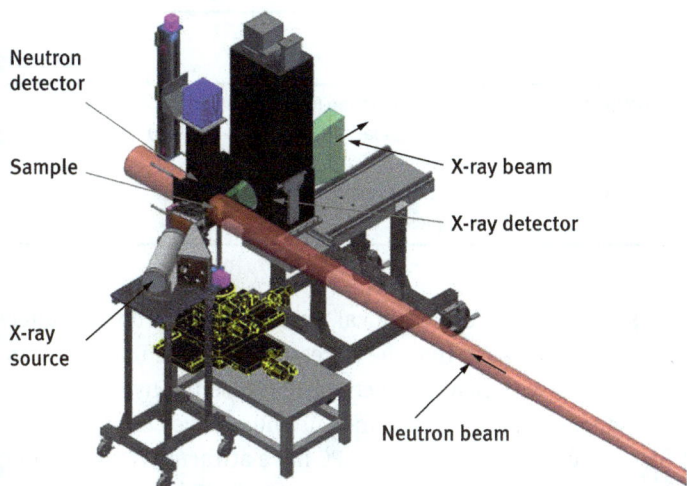

Figure 4.8: Schematic of combined neutron and X-ray imaging system at NIST. The neutron beam comes from the NIST neutron accelerator facility. The X-rays come from a laboratory X-ray source. Reproduced with permission from [Lamanna 2015] Electrochemical Society (ECS): 2015; Vol. Abstract MA2015–02, 91. Copyright 2015, The Electrochemical Society. LaManna, J., Hussey, D. S., Jacobson, D. L., Baltic, E., Simultaneous Neutron and X-Ray Tomography for Advanced Battery Research. The resolution image was received from Dr. Jacob LaManna, NIST.

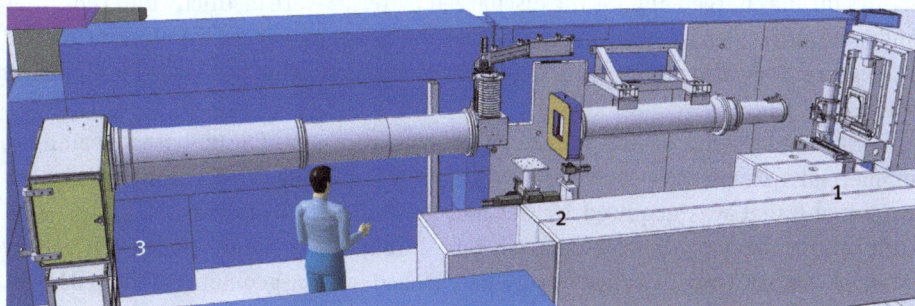

Figure 4.9: Sketch of NEUTRA beamline at SINQ Switzerland. The measurement positions 1,2,3, are explained in the table (Credit: SINQ, PSI).

[Lehmann 2001]. NEUTRA has been instrumental in developing neutron tomography for PEM-FC research. To date, NEUTRA has the following specifications. The beamline is made from evacuated neutron flight-tubes going through a 20-mm diameter pinhole into a heavy water (D_2O) SINQ moderator tank. Outside the SINQ shielding wall at position 1, an optional 320-kV X-ray source can be moved into the irradiation position. The beamline has three measurement positions.

Position	Main feature	Options	Maximum field of view
1	Sample irradiation by neutrons	320-kV X-ray tube (optional)	–
2	Combined neutron and X-ray measurements	Highly radioactive samples in shielded container	15×15 cm^2
3	Radiography/tomography of large samples	Vertical and horizontal scanning of samples	30×30 cm^2

The majority of neutron studies on electrochemical systems was probably done on PEM-FC fuel cells. The reason for this is that the necessary water management in such PEM-FC can be nicely imaged with neutron methods because the hydrogen ions (protons) in water absorbs neutrons very well, and thus provides a welcome chemical contrast. As a matter of fact, similar to batteries, PEM-FC have attracted interest from many different analysis and diagnostics communities [Zhang 2013]. The number of battery studies in neutron radiography and tomography is in the meantime increasing. One reason for this delay is that the organic electrolytes in batteries contain much hydrogen that adds a not welcome contrast with neutron methods. Butler et al. [Butler 2011] have looked into a prismatic shape lithium ion polymer battery with neutron radiography with neutron wavelengths close to a spectral feature characteristic of the LiC$_6$ negative electrode material. Over the wavelength range of 3–4 Å, the neutron attenuation characteristics for the charged and for the discharged batteries were distinctly different. They conducted also *operando* studies and observed the disappearance of the aforementioned LiC$_6$ feature. They also attempted a 3D image reconstruction but this was of preliminary nature.

X-ray tomography is not only done at synchrotron radiation-based beamlines. Commercial systems were pioneered by Xradia, a California (Pleasanton)-based start-up company that was acquired by German Zeiss in 2013 [Nitschke 2013]. Yufit et al. [Yufit 2011] made a *postmortem* analysis on a lithium ion polymer pouch cell battery with a laboratory X-ray µCT. Specifically, they looked into the geometry of the cell layers and found mechanical distortions and concluded that the deformation was caused by pressure build up as a result of gas generation. This study is similar to the investigation of an SOFC stack with neutron tomography. Here, it is not necessary to disassemble the stack. Rather, changes in the stack which may have occurred during operation can be visualized without opening the stack.

4.4 Tomography

Tomography is an imaging method where a 3D representation of an object is created (image reconstruction) by a stack of projected image sections. This is a more complex methodology than mere radiography. The necessary image sections are obtained by irradiating the body with probes (neutrons, electrons, X-rays, other radiation), as illustrated

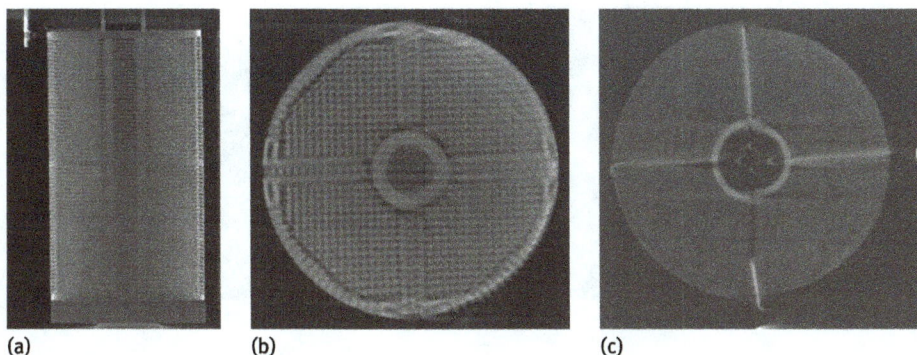

(a)   (b)   (c)

Figure 4.10: (a) Frontal/coronal section and two transversal (b)/axial (c) sections of the reconstructed images of the SOFC stack from Hexis. Data recorded at NEUTRA beamline at SINQ.

in Figure 4.10, so that one image of the entire body is projected on the probe detector. The object is then stepwise incrementally rotated on a turntable on an axis perpendicular to the irradiating beam, and another image is recorded before the object is rotated by another angular increment for the next image recording. With a sufficient number of projected images, it is possible to mathematically reconstruct the 3D structure of the body [Herman 2006, Kak 2001] and arrive at impressive 3D images, which can provide very good features of the microstructure of porous bodies or the architecture of complex components and devices.

4.4.1 Neutron tomography on SOFC

The aforementioned SOFC stack was subject to neutron tomography at NEUTRA with the stack on a rotating table. Figure 4.10 shows image sections from the reconstructed tomography data. On the left, we see a frontal view section of the entire stack height in the orientation how the images were acquired. The image reconstruction allows calculation of axial sections of the stack, which are shown for a top end plate and one electrode assembly from the middle range of the stack. We can see clearly the structural integrity of the electrode assembly on the submillimeter size range.

Depending on the efforts taken during tomography (spatial resolution, angular resolution, and time resolution), it is possible to have image stacks of hundreds or thousands of images through which one can move (on the computer) in all spatial directions in a movie in order to inspect the entire SOFC for structural damages. The next Figure 4.11 shows the tomography image from outside, and then two more images that show coronal sections, including a section highlighting the center axis of the SOFC. The fourth image shows a tilted cross-section of the SOFC. To the best my knowledge, a SOFC stack has not yet been investigated with neutron tomography when under electrochemical operation. However, there is no fundamental reason that prevents us from making such

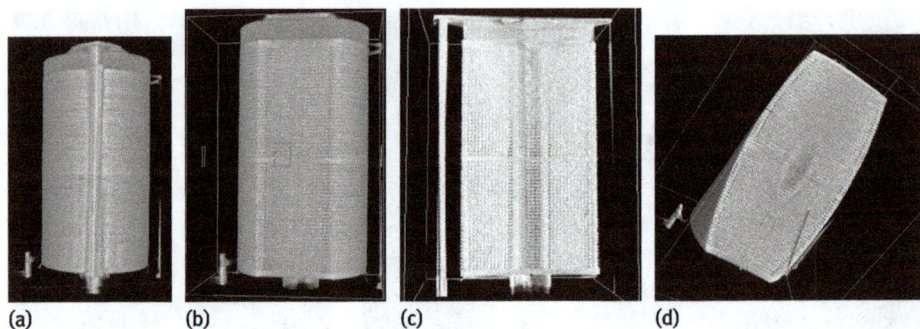

Figure 4.11: Neutron tomography images of SOFC in coronal direction showing the stack from outside and from two sections inside the SOFC. Data recorded at NEUTRA by George Necola.

experiment. NEUTRA, e. g., is able to look into combustion engines while under operation, and with lock-in technology, it can make snapshots of particular piston situations during operation with a high time resolution although the acquisition of one image takes 1000 times more than one engine cycle [Grünzweig 2012]. They have demonstrated that this in an impressive movie that can be viewed on YouTube [Grünzweig 2012].

The spatial resolution of neutron radiography and tomography is currently around 20 micrometers, to the best of my knowledge. Certainly, in every method, resolution is constantly being pushed to the maximum [Tremsin 2011]. As we will see later, spatial resolution in scanning transmission X-ray microspectroscopy (STXM) is by a factor 1000 better, i. e., 20 nm. Therefore, neutron imaging resolution is easily outperformed by other methods. As of yet, it is therefore impossible to look into the microstructure of materials with neutron imaging. With small-angle neutron scattering, it is possible to investigate the microstructure, but this is not an imaging method. The complementarity of small-angle scattering and tomography methods has certainly be realized and applied to SOFC [Allen 2004].

4.4.2 X-ray tomography on SOFC

X-ray tomography can provide spatial resolution in the submicrometer range. The study by Chen-Wiegart et al. [Chen-Wiegart 2012b] deals with the X-ray tomography of a SOFC electrode assembly. Similar to the SAXS work in Chapter 5.4.3, we see an assembly of cathode (LSM-YSZ), electrolyte (YSZ), and anode (Ni-YSZ). Note that this assembly has a fairly thin electrolyte layer of only 10-micrometer thickness. Although the title of their study suggests this is a nondestructive work, it is clear that a real SOFC has to be dismantled and that the electrode assembly has to be broken and polished to make it fit into the measurement apparatus. The result of their analysis shows a 3D reconstruction of the anode layer with color code distinguishing the Ni phase from the YSZ phase and from the pores (which is also considered a phase). The resolution is clearly better than

1000 nm. The authors advertise their study with the statement that they have measured an unusual large volume of 3600 μm^3 at high resolution, which would be a sampled size of 6 μm × 10 μm × 60 μm, for example. On the basis of the obtained microstructure parameters and the McLachlan model [Mclachlan 1989], they predicted an electrochemical anode polarization resistance that was in good agreement with impedance data.

The resolution and the probed volume that can be achieved with this X-ray tomography method can also be obtained with SAXS [Allen 2009]. Moreover, ASAXS can provide the same chemical contrast like in the study shown here [Allen 2014]. However, as of yet, it appears that SAXS is unable to determine the tortuosity of porous material. In chemical and electrochemical technology, the tortuosity of pore channels in materials and electrodes is however of fundamental importance. The tortuosity "τ" is a topological parameter and can be written in its simplest definition as the ratio of the length L of a curved segment of a pore channel versus the distance C of the two end points C ($\tau = L/C$) and is closely related to the connectivity of pores and particles and governs the mass transfer in porous media and also the charge transport by electrons and ions in electrodes.

Geometrical analysis of sections allows for derivation of the tortuosity in porous electrodes and catalyst supports and has been exercised for SOFC and lithium ion battery cathodes using X-ray tomography, e. g., by Chen-Wiegart et al. [Chen-Wiegart 2014]. They developed a so-called distance propagation method that computes the tortuosity from 3D tomography data. They formed also a concept of tortuosity distribution that shows the spatial tortuosity along with a statistic histogram. The authors tested their new approach on a lithium ion battery system ($LiCoO_2$) and on a SOFC system (LSM-YSZ) based on FIB images. The outcome confirmed good agreement with effective diffusion-based tortuosity values. Related electrode processing studies are provided by Chung et al. and Ebner et al. [Chung 2013, Ebner 2013].

When the energy resolution is large enough, it is possible to investigate the fine structure of the X-ray absorption or emission features and then to arrive at additional chemical information that goes beyond mere geometrical and elemental analysis. For example, it is possible to determine the oxidation states of particular regions in the materials. Advanced Photon Source beamline 32-ID-C and Stanford Synchrotron Radiation Lightsource beamline BL6–2c allow for X-ray tomography [Nelson 2011]. Nelson et al. investigated Ni/NiO samples for the method development for fuel cell, battery, and supercapacitor material analyses (Figure 4.12). Chen-Wiegart et al. [Chen-Wiegart 2012a] used a similar beamline at NSLS in Brookhaven and investigated actual SOFC electrode samples at the Ni K-edge.

4.4.3 Tomography on PEM-FC

An X-ray tomography study on a PEM-FC has been conducted by Sinha et al. [Sinha 2006]. The rotating sample table for the tomography is sketched in Figure 4.13. The fuel cell

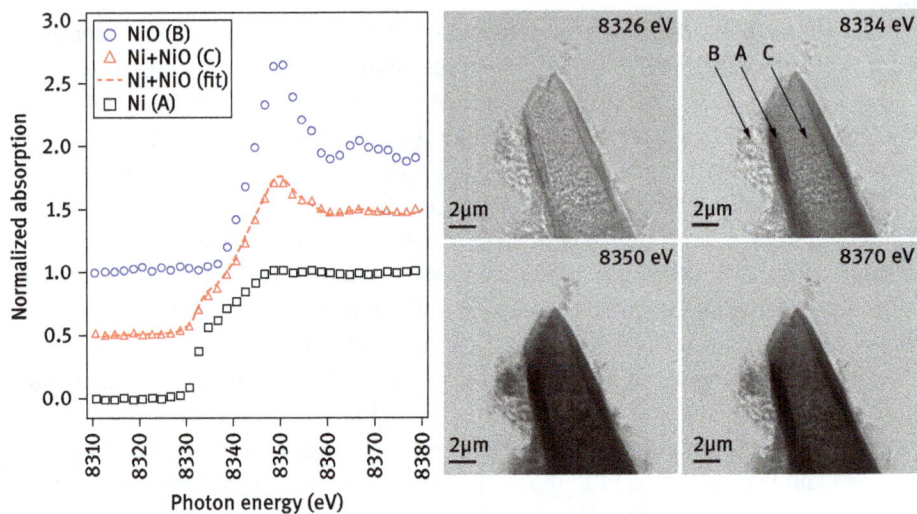

Figure 4.12: The distinction between oxidation states is enabled by image subtraction at the energy levels associated with the primary features in the spectra. Transmission images taken at X-ray energies of 8326, 8334, 8350, and 8370 eV illustrate the capability to distinguish Ni oxidation states. Representative regions of each material Ni (A) and NiO band overlapping Ni–NiO C are highlighted at 8334 eV. A linear combination of the Ni and NiO spectra, shown as the red dashed line, reproduces the spectrum for the region of overlapping Ni–NiO and corroborates the spectra obtained. [Nelson 2011] Reprinted with permission from [Nelson et al., Applied Physics Letters 98, 173109, 2011]. Copyright [2011], AIP Publishing LLC.

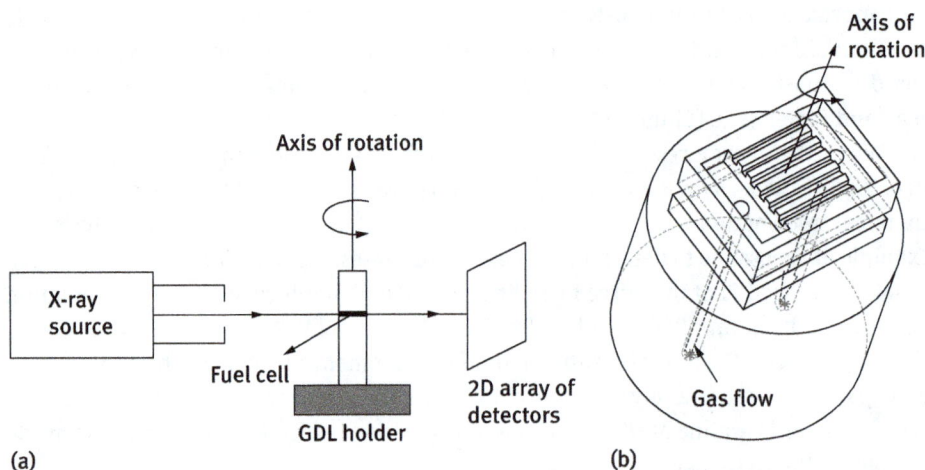

Figure 4.13: (a) Schematic of an experimental setup for X-ray tomography and (b) a PEM-FC with the lower part of GDL holder where the flow field is composed of parallel gas channels [Sinha 2006]. Reproduced with permission from Electrochemical and Solid-State Letters, 9, 7, A344–A348, 2006. Copyright 2015, The Electrochemical Society.

is mounted on a GDL plate that receives the gas flow from underneath. The plate is rotated on a turntable step by step while images are recorded on a 2D array detector [Sinha 2006]. For a recent PhD thesis on tomography of PEM-FC, see Eller [Eller 2013].

Cheng et al. [Cheng 2014] reported an approach of solving the phase problem in tomography; see Figure 4.14. They applied this method on a polymer sample, which was partially soaked with water, in order to investigate polymer-water interfaces. They rotated the sample on a turntable but also varied the distance between sample and detector. With this method, they could reconstruct the 3D X-ray refractive index of the sample.

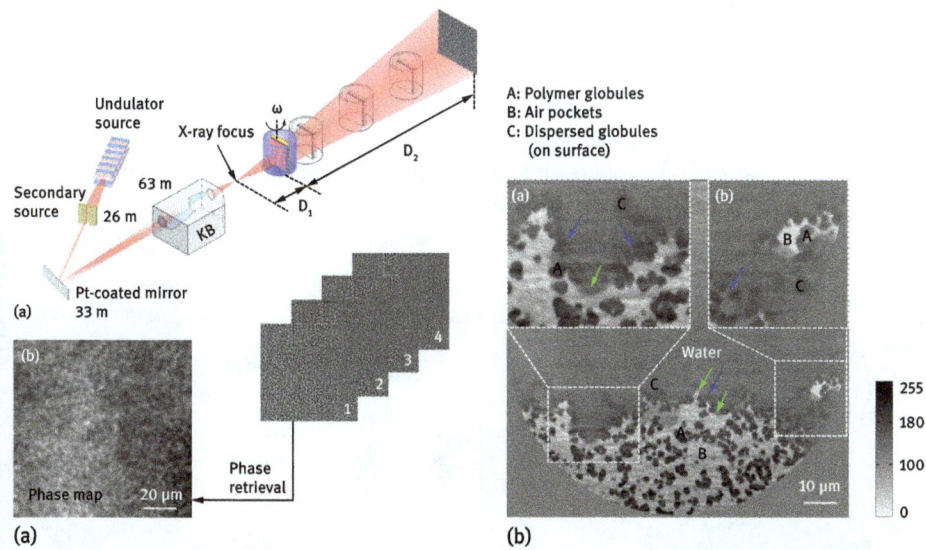

Figure 4.14: (Left) Principle of holotomography. (a) Sketch of the experimental setup at ESRF beamline ID22NI. (b) Radiography images were recorded at four focus-to-sample distances over a complete tomographic scan with 1199 projection angles. The four images were used to reconstruct the phase shift at each angle (phase retrieval). The retrieved phase maps were then used in a tomographic reconstruction step to reconstruct the 3D refractive index decrements of a specific region of interest the sample. (Right) High-resolution reconstructed 2D cross-section of the microporous BMA-EDMA surface on a glass substrate (sample 2). Effective pixel size was 50 nm. Insets (a) and (b) are zoom-in images of the white dashed selections. Red color labels A, B, and C denote polymer globules, air pockets, and dispersed globules, respectively. Air was trapped by the structured BMA-EDMA polymer. The coexistence of the Wenzel state (blue arrows) and Cassie–Baxter state (green arrows) when water interacting with the rough surface was observed. A volume of $0.1 \times 0.05 \times 0.01 \, \mathrm{mm}^3$ was rendered in 3D via gray level segmentation. [Cheng 2014] – Published by The Royal Society of Chemistry.

The result is an image as shown in the right panel of Figure 4.14. It provides a chemical contrast that is not based on varying the X-ray energy. The dark spherical objects are the polymer, and the bright areas designate the air pockets, which are not filled with water. The compact homogeneous gray-shaded area occupies most of the image area and is made from the water. There are some isolated polymer globuli swimming in the water layer (dispersed globules).

The main purpose of microscopy and tomography is to provide a visual impression of the geometrical structure of the samples. When X-rays are used as probes, it is possible, as we have shown already, to provide images of this geometrical structure with an elemental (chemical) contrast. This is, e. g., relevant for catalyst research [Meirer 2015]. Garzon et al. [Garzon 2007] have used *ex situ post mortem* high resolution with high contrast X-ray computed tomography (XCT) for the imaging of the morphology, platinum distribution, and carbon content of fuel cell membrane/electrode assemblies after various cycling stages. Their studies brought about information on the redistribution of platinum in the cathode and the migration of platinum into the electrolyte membrane and corrosion of the carbon support.

Neslon et al. showed how a tomography on a rotating sample can be done with an X-ray beam that passed a capillary condenser and a pin hole. A zone plate behind the rotating sample projects the image on a CCD camera. This setup allows for combination of tomography with XAS (Figure 4.15) [Nelson 2011].

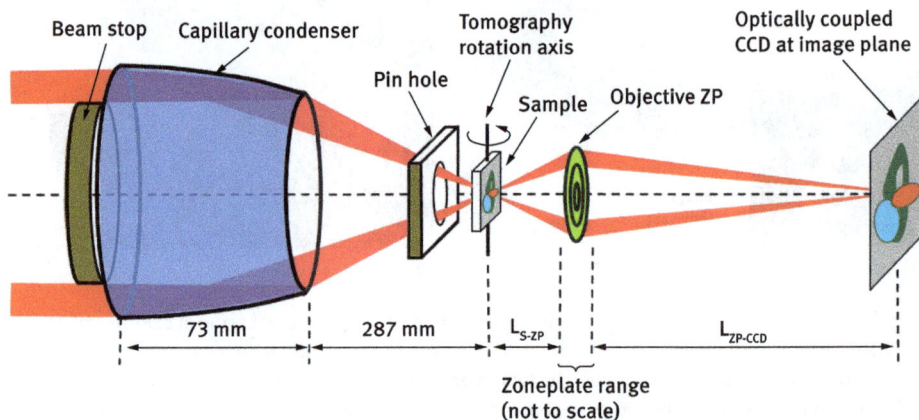

Figure 4.15: Schematic of the X-ray optical setup for transmission X-ray tomography with hard X-rays at SSRL and APS. Tunable monochromatic X-rays pass through a capillary condenser through a pinhole and hit the sample which can be rotated. The transmitted X-rays hit a zone plate and are then focused on a CCD camera for image acquisition. Rotation angle and X-ray energy are the functional parameters for combining XANES with tomography [Nelson 2011]. Reprinted with permission from [Nelson et al., Applied Physics Letters 98, 173109, 2011]. Copyright [2011], AIP Publishing LLC.

A not so well-known type of fuel cell is the phosphoric acid fuel, where a high concentration of H_3PO_4 is used as electrolyte at 150 °C–200 °C operation temperature. The phosphoric acid fills the pores in the porous components of the fuel cells, but it has always been problematic to find out to which extent these pores are efficiently filled. Eberhardt et al. [Eberhardt 2014] used a propagation-based phase X-ray contrast imaging and a referencing method. They developed experimental calibration curves and correlated gray-scale values and phosphoric acid concentrations, which they established for the GDL, catalyst layer, and electrolyte separator membrane. Finally, they made a comparison of image-based analyses and chemical analysis.

4.4.4 Tomography on batteries

Shearing et al. pointed out in their review article [Shearing 2013] that the improvement in microscopy methods allows more and more for the derivation of direct relationships between the microscopic structure of electrode (and in my opinion certainly also electrolyte materials; Artur Braun) and the performance of the components that constitute the device, and the device itself. In their article, they present multilength scale X-ray tomography results with case studies on SOFC and lithium battery materials.

The kind of tomography shown here is a basically nondestructive transmission-based method. The image depends on the intensity of the radiation arriving at the detector. Therefore, "thick" objects are not suited for tomography, and the critical thickness beyond which no tomography is possible depends on the absorption cross-section of the chemical elements in the object and the optical thickness of the object and the energy or wavelength of the probes.

Shearing et al. [Shearing 2010] have looked into the microstructure of a porous lithium-ion battery with X-ray tomography. They reconstructed a $43 \times 348 \times 478$ μm^3 sample volume, and the volume per voxel was 480 nm, around 0.5 micrometers. They showed that the minimum size for a representative volume element was about $43 \times 60 \times 60$ μm^3 for volume-specific surface area, "but as large as the full sample volume for porosity and tortuosity."

The main purpose of this book is how to learn about the electrochemical properties of materials and their assessment with synchrotron-based methods. As already declared in the Introduction, not only charge transfer but also heat and mass transfer are important properties for materials architectures in electrochemical energy converters. When we come to the device level, heat generation becomes an issue that needs to be taken care of in the engineering of device architectures. Bo Yan et al. determined the heat generation in a lithium ion battery and made a parallel computational modeling of the heat generation. Their 3D model predicted higher heat generation than an alternative porous electrode theory model [Yan 2013]. The same group investigated also the topological parameters [Yan 2012] of a lithium ion battery with X-ray computer tomography and used the data for the simulation of the electrochemical performance, i. e., OCV, current density, spatial distribution of reactants, etc. The outcome of this study was different from the outcome when they applied a simple pseudo 2D model.

Harris [Harris 2013] finds that most Li-ion battery failure can be ascribed to the presence of nano and microscale inhomogeneities that interact at the mesoscale, as is the case with almost every material, and that these inhomogeneities act by hindering Li transport. This means the Li ions do not get to the right place at the right time. They therefore define "inhomogeneities" as regions or objects with "sharply varying properties." This includes interfaces, irrespective whether they are made by design or by accident. They applied X-ray tomography and digital image correlation along with several other analytical methods in order to identify and quantify such inhomogeneities.

It is certainly of interest whether battery studies (and studies on any electrochemical energy converters [Andrews 2013, Nelson Weker 2015]) are conducted not only *ex situ* but also *in situ* or *operando*, which means that the battery is investigated while under operation, either in charging or discharging mode. Kallfass et al. [Kallfass 2012] applied X-ray tomography *ex situ* and *in situ* on commercial type lithium batteries and studied the response of the battery on electrical and thermal short circuit and on overcharging. Morphological changes could be directly correlated with current, voltage, and temperature characteristics. For 26650-type cells, they observed destruction of the metal containers during short circuit and overcharge, whereas on the 18650-type cells "with identical chemical components," no damage was found but the safety valve opened. They concluded it was necessary to determine the short circuit and overcharge of every single lithium ion battery type in detail before its technical application.

Tomography can be done with electron microbeams, such as provided by a transmission electron microscopy. Yoshizawa et al. [Yoshizawa 2006] have looked into the structure of carbon spheres with 100–200 nm diameter, which are used in lithium ion battery anodes (negative electrodes). The tomography image clarified the morphology of the carbon particles, which had been graphitized at 2800 °C. Specifically, the image obtained from the 3D reconstruction exhibited the inner texture more clearly than the usual TEM bright-field picture.

Genc et al. [Genc 2013] combined electron and X-ray tomography to determine the distribution of Ni, Mn, and O in a $Li_{1.2}Ni_{0.2}Mn_{0.6}O_2$ nanoparticle in lithium batteries. Simultaneously acquired tomography results provide additional 3D chemical information of the material especially when there is low atomic number (Z) contrast in the material of interest.

Shui et al. combined X-ray tomography with X-ray diffraction at the synchrotron with an operating lithium air battery [Shui 2013]. They discovered a partial recovery or metallic lithium during the charging half cycle and constant accumulation of LiOH under the charging and the discharging cycle. Specifically, a LiOH layer forms on the negative electrode that separates it from the separator. This layer is porous and provides charge and mass transport.

A very extensive study on a (vanadium) redox flow battery system was conducted by Qiu et al. [Qiu 2012]. They used XCT in order to determine the structure parameter (surface area, pore volume, pore size, and particle size), which were then used as input data for multiphase flow modeling with the Lattice Boltzmann method and electrochemical reaction modeling. Thus, they obtained results for the distribution of the cell potential, overpotential, current density profile, and distribution of the electrochemical species.

Fukunaga et al. [Fukunaga 2005] prepared nickel/metal-hydride (Ni/MH) batteries with a quasi-3D substrate and CoOOH-coated $Ni(OH)_2$ as active material. Their swelling and distribution of electrolyte were investigated with XCT and atomic absorption spectroscopy after cycling. They found that the degradation in cycling was due to the increase in cell resistance associated with the transfer of electrolyte from the separator to the positive electrode.

We have learned now about the utility of X-ray and neutron radiography for the imaging of particles and structures and pore spaces. The focus was laid on the quantification of topological characteristic, i. e., structural parameters in the two dimensions projected. Some studies even showed how a chemical contrast can be obtained with radiography. In this chapter, we learn how transmission X-ray microscopy becomes the extension of the radiography in the sense that we achieve also molecular structure and electronic structure information that goes beyond the mere chemical element contrast in structures.

Operando and *in situ* experiments require more elaborate approaches but are also more rewarding [Shearing 2011]. A study of transmission X-ray microscopy on a battery system is provided in Wang et al. [Wang 2013e]. Figure 1 in their paper shows the principle of the data structure of the TXM-XANES experiment for an *in situ* study. Stacks of images with a size of, e. g., 1024 × 1024 pixels are recorded for a range of X-ray energies, in this specific case covering the copper K-edge from 8960 to 9040 eV. Thus, every pixel can be assigned a full XANES spectrum. Using conventional XANES analysis, the chemical composition (within the limits of accuracy of the XANES analysis) of the material at the corresponding pixel position can be determined. The image on the upper right shows a particle arrangement in the battery cathode. The different colors reflect the chemical heterogeneity. When looking at Figure 1 in their paper [Wang 2013e], we identify a large particle core (Cu_2O blue) and a particle shell (Cu red). The greenish region is assigned CuO. For this particular sample, the composition was to 35 % Cu, 40 % CuO, and 25 % Cu_2O.

Figure 3 of Wang et al. [Wang 2013e] shows how the particle is scanned versus the battery charging time. Let us pick a particular image in the stack that corresponds to a particular time of the charging and that therefore constitutes a particular charging state. In stage "a," the particle has a blue color, which the initial stage indicating CuO (Cu^{2+}) in the bulk and at the surface. However, there is a greenish corona visible on the boundary of the particle, indicative of Cu_2O (Cu^+). Hence, at stage "a," the electrode is becoming reduced and the Cu^{2+} in CuO becomes reduced to Cu^+ at the boundary. In stage "b," the surface becomes reddish, which is indicative of metallic copper (Cu^0). The line scan for phase b shows a high concentration of CuO in the particle bulk and a reddish corona at the boundary, consequently. With increasing charging of the battery, the thickness of the Cu metal film increases and propagates into the particle interior. Now, there is an interface between the CuO core and the Cu film, which is Cu_2O. It appears therefore that Cu_2O is an intermediate phase. A color code is used for the triplot phase diagram of Cu, Cu_2O, and CuO.

Figure S1 in the Electronic Supplementary Information in [Wang 2013e] shows two photos of the actual transmission X-ray microscope at beamline X8C at NSLS. The X-rays arrive from the monochromator on the right and hit the sample, which is mounted on a rotatable sample holder that can be moved in all three space directions with a stepper motor. An extensive open access review on hard X-ray scanning X-ray microspectroscopy is given by Somogyi and Mocuta [Somogyi 2015].

In an *ex situ* study [Chen-Wiegart 2013], the same research group investigated a composite battery cathode containing $LiCoO_2$ and $Li(Ni_{1/3}Mn_{1/3}Co_{1/3})O_2$. The latter phase exhibited more cracking. In addition to chemical contrast by X-ray absorption, they used also Zernicke contrast generation [Chen 2015a, Holzner 2010], which enabled them to look into the carbonaceous phases, such as binder and carbon black and graphite in the electrode assembly [Chen-Wiegart 2013]. They used the data also for a complete microstructure analysis and for vague conclusions on the origin of capacity loss. Liu et al. [Liu 2013] studied the same material at the Advanced Light Source with spatially resolved X-ray spectroscopy and arrived at conclusions about the charge carrier dynamics in the battery during operation. Chen-Wiegart et al. [Chen-Wiegart 2012c] showed in a follow-up *operando* study a sequence of $LiVO_2$ particle images that undergo a discharging (reduction) from 0 to 1116 mAh/g and subsequent charging (oxidation) from 30 to 760 mAh/g, where the particle undergoes a core-shell chemical reaction, mechanical fracturing, and redensification. The $LiVO_2$ battery cathode particles were of around 20 micrometers each.

As I have mentioned in Part 2, the battery manufacture may incur already changes in the electrode material before any electrochemical treatment. Adding graphite, carbon black, and polymer binders and carbonaceous solvents creates a local reducing environment at the surface of the active metal oxide (spinel) particles because carbon is the oldest known reducing agent [Braun 2002d]. This effect was observed with X-ray spectroscopy [Braun 2002d]. Wang et al. [Wang 2013e] investigated this effect recently with X-ray transmission microscopy for the case of carbon coating of lithium iron phosphate particles.

4.5 Scanning transmission X-ray microspectroscopy (STXM)

Spectroscopy experiments and diffraction experiments sometimes give hints that a sample is not entirely phase pure. However, without further inspection with other methods, the overall structure of the probed material, or the sample, may remain elusive. Pecher et al. [Pecher 2005] have put it at a Goldschmidt Conference 2005 as follows:

> "We've used a suite of surface and bulk sensitive microscopic and spectroscopic techniques (TEM, HREELS, XPS, XRD and synchrotron L-edge STXM) to study nanomagnetite particles coming from different sources (biosynthesis, wet chemical synthesis routes (Gee et al.)). We have found that the application of nonmicroscopic techniques, such as XPS or conventional powder XRD can yield incomplete information because of particle size dependant variations in structural and electronic composition and because of effects of aggregation."

From this perspective, it is sometimes necessary to inspect the chemistry of a material with a very high spatial resolution.

Cody et al. [Cody 2003] had a different rational for using a chemically sensitive probe with very high spatial resolution:

> "It is unlikely that we will ever have sufficient mass of organic (cosmic) matter from these sources to apply solid state NMR to solve their chemical structures."

The amount of available sample material may have a significant effect on what kind of analytical techniques come to application. Because environmental and atmospheric scientists typically rely on limited amounts of sample on filters, spectromicroscopy techniques are necessary because focused beams allow direction of the radiation on single particles. Cody was probably referring to NASA's Stardust Space Mission, which returned cosmic dust sample trapped in an aerogel receiver after it had been in space for 6 years. It contained 145 dust particles larger than 10 micrometers, some of which were sliced and distributed in the scientific community for analyses. For sure, this was not enough material for one Nuclear Magnetic Resonance (NMR) analysis. However, one particle only with 1 micrometer is sufficient for a STXM analysis [Sandford 2006, Sandford 2007].

STXM allows for the chemical mapping of thin films and particles and the imaging of small structures. The samples are typically deposited on lacy carbon or copper TEM grids, or on silicon wafers with a very thin silicon nitride window (Silson Ltd, Insight Park, Welsh Road East, Southam, Warwickshire, CV47 1NE, England [Anastasi 2008]). The wafer size can range from 5×5 mm to 10×10 mm, and then sample silicon nitride Si_3N_4 window with a thinness of 50 to 100 nm can be 1×1 mm. A typical maximum sample thickness lime a film or particle is less than 1 micrometer. Thicker films absorb too much of the X-rays, and no image can be obtained. The wafer is then put on a sample holder that is scanned in lateral direction in x-y increments of say 100 nm by a stepper motor. The X-ray beam is focused on the sampling position with a so-called zone plate, the foundations of which have been laid in the pioneering paper by Janos Kirz in the early 1970s [Kirz 1974]. In the early 1990s, instrumentation was built to allow for practical application of this technology at the NSLS.

The first report on STXM which I found dates back to 1990, where X-ray spectromicroscopy was conducted on a model sample and explicitly mentioned in the publication title [Ade 1990]. Two years later, Ade et al. came up with a paper in Science [Ade 1992] where they had looked at carbon contrast in organic polymer samples and also in a DNA sample, the latter of which may have been the first application of STXM to biology.

Maybe the first application of STXM to energy and environment sciences comes from Cody et al. [Cody 1995a, Cody 1995b, Cody 1995c], who reported a study on a natural cooked coal sample (Pennsylvania) and a laboratory-derived sample (carbonized pyrene, 1000 °C). They showed spatially resolved C 1s Near-edge X-ray absorption fine structure (NEXAFS) spectra with an energy resolution of 0.3 eV and a spatial resolution of 55 nm, expressing confidence that a 10-nm resolution could be achieved in future. I am listing here somewhat the chronology of STXM development in order to show what a long time can pass before an analytical method becomes established as such and becomes established in a particular field of application. During my time at Berkeley from 1999 to 2001, I had the opportunity to access Beamline 5.3.2, where I planned a "battery experiment" with Dr. Klaus Pecher, who was beamline scientist back then at the Advanced Light Source. The goal was to make a chemical mapping of spinel particles with respect to the Mn 2p absorption threshold. My assumption was that it should be possible to identify chemical heterogeneities such as Mn^{2+}, Mn^{3+}, and Mn^{4+} in spinel particles form partially lithiated or delithiated lithium battery electrodes.

During running STXM scans, it turned out, however, that the particles from my cathode were absorbing too much X-rays. Hardly any X-rays reached the detector. The particles were simply too thick for STXM. As a rule of thumb, I would say particles of 3D metal oxides with 1-micrometer thickness rule out a STXM experiment with metal L-edges or oxygen K-edge. The transmission is too low. However, a 100-nm thin electrode film is just fine for STXM, which can easily be tested when using the convenient web-based tool at the Center for X-ray Optics at Berkeley National Laboratory [Gullikson 1995]. What I learned from this experiment was that "sample prep" is of utmost importance for complex microscopy experiments such as STXM. Conventional spectroscopy measurements may be possible under much more relaxed conditions.

Progress in X-ray optics made that lateral resolutions of 50 nm and better are feasible today. By scanning these increments step-by-step, an image can be stored. Elemental contrast variation is obtained by tuning the X-ray energy to an absorption edge and then scanning around this absorption edge. For every energy increment, e. g., 0.1 eV, the sample holder has to be moved again in the x-y plane for every pixel. The result is a voxel space where every pixel can be assigned one full NEXAFS spectrum per scanned element. The X-ray optical or elemental contrast can be visualized in images. This is not a 3D reconstruction method. The images are summed projections in the z-direction. The acquisition of the images can take several hours. The wider the scanned area is, the longer the necessary time in the experiment. The STXM method has been applied in environmental soil science [Kinyangi 2006, Lehmann 2005] and environmental aerosol science [Braun 2002c, Braun 2004b, Braun 2005a, Braun 2006d, Maria 2004, Russell 2002], to name a few examples only.

4.5.1 STXM on PEM fuel cells

Electrochemical *in situ* studies with STXM have been attempted at the ALS in the early 2000s, as the author of this book has witnessed occasionally during his own NEXAFS campaigns at the ALS. For a review on STXM work done at ELETTRA in Italy, see the short review and papers [Bozzini 2012a, Bozzini 2012b, Bozzini 2013]. The first reported electrochemical *in situ* work with STXM on an electrochemical cell was by the Hitchcock group in 2005 [Guay 2005, Hitchcock 2005]. The electrochemical cell is shown in

Working electrode

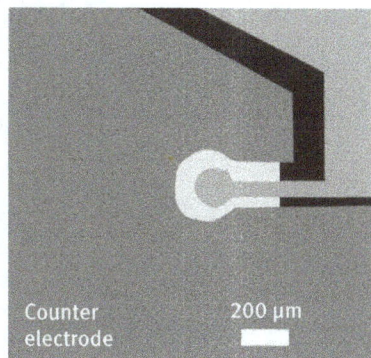

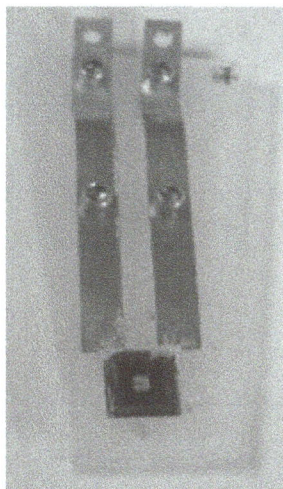

(a)   (b)

Figure 4.16: (a) Details of the 2-electrode cell used for STXM with *in situ* potential control. The chemical state of polyaniline electrodeposited on the working electrode is varied by potentials applied via the spring contacts (b). Reproduced after [Guay 2005]. Reprinted with permission from (Guay, D., Stewart-Ornstein, J., Zhang, X. R., Hitchcock, A. P., *In situ* spatial and time-resolved studies of electrochemical reactions by scanning transmission X-ray microscopy. Analytical Chemistry 2005, 77 (11), 3479–3487). Copyright (2005) American Chemical Society.

Figure 4.16. They coated on a thin Au current collector film by electrodeposition a thin polyaniline film. This film was covered with an aqueous overlayer of 1 molar HCl electrolyte solution. This arrangement was assembled between two thin Si_3N_4 windows and then subject to cyclic voltammetry, while at the same time STXM scans were recorded over the electrode area along the C 1s and N 1s X-ray absorption threshold. The polyaniline film was converted to leucoemeraldine as the reduced state to emeraldine chloride in the oxidized state (Figure 4.17). The experiment was conducted at Beamline 5.3.2 at the ALS. They obtained electrochemical contrast for carbon and nitrogen with a spatial resolution better than 50 nm and conducted kinetic studies with subsecond time resolution.

With such fundamental electrochemical study accomplished, the researchers could move on and develop the know-how for looking into applied electrochemical systems. Berejnov et al. conducted a study [Berejnov 2012] where they looked into the "platinum degradation" in PEM-FC fuel cells. An operating fuel cell poses actually a quite harsh environment on its components. This is not surprising. The same applies to any electrochemical energy converter. The studies that we are nowadays able to make show us that the electrochemical environment causes dissolution, mass transport at the nanoscale, which over time accumulates into disintegration of electrodes and electrolytes. We know from Chapter 4.1.2 how Zn dissolves in a Cu-Zn battery, for example. There is not even one material in energy converters that is not fundamentally prone to degradation.

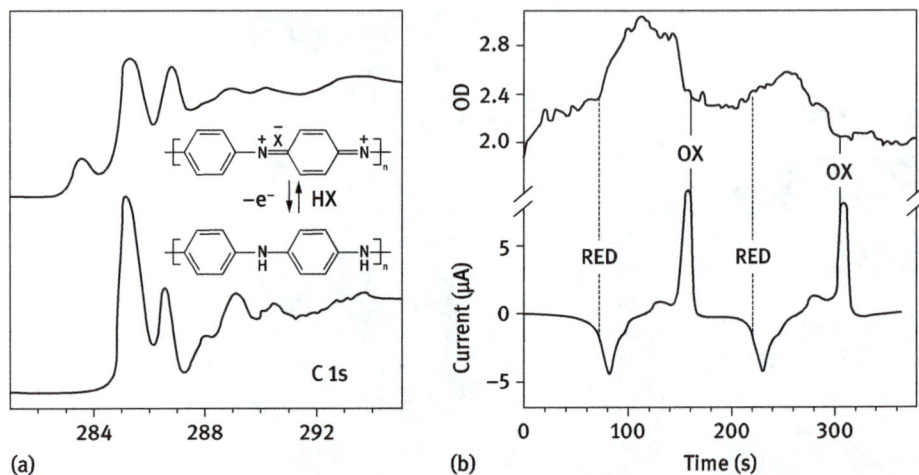

Figure 4.17: (a) C 1s NEXAFS spectra of oxidized (upper) and reduced (lower) polyaniline film obtained from STXM. (Right, upper) Optical density changes at 283.6 eV (quinoid π^*) recorded in parallel with cyclic voltammetry. (b, lower) Current response in CV. Reproduced after [Guay 2005]. Reprinted with permission from (Guay, D., Stewart-Ornstein, J., Zhang, X. R., Hitchcock, A. P., *In situ* spatial and time-resolved studies of electrochemical reactions by scanning transmission X-ray microscopy. Analytical Chemistry 2005, 77 (11), 3479–3487). Copyright (2005) American Chemical Society.

Therefore, we frequently find publications, projects, and even entire research programs on degradation in fuel cells, batteries, supercapacitors, photo-electrochemical cells, and solar cells.

Berejnov et al. [Berejnov 2012] investigated a platinum catalyst-coated polymer membrane. Two Pt precursors with and without nitrogen containing ligands were used in the processing of the cathode catalyst layer. The local environment of the assembly, as investigated with STXM-based NEXAFS spectroscopy (Figure 4.21), was found to be directly related to the Pt precursor used in catalyst composite membrane fabrication. The STXM maps at the nitrogen edge show spectral characteristics around the dendritic Pt particles in those membranes, which are made with platinum grown from nitrogen containing precursors. During degradation, the platinum is carried away from its original position, along with the nitrogen containing ligands.

Figure 4.18 shows an illustration of STXM data processing for nitrogen 1s image sequence applied to a microtome cross-section of the catalyst-coated membrane (CCM). The X-ray energy is scanned across the nitrogen 1s edge while the images are recorded. Shown are the images for 440, 405, and 390 eV. The layers are polystyrene support, cathode, continuous, and porous parts of the polymer electrolyte membrane. The right panel shows nitrogen 1s spectra of the *postmortem* membrane in the regions of the Pt core (second from top, red), edge of Pt particles (third, green), and membrane (fourth, blue). The inset is the average of 53 images of the N 1s STXM image sequence in the region of the Pt band. The N 1s spectrum of the cathode of the same section of the EOL sample A is

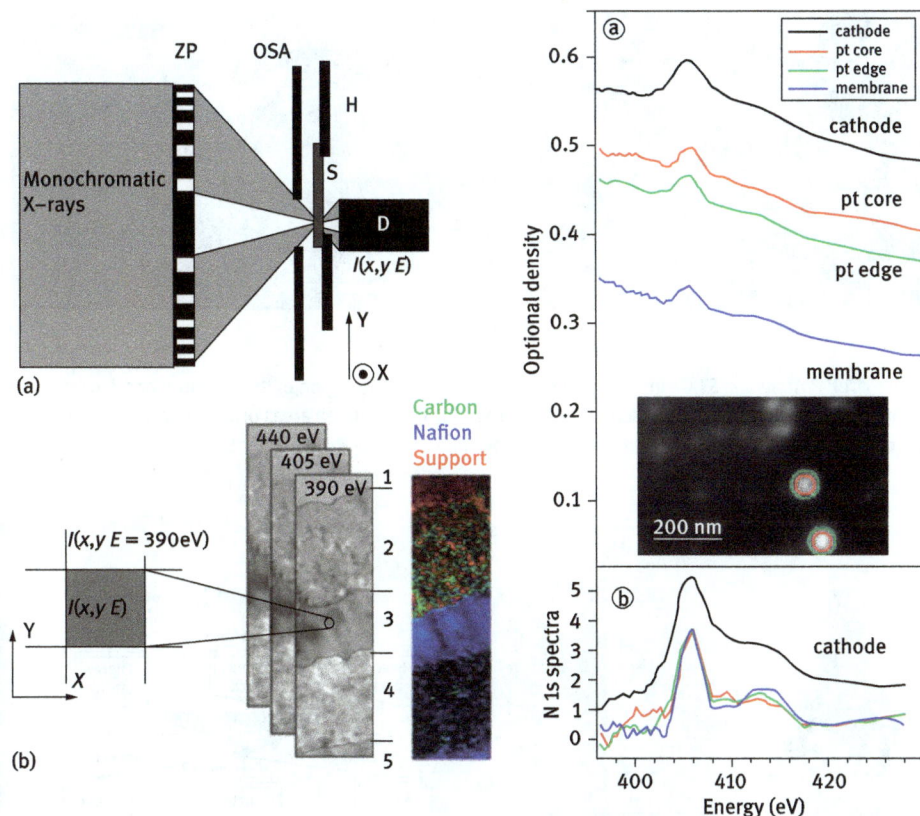

Figure 4.18: (a, upper left) Simple sketch of STXM image acquisition and processing principle. (a) Monochromatized X-rays are focused by the zone plate ZP and pass through an order selecting aperture OSA on the sample S on the sample holder H, which can be moved by a stepper motor in (x, y) direction. The X-rays hit the detector D after passing through the sample. Reproduced from [Berejnov 2012] with permission of the PCCP Owner Societies.

also shown (first, black). (b) N 1s spectra of the membrane, Pt-edge, Pt-core, and cathode after subtraction of a curved background extrapolated from the pre-N 1s region and unit normalization in the continuum. The spectrum of the cathode is offset by 1 unit.

Figure 4.19 shows two color-coded STXM images from the catalyst-coated membrane with platinum with a nitrogen ligand precursor (left image) and with no nitrogen ligand precursor (right image). The red dots represent the platinum catalyst particles. The green boundaries around the platinum in the left image is the spectral signature of the nitrogen from the precursor.

It is possible to extend the conventional STXM into "STXM tomography" by making angle scans. This has been realized before 2007 at the Advanced Light Source on a microcapillary of 4-micrometer diameter, which contained water, polystyrene microspheres, and polyelectrolyte (setup in Figure 4.20). The microstructure of this dispersion shows

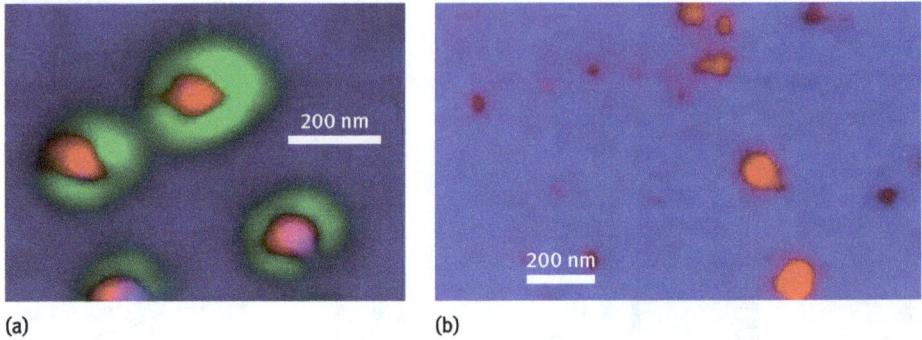

(a)   (b)

Figure 4.19: *Postmortem* STXM images of the CCM, recorded in the region of the platinum band for the nitrogen edge. The red dots represent the platinum particles; the green areas originate from the nitrogen precursor. Reproduced from [Berejnov 2012] with permission of the PCCP Owner Societies.

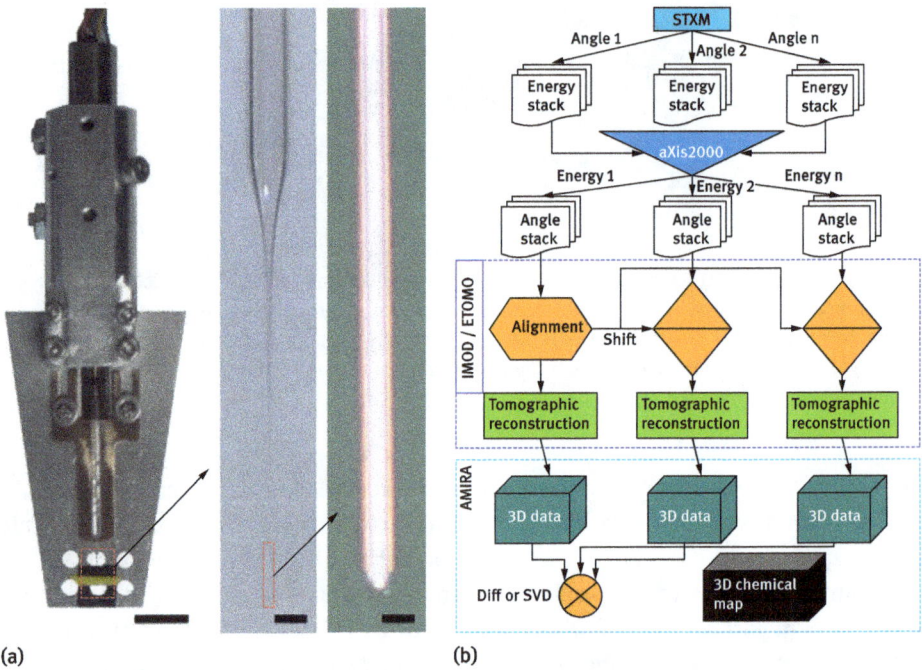

(a)   (b)

Figure 4.20: (Left) Fine tip of an empty capillary (scale bar = 5 mm), along with a close-up of the glass capillary (scale bar 1 mm). (Center) Photograph of the tomography rotation stage with plate used to support the rotation structure in the STXM kinematic mount (scale bar 10 mm). (Right) Schematic/flow chart for the acquisition and processing of STXM tomography data. Reproduced from [Johansson 2007] with permission of the International Union of Crystallography.

peculiarities in STXM tomography, which had not been known before [Johansson 2007]. It should be possible to conduct such STXM tomography also under electrochemical conditions, e. g., in a redox flow battery.

The Hitchcock group investigated further PEM-FC problems of industrial relevance with respect to synthesis, processing, and fabrication of components and assemblies. The preparation of such sample for STXM analysis is actually quite complex and laborious when compared with other synchrotron experiments. They looked at nanostructured thin film (NSTF) catalysts, which served also as current collectors or electrodes in PEM-FC [Lee 2014]. Perylene PR149 was the precursor material that was formed into whiskers and then coated with platinum catalysts. In the end, one obtains an ideal model catalyst CCM; see Figure 4.21.

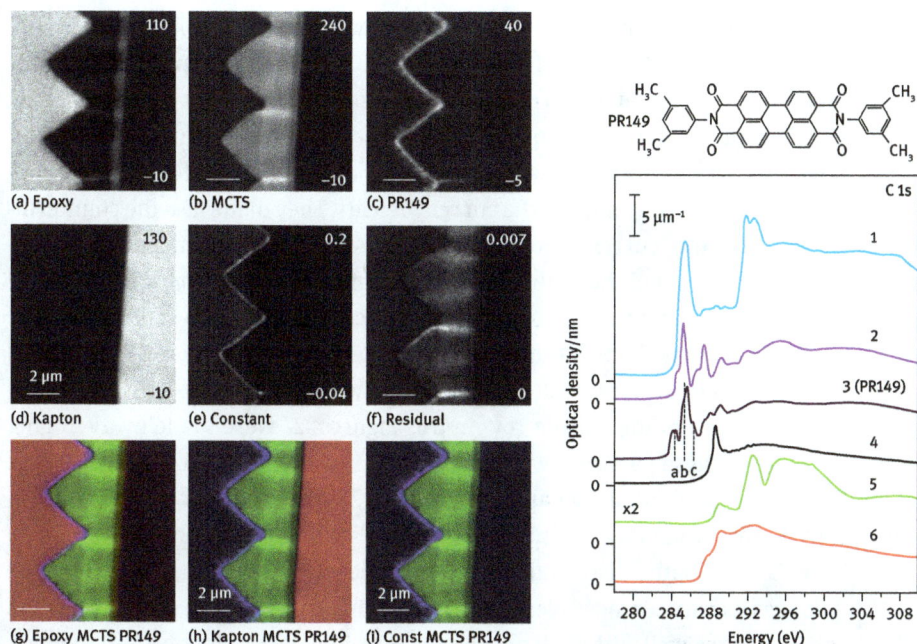

Figure 4.21: (Left). Component STXM maps for (a) epoxy, (b) MCTS, (c) PR149, (d) Kapton, and (e) constant (a metal release layer below the PR149 whiskers), derived from a stack fit analysis of a carbon 1s image sequence measured from an uncoated NSTF whisker sample, grown on MCTS and attached to Kapton. Numerals on lower and upper left side of epoxy, MCTS, PR149, and Kapton images indicate minimal and maximal thickness in nm. For constant and (f) residual images, the numerals denote minimal and maximal OD values. (g), (h), and (i) are color-coded composites (rescaled) of the indicated component maps. (Right) C 1s spectra of carbon support e1, Kapton e2, perylene red powder e3 (PR149, structure presented at the top of this figure), microstructured catalyst transfer substrate (MCTS) e4, perfluorosulfonic acid (PFSA) e5, and embedding epoxy e6. The intensity scale is absolute optical density per nanometer thickness. Offsets are used for clarity. The characteristic C 1s/π^* peaks in PR149 labeled e, a, b, c, and e occur at 284.2, 285.3, and 286.4 eV. See the original publication for spectral assignments. The figures were provided to the author by A. P. Hitchcock, McMaster University.

These CCM samples were examined before the operational testing and after a number of different protocols: start-up/shut-down, and also reversal tests. They could show that the perylene support is present in the pretest materials but completely absent in the post-test materials. The authors believed that this loss of perylene could be attributed to cracks in the catalyst, paralleled by hydrogenation at the anode during device operation. Despite the loss of the perylene, the Pt shells forming the NSTF anode catalyst layer performed well (Figure 4.21).

4.5.2 STXM on lithium ion batteries

Li et al. [Li 2014] looked with STXM into a $LiFePO_4$ cathode during electrochemical cycling. In particular, they scanned an area with an overwhelming number of 3000 individual particles in the electrode film and assigned via the Fe 2p NEXAFS spectra particular states of charge for every particle. This approach produces data with some statistical significance. This can be done with some ease with the automated data acquisition and image analysis software, which is nowadays routinely available. They find that the population of active particles depends on the current density. They made also the counterintuitive discovery that the "current-per-active-internal surface area" is virtually independent from the cycling rate of the battery, potentially giving rise for a paradigm change in battery technology. Figure 4.22 shows two reference NEXAFS Fe 2p spectra for Fe^{2+} and Fe^{3+} with the obvious chemical shift, and on the center and right side STXM images with color code highlighting Fe^{2+}-rich and Fe^{3+}-rich and mixed valency-rich particles of $Li_{1-x}FePO_4$. The spectra at the bottom row in this Figure 4.22 were made by averaging the spectra from all those particles, which have the comparable state of charge. This is also the task of the data and image analysis software, but it depends on the actual operator where to set reasonably the limits. The Fe^{3+} is set to green color and the Fe^{2+} to red color. Hence, the spatial distribution of the Fe valence is clearly visible by color in the STXM images. Because images were recorded during charging and discharging, the color changes occur as well, although not shown here. The data reduction and the data analysis are laborious but also rewarding.

The relative fraction of active particles, i. e., particles that immediately respond to the charge and discharge process, can thus be determined by counting particles area in the probed STXM window and related to the state of charge or speed of charging and discharging. This is further demonstrated in Figure 4.23, which shows the active particle fraction as a function of charging rate for a few data points for charging and discharging. The figure shows also the simulated active particle fraction based on the Cahn–Hilliard model with Butler–Vollmer kinetics, as is shown in the equation below with rate constant k_0 (is being used as fit parameter), activities in solid particle $a_{Li,s}$ and electrolyte $a_{Li,yte}$, elementary charge e, and reaction overpotential η at the triple-phase

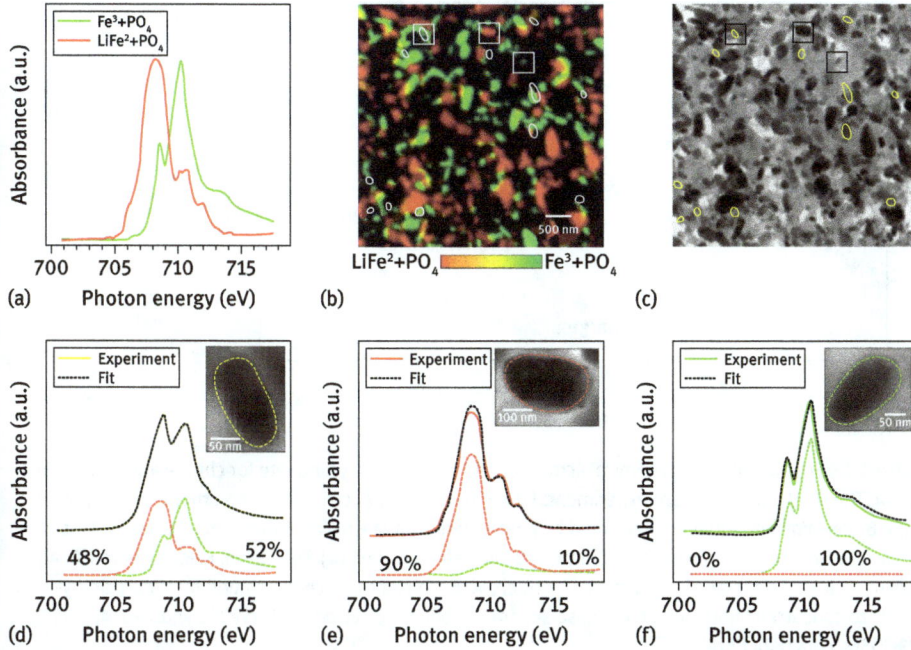

Figure 4.22: (a) Fe 2p NEXAFS spectra from $Fe^{3+}PO_4$ and $LiFe^{2+}PO_4$ for reference. (b) STXM image of cathode with color code, highlighting differences in Fe oxidation state. (c) Image with highlighted particles of mixed valency. (d) Fe 2p NEXAFS spectrum averaged from particles with mixed oxidation state. (e) Fe 2p NEXAFS spectrum averaged from particles with predominantly Fe^{2+} oxidation state. (f) Fe 2p NEXAFS spectrum averaged from particles with only Fe^{3+} valency. Reprinted by permission from Macmillan Publishers Ltd: Nature Materials (Li, Y., El Gabaly, F., Ferguson, T. R., Smith, R. B., Bartelt, N. C., Sugar, J. D., Fenton, K. R., Cogswell, D. A., Kilcoyne, A. L. D., Tyliszczak, T., Bazant, M. Z., Chueh, W. C., Current-induced transition from particle-by-particle to concurrent intercalation in phase-separating battery electrodes. *Nature Materials* **2014**, *13* (12), 1149–1156), copyright (2014).

boundary, and TS is the transition state activity coefficient. α is a phenomenological fit parameter for the experimentally observed asymmetry between charge and discharge:

$$J = k_0 \frac{a_{Li,yte}^{1-\alpha} a_{Li,s}^{\alpha}}{\gamma_{T_s}} \left[\exp\left(-\frac{\alpha e \eta}{k_B T} \right) - \exp\left(\frac{(1-\alpha) e \eta}{k_B T} \right) \right]$$

The mobility of the lithium in the electrode and electrode is given by

$$\mu_{Li,LFP} = \Omega_a (1 - 2\tilde{c}_s) + k_B T \ln\left(\frac{\tilde{c}_s}{1 - \tilde{c}_s} \right) - \frac{\kappa}{\rho_s} \nabla^2 \tilde{c}_s + \frac{B_0}{\rho_s} (\bar{c}_s - \tilde{c}_s)$$

with the regular solution parameter Ω_α, the gradient penalty κ from the Cahn–Hilliard equation, B_0 is a coherent stress penalty, ρ_s is the lithium site concentration, and c tilde and c bar are the lithium concentrations relative to the number of lithium sites and the

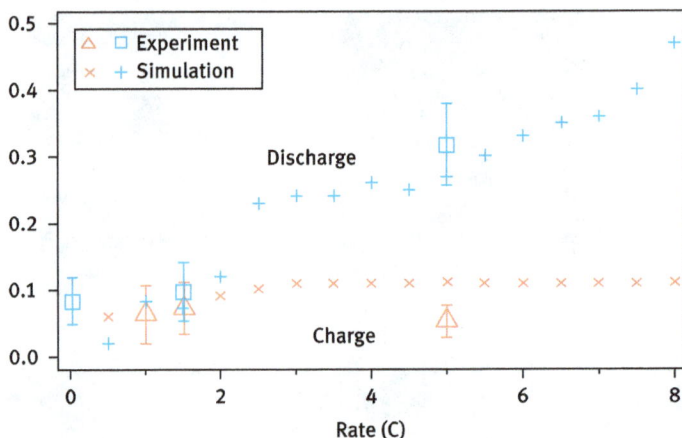

Figure 4.23: Fraction of electrochemical active particles versus the cycling rate for charge and discharge direction. The solid data points are experimental data from the STXM campaign. The other data are modeled by the Butler–Vollmer equation. Reprinted by permission from Macmillan Publishers Ltd: Nature Materials (Li, Y., El Gabaly, F., Ferguson, T. R., Smith, R. B., Bartelt, N. C., Sugar, J. D., Fenton, K. R., Cogswell, D. A., Kilcoyne, A. L. D., Tyliszczak, T., Bazant, M. Z., Chueh, W. C., Current-induced transition from particle-by-particle to concurrent intercalation in phase-separating battery electrodes. *Nature Materials* **2014**, *13* (12), 1149–1156), copyright (2014).

average filling fraction, respectively. The few experimental data points demonstrate the agreement of experiment with the Cahn–Hilliard model and the Butler–Volmer equation, which were least squares fitted to the data. It is rewarding to see that this experiment validates the model.

Li et al. [Li 2014] have plotted the fraction of active particles as determined from the STXM analysis and their, and only their corresponding current density, both as a function of the discharge rate (Figure 4.23). It is indeed amazing how well-simulated Butler–Volmer kinetics reproduces the experimental data and predicts a solid-solution filling of the particles. We observe that with discharge rates up to 5 C, the active particle fraction increases with increasing rate, which means the material handles the increased current by increasing the active particle fraction. At 10 C, a plateau is reached, and the authors associate the fact that incomplete (smaller 100 %) participation of filling is due to a miscibility gap.

4.5.3 STXM on biological samples

Two works on biological systems shall be mentioned here in view of anticipated future work on bioelectrochemistry, where STXM and similar spatially resolving spectroscopy could be of interest [Miot 2014]. This could hold, e. g., to bio fuel cells, or bio electrolyzers. Biological materials contain substantial amounts of carbon. There is firm prejudice in some communities that it makes no sense to look with X-ray spectromicroscopy into cells and biofilms because the elemental carbon abundance would not allow for discrimination of any molecular contrast. Such prejudice has been, e.g., the case for the toxicological interaction of soot with living cells. As we know by now, STXM allows you to discriminate in a fully carbonaceous sample aliphatic from graphitic carbon, for example. The carbon STXM papers have clearly illustrated this for soot, coal, and also for polymer blends, to name a few.

Toner et al. [Toner 2005] looked at the ALS into the oxidation of manganese by bacterial *Pseudomonas putida* biofilms. Its strain MnB1 oxidizes Mn^{2+} (which we know from the battery works that it can dissolve in some solvents, Chapter 8.1.2) in freshwater and soils. The researchers exposed the biofilm to aqueous Mn^{2+} containing environment and conducted Mn 2p NEXAFS with the STXM instrument at a 40-nm lateral resolution. The manganese became oxidized from Mn^{2+} to Mn^{3+} and Mn^{4+}, became precipitated, and was completely phagocytosized by the bacterial biofilm material. They conclude that STXM is a promising tool for researching on hydrated interfaces between bacteria and minerals, "particularly in cases where the structure of bacterial biofilms needs to be maintained."

The final contribution is by Leung et al. [Leung 2012] with a work about X-ray photoemission electron microscopy (X-PEEM) on protein adsorption on polymers. X-PEEM stands for X-ray photoelectron emission microscopy and is particularly surface sensitive. Its spatial resolution is around 80 nm. It is not a transmission method and, therefore, does not integrate over the entire probed thickness, in contrast to STXM. X-PEEM does however require an ultrahigh vacuum environment and is thus not (not yet) suited for experiments with samples in a liquid environment. At this time, *in situ* biological experiments may be ruled out.

Recent synchrotron-based soft X-PEEM studies of protein and peptide interaction with phase segregated and patterned polymer surfaces in the context of optimization of candidate biomaterials are reviewed, and a study of a new system is reported by [Leung 2012]. X-PEEM and atomic force microscopy were used to investigate the morphology of a phase-segregated thin film of a polystyrene/poly(methyl methacrylate)-b-polyacrylic acid (PS/PMMA-PAA) blend, and its interactions with negatively charged human serum albumin (HSA) and positively charged SUB-6 (a cationic antimicrobial peptide, RWWKIWVIRWWR-NH2) at several pHs. At neutral pH, where the polymer surface is partially negatively charged, HSA and SUB-6 peptide showed contrasting adsorption behavior, which is interpreted in terms of differences in their electrostatic interactions with the polymer surface.

Leung et al. [Leung 2012] stated, by referring to Stewart-Ornstein et al. [Stewart-Ornstein 2007], that it is currently difficult to use NEXAFS spectroscopy for the identification of various proteins because of the similarity of amino acid residues. Yet, they have had success in studying a difficult case of protein–peptide adsorption [Leung 2010]. They speculate on whether labeling with metal ions or quantum dots could be helpful for further progress in this analysis method.

4.5.4 STXM on soot from combustion processes

Particulates sampled from the ambient environment originate from numerous sources. Source apportionment is an important task in environmental regulation and policy. During my time in Kentucky, I mainly worked on the molecular speciation of airborne particulate matter and its method development with X-ray spectroscopy. If the particulates emitted from particular sources also contain particular spectroscopic fingerprints, a source apportionment should to some extent be possible. I therefore made many relatively straightforward carbon C 1s NEXAFS experiments on soot samples, soot from diesel engines, wood fire places, gasoline engine soot, helicopter turbine soot, and many more. Historically, it was the civil engineers who were interested in the particular matter (PM) burden. They wanted to know how much particulate matter in microgram/m^3 would be around in the environment or in buildings, for example. Toxicologists and epidemiologists, too, were interested in this because of the suspected correlation of mortality and PM concentration. The famous Six Cities Study gives empirical statistical account—and a strong message—on how hospitalization and mortality increase in urban areas when the PM concentration to which humans are exposed increases [Dockery 1993].

In the early studies, there was no discrimination between the particulates. Only their relative mass versus the sampled ambient air volume was counted. With increasing awareness, researchers became interested in further analyzing the PM. One criterion was the particle size distribution. There was indication that smaller particles could cause more adverse health effects (cardiac and pulmonary diseases, for example) than large particles. Particle size determination became therefore important. At some point, therefore, attention was paid to the particle size distribution. Particles smaller than 10 micrometers and smaller than 2.5 micrometers were considered relevant.

At some point, it turned out that the chemical composition had some relevance to adverse health effects, in addition to the size. Hence, molecular speciation became important. Particles emitted from coal power plants as fly ash may contain carbon, sulfur, chromium, and arsenic, for example. The two latter elements may cause toxic responses in human tissue. It may therefore be important to discriminate in particles what elements they contain and also there these elements that are located in or on the particle.

PM researchers are particularly interested in the black carbon portion in PM because it is supposedly a marker for health related issues. At the Consortium for Fossil Fuel Sciences at the University of Kentucky, I specifically looked into the carbon molecular signature of PM. Carbon has a very rich chemistry, which can be probed with infrared spectroscopy and Raman spectroscopy, NMR, and with X-ray spectroscopy.

The C 1s NEXAFS spectra of PM typically contain a nice manifold of molecular species, which also allow for some toxicological insight [Bolling 2012, Totlandsdal 2012]. At Brookhaven National Laboratory (National Synchrotron Radiation Light Source, Beamline X1 A), we conducted a STXM study on diesel soot particles. Without any analytical instrument other than your six senses, you can make out differences in the structure of diesel soot depending on its engine operation conditions. Diesel soot produced under lean fuel engine conditions happens to be dry and very dielectric. The particle agglomerates repel one another. Soot obtained from a fat mixture of fuel and air tends to be heavier and wet. The latter can be handled with ease.

Facing the question how to distribute the soot particles properly on the silicon nitride sample holder window, I decided then to disperse the soot particles in acetone and put a drop of this solution with a micropipette on the window. The solvent eventually evaporated, and a stain of soot samples was well dispersed on the silicon nitride window. Figure 4.24 shows a roughly 10 × 10-micrometer sample region with STXM images recorded at 282.59 and 284.87 eV X-ray energy. The black color contrast between both images indicates the graphite-like carbonaceous nature of the soot particle, which becomes particularly pronounced because of the chemical absorption contrast for graphite at around 285 eV. This is literally in contrast to the lower X-ray energy image. Figure 4.25 shows a somewhat more sophisticated contrast, which distinguishes between the graphite-like core of the soot agglomerate and a stain that is built from the washed out volatiles that were adsorbed on the soot particles, such as residual lubricant oil. C 1s NEXAFS on soot thus allows for the discrimination of aliphatic and aromatic carbon, for example.

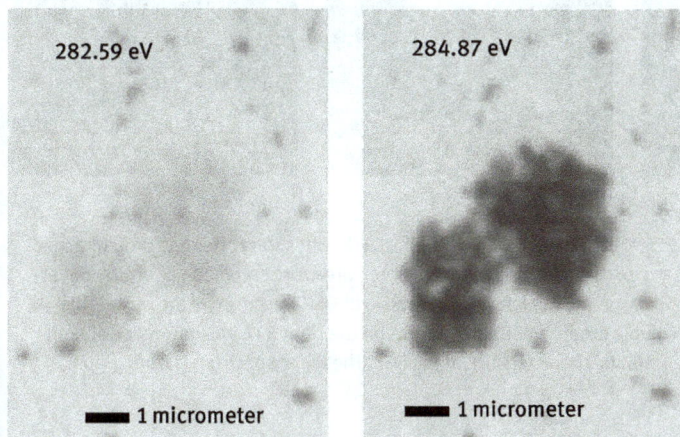

Figure 4.24: STXM images of diesel soot particles dispersed over Si_3N_4 windows. The left image is recorded with 282.59 eV X-ray energy, this is the C 1s pre-edge. The right image shows the same sample recorded at 284.87 eV, where the C=C bond of graphite shows a strong resonance and hence absorption. The solid graphite-like core of the sot becomes thus visible in the STXM image and allows for chemical speciation.

Figure 4.26 shows the inspected area with dispersed soot particles at three different X-ray energies [Braun 2008a]. The green dots denote the region where pixels were summarized for spectra. The red regions denote the pixels where background spectra were recorded. In the STXM image on the left, the green regions denote solid carbon

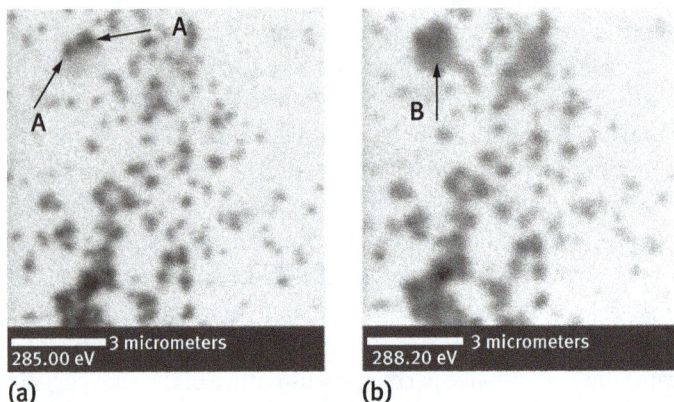

Figure 4.25: STXM image (10 × 10 micrometers) of soot particles at 285 (a) and 288.20 eV (b) photon energy. "A" and "B" positions denote selected areas where NEXAFS spectra are presented.

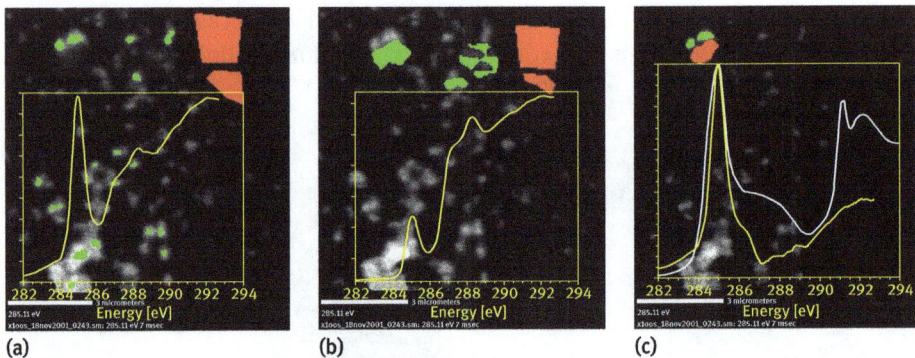

Figure 4.26: STXM images at 285.11 eV (due to C=C bonds) of DEP (bright spots) after one drop of added acetone was evaporated, causing extraction of the solubles. Green areas are summed up for I(E); red areas are summed up for Io(E). Spectrum on the left (a) is indicative of DEP. Spectrum in the center (b) is representative of an acetone-soluble organic matter. Spectrum on the right (c) is the difference of both spectra, highlighting the semigraphitic characteristics of the DEP. The white spectrum is from graphite. [Braun 2008a].

particles cores from, as we see later, graphite-like carbon with C=C double bonds. The corresponding spectrum from the summarized intensity of the green pixels is pasted in yellow color in the STXM image. The strong peak at 285 eV is from C=C double bonds indicative of graphite-like structures.

The image in the middle was used to collect pixels from the surrounding stained area, which is likely the washed out solvents and volatiles from lubricant. The peak at 285 eV from C=C is present, but not dominating the spectrum. Instead, structures from aliphatic carbon are dominating the spectrum. The right image compares the difference between both regions. In addition, I have plotted a C 1s NEXAFS spectrum from

graphite so as to highlight spectroscopic similarities between graphite and soot particle core. We will see a STXM soot extract analysis in Part 10 in this book on radiation
damages [Braun 2006d].

Soot has taken center stage also with respect to climate change [Chameides 2002].
More papers on C 1s STXM can be found with Lynn Russell from the Scripps In-
stitute [Bahadur 2010, Day 2009, Liu 2009, Maria 2004, Schwartz 2010, Shakya 2013,
Takahama 2007, Takahama 2008, Takahama 2013]. There are also numerous papers from
Johannes Lehmann at Cornell University [Heymann 2014, Kinyangi 2006, Lehmann 2005,
Liang 2006, Liang 2008, Solomon 2012]. It was predominantly the STXM instruments at
Beamline X1A at NSLS and 5.3.2 at ALS that were used for their studies.

4.5.5 STXM on Liesegang rings

We have used STXM recently for the chemical speciation of so-called Liesegang rings.
It would be interesting to deposit, e. g., electrocatalysts in an ordered manner on substrates by using the Liesegang phenomenon [Braun 2011c, Toth 2011, Walliser 2015a,
Walliser 2015b]. We heard of Liesegang rings already in the SAXS chapter in relation
with Fischer–Tropsch catalysts. In the particular case here, we deal again with a catalyst problem. Liesegang rings form when ions in gels diffuse to one another and can
chemically react with precipitation of solid phases [Liesegang 1906, Liesegang 1914,
Liesegang 1915, Liesegang 1939]. Liesegang rings are believed to be a macroscopic manifestation of quantumphysical phenomena [Furth 1933, Hermann 1935]. Under particular
conditions, the "concert" of reaction and diffusion causes precipitates, which arrange
in an oscillatory pattern. Here, we had to distinguish chromate from dichromate and
from silver on the micrometer size range. Both species are supposed to be formed in
the rings, but we cannot distinguish them on this narrow size scale.

We therefore need an analytical method for chemical speciation with high spatial
resolution, which in general can be performed with STXM [Toth 2011]. Figure 4.27 shows
the STXM experimenters at the POLLUX Beamline at the Swiss Light Source, Villigen PSI.
Figure 4.28 shows a micrograph with Liesegang rings where we need to specify whether
the rings are formed by chromate or by dichromate, and whether the flat range in between the rings forms chromate and/or dichromate. We have therefore recorded STXM
images at the chromium L-edges, oxygen K-edge, and silver M-edge. We then have inspected the STXM stacks spatially and mapped spectra for the ring ranges and the flat
ranges in between the rings.

There is one interesting observation in the silver chromate system. We discover that
the submillimeter Liesegang rings are actually divided into further subrings, secondary
rings in the micrometer range. The center panel of Figure 4.28 shows a 100-micrometer
thick Liesegang ring sector with approximately 20 secondary ring ripples. The two panels on the left show the areas where we have extracted STXM spectra. This includes the
dark rings for one spectrum type A and the bright range between the dark rings, which

Figure 4.27: POLLUX beamline control room for STXM experiment at Swiss Light Source, PSI, during experiment campaign. Artur Braun (Empa), Istvan Lagzi (Budapest University of Technology and Economics), Rita Toth and Florent Boudoire (Empa), and Roche Walliser (U Basel). Photo courtesy Benjamin Watts, PSI.

is assigned spectrum type B. The two panels on the right show the flat range in between the primary Liesegang rings where we also identify secondary rings. Here, we sample at the secondary ring location spectrum type C and the "empty" flat range in between with spectrum D. We have visualized this situation in Figure 4.29 and shown the regions in and in between the Liesegang rings where we anticipate chromate and dichromate.

When you consider the molecular structure of chromate and dichromate, you notice that in both cases chromium persists as Cr^{6+}. The coordination of the Cr^{6+} by the oxygen ions is in both cases four. However, in the dichromate, two such coordination molecules are joined by one oxygen ion, plus the bonding lengths are slightly different. It is therefore not at all trivial to make out the differences of both species in the NEXAFS spectra, which we will derive from STXM stacks.

Figure 4.30 shows a series of NEXAFS spectra recorded at positions A–D, which are depicted in the images in Figure 4.28. We recorded STXM images at the Cr L3,2 edge energies (576.2 and 584.9 eV) (Figure 4.28). Cr L edge, O K edge, and Ag M edge spectra were acquired over a 200 × 200-nm area, representative of the secondary bands in and between the primary bands (Figure 4.28, positions A and C), in the area between the sec-

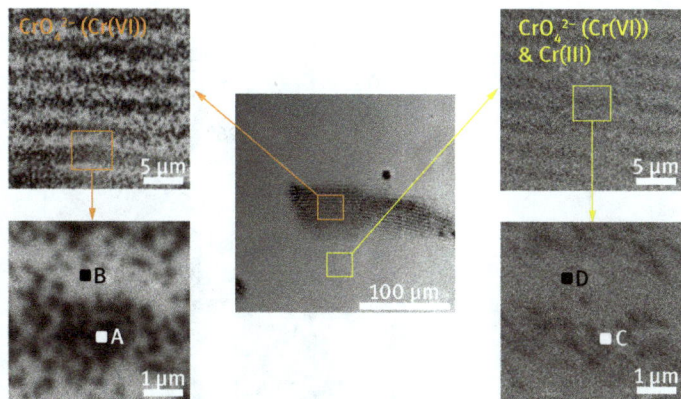

Figure 4.28: STXM images of the precipitation bands. The images were measured at 576.2 eV (corresponding to absorption by Cr). In the middle: Image of a primary band and its surrounding area (1.25 μm pixels). Left side: Images of the fine detail of primary bands (top: 125 nm pixels, bottom: 50 nm pixels). A: a secondary band in a primary band, B: background in a primary band. Right side: Images of the fine detail of an area between two primary bands (top: 125 nm pixels, bottom: 50 nm pixels), C: secondary band, D: background. The precipitates in the secondary and primary bands are Ag_2CrO_4 and $Ag_2Cr_2O_7$, respectively.

ondary bands at these two locations (Figure 4.28, positions B and D), and in and between primary bands when secondary bands were not present (image is not shown).

The Cr L3,2 edges spectra present two sets of features, one in the L3 region (570–579 eV) and one in the L2 region (580–590 eV) (Figure 4.28). The similarity of spectra A and E with the Cr^{6+} spectrum from [Grolimund 1999] shows that in the precipitation bands, the chromium is present as Cr^{6+}. In spectra B, C, and D, the L2 peak (~584 eV) is slightly shifted in the direction of lower energies. These spectra are similar to the Cr^{3+} spectrum from Grolimund et al. [Grolimund 1999], which means that between the precipitation bands (background), Cr^{3+} also has an influence. Cr^{3+} is present in the beamline optics, and the gelatin reduces some of the Cr^{6+} to Cr^{3+}. In the background, a very small amount of precipitate is present, and because we measure a few hundred nanometer thin films, in the ~30-nm high secondary bands between two primary bands (location D), also very little precipitate can form. We can conclude that the precipitate at each location is Cr^{6+}, and when a very small amount of precipitation is present, Cr^{3+} can also be detected.

The oxygen K shell pre-edge spectra (Figure 4.30 center panel) provide further information. All chromates have structures made up of tetrahedral CrO_4^{2-} units sharing an O corner. In spectra A, B, and E, the presence of two peaks suggests that the precipitate in the primary bands (regardless of containing secondary bands or not) contains $Cr_2O_7^{2-}$ ions. The peaks are narrower compared with the $K_2Cr_2O_7$ reference spectrum, possibly because of bonding with Ag^+ ions. Spectra C and D contain only one peak, and its position matches with the CrO_4^{2-} ion peak in K_2CrO_4 reference spectrum. The right panel of Figure 4.30 shows the Ag M4,5 edges spectra. Although the peaks corresponding to the M4,5 edges at 368 and 374 eV are clearly visible on the Ag reference spectra, they

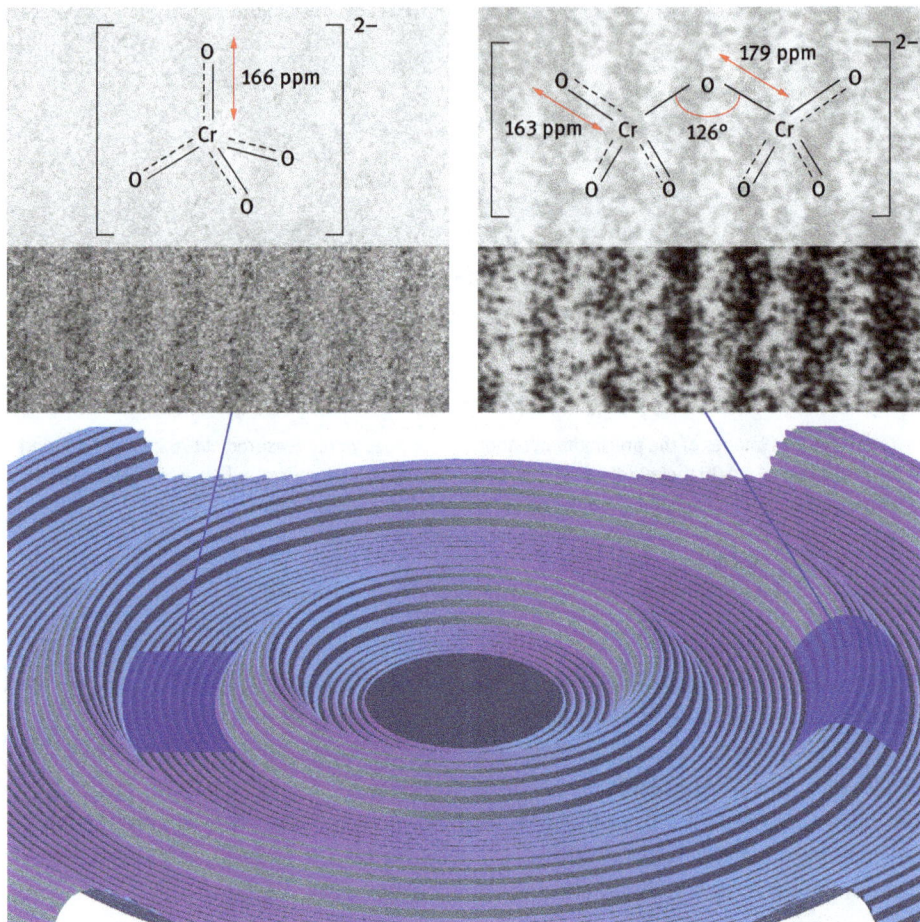

Figure 4.29: Computer visualization of Liesegang primary rings with distinct height profile, convoluted with finer secondary rings that are in the flat region between primary rings and also on top of the primary rings. The locations of the found molecular structures of chromate and dichromate are depicted accordingly.

are not discernible on any of the precipitate spectra. Yet, the broad, flat peak at the range of the Ag M4,5 edges suggests that silver is present throughout the entire pattern.

Analysis of the Cr 2p multiplet shows that the secondary and primary bands consist of Cr^{6+}. The O K pre-edge spectra reveal that the secondary bands contain CrO_4^{2-} ions, whereas the primary bands contain $Cr_2O_7^{2-}$ ions (sketched in the schematic of Figures 4.29 and 4.30). Because the Ag M edge spectra show that silver is present in both precipitate bands, it is relatively safe to say that the two precipitates in the secondary and primary bands are Ag_2CrO_4 and $Ag_2Cr_2O_7$, respectively.

The precipitation process was thus proposed to be governed by the chromate-dichromate balance reaction, which is dependent on the pH. Because the gelatin films

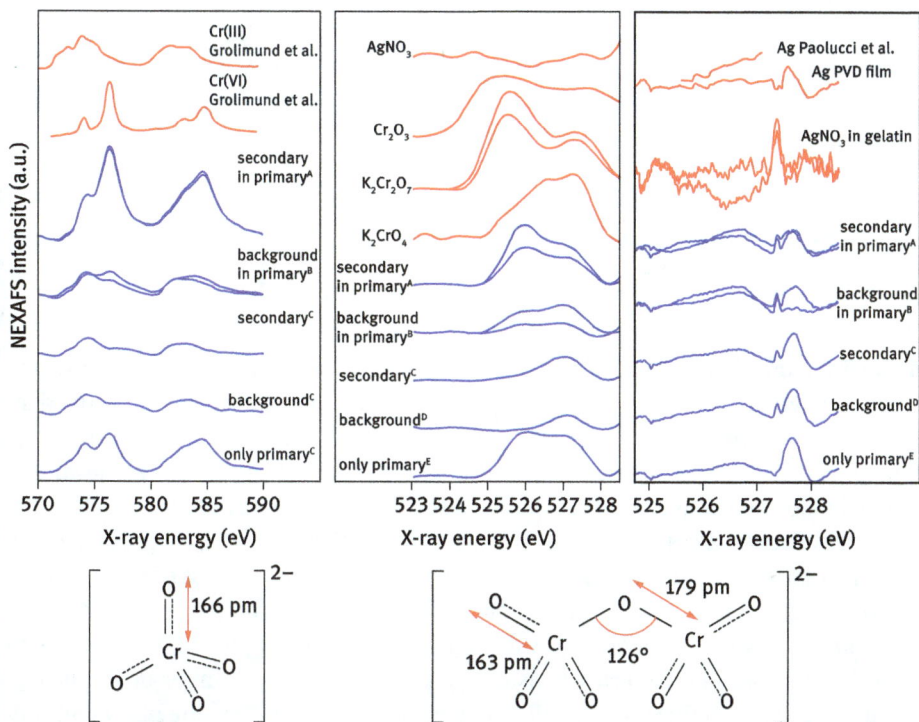

Figure 4.30: STXM spectra. (a) Cr $L_{3,2}$ edges, (b) oxygen K edge, and (c) Ag M edge average spectra of 10 square areas of 20 × 20 nm in a secondary band in a primary band (A), the spacing between these secondary bands (B) at two different positions, a secondary band between two primary bands (C), the spacing between these secondary bands (D) at two different positions, a primary band (E) when secondary bands are not present, plus (a) the Cr(III) and Cr(VI) spectra [Grolimund 1999], (b) our $K_2Cr_2O_7$, K_2CrO_4 and $AgNO_3$ reference spectra and Cr_2O_3 spectrum [Chiou 2011], and (c) spectra of $AgNO_3$ in a gelatin film, our physical vapor deposited Ag film and a polycrystalline Ag sample [Paolucci 1991]. [Toth 2016] Toth R., Walliser R. M., Lagzi I., Boudoire F., Duggelin M., Braun A., Housecroft C. E., Constable E. C.: Probing the mystery of Liesegang band formation: revealing the origin of self-organized dual-frequency micro and nanoparticle arrays. *Soft Matter* 2016, 12:8367–8374. – Published by The Royal Society of Chemistry.

contain a certain amount of NH_4OH in addition to the $K_2Cr_2O_7$ salt, the initial equilibrium is shifted to the right-hand side of equation toward the chromate ions:

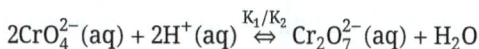

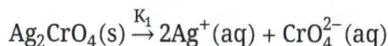

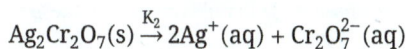

$$2CrO_4^{2-}(aq) + 2H^+(aq) \overset{K_1/K_2}{\Leftrightarrow} Cr_2O_7^{2-}(aq) + H_2O$$

$$Ag_2CrO_4(s) \overset{K_1}{\rightarrow} 2Ag^+(aq) + CrO_4^{2-}(aq)$$

$$Ag_2Cr_2O_7(s) \overset{K_2}{\rightarrow} 2Ag^+(aq) + Cr_2O_7^{2-}(aq)$$

Therefore, the secondary bands must form first as Ag_2CrO_4 precipitate bands, and as the pH increases (depleting the OH^- ions), the $Ag_2Cr_2O_7$ primary precipitation bands also form. As the Ag_2CrO_4 precipitate is more stable than $Ag_2Cr_2O_7$ (the molar solubility values of Ag_2CrO_4 and $Ag_2Cr_2O_7$ in water at 25 °C are 7.8×10^{-5}M and 3.7×10^{-3}M, and their K_{sp} values are 8.83×10^3 m/kg and 2.0×10^{-7} m/kg, respectively), the two precipitates can form consecutively and both of them can be present at the same time.

4.6 X-ray ptychography

4.6.1 Concept

The next Chapter 5 will cover the resonant methods, but I want to cover the X-ray ptychography here in the chapter on imaging. The resonant methods are based on the scattering contrast, which is obtained by tuning the primary X-ray energy across an absorption threshold of an element. This allows in general for the phase[1] retrieval. Let us shortly move back to crystallographic phase[2] analysis by X-ray diffraction. The theory of X-ray data analysis goes back to Bragg [Perutz 1990, Thomas 2012]. X-rays strongly interact with the electrons. The electron density distribution $\rho(x, y, z)$ is basically a manifestation of the atomic positions h, k, and l in a cartesian coordinate system with x, y, and z axes. The X-ray detector can technically measure the amplitude if the structure factor F_{hkl}, but not the scattering phases a_{hkl}. Measured is the modulus of F, which is the square of F. With this mathematical square building operation, the phases annihilate each other; they get lost. It is therefore *a priori* impossible to reconstruct a real space image from a scattering pattern. This is known as the phase problem [Taylor 2003], and the recovery of the phase for the reconstruction of the real space from the reciprocal space data has been of fundamental and practical interest since.

One of the pioneers in phase retrieval was Walter Hoppe [Cosslett 1982], whose illustrious career covered research on X-rays, electron microscopy, photochemistry, and also the design of microcomputers. The latter turned out to be helpful for the full scale utilization of microscopy images, diffraction pattern, and scattering curves toward phase retrieval and real space image reconstruction from otherwise hardly legible raw data. This included the design of an analog computer, which could carry out virtually in real time a Fourier transform that could be used during the experiment to control and optimize the operation parameters.

There are cases where at least the sign ($\pm$) of the scattering phase can be retrieved, such as from the zeros of the distribution of the scattering intensities. This aids in carrying out the Fourier transform toward the real space electron density distribution. Hose-

---

1 The phase in this context is the phase angle in the argument of the exponential of the wave function, the scattering phase.

2 The phase in this context is not the scattering phase, but the crystallographic phase, the modification of the material, such as cubic phase, tetragonal phase, perovskite structure, and the like. It is not a term of scattering physics, but a term of materials chemistry.

mann and Bagchi [Hosemann 1952] have shown that a unique crystallographic structure can be resolved from an X-ray diffraction experiment. This is based on the condition that (i) the fine structure of the scattered intensities in the Fourier space is known, and (ii) the crystal is finite and centrosymmetric. Under this condition, the scattered intensity is subject to (inverse) Fourier transform, which yields a continuous real space function which Hosemann and Bagchi name "Q." This function is sectioned into lattice steps, which then is mathematically "folded" with a folding polynomial into the electron density distribution. Here is also the etymologic origin of the use of the green work "ptycho," which means "to fold."

Instructive for the mathematical treatment of X-ray intensities is the Patterson function [Patterson 1935], which is related with the electron density $\rho(xyz)$ probed by an X-ray beam:

$$\rho(x,y,z) = \frac{1}{V} \sum F_{hkl} \exp(ia_{hkl}) \exp(-2\pi ihx + ky + lz) \tag{4.1}$$

Because the Patterson function is a folding product of a function and its conjugated complex function, it is suitable to treat the scattering problem along the mathematical folding theorem. The electron density $\rho(xyz)$ in a crystal measured at a point (xyz) can be written as a summation of Fourier series [Patterson 1935] and contains the structure factor $F(hkl)$:

$$\rho(xyz) = \sum_{hkl=-\infty}^{\infty} \sum \sum F(hkl) \exp 2\pi i(hx/a + ky/b + lz/c) \tag{4.2}$$

But the experiment yields only the quantity F^2.

In return, the structure factor is (compare [Bragg 1929], p. 537, and [Ewald 1921]):

$$F(hkl) = |F(hkl)| \exp i\theta(hkl)$$

$$= \frac{V}{abc} \int_{-a/2}^{+a/2} \int_{-b/2}^{+b/2} \int -c/2^{+c/2} \rho(xyz) \exp 2\pi i(hx/a + ky/b + lz/c)dxdydz \tag{4.3}$$

The convolution of the electron density with its inverse electron intensity, $\rho(\vec{r}) * \rho(-\vec{r})$, is equivalent to the Patterson function $P(\vec{u})$. The fact the Patterson function is a convolution, a "folding" ("Faltungsprodukt"), is the reason why the method of phase retrieval using this property is called ptychography. Because ptycho is Greek for folding.

4.6.2 Application

Synchrotron sources of the fourth generation contain those, where the emittance in the storage ring is smaller than the emittance after the insertion device. Those operate near the diffraction limit (Diffraction Limited Storage Ring DLSR) and allow for ptychography, provided the beam optics have sufficient high accuracy and precision. Along with a STXM-type X-ray microscopy setup, one can record diffraction patterns and reconstruct by phase recovery a real space pattern.

Nelson-Weker and Toney summarize in their paper [[Nelson Weker 2015], and references therein]: "Ptychography combines point scanning with 2D detection of the coherently scattered X-rays [Thibault 2008]. Overlapping spots on a sample are illuminated with a coherent X-ray beam, and extended structures can be imaged with contrast of elemental and chemical composition [Shapiro 2014]."

A very good explanation of the technicalities is provided by Franz Pfeiffer [Pfeiffer 2017]. There is a complementarity between the XFEL and the DLSR and it has been suggested to build large scale facilities with both, the first for the time resolution and the second for the spatial resolution. But there are additional goodies that come with DLSR; see [Eriksson 2014].

Hitchcock and Toney [Hitchcock 2014] provided a perspective on their "dream experiments" on energy related materials once the diffraction limited storage rings with their dramatically increased brightness become available. These include experiments on batteries, electrocatalysis and artificial photosynthesis, and fuel cells, for example.

Wise et al. [Wise 2015] reported 10 years ago on the development of a soft X-ray ptychography beamline at SSRL specifically for the investigation of energy storage materials. They mean batteries, of course, and suggest an *operando* study on a battery cathode assembly with NEXAFS.

Bozzini et al. [Bozzini 2011] applied ptychography on an air electrode for the Zn air battery. The air electrode is a porous manganese oxide or lanthanum manganese oxide assembly, which allows for the gas diffusion.[3] The Mn is the electrocatalyst for oxygen reduction reaction (ORR). Bozzini et al. added Ni as further metal catalyst to the gas diffusion electrode and recorded electron micrographs, STXM images, and ptychography images, in addition to Ni L-edge spectra. This provided information on the distribution of the oxidation states of the Ni in the electrode.

---

3 You can read about this type of battery in my book [Braun 2019].

5 Resonant methods and chemical contrast variation

Although the title of this book does not explicitly say so, the book does feature synchrotron radiation-based X-ray methods. The advantage of using synchrotrons is that they provide a high photon flux with high brilliance with tunable X-ray wavelengths. However, a synchrotron is not so very distinct from a conventional laboratory X-ray tube. For a review article on the history of X-ray, see Assmus [Assmus 1995] in *Beam Line—A Periodical of Particle Physics*, which is meanwhile retired.

5.1 X-ray tube (Kathodenstrahlen)

In an X-ray tube, electrons are produced by an emitter (hot or cold) cathode (Kathodenstrahlen), and the electrons are accelerated by an electric field onto a positively charged anode target, which usually is a copper metal block, or some other well-conducting metal element that does not melt so easy. The electrons hitting the target constitute an electric current that typically amounts some 10 mA. The acceleration voltage is typically some 10 kV. Figure 5.1 shows a modern X-ray tube for XPS instruments from SPECS Surface Nano Analysis GmbH, for example.

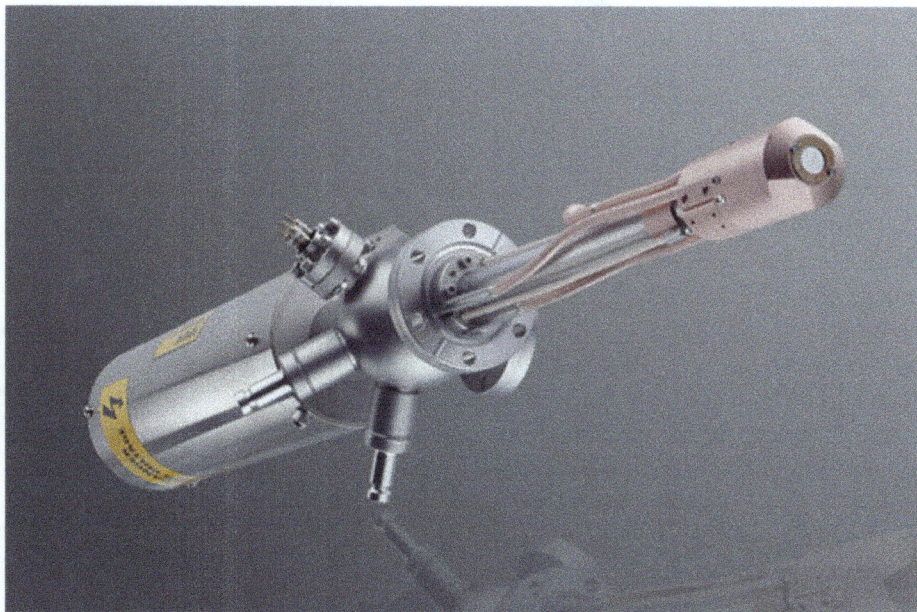

Figure 5.1: Modern X-ray source XR50 for XPS instruments from SPECS Surface Nano Analysis GmbH, Berlin, Germany. Source for characteristic X-ray emission lines Mg at 1253 eV, Al at 1487 eV, Ag at 2984 eV, Zr at 2042 eV. The anode base is made of silver to avoid any CuL_α stray radiation.

https://doi.org/10.1515/9783110794038-005

When arriving at the target, the high-speed electrons are decelerated by the impact, upon which they release their kinetic energy $E_{el} = q \cdot U \approx E_{kin} = \frac{1}{2}mv^2$ as electromagnetic radiation in the so-called "Bremsstrahlung" (German "bremsen" means "to break," to decelerate, to stop; German "Strahlung" means "radiation"). The spectrum of this Bremsstrahlung is recorded as a very broad distribution of electromagnetic waves in the X-ray range. In parallel, the target material will emit its element-specific, sharp X-ray lines with a high intensity. Moseley [Moseley 1913a, Moseley 1913b, Moseley 1913c] has investigated "the high-frequency spectra" of numerous elements over 100 years ago and found in his radiation experiments an inverse parabolic relationship between the number of electrons Z (atomic number) in the elements and the characteristic wavelengths. When the atomic number of a target material increases, then the wavelength λ of the characteristic X-ray radiation decreases, which is known as Moseley's law:

$$\frac{1}{\lambda} = c(Z - \sigma)^2$$

with some constants c and σ. Typical target materials are listed in Table 5.1 [Cockcroft 1997].

Table 5.1: Typical metals that can be used as targets for X-ray tubes. Based on information from [Cockcroft 1997].

Target metal	Ti	Cr	Fe	Co	Ni	Cu	Mo	Ag	W	Au
Atomic number Z	22	24	26	27	28	29	42	47	74	79
Kα Wavelength λ (Å)		2.29	1.94	1.79		1.54	0.71	0.56		

I refer here to some explanations on X-ray sources in the laboratory made by J. K. Cockcroft (UCL) [Cockcroft 1997]. Note the large gap in the values between Cu and Mo and the fact that practically only discreet wavelengths are available in the laboratory, unless the Bremsstrahlung with its wide wavelength range is used. The latter has a very low intensity and is therefore normally not used. Extensive lists of characteristic lines are given in the *International Tables for Crystallography* published by the International Union of Crystallography (IUCr).

As Cockcroft correctly writes, in many cases, a choice for a particular anode may practically not exist, either due to cost (every extra target costs extra money), or due to the loss of beamtime when the diffractometer needs to be shut off because of changing the X-ray target. Those who frequently look in the literature will notice that the overwhelming majority of X-ray diffractograms were generated with diffractometers equipped with a copper target. In line with Moseley's law, copper gives the shortest wavelength above 1 Å. Copper is so much common as a target that it is not in every single case disclosed in literature. It can cause confusion when this information is missing because certainly a different target than copper may have been used. You may notice this when you compare your diffractogram from your material with a diffractogram in literature from a comparable material. The Bragg reflections may be then at different 2θ positions.

The wavelengths provided, e. g., by molybdenum or silver targets are normally too short, and thus not suitable for most powder diffraction work in the laboratory. Short wavelengths scatter weakly and compress the diffraction pattern toward low Bragg angles with consequent loss of d-spacing accuracy and resolution.

For some studies, as Cockcroft continues, copper anodes are clearly not the best choice as illustrated by the two diffractograms that follow. He takes an English penny coin (modern, from 1998) and measures it with a diffractometer equipped with a copper target; see Figure 5.2. The top panel diffractogram has a very good quality with sharp peaks. Then Cockcroft removes the top surface of the penny coin (possibly by grinding with sand paper). After that treatment, the diffractogram, shown on the bottom panel of Figure 5.2, looks much different. The background intensity is high, and the signal intensity is comparably low.

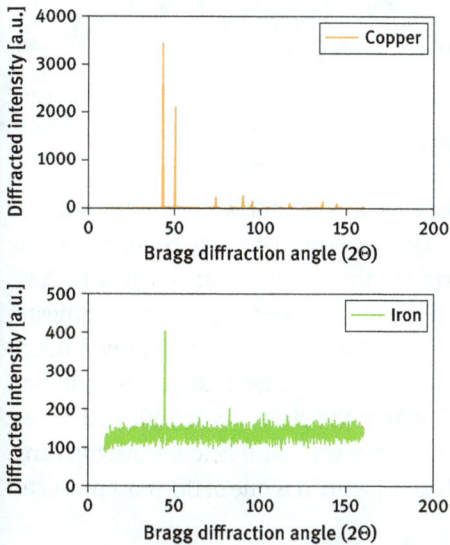

Figure 5.2: X-ray diffractogram of copper (top) and "copper penny" with copper surface removed. Photo by courtesy of Jeremy Karl Cockcroft, University College London [Cockcroft 1997].

The high background originates from the fluorescence of iron atoms in the penny, which are excited from the copper radiation of the X-ray target. Iron atoms fluoresce in the X-rays produced by a copper target. We also learn that the new pennies are actually only copper-coated iron coins. Even larger fluorescence for copper X-rays is found on cobalt samples, and to a lesser extent the manganese fluoresces. If this is a problem for studies on, e. g., iron samples, then one should switch to an iron or a cobalt anode.

Sometimes it helps to just use a graphite monochromator that is placed after the sample. An alternative "poor man's" solution is to use a post-sample graphite [Laing 1981].

5.2 Synchrotron storage rings

The X-ray tube discussed is basically a linear particle accelerator, with electrons being the particles. A large-scale linear accelerator, i. e., the Stanford Linear Accelerator Center at Menlo Park in California, accelerates electrons over an impressive distance of over 2 miles at relativistic speed (Figure 5.3).

Figure 5.3: Aerial photograph of Stanford Linear Accelerator Center in Menlo Park, California. Photo credit Peter Kaminski, USGS [Kaminski 2004].

A synchrotron storage ring is a charged particle accelerator where electrons are accelerated to velocities close to the speed of light by repeated acceleration on a ring geometry over a smaller geometric area. Therefore, the electrons carry a very high kinetic energy. It needs magnets in order to bend the electrons on their circular trajectories. This bending and this deviation from their straight trajectory constitutes a deceleration that is met by a loss of kinetic energy in the form of electromagnetic radiation [Iwanenko 1944]. At the spots on the trajectory where the electrons are allowed to "escape" tangentially, they emit electromagnetic radiation comparable to Bremsstrahlung. Figure 5.4 shows an aerial view of SPring8 in Hyogo, Japan. It is one of the most powerful synchrotrons on the globe.

This synchrotron radiation can be considered similar to the "white" X-ray light of an X-ray tube. Because it has very high intensity, it is not necessary to hit a target and create characteristic radiation. The white X-ray light can be guided on a monochromator, which selects out X-rays with a finite wavelength, which then can be used for the various X-ray analytical methods, which this book is about.

We have seen in the previous section how the fluorescence of one element can virtually contaminate an X-ray diffractogram to an extent that a measurement is not meaningful anymore. If we can select and "move away" the X-ray wavelength from a fluorescence effect, we get a better X-ray diffraction (XRD) signal. We have more freedom to do so at synchrotron beamlines where we have typically X-rays with tunable wavelength.

Figure 5.4: Aerial photograph of SPring8 (Super Photon ring-8 GeV) synchrotron center in Hyogo, Japan. With a storage ring circumference of over 1400 m and maximum electron energy of 8 GeV, Spring-8 is one of the most powerful synchrotrons on the globe. Fifty-six beamlines take advantage of photons ranging from 12 meV to 2.9 GeV energy. Photo credit: http://acc-web.spring8.or.jp/~/SP8photo.png Akihiro Yamashita, Spring8.

We can do also the opposite and move the X-ray wavelength close to an X-ray absorption threshold of a particular element in the sample. We thus obtain an X-ray optical contrast, which helps to enhance the spectroscopic response of a particular element in a complex sample. This is called resonant scattering or resonant spectroscopy, or resonant (anomalous) diffraction. We have already learned about resonant methods when we studied the X-ray wavelength dependent microscopy and tomography methods, most notably scanning transmission X-ray spectromicroscopy.

5.3 The X-ray Free Electron Laser (XFEL)

The outstanding benefit of using an X-ray Free Electron Laser XFEL instead of conventional X-ray sources, even synchrotrons, is the short X-ray pulse duration coming from XFEL. This allows, e. g., to perform X-ray diffraction of materials and molecules, and X-ray spectroscopy at a snapshot time increment. Read, e. g., [Leach 1988]. The number of XFEL installations is growing worldwide. For a recent review on the current status and future XFEL, see [Huang 2021].

There are instances in research where such snapshot experiments are necessary, because the processes that we want to observe are so short lived, such as some excited states in photochemistry. The creation and annihilation of charge carrier pairs is important in semiconductor photo-electrochemistry, where light is absorbed by photo-electrodes. The photons of energy $h\nu$ may create electron and holes pairs, which may diffuse to the electrode surface and perform there chemical work, e. g., water oxidation (photo-electrochemical water splitting for solar hydrogen production), carbon dioxide reduction or nitrogen reduction for electrochemical ammonia synthesis.

Unwelcome is a premature recombination of electrons and holes, which may happen at defect sites in the photoelectrode material. The speed at which creation and annihilation of charge couples take place is very fast and happens in the picosecond (ps) time range. You cannot detect this domain with a conventional photo-electrochemical setup and also not with a conventional optical or X-ray optical experiment. Since the process is ultrafast, the corresponding detection method is called ultrafast spectroscopy. This line of research is done with ultrashort laser light pulses. For a review on the dynamic aspects of photo-electrochemistry, see the work by Lawrence "Laurie" Peter [Peter 1990].

However, for electronic and molecular structure studies with X-rays at the ultrafast time scale, X-rays with a high time coherence are necessary. This requires an X-ray laser. Unlike conventional optical laser, which require a laser active medium such as a rubin crystal[1] or other laser media, like krypton gas, CO_2 gas, and others.

The synchrotrons provide pulsed X-ray bunches and insofar allow for experiments with a particular time resolution by using lock-in techniques. Most of the time, such experiments are not necessary for the majority of user community, and to the best of my experience, the synchrotron beamlines are then practically not tuned for this purpose. If you want to use the pulsed mode, likely the beamline scientists will give you, with proper advanced notice and necessary preparation.

I wanted this once at the Advanced Light Source, but I became overwhelmed with other projects and was running short on staff and students, and thus had to focus on other, more essential projects.

One new application of the X-ray free electron laser XFEL is the diffractive imaging of small particles. This method can be useful for the investigation of such particles while they are being coated with some overlayer. Bozzini et al. present an electrochemical sys-

---

1 The first optical laser ever was based on the stimulated photo emission by chromium doped rubin crystals, Al_2O_3, which were single crystals with long shape and finely polished front and end planes. The material has three energy levels. The optical pumping is done with a xenon lamp, which populates the energy levels 4F1 and 4F2, part of which relaxes by spontaneous fluorescence to ground state 4A2. A large portion of electron excited states relax to a metastable energy level 2E, which is the laser level. To the extent that there are more electrons in excited state 2E than in ground state 4A2, which slowly relax by spontaneous emission of photons and have a wavelength of 694.3 nm. Those photons can cause stimulated emission in the excited crystal, which constitutes coherent laser light. Later, other media were developed, which could produce laser light.

tem, which allows for the oxidation and reduction of nanoparticles and microparticles that are injected through a liquid jet into the XFEL X-ray beam [Bozzini 2011].

One recent study using an XFEL is the X-diffraction for the structure resolution of the oxygen evolving complex in photosystem II [Askerka 2017]. A strong synchrotron beam can cause radiation damage specifically in molecules, and the diffractogram or spectrum does not necessary constitute the true structure of the molecule.[2]

Another application of the XFEL involves ultrafast Kβ-spectroscopy and ultrafast optical absorption spectroscopy on iron complexes in order to learn about the metal-to-ligand charge transfer (MLCT) and spin crossover [Zhang 2017]. Such iron complexes are important in photosynthesis, e. g., as hydrogenases, and in the heme porphyrins such as hemoglobin. Prussian blue is sometimes used as model compound for related studies.

The beam from an XFEL is so ultrashort that the spectrum or diffractogram is recorded before the sample is damaged by the high intensity beam. This is one major instrumental advantage of the XFEL. It seems researchers are not always looking for very short X-ray pulses. When reading the status reports of the XFEL at the so-called Koffler accelerator at Weizmann Institute in Rehovot, Israel, they were interested in extending the duration of the pulses. Note, however, that this accelerator is not a synchrotron, but more like a Van de Graaf-type accelerator [Ben-Zvi 1988, Ben-Zvi 1990].

5.4 Anomalous X-ray diffraction

As already mentioned in this book, X-ray diffraction (XRD) is still the most widely known method in materials characterization, analysis, and diagnostics. X-ray diffractometers are available at many laboratories in companies and universities and facilities. The diffractometers require an X-ray source, which is typically an X-ray tube in which electrons are produced in an emitter filament and then accelerated with a high voltage of 10 or 20 keV, typically on a metal target. The high energy electrons are stopped by this target, and Bremsstrahlung over a wide energy range is produced. Also, the target material emits a sharp emission line, which is characteristic of the target material. Copper targets then emit Cu emission lines for the K-shell, L-shell, and M-shell. A monochromator is used for the selection of the Cu K lines. A filter is then used in order to select the highest intensity line, the Cu Kα radiation. If there is still radiation with other wavelengths in the beam, these wavelengths will also cause diffraction. However, according to Bragg's law, potential Bragg reflections will be shifted on the 2θ scale. Therefore, it will be difficult to tell apart whether a Bragg reflection originates from one particular wavelength or from another one. There exist other metal targets, such as Fe and Mo, for example. It may be useful to vary the targets if the sample contains elements identical

---

2 It may be even worse with electron microscopy, because energetic electron beams, too, involve a considerable heat dissipation.

with the target element. In this case, the fluorescence of the elements may cause a too large scattering background that makes XRD analysis more difficult.

This is the first reason why it is worthwhile to have a variable but monochromatic X-ray source available, such as from synchrotron radiation sources. Note, however, that the "white radiation" from an X-ray tube in principle can be used with any monochromator to tune to any desired X-ray wavelength. It is only the limited photon flux that poses a problem when compared to synchrotron radiation centers, where typically huge flux is available.

When "playing around" with variable X-ray wavelength, it turns out that the scattered or diffracted X-ray intensity is particularly large when the probed material contains a chemical element, which has an absorption threshold just slightly above this wavelength. This can be used for so-called X-ray contrast variation, which then allows producing X-ray diffractograms that are particularly representative to one particular element in the probed sample. This is then called anomalous X-ray diffraction because of the underlying anomalous dispersion of the scattering cross-section.

5.4.1 Distinguishing W and Cu in $CuWO_4$ photoanodes

Copper tungsten oxide is a mineral and prospective water splitting photoanode material [Baer 2010]. To the best of the author's information, it was the first time mentioned as water splitting photoanode material by Benko et al. [Benko 1982]. In a more recent study by Pandey et al., they use a spray deposition process [Pandey 2005]. The X-ray diffractogram right after deposition shows no Bragg peaks, but only a very broad hump extending from 15° to 45° on the 2θ scale, suggesting that the material is amorphous after deposition. Upon thermal treatment at 450 °C, Bragg reflections evolve and sharpen in the aforementioned 2θ range.

The rationale for moving from WO_3 (yellow color) to $CuWO_4$ (orange-red color) is its band gap energy and moving this to an energy range, which increases its efficiency as a photoelectrode for solar applications. By bringing Cu as a second cation into play, we so to speak make a doping or even substitution, given the large Cu concentration in $CuWO_4$. The crystallographic difference between WO_3 and $CuWO_4$ is easy illustrated by just looking at the diffractograms of both in Figure 5.5 of the paper by Yourey and Bartlett [Bartlett 2011, Yourey 2011].

Lately, $CuWO_4$ has found the interest by several researchers. Lalic et al. [Lalic 2011] and Rajagopala et al. [Rajagopal 2010] made *ab initio* and first principle DFT studies and determined by total energy calculations the optimum crystallographic structure of $CuWO_4$. X-ray and electron spectroscopy studies were performed by Atuchin et al. [Atuchin 2011a, Atuchin 2011b] and Khyzhun [Khyzhun 2000], for example. Ruiz-Fuertes et al. [Ruiz-Fuertes 2008] investigated the optical properties of $CuWO_4$ at high pressure.

As we know already from previous chapters in this book, the functional properties of materials depend on their structure, which in turn, depend on their synthesis and

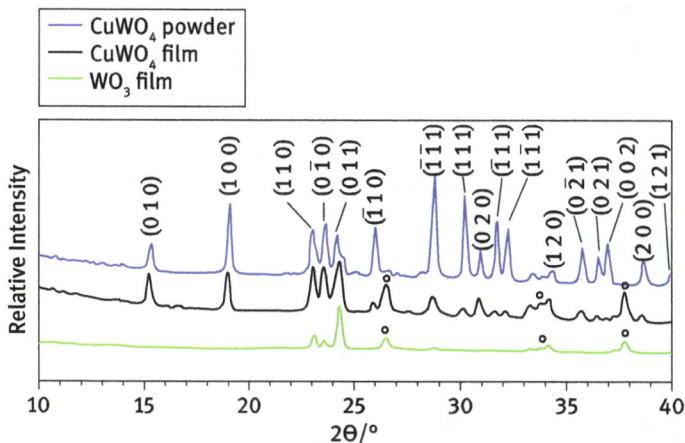

Figure 5.5: X-ray diffraction (XRD) pattern of the crystalline CuWO$_4$ electrode (black) after annealing at 500 °C and a CuWO$_4$ (blue) powder made by solid-state synthesis, with their corresponding Miller indices. The peaks with circles on top are reflections associated with the FTO substrate. Reproduced from [Yourey 2011] with permission of The Royal Society of Chemistry.

processing conditions. This holds also for copper tungstate. Below is a photograph of a collection of copper oxide, tungsten oxide, and copper tungsten oxide films deposited on indium tin oxide glasses and the subject to thermal after treatment at around 500 °C in strong reducing Argon atmosphere, and in oxidizing mixture atmospheres with 10 % and 50 % oxygen added. The "chromatic" appearance of the samples with different stoichiometry is obvious. Reduced copper oxide is yellowish, reduced tungsten oxide is bluish, and their oxidized forms become pale.

Copper tungsten oxide looks robust in terms of color, which is reddish brown for all processing gas conditions in this simple study. This is shown in Figure 5.6, which shows a collection of CuO, WO$_3$, and CuWO$_4$ samples processed under different thermal gas phase conditions. A simple visual inspection of samples can already be helpful because a color of a material may tell you already something about its chemical constitution. We have seen before that minute changes in the oxygen stoichiometry in WO$_3$ can make the sample blue or yellow. We see here in the samples of Figure 5.6 finer shades of coloring depending on the process conditions and stoichiometry.

When you carry out *in situ* studies where you expect chemical changes to happen, a corresponding color change should be noticeable, sometimes. If it is not noticeable, this may be an indication, e. g., that the targeted temperature on the sample was not reached. We experienced this with *in situ* AP XPS measurements on the samples mentioned here.

We conducted on this material photoelectrochemical studies [Chang 2011, Gaillard 2012. Gaillard 2013] and made also structural studies. We looked into films deposited on fluorine-doped tin oxide (FTO) glass as shown in Figure 5.7. The choice of reactive gas and reaction temperature can be decisive for the later function of the materials as pho-

Figure 5.6: Chromatic appearance of CuO, WO$_3$, and the mixed compound CuWO$_4$ prepared subject to different processing conditions, i. e., 500 °C under Argon, 10% O$_2$, 50% O$_2$, and air. Samples courtesy Yuanchen Chang and Nicolas Gaillard, University of Hawai'i at Manoa.

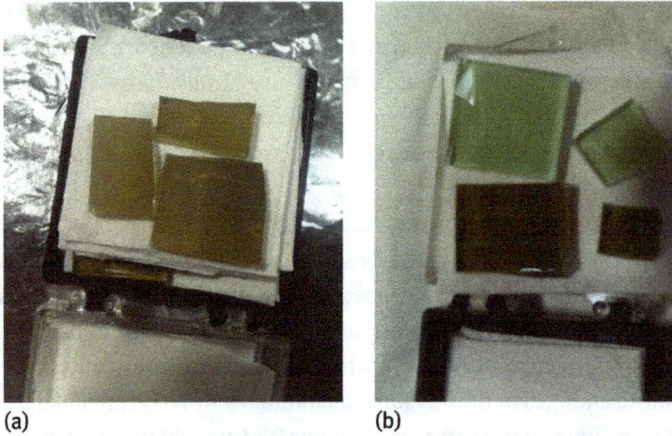

(a)　　　　　　　　　　　　　　(b)

Figure 5.7: Copper tungstate samples for anomalous XRD experiments at SSRL. Note the difference in color between (a) and (b) depending on how the samples were treated by using different gas atmospheres. Samples courtesy Yuanchen Chang and Nicolas Gaillard, University of Hawai'i at Manoa.

toelectrode. Often, like in this case here, the color of the material allows for a quick assessment of the structure of the material. We have therefore prepared samples already at our home lab in Hawai'i (University of Hawai'i at Manoa, School of Ocean and Earth Science and Technology, Hawai'i Natural Energy Institute, Thin Film Laboratory).

However, we also brought samples to Stanford Synchrotron Radiation Lightsource (SSRL) that should be processed at the beamline. The Wiggler Insertion Device Experimental Station 7-2 at SSRL has a diffractometer instrument that allows for heating of the sample under gas atmosphere while X-ray spectra and diffractograms can be recorded. It is therefore possible to make *in situ* gas phase processing experiments. To allow for such extreme environment, a half dome from polyimide is put over the sample and sealed so as to provide a controlled atmosphere and temperature around the sample during measurement. The sample holder has a heating stage and a gas inlet and outlet. The maximum temperature is limited so as to avoid that the polyimide dome becomes burned. Polyimide was chosen because it allows for a good X-ray transmission. Figure 5.8 shows the author of the book in the experimental hutch of Beamline 7-2 at SSRL at the diffractometer instrument, preparing an XRD experiment *in situ* at high temperature and under argon gas flow. His left hand with plastic glove points to the polyimide half dome.

Figure 5.8: The author at beamline 7-2 at SSRL diffractometer. Near my left hand is the polyimide half dome, which allows feeding the sample in the diffractometer X-ray optical path being subject to various gas exposures while being heated.

One question that arises when dealing with a compound like $CuWO_4$ is whether the Cu and W atoms do order, or not [Doumerc 1981].[3] We therefore conducted anomalous XRD with respect to the Cu and W X-ray K-shell absorption edges. For the preparation of the anomalous XRD, it was thus necessary to determine the necessary energies by running a XANES spectrum. Figure 5.9 shows in one scan the Cu spectrum ranging from 8600 to 9250 eV, and the W spectrum from 9400 to 13500 eV. The red rings in the spectra signify the X-ray energies, which we chose for the subsequent anomalous XRD at the same beamline for the contrast variation with respect to Cu and W.

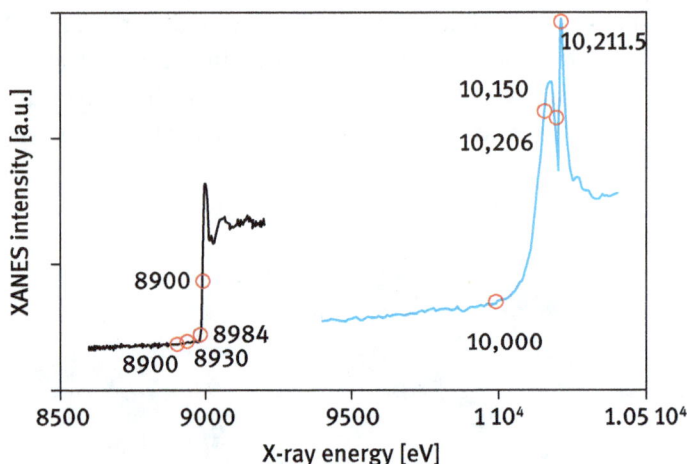

Figure 5.9: Energy values at the Cu and W absorption edge, which were selected for subsequent anomalous XRD at beamline 7-2 at SSRL.

After the XANES measurements, the sample position was readjusted in the X-ray optical path for the diffraction experiment. The selected X-ray energies were 8900, 8930, 8984, and 8990 eV for the copper. For the tungsten, we selected X-ray energies 10000, 10206, 10150, and 10211.5 eV.

Figure 5.10 below shows a set of X-ray diffractograms recorded at different temperatures and different X-ray energies. Obviously, this data set is at first glance not only complex but confusing. But I am showing you this for pedagogical reasons. We notice Bragg reflections, some of which do overlap. We notice differences in the relative intensity of the peaks, and peaks that cannot be associated with other diffractograms. The

---

3 The ordering of cations is a frequently heard issue. All additional ordering in a material should have influence on its entropy, and thus affects its total energy, and likely then the synthesis conditions, and maybe the phase diagram. The ordering of oxygen vacancies is interesting in solid electrolytes, and so is the ordering of dopants and substitution elements like Y in the BaCeY-oxide and BaZrY-oxide. Sometimes, this needs to be found out with high-resolution neutron diffraction.

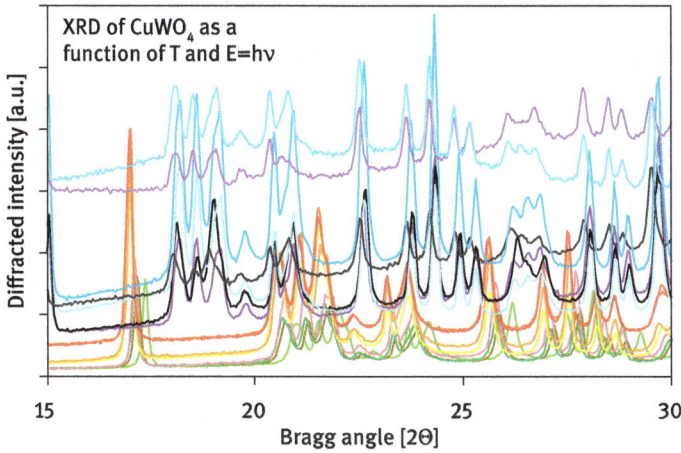

Figure 5.10: Cu resonant and W resonant X-ray diffractograms of a copper tungstate film measured at ambient temperature around 25 °C. Because the diffractogram were recorded with different X-ray wavelengths, their assessment is difficult when using the Bragg angle as the coordinate.

first problem we have here is that diffractograms recorded at different X-ray energies are plotted on the 2θ scale. With increasing X-ray energy, the diffractograms are shifted toward lower diffraction angles. Therefore, as we have explained in Chapter 3, it is necessary to find a coordinate, which accounts for this shift. The physical correct metrics for this is the momentum transfer, which is expressed by the wave vector k (or, K, q, Q), which basically reads $k = 4\pi \sin \theta / \lambda$, with X-ray wavelength λ. This coordinate transformation will immediately shift the diffractograms to the proper and aligned "angular" position. What we notice then is that the relative intensity will decrease substantially when we are beyond the X-ray absorption edge. This is an effect that we will have to accept.

Figure 5.11 shows how we have made this transformation from Bragg angle 2θ to momentum transfer k. The patterns are now obviously aligned with respect to peak positions. The red vertical lines on the bottom denote the Bragg reflections of the reference material JCPDF-21-0307.

We have kept the sample at 100 °C under argon atmosphere and recorded diffractograms over time (Figure 5.12). The diffractograms are shifted on the intensity scale for better discrimination. The position of the peak at around $k = 0.2716$ is shifting as we notice from scan 1 to scan 2. The peak at 0.295 in the final, top scan decreases significantly in intensity when compared to its neighboring peak at 0.290. I am not commenting further on how long the time was to incur such changes. Relevant is that changes happen over time, and these can be monitored with repeated scans. It would be even possible to redo the entire experiment with new samples and focus the study on those few peaks that exhibit noticeable changes. This will save beamtime and allow for a better count-

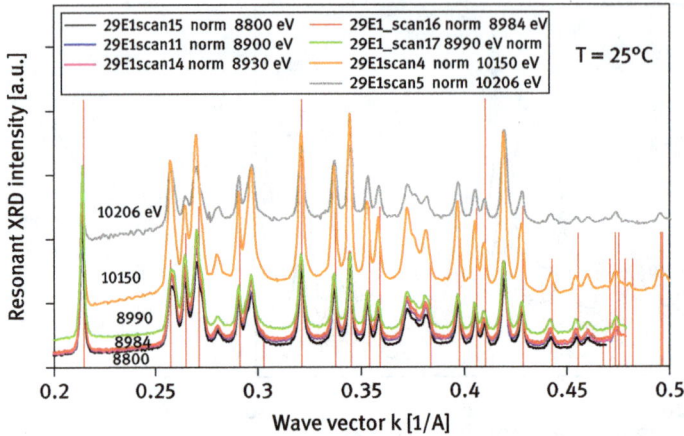

Figure 5.11: Cu- and W-resonant XRD for sample 29E1 (not further specified). The incoherent scattering background increases with increasing energy, which is a correct behavior. I would call this the anomalous effect of 0th order.

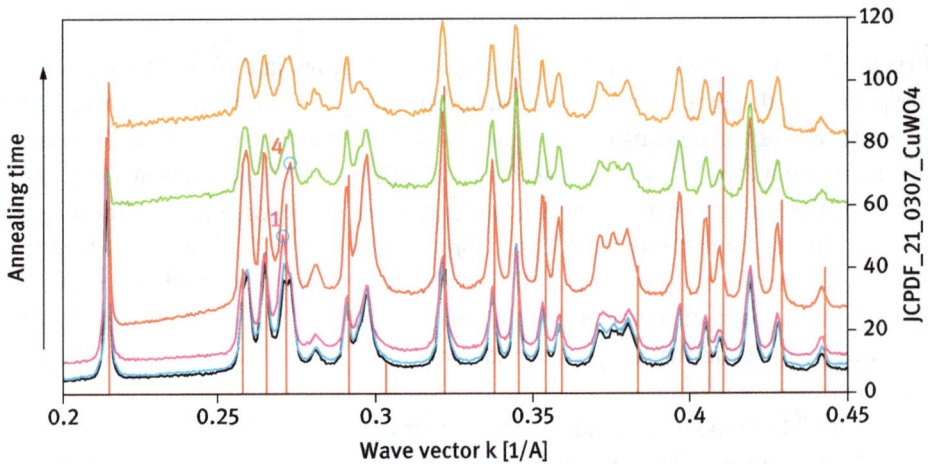

Figure 5.12: Cu- and W-resonant XRD at 100 °C. Systematic changes over this peak happen not over the resonant energies, but over the annealing time, suggesting the changes in the relative peak heights are from structural reorganization during heating.

ing statistics and subsequent structure refinement with kinetic analyses of the structural changes, which happen while the sample is annealed at 100 °C.

The next Figure 5.13 shows the diffractograms recorded at different annealing temperatures under argon atmosphere. The diffractograms are aligned on their baselines for better comparison. The practitioner of the data analysis has to find out sometimes by multiple looking at the data under various perspectives how to normalize the curves properly so that they serve their purpose of true data analysis. You notice that in this

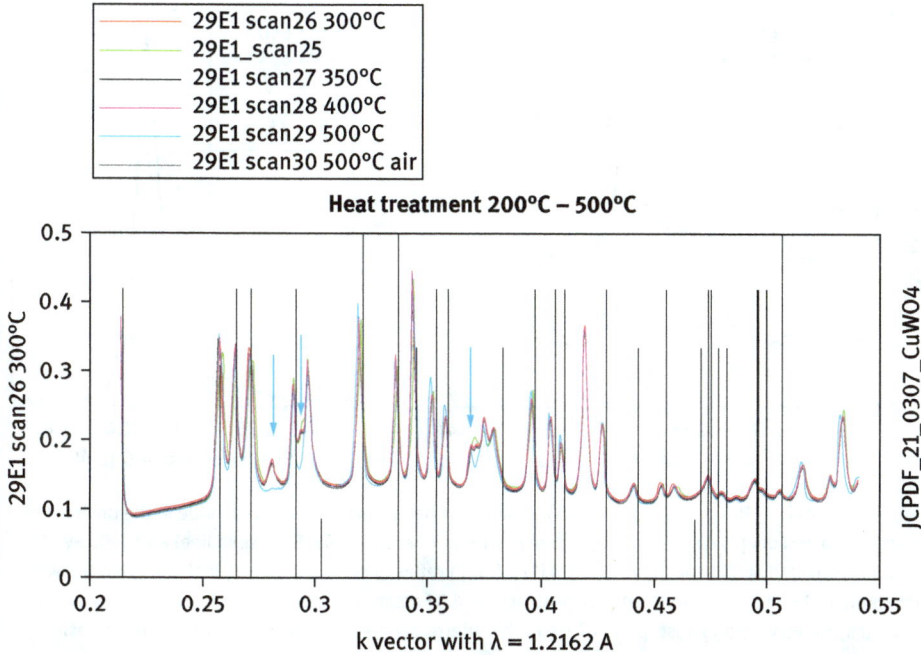

Figure 5.13: Cu- and W-resonant XRD at 100 °C for sample 29E1 (not further specified). Systematic changes over this peak happen not over the resonant energies, but over the annealing time, suggesting the changes in the relative peak heights are from structural reorganization during heating.

particular set of diffractograms, no background subtraction was made. This was not necessary but also not advised at this very early stage of analysis. After all, the background intensity reflects part of the structure of the material and can provide insight into the function—and failure—of the material [Gray 1998].

We see in Figure 5.13 that 500 °C is the temperature where structural changes in the films occur fast, as evidenced by the diminishing of the peak at $k = 0.28$, 0.29, and $0.37\,1/\text{Å}$.

5.4.2 Separation of manganite spinel diffractogram from other battery components

An early example of resonant XRD on battery cathodes is shown in Figure 5.14. Shown are the [111] Bragg reflections from a $LiMn_2O_4$ battery cathode recorded *operando* at BL 2-3 SSRL during electrochemical lithiation at 2.97 V, which was the open circuit voltage, and 1.54 polyimide a highly discharged state (Q-values of 1.392 and $1.395\,\text{Å}^{-1}$). At open potential, the [111] peak is relatively sharp, despite that it was recorded from the *in situ* spectroelectrochemical cell.

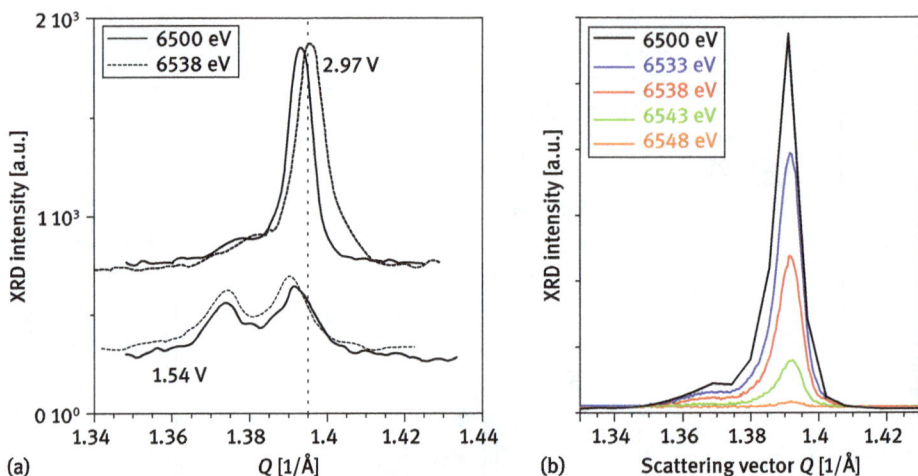

Figure 5.14: (a) [111] Bragg reflections from a lithium ion battery with $LiMn_2O_4$ cathode in the uncharged (single peak) and deeply discharged (split peak) state for energies at 6538 eV (solid line) and 6500 eV (dotted line). Recorded at SSRL beamline 2-3. (b) [111] Bragg reflex of cubic $LiMn_2O_4$ spinel phase (Fd3m) with maximum at $Q = 1.39\ 1/Å$. Small satellite peak at $Q = 1.37$ assigned to tetragonal phase of $LiMn_2O_4$ with I41/amd symmetry. Strong absorption of Mn in this energy range—peak heights shrink with increasing photon energy.

The dotted spectrum was recorded at the Mn K pre-edge of 6500 eV and the solid line spectrum of 6538 eV. Both curves are shifted by 500 a. u. upward on the intensity scale for better distinction from the two lower diffractograms. The peaks are shifted, but this shift is not a result of a resonant effect around the K-edge. Rather, the voltage (potential) of the battery has changed while we recorded the diffractograms, and the peak shifted toward smaller Q-value, suggesting a contraction of the lattice with respect to the [111] reflection. It seems, however, that the shoulder in the high Q flank of the curve measured with 6500 eV disappears either because of moving to higher X-ray energy or because of structural changes that take place in the sample. Upon extensive lithiation to 1.54 V, the low Q flank of the [111] reflection evolves into a strong peak, which has comparable height to the original [111] peak. This holds for the off-resonance and on-resonance energies of 6500 and 6538 eV.

During discharging, the [111] reflection splits in two peaks at 1.37 and 1.39 $Å^{-1}$. This cubic → tetragonal phase transition is known as the Verwey transition and supposedly caused by the Jahn–Teller effect; see Dunitz and Orgel [Dunitz 1957a, Dunitz 1957b] and Yamada et al. [Yamada 1999]. The diffractogram of the open cell shows a slight densification of diffracted intensity at $Q = 1.38\ A^{-1}$, revealing that the spinel as prepared is not entirely a pure cubic phase. The presence of two phases may be due to the fact that $LiMn_2O_4$ undergoes a phase transition from cubic to tetragonal at approximately 300 K (Verwey transition, [Garcia 2004, Shimakawa 1997]), and our experiments were conducted at room temperature. The critical balance of parameters that influences phase changes and phase purity in lithium manganese oxides was a topic that received much attention around 20 years ago ([Mishra 1999, Takano 1999], and references therein).

When we plot the two diffractograms, obtained at 6500 and 6543 eV, in the classical way versus the 2θ-angle, the latter diffractogram will be shifted toward smaller angles. We can compensate for this effect by using the scattering vector Q instead: $Q = 2 \sin \theta / \lambda$. The X-ray wavelength λ is related with the X-ray energy E via $E = hc/\lambda$, with h being Planck's constant and c being the speed of light. Yet, we see in Figure 5.14 that corresponding Bragg reflections are shifted, even if we plot them versus the scattering vector Q. Thus, the shift is a result of structural changes in the electrode material during discharge.

A systematic multianomalous XRD study was done on $LiMn_{1/3}Ni_{1/3}Co_{1/3}O_2$ at the 3d metal edges, in addition to high-resolution neutron diffraction by Whitfield et al. [Whitfield 2005a, Whitfield 2005c]. There, the diffractograms were obtained with different X-ray wavelengths and plotted versus the technical 2θ Bragg diffraction angle. When we put these diffractograms on the same 2θ scale, we would notice that they do not match. Because of the reciprocal space relation in Bragg's law,

$$n \cdot \lambda = 2 \cdot d \cdot \sin\left(\frac{2 \cdot \Theta}{2}\right)$$

a decrease in the X-ray wavelength causes a shift of the diffraction peaks toward lower 2θ angles. This is why diffractograms recorded with different wavelengths λ do not match when plotted versus the diffraction angle 2θ. A coordinate transformation $\theta \rightarrow (\sin \theta)/\lambda$ renormalizes the position of the Bragg reflections so that diffractograms become directly comparable when the intensity is plotted versus the scattering vector (momentum transfer)

$$2\Theta \rightarrow q = 4\pi \sin \frac{\Theta}{\lambda}$$

A complete and comprehensive data treatment is exercised in [Xiao 2004]. See also Nakayama et al. [Nakayama 1988].

With another multianomalous XRD study, the architecture of the oxygen evolving complex in photosystem II has been resolved. For example, diffractograms were recorded at the Mn K-edge and the Ca-edge, where Ca^{2+} has a critical "f" feature, which helped to establish X-ray contrast between both metals, and thus allowed to distinguish the positions of the Mn atoms from Ca^{2+} [Ferreira 2004].

The use of the anomalous scattering properties allows not only for the distinction of different chemical elements. When the necessary energy resolution is given, we can also distinguish different oxidation states of ions and even different orbital origin of scattering, as we will see later in resonant photoemission spectroscopy. Xiao et al. [Xiao 2004] have investigated the cation distribution Mn^{2+} and Mn^{3+} in the Mn_3O_4 spinel by anomalous XRD. A good rationale for when it is necessary to combine XRD, ND, and anomalous XRD for more precise information on where particular atoms are located in a crystal lattice is provided by Whitfield et al. [Whitfield 2005a, Whitfield 2005c]. Suzuki et al. [Suzuki 2006] have investigated with XRD and anomalous XRD and reverse Monte Carlo simulations how Cr in chromium doped β-FeOOH affects the overall crystal structure.

5.4.3 Diffraction on substituted protonated ceramic proton conductors

Transport properties are often influenced by ordering phenomena. It has been postulated for some materials that their oxygen vacancies may order at some particular temperature and then impede ionic transport. Let us discuss this for yttrium-substituted barium zirconate. The proximity of the Y^{3+} dopant and the oxygen vacancy (for an example of crystal lattice, see Figure 2.8) suggests that oxygen vacancy ordering might actually go along with Y^{3+} ordering. At least from the experimental point of view, and because of the peculiar role which Y^{3+} plays in proton conduction, it is then worthwhile to study specifically the Y^{3+} positions, which permits to establish whether Y forms a solid solution with random statistical distribution, or whether they are ordered in the lattice. We thus conducted XRD at the Y K-shell absorption edge by taking advantage of the anomalous dispersion of Y and Zr at 17 and 18 keV X-ray energy, respectively. This experiment was done at the Swiss Light Source. X-ray powder diffractograms of dry $Ba_{0.9}Y_{0.1}ZrO_3$ (BZY10) at 340 K were obtained with different photon energies (Figure 5.15, top left panel). At about $Q = 4.49\,\text{Å}^{-1}$, we notice a peak which splits at 16 keV, but not at the higher energies. At about this Q value, the (300) and (221) Bragg reflections coincide. Close inspection shows that this peak has actually two shoulders at about $Q = 4.50$ and $4.51\,\text{Å}^{-1}$, the latter of which is a separate peak at 16 keV, around 1 keV below the yttrium absorption edge (inset in Figure 5.15, top).

The intensity in this Q-range has been deconvoluted for the diffractograms of all energies (right panel, Figure 5.15) in order to illustrate the three peak positions and the change of their relative intensities. BZY10 actually represents two isostructural phases, α and β, with similar lattice constants [Azad 2008]. Two inequivalent peaks of each phase coincide. Bragg reflections of the cubic ABO_3 perovskites with one even and two odd indices have structure factors proportional to

$$F_{003} = f_B + f_O - f_A,$$

which is very small for the main phase β in the condition of normal scattering. The multiplicity of equivalent reflections is 6; thus, the intensity for the (003) reflection reads

$$I_{003} = 6 \cdot |F_{003}|^2.$$

Anomalous factors near the absorption edge may have a large effect on the intensity. I briefly exercise this (following the work from Antonio Cervellino/SLS and Vladimir

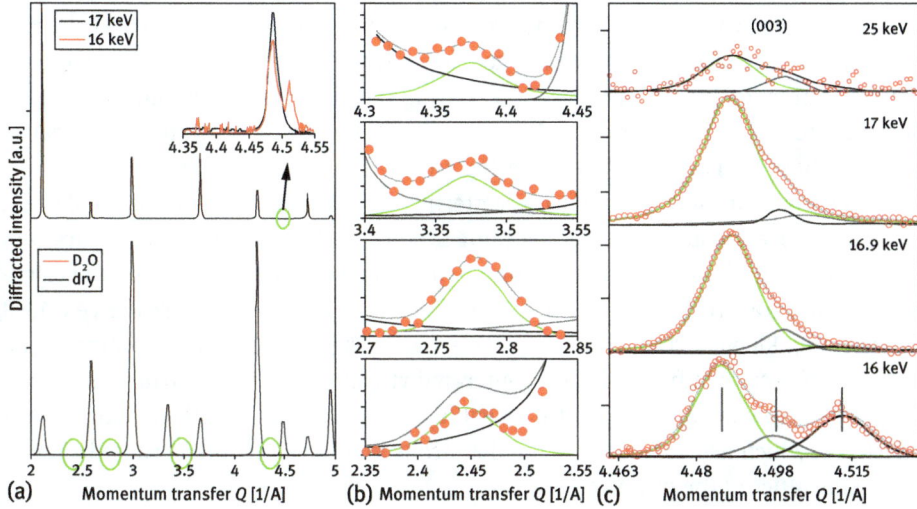

Figure 5.15: Synchrotron X-ray diffractograms recorded at 16 and 17 keV of dry BZY10 ((a) upper left panel) and neutron diffractograms (ND) of deuterated and dry BZY10 ((b) lower left). Green circles in the ND denote extra reflections from deuteration, magnified in the middle panel (b). Green circle in upper left panel denotes the (300) reflection, as magnified therein for 16 and 17 keV. Right panel (c) shows evolution of (300) intensity and shoulders as a function of photon energy [Braun 2009g, Braun 2009l].

Pomjakushin/SINQ, whose contributions I gladly want to acknowledge here) for the ABO_3-type material BZY10 with following sites and compositions A: (1/2,1/2,1/2), Ba_{1-y}, Y_y; B: (0,0,0), $Zr_{0.9+y}$, $Y_{0.1-y}$; and O: (1/2,0,0), $O_{2.95/3}$. The site form factors then read

$$f_A(Q) = (1-y) \cdot f_{Ba}(Q) + y \cdot f_Y(Q)$$
$$f_B(Q) = (0.9+y) \cdot f_{Zr}(Q) + (0.1-y) \cdot f_Y(Q)$$
$$f_O(Q) = (2.95/3) \cdot f_O(Q)$$

With the help of the tabulated Q-dependent form factors for the particular elements, we obtain for the (300) Bragg reflection relative intensities of 104 a. u. at 16 keV, 117 a. u. at 16.9 keV, 121 a. u. at 17 keV, and 59.2 a. u. at 25 keV. This variation of intensities agrees with our empiric observed (300) intensities in the right part of Figure 5.15 (green dominant Voigt distribution). Thus, Y atoms occupy particular lattice positions and probably form an ordered solid solution. The peaks persist at higher temperatures. Although resolution and statistical robustness of the diffraction data shown here are not sufficient to pose one particular unambiguous crystallographic model with unique lattice positions for Y and H/D, the observations provide evidence for ordering of these species [Braun 2009j].

A common way for contrast variation in neutron scattering and neutron spectroscopy is by substituting particular elements with isotopes that cause a larger contrast with the other elements than their original constituents. This is well known for hydrogen-containing materials, where protons can be replaced with deuterons, or water H_2O with heavy water D_2O. Depending on whether we are interested in the coherent or incoherent scattering, we may have to choose for one or for the other substituent. Neutron diffraction is a use of coherent scattering and, therefore, we replace water H_2O for the proton loading of our ceramic proton conductors by heavy water D_2O because deuterons have a larger coherent scattering cross-section than protons.

When doing quasi-elastic neutron scattering, we take advantage of the incoherent scattering process and, therefore, load out proton conductors with H_2O. Figure 5.16 shows a neutron diffractogram of a deuterated proton conductor, along with the calculated diffractogram from Rietveld refinement [Rietveld 1966, Rietveld 1967, Rietveld 1969, Rietveld 2010, Rietveld 2014]. The lower left panel of Figure 5.15 shows a section of the neutron diffractogram from dry and deuterated yttrium-substituted barium zirconate. The green circles denote small yet noticeable and statistically significant extra Bragg reflections, which are attributed to the deuterons that have occupied ordered sites in the crystal lattice of the proton conductor.

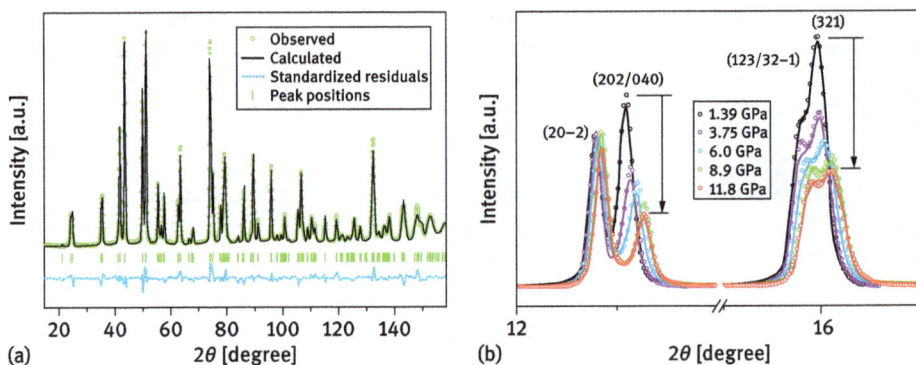

Figure 5.16: Neutron diffraction pattern for deuterated $BaCe_{0.8}Y_{0.2}O_{3-\delta}$ under ambient condition (a) and fragment of XRD patterns and refinements for hydrated $BaCe_{0.8}Y_{0.2}O_{3-\delta}$ under various pressures (b) [Chen 2011d]. Reprinted with permission from (Chen, Q., Huang, T.-W., Baldini, M., Hushur, A., Pomjakushin, V., Clark, S., Mao, W. L., Manghnani, M. H., Braun, A., Graule, T., Effect of Compressive Strain on the Raman Modes of the Dry and Hydrated $BaCe_{0.8}Y_{0.2}O_3$ Proton Conductor. Journal of Physical Chemistry C 2011, 115 (48), 24021–24027). Copyright (2011) American Chemical Society.

We have now learned how the positions and the intensities of Bragg reflection may change and vary depending on the choice of the X-ray wavelength. Now I want to demonstrate another experiment where the same changes happen but because of other processes. The right panel of Figure 5.16 shows a set of X-ray diffractograms from ceramic proton conductor $BaCe_{0.8}Y_{0.2}O_{3-\delta}$ powder, which was subject to high mechanical pressure in a diamond anvil cell. The experiment was conducted at Beamline 12.2.2 [Kunz 2005] at the advanced light source (ALS) with an X-ray wavelength of 0.496 Å [Chen 2011d]. The pressure on the anvil cells and powder was increased from 1.39 to

11.9 GPa while X-ray diffractograms were recorded. This pressure compresses the crystal lattice, which manifests in an overall shift of the Bragg reflections toward lower 2θ angles. With the Rietveld structure refinement, we can determine the variation of the unit cell volume V and the superexchange angle β versus the external pressure, as is shown in Figure 5.17.

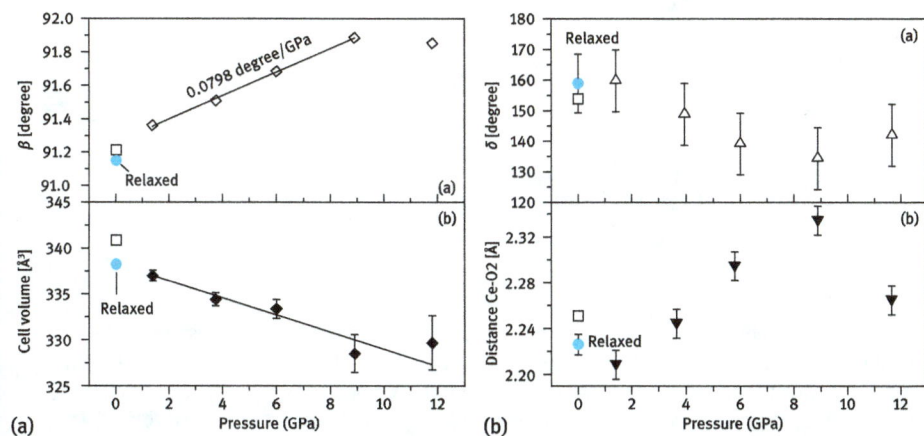

Figure 5.17: Evolution of the angle between Ce-O2-Ce (γ) for hydrated BaCe0.8Y0.2O3δ (a) and distance between Ce and O2 (b) as a function of pressure. Reprinted with permission from (Chen, Q., Huang, T.-W., Baldini, M., Hushur, A., Pomjakushin, V., Clark, S., Mao, W. L., Manghnani, M. H., Braun, A., Graule, T., Effect of Compressive Strain on the Raman Modes of the Dry and Hydrated BaCe0.8Y0.2O3 Proton Conductor. Journal of Physical Chemistry C 2011, 115 (48), 24021–24027. Copyright (2011) American Chemical Society.

In addition, we notice that, e. g., the intensity of the (202) and (040) pair is steadily decreasing upon pressurizing. The same is observed for the (321) peak. This decrease can be rationalized and simulated by local movement of oxygen ions occupying the crystallographic O2 site. Refer to Figure 5.18, where we have plotted the positions of the Ba, Ce, and O atoms for better visualization. However, the intensity of the (20-2) reflections is invariant versus pressure change. The reason for this invariance is that the O2 site with miller indices (0, b, 0) is a Bragg forbidden peak intensity contribution if O2 displaces only along the b-axis direction when the pressure is increasing.

Furthermore, in this assumption, all atoms other than O2 should keep the fractional coordinates fixed during the lattice changes as a function of pressure. Because the constant intensity of the (20-2) reflections upon increasing pressure implies that all other atoms—with the exception of O2—keep the fractional coordinates fixed, we may ignore the intensity of the O3 contribution in the Raman spectra, the O3 site of which is the other oxygen site linking two Ce centered octahedra in the b-axis.

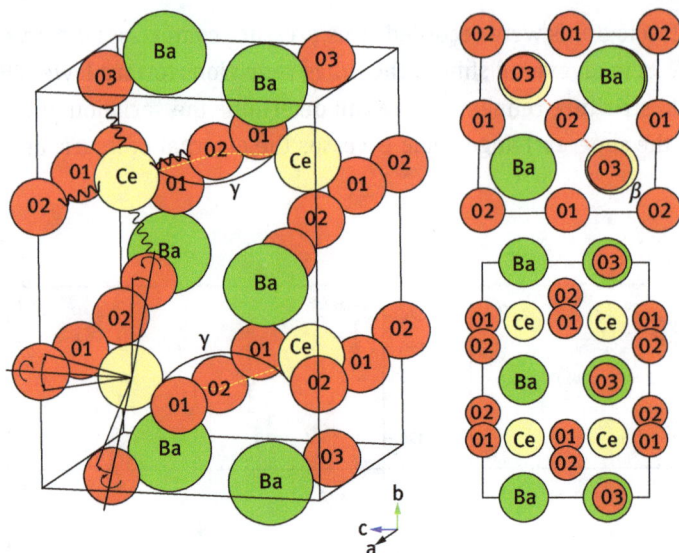

Figure 5.18: Schematic of the crystal structure for hydrated BaCe0.8Y0.2O3-d at a pressure of 1.39 GPa. Reprinted with permission from (Chen, Q., Huang, T.-W., Baldini, M., Hushur, A., Pomjakushin, V., Clark, S., Mao, W. L., Manghnani, M. H., Braun, A., Graule, T., Effect of Compressive Strain on the Raman Modes of the Dry and Hydrated $BaCe_{0.8}Y_{0.2}O_3$ Proton Conductor. Journal of Physical Chemistry C 2011, 115 (48), 24021–24027). Copyright (2011) American Chemical Society.

5.5 Anomalous small-angle X-ray scattering

5.5.1 *In situ* ASAXS on lithium batteries

We saw in Chapter 3 how the microstructure of a chemically homogeneous material like glassy carbon can be well studied with small-angle X-ray scattering (SAXS). When the sample is chemically heterogeneous, when it is a multiphase material from phase A and phase B, it can be difficult to tell apart whether a particular microstructure feature originates from phase A or from phase B. A lithium battery is a very good example for such case. The battery cathode is typically a blend of lithium 3d metal oxide ($LiMn_2O_4$, $LiCo_2O_4$, $LiFePO_4$, ...), carbon black, graphite, and polymer binder cast on an aluminum foil as current collector. The separator can be made from glass fiber, polymer, or some other electronically insulating porous foil. The anode can be lithium metal, lithium carbon, or some different lithium source. The electrolyte can be $LiPF_6$ salt, dissolved in some organic liquid electrolyte. These components are then assembled in a cell, wired to a potentiostat or battery cycler and can be charged and discharged. When this cell is subject to SAXS, a potential Guinier shoulder in the scattering curve may be attributed to the carbon black or the metal oxide, for example. We can avoid this problem of identification when we select the X-ray energy for the SAXS experiment such that it is near an absorption threshold of the 3d metal in the battery cathode.

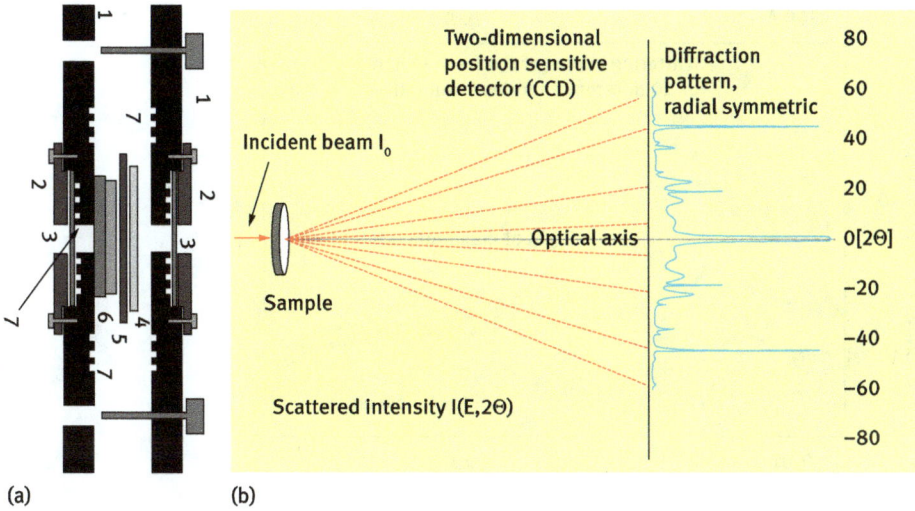

(a)   (b)

Figure 5.19: Schematic showing the geometrical relation between cell, optical path, scattering direction, and detector. (a) Electrochemistry X-ray *in situ* cell based on (1) stainless steel plate, (2) *aluminum* metal washer plate to fix Be X-ray windows, (3) beryllium windows with 380-micrometer thickness, (4) lithium foil with 125-micrometer thickness, (5) Celgard® foil polymer separator with 25-micrometer thickness, (6) spinel electrode assembly of 40-micrometer thickness casted on 25-micrometer aluminum foil, (7) BUNA® O-rings with groves. Cell manufactured by Eric M. Granlund, Machine Shop College of Chemistry, UC Berkeley. (b) Schematic of the spectro-electrochemical cell diffraction process for *operando* battery studies. The X-rays penetrate cell and sample coming from the left through the first beryllium X-ray window at position 3, passing through the aluminum metal current collector with lithium battery cathode on it (6), then through the electrolyte filled polymer separator (5), then through the lithium metal counterelectrode, (4) and finally through the second beryllium X-ray window on the right (3). The inner sides of the Be windows were coated with polyimide KaptonR foil to protect them from corrosion by $LiPF_6$ electrolyte dissolved in dimethyl carbonate. All these materials, including the air, are scattering centers for the X-ray beam. The scattered X-rays will be projected on the two-dimensional position sensitive detector where they mark the scattering pattern, which typically is radial symmetric, Figure 3.15 in Chapter 3.3, unless the sample has anisotropy. The radial averaged pattern is an X-ray diffractogram or SAXS curve, depending on the distance of the detector from the sample. The right side of the yellow panel shows the azimuthal averaged XRD pattern.

Figure 5.19 illustrates how an *in situ* battery cell with numerous components is exposed to an X-ray beam and then scatters information, e. g., on a CCD detector as a diffraction pattern or scattering pattern. We have recorded scattering curves from this cell at 20 different X-ray energies covering the Mn K-edge from 6350 to 6550 eV, as shown in Figure 5.20. These contain the structure information from all matter probed by the X-ray beam. We note strong intensity differences in the Q-range $> 0.06\,Å^{-1}$, whereas the scattering curves for very small Q virtually overlap. The range where the scattering curves do not overlap is where we can hope to obtain some structure information. The differences in intensity originate from the differences in the scattering factors f', and thus

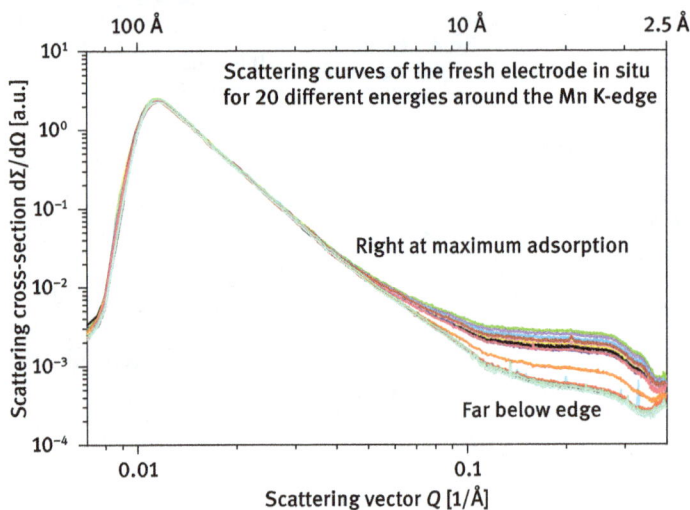

Figure 5.20: Small-angle scattering curves recorded from a lithium ion battery *in situ* cell for 20 X-ray energies around the Mn K-shell absorption edge. The different scattering intensity for high Q scattering vectors is a manifestation of the anomalous small-angle scattering (ASAXS).

yields the anomalous, element-specific scattering, anomalous small-angle X-ray scattering (ASAXS) [Haubold 1994].

In Figure 5.21, we plot the ASAXS curves obtained at 6400 and 6545 eV. These energies are far away from the Mn K-edge and just about before the Mn K-edge. The X-ray scattering contrast originates from the variation of the atomic form factor f' of manganese. The atomic form factors are a function of X-ray energy. Therefore, Goerigk et al. [Goerigk 2003] describe anomalous scattering as a "labeling technique." Because of the resonance between X-ray absorption and elastic scattering near the X-ray absorption edge of a particular chemical element, strong variation in its atomic scattering factor $f(E)$ is observed:

$$f(E) = f_0 + f'(E) + if''(E)$$

f_0 is the atomic number Z of the particular element. The values for the atomic form factors f' and f'' are the ones calculated by Cromer and Liberman [Cromer 1970]. If the calculated, tabulated numbers are not accurate enough, the form factor $f''(E)$ can be determined by an experimental XANES spectrum [Braun 2001a, Braun 2001f]. Using the Kramers Kronig relation, f' can be determined from f''. This can be done with the DIFFKK computer code published by Cross et al. [Cross 1998] and which available at the CARS9 Beamline website [Newville 2023].

We have subtracted the ASAXS curves by using a weighting factor of 1.356, which results in the light green scattering curve in Figure 5.21 bottom. This ASAXS curve contains now specifically the scattering from the Mn in the beam, and thus the corresponding mi-

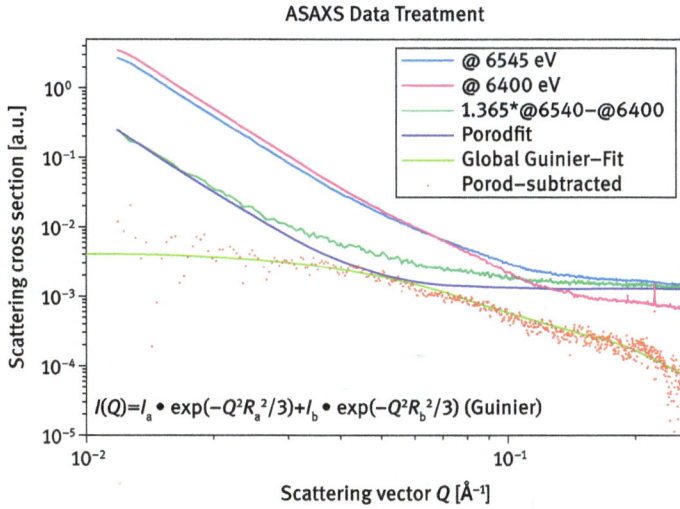

Figure 5.21: SAXS curves recorded at 20 different X-ray energies from 6350 to 6550 eV.

crostructure information. With this curve, we can now perform further processes, such as the Porod subtraction. This is done by fitting a Q^{-4} profile (blue) to the green curve. Subtraction of this Porod scattering background yields the red data points, which were then fitted with two exponentials, i. e., two Guinier ranges. This is a simple ASAXS data analysis. This experiment was done in the year 2000 at the APS BESSRC-CAT Beamline. The time for acquisition of one 2D ASAXS curve with significant statistical quality took 0.1 seconds. The photon flux of the APS was high enough to warrant a statistically robust SAXS pattern in a time much shorter than 0.1 seconds. This was the minimum time for one scattering image because of the slow, i. e., long beam shutter response time. An impressive number of patterns were recorded during charging and discharging of the *in situ* cell and azimuthal averaged and further analyzed as shown here.

Figure 5.22 shows three ASAXS curves recorded from the *in situ* battery cell when it was uncharged, charged, and then again discharged. Thus, out of several thousands of ASAXS pattern have analyzed only a few. For a complete analysis of all scattering data, it would be necessary to develop computer macros that can handle this huge amount of data based on comparable references. Quick visual inspection of the three ASAXS curves in Figure 5.22 shows that two size ranges are subject to changes, depending on the state of charge. Note that these three ASAXS curves show the chemical contrast with respect to manganese. Therefore, changes, e. g., only in the carbon matrix, or polymer separator, are by intention not visible and can be ignored. The feature size that we identify has a range of around 10 nanometers and around 20–30 nanometers. We have inspected several dozens of ASAXS curves recorded during the entire charging and discharging process with respect to these two features, plotted in the right panel of Figure 5.22. The 10-nm feature is slightly increasing from 8 to 10 nm when the lithium is extracted during the

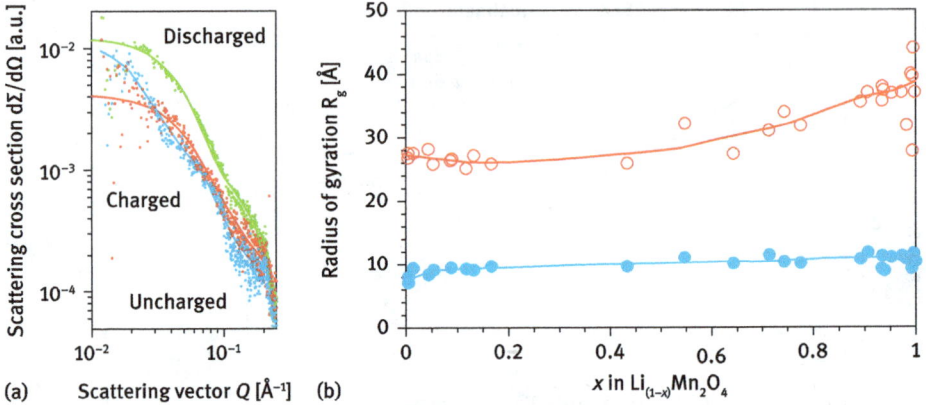

Figure 5.22: (a) ASAXS curves recorded from the *in situ* battery cell for the uncharged, charged, and discharged state, after contrast variation with respect to 65400 and 6545 eV, and Porod subtraction. The solid lines through the data points are two Guinier fits. (b) Variation of the radius of gyration determined from the Guinier fits in the left panel, as a function of the lithium content $(1-x)$ during charging and discharging.

battery charging. Note that x in this graph refers not to the lithium concentration, but to x in $Li_{(1-x)}Mn_2O_4$. The second feature that ranges around 30 nm starts out at around 26 nm, then slightly decrease to 24 nm during lithium extraction, and then increases to around 40 nm while further lithium is extracted.

We have thus obviously changes in the microstructure of the spinel particles going on during delithiation and relithiation of the battery, as evidenced with ASAXS.

When we look at the wider scattering angles, we notice a change at around $Q = 0.28 \, Å^{-1}$. This observation was not done during the ASAXS measurement at APS, but during an *in situ* XRD campaign at SSRL. Figure 5.23 below shows the diffracted intensity obtained in θ-2θ mode plotted as $I \cdot Q^2$ versus Q. In the uncharged stage "A," we see a peak intensity at around $Q = 0.28 \, Å^{-1}$, which disappears upon delithiation. At the fully delithiated stage "B," we can interpret the diffractogram as undergoing a split from one peak at $Q = 0.28 \, 1/Å$ to a two-peak structure at $Q = 0.28 \, Å^{-1}$ and $Q = 0.337 \, Å^{-1}$. These Q-values correspond to real-space distances of 18.6 and 22.4 Å, which ranges in between the two aforementioned microstructure change ranges.

One practically important consequence of the ASAXS principle is that any scattering signal, which is independent from the ASAXS signal can be treated as a simple scattering background. Goerigk et al. [Goerigk 1997] have demonstrated this for Co particles in a Cu matrix with the scattering contrast $(f_{Cu} - f_{Co})^2$:

$$f(E) = f_o + f'(E) + if''(E)$$

$$q = \frac{4\pi}{\lambda} \sin(\theta)$$

$$\frac{d\sigma}{d\Omega}(q) = (\varrho_{Cu}(E) - \varrho_{Co}(E))^2 S_\phi(q)$$

Reversible "Flip" of Bragg-Reflex

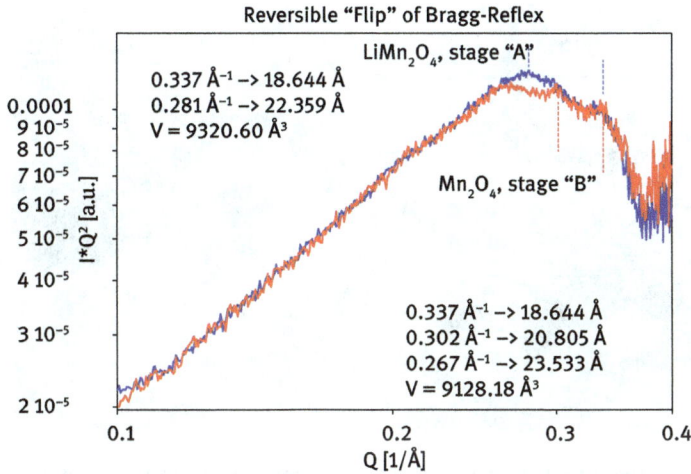

$LiMn_2O_4$, stage "A"

0.337 Å^{-1} –> 18.644 Å
0.281 Å^{-1} –> 22.359 Å
V = 9320.60 Å^3

Mn_2O_4, stage "B"

0.337 Å^{-1} –> 18.644 Å
0.302 Å^{-1} –> 20.805 Å
0.267 Å^{-1} –> 23.533 Å
V = 9128.18 Å^3

Figure 5.23: XRD intensity from the uncharged (blue curve) and completely charged (red curve) $LiMn_2O_4$-based battery. The data were recorded *in situ* at SSRL.

5.5.2 *In situ* ASAXS on Pt-based PEM-FC assemblies

Early pioneering ASAXS work can be found by Haubold and Wang [Haubold 1995], who investigated platinum electrocatalysts for polymer electrolyte membrane fuel cell assemblies. This research was done at the JUSIFA beamline at HASYLAB in Hamburg, Germany [Haubold 1989]. JUSIFA ASAXS beamline, Jülich User dedicated Small-Angle Scattering Facility at Hasylab (DORIS), E = 4.5–35 keV, $\delta E/E$ = 1×10^{-4}, Q = 0.006 … 1Å^{-1}. This beamline was funded upon the proposal by the Jülich IFF by the German Department for Education and Research in 1985 with an amount of comparable to 750,000 Euro in 2009. Figure 5.24 shows the JUSIFA beamline in Hamburg, with Dr. Haubold inspecting the sample position. We realize that a SAXS beamline is "long." The left side shows a prismatic box, which is the wire proportional counter detector. Left from the left hand of Haubold is a large beam shutter. The metal tube beamline can be operated in a long and in a short version so as to cover two different Q-ranges. Sometimes, it is necessary to measure over both ranges, and thus run in the long and in the short measurement geometry. Practically, you measure all your samples first in one geometry then change the geometry (this can take a while) and measure all samples again in the other geometry. In the end, you have to merge for every sample the two SAXS patterns from the different Q-range. This was the technical status until late 1990s. Meanwhile, doing SAXS has become more comfortable because of technical progress in instrumentation and computer software.

Haubold et al. used a spectro-electrochemical cell, which was made from polycarbonate Plexiglas®. They investigated carbon supported Pt electrocatalysts [Goerigk 2003, Haubold 1995, Haubold 1996, Haubold 1999, Vad 2002]. They used different Pt loadings

Figure 5.24: H.-G. Haubold inspecting sample position at JUSIFA ASAXS beamline, HASYLAB. Haubold et al. Rev. Sci. Instrum. 60(7), July 1989 [Haubold 1989].

ranging from 5 % to 80 %. Figure 5.25 shows a brief summary of their experiment, which was published in [Haubold 1997]. The cyclic voltammogram shows two potentials (0.25 and 1.10 V vs. $Ag^+/AgCl$ reference) where oxidation state was kept constant and then ASAXS curves were recorded at 10350 and 11546 eV for Pt L-edge contrast variation. Note that these two scattering curves are marked by dots and by circles that virtually overlap. Only very close inspection shows that there is an anomalous effect that can be utilized by the weighted subtraction of either curve, which yields the ASAXS difference curve shown at the bottom of the top right panel of Figure 5.25, shown with open squares. This can be further modeled with a Guinier fit of with a sphere geometry particle fit. It makes a difference whether these ASAXS curves are obtained from the reduced Pt particles at 0.25 V, or in the oxidized state at 1.1 V. This difference becomes clear by looking at the lower left panel graph of Figure 5.25. This curve is further modeled with a Rayleigh sphere geometry shape factor function, multiplied with a log-normal distribution and then integrated from $R = 0$ to R-> infinity. The lower right panel of Figure 5.25 shows the thus obtained particle size distribution for the reduced and oxidized Pt electrocatalysts for the electrode with 10 % loading and 5 % loading [Haubold 1997].

Bota et al. [Bota 2002] have used ASAXS at JUSIFA for the characterization of the nickel in a Raney-type catalyst was characterized by using anomalous SAXS. The particle size distribution was in agreement with the crystallite size obtained from Scherrer analysis of the Bragg peak width from XRD and with the size determined by TEM. They found that the Ni particles had cylindrical shape. ASAXS was also used to characterize nanostructured metal/organic networks, which were synthesized by cross-linking aluminum-organic-stabilized platinum nanoparticles in a solution with organic spacer molecules [Bonnemann 2002].

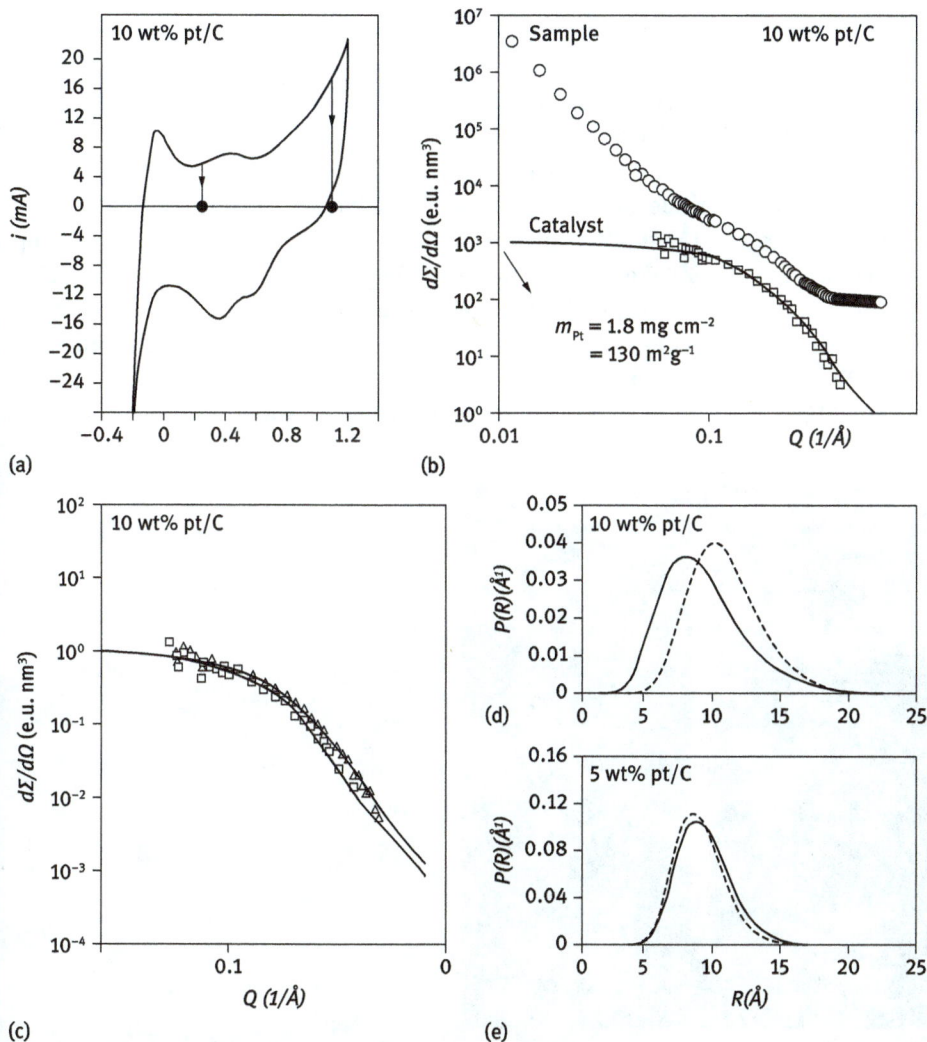

Figure 5.25: (a) Cyclic voltammogram of dispersed Pt/C electrodes as prepared from 10 wt% Pt/C E-TEK catalyst powder, sweep rate 10 mV/s. 0.25 and 1.1 V versus Ag/AgCl are the potentials for potentiostatic *in situ* ASAXS measurements. (b) Small-angle scattering cross-sections at two X-ray energies E_1 (dots) (10350 eV) and E_2 (circles) (11546 eV) and difference (squares) for 10 wt% Pt/C electrode in the reduced state at 0.25 V versus Ag/AgCl. The differences arise from the Pt particles and are fitted by log-normal particle size distributions (solid line). (c) Separated particle scattering and fits (solid lines) with log-normal size distributions for the reduced (triangle) and oxidized (square) particles at 0.25 and 1.1 V versus Ag/AgCl, respectively. (Bottom right) Size distributions of the Pt catalyst particles in the reduced (solid line, (d)) and oxidized (dashed line, (e)) state at 0.25 and 1.1 V versus Ag/AgCl, respectively. Reproduced from [Haubold 1997] with permission of the International Union of Crystallography http://journals.iucr.org.

5.5.3 ASAXS on SOFC assemblies

We continue this ASAXS chapter with a study we did on solid oxide fuel cell (SOFC) electrode assemblies. As we know, SOFC electrode assemblies are composed of a ceramic cathode, ceramic electrolyte, and an anode "cermet," which is a porous ceramic matrix with metal catalysts inside. SAXS is a well-suited method for the microstructure determination of ceramics [Allen 2004, Allen 2005, Wong-Ng 2005]. Figure 5.26 shows an optical micrograph of a SOFC electrode from a HEXIS fuel cell stack [Mai 2013, Mai 2009, Mai 2011]. We were interested in the microstructure and porosity of the electrodes and electrolytes with respect to operation under sulfur-containing fuel. It is difficult to tell apart either SOFC components from a conventional SAXS study. We did such SAXS study recently [Allen 2009]. In an extension of that conventional SAXS study, we conducted

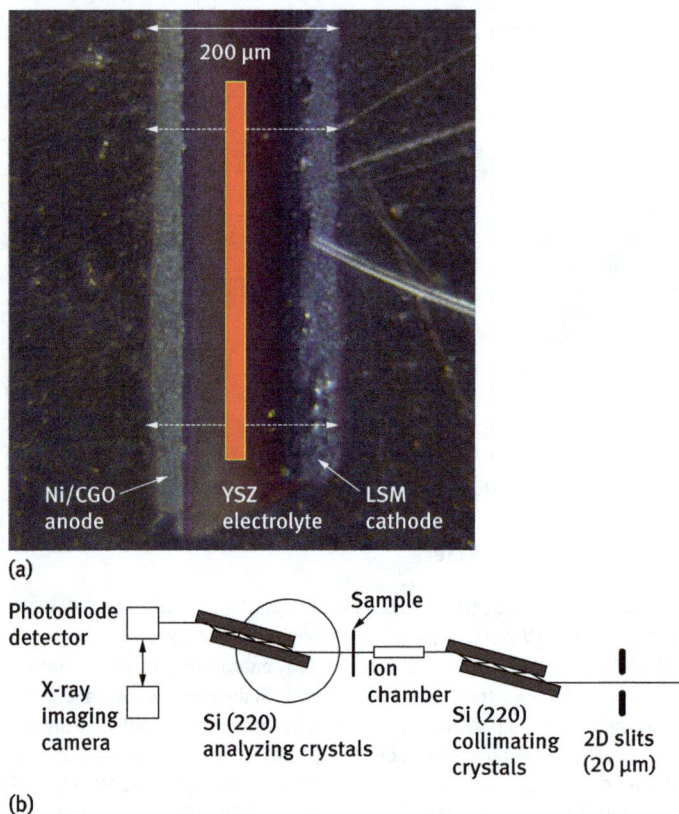

(a)

(b)

Figure 5.26: (a) Optical micrograph of a 200-micrometer thick SOFC assembly with a Ni/CGO anode, YSZ electrolyte, and LaSrMn-oxide cathode. The red bar indicates the rectangular shape and size of the X-ray beam, which scans the sample region step-by-step. (b) Schematic of USAXS instrument at APS sector 15-ID ChemMatCARS. A. J. Allen, J. Ilavsky, P. R. Jemian, and A. Braun, *RSC Adv.*, 2014, 4, 4676. Reproduced by permission of The Royal Society of Chemistry.

ASAXS at the Ni K-edge and Zr K-edge energies [Allen 2014]. Because the Ni K-edge is at the relatively low X-ray energy of 8333 eV, it was necessary to thin down the specimens so as to warrant sufficient X-ray transmission through the entire SOFC assembly. This is quite difficult because of the brittleness of the ceramic samples. It was therefore necessary to fix the ceramic assemblies in a solid resin to make sure that breaking of the sample did not occur while thinning it down.

The technical specifications of the SOFC electrode assembly are summarized in Table 5.2. It turned out during our first SAXS study on these materials that the anode and the cathode were made of two layers each [Allen 2009]. We investigated a pristine sample, a sample operated from an SOFC stack that was propelled with sulfur containing natural gas (sample B), and a stack which was operated with the same fuel, but with a sulfur filter attached (sample A).

Table 5.2: Summary of SOFC samples studied with SAXS at APS UNICAT. A. J. Allen, J. Ilavsky, P. R. Jemian, and A. Braun, *RSC Adv.*, 2014, 4, 4676. Reproduced by permission of The Royal Society of Chemistry.

Sample	A filter	B sulfur	C pristine
Outer anode	≈ 25 μm thick; 60 % mass Ni, 40 % mass CGO; ≈ 20 % porosity; [CGO = $Ce_{0.9}Gd_{0.1}O_{1.95}$]		
Inner anode	≈ 7 μm thick; 60 % mass Ni, 40 % mass CGO; ≈ 10 % porosity; [solid = 55 % volume Ni, 45 % volume CGO]		
Electrolyte	≈ 140 μm thick; 3 % mass Y_2O_3, 97 % mass ZrO_2 (YSZ); dense		
Inner cathode	≈ 10 μm thick; 50 % mass LSM, 50 % mass YSZ; ≈ 15 % porosity [solid = 49 % volume LSM, 51 % volume YSZ]		
Outer cathode	≈ 20 μm thick; 100 % LSM; ≈ 30 % porosity [LSM = $La_{0.8}Sr_{0.2}MnO_3$]		
Service time	1000 h	1000 h	None
Service temperature	950 °C	950 °C	N/A
Fuel	Desulfurized (no sulfur in fuel)	Natural Gas (sulfur in fuel)	N/A

Let me now explain the data structure we want to obtain. The entire analysis strategy was made by Andrew Allen from NIST. We have a nominally three-layer system with a supposed porosity gradient. The electrolyte should be by default without any porosity. Anode and cathode should be porous with high internal surface area per volume so as to warrant a large electrochemical interface between electrode and gas for electrochemical conversion at the triple-phase boundary. Possibly there is a gradient in porosity when we scan the assembly from the surface to the anode to the interface of anode and electrolyte. The same should hold for the cathode. We therefore would like to scan their sample along its "z-direction." Thus, we rotate the cell assembly in Figure 5.26 by 90° counterclockwise and arrive at the situation in Figure 5.27. We maintain a sample height of 200 micrometers. This is the height of the repeating unit in the SOFC stack. We reduced the thickness of the assembly down to 40 to 80 micrometers so that we have

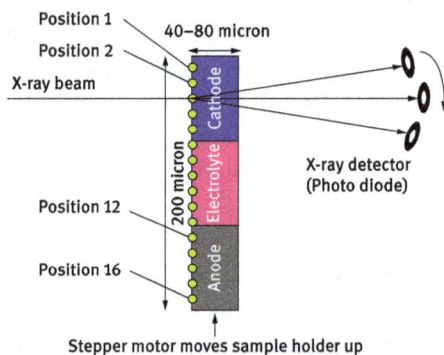

Figure 5.27: Schematic for USAXS scattering curve recording with high spatial resolution for porosity gradient profiling on a sample. The stepper motor moves the sample in increments of 10 to 20 micrometers through the X-ray beam. At every position (here 1–16), a USAXS curve can be recorded with selected X-ray energies.

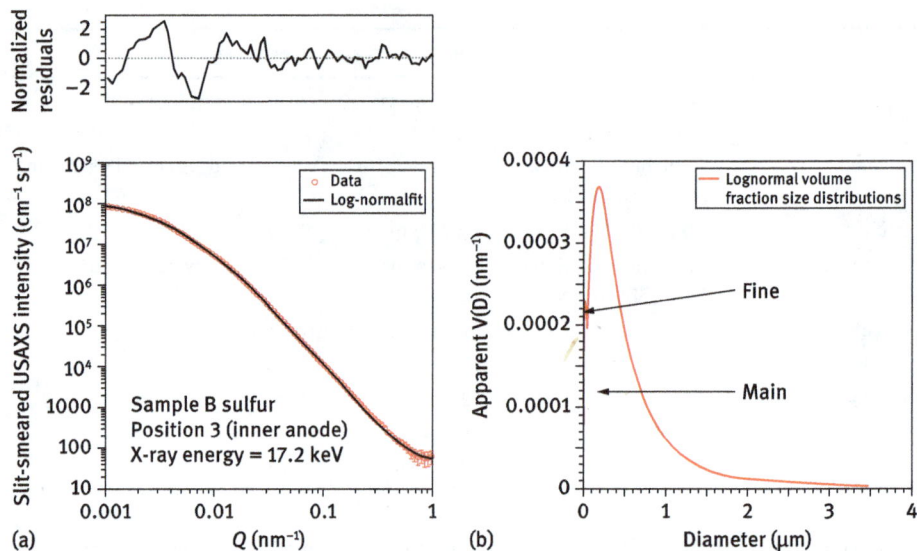

Figure 5.28: Typical reduced and calibrated USAXS curve (a) with two-component log-normal size distribution fit and (b) apparent volume fraction size distributions obtained. [A. J. Allen, J. Ilavsky, P. R. Jemian, and A. Braun, *RSC Adv.*, 2014, 4, 4676]. Reproduced by permission of The Royal Society of Chemistry.

the cathode, electrolyte, and anode layers thin enough so that X-rays with energies corresponding to the Ni K-edge and Zr K-edge can pass and make a scattering pattern on the detector. Figure 5.28 shows one such typical USAXS curve with a two-component size distribution. The beamline at UNICAT at the APS in Argonne National Laboratory is an ultrasmall-angle X-ray scattering (USAXS) beamline, which can resolve structures from 1 nm to several micrometer size in real space. The X-ray beam is so thin that stepper motors can yield a spatial resolution across the sample of around 10–20 micrometers. We thus can record a USAXS profile with a resolution of 10- to 20-micrometer steps, as is indicated in the schematic in Figure 5.27. There we moved the sample from position 1 to position 16, and thus took 16 USAXS curves along the anode, electrolyte, and cathode layer.

We must select the X-ray energies for every layer properly so that we can carry out ASAXS at the anode with respect the Ni metal catalyst, and at the electrolyte with respect to the Zr, for example. We did not make ASAXS for the Mn K-edge because the X-ray energy of 6500 eV would have required even more critical thinning of the brittle sample. Note that we must do these experiments for three different samples, i. e., the pristine samples and samples A and B.

Figure 5.29 shows the anomalous scattering contrast factors for the Ni/void and Zr/void interfaces close to the respective absorption edge energies, along with linear fits of $|\delta\rho(E)|^2$ versus $\ln((E_0 - E)/E_0)$ for $E < E_0$. For $|\delta\rho|_E^2 > 0$, $|\delta\rho(E)|^2$ decreases when E increases toward E_0. Shown are also the equivalent plots for the YSZ/LSM interface at the Zr absorption edge. For this interface on the cathode side of each SOFC, $|\delta\rho(E)|^2$ increases as E increases toward E_0 from below with both $|\delta\rho|_0^2$ and $|\delta\rho|_E^2$ negative (see Table 5.3). Allen et al. call this a negative anomalous effect [Allen 2014]. Considerable loss of linearity is obvious in Figure 5.29 next to the absorption edge. This is because of $|\delta\rho|$ being small so that the requirements for linearity are violated. A corrected fit is shown, and corrected fit parameters are included in Table 5.3. Tables 5.3 and 5.4 show the scattering contrast factors for interfaces in the SOFC assembly with respect to the Ni (Table 5.3) and Zr (Table 5.4) K-shell absorption edge.

Interfaces labeled a, b, c, etc., are discussed later in the text.

Interfaces labeled a, b, c, etc. are discussed later in the text [A. J. Allen, J. Ilavsky, P. R. Jemian, and A. Braun, RSC Adv., 2014, 4, 4676]. Reproduced by permission of The Royal Society of Chemistry.

In order to increase the X-ray transmission, it was necessary to thin down the samples to around 30 micrometers. Yet, substantial multiple scattering corrections were

Table 5.3: Scattering contrast factors for SOFC interfaces near Ni X-ray K-absorption edge.

Interface	Fixed contrast factor, $|\delta\rho|_0^2$ (10^{28} m^{-4})	Anomalous contrast factor, $|\delta\rho|_E^2$ in $|\delta\rho|_E^2 \ln((E_0 - E)/E_0)$ (10^{28} m^{-4}) ($E_0 = 8.333$ keV)
YSZ/void (f)	2096.3	0
YSZ/CGO	29.240	0
CGO/void (c)	2614.3	0
LSM/void (g)	2043.7	0
YSZ/LSM (h)	9.3060	0
Ni/void (a)	5062.4	302.94
Ni/YSZ	601.74	82.309
Ni/CGO (b)	373.31	61.161
Ni$_3$S$_2$/void (x)	2271.8	105.14
YSZ/void (f)	2096.3	0

Table 5.4: Scattering contrast factors for SOFC interfaces near Zr X-ray absorption edge. A. J. Allen, J. Ilavsky, P. R. Jemian, and A. Braun, RSC Adv., 2014, 4, 4676. Reproduced by permission of The Royal Society of Chemistry.

Interface	Fixed contrast factor, $\|\delta\rho\|_0^2$ (10^{28} m^{-4})	Anomalous contrast factor, $\|\delta\rho\|_E^2$ in $\|\delta\rho\|_E^2$ $\ln((E_0 - E)/E_0)$ (10^{28} m^{-4}) ($E_0 = 17.998$ keV)
Ni/void (a)	5241.9	0
Ni/CGO (b)	413.35	0
CGO/void (c)	2711.5	0
LSM/void (g)	2075.0	0
Ni$_3$S$_2$/void (x)	2275.9	0
YSZ/void (f)	2111.8	61.44
YSZ/CGO	20.07	−14.45
Ni/YSZ	684.0	−43.53
YSZ/LSM (h)	−15.45	−5.03

needed for data measured near the Ni K-absorption edge. As a precaution, in comparing the predicted $|\delta\rho(E)|^2$ with those measured experimentally, only relative comparisons were explored among the various phases for the Ni K-absorption edge measurements. Absolute comparisons were made using data measured near the Zr K-absorption edge, where little or no significant multiple scattering occurred.

Figure 5.29 shows the experimental obtained X-ray scattering contrast (for the anomalous form factor) for Ni and for Zr with respect to the interfaces of Ni metal to air, and Zr to air and Zr to manganese. We clearly see the resonance effect for the Ni edge and Zr edge energy position. The left panel of Figure 5.30 shows the X-ray transmission data profiles across the thickness of the three measured samples. Cathode and anode have high transmission, the electrolyte absorbs much, and thus has low transmission, and there appear to be transmission minima at the interfaces of electrolyte with the two electrodes. With a two-component log-normal size distribution model applied to the SAXS curves, we obtain the apparent volume fraction size distributions, as shown in the right panel of Figure 5.30. The scattering contrast factors for the Ni K-edge and the Zr K-edge are determined for all potential interfaces, which may occur in the electrode assembly and listed in Table 5.3 and Table 5.4, respectively.

With these quantitative and qualitative consistencies, we are able to carry out a full quantitative assessment of the AUSAXS spectra, which is exercised step-by-step by Allen et al. [Allen 2014]. We obtain, e. g., the surface areas for the interfaces between the various phases in the SOFC assembly, such as the interface between LSM and YSZ. Note that this chemical distinction cannot be made with a conventional SAXS analysis. Only the contrast variation permits this distinction. Moreover, we obtain this information for a specific point across the thickness of the assembly because of the microfocus of the beam. This is demonstrated for the pristine assembly and the two operated assemblies in the left panel of Figure 5.30. The surface areas for the YSZ/void interface and LSM/void

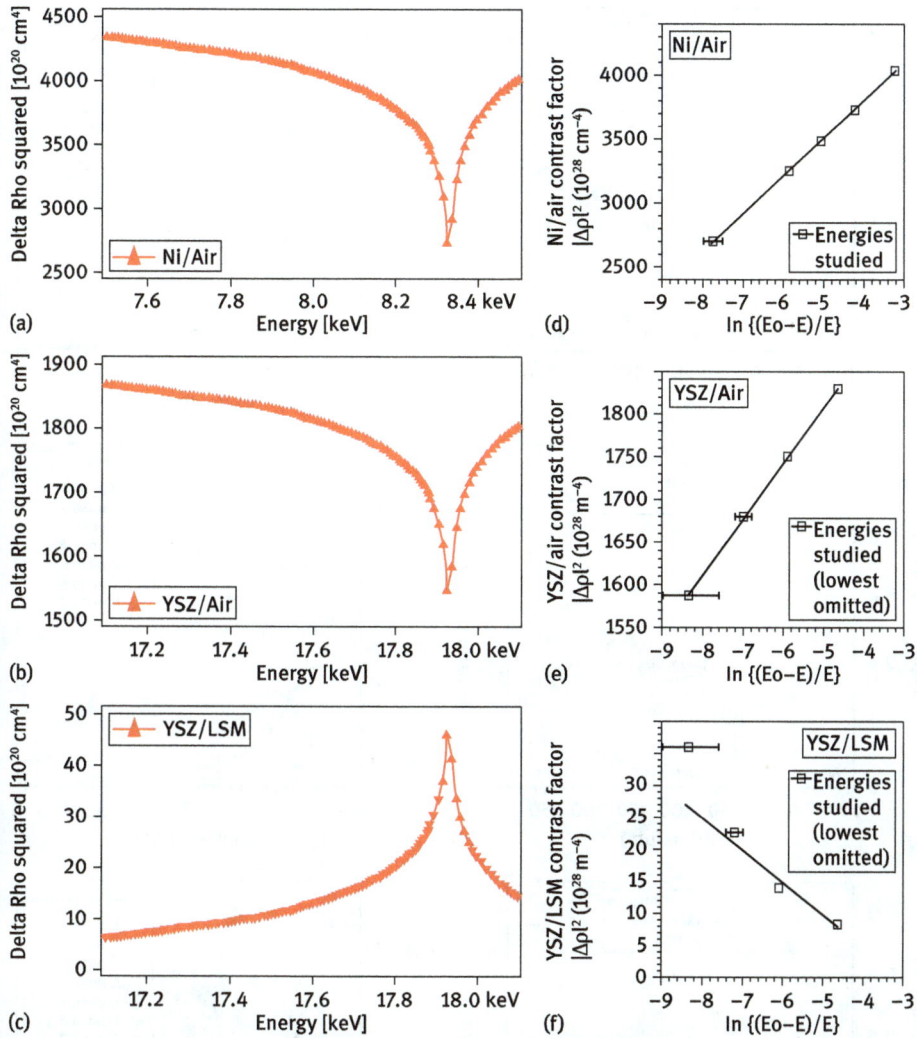

Figure 5.29: Real part of the calculated scattering contrast factor, $|Dr|2$, versus X-ray energy for (a) the Ni/void interface near the Ni X-ray absorption edge; (b) the YSZ/void interface; and (c) the YSZ/LSM interface near the Zr X-ray absorption edge; (d–f) corresponding straight line fits for (a), (b), and (c) of $|Dr|2$ versus $\ln((E_0 - E)/E_0)$ for $E < E_0$. Here, the horizontal bars reflect the precision of the tabulated X-ray energies [A. J. Allen, J. Ilavsky, P. R. Jemian, and A. Braun, *RSC Adv.*, 2014, 4, 4676]. Reproduced by permission of The Royal Society of Chemistry.

interface are shown in the right panel of Figure 5.30. Note that we now basically map out the electrochemically active interfaces in the electrode assembly. The data are summarized in Table 5.5 and Table 5.6 for the interface areas for anode and for cathode, respectively.

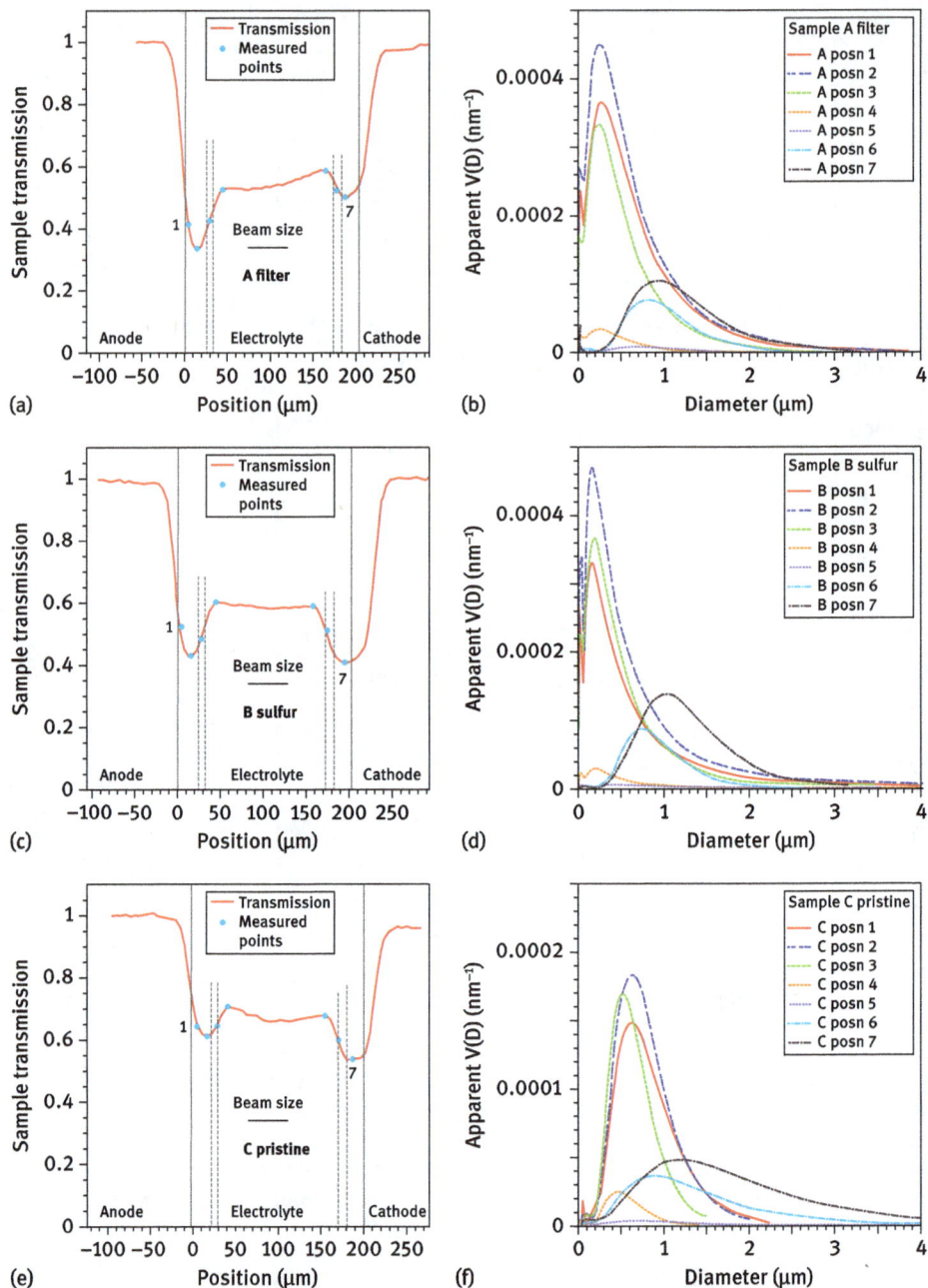

Figure 5.30: X-ray transmission profiles of SOFC sections studied showing measurement positions within the anode, electrolyte, and cathode for each of the three SOFC samples studied: A filter, B sulfur, and C pristine; and (b) corresponding apparent volume fraction size distributions obtained using the 2-component log-normal model. In (a) and in following figures, the vertical dashed lines indicate the inner anode and cathode layers [A. J. Allen, J. Ilavsky, P. R. Jemian, and A. Braun, *RSC Adv.*, 2014, 4, 4676]. Reproduced by permission of The Royal Society of Chemistry.

Table 5.5: Interfacial surface areas (in $m^2 \, cm^{-3}$) on the anode side of each SOFC sample.

Position Ni or Ni_3S_2	Sample A filter Ni/void (a)	Sample B sulfur Ni/void (a)	Sample B sulfur Ni_3S_2/void (x)	Sample C pristine Ni/void (a)
Position 1				
a or x:	0.370 ± 0.003	0.756 ± 0.016	1.935 ± 0.046	0.209 ± 0.003
	0.893 ± 0.014	0.693 ± 0.076	1.183 ± 0.076	0.209 ± 0.013
b:	2.495 ± 0.019	1.605 ± 0.050	1.368 ± 0.046	0.314 ± 0.007
c:				
Position 2				
a or x:	0.924 ± 0.013	1.463 ± 0.035	2.464 ± 0.091	0.299 ± 0.011
	2.230 ± 0.063	1.341 ± 0.168	4.263 ± 0.322	0.298 ± 0.051
b:	2.125 ± 0.034	0.953 ± 0.055	1.241 ± 0.054	0.190 ± 0.017
c:				
Position 3				
a or x:	1.070 ± 0.026	0.865 ± 0.022	1.760 ± 0.065	0.168 ± 0.005
b:	0.406 ± 0.031	0.793 ± 0.106	2.115 ± 0.106	0.454 ± 0.031
	2.583 ± 0.141	1.063 ± 0.048	1.056 ± 0.045	0.055 ± 0.006
c:				
Position 4				
a or x:	0.090 ± 0.006	0.072 ± 0005	0.100 ± 0.035	0.028 ± 0.003
	0.344 ± 0.032	0.051 ± 0.010	0.236 ± 0.060	0.122 ± 0.011
b:	0.011 ± 0.011	0.042 ± 0.003	0.039 ± 0.039	0.001 ± 0.004
c:	0.023 ± 0.023	0.030 ± 0.009	0.079 ± 0.079	0.076 ± 0.076
f:				
electrolyte				
f:	0.021 ± 0.002	0.026 ± 0.003	0.026 ± 0.003	0.024 ± 0.002

a = Ni/void, x = Ni_3S_2/void, b = Ni/CGO, c = CGO/void, f = YSZ/void. [A. J. Allen, J. Ilavsky, P. R. Jemian, and A. Braun, *RSC Adv.*, 2014, 4, 4676 DOI: 10.1039/C3RA46886K]. Reproduced by permission of The Royal Society of Chemistry.

Table 5.6: Interfacial surface areas (in $m^2 \, cm^{-3}$) on the cathode side of each SOFC sample.

Position	Sample A filter	Sample B sulfur	Sample C pristine
electrolyte			
f:	0.039 ± 0.004	0.024 ± 0.002	0.021 ± 0.002
Position 5			
f:	0.084 ± 0.020	0.056 ± 0.013	0.038 ± 0.002
g:	0.001 ± 0.016	0.000 ± 0.011	0.003 ± 0.002
h:	0.011 ± 0.213	0.006 ± 0.138	1.281 ± 0.318
Position 6			
f:	0.296 ± 0.139	0.444 ± 0.015	0.353 ± 0.018
g:	0.197 ± 0.114	0.141 ± 0.014	0.016 ± 0.017
h:	2.220 ± 1.498	7.301 ± 2.538	8.226 ± 2.650
Position 7			
f:	0.326 ± 0.284	0.081 ± 0.231	0.288 ± 0.108
g:	0.387 ± 0.226	0.729 ± 0.192	0.129 ± 0.089
h:	0	0	0

f = YSZ/void, g = LSM/void, h = YSZ/LSM. [A. J. Allen, J. Ilavsky, P. R. Jemian, and A. Braun, *RSC Adv.*, 2014, 4, 4676 DOI: 10.1039/C3RA46886K]. Reproduced by permission of The Royal Society of Chemistry.

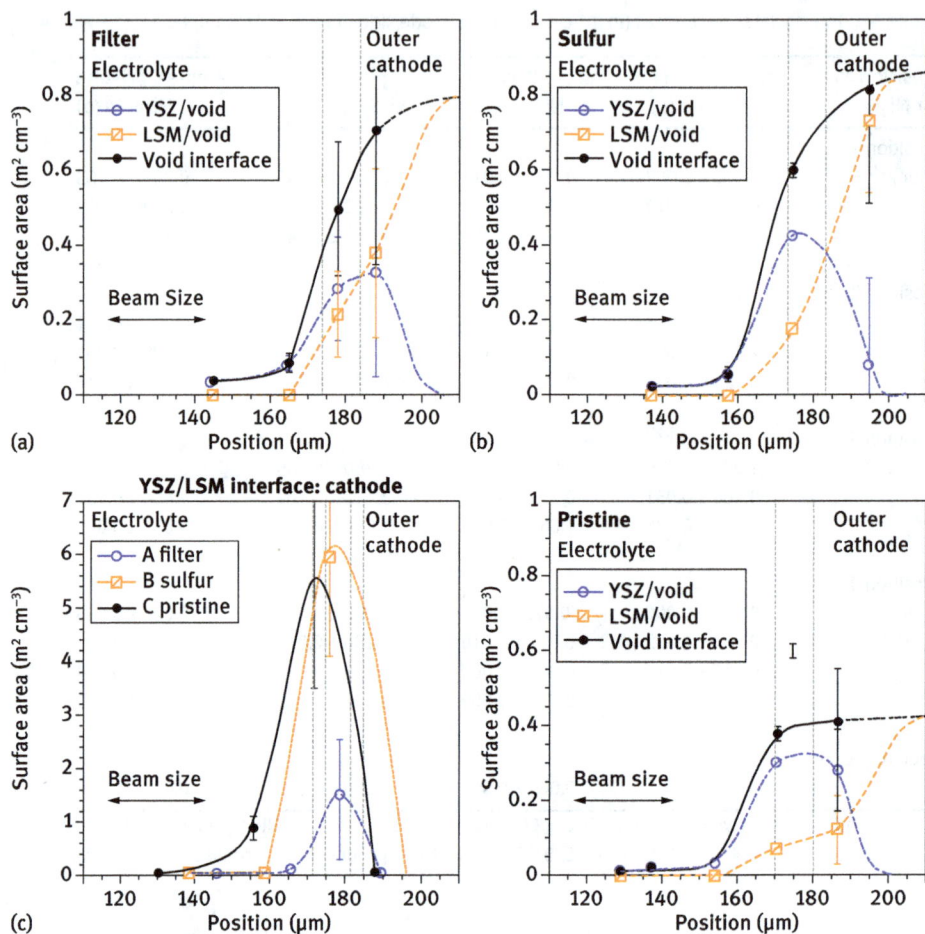

Figure 5.31: (Left) Comparison of the interface area between YSZ and LSM determined across the SOFC electrode assembly by AUSAXS scans for three conditions, i. e., pristine, operated with sulfur containing fuel and sulfur filtered fuel. (Right) Surface areas of the YSZ/void interface, LSM/void interface, and void interface for the three aforementioned samples, obtained from the same AUSAXS scans [A. J. Allen, J. Ilavsky, P. R. Jemian, and A. Braun, *RSC Adv.*, 2014, 4, 4676 DOI: 10.1039/C3RA46886K]. Reproduced by permission of The Royal Society of Chemistry.

Let us look at the surface areas at the interface between the YSZ electrolyte and the LSM cathode in Figure 5.31. The left part in the lower left panel of Figure 5.31 represents the electrolyte side. The pristine sample has a maximum surface area for the YSZ/LSM interface of around $5\,m^2/cm^3$. The surface area distribution is further centered on the electrolyte side, but the two operated cells have the corresponding surface area centered on the cathode side. The SOFC operated with the sulfur filter has a very low surface area of below $2\,m^2/cm^3$. The sample operated with sulfur containing fuel has the highest surface area, specifically around $6\,m^2/cm^3$. Now let us look at the interfaces of YSZ versus void, LSM versus void, and the void interface for all three samples. These data are plotted in the right panel of Figure 5.31. The void interface is large for both operated samples, around $0.8\,\text{m}^2/\text{cm}^3$, whereas the pristine sample has only half this value, $0.4\,\text{m}^2/\text{cm}^3$. This holds for the outer part of the LSM cathode. The void interface is therefore growing during SOFC operation. The same holds for the surface area between LSM cathode and voids. The YSZ/void interface area, however, shows not any substantial difference for any of the three samples. Noteworthy, though not surprising, is that the surface areas in the electrolyte are all substantially low and close to 0.

With the imagination of the microstructure of the SOFC electrode assembly, which is built from different phases for m different interfaces with different functionality, we can now speculate which processes take place during SOFC operation depending on whether the fuel gas contains sulfur or not. Figure 5.32 shows a simplistic schematic of the anode, which may be built up from, say, a CGO electrolyte, nickel catalyst particles and voids. Exposure to sulfur at high temperatures during SOFC operation makes that sulfur reacts with nickel to, e. g., Ni_3S_2, which causes structural reorganization such as pore widening and formation of new phases.

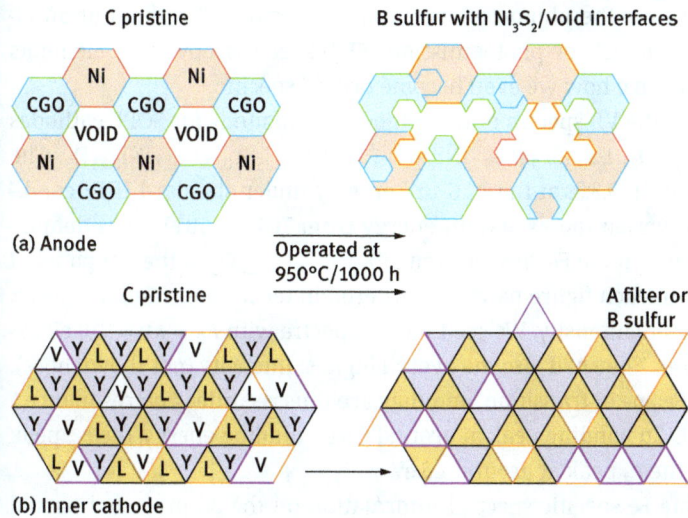

Figure 5.32: Idealized schematics illustrating generic microstructural changes in (a) anode and (b) inner cathode layers during SOFC service life. In anode: Ni_3S_2 formation is highlighted by a glowing Ni/void interface; for boundaries: red = Ni/void, green = CGO/void, blue = Ni/CGO. In cathode: L = LSM, Y = YSZ, V = voids; for boundaries: orange = LSM/void, purple = YSZ/void, black = YSZ/LSM. [A. J. Allen, J. Ilavsky, P. R. Jemian, and A. Braun, *RSC Adv.*, 2014, 4, 4676 DOI: 10.1039/C3RA46886K]. Reproduced by permission of The Royal Society of Chemistry.

5.6 Resonant photoemission spectroscopy

5.6.1 SOFC cathodes

We have seen in Chapter 2 the various applications of valence band (VB) XPS. There is sometimes confusion about which terms are used for which techniques. XPS stands for X-ray photoelectron spectroscopy, whereby the X-rays excite the atoms and the photoelectrons are being detected by a channeltron. Thus, the electrons are the probed species.

Photoemission spectroscopy and X-ray photoelectron spectroscopy is often used synonymously. When ultraviolet radiation is used, rather than X-rays as excitation source, then we speak about UPS, and this still is a method of photoemission spectroscopy.

The difference between X-ray core level spectra and VB spectra is that the former are specific to the probed element, whereas the latter are substantially hybridized. When you look at the VB spectrum of a compound, it is beforehand not possible to tell which spectral details originate from which particular transitions. This is an unfortunate detail of VB spectroscopy because the VB contains almost all information on the functionality of a material, such as electric and magnetic and optical properties.

However, when we tune the photon energy to an absorption threshold of an element, we can incur also in XPS or photoemission (PES) spectroscopy some elements specificity. Later we will learn how we even become orbital specific.

We see in Figure 5.33 the VB spectra of three model compounds for SOFC cathodes [Erat 2010h], $LaFe_{0.75}Ni_{0.25}O_3$, $La_{0.5}Sr_{0.5}Fe_{0.75}Ni_{0.25}O_3$, and $La_{0.25}Sr_{0.75}Fe_{0.75}Ni_{0.25}O_3$. The spectra were recorded at BL 9.3.2. at the ALS in Berkeley under standard UHV conditions. As you notice, we varied the excitation energy from 704 to 716 eV, as is demonstrated by the vertical lines in the Fe 2p spectrum of $LaFe_{0.75}Ni_{0.25}O_3$ in the left panel of Figure 5.34. The spectra in each figure panel are ordered in terms of excitation energy, and not in order of absolute intensity. We see that the spectra with $E_0 = hv = 710$ eV exhibit the sharpest features. They had also the overall highest intensity (not shown here). At 710 eV, we excite the Fe 2p–3d transition, and they are thus recorded "on resonance." We clearly identify the O 2p bonding peak at around 6.5 eV binding energy, the t_{2g} peak at around 3.8–4 eV, and the e_g peak at around 1.5 eV.

We thus obtain some Fe-specific spectral information on the 3d projected partial density of states in the VB spectra. The rationale behind this study was to find out about the origin of those electron holes, which are important charge carriers for conductivity. When we compare the above spectra with spectra in literature from comparable compounds that contain no nickel, then we conclude that substitution of Fe by Ni forms electron holes which are mainly O 2p character. The substitution of La by Sr increases the hole concentration to an extent that the e_g structure vanishes. The variation of the e_g and t_{2g} structures, not shown here, is paralleled by the changes in the electrical conductivity.

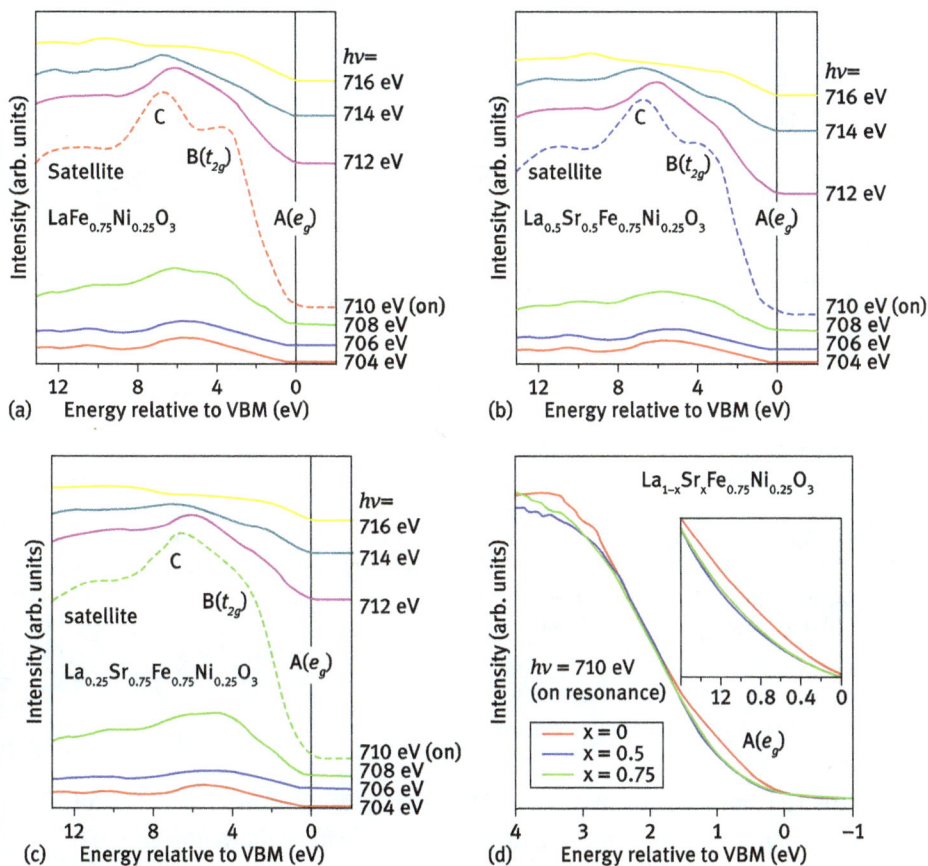

Figure 5.33: Fe 2p–3d resonant valence-band photoemission spectra of $La_{1-x}Sr_xFe_{0.75}Ni_{0.25}O$. There is on-resonance condition at 710 eV. (a) $x = 0$, (b) $x = 0.50$, (c) $x = 0.75$. (d) Expanded valence-band spectra at Fe 2p–3d on-resonance (710 eV), with an inset showing that the intensity for $x = 0.75$, is in between those for = 0 and = 0.5. Reprinted from [Erat, S., Wadati, H., Aksoy, F., Liu, Z., Graule, T., Gauckler, L. J., Braun, A., Iron-resonant VB photoemission and oxygen near edge X-ray absorption fine structure study on $La_{1-x}Sr_xFe_{0.75}Ni_{0.25}O_{3-\delta}$. Applied Physics Letters 2010, 97 (12)] with the permission of AIP Publishing.

We have performed resonant VB PES and NEXAFS at O K edge and Fe L2,3 edges on $La_{1-x}Sr_xFe_{0.75}Ni_{0.25}O_{3-\delta}$ ($x = 0.0$, 0.5, and 0.75). The spectral weight of occupied e_g and t_{2g} states of Fe is reduced upon electron hole doping up to 50 % and then starts to increase, which is parallel to changes in electrical conductivity. The spectral weight of the pre-peak in the oxygen NEXAFS spectra due to the p-type electron holes created on Ni 3d increases with increasing x. This is clear from the right panel of Figure 5.34, which shows the oxygen NEXAFS spectra. With respect to $La_{1-x}Sr_xFeO_3$, it is observed that the spectral weight transfers from below EF to above it across the gap. However, here in $La_{1-x}Sr_xFe_{0.75}Ni_{0.25}O_{3-\delta}$, it is difficult to conclude this because the Ni 3d closer to EF than Fe 3d and we do not consider any metal (Fe)-metal (Ni) electron hole transfer.

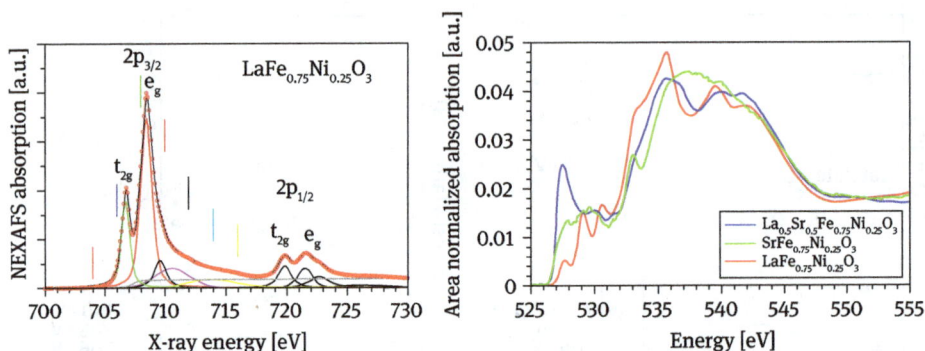

Figure 5.34: (Left) Normalized Fe L2,3 edge X-ray absorption spectra of $La_{1-x}Sr_xFe_{0.75}Ni_{0.25}O$ for $x = 0$, and $x = 0.50$. Vertical lines denote the energies where VB PES spectra were collected. Bottom panels— Atomic multiple simulation spectra for the same stoichiometries. O K edge X-ray absorption spectra of $La_{1-x}Sr_xFe_{0.75}Ni_{0.25}O$ for $x = 0$, and $x = 0.50$ with suggested peak deconvolution into Voigt functions. Arrow in top panel spectrum shows the difference between $e_g\uparrow$ peak intensity and normalization level (dashed horizontal line). Spectrum in bottom panel has additional peak at ~528 eV. Reprinted from [Erat, S., Wadati, H., Aksoy, F., Liu, Z., Graule, T., Gauckler, L. J., Braun, A., Iron-resonant valence band photoemission and oxygen near edge X-ray absorption fine structure study on $La_{1-x}Sr_xFe_{0.75}Ni_{0.25}O_{3-\delta}$. Applied Physics Letters 2010, 97 (12)] with the permission of AIP Publishing.

5.6.2 Mn 2p resonant photoelectron spectroscopy for SOFC chromium poisoning

Most of the functional components of SOFC are from ceramic material, except for the metallic interconnect plates that combine the electrode assemblies into an SOFC stack. The high operation temperature of SOFC (600 °C–1000 °C) makes that chromium in the interconnect plates becomes evaporated and reacts malign with the cathode material. This chromium poisoning causes deleterious degradation of the electric conductivity of the cathode. We looked with X-ray spectroscopy into the electronic structure of the $(La_{0.8}Sr_{0.2})_{0.98}Mn_{1-x}Cr_xO_3$ compound with $0 \leq x \leq 0.1$ with soft X-rays at ambient and elevated temperature [Tsekouras 2015]. The oxygen NEXAFS spectra suggest there is no hybridization between Cr 3d and O 2p states for lowest level substitutions. Mn $2p_{3/2}$ resonant PES measurements pointed to a decreased Mn 3d–O 2p hybridization upon chromium substitution. The deconvolution of the oxygen spectra took into account the effects of exchange and crystal field splitting, whereby we included a new approach where the pre-peak region was described using the nominally filled $t_{2g\uparrow}$ state. Substitution by chromium to 0.1 relative concentration level results in the lowering in the energy of the $t_{2g\uparrow}$ state by 0.17 eV. This fits surprisingly well with the 0.15-eV increase in activation energy for the oxygen reduction reaction, while decreased overlap between hybrid O 2p–Mn 3d states was in qualitative agreement with previously observed decrease in electronic conductivity. Cr poisoning in SOFC appears thus to be genuinely a consequence of altered electronic structure, rather than the generally accepted effect of microstructure. We will see this in the final part of this book in Chapter 9.2 where we look into the correlation of electronic structure and conductivity.

In order to preliminary probe the electronic structure of the $(La_{0.8}Sr_{0.2})_{0.98}$ $Mn_{1-x}Cr_xO_3$ model series, Mn 2p NEXAFS spectra were recorded (compare [Cramer 1991]). Figure 5.35 shows the Mn 2p NEXAFS spectrum of $(La_{0.8}Sr_{0.2})_{0.98}MnO_3$ along with an energy diagram, illustrating the effects of spin-orbit splitting of the Mn 2p core-hole and octahedral field splitting, explaining the orbital origin of fitted peak functions. A meaningful fit is possible by employing Gaussian functions for the $2p_{3/2}$ t_{2g}, $2p_{1/2}$ t_{2g}, and $2p_{1/2}$ e_g orbitals plus a pseudo-Voigt function for the $2p_{3/2}$ e_g orbital. We applied the same procedure for all Mn 2p NEXAFS spectra. Spin-orbit coupling and octahedral field splitting values of around 10.7 and 1.7 eV were obtained, respectively. The Mn 2p NEXAFS spectra were necessary for the selection for X-ray energies resonant with the $2p_{3/2}$ absorption peak to use during resonant PES measurements.

The results of Mn 2p NEXAFS measurements allow us to select photon energies corresponding to the $2p_{3/2}$ feature for resonant photoelectron spectroscopy. By this, we can enter better into understanding the VB features, which are hybridized *per se* and which we can tell apart by going resonant. The right top panel of Figure 5.35 shows the Mn 2p resonant PES spectra of $(La_{0.8}Sr_{0.2})_{0.98}MnO_3$ as the incident photon energy was swept through the $2p_{3/2}$ absorption peak of the corresponding Mn 2p NEXAFS spectrum. The selected energies are shown in the middle panel on the right and indicated with roman numerals.

Following the work by [Agui 2005, Saitoh 1993a, Saitoh 1993b, Saitoh 1995, Saitoh 1999], the features in Figure 5.35 top right panel VB spectra are assigned to Mn 3d e_g and t_{2g} orbitals (ca. 1.0–4.2 eV), hybrid O 2p–Mn 3d orbitals (ca. 4.2–9.4 eV), and to a satellite feature above ca. 9.4 eV. The bottom panel of Figure 5.35 shows the on-resonance Mn $2p_{3/2}$ spectrum of $(La_{0.8}Sr_{0.2})_{0.98}MnO_3$ measured using 643.5 eV incident photon energy and its deconvolution into a Shirley background and Gaussian functions corresponding to Mn 3d e_g (at ca. 2.1 eV) and t_{2g} (at ca. 3.7 eV) orbitals, hybrid O 2p – Mn 3d orbitals ("C" at ca. 4.7 and "D" at ca. 7.7 eV), and a satellite feature (at ca. 10.6 eV).

For the visualization of the effect of annealing and chromium substitution, we show in Figure 5.36 the normalized, on-resonance Mn $2p_{3/2}$ VB spectra of $(La_{0.8}Sr_{0.2})_{0.98}MnO_3$ at 300 K and at higher temperature, and $(La_{0.8}Sr_{0.2})_{0.98}Mn_{0.9}Cr_{0.1}O_3$ at elevated temperature, measured at 643.5 eV photon energy. Visual inspection reveals that heating $(La_{0.8}Sr_{0.2})_{0.98}MnO_3$ causes higher signal intensity in the part of the spectrum that we attribute to O 2p–Mn 3d hybrid states C and D, in comparison to Mn 3d e_g and t_{2g} orbitals. That is, O 2p–Mn 3d hybridization within $(La_{0.8}Sr_{0.2})_{0.98}MnO_3$ increased upon heating. For the quantification of this increase in hybridization, we define the intensity ratio $(C + D)/(e_g + t_{2g})$ of integrated peak areas, as shown in [Tsekouras 2015]. Accordingly, heating of $(La_{0.8}Sr_{0.2})_{0.98}MnO_3$ results in an increase in the $(C + D)/(e_g + t_{2g})$ ratio from 3.17 to 3.73, which is consistent with the increase in electronic conductivity from 131 S cm^{-1} at room temperature to 174 S cm^{-1} at 850 °C for this material. In contrast to the

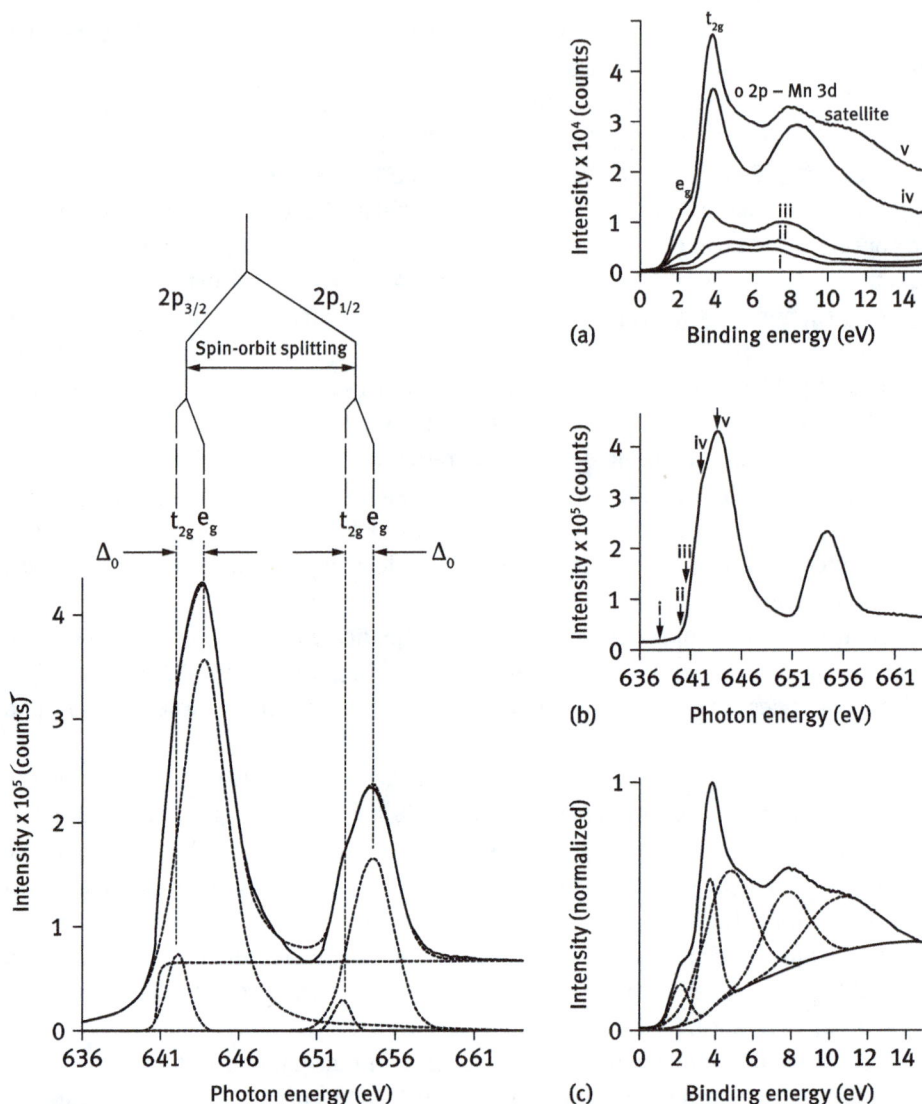

Figure 5.35: (Left) (−) Mn 2p near-edge X-ray absorption fine structure spectrum of $(La_{0.8}Sr_{0.2})_{0.98}MnO_3$. (—) Fit components. (•••) Sum of fit components. Mn 3d energy level diagram showing effects of spin-orbit and octahedral field (DO) splitting superimposed to indicate origin of fitted peak functions [Cramer 1991]. (Right) (a) Mn $2p_{3/2}$ resonant photoelectron spectra of $(La_{0.8}Sr_{0.2})_{0.98}MnO_3$ as incident photon energy increased through (i) 638.0, (ii) 640.0, (iii) 640.5, (iv) 642.0, and (v) 643.5 eV. (b) Room temperature Mn 2p near-edge X-ray absorption fine structure spectrum of $(La_{0.8}Sr_{0.2})_{0.98}MnO_3$ with incident photon energies used in (a) indicated by arrows. (c) Deconvolution of on-resonance Mn $2p_{3/2}$ resonant photoelectron spectrum of $(La_{0.8}Sr_{0.2})_{0.98}MnO_3$ with a Shirley background and Gaussians corresponding to Mn 3d e_g (ca. 2.1 eV) and t_{2g} (ca. 3.7 eV) orbitals, hybrid O 2p–Mn 3d orbitals (ca. 4.7 and 7.7 eV), and a satellite feature (ca. 10.6 eV); incident photon energy = 643.5 eV. Reprinted from [Tsekouras, G., Boudoire, F., Pal, B., Vondracek, M., Prince, K. C., Sarma, D. D., Braun, A., Electronic structure origin of conductivity and oxygen reduction activity changes in low-level Cr-substituted (La, Sr)MnO3. Journal of Chemical Physics 2015, 143 (11)] with the permission of AIP Publishing.

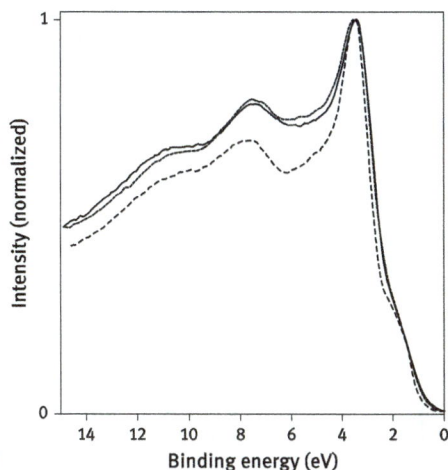

Figure 5.36: Mn 2p resonant photoemission spectra of (—) $(La_{0.8}Sr_{0.2})_{0.98}MnO_3$ at 25 °C, (–) $(La_{0.8}Sr_{0.2})_{0.98}MnO_3$ at 300 °C, and (•••) $(La_{0.8}Sr_{0.2})_{0.98}Mn_{0.9}Cr_{0.1}O_3$ at 300 °C. Incident photon energy = 643.5 eV. Reprinted from [Tsekouras, G., Boudoire, F., Pal, B., Vondracek, M., Prince, K. C., Sarma, D. D., Braun, A., Electronic structure origin of conductivity and oxygen reduction activity changes in low-level Cr-substituted (La, Sr)MnO$_3$. Journal of Chemical Physics 2015, 143 (11)] with the permission of AIP Publishing.

overt effect of heating, the effect of low-level chromium substitution is not ascertainable by simple observation [Tsekouras 2015]. However, quantification of the extent of hybridization reveals a decrease in the $(C + D)/(e_g + t_{2g})$ ratio at elevated temperature from 3.73 to 3.57 upon low-level chromium substitution, indicative of lowered O 2p–Mn 3d hybridization, and qualitatively consistent with the previously observed lower electronic conductivity at 850 °C of $(La_{0.8}Sr_{0.2})_{0.98}Mn_{0.9}Cr_{0.1}O_3$ (89 S cm^{-1}) compared to $(La_{0.8}Sr_{0.2})_{0.98}MnO_3$ (174 S cm^{-1}) [Tsekouras 2014].

5.6.3 VB PES on photoelectrodes

Below we have recorded VB photoemission spectra from an iron oxide photoelectrode (pristine α-Fe$_2$O$_3$), dip coated on FTO glass. The spectra were recorded at ALS BL 9.3.2 with an ambient pressure XPS instrument. We have varied the excitation energy from 700 to 712 eV and then recorded VB spectra. Figure 5.37 below shows eight spectra recorded with eight different energies. The PES intensities have not been manipulated with respect to each other. We notice that the overall spectrum recorded for 700 and 702 eV is very low, whereas the intensity of the spectrum recorded at 706 eV is very high. The intensities resulting from the other excitation energies range in between.

Closer inspection of the set of spectra shows that the state with t_{2g} orbital symmetry looks particularly enhanced. The same holds for the O 2p bonding peak. The spectrum measured with 706 eV photon energy is thus called "on resonance," whereas the spectra

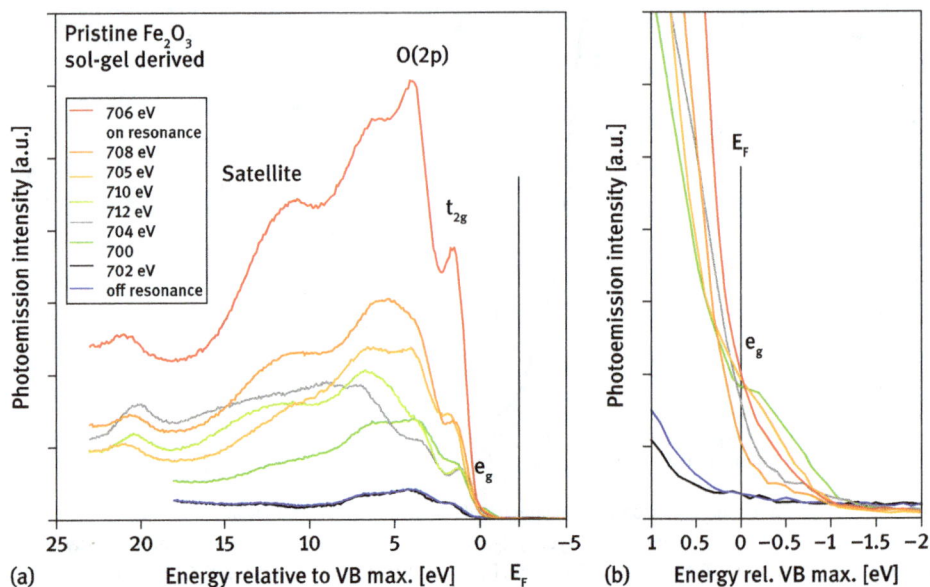

Figure 5.37: Resonant photoemission spectra in the VB region obtained with excitation photon energies from 700 eV to 712 eV. The panel on the right shows a magnification of the energy range around the e_g transition.

with a showll low profile and low intensity at 702 eV and below are called "off-resonance." The right panel of Figure 5.37 shows a magnification of the intensity shoulder around the e_g peak near the Fermi level. The spectra for photon energy 704 and 705 eV are drawn in thick lines in order to highlight that these have the highest intensity for the e_g state. This is very interesting because the highest relative intensity for the t_{2g} peak was found at a different photon energy excitation, i. e., 706 eV. The e_g peak for 706 eV has clearly lower intensity in the e_g region. Therefore, we can say that we have a fine structure in the photon energy dependent VB spectra, which goes beyond the mere distinction between on resonance and off resonance. Rather, every peak in the PES spectrum may have its particular resonant excitation energy.

How do we look at this? Let us look at the Fe 2p spectrum of a comparable sample in Figure 5.38, which was measured at BESSY in Berlin. Observe the "rings" drawn on the spectrum, which match the color of the spectra in Figure 5.37. We realize now at which energetic fine levels we have excited the iron cation in the α-Fe_2O_3 with orbital symmetry sensitivity.

To some electrochemists, this approach may resemble some similarity with the preparation of a Mott–Schottky plot. First, you run a cyclic voltammogram (CV), basically an I-V curve in an electrochemical system where you polarize the working electrode between an upper and a lower electrochemical potential. When an electrode is inserted in an electrolyte, its Fermi level will try to adjust and match the RedOx potential of the electrolyte. This works only by bending of the energy bands in the electrode.

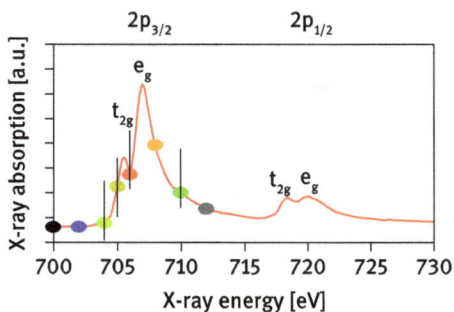

Figure 5.38: Fe 2p NEXAFS spectrum of a-Fe_2O_3. The filled dots on the spectrum denote the X-ray energies, which were selected for subsequent resonant photoemission experiments.

The energy bands are bent in either direction during running the I-V curve. The CV reflects to some extent the electronic flat band profile.

For the investigation of the charge carrier dynamics, it is therefore advised to do impedance spectroscopy at particular positions of the electrochemical potential. One therefore scans impedance spectra along various external DC bias potentials and chooses the potentials at positions of the CV, which are considered particularly relevant. Thus, in electrochemistry, we are guided by the CV for choosing the points of interest for running the impedance spectra with subsequent Mott–Schottky analysis. And in resonant PES we are guided, e. g., by the L-edges of the metal ion, which we want to use for a deeper investigation of the electronic structure of a compound. It is important for a scientist to be able to see and to connect various seemingly unrelated methods and combine them for an exhaustive investigation of a system. In Part 8 (correlation), we will learn how we widen this horizon of conceptual approaches to combine different fields of scientific research.

Let us come back to the resonant PES on photoelectrodes. In Chapter 2.3, we have looked into the VB of iron oxide photoelectrodes, which were differently oxidized in electrolyte by anodization. Figure 5.39 shows the four XPS spectra of different anodized iron oxide electrodes recorded with a laboratory-based XPS spectrometer with Mg Kα 1253 eV X-ray energy. We do identify the shallow shoulders of e_g and t_{2g} peaks, which are indicative to Fe^{3+} and Fe^{2+} at the surface, respectively. Moreover, we see the broad oxygen O 2p bonding peak. On the right panel of Figure 5.39, we see the spectra from the same samples, but measured at beamline 9.3.2 at the ALS. The photon energy was 705 eV, and we see now much sharper t_{2g} peaks. Moreover, we notice a sharper O 2p peak, whereas the e_g peak is relatively diffuse as a shoulder. With 705 eV, we are close at the on-resonance condition for the Fe 2p contrast variation. The samples that we subject to such spectroscopy experiment can be the same like the one we use for the photo-electrochemical experiment. Figure 5.40 shows two metal sample holders from BESSY II at HZB Berlin on which two photoelectrodes are mounted with double sticky carbon take. The photoelectrodes are glass slides coated with a transparent conducting oxide

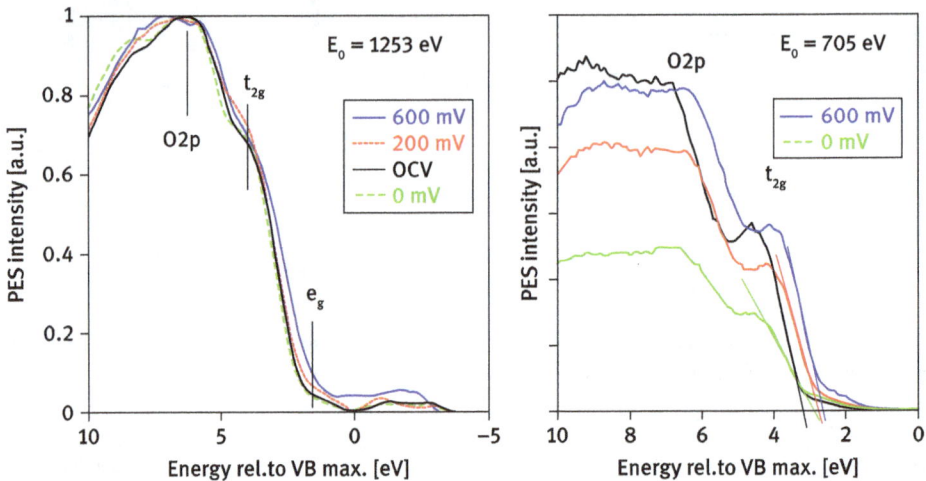

Figure 5.39: (a) XPS spectra of iron oxide photoelectrodes that were subject to electrochemical oxidation at 600 mV, 200 mV, 0 mV, and open circuit potential only. Spectra were recorded with a laboratory-based X-ray source with 1253 eV (Mg lamp) photon energy. (b) XPS spectra of the same samples, measured with 705 eV X-ray energy at the ALS in Berkeley.

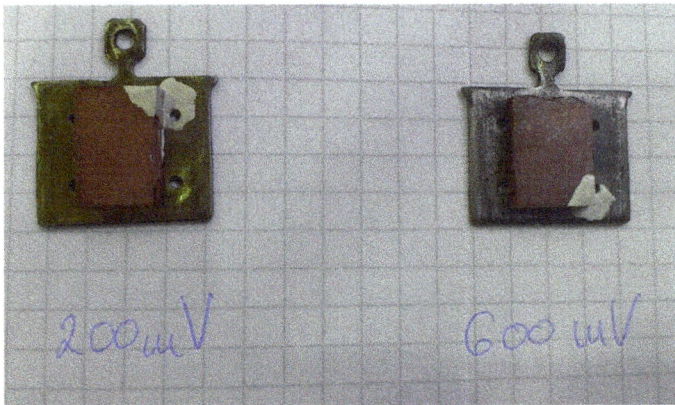

Figure 5.40: BESSY-II sample holders with iron oxide films deposited on FTO glass after oxidation at 200 and 600 mV in a KOH electrolyte. The samples are fixed with double sticky carbon tape on the metal sample holder. Silver paint is used to provide an electric conducting path from the iron oxide film to the sample holder because the FTO glass would not provide electric contact directly to the sample holder.

such as FTO, on which we have coated the iron oxide film, as can be seen from typical reddish brown rust color. We have cut off the free FTO layer on glass in order to match the sample size with the sample holder size. In order to provide electric contact between the iron oxide film and the sample holder, so as to avoid disturbing electric charging of the sample when exposed to the X-rays, we have painted a thick silver paint path. This is visible as a silver stain in the upper right and lower right sample corner of the left side and right side sample in Figure 5.40. These samples have been oxidized at 200 and 600 mV, respectively. After this electrochemical experiment, they were rinsed, dried, and then transferred into the UHV chambers for XPS.

Figure 5.41 shows a cyclic voltammogram of hematite, which we have reproduced from a publication by Dare-Edwards et al. [Dare-Edwards 1983]. The rings indicate the potentials at which we have anodized our hematite samples before we measured them with VB PES. At the higher potential, we carry out water oxidation, either in dark or under illumination. Researchers typically disclose only the photocurrents of their systems, *in lieu* of the produced oxygen or hydrogen gas. In Figure 5.42, we show the actual evolved oxygen gas at the photoanode, which we measured with a gas chromatograph. The panel on the left shows the GC detector signal (chromatogram) for 400 and 500 mV bias potential versus the retention time. The right panel shows the accumulated oxygen volume versus time, as a function of the applied bias potential. Even at 200 mV, we measured already some oxygen evolution, but 400 mV gave an appreciable gas evolution.

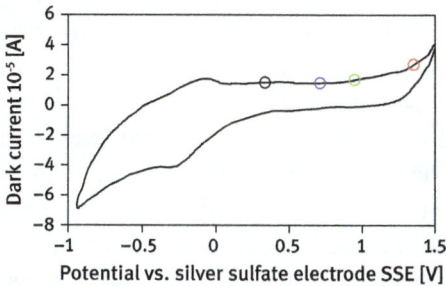

Figure 5.41: Cyclic voltammogram of a-Fe$_2$O$_3$ reproduced from Dare-Edwards et al. [Dare-Edwards 1983] with permission of The Royal Society of Chemistry. The four color bullets indicate roughly the potential positions that are chosen for anodization of samples for follow-up VB PES measurements.

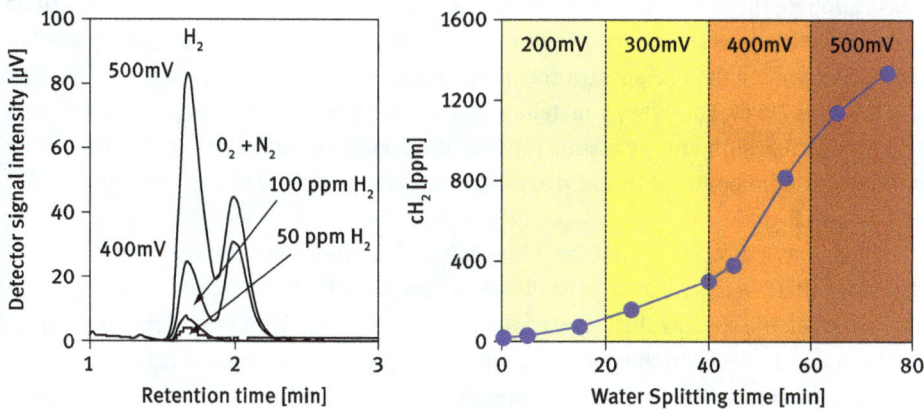

Figure 5.42: (Left) Gas chromatograph (GC) detector signals for solar photoelectrochemical water splitting at 400 mV and 500 mV versus Ag/AgCl reference, and (right) evolution of hydrogen concentration during a water splitting experiment with changes of bias potential from 200 to 500 mV according to GC data.

We therefore had an interest in the electronic structure of the electrode surfaces, which had been operated at various potentials where we have different extent of oxygen evolution, including the reference values where we anticipate no oxygen evolution, such as open circuit potential (OCV) and 0 mV bias.

Upon closer visual inspection of the four spectra measured at 705 eV X-ray energy (Figure 5.39 right panel), it becomes obvious that the spectra are shifted on the energy axis. In order to determine the position of the spectrum, the flank with increasing intensity after the VB top near E_F was extrapolated to zero intensity and then the intercept with the ordinate was determined. This makes that the spectrum of the sample, which was kept only under OCV in the electrolyte, has the highest energy position at around 3.3 eV (not accurately the binding energy, but the energy relative to the VB maximum), whereas the sample anodized at 0 mV has the position at 3.0 mV. The spectrum of the sample anodized at 200 mV has the intensity intercept at 2.8 eV, and the one anodized at 600 mV (this is beyond the water splitting onset potential) has the position at around 2.5 eV.

The iron oxide (α-Fe_2O_3) has a particular electric surface charge after processing and exposure to the ambient atmosphere. Upon insertion into the 1-molar KOH electrolyte, which we typically use for PEC experiments, we measure a potential of some 10 mV in the open circuit mode. Open circuit mode means that no external electric potential is applied to the working electrode and counterelectrode in the photoelectrochemical cell. Then the sample was removed from the electrolyte, rinsed with distilled water, dried, and stored for later analyses. Another aliquot electrode was inserted, and the potentiostat was switched on to apply a 0-mV bias, upon which a net current can flow. After some time, the electrode was removed, rinsed in distilled water, dried, and stored for later analyses. We then went ahead and prepared the 200- and 600-mV bias electrodes. Anodization means electrochemical oxidation. When a positive bias is applied, the electrode will be oxidized. Any Fe^{2+} present in the electrode will be subject to oxidation. The question is: at which potential can this process take place? In Figure 5.43, we see the PES spectra recorded at four different photon energies, i. e., 704, 705, 706, and 710 eV. For every constellation, we have determined the spectral shift and then plotted this shift energy versus the four different bias potentials. It has turned out that the four data points in this plot are on a smooth line for the photon energy of 705 V, as shown in Figure 5.44.

What does it mean then that the PES spectra are shifted on the binding energy scale? Obviously, this is a chemical or even electrochemical shift. With increasing anodization bias potential, the spectra shift toward the Fermi energy. This is a very strange observation because anodization means oxidation, and typically an oxidation causes a shift of the spectrum toward higher binding energies—away from the Fermi level. This sometimes goes along with a decreased quality of the spectrum. Spectra may become noisy when the sample becomes oxidized and less conducting. Here, we observe the opposite, which means that we are making hole doping upon anodization. This could be a

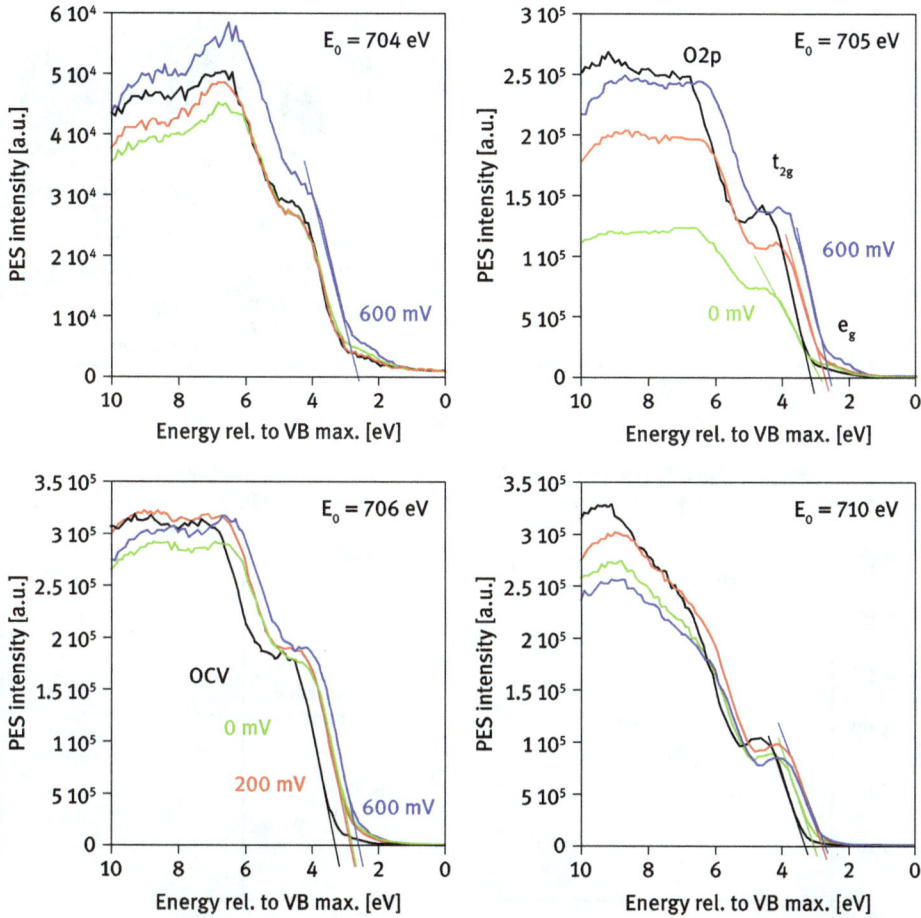

Figure 5.43: Comparison of the resonant PES spectra for four different anodized hematite photoelectrodes (OCV, 0 mV, 200 mV, 600 mV) for four different photon energies 704, 705, 706, and 710 eV.

manifestation of highly oxidized iron species on the electrode surface. We had speculated over this already after a previous investigation with oxygen NEXAFS spectroscopy [Bora 2011a]. The Pourbaix diagram of iron oxide [Li 2012] suggests that under high potential and high pH, such as in 1 molar KOH, we would expect iron species with a superficial high oxidation state of 4, 5, or even 6. To a chemist, such high iron oxidation state sounds superficial.

We know from the studies on iron perovskites in SOFC cathodes that we have to deal with a species, which can be interpreted as Fe^{4+}, whereas it is spectroscopically $Fe3d^5$ with L as an oxygen 2p ligand hole [Braun 2009a]. The perovskite structure can stabilize species that are not necessarily stable in ambient conditions in nature. Moreover, surfaces in electrochemical environment can favor even more exotic species, which only under particular analytical circumstances as shown here become apparent.

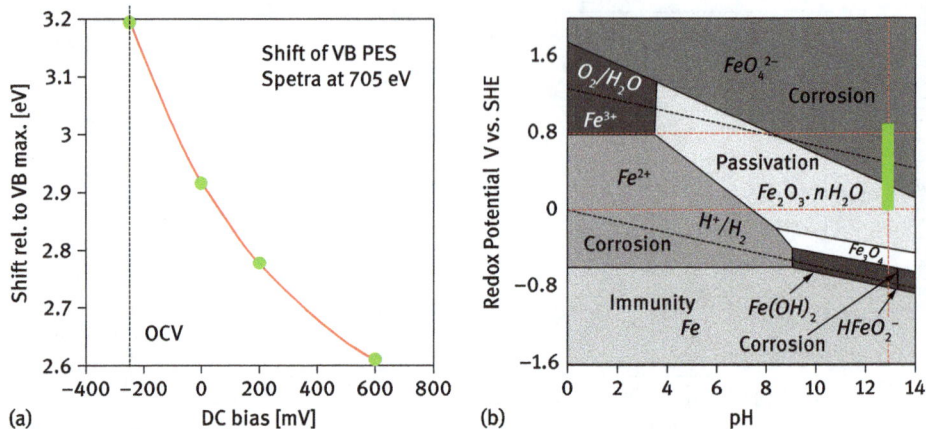

Figure 5.44: Variation of the shift of the VB spectra from Figure 5.43 with increasing anodic bias on the hematite photoelectrode. Pourbaix diagram of iron oxide [Gajda-Schrantz 2013].

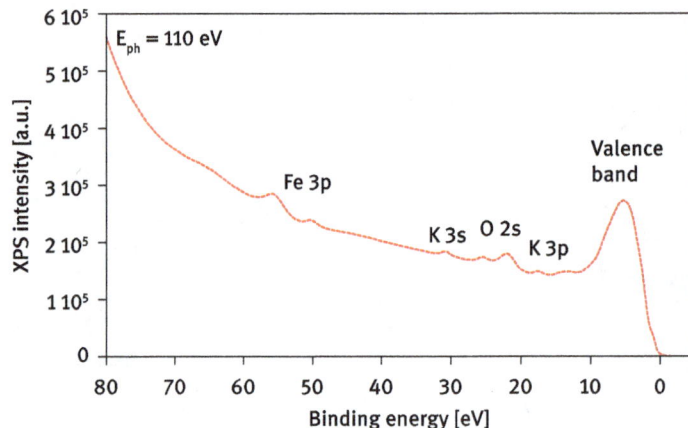

Figure 5.45: Survey spectrum of α-Fe$_2$O$_3$ photoelectrode with photon energy of 110 eV recorded at BESSY-II Berlin, SurICat beamline.

As for now, we have investigated the photoelectrode with Fe 2p resonant PES. We certainly can select other resonant X-ray excitations such as the Fe 3p transitions, provided the necessary photon energy range is available at the beamline. Figure 5.45 shows the survey spectrum of a hematite sample obtained with relatively low photon energy of 110 eV. The spectrum was recorded at BESSY-II in Berlin at the SurICat beamline (Figure 5.48). We notice the shallow Fe 3p structure at around 55 eV, the O 2s structure at around 24 eV, and the structure from residual potassium from the KOH electrolyte, i. e., K 3s and K 3p at 30 and 18 eV, respectively.

We then took a hematite electrode and anodized it at 500 mV in KOH electrolyte for 10 min and recorded resonant VB PES spectra with photon energies from 50 eV to 60 eV

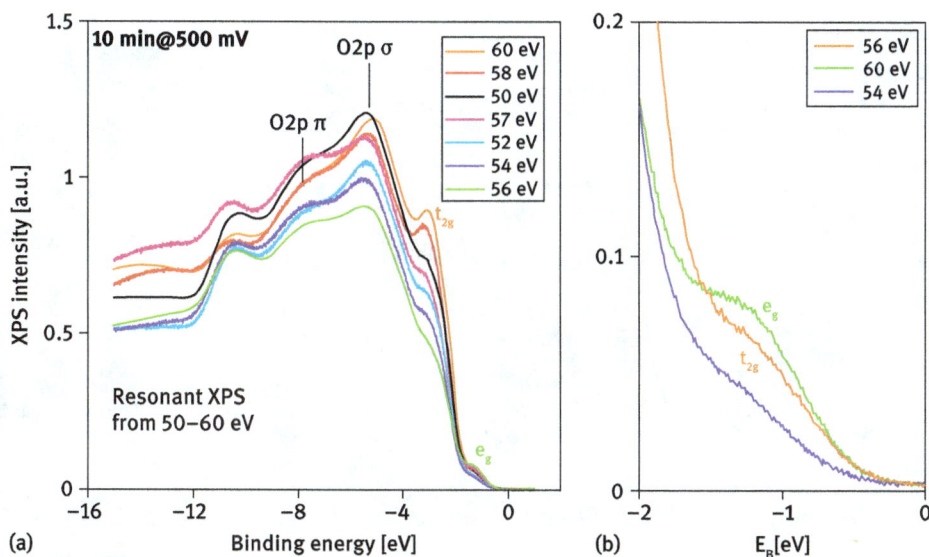

Figure 5.46: (a) Resonant PES spectra of Fe_2O_3 with photon energies from 50 to 60 eV. (b) The t_{2g} structure is resonant enhanced at 60 eV, whereas the e_g structure is resonant enhanced at 56 eV.

in 1 eV steps, as shown in Figure 5.46. We see how nice the relative spectral weight of the t_{2g} and e_g peaks are changing with the resonant excitation. The overall highest intensity is obtained with 50 eV for the O 2p σ peak at a binding energy of 5.5 eV. However, the highest intensity for the t_{2g} peak is at X-ray excitation energy of 60 eV, and the lowest intensity at 56 eV. When we look into the e_g structure, which is magnified on the right panel of Figure 5.46, e_g has highest intensity at 56 eV. Therefore, if we are interested in the e_g and t_{2g} structures and how they vary during synthesis, processing, or function or degradation of materials and devices, we should investigate the samples by selecting 56 and 60 eV resonant excitation, provided we have these energies available at the synchrotron beamline. As an alternative, we may look at the Fe 2p resonant excitation, which is around an X-ray energy of 700 eV.

Resonant excitation is therefore not just among different elements, but also among different orbitals, obviously.

It is possible to do the necessary electrochemical treatment at the SoLIAS endstation at BESSY-II in their "Glaskreuz" glass cross-tubing preparation chamber, which is shown in Figure 5.47. The SoLIAS end station used to be one of the few electrochemistry facilities at a synchrotron [Mayer 2005, Schwanitz 2007a, Skorupska 2005]. Another end station for surface and interface science and catalysis, but not for electrochemistry is the SurICat beamline at BESSY-II, as shown in Figure 5.48.

We have looked into samples with Fe 3p resonant PES, which were anodized at different bias potentials, 500, 600, and 700 mV. Figure 5.49 compares the resonant PES spectra of two samples anodized for 0–1440 min and 500 and 700 mV. Close inspection

Figure 5.47: "Glass Cross" electrochemical cell for *in situ* electrochemical preparation of samples prior to synchrotron XPS measurements at SurICat beamline at BESSY-II Berlin. The glass cell is hermetically sealed and has ports for electrolyte supply and electric feedthroughs for a three-electrode setup.

of the e_g peak shows that its intensity—for both samples left and right panels in Figure 5.49—decreases with increasing anodization time. Because the e_g feature is caused by iron atoms with Fe^{2+} valence, we conclude that there is an abundant concentration of Fe^{2+} at the pristine electrode, which becomes depleted upon anodization. At the same time, the t_{2g} feature indicates that Fe^{3+} is increasing. Hence, the anodization oxidizes the Fe^{2+} into Fe^{3+}.

Figure 5.48: SurICat beamline and end station at BESSY-II in Berlin. Some end stations at BESSY are mobile. Depending on the experimental needs of the research proposals, they can be transported to different beamlines within the BESSY-II building. Not every synchrotron facility has this philosophy of moving end stations to different beamlines.

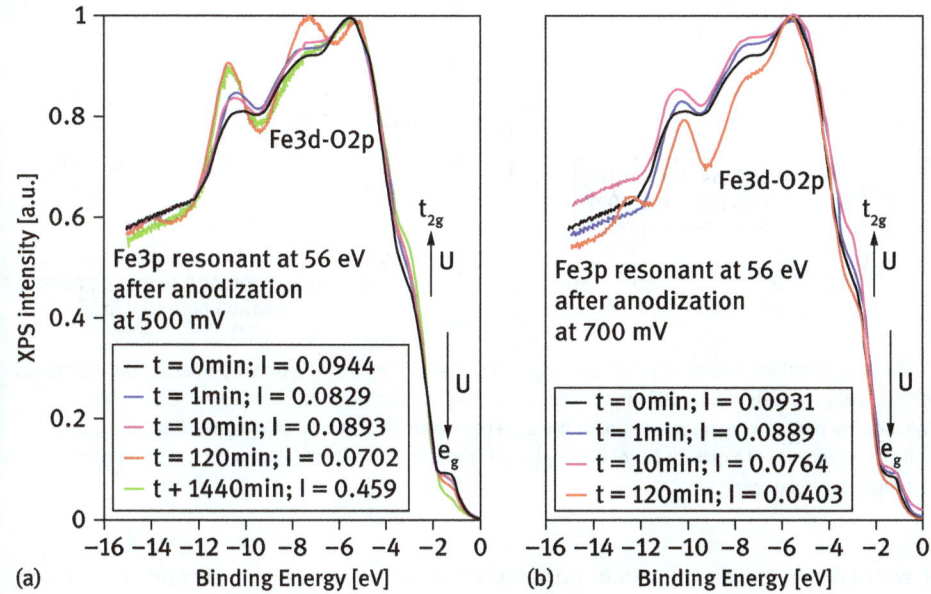

Figure 5.49: Resonant VB PES spectra of hematite measured with 56 eV photon energy. The samples were anodized at 500 mV (a) and 700 mV (b) for $t = 0$ min, 1, 10, 120, 1440 min.

We observe the same for the sample anodized at 600 (not shown here) and 700 mV. In order to better quantify this systematic behavior, we have simply measured the height of the e_g peak for the various anodization times and plotted it versus the anodization time. In anticipation that the anodization process on the electrode surface could be diffusion controlled, we plotted the peak height versus the square root of the anodization time in a Cottrell plot. For the explication of a Cottrell plot in electrochemistry, refer to Morita et al. [Morita 1987].

It turns out, as we see from Figure 5.50 left panel, that the behavior is linear in this Cottrell plot and, therefore, we have an indication for a diffusion controlled growth process of Fe^{2+} becoming oxidized to Fe^{3+}, which grows with a diffusion front likely into the electrode interior. The steeper the slope, the higher the anodization potential. From the slope, we can determine some diffusion constant "k", which we interpreted as a pre-factor for a thermodynamic activated growth process with an exponential behavior. We have plotted this growth prefactor k versus the anodization potential and forced an exponential fit through the few only three data points that we have for Figure 5.50.

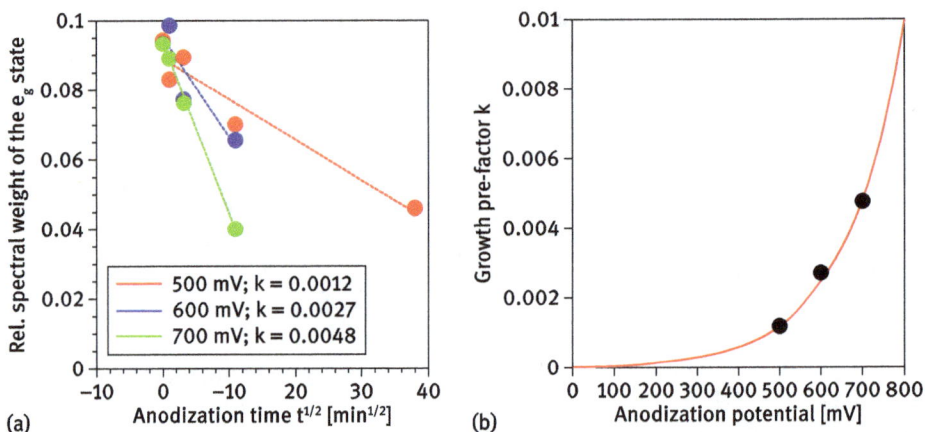

(a)   (b)

Figure 5.50: (Left) Relative spectral weight (peak height) of the e_g orbital symmetry peak for the electrodes anodized at potentials of 500, 600, and 700 mV versus $Ag^+/AgCl$ reference for times $t = 0$ to $t = 1440$ min. The data are plotted versus square root of the time t (a parabolic Cottrell plot). (Right) Slope of the Cottrell plot is considered a growth factor k and plotted versus the anodization potential and fitted with an Arrhenius-type exponential.

If we had performed this study only with laboratory-based nonresonant photon energy, we would not have identified and quantified the small yet systematic and significant changes in the spectral weight of the relevant e_g and t_{2g} peaks in the spectra.

We have another example of the great utility of soft resonant PES to show. We took another hematite photoelectrode and ran a Fe 3p resonant study. This time, we too a solar light simulator and pointed it from outside the XPS chamber at BESSY-II SurICat into the chamber onto the sample, while recording spectra. We repeated the study with the light source removed and darkened all windows on the chamber. Therefore, we had a dark and a light dependent resonant VB PES study on a photoelectrode, although without electrochemical polarization. The left panel of Figure 5.51 shows the sample glued on the sample holder and in the right panel the sample holder plus sample through the window in the UHV chamber. Figure 5.52 shows the solar simulator lamp mounted in front of the chamber, pointing the 1.5 AM solar simulated light on the sample in the chamber. In Figure 5.53, we compare the VB PES spectra recorded at 57.5 photon energy when the sample was in the dark (upper spectrum, blue) and under illumination with the solar simulator (lower spectrum, green). First of all, we notice that the overall inten-

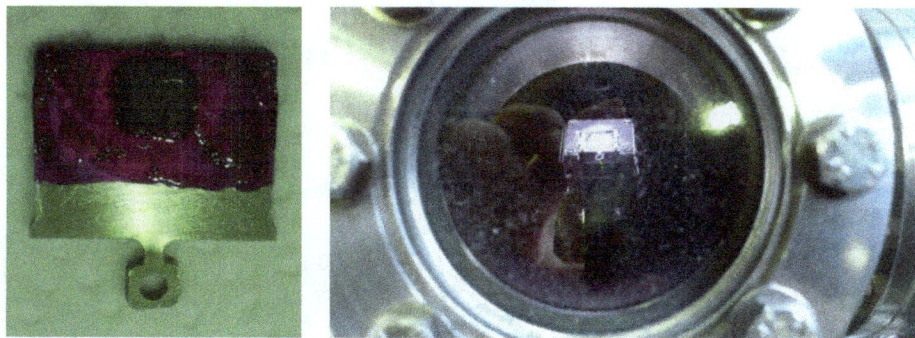

Figure 5.51: (Left) Standard sample holder from BESSY-II with hematite single crystal glued on top, partially coated with Lacomite varnish. (Right) Sample holder in measurement position in the SoLiAs end station measurement chamber at BESSY-II in Berlin.

Figure 5.52: SoLiAs beamline for resonant photoemission on iron oxide films. The solar simulator was mounted at the UHV chamber and then illuminated the sample during VB PES spectra recording.

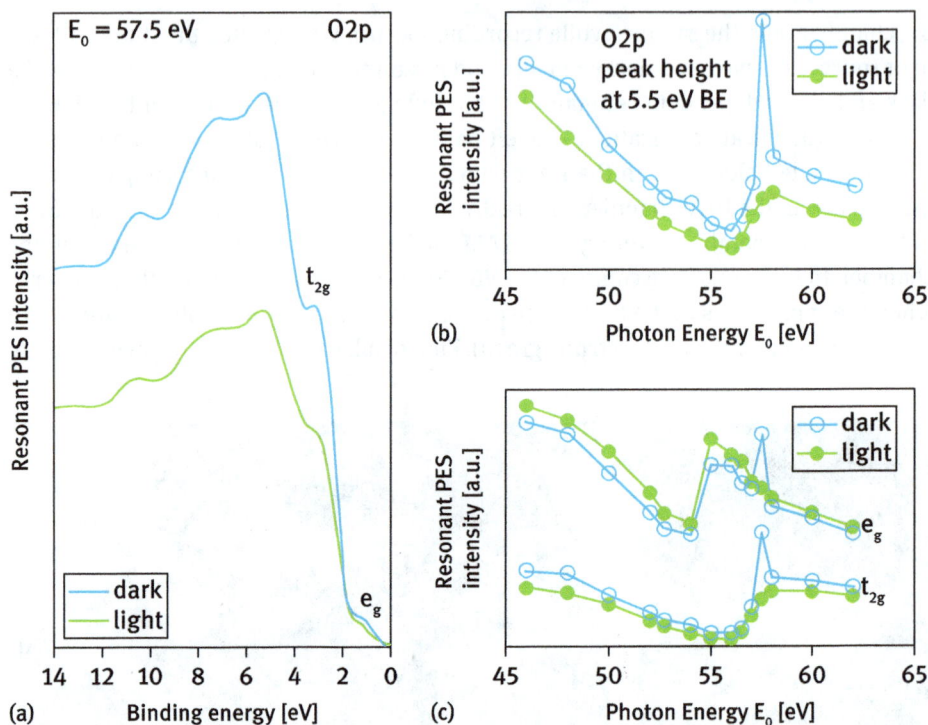

Figure 5.53: (a) Fe 3p resonant PES spectra resonant photoemission intensity at the O 2p bonding peak position, t_{2g} peak (b), and e_g peak (c) position versus the excitation energies ranging from 46 to 62 eV during dark and light condition. Note that in the dark condition, the intensity has a spike at 57.5 eV excitation energy. This spike is attributed to a light excited state upon illumination, which accounts for the photoconductivity of the iron oxide. When the material is illuminated with visible light from the solar simulator, this excited state is discharged and the spike of peak intensity disappears again.

sity is higher for the dark experiment. We made the same experiments also for the other photon energies ranging from 46 to 62 eV, thus covering the entire Fe 3p core level range. Close inspection and comparison of the heights of the O 2p bonding peak, t_{2g} peak, and e_g peak of these spectra reveals an overall parallel trend, i. e., the intensity under dark is higher than the intensity under illumination with visible light from the solar simulator. However, at 57.5 eV, we notice a sharp spike in the intensity for all three prominent structures, when the sample was dark. This spike is "quenched" when the light is turned on. Probably, the light is generating electron-hole pairs and improving the photoconductivity of the hematite, which then manifests in quenching of the state.

The solar simulator mounted to the UHV chamber shown in Figure 5.52 was not part of the end station at BESSY II. It was part of our photo-electrochemical laboratory equipment at Empa, which we have shipped in a crate from Switzerland to the BESSY II synchrotron in Berlin. The shipment included the potentiostat, solar simulator, preparation cells, PC for data acquisition, control units, vials, chemicals, and so on, everything that we needed at the synchrotron and that was not provided by the synchrotron for our experiments at that time. Often it needs a written approval by the synchrotron administration before external equipment and chemicals may be brought to the synchrotron. The clearance procedure is typically part of the proposal and beamtime scheduling procedure. When the synchrotron is in a foreign country, the equipment needs shipping documents for the customs authorities. Sometimes an A. T. A. Carnet is sufficient for this purpose.

Figure 5.54 shows a photo of a quarter segment of the experimental hall at BESSY II. As I mentioned already, some of the end stations at BESSY II are mobile and can be moved to various beamlines. Technicians, engineers, and scientists prepare the scheduled experiments in advance, based on the experiment specifics declared in the beamtime proposals.

Taking advantage of the beam was then up to the experimenters, including the provision of other infrastructure such as out electrochemical equipment. There may be even situations where no experiment chamber is available. Then the user has to provide such chamber, e. g., a UHV chamber plus an XPS spectrometer. An end station such as the one shown in the bottom photo in Figure 5.54 is basically "mobile." It can be disconnected from the beamline tubes and then moved in the experimental hall (upper photo in Figure 5.54) to another suitable beamline for some time.

As a general rule, the beam at synchrotrons, neutron sources, and other facilities is free of charge for internal and external users. This is with the provision that the scientific data produced at these facilities are for the public domain. Synchrotrons and neutron sources do welcome users from industry. If their work is confidential, the data are certainly not for the public domain. Then the facilities typically reserve the right to charge the user from industry for the costs for the operation of the beamline for their beamtime.

Not all beamlines are fully funded and operated by the synchrotron. If you have the necessary funding and when the space at the synchrotron is available, you may propose a new beamline for your own purposes and build it together with the synchrotron. You may have to hire your own staff, which operates the beamline or end station. If you do not have all the necessary funds, which typically come from your university or research lab, from funding agencies or from private foundations, you may team up with other researchers and form a consortium or a participating research team (PRT; groups of researchers from one or more institutions). The PRT construct and operate beamlines at synchrotrons and have principal responsibility for the experiment end station equipment. They are entitled to a percentage of the beamline operating time commensurate with the resources the PRT contributes to the beamline.

Some beamlines are therefore run by external groups who do not belong to the synchrotron but to a university, a government research lab, or a private company. These groups may have a particular interest in a specific use of the beamline and end station,

Figure 5.54: Experiment hall at BESSY-II in Berlin. Notice the curved architecture of the building adjusted to the storage ring technology. (Bottom) Beamline end station at BESSY-II. The metal tube on the upper right is the evacuated X-ray beamline ("Strahlrohr"), guiding the X-rays into the measurement chamber. The metal tube on the left is the sample feeding tube for manipulation of the sample position in the evacuated measurement chamber in the center. The white polystyrol container above chamber is used to hold Dewar with liquid nitrogen for cooling for low temperature measurements in the chamber.

and thus may provide extra equipment so as to provide a particular sample environment. At BESSY-II, you may find beamlines associated with the Max-Planck-Gesellschaft, with Humboldt Universität zu Berlin, or with the Physikalisch-Technische Bundesanstalt (PTB). PTB operates a sector at BESSY-II in Adlershof in Berlin and has in addition its own small synchrotron across the street in the Willy-Wien Laboratorium.

It is becoming more common that synchrotron centers invest funds in the electrochemical sample environment, and thus make it suitable for their users to not ship their own equipment to the synchrotron. Several beamlines at the ALS in Berkeley and SSRL in Stanford provide, e. g., potentiostats for electrochemical *in situ* and *operando* X-ray experiments. Many synchrotrons and neutron sources have a glove box for the chemical preparation of samples. Some synchrotrons have well-equipped chemistry labs or biology labs or film deposition equipment such as PLD chambers.

Some beamline operations are organized and scheduled that the beamtime is shared by two different user groups in parallel. This can mean that two users are invited for 2 weeks of beamtime. The users have to share this beamtime proportionally. One user can begin and measure samples for say 24 hours. It would require two or three researchers for one group to work in shifts for this beamtime. After 24 hours, the second group takes over the beam and uses it for own experiments, but maybe in a second measurement chamber next to the first chamber and coupled with the same beamline. Meanwhile, the first group can analyze their data from the first 24 hours and prepare the next sample, e. g., sample heating, sputtering, the next deposition of another film, and so on. When the second group is done with their first 24 hours of beamtime, the first group takes over and continues for the next 24 hours with measurements.

5.6.4 Lithium ion battery cathodes

In this section, we look at lithium ion battery cathodes, specifically lithium cobalt oxide and the model compound sodium cobalt oxide $Na_{0.69}CoO_2$, $Na_{0.69}CoO_2$, and $Li_{0.9}CoO_2$. This study was part of a Swiss-Polish research project on batteries for automotive application [Braun 2012i]. We have subjected the compounds $Na_{0.69}CoO_2$ and $Li_{0.9}CoO_2$ to resonant photoemission spectroscopy at the Co 3p X-ray absorption thresholds. With a precise data reduction and analysis, we were able to derive the band positions of the O 2p bonding peak and the Co3d orbitals with t_{2g} and e_g orbital symmetry. Close inspection of the refined spectra shows the existence of a minute structure in the density of states near the Fermi energy, which could be interpreted as a $3d^6L\,e_g$ hole state that likely accounts for the observed electronic conductivity of the battery electrode. The spectra were recorded at Beamline 8-1 at SSRL, Menlo Park, CA, under ultrahigh vacuum conditions as typical for conventional XPS experiments.

Figure 5.55 shows Beamline 8-1 at SSRL in Menlo Park, CA. The end station is designed for photoemission spectroscopy with soft X-rays from 15 to 175 eV. The experimental station looks less complex than some of the other beamlines shown in this book.

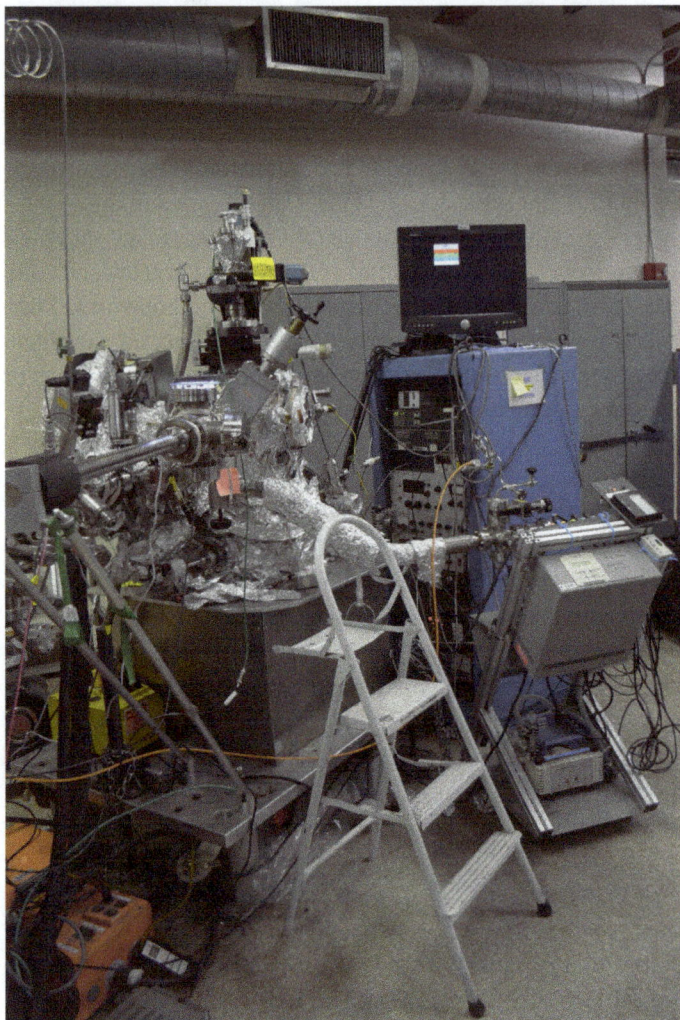

Figure 5.55: Beamline 8-1 at Stanford Synchrotron Radiation Laboratory (SSRL).

However, this is a very good instrument. The photoemission spectra were obtained under Fe 3p resonant condition. This means that we have excited the materials with X-ray energies located around the Co 3p X-ray absorption threshold. This allows for the spectroscopic enhancement of electronic structure features not only with elemental resolution but also with orbital resolution. Figure 5.56 shows the spectra of $Na_{0.72}CoO_2$ that were obtained with photon energies from 55 to 65 eV. Because no calibration for the energy axis was made, the spectra are shown with reference to the VB maximum, and not with reference to the Fermi energy. Binding energy calibration can be made by measuring the core level spectra of a gold foil, whose energy levels are certainly known.

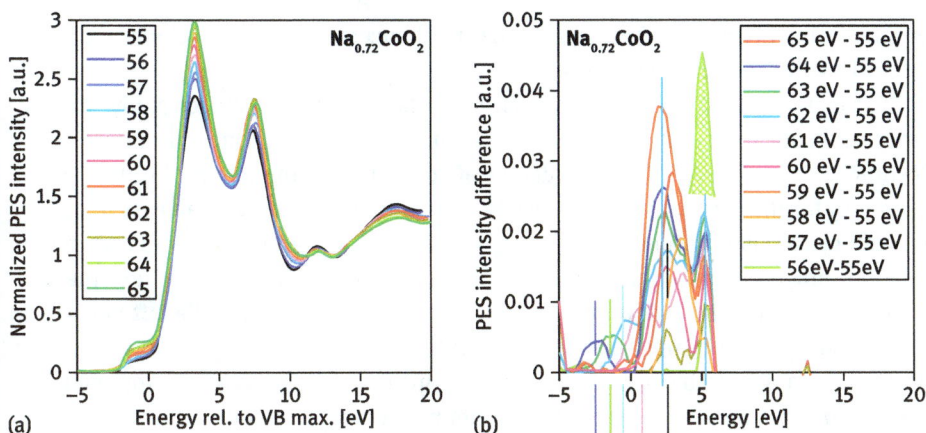

Figure 5.56: (a) Co 3p resonant photoemission spectra of $Na_{0.72}CoO_2$ recorded with photon energies from 55 to 65 eV. (b) Relative resonant PES intensities obtained by subtracting the 55-eV spectrum from all other spectra. The shift of the hump at -2.5 eV by around 1 eV steps is a spectroscopic artifact and originates not from the sample property. The peak at 5 eV binding energy is not shifting, and thus is an original Co 3p resonant peak, which is to be used in the derivation of the DOS.

Because gold is not much subject to oxidation or other chemical changes in ambient environment, its core level positions can be taken as face value for gold. The difference between the measured core level position and the tabulated core level position in literature is then the energy shift, which must be applied to the spectrum for calibration.

We notice in Figure 5.56 that the intensity of the prominent peaks change systematically with increasing photon energy. We have used the spectrum recorded at 55 eV and set the intensity in the intensity minimum at 10 eV with reference to VB maximum to unity intensity. You can consider this as a technical normalization procedure for further analysis.

The next step in the quantitative data analysis is the direct subtraction of one deliberately selected spectrum from another selected spectrum. The rationale for selecting the spectrum from 55 eV photon energy was that this one has the overall lowest intensity. The difference spectrum would therefore have no negative intensity. This is exercised in the right panel of Figure 5.56, where we have subtracted the 55-eV spectrum from the 59 eV spectrum, without any weighing factor. This method causes an enhancement of some spectral details, such as the e_g structure at around 0 eV, two sharp peaks at 3.5 and 7.5 eV, and a shoulder in the intensity minimum at before 10 eV. This has been done for all the spectra with all the different excitation energies, as is shown in the right panel of Figure 5.55. Correspondingly, the data structure becomes very complex. The position of the small peaks between -5 and $+1$ eV is shifting toward larger binding energies with decreasing excitation energy. This is a spectroscopic artifact that has no relation with the investigated material. The position of the peak at 2 and 5 eV is however not shifting, but increasing in intensity. This is a resonant enhancement effect and shows that there are peaks with Co 3p resonant peak.

The experimental work at the beamline is relatively trivial for such study. We have several specimen to study in *ex situ* condition. We are at this time only interested in the differences in the electronic structure of comparable compounds of different stoichiometry for the very same purpose of battery cathode studies. We can easy distribute the battery cathode powder on sticky carbon tape, or we can press it into soft indium foil. The samples are then mounted on the sample holder and inserted into the UHV chamber with the load lock tube, which is visible in the photo (middle left portion) in Figure 5.56. The end of this tube has a black thick cylinder which contains a magnet, which acts on a movable sample holder stick inside the tube. With this method, it is possible to move the inner stick, which contains the sample holder without and direct mechanical connection. This warrants that no hole with a movable part with a necessary sealing is required. Figure 5.57 shows Empa/EPFL PhD student Mr. Yelin Hu carrying out the experiment at Beamline 8-1. His left hand holds the black cylinder with the magnet. Mr. Hu is about to move this part forward, and thus the magnet will move the sample holder inside the tube forward as well. In the upper right photo of Figure 5.57, we look through the vacuum window at the inserted sample holder. The lower left photo in Figure 5.57 shows on the right the sample holder connected at the sample (holder) manipulator, with which we can move the sample holder up and down, left and right, and back and forward. We also can rotate the sample manipulator. This is necessary in order to move the sample on the sample holder properly in the X-ray beam and rotate it toward the detector. On the left, we see the "grip" end, the clamp of the push rod. It has a mechanism with which we can push the sample holder on the sample manipulator and also remove it later when we are done with the experiment. In the lower right panel of Figure 5.57, we are looking at the sample holder fixed on the sample manipulator. The manipulator was rotated so that the photoemission of the samples, which we see lined up in two rows, will be pointed to the detector inlet on the right side of the photo. The movements on the sample (holder) manipulator are done with micrometer gauge screws, either manually or with electric motors.

The next system we are looking at is $Na_{0.69}CoO_2$, with the spectra in Figure 5.58. I have typed in the lowest energy core level positions for the relevant constituents in the sample: Na 2p at around 31 eV, O 2s at around 41.5 eV, Co 3p around 59 eV, and Na 2s at around 63 eV. We find these energies in the literature [Thompson 2009]. In this energy range, we have to be prepared for core level intensity. The right panel of Figure 5.58 shows the off-resonant spectra recorded at 90 and 100 eV. The Na 2p is clearly pronounced. We begin with the not normalized spectra and read their peak height at the binding energies of 5.8 and 9.8. We notice that substantial changes occur in the resonant range with X-ray energies around 60 eV. The two spectra recorded with X-ray energies of 90 and 100 eV have a much lower intensity.

For the purpose of getting better insight in the spectra, it may be necessary to plot the spectra in various ways. We demonstrate this in Figure 5.59. For example, by normal-

Figure 5.57: EPFL/Empa PhD student Yelin Hu at Beamline 8-1 of Stanford Synchrotron Radiation Laboratory after preparation for resonant photoemission studies on lithium batteries, photo-electrochemical cells, and bioelectrodes. His left hand holds the load lock shift (black), which is connected to the UHV measurement chamber. The load lock tube can be attached to a glove box where the sample can be prepared in protective environment. After disengaging from the glove box, the sample is kept safe in the load lock on a clamp and can be brought to the beamline. Attached to the flange, the sample can be pushed into the chamber onto the sample holder on another clamp. The load lock stick is removed and the sample holder is rotated so that the samples can be exposed to the synchrotron X-rays, yielding photoelectrons or X-rays to the detectors. View into the measurement chamber with sample holder at sample manipulator rod (center). The rod on the left is the sample deliver/removal stick, which is used to bring the sample through a load lock onto the sample holder by firm pushing.

izing the spectral intensity to unity at the binding energy of 7.5 eV, we have confirmation that the intensity onset near the Fermi energy is not evolving homogeneous with photon energy. Hence, we have there a resonance effect. The "white line" at around 5 eV is increasing homogeneously with increasing photon energy, as indicated by the vertical arrow. When we zoom into the intensity onset region around 0 eV, we notice that this structure is not increasing homogeneously.

The highest intensity in this energy range is obtained at the photon energy of 62 eV, as indicated by the two 90° pointed arrows in the right panel of Figure 5.59. We notice that the position of the energy minimum at around 13 eV shifts toward higher binding energies, as indicated by the horizontal arrow. The right panel of Figure 5.59 shows that the intensity near the Fermi level is not simply one shoulder or one peak. Rather, it looks almost like a convolution of a shoulder with a plateau, the position of which is shifting. This intensity originates from a Co 3p-3d resonant excitation and contains interesting structural information.

8 O 543.1* 41.6*
11 Na 1070.8† 63.5† 30.65 30.81
27 Co 7709 925.1† 793.2† 778.1† 101.0† 58.9† 59.9†

E_{photon} = 100 eV

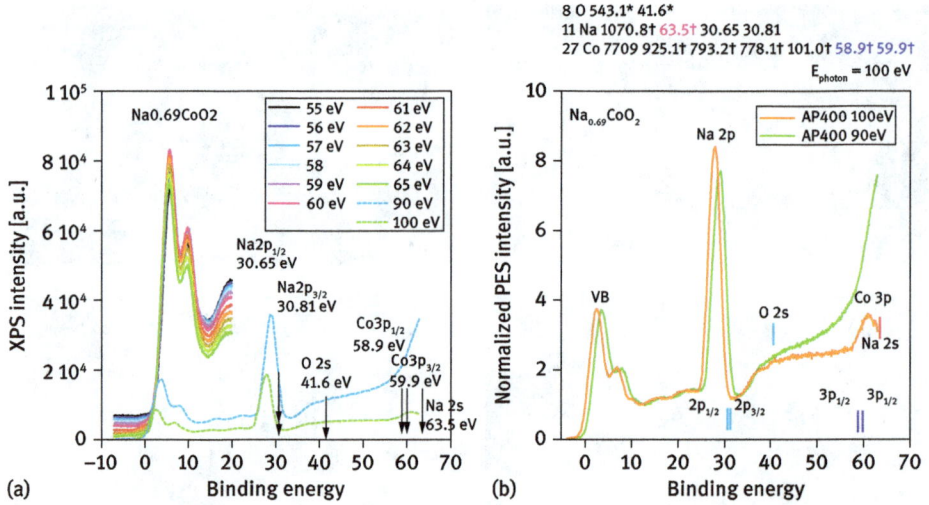

(a) (b)

Figure 5.58: (a) Comparison of PES spectra of $Na_{0.69}CoO_2$ recorded off-resonance with 90 and 100 eV photon energy, and on-resonance around 55 to 65 eV. (b) The two latter spectra are survey spectra and show the core level peaks from Na, O, and Co. The slight shift of the spectra has no origin in the materials structure.

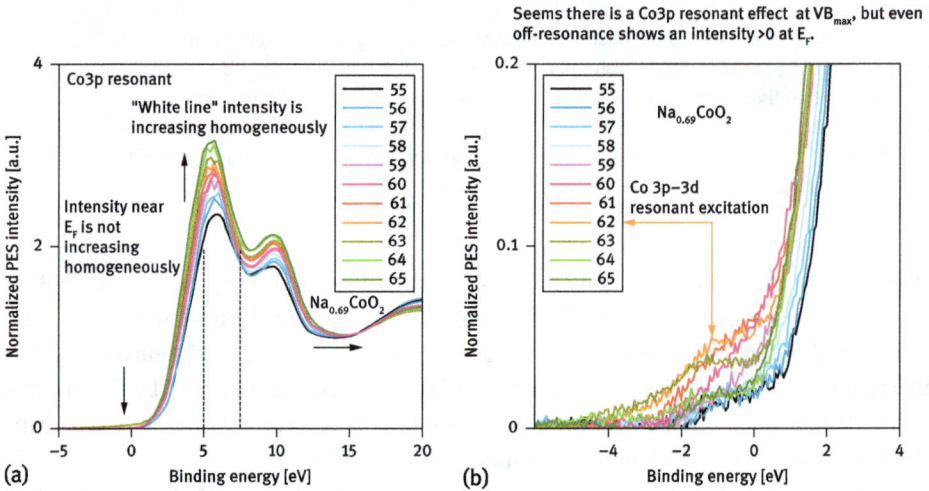

(a) (b)

Figure 5.59: (a) Co 3p resonant photoemission spectra of $Na_{0.69}CoO_2$ with photon energies from 55 eV to 65 eV. Peaks at 5 and 10 eV grow homogeneously with photon energy. Intensity at $E_B > 12$ eV seems shifting to larger E_B. (b) Magnification of the range -4 eV$<EB<2$ eV shows a "pre-edge" structure from Co 3p-3d excitation, which shows resonant enhancement around 60–63 eV.

I have plotted the determined intensities of the peaks at the binding energy of 5.8 and 9.8 eV versus the X-ray energy in the left panel of Figure 5.60. This plot helps us to select the X-ray energies where we will perform spectral intensity subtractions. It turns out that for both peaks, the maximum intensity versus the photon energy is around 60 eV. This is gratifying because then we need not make a decision among potentially two different photon energies, which serve as basis for spectra subtraction.

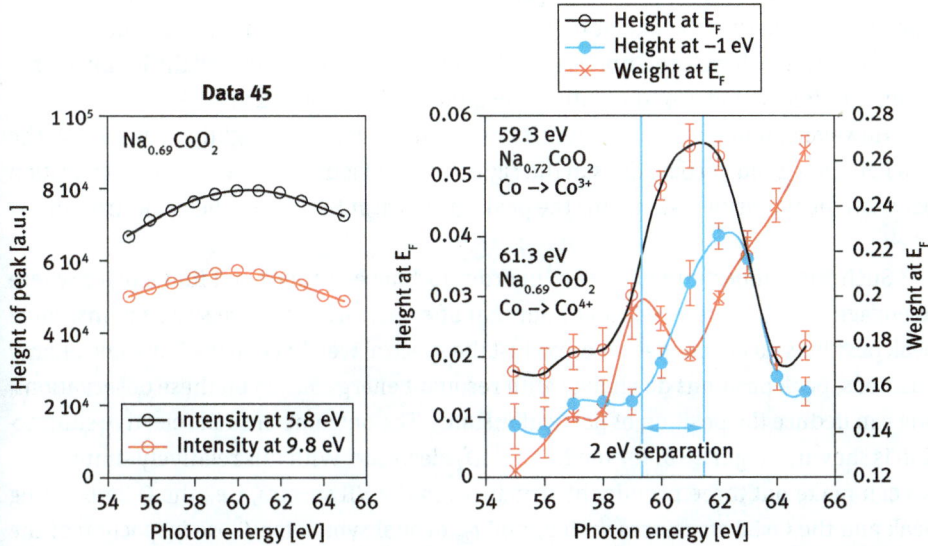

Figure 5.60: (Left) Comparison of the spectral weight variation for the two prominent peak positions at 5.8 and 9.8 eV from $Na_{0.69}CoO_2$ for photon energies from 55 to 65 eV. The highest intensities are achieved at around 60 eV. (Right) Variation of the spectral weight of peaks at EF and −1 eV from for $Na_{0.69}CoO_2$ (black and blue) and $Na_{0.72}CoO_2$ (red).

The right panel of Figure 5.60 compares the spectral intensity measured at the Fermi energy EF and at the energy position −1 eV (compare with the spectra in Figure 5.59, where a preedge peak shows up at −1 eV). We see here at which excitation energy the largest difference is located. By subtracting spectra recorded at one particular photon energy from spectra recorded at a "reference" energy, we can bring forward particular structures that are hidden in conventional PES spectra from laboratory sources.

Let us first look at the black curve, which describes the variation of the intensity from all recorded spectra shown in Figure 5.56 at the Fermi energy (resp., the VB maximum). From 55 to 58 eV, the intensity is enhanced but constant. At 59 eV photon energy, the intensity is increasing. At around photon energies of 60 eV to 62 eV, we obtain the highest intensity. Beyond 62 eV photon energy, the intensity decays rapidly.

Now we inspect the intensity of the peak plateau structure around −1 eV (blue curve). The intensity is lower than the intensity from 55 to 60 eV for the intensity measured at EF. The intensity maximum is found slightly above 61 eV photon energy. The separation between both maxima is 0.8 eV in photon energy.

Let us do the same procedure with the previous battery electrode material $Na_{0.72}CoO_2$. The data for this compound are shown by the red line in the right panel of Figure 5.60. The separation of the intensity maxima is 2 eV between 59.3 and 61.3 eV. This indicates that the cobalt in $Na_{0.72}CoO_2$ has the oxidation state Co^{3+} (should be nominal $Co^{+3.28}$), whereas the oxidation state for the $Na_{0.69}CoO_2$ is Co^{4+} (should be nominal $Co^{+3.31}$). We have therefore a quite large shift of spectral weight toward the Fermi energy when the stoichiometry of the sodium cobaltate shifts from 0.72 to 0.69.

However, an intermediate intensity maximum is found at slightly above 59 eV. The two light blue solid vertical lines in the right panel of Figure 5.60 show a 2-eV separation between the intensity maxima for the peak structure at EF and for the peak structure at −1 eV.

Such "prestudies" are necessary in order to "move around" and find a place where to engage with the spectra. It turns out that after performing these subtractions, some peak positions do not change whereas just the spectral weight changes. For some energy positions, peak positions do change with resonant energy. Based on these observations, one can deduce the position of particular states. The outcome of such selective subtraction is shown in Figure 5.61 for the $Na_{0.69}CoO_2$ electrode. With this relatively simple way, we can make out three prominent structures in the VB spectra, i. e., the O 2p bonding peak and the Co 3d structures with e_g and t_{2g} orbital symmetry. Close inspection of the subtracted spectra shows a prevalent intensity at around +1 eV, which we can interpret as a defect state, possible with the notation $3d^6L$ for a ligand hole state.

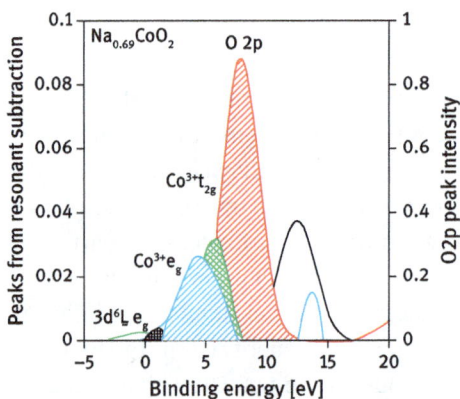

Figure 5.61: Resonant subtracted density of states from $Na_{0.69}CoO_2$. The blue shade area is the result of subtraction of the 60-eV spectrum minus the 56-eV spectrum. The green shade area is the result of the 64-eV spectrum minus the 56-eV spectrum.

It is interesting to notice that in the standard literature on battery research only few papers consider the electronic structure of electrodes as it is known from condensed matter physics. Condensed matter physics provides very good models for the description of electrode materials, including the very important metal oxides. However, these models are rarely used by battery researchers. One of the few papers that takes explicitly reference to the density of states is found by Tarascon et al. [Tarascon 2004], who sketch the DOS of the $Li_{1-x}CoO_2$ system, as shown in Figure 5.62. It shows filled relatively broad and wide oxygen sp orbitals in the neighborhood of a partially filled cobalt 3d state—with t_{2g} orbital symmetry and e_g, respectively. The e_g orbitals are somewhat broader and less intense than the t_{2g}. They are empty and also more remote from the Fermi level. Depending on the lithium concentration in the electrode, the Fermi level moves through the Co $3d$ t_{2g} state, while the hybridization between the Co $3d$ t_{2g} is changed as well. The intuition of the chemist is correct, but it would be nice if we had direct empirical evidence. I have shown this evidence with the lithium cobaltate materials in this section.

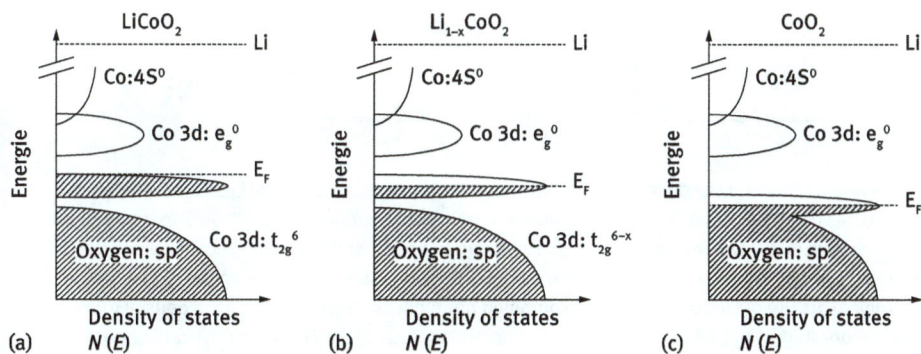

Figure 5.62: Schematic of the band structure evolution of $LiCoO_2$ as Li is removed from the material; the main feature being the merging of the Co 3d orbitals and O 2p orbitals and the corresponding shift of the Fermi level. Reproduced from Tarascon et al. by kind permission from RSC [Tarascon 2004].

Figure 5.63 shows a recently published version [Ensling 2006, Ensling 2010] for the experimental determination of the electronic density of states (DOS) of the lithium battery positive electrode material $LiCoO_2$ based on oxygen NEXAFS and VB PES spectra, along with a calculated DOS based on DFT+U methods. To the best of my knowledge, this is the first paper where a real DOS of a battery is shown. The NEXAFS spectrum is transformed with the PES spectrum on a common scale for ease of comparison. The agreement of experimental and computed spectra is gratifying.

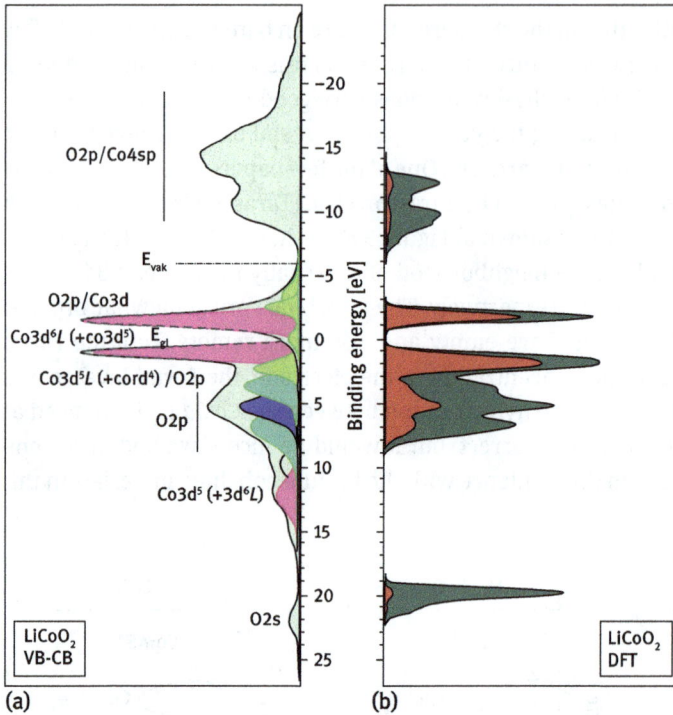

Figure 5.63: Comparison of the experimentally obtained electronic structure of $LiCoO_2$ and the density of states (DOS) based on DFT + U calculations. For this illustration, the O 1s NEXAFS spectra and the VB PES spectra ($h\nu$ = 175 eV) were aligned on a common energy scale. [Ensling 2006, Ensling 2010]. Reprinted this figure with permission from Ensling D., Thissen A., Laubach S., Schmidt P. C., Jaegermann W.: Electronic structure of $LiCoO_2$ thin films: a combined photoemission spectroscopy and density functional theory study. Physical Review B 2010, 82. Copyright (2010) by the American Physical Society." http://journals.aps.org/prb/pdf/10.1103/PhysRevB.82.195431.

5.6.5 Resonant inelastic X-ray scattering

We have seen that we can concentrate on particular features in spectra by varying the excitation energy. Practically, we do this by collecting scans over a particular energy range while we vary the excitation energy. We measure in this case the energy change and the momentum change of the scattered photon. This is typically a photon-in–photon-out experiment that can be done with hard X-rays and with soft X-rays. The method is called resonant inelastic X-ray scattering (RIXS) and has become increasingly popular following the seminal work by Gelmukhanov and Agren [Gelmukhanov 1994]. Frequently, RIXS is also called resonant X-ray Raman spectroscopy. Caliebe et al. [Caliebe 1998] investigated hematite with Fe K-edge XANES and 1s2p RIXS and supported their experimental data with mathematical modeling. They show that the double peak in the pre-edge of the Fe K-edge XANES spectrum originates from the crystal field splitting in a cubic symmetry setting. This double peak has components from orbitals with e_g and t_{2g} symmetry, whose intensity in the RIXS spectra is enhanced when the excitation energy is tuned across the absorption threshold.

RIXS can be used for the investigation of a manifold of energy relevant systems, such as metal oxides, coordination compounds, and even complex metaloproteins. Glatzel et al. studied Mn oxides, coordination compounds including PS II with RIXS [Glatzel 2004, Glatzel 2005]. An overview of the method including the instrumentation is provided in the conference paper by Glatzel et al. [Glatzel 2009]. Although this book shows already the complexity of experiments when combining electrochemistry with X-ray (and neutron) methods, the latter becomes very complex by themselves, e. g., by combining XAS and XES and RIXS and thus deserve new protocols for wise application of these methods [Glatzel 2013].

An early soft X-ray Raman study was performed by Kuiper et al. [Kuiper 1998]. Liu et al. outline in their perspective paper the potential of RIXS for catalysis studies [Liu 2015]. An *in situ* reaction cell for catalysis studies with soft X-ray RIXS, e. g., for oxygen, is provided by Kristiansen et al. [Kristiansen 2013]. Guo et al. have studied the RIXS of nitrogen in Deoxyribonucleic acid (DNA) duplexes [Guo 2010]. Battery materials [Augustsson 2004, Augustsson 2005] and hydrogen storage materials [Guo 2006] were investigated with RIXS, for example.

A very rare case of ultrasoft RIXS with energies below 100 eV can be found in Wang et al. [Wang 2013a], who studied the M2,3-edge RIXS on NiO, two NiF_2, and two other covalent complexes, in which they observed different d-d transition patterns.

Figure 5.64 shows seven M-edge RIXS spectra of NiO, NiF_2, $NiCF_2$, and $NiCF_3$ with labeling letters a–f. The spectra a–c from NiO show two peaks at −1.1 and −1.8 eV. There is a weak feature at −3 eV when excited with 66, 68, and 70 eV photon energy. These peaks originate from the d-d transitions of Ni inside the NiO coordination cage. At different excitation energies, the NiO M-edge RIXS reproduce the d-d transition peaks, although minor differences are noticeable. We anticipate that in samples with more structural complexity different excitation energies could produce slightly different RIXS features. Therefore, the RIXS method might provide better selectivity and deeper understanding of the specimen when measured in comparison to conventional XAS.

Wang et al. observed also the RIXS on NiF_2, $NiCF_2$, and $NiCF_3$ (spectra d, e, and f, respectively). For these spectra as common ground for reference, the photon excitation energy of 70 eV was used. Nickel in NiF_2 is a high spin Ni(II) ion with a structure similar to NiO. Its RIXS feature shows the d-d transitions at 0.8 and 1.5 eV, unlike NiO. The difference in the RIXS between NiO and NiF_2 is not surprising. In complexes with comparable ionicity and geometry, the d-d structure can show differences.

Ni(II)CF_3 is a covalent complex. It shows virtually no RIXS feature; see spectrum (e). The absence of RIXS features is consistent with its low spin Ni(II) nature. It usually has a large d-d transition energy difference and could have no feature in the probed region of

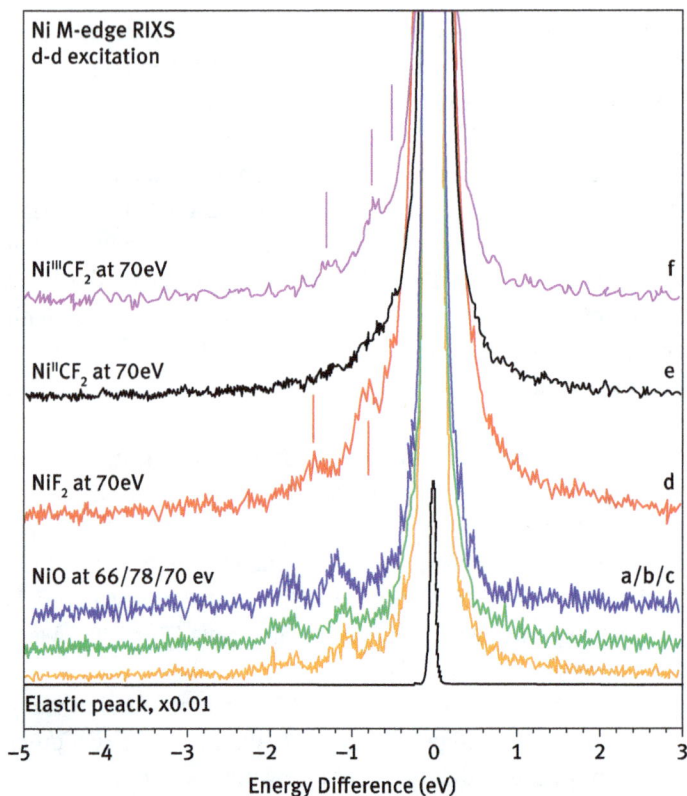

Figure 5.64: $M_{2,3}$-edge RIXS (the energy difference spectra) of NiO, excited at 66 (a, yellow), 68 (b, green), 70 eV (d, blue); and of NiF_2 (d, red), Ni(II)CF$_3$ (e, black), and Ni(III)CF$_3$ (f, purple) at 70 eV. The black curve at the bottom illustrates the elastic peak position (x0.01). Vertical bars correspond to the possible d-d transition peaks. Reproduced from [Wang 2013d] with permission of the International Union of Crystallography. (http://journals.iucr.org/).

several eV. Ni(III)CF$_3$ is also a covalent complex, yet it shows a feature at 0.7 and 1.3 eV. As Wang et al. put it:

> "The marginal spectral features of $M_{2,3}$-edge RIXS on this covalent complex provides hope that future RIXS studies could be extended to more biologically relevant complexes if beamlines and spectrometers continue to improve. It also motivates the exploration of more sensitive and more effective detection methods for a wider application."

5.6.6 Nuclear resonance vibrational spectroscopy

A not so well-known and new emerging technique is nuclear resonance vibrational spectroscopy (NRVS) at synchrotrons [Wang 2021]. This is a synchrotron-based vibration spectroscopy method, which requires Mössbauer typical isotopes, such as ^{57}Fe. You excite the sample with an X-ray energy that is equal to a nuclear Mössbauer energy level, such as 14.4 keV in the case of iron. The thermal vibrations of the Mössbauer element cause then their spectroscopic signature in the X-ray response. Changing the molecular environment of this Mössbauer element adds an additional spectroscopic influence, which can be interpreted by comparison with the pristine sample. Adding protons or deuterons can have already an influence.

The benefit of NRVS is that you obtain vibration spectra, like from infrared and Raman spectroscopy. But the NRVS spectra are element specific. You cannot have this with the conventional vibration spectroscopy methods. Hongxin Wang in the Cramer group and colleagues used this NRVS to study the Fe site in the Fe-S cluster-free hydrogenase from *Methanothermobacter marburgensis* [Guo 2008]. They also looked into the nitrogenase protein with, which is important for nitrogen fixation. The title of their manuscript, which appeared in JACS [Xiao 2006], begins with "How nitrogenase shakes - ..." This sounds certainly exciting.

I have become interested in the vibration structure of materials and molecules, and their influence on the charge and mass transport and how those overcome kinetic barriers. Therefore, I wanted to look into our ceramic proton conductors with NRVS and teamed up again with my colleagues from the Cramer group. The problem is that my BCY and BZY proton conductors do not contain Mössbauer elements. I therefore had to find a materials system of proton conductors that would fit. $BaSnO_3$ has been investigated for proton conductivity, and so went my choice and I got beamtime at SPring8

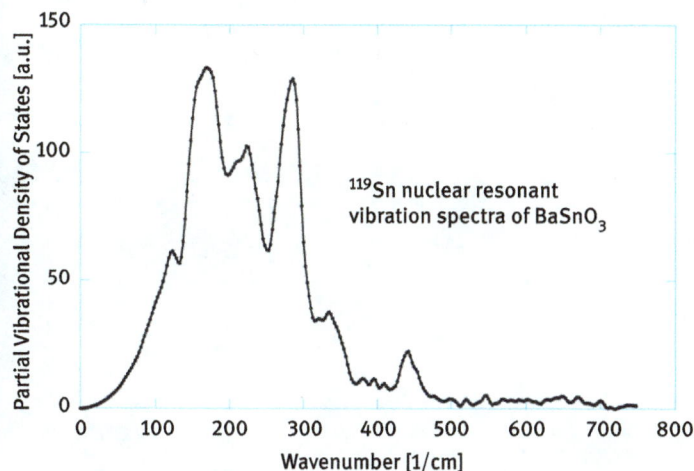

Figure 5.65: Vibrational density of states ("vibration spectrum") of $BaSnO_3$ with isotope ^{119}Sn obtained from NRVS spectra with 23.88 keV transition energy at BL 35XU at SPring8 in Japan; Artur Braun (Empa), Yoshitaka Yoda, Nobumoto Nagasawa (SPring8), Hongxin Wang, Stephen P. Cramer (SETI), Leland B. Gee (SSRL).

in Japan 2022. [119]Sn tin is a Mössbauer isotope and is suitable for NRVS. The excitation energy for [119]Sn is 23.88 keV.

A particular protocol is required for the conversion of NRVS spectra to the vibration density of states. Note that the vibration density of states from NRVS is specific to the relevant isotope. Figure 5.65 shows the such obtained vibration density of states of tin in the $BaSnO_3$ pellet versus the wavenumbers.

Given that for this method isotopes are required, the samples may become quite costly. You must purchase the isotopes form a specialized vendor and then often prepare the samples entirely by yourself.

6 Surface-sensitive and volume-sensitive methods

The purpose of this book is to showcase the utility of X-ray methods for analyses of electrochemical energy systems. It is not so straightforward to tell apart the interrelations between structure and functionality in this context. Chemical reactions in electrochemical systems typically take place at the electrode surfaces. For intercalation batteries, of course, an ionic front propagates from the surface into the electrode interior and the reactions take place therefore also in the bulk of the electrode. The electrode is also an *absorber*, and the charge capacity of the battery scales linear with the volume of this electrode (charge absorber) so to speak, whereas historically, electrochemistry on electrodes is considered surface science.

The next thing is that in electrochemical energy systems, double-layer capacities may play a role, although these are not considered due to electrochemical reactions. Rather, they are of an electrostatic nature. They do not scale with the bulk volume of the electrode but with the internal surface area of the electrode. Yet, because the internal surface area scales with the volume of the porous electrode, this surface sensitive property is in practice then a volume effect.

This chapter will address the difference between surface-sensitive and bulk-sensitive measurements and methods. Surface and bulk is an oversimplification, which we adopt from simplistic sketches of an ideal single crystal that is typically used for conceptual purposes, such as for the introduction of Tamm surface states [Tamm 1932]. However, surface and bulk do have an *"in between,"* which can extend from the nanometer to the submicrometer range in the interior of the material. In other chapters in this book, I have shown how, e. g., the XPS spectra of metal oxides may vary despite being recorded from the same sample. The penetration depth of the X-rays varies with the X-ray energy, and the escape depth of the probed electrons or photons is also a function of energy. I have met many researchers, junior and senior, who could hardly handle this apparent discrepancy. This is not an analytical *problem*. The opposite is true. It is an analytical *opportunity* for making depth resolved materials studies. For electrodes and for electrolytes, this is a very important issue. In semiconductors, this can be the charge carrier accumulation or depletion layer. We have also seen that the oxygen substoichiometry of metal oxides is often concentrated at or near their surface [Chapters 2.3 and 3.5]. Then, e. g., the surface of iron oxide or tungsten oxide is terminated with more reduced cation species, and the average oxidation state gradually increases the bulk toward the nominal oxidation state.

A conventional laboratory-based X-ray diffractogram recorded with Cu Kα radiation of 1.54056 Å wavelength and X-ray energy of 8027.84 for Kα1 and 8047.82 for Kα2 has an X-ray attenuation of around 8–9 micrometers for Fe_2O_3 with 5.24 g density. A ceramic pellet of, say, 5 mm thickness will therefore be probed only at the top 10-micrometer surface. To a surface scientist, a layer of 10 micrometers is already the bulk. The subsequent phase analysis with the diffractogram is only representative of these few micrometers. The X-ray diffractogram does not show what is underneath this thick top layer. This is

https://doi.org/10.1515/9783110794038-006

one of the reasons why often a very thin fine dust layer of powder from the sample is sufficient for obtaining a diffractogram. You can go and scrap with a doctor blade powder from the sintered pellet surface and make the first powder XRD sample, and then you dig deeper in the pellet and scrape another layer from the pellet from the deeper region. A simple ruler or microgauge may help with measuring the thickness level. With one or two handful of such obtained powder samples, it is thus possible to make a simple depth resolved X-ray phase analysis of the pellet. Certainly, using the blade means we are employing ultimately a destructive method, although the use of X-rays *per se* is not considered a destructive method.

When we increase the X-ray energy—by whatever means—to 30,000 eV, then the attenuation depth is around 300 micrometers. Even with this extremely high X-ray energy (truly hard X-ray range from my perspective but there are engineers who use 100 keV and above), we would not be able to probe 1 mm of the 5-mm disk. Likely, even if we could penetrate the entire 5-mm ceramic disk with X-ray, and thus perform a transmission XRD experiment, the diffractogram would be a convolution of the potential structural gradient that extends over the entire sample. This becomes particularly obvious when doing a Rietveld structure refinement on the data. It then turns out that the diffractogram cannot be fitted with one crystallographic phase only. This gives a hint that should be taken seriously.

We have looked into a $(Sc_2O_3)_x$ $(ZrO_2)_{1-x}$ solid electrolyte (6ScSZ, 6 % scandium stabilized zircon oxide) with neutron and X-ray diffractometry. The samples had been subjected to an ageing study and compared with the conductivity change over time. This was part of a research project on solid oxide fuel cell degradation [Atkinson] funded by the European Commission (SOFC-Life). Figure 6.1 shows the evolution of the area serial resistance (ASR) of a Ni-CGO anode (top, blue) and the 6ScSZ electrolyte (bottom, red). The resistance is homogeneously increasing over time for both components.

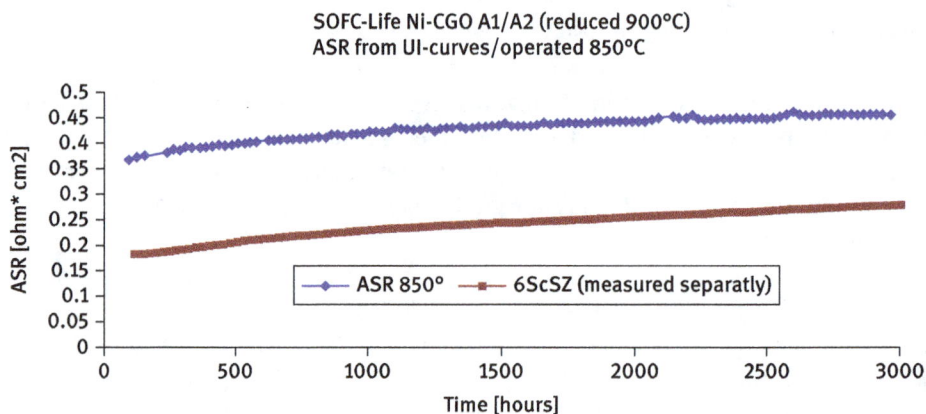

SOFC-Life Ni-CGO A1/A2 (reduced 900°C)
ASR from UI-curves/operated 850°C

Legend: ASR 850°   6ScSZ (measured separatly)

X-axis: Time [hours]
Y-axis: ASR [ohm* cm2]

Figure 6.1: Area serial resistance (ASR) of a Ni-CGO SOFC anode (top blue curve) and a 6ScSZ electrolyte (bottom, red). Data courtesy of Dr. Tzu-Wen Huang, Empa.

X-ray diffraction was the first method of choice for structure analysis, but we also had neutron diffraction available. The comparison provided an interesting insight. Figure 6.2 shows the diffractograms of 6ScSZ, which was aged for various times. The diffractograms were recorded with Cu Kα X-rays (bottom in left panel) and with neutrons (top in left panel). The phase composition differs in the different aging samples. After 1250 hours of aging, the amount of tetragonal phase grows as cubic phase decreasing, which is due to a local ordering effect toward the formation of the tetragonal phase in the cubic matrix, makes this material undesirable for a SOFC. For this, we found that we need more neutron beam time to investigate the phase information and short range ordering the effect of oxygen on this topic.

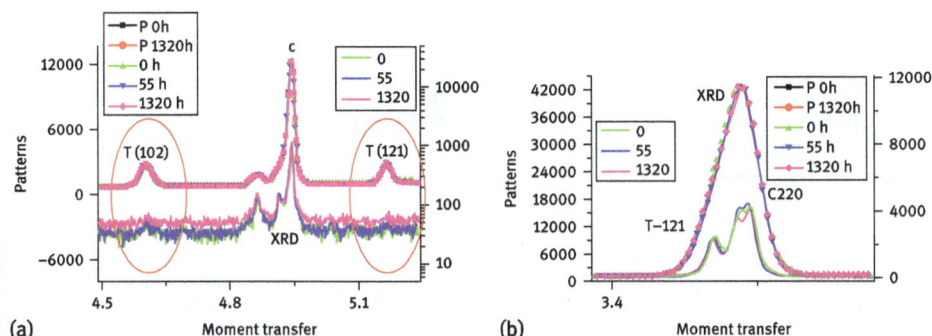

Figure 6.2: Comparison of X-ray (bottom left diffractograms) and neutron (top right diffractograms) diffraction data for the 6ScSZ SOFC electrolyte layer model compound. The right panel shows magnified a neutron diffraction peak on top and the XRD peak underneath. The green curve is form the pristine 6% Sc substituted zircon oxide (6ScSZ). The blue curve is from the 55-hour-aged sample. The purple curve is from the 1320-hour-aged sample. Samples and data courtesy of Dr. Tzu-Wen Huang, Empa.

Let us look at the peak at $3.5\,\text{Å}^{-1}$ momentum transfer in the right panel of Figure 6.2. The XRD method (three overlapping lower curves) resolves this one as a nice double peak with tetragonal phase (−121) and cubic phase (121) indexing. Now observe the ND peak above. It is very broad and shows a convolution of the two aforementioned peaks. We would guess at first glance that the X-ray diffractometer had a better resolution than the neutron diffractometer, simply because the X-ray diffractograms show sharper peaks. But this is not the case. Rather, the X-ray beam probes only the 10-micrometer top surface layer of the sample, whereas the neutron beam probes the entire sample thickness, including the 10-micrometer top surface layer of the sample.

Note that the neutron scattering factors do not depend on the diffraction angle. Figure 6.3 shows the diffractograms of the 6ScSZ sample over the wide range from 40° to 100°. The blue curve is the neutron diffractogram and the green curve the X-ray diffractogram. The top panel shows with the blue dashed line the variation of the neutron scattering factors with diffraction angle. It is constant over 2θ.

(a)

(b)

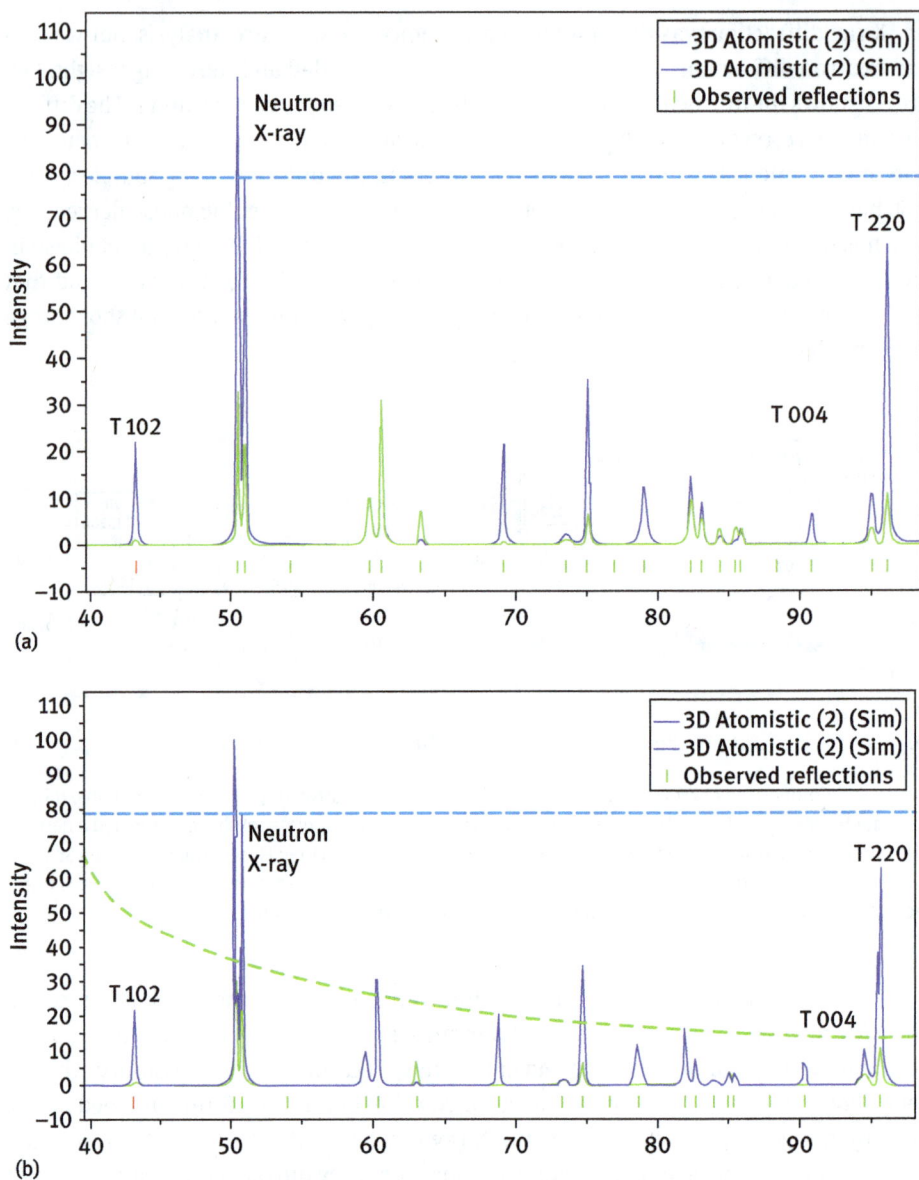

Figure 6.3: Neutron diffractogram (blue) and X-ray diffractogram (green) of $(Sc_2O_3)_x$ $(ZrO_2)_{1-x}$ solid electrolyte (6ScSZ, 6% scandium stabilized zircon oxide). The thick dashed lines indicate the variation of neutron and X-ray scattering factors with the diffraction angle 2θ. Data courtesy of Dr. Tzu-Wen Huang, Empa.

In the bottom panel, we see the two diffractograms plus the variation of the X-ray scattering factors over 2θ. It is rapidly decaying with the diffraction angle.

We have now compared two very similar structure determination methods. Another comparison is made when you run an XRD and make the phase analysis, but the

XPS spectrum tells you something different. We first have to realize that XRD investigates the crystallographic structure for determination on the unit cell. X-ray spectroscopy is not designed for this task. As we know by now, with XPS or PES, we can determine the elemental composition at the surface, including the oxidation state and further molecular structure information, such as hybridization. When the XRD tells us that the sample is hematite α-Fe_2O_3 with Fe^{3+} iron oxidation state, it is possible to detect strong Fe^{2+} signatures in the PES spectra of the same sample. This is because of the oxygen deficiency at metal oxide surfaces.

Nuclear magnetic resonance (NMR) spectroscopy is considered a bulk-sensitive method. Similar to neutron diffraction, NMR needs relatively large quantities of material for one experiment. Similar to Mössbauer spectroscopy, NMR is a method with extremely high energy or wave number resolution, and thus well suited for molecular structure measurements. It is therefore well established in analytical chemistry and also in some fields of materials science. We used NMR for the study of lithium ion battery cathode materials [Tucker 2000, Tucker 2002a, Tucker 2002b, Tucker 2002c, Tucker 2002d]. In the course of the study of cycled battery cells, some findings in comparison of cycled and uncycled electrodes were not conclusive. As we had an opportunity to look at these and other samples with hard and soft X-ray spectroscopy, we conducted a Kβ study at the Advanced Photon Source (APS) Bio-CAT and at SSRL, and a NEXAFS study at the Advanced Light Source. The studies were done on powder samples, which were spread over Kapton® tape. We have seen this study already in Chapter 2.2 in this book.

At the time when this study was conducted, in the late 1990s, only the cycled and noncycled electrodes were usually investigated. But we chose to look at the materials during all stages of synthesis and processing and operation in order to understand the influence of these steps on the electronic structure. We therefore looked at the $LiMn_2O_4$ powder after synthesis, then a dry electrode that had been casted on aluminum foil also contained carbon, graphite, and binder, and such electrode soaked in $LiPF_6$ dissolved in dimethyl carbonate (DMC) battery electrode, and three electrodes cycled for 1, 10, and 100 times.

The top row in Figure 6.4 shows Kβ photoemission spectra of one cycled and one uncycled battery $LiMn_2O_4$ cathode. Even the magnification of the $K_{\beta 1,3}$ peak in the right panel shows hardly any differences in spectra. However, when forming the first statistical moment $\langle E \rangle$ of the peak, the values differ from each other systematically. This first statistical moment defines the center of gravity of the spectral weight and can be used as a metric for the chemical shift, which in turn can be translated into formula units for the oxidation state of the probed element.

$$\langle E \rangle = \frac{\int_{6485\,eV}^{6459\,eV} I(E) \cdot E \cdot dE}{\int_{6485\,eV}^{6459\,eV} I(E) \cdot dE}$$

We can apply the same formula for obtaining $\langle E \rangle$ for the L-edge NEXAFS spectra, which are shown in the bottom panel of Figure 6.4. Here, we do note substantial differences

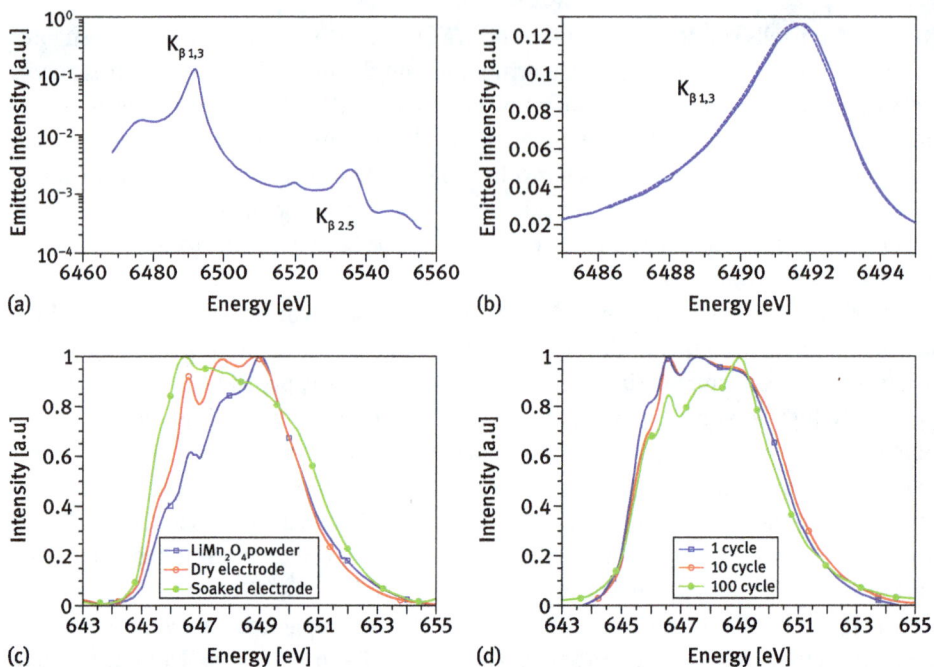

Figure 6.4: (top)Manganese Kβ emission spectra of a cycled and uncycled electrode. The right panel shows the magnified manganese Kβ1,3 emission spectra of a cycled (dashed line) and uncycled. (bottom) Mn 2p NEXAFS spectra of the uncycled electrode materials (left) and 1, 10, and 100 times cycled electrodes (right).

in the spectra of the samples at first glance. We now can compare the chemical shift of the Kβ and NEXAFS spectra, when we plot them versus the number of battery cycles, as shown in Figure 6.5. LiMn$_2$O$_4$ powder, dry electrode, and soaked electrode are materials from noncycled state, and thus are assigned the cycle number "0." Shift toward higher $\langle E \rangle$ means reduction of the Mn. Obviously, the Mn in the spinel powder becomes

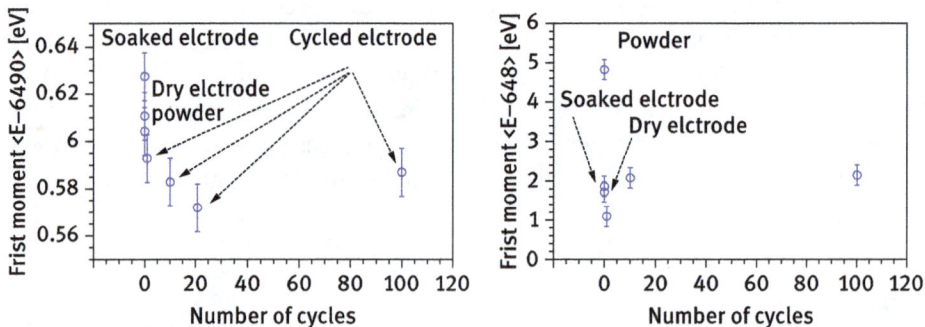

Figure 6.5: Comparison of the first moment variation with electrode processing and cycling for hard and soft X-ray spectroscopy. (left) First moment $\langle E \rangle$ of electrode samples as obtained from manganese Kβ spectra. (right) First moment $\langle E \rangle$ as obtained from Mn 2p NEXAFS spectra.

reduced when it is mixed with carbon and binder during the electrode processing and more reduced when it is soaked with the organic electrolyte. This is not surprising, as carbon and organic electrolyte are soft reducing agents. Upon cycling, $\langle E \rangle$ decreases, which means the Mn becomes oxidized. The changes in $\langle E \rangle$ are more pronounced for the surface sensitive NEXAFS spectra than for the bulk sensitive Kβ spectra.

We have here the spectral signatures from two different physical processes: one from the X-ray emission and one from the X-ray absorption, hence probing unoccupied and occupied states, respectively. But the information on the valence state of the particular element under consideration is the same.

Let me briefly give the theory of the Kβ emission process in the paper published by Glatzel et al. [Glatzel 2001]. Upon photoionization, such as photoabsorption, or following a capture process, an excited *intermediate* state $1s3d^5$ is created from the $1s$ electron state. This process is paralleled by the creation of the $1s$ hole, which in turn will be filled by a $3p$ electron. The filling by this $3p$ electron is accompanied by the emission of the Kβ fluorescence lines, as shown in Figure 6.6. Filling the $3p$ electron leaves a *final* state $3p^53d^5$ in the $3p$ shell. The intermediate state $1s3d^5$ and the *final* state $3p^53d^5$ are subject to the Coulomb interaction in their molecular environment as well as to the spin-orbit coupling. These two interactions cause a splitting of the energy levels into sublevels, which are spectroscopically observable. When we look at Mn, which has the electron structure [Ar] $3d^5\,4s^2$, we see Mn^{2+} has the structure $3d^5$ with the $3d$ shell half-filled. This ion has ^{6}S symmetry and there is *no* spin-orbit coupling. During the aforementioned *intermediate* state, there is a coupling of the $1s$ hole with the $3d$ electrons, which makes ^{5}S and ^{7}S terms that are split and separated by the exchange interaction. The $3p$ hole is coupled with the $3d$ electrons in the *final* state.

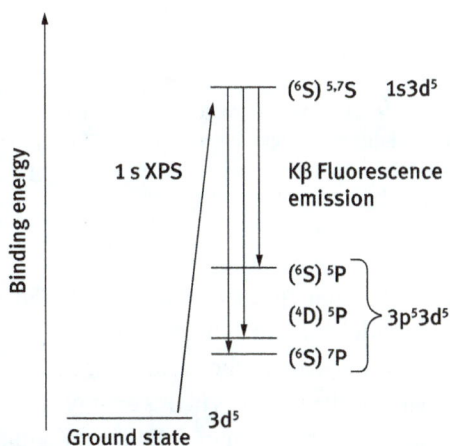

Figure 6.6: Schematic for the 1s photo-ionization and Kβ fluorescence emission. Energy terms parentheses stand for the 3d parent. Only the three final states with the strongest intensities are shown (Figure adapted with permission from [Glatzel 2001]). Copyrighted by the American Physical Society.

6.1 X-ray Raman spectroscopy

It would be interesting if we had either surface-sensitive Kβ method to compare with the conventional Kβ or a bulk-sensitive NEXAFS method to compare with the conventional surface sensitive soft X-ray NEXAFS method. We probably would favor soft X-ray spectra over hard X-ray spectra because the energy resolution in soft X-ray experiments is typically better than in experiments with hard X-rays. The left panel of Figure 6.7 illustrates how spectral features may come out better in a NEXAFS spectrum than in a hard X-ray XANES spectrum [Cramer 1991]. Cramer et al. have recorded a Mn K-edge XANES spectrum around 6500 eV and rescaled it on the energy axis to match it with the energy axis of a soft X-ray L-edge spectrum of manganese. In older literature, you can sometimes observe how the absorption edge position in a K-edge spectrum; let us say E_E = 6540 eV is arbitrarily set to $E_{E'}$ = 0 by subtracting the value of E_E from the entire energy column in the spectrum. This is a simple coordinate transformation, which facilitates comparison of different spectra.

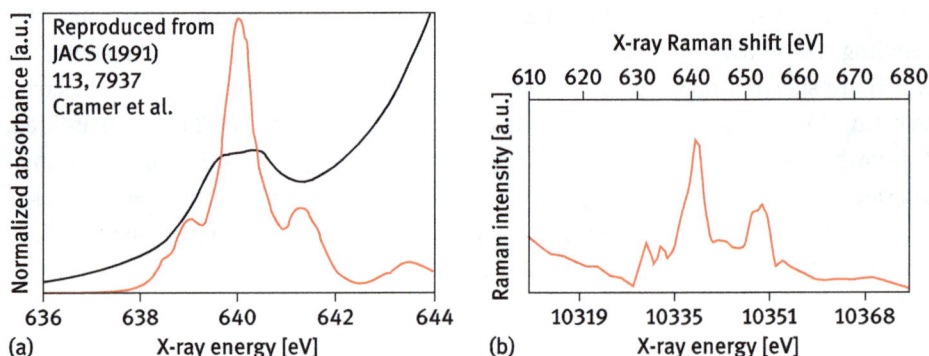

Figure 6.7: (a) Comparison of K-edge (black diffuse spectrum) and L3-edge (red peaked spectrum) of $MnCl_2$, as digitally reproduced from [Cramer 1991]. Spectra are arbitrarily aligned according to [S. P. Cramer et al., Ligand field strengths and oxidation states from manganese L-edge spectroscopy, J. Am. Chem. Soc. 113 (21) (1991) 7937–7940.]. (b) X-ray Raman spectrum of Mn in the lithium battery studied in this paper, recorded around 10,300 eV, which includes the Mn L2,3 spectrum. The top axis shows the corresponding energy range for a typical soft X-ray NEXAFS spectrum.

The right panel of Figure 6.7 shows part of an emission spectrum obtained at SSRL with 11,000 eV X-ray excitation energy from an operating lithium battery cell with a manganese oxide-based positive electrode. The part of the spectrum shown in Figure 6.7 extends from around 10310 eV to around 10375 eV. The spectrum shows in this energy range a multiplett like structure. On the top axis, you can see the Raman shift energy from 610 eV to 680 eV, which visually confirms that we are looking actually at a Mn multiplett spectrum. See Section 8.1.4 for a detailed explanation of the experiment.

We have therefore a solution for the aforementioned problem, for which we need soft X-ray information with a bulk-sensitive method. The X-ray Raman process can be used to extract a soft X-ray spectrum from a hard X-ray excitation. Typically, one can excite the sample with say 10,000 eV, e. g., for fluorescence. The inelastic scattered intensity contains also the spectroscopic signature from processes, which are equivalent to an X-ray absorption process [Cohen 1973]. This information is basically sitting like a piggyback on the fluorescent radiation, and thus overcomes the limited attenuation depth and information depth of soft X-rays.

I will be presenting a complete X-ray Raman *operando* study on a working lithium battery cell later in Section 8.1.4. Some basics of the method shall be shown here. X-ray Raman spectroscopy is a *photon-in–photon-out* method as simply illustrated in Fig-

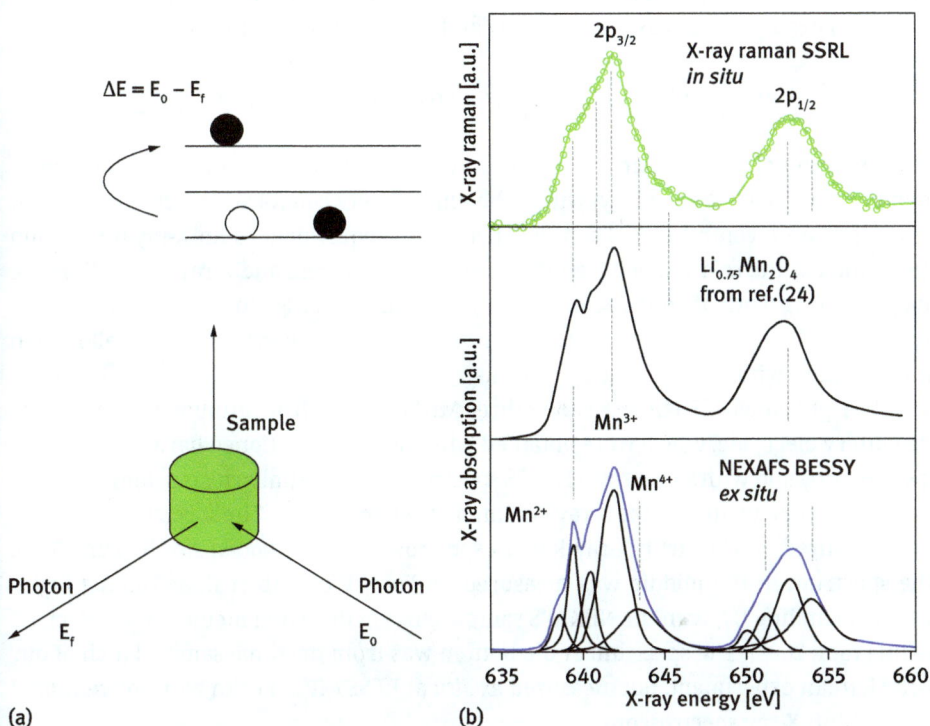

(a) (b)

Figure 6.8: (a) Simplistic sketch of the X-ray Raman method. An X-ray photon with energy E_0 is incident on the sample (green cylinder pellet). An electron will be changed into an excited state, and a flourescent photon with energy E_f will be emitted. The energy difference δE is an energy loss, which contains the Raman scattering signature. Reproduced from [Bergmann 2002]. (b) Comparison of *in situ* X-ray Raman and ex situ NEXAFS Mn 2p spectra. (right, top) X-ray Raman derived Mn 2p spectrum of our sample measured *in situ* prior to charging. (middle) Mn 2p NEXAFS spectrum of $Li_{0.75}Mn_2O_4$ from [Yoon 2003]. (bottom) Mn 2p NEXAFS spectrum of our cathode assembly recorded in dry state *ex situ* at BESSY. The thin vertical lines are guides to the eye to illustrate deconvoluted peaks. The peak positions for Mn^{2+}, Mn^{3+}, and Mn^{4+} are adopted from [Piper 2011].

ure 6.8. The incoming photon with energy E_0 is scattered in the sample and has an energy loss δE and the sample emits another (a secondary) photon with final energy E_f. When we vary this energy loss by changing the primary photon energy E_0 with a fixed analyzer energy which is set to E_F, we obtain a spectrum, which shows three prominent spectroscopy events. This is the elastic peak at/with E_0, a very broad Compton scattering intensity, and the X-ray Raman scattering spectrum of the probed sample with a very low intensity.

The transition probability w for the X-ray scattering process [Tohji 1989a, Tohji 1989b] is written as

$$w = \left\{ (4\pi^3 e^4 h) \big/ (m^2 \nu_i \nu_j) \right\} \cdot (1 + \cos^2 \theta) \times |\langle f| \exp(iqr)|i \rangle|^2 \cdot \delta(E_f - E_o - h(\nu_i - \nu_j))$$

where f and i are the *final* and the *initial* states, respectively. When the product $qr \ll 1$, the scattering dipole approximation holds and the equation simplifies to

$$w = \left\{ (64\pi^5 e^4 h) \big/ (m^2 c^2) \right\} \cdot (1 + \cos^2 \theta) \cdot \sin^2(\theta/2) \cdot |\langle i|r|f \rangle|^2$$

This condition of dipole approximation can be obtained by selecting the right experimental measurement geometry, where the angle θ between X-ray source, sample, and detector plays a central role. Then the spectrum is equivalent to an X-ray absorption spectrum. For the relation between the X-ray Raman process and X-ray absorption, see the pioneering work of Cohen et al. [Cohen 1973]. Superimposed to the Compton scattering, we then have the absorption spectrum. We identify in the spectrum the well-known Mn multiplett with the $2p_{3/2}$ and $2p_{1/2}$ peaks.

The right panel of Figure 6.8 shows three Mn 2p spectra from lithium manganese oxide battery electrode, which were obtained with the aforementioned hard X-ray Raman spectroscopy and with soft X-ray NEXAFS spectroscopy. The similarity is astonishing. We even find fine structures in the X-ray Raman-derived multiplets. The spectrum on the top was measured at SSRL with X-ray Raman spectroscopy *operando/in situ* [Braun 2015j]. The spectrum in the middle was measured by Yoon Won-Sub et al. at Pohang Lightsource, Beamline U7, with the NEXAFS sample current detection method (total electron yield) [Yoon 2003]. The spectrum at the bottom was from the same sample batch of our X-ray Raman experiment, but measured *ex situ* at BESSY II in Berlin with conventional NEXAFS soft X-ray spectroscopy.

Note that we have here one spectrum obtained with around 10 keV excitation energy. These X-rays have penetrated the entire battery electrode and probe thus the bulk and surface of all spinel particles in the X-ray beam and produce their corresponding Mn multiplet. This is not to be underestimated, because we have the bulk-sensitive electronic structure information from the intercalation process, for which we have in parallel the surface-sensitive multiplet from NEXAFS spectroscopy, and we can compare both at equal level, more so than comparing the Kβ XES spectrum with the NEXAFS spectrum. Certainly, this approach allows also for the derivation of soft X-ray spectra from samples, which are inside a massive *in situ* spectroscopy battery cell, as is shown in Chapter 8.1 in Figure 8.11 in this book. Another motivation for performing such inelastic X-ray scattering experiments is the acquisition of X-ray spectra from low Z elements [Bergmann 2002], whose absorption and emission energies are very low, such as sulfur with 2400 eV in the tender X-ray range, oxygen at 525 eV, carbon at 285 eV, and certainly lithium with 54 eV. Here, too, the low necessary primary energy mandates that the spectra are only surface representative. In order to prepare (1999–2001) for further method development for lithium ion battery research, we tried to collect lithium spectra with X-ray Raman spectroscopy. We did so *ex situ* with tetrahedrally coordinated model compounds, specifically Li_2CO_3, Li_2SO_4, and LiOH, the spectra of which are shown in Figure 6.9. The left panel compares the lithium orthosilicate (powder sample) spectra recorded with soft X-ray NEXAFS at Beamline 4.0.2 at the Advanced Light Source, and the same material (pressed pellet) recorded with 10 keV X-ray Raman at BioCAT Beamline at the Advanced Photon Source. In the right panel of Figure 6.9, we compare the X-ray Raman spectra of three different lithium compounds.

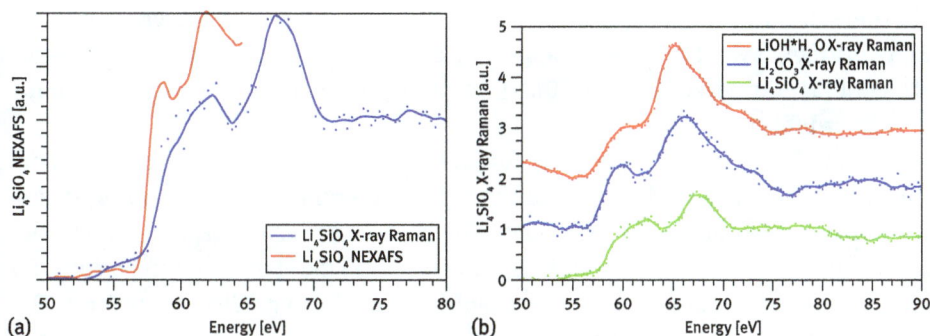

Figure 6.9: (a) Comparison of a NEXAFS spectrum and an X-ray Raman spectrum from Li_4SiO_4 powder and pressed pellet, respectively. The NEXAFS spectrum was recorded at BL 4.0.2 at the ALS (Artur Braun and Hongxin Wang). The X-ray Raman spectrum was recorded at Bio-CAT at the APS. (b) X-ray Raman spectra (measured at APS Bio-CAT by Uwe Bergmann and Pieter. Glatzel) recorded from dense, tightly pressed pellets of $LiOH \cdot H_2O$, Li_2CO_3, Li_4SiO_4. See also [Bergmann 2002].

What we learn from this work on lithium is that we can resolve the basic structure of the element in various compounds. We see the absorption edge and also finer structures. Note that these X-ray Raman spectra reflect the bulk of the entire probed sample, whereas the NEXAFS spectrum is obtained with around 50–60 eV photons, which have a low penetration depth. A "sharper" NEXAFS spectrum may therefore originate from the fact that only a thin surface region is probed. And this surface region may have a relatively "sharp" or well-defined electronic or molecular structure with low diversity in species. A thick sample is comprised of surface and bulk and regions in between. They all may have slightly different spectral signatures which add up in an actual spectrum, and thus appear more diffuse with broader peaks.

6.2 Hard X-ray XPS and hard X-ray PES (HAX-PES)

The surface sensitivity of an analytical technique becomes a problem, as we know by now, when we are interested in bulk properties or properties of regions beneath the surface of a material or component. While XPS is *the* surface method, researchers wanted to use it also for subsurface studies for a long time. For several years, developers have made progress in making hard X-ray XPS possible. Hard X-rays in the keV allow for accessing energetically deep core levels and generate photoelectrons with high kinetic energy and, therefore, longer inelastic mean free paths. Therefore, we can probe also somewhat deeper underneath the surface. Note than when the term "deep" is used in X-ray spectroscopy, it typically refers to the energetic position of an electronic state with respect to the Fermi energy, not to any spatial depth in the sample.

We remember the problems we had with tungsten oxide films deposited on TiO_2 substrates. Diffusion of Ti from the substrate to the film surface was observed with the (110) substrate, as was evidenced by appearance of the Ti 2p core level signature between 623 and 673 K [Braun 2012a, Chen 2013d]. The film grown on the (001) surface did not show interdiffusion.

Looking back to the previous section, we remember that electron hole pairs are generated in the bulk of the material throughout the range where the absorption of photons with relevant energy takes place. And the structure of a surface and subsurface can significantly differ from the bulk. This can be due to the elemental composition, for example. During synthesis and processing of materials, such gradients may form. This can have a malign or benign influence on the material, but it is not always possible or easy to elucidate such gradients.

Hence, conventional surface-sensitive XPS does not necessarily probe the sample depth regions characteristic for device functionality. However, the tunability of the mean escape depth of photoelectrons in photoemission allows for the determination the profile of various species nondestructive as a function of the depth from the surface [Nanda 1999, Nanda 2001] and study the interface structure of aforementioned films.

Combination of high-brilliance synchrotron X-ray sources and efficient detection techniques for energetic electrons makes it now possible to vary the effective escape depth (λ_{eff}) up to some 10 nanometers and the elegance of this technique has already been proven in many contexts ([Claessen 2009, Granroth 2011, Santra 2009, Sapra 2006, Sarma 2010]). We have therefore performed hard X-ray photoemission (HAXPES) measurements on one as-deposited oxygen-deficient film and on one subsequently oxidized WO_3 films grown on a TiO_2 substrate to investigate the chemical composition across the film and at the interfaces and the effect of the annealing. Our experiment was performed using the HAXPES instrument at an undulator beamline P09 of PETRA III (DESY, Germany [Gloskovskii 2012]).

Figure 6.10 shows the spectra obtained for W 4d and Ti 2p (inset) at three different photon energies from 4500 eV to highly bulk-sensitive 8000 eV for the oxidized yellow (a) and the as-deposited blue films (b). The position of the Ti 2p peaks obtained at 458.5 and at 464.2 eV with a characteristic spin orbit separation of 5.7 eV is easily associated with Ti^{4+} in TiO_2. The W 4d peaks of WO_3 are measured at 247.1 and 259.5 eV with a spin orbit separation of 12.4 eV.

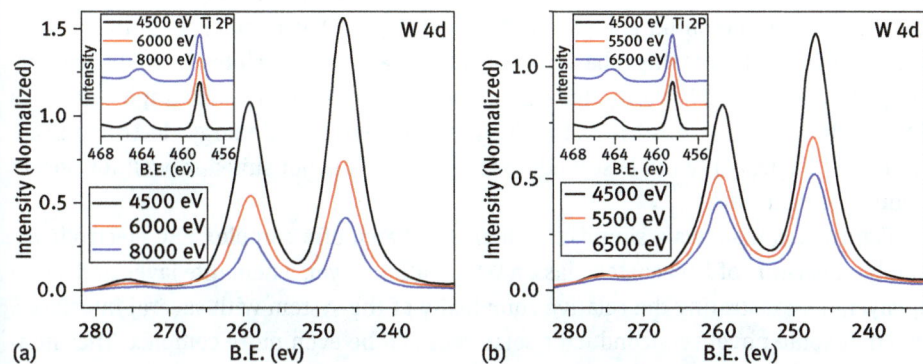

Figure 6.10: Photoemission spectra of W 4d and Ti 2p (inset) core levels obtained at three different photon energies from the annealed yellow sample (a) and the unannealed blue sample (b). The W 4d spectra are normalized with respect to the intensity of the Ti 2p core levels obtained at the same energy. Reprinted from [Bora 2013b], Bora D. K., Hu Y., Thiess S., Erat S., Feng X., Mukherjee S., Fortunato G., Gaillard N., Toth R., Gajda-Schrantz K., Drube W., Graetzel M., Guo J., Zhu J., Constable E. C., Sarma D. D., Wang H., Braun A.: Between photocatalysis and photosynthesis: Synchrotron spectroscopy methods on molecules and materials for solar hydrogen generation. JOURNAL OF ELECTRON SPECTROSCOPY AND RELATED PHENOMENA 2013, 190:93–105, Copyright (2013), with permission from Elsevier.

The sharp peak shape obtained at all different photon energies clearly indicates the presence of only one kind of oxide species for both W and Ti in these samples. To compare the intensity ratios of these two core levels, we normalized all spectra with their corresponding photo ionization cross-sections [Granroth 2011]. The W 4d peaks were then normalized relative to the intensity of the Ti 2p core levels at those photon energies shown in Figure 6.10(a) (annealed sample) and (b) (unannealed sample). For both samples, the normalized W 4d intensity decreases with increasing photon energy because WO_3 is on top of TiO_2, and with increasing photon energy, we are probing more of the TiO_2 layer across the interface. We have estimated the thickness of the WO_3 layer by fitting [Claessen 2009] the intensity ratio variation yielding a thickness around 12 nm for the oxidized film. Besides this, no changes of the peak profiles are observed, which is indicative of a homogeneous chemical composition throughout the layer. This suggests presence of a relatively "sharp" WO_3-TiO_2 interface in this thin film, which was not affected by the annealing in O_2.

Of interest in semiconductor photo-electrochemistry and many other fields are surface states and surface defects. The concept of surface states originates from Tamm [Tamm 1932]. At crystal surfaces, unsaturated bonds cause energetic states, which are not present in the bulk of the crystal. These states may have particular influence on the surface properties of the crystal or material. This is related with the formation energy of lattice defects in the crystal. One can compute the energy, which is necessary to form such defect, e. g., an oxygen vacancy. It turns out that this energy is very low when a vacancy is to be created at the crystal surface, and it is somewhat higher in the bulk. There is a maximum energy for the third oxygen layer underneath the surface [Warschkow 2002]. Now this is based on calculus from first principles. It would be interesting to experimentally verify this claim. It would require a crystallographic analysis of the surface and the subsurface, but this is not so trivial. Low energy electron diffraction (LEED) is probably the best method for this, but it is not suited so well for poorly conducting materials.

Figure 6.11(a) shows a simplistic schematic sketch of a complex PEC electrode assembly, consisting of FTO-coated glass, a WO_3 scaffold layer, a hematite layer, and a NiO top layer, demonstrating the relative complexity of the system with several interfaces. Note that some novel semiconductor solar cells can be even more complex. The situation is further complicated by the possibility that the defect concentration is not only a matter of stoichiometry, but also a matter of symmetry breaking at surfaces and interfaces, providing that the defect concentration at surfaces and interfaces may differ from the defect density in the bulk. The right panel of Figure 6.11 shows the computed defect energy in hematite as a function of depth in the crystal, having a minimum in the second oxygen layer and a maximum in the third oxygen layer. In particular, the defect energy for an oxygen vacancy in the third oxygen layer of TiO_2 is by around 0.3 eV higher than in the bulk [Warschkow 2002].

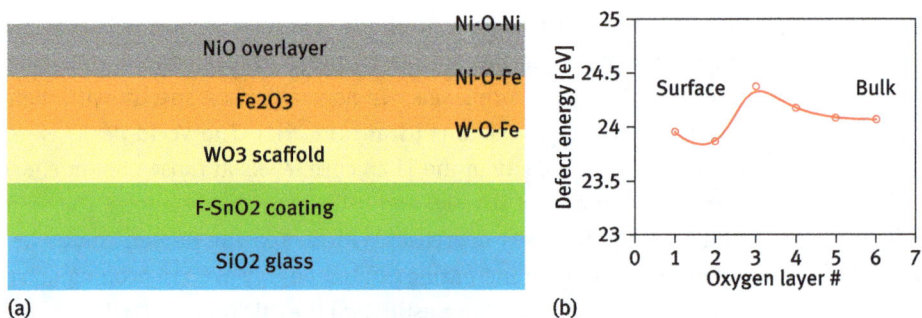

(a) (b)

Figure 6.11: (a) Simplistic schematic of a complex superlattice structure with several different heterojunctions. (b) Computed defect energy in hematite as a function of the oxide layer position (depth) in the near surface region [reproduced after Warschkow et al. [Warschkow 2002]].

I have tried to illustrate in Figure 6.12 how a sample with a smooth surface and a rough surface might cause different X-ray spectra. The left panel of Figure 6.12 stands, e. g., for a metal single crystal or polycrystal (blue color, say cobalt, e. g., density ρ = 8.9 cm^3) with a highly polished surface with little or "absent" surface roughness. The gray film on the top of the metal could be a metal oxide film, of say, 1 nm thickness. When you record X-ray or XPS spectra from this sample, the beam (yellow color) may penetrate the sample some nanometers or some micrometers and return the spectroscopic information correspondingly. A highly surface sensitive method like XPS with a penetration depth of 1 nanometer will provide an XPS spectrum with predominantly metal oxide signature. If we run a Co 2p NEXAFS spectrum on this sample, the attenuation length will vary from 600 nm to 80 nm across the absorption threshold. Therefore, even after the absorption threshold, we will probe a range with a depth of up to 80 nm. The 1 nm film on the surface of the cobalt is insignificant and likely not noticeable. If the cobalt oxide film was 10 nm thick, then we might see the oxide signature in the NEXAFS spectrum.

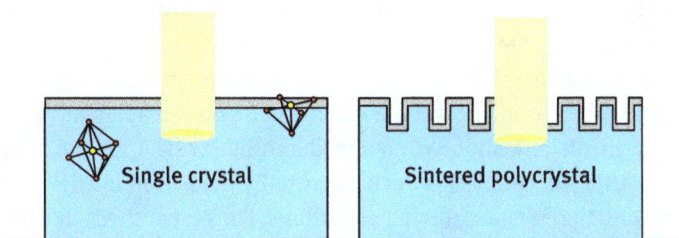

Figure 6.12: Schematic representation of a sample with smooth surface (left) and rough surface (right). The blue color shall signify a metal block with a gray oxide layer on the top. The yellow beam is the X-ray beam.

Let us now look at the right panel of Figure 6.12. The pillar structure shall resemble the surface roughness. When we compare the yellow beam over the rough surface in the right with the smooth surface on the left, it becomes clear that we have more "gray" film material probed by the beam when the surface is rough. The probed film volume is the product of the cross-section of the X-ray beam times the film thickness for the smooth film. For the rough film, we have in addition the vertical components, which contribute to the spectrum as film material. This helps to understand how spectra from smooth single crystals and rough powder materials may differ. I have been frequently asked by juniors whether it is possible or allowed to XPS measurements on powder materials.

It is certainly possible to run XPS spectra on powders. When you are a manager or operator of an XPS laboratory, you likely may have received your initial training as a diploma student in a physics institute at some university where single crystal studies were the normal business. There it was likely a normal routine to first sputter away all surface adsorbates before doing any measurement on the single crystal surface. In short: the ideal single crystal was the study subject. With this education and training, and with this ideal crystallographic clean picture in mind, there is probably no worse sample than some "ill-defined" high-surface-area powder with all kinds of surface absorbates. In addition, the vacuum pumps will make that the powder, once distributed on the sample holder, may swirl away the powder when they begin to pump after sample insertion into the chamber. Therefore, chances are that you will be rejected with your proposal to measure powder samples.

When you fix and stick the powder on a sticky carbon tape or press it onto a soft indium metal foil, there is less chance that powder particles become sucked and spread around and contaminate the UHV chamber. The carbon tape, however, may contain volatile organics, which will make the pressure in the vacuum chamber slightly or heavily increase. Some XPS lab managers do also not want this because of the possibility of contamination by such organics.

From my own experience that I can tell, statistically, the most productive and successful researchers are those who do not focus on cleanliness of their equipment but on readiness for a new collaboration. The contamination issues may be prevalent, but if you do not want your equipment to become contaminated, or dirty, then you simply better not use it at all. Fortunately, more and more researchers realize that samples from real applications simply are not "academically" clean. You have to cope with all the shortcomings that arise from looking into realistic sample scenarios. If you can avoid adverse contamination, avoid it. But if it is unavoidable, go ahead and live with it for the sake of a successful experiment. After all, catalysts, which are usually measured in UHV, are not being used in UHV but under realistic catalytic conditions. For researchers like me who are working with chemical engineers, I have to include the industrial environment and the user environment. This makes the research much more interesting.

We will see later in this book experiments where we looked into photo-electrochemical cells for water splitting, which were inserted in UHV chambers. On some occasions, protective windows would break and the KOH electrolyte would pour into the UHV chamber and trigger all protective valves—and certainly contaminate the UHV chamber inside. After a while, the chamber would be opened and cleaned and readied for new campaigns. If the beamline scientist would have had the attitude of only protecting the instrument, then a large number of pioneering experiments would never have been made.

In other words: if you want to bring a satellite into space, you will have to sacrifice the rocket. We will come later in this book in Part 8 to *operando* studies on water splitting, where I will give a detailed account on processes taking place at the photoelectrode surface. But here I want to focus briefly on the surface, subsurface, and bulk question.

We learned from initial *ex situ* NEXAFS studies that operation of a PEC electrode forms particular hydroxyl chemical species on the electrode surface, as indicated by a new intensity in an intensity minimum in the oxygen NEXAFS spectra of iron oxide [Bora 2011a]. See the energy range with a circle at around 532 eV in Figure 6.13. This conclusion was based on observations, which we had made with films that had a different

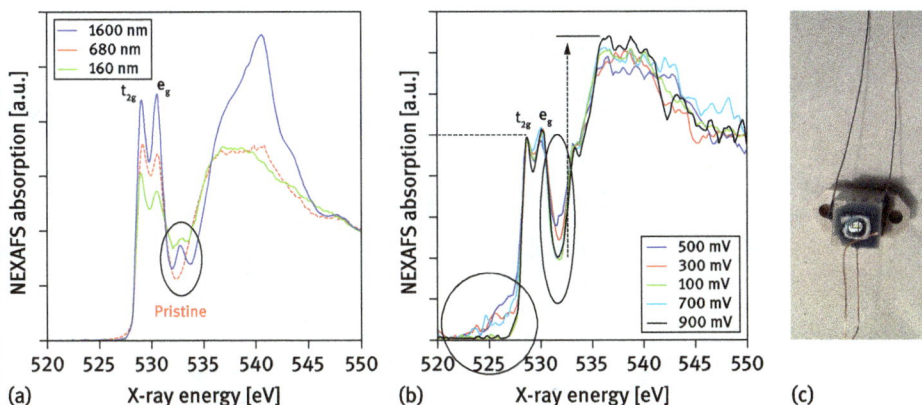

Figure 6.13: (a) Oxygen NEXAFS spectra of an iron oxide photoelectrodes with three different thicknesses (160, 680 and 1600 nm), which was pristine (680 nm) or subject to a DC bias of 600 mV versus $Ag^+/AgCl$ reference (160 and 1600 nm). (b) In situ oxygen, NEXAFS spectra recorded under illumination and potentials from 100 mV to 900 mV. (c) Miniature x-ray spectroelectrochemical cell from PEEK plastic, with three copper wires connecting to working electrode (shiny Si_3N_4 membrane on top, underneath blue glue), reference electrode and counter electrode. The photoelectrode layer is inside the cell, in contact with electrolyte. Available at Advanced Lightsource, Berkeley.

thickness, i. e., 160, 680, and 1600 nm. The relative spectral weight of the signature from the oxygen of the FTO layer underneath helped to get a confirmation on the thickness of the films.

In the subsequent *operando* study at the ALS, we learned that under illumination and 500 mV bias that we could see two hole states, which we could "switch off" by either reducing the bias potential or by switching off the light [Braun 2012]. The middle panel of Figure 6.13 shows the onset of two new peaks at the pre-edge of the oxygen spectra for DC bias of 300 mV and above in the energy range around 525 eV. At the same time, we see also the hydroxyl or oxy-hxdroxyl feature at around 532 eV increasing and, at a too high potential, also again decreasing. The miniature x-ray spectroelectrochemical cell is shown in the right panel of Figure 6.13.

Prior to running the *operando* experiment, we measured the iron oxide photoanode film in the cell when no electrolyte was filled. This spectrum is shown in the left panel of Figure 6.14 with red color (dry hematite). After filling with electrolyte, we obtain the spectrum, which is drawn in the left panel in Figure 6.14 with a brown color. This is the oxygen spectrum resulting from the oxygen in the iron oxide electrode and the oxygen in the water and the electrolyte. It is interesting to observe how the peak in the pre-edge with t_{2g} orbital symmetry at around 529 eV has low intensity when the iron oxide is dry. Upon exposure to the KOH electrolyte, the peak intensity becomes equal with the e_g peak. When we subtract the spectrum from the dry electrode from the spectrum from the wet electrode, we should obtain in principle the spectrum of the electrolyte. At large, since the electrolyte is diluted KOH, we obtain a water spectrum by this difference.

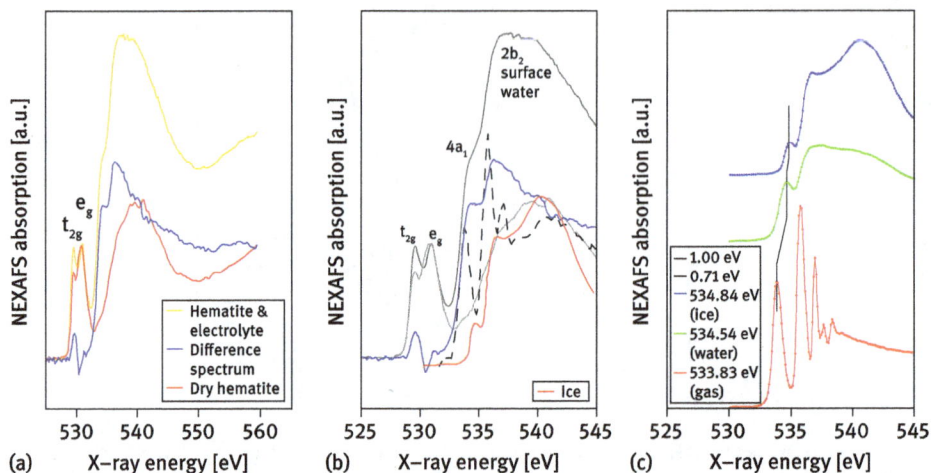

Figure 6.14: (a) Experimental NEXAFS spectra of iron oxide photoanode in dry (red spectrum) and KOH-wet (light brown spectrum) condition, along with their difference spectrum (blue). (b) Same experimental spectra like before in dim colors, with the blue difference spectrum plus reference spectra from ice (solid red) and from surface water (dashed blue), reproduced from [Wilson 2001]. The assignment of internally antibonding OH orbitals was taken from [Parent 2002] and [Cavalleri 2002]. (c) Shift of the "hydrogen bond" peak toward lower X-ray energies when the aggregate state changes from ice to water to gas. The spectra are reproduced from [Myneni 2002].

The middle panel of Figure 6.14 shows the aforementioned dry, wet, and different spectra along with two spectra from ice and from surface water, which were taken from [Wilson 2001]. It is hard to tell to what extent our difference spectrum, which should be representative to water, corresponds with the spectrum from surface water or from ice. Surface water shows two sharp peaks at around 534 and 536 eV with spectroscopic assignment $4a_1$ and $2b_2$ [Parent 2002, Cavalleri 2002]. Peak $4a_1$ is considered a hydrogen bond breaking peak, which shifts toward lower X-ray energies when then number of hydrogen bonds is decreasing. Ice has a large concentration of hydrogen bonds that literally melt away on the transition to water and transition to vapor. We notice such a shift when we compare with our difference spectrum, which should be indicative to the water in the electrolyte [Braun 2016c]. This shift is further illustrated in the right panel of Figure 6.14 [Myneni 2002].

6.3 X-ray standing waves for electrochemical double layer studies

I want to present a not very well-known method, the X-ray standing wave (XSW) technique. This is a highly surface-sensitive method, which can be also applied to buried interfaces. It requires basically very smooth surfaces and is therefore probably not suited for conventional energy converter devices with very rough and porous components or powder materials. It is well suited for model system studies and has gained more recognition recently. The XSW method as such has been proposed over 50 years ago by Batterman [Batterman 1964] and is well described in reviews [Vartanyants 2001, Zegenhagen 2013]. XSW has been applied, e. g., for the study of the structure of the electrochemical double layer on a metal oxide electrode [Fenter 2000, Zhang 2004].

Recently, the XSW method was combined with near ambient pressure XPS (NAP-XPS) [Nemsak 2014] for the determination of the spatial distribution and molecular structure of solvent atoms on the surface of a metal oxide semiconductor, including the electrochemical double layer. This is because tuning the field of the exciting X-ray wave into a standing wave allows for greater depth sensitivity in photoemission [Fadley 2013].

Nemsak et al. [Nemsak 2014] have investigated a polycrystalline smooth iron oxide film, which was immersed in a mixture of NaOH and CsOH electrolyte. They could determine chemical-state resolved depth profiles of every chemical element in the sample with sub-nm spatial depth resolution. The electrolyte cations Na and Cs, e. g., have different distributions across the depth. For example, they show that the Na and Cs ion have distinctly different depth distributions. In the planning of their experiment, it was necessary to build a standing wave over the iron oxide layer.

They deposited the iron oxide film on an X-ray mirror multilayer heterostructure assembly from silicon and molybdenum, which provided with 14 and 42 electrons per atom (see Appendix with periodic table of elements), respectively, a large scattering contrast. Note that the sample under study is the liquid electrolyte and its ions. An iron oxide layer was needed as a "substrate," which had a thickness of around 37 Å. The Mo/Si multilayer underneath of around 250 nm thickness was the necessary X-ray optical infrastructure for the generation of the standing wave. The wavelength is determined by the superlattice spacing of Mo and Si layers, 34 Å. The adsorbed water layer was only 10 Å thin. The temperature of the system around the water was only 2.5 °C, slightly above freezing point but still in vapor form. This is the basic geometrical description of the sample environment in Figure 6.15 (panel a). The oxygen core level spectrum in Figure 6.15 shows the oxygen in crystalline form from the iron oxide substrate, then OH groups from the electrolyte and the oxygen in the liquid and vapor forms of water in the electrolyte (panel b). Dependent on the angle of incidence, the relative peak heights

▶ **Figure 6.15:** The sample configuration, together with some key rocking curves for the various elements in the sample. (a) The sample configuration, with some relevant dimensions noted. (b) An O 1s spectrum, resolved into four components by peak fitting. The component labeled OHmay also contain contributions from carboxyl and bicarbonate species. (c) The rocking curves derived from the four components of (b). (d) A Cs 4d spectrum, including peak fitting. (e) Analogous overlapping Na 2p and O 2s spectra, with peak fitting. (f) The rocking curves for Cs 4d and Na 2p derived from spectra such as those in (d) and (e). (g) A typical C 1s spectrum, showing the two components, one at low binding energy (LBE) and one at higher binding energy (HBE). (h) The rocking curves for the two C 1s components. Reprinted by permission from Macmillan Publishers Ltd: [Nature Communications], Nemsak, S., Shavorskiy, A., Karslioglu, O., Zegkinoglou, I., Rattanachata, A., Conlon, C. S., Keqi, A., Greene, P. K., Burks, E. C., Salmassi, F., Gullikson, E. M., Yang, S.-H., Liu, K., Bluhm, H., Fadley, C. S., Concentration and chemical-state profiles at heterogeneous interfaces with sub-nm accuracy from standing-wave ambient-pressure photoemission. Nature Communications 2014, 5. Copyright (2014) [Nemsak 2014].

$H_2O(g){:}P_{H2O} = 0.4$ Torr, 2.5°C

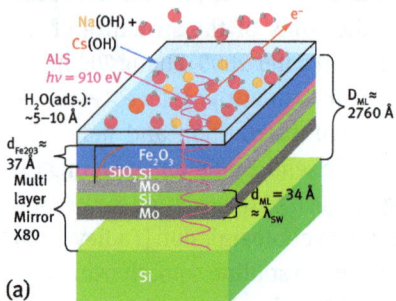

(a)

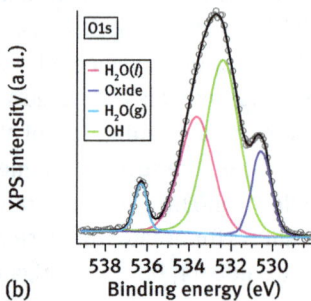

(b)

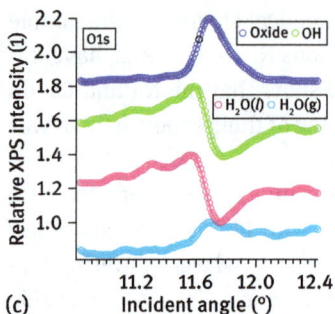

(c)

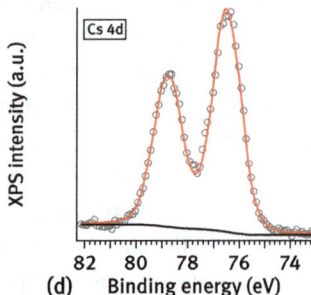

(d)

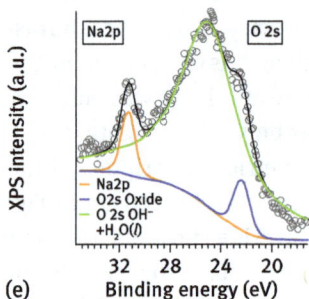

(e)

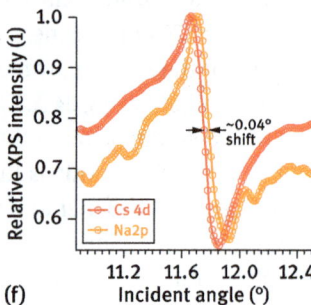

(f)

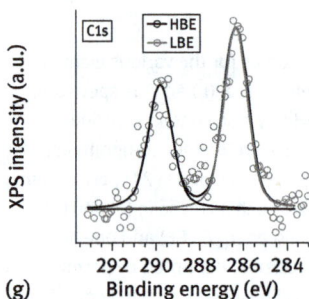

(g)

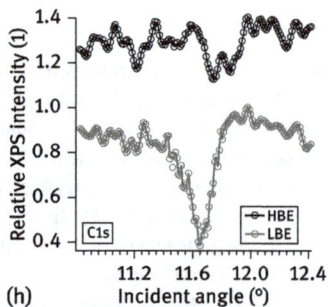

(h)

of these four species vary, as we see from panel c of Figure 6.15. The Na 2p peaks are at 30.65 eV (Na $2p_{1/2}$) and 30.81 eV (Na $2p_{3/2}$). Panel e shows how the Na 2p peak is proximal with the O 2s peak at 41.6 eV. They are able to assign features from the liquid water and hydroxyl groups and the crystalline oxygen from the substrate. The Cs 4d peaks are at 79.8 eV and 77.5 eV. The interferometric nature of this experiment, combined with spectroscopy, allows for detection of molecular structure with in the Ångstrom range thickness resolution.

6.3.1 Resonant photoemission spectroscopy on thin thermochromic window VO$_2$ films

Vanadium dioxide has a metal insulator transition at 68 °C, which is accompanied by changes in the optical properties. This effect can be used as thermochromic windows in terrestrial and outer space applications. This was a project between Empa, iThemba LABS and ETHZ and University of South Africa with doctoral student Itani G. Madiba [Braun 2015h]. In outer space, the vanadium dioxide will be exposed to cosmic radiation. Such cosmic radiation can severely impact the structure and function of materials. We have therefore deposited VO$_2$ films on silicon wafers and exposed them to γ-radiation of doses up to 100 kGy by using a ^{60}Co nuclide source with 1.17 and 1.33 MeV photon energy at iThemba LABS. We anticipate that the γ-radiation causes local structural perturbations, which can amount to defects with a corresponding change in electronic structure and function. Figure 6.16 illustrates the vanadium oxide film assembly. We can simulate the X-ray transmission with the software, which is available at the Center for X-ray Optics in Berkeley [Gullikson 1995]. With a density of 4.57 g/cm^3 for VO$_2$ and a thin film thickness of 200 nm, we get the following output profile for the X-ray transmission T in the photon energy range from 20 to 50 eV, as shown in the right panel in Figure 6.16.

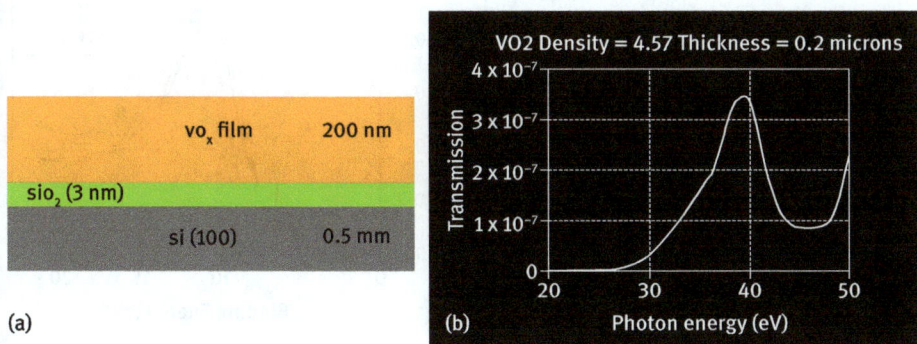

(a) (b)

Figure 6.16: (a) Schematic of the 200 nm thick vanadium oxide film deposited on a silicon (100) single crystal wafer. The wafer has a very thin silicon oxide layer on top. (b) X-ray transmission for X-rays with energies from 20 to 50 eV for VO$_2$ with 4.57 g/cm^3 density and 200 nm thickness. The transmission profile is taken from the webpage online simulation output from [Gullikson 1995, Henke 1993a, Henke 1993b].

T is 3×10^{-7} in this energy range. The software returns an attenuation length of 14 nm at 40 eV. The attenuation length is the path after which the beam intensity has decreased to 1/e of its original intensity. Hence, only the top surface of the film is excited by the X-rays. The O 2s state is absorbing at 41.2 eV, and from vanadium the M2 $3p_{1/2}$ and M3 $3p_{3/2}$ at 37.2 eV [Thompson 2009], respectively. Silicon has the lowest energy core levels at 99 eV and is therefore not of interest. We are measuring therefore at 40 eV resonant on oxygen and vanadium.

We carried out photoemission spectroscopy at the synchrotron storage ring of the Physikalisch-Technische Bundesanstalt (PTB) in Berlin. Excitation energies (I am using the terms *excitation energy, X-ray energy, and photon energy* synonymously in this book) were 40, 60, 80, and 125 eV. Quick comparison of the spectra in Figure 6.17 shows that the as-deposited vanadium oxide film shows no considerable spectral changes for any

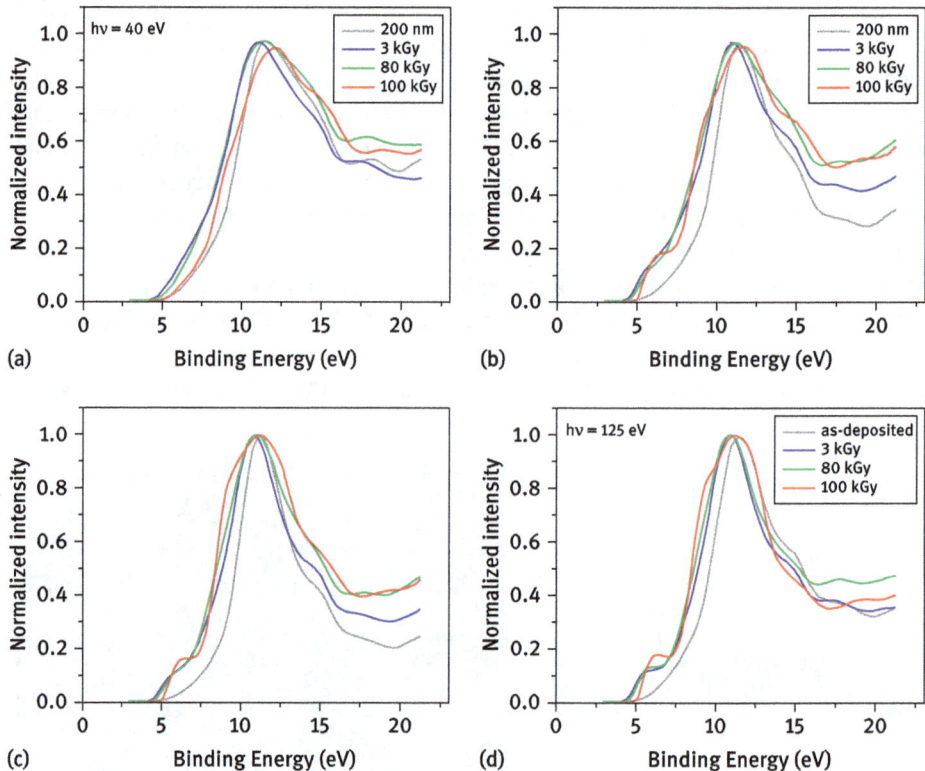

Figure 6.17: Valence band photoemission spectra of an as deposited vanadium oxide film (gray spectrum), and three such films exposed to γ-radiation for 3 kGy (blue spectrum), 80 kGY (green spectrum), and 100 kGy (red spectrum). The excitation energies were 40 eV (a), 60 eV (b), 80 eV (c), and 125 eV (d). Films were deposited by Nicolas Emond and Mohamed Chaker (INRS Univ. Quebec, Canada), and γ-irradiation was done by Philip Beukes, iThemba LABS, South Africa.

of the four different excitation energies. Maybe for the spectrum recorded with 125 eV, we see a slight shoulder at 7–8 eV binding energy.

When we look at the films irradiated with y-rays, a new peak at 5 eV shows up, plus a strong shoulder at around 8 eV. These two structures are particularly pronounced for the excitation energies 60, 80, and 125 eV. The sharpness of the peak at 5 eV increases with increasing excitation energy. We can therefore assume that we are here not looking at a resonant effect, but at an effect between surface sensitivity and bulk sensitivity. The attenuation depths increase with increasing excitation energy, as we can check online with the tool from Eric Gullikson at Berkeley Lab [Gullikson 1995], from 12 to 65 nm. When we compare this with the mean free path of electrons in solids [Seah 1979] as shown in Section 2.3.1, electrons with a kinetic energy between 10 and 100 eV escape from a thin layer of 1 monolayer to 10 monolayer. We have also spectra (measured by Ulrich "Uli" Müller from Empa, not shown here) recorded with an XPS spectrometer operated with Al Kα radiation ($h\nu$ = 1486.7 eV). These spectra do show the strong peak at the valence band onset, which we see at 5 eV. For this high excitation energy, the attenuation length in VO_2 is 900 nm. Hence, the entire VO_2 film is affected. We can therefore reasonably assume that the γ-rays with larger than 1 MeV energy will fully penetrate the

Figure 6.18: MSc Itani G. Madiba (Empa and University of South Africa), Dipl.-Ing. (FH) Hendrik Kaser, and Dr. Michael Kolbe (both PTB) finalizing the sample manipulator position and preparing for the resonant PES scan program at the PTB synchrotron storage ring in Berlin.

200 nm VO_2 film and cause structural changes in the entire sample. The surface remains interestingly unaffected by this irradiation, because the surface-sensitive low photon energy measurements at X-ray energy of 40 eV produce VB PES spectra, which do not show significant differences. Within increasing X-ray energy, and thus increasing penetration by X-rays, the features at 5 and 7–8 eV become stronger; we are hence probing more of the bulk. Consequently, we are looking at a bulk effect. Figure 6.18 shows the experimental crew at PTB in Berlin carrying out the PES measurements.

7 Organic and bioorganic samples

"You see? That is why scientists persist in their investigations, why we struggle so desperately for every bit of knowledge, stay up nights seeking the answer to a problem, climb the steepest obstacles to the next fragment of understanding, to finally reach that joyous moment of the kick in the discovery, which is part of the pleasure of finding things out." (Freeman J. Dyson in his Foreword to "The Pleasure of Finding Things Out" [Feynman 2005]).

To the nonscientists, "X-raying" is usually associated with the medical imaging of the human or animal body. This is for the orthopedic detection of fractions in bones or anomalies in the lungs, and so on [Dommann 2003]. The first application of X-rays was actually an application in the biological, i. e., medical field [Röntgen 1901]. However, it has no direct electrochemical application. This part of the book deals not only with organic but also with bioorganic samples that are studied with X-rays. Not all of them have direct or obvious relation to electrochemistry.

The majority of samples in X-ray and neutron studies are from inorganic materials. This part of the book is dedicated to organic and bioorganic samples and studies. Their presence in electrochemical energy conversion and storage systems is limited, but the importance of the organic and bioorganic materials will be steadily increasing in the future.

A large portion of the bioorganic samples are for protein crystallography studies. Protein crystallography is one of the best established routines in X-ray research. The great interest from pharmacy and life sciences has pushed the field so much that high throughput technology for sample preparation, measurement, and analysis has been developed quite early. The protein crystallography beamlines are possibly among the most productive. Multiple-wavelength anomalous dispersion (MAD) phasing was substantially developed for protein crystallography [Karle 1980, Hendrickson 1979].

The field of small-angle scattering benefited also a lot from protein structure and folding studies. These studies typically take place with proteins in solution, whereas protein crystallography is conducted on well-prepared single crystals. Nowadays, very complex structure factors are calculated and tested against experimental SAXS curves.

Protein spectroscopy has been a field of great interest for many decades. In Part 2, I have presented a $K\beta$ study on photosystem II. Metaloproteins from photosynthesis research have been investigated with EXAFS and with molecular and spin structure methods. As these proteins are material for energy conversion in photosynthesis and artificial photosynthesis and have catalytic functions, they have a place in this book.

Related with these biocatalysts are macromolecular assemblies with particular light harvesting properties, such as light antenna proteins, porphyrins, and other types of dyes that have found a steady application in dye sensitized solar cells (DSSC), for example. Such assemblies have been studied with X-ray scattering and spectroscopy already.

I have not covered the field of ionic liquids in this book. These are organic materials and can be used as electrolytes in batteries and in supercapacitors, for example. They

https://doi.org/10.1515/9783110794038-007

have no vapor pressure, so to speak. Therefore, they can be used in conventional UHV based methods for *in situ* and *operando* experiments.

Organic electrolytes are used in batteries and supercapacitors and, therefore, generally fall in this part of the book. What other materials and components in electrochemical energy converters and storage media are organic? Certainly, the proton exchange membrane in PEM-FC fuel cells have been studied with X-ray scattering and spectroscopy.

Organic membranes are used as proton conductors in PEM-FC fuel cells and constitute an integral part in electrochemical energy conversion. Polymers have gained an increasing interest in the 1990s with respect to X-ray analysis. The book "NEXAFS Spectroscopy" by J. Stöhr [Stöhr 1992] features particularly the samples with a substantial amount of lightweight elements, i. e., carbon and oxygen, and to a lesser extent nitrogen, all of which are important for polymers and biological samples.

In X-ray and photoelectron spectroscopy (XPS), the signals from carbon and oxygen are normally attributed to impurities from ambient environment, particularly CO_2 or water. Hence, the pronounced interest in the relative abundance of these elements in standard XPS analytical service.

The application range of organic materials in high temperature electrochemical energy converters is quite limited. They are not compatible with the high temperature environment of SOFC because organic materials would get combusted. However, the fuel gas can be considered an organic component. Our SOFC sulfur studies thus fall into this range. I have presented the sulfur XANES work in this book. Also, the SAXS work on the SOFC electrode assemblies had a relation to sulfur. A very early sulfur resonant protein crystallography work can be found in the work of Hendrickson and Teeter [Hendrickson 1981].

With few exceptions [Vyalikh 2004, Vyalikh 2005, Vyalikh 2006, Vyalikh 2009a], there are virtually no VB XPS studies on bioorganic materials interfacing inorganic materials. Figure 7.1 below shows a schematic of a bacteria with an inner and outer cell membrane, inside which a natural cell metabolism takes place. This includes a manifold of chemical reactions that we may want to be interested in and like to observe, e. g., the oxidation and reduction of iron to Fe^{2+} and Fe^{3+} in an iron hydrogenase. The previously mentioned oxygen-evolving complex in PS II measured with XES certainly falls in the category of bioorganic samples (Chapter 2.2.1). When we consider *Shewenella* bacteria, which have iron oxide pili for charge transfer [Qian 2014], it would be interesting to investigate this with valence band spectroscopy. Samples as thick as those shown in the right panel of Figure 7.1 (submillimeter range thickness of blue-green algae film) would not allow for probing the interface of algae and iron oxide electrode with XPS, for example. It is therefore important to prepare samples properly for the targeted analytical method.

There are few XPS core level studies, which are only of limited assistance when it comes to understanding the electric and electronic properties of biointerfaces. "The function and efficiency of most organic electronic and opto-electronic devices greatly

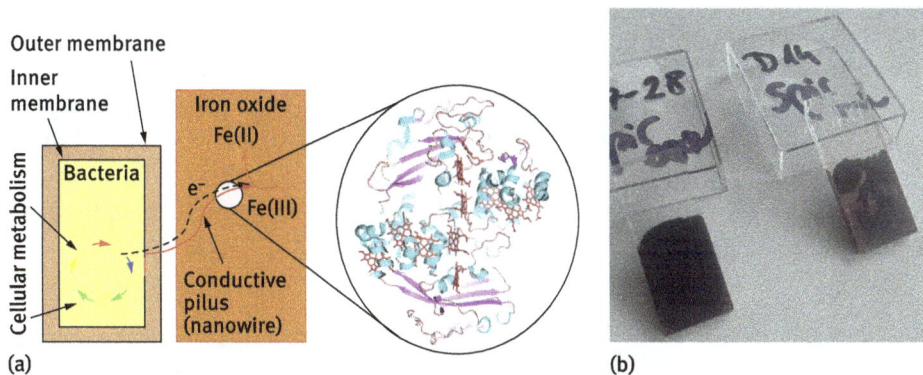

Figure 7.1: (Left) Schematic representation of the dissimilatory reduction of iron oxide by bacteria such as some *Geobacter* and *Shewanella* species that produce conductive nanowires. Electrons produced by the cellular metabolism are transported through the outer bacterial membrane along the conductive nanowire that contains multiheme cytochrome-type proteins. As an example, the extracellular 11-heme cytochrome UndA from *Shewanella* sp. HRCR-6 (PDB ID:3UCP) is reported in the inset that shows heme molecules in red, a-helices in blue, and b-sheets in purple. (Right) Photos of two hematite film photoelectrodes coated with a film of *Spirulina* cell extracts of several 100 mm thickness [Braun 2015b]. Reproduced from [Braun 2015b] with permission of Wiley. © 2015 Wiley-VCH Verlag GmbH & Co. KGaA, Weinheim.

The function and efficiency of most organic electronic and opto-electronic devices greatly depend on the electronic structure of the interfaces therein. Charge injection from electrical contacts into the organic semiconductor, charge extraction, or exciton dissociation at organic semiconductor heterojunctions are crucial processes that must be optimized for high device efficiency. Consequently, the energy levels at these interfaces must be matched to allow for optimal performance. The key mechanisms that determine the energy level alignment at organic/electrode and organic/organic interfaces are reviewed, and methods to adjust the levels at such interfaces are presented.

Figure 7.2: Abstract of Norbert Koch article [Koch 2012] underscoring the importance of electronic structure of interfaces in photoelectrode assemblies. © 2012 WILEY-VCH Verlag GmbH & Co. KGaA, Weinheim.

depend on the electronic structure of the interfaces therein", see Figure 7.2 [Koch 2012]. There exist, however, VB XPS studies on polymers interfacing electrodes, which point in a direction that we actually want to follow. Norbert Koch in Berlin seems to be one of the pioneers in this field [Koch 2012].

7.1 X-ray studies on polymer electrolyte membranes

The best known fuel cell is probably the polymer electrolyte membrane fuel cell (PEM-FC), which converts oxygen and hydrogen at ambient conditions in water and electricity. We have dealt with that one already in Part 4 for X-ray and neutron imaging. PEM stands synonymously for proton exchange membrane and, in the German language, polymer electrolyte membrane. Both terms are correct. The membrane material is the fluoropolymer NAFION® and comes as solid polymer material or as liquid for laboratory use. As we see from its structure formula as shown in Figure 7.3, the NAFION® molecule is built up from carbon, fluorine, oxygen, and one sulfur and one hydrogen atom. The membrane provides the electrochemically necessary proton transport from anode to cathode and prevents electronic shortcut between both electrodes. The membranes work only when a minimum humidity is available, and drying out can severely harm to the structure and function of the membrane. NAFION® is also available as liquid solution for drop casting, spin coating, and dip coating (Figure 7.3).

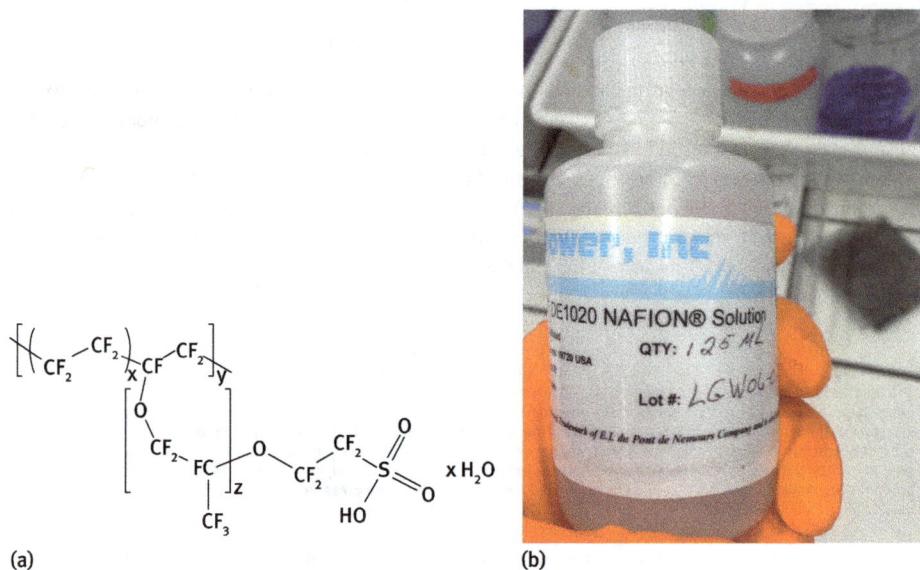

(a)   (b)

Figure 7.3: (a) The structure formula of fluoropolymer NAFION [Roland 1952]. Note that water, like crystal water, is actually an integral part of NAFION. (b) NAFION® solution (Photo courtesy Yelin Hu, Empa).

Xiao et al. [Xiao 2014] have made a mechanical testing of NAFION N117 and included an XRD study on pristine and cycled membranes. Memioglu et al. [Memioglu 2014] have investigated polymer composites (poly(3,4-ethylene dioxythiophene) PEDOT) with thermogravimetry, cyclic voltammetry, infrared spectroscopy, and X-ray diffraction. The phosphoric acid fuel cell operates above ambient temperature, at around 160 °C. Che et al. [Che 2015] looked into a polymer membrane for this purpose with XRD. When we look into the diffractograms of the aforementioned polymers, it is obvious that these are not suited for a conventional crystallographic analysis. The polymers are hardly "crystalline" in the conventional manner. However, this is no problem when we just want to understand the correlation between structure and function of a material and component, as is demonstrated by Li et al. [Li 2010]. X-ray scattering backgrounds, small-angle scattering ranges, or peak shifts or redistribution of diffracted intensity may allow for a very good semiquantitative analysis and an explanation of what is going on in a material during operation or processing. Since conventional XRD is typically not applicable in poorly ordered materials like polymers, researchers frequently prefer to call the method wide-angle X-ray diffraction or scattering (WAXS), and not XRD. In the extreme case, emphasis is then not put on angular resolution, but on high counting statistics for flat backgrounds. A good brief summary of the use of this method for PEM is provided by Blanton and Koestner [Blanton 2015]. Mauritz and Moore have extensively elaborated on the NAFION structure and function in their review article and specifically show XRD, WAXS, and SAXS experiments [Mauritz 2004]. Sulfur and vanadium XANES was conducted on sulfonated membranes by Vijayakumar et al. [Vijayakumar 2015]. NAFION membranes are also of interest for the redox flow batteries [Schwenzer 2011]. Derivates of polymer membranes, such as those combined with metal ions, can also be investigated, e. g., with EXAFS [Pan 1983]. Every component from solid carbon in battery electrodes or graphite end plates in PEM-FC certainly can be investigated with XRD and SAXS and WAXS.

7.2 Organic solar cells

Functionality and performance of organic solar cells depend at large on the structure of their polymer films. Note that the structure (the architecture) of the polymer film is different from the molecular structure of the organic polymer material that constitutes the film. This is sketched in Figure 7.4 after Hiszpanski and Loo [Hiszpanski 2014]. Crystal structure, size, and portion and orientation of crystals determine the film structure, and these can be determined with the aforementioned XRD and SAXS methods.

The analysis of polymers can be difficult because of the manifold of complex chemical groups that constitute the samples. Primary and secondary shifts of binding energy may cause relatively broad peaks or featureless photoelectron humps. Gengenbach et al. [Gengenbach 1996] have developed an analytical protocol for the XPS investigation of ageing of plasma deposited polymer films. Their protocol is based on least square fits of a series of spectra, which they obtain by monitoring the surface composition of a plasma polymer film over some time after deposition. The information that they obtain from the first round of curve fitting is then used as additional constraints for the following round of least square fits. This is basically a self-consistent procedure.

Organic solar cells are certainly a subject of interest in this part of the book. We have yet to mention the use of NEXAFS spectroscopy, particularly at the carbon, oxygen, and nitrogen edges. Their absorption cross sections are very high, and it is therefore throughout possible that we incur large radiation damage effects (see the section on radiation damage). A very illustrative review on the use of soft-ray absorption spectroscopy for the characterization of polymer films and blends is provided by McNeill and Ade [Mcneill 2013].

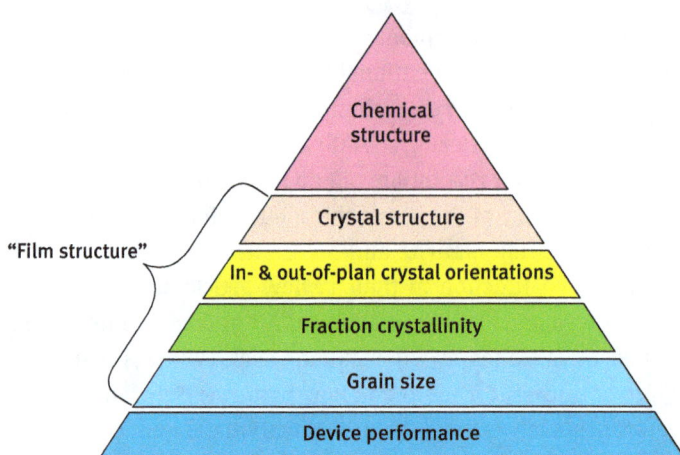

Figure 7.4: The performance of devices composed of organic semiconductors is related to the molecular structure of the organic semiconductor and the ordering of the molecules in the solid state, typically referred to as the film structure. Elucidating the effect of the film structure on device performance may be enhanced by considering individually the various aspects of ordering that may occur in the organic semiconductor thin film across several length scales. Reproduced from [Hiszpanski 2014] with permission of The Royal Society of Chemistry.

It is an established fact that the orientation of anisotropic molecules residing on electrode surfaces has a noticeable or significant influence on the charge transfer, energy transfer, and light absorption, or energy level alignment. It has turned out that the relative orientation of such molecules has also an influence on the spectral characteristics of its NEXAFS spectra. The spectra give therefore account of the orientation and the functional performance, e. g., in organic solar cell materials. Zhong et al. [Zhong 2015] have investigated the influence of the molecular orientation at the organic-organic heterojunction (OOH) interface with NEXAFS spectroscopy. They investigated also the gap state-mediated, orientation-dependent energy level alignment at OOH interfaces. Peisert et al. [Peisert 2015] have investigated the orientation-dependent NEXAFS and PES spectra of transition metal phthalocyanine molecules. They specifically looked into the role of the central 3d metal atom in interfacial charge transfer processes. Roth et al. [Roth 2015] have used EELS for the investigation of phenanthrene-type hydrocarbons organic films doped with K and Ca. Note that their EELS spectrometer was not a so-called "TEM-EELS," but a dedicated spectrometer [Fink 1989]. Anselmo et al. [Anselmo 2013] used C1s NEXAFS spectroscopy for orientation-dependent molecular structure studies on polymer films blended with fullerene.

Stadtmueller et al. [Stadtmueller 2015] have investigated organic films adsorbed on metal single crystal substrates, particularly Ag(111). The films were from homomolecular adsorbate systems, perylene, and naphtalene derivates, in particular 3,4,9,10-perylene-tetracarboxylicacid-dianhydride (PTCDA) and 1,4,5,8-naphthalin-tetracarboxylicacid-dianhydride (NTCDA). The other system includes different metal-phthalocyanines. They investigate the organic-inorganic heterostructures with X-ray standing wave spectroscopy and conventional XPS. Very interesting are their valence band investigations and their studies on the bond breaking at the interface and the influence on the charge balancing. Although it appears that the link between geometric structure and electronic structure was lost, substrate-mediated charge transfer from donor to acceptor was seen as the rational for the experimental observations. They conclude that this effect is of general validity for π-conjugated molecules adsorbing on the surfaces of noble metals.

Light-emitting diodes (LED) belong also to the class of organic solar cells, as they are operated in reverse mode, so to speak. Ou et al. [Ou 2015] have used photoemission spectroscopy in order to probe the alignment of the energy band levels in tandem organic LEDs. An introduction article into the electronic properties and structure of organic films and their determination with photoemission spectroscopy (PES) is provided by Amsalem et al. [Amsalem 2015] and Koch [Koch 2012] and by the dissertation of Frisch [Frisch 2014].

7.2.1 Organic dyes for photosensitization

Metal oxides for photoelectrodes that are stable in electrolytes have usually large band gaps. This is because the bond strength scales usually with the band gap energy. Such photoelectrodes are not suitable for solar applications because they absorb light predominantly in the ultraviolet range. Dye molecules can absorb light and form photoelectrons that can diffuse to the conduction band of a semiconductor in intimate contact with the dye. The dye can relax into its ground state by oxidation of a water molecule. This is called dye sensitization. Early works (compare [Allisson 1930a, Allisson 1930b]) on dye sensitization on metal oxides can be found in the PhD thesis and publications by Helmut Tributsch [Tributsch 1968, Tributsch 1969a, Tributsch 1969b, Tributsch 1969c, Tributsch 1971], where ZnO was the semiconductor. Tributsch received these zinc oxide substrates from Gerhard Heiland [Tributsch 2015], who had made many contributions to the surface electronic transport properties of ZnO with 4-point DC methods [Bauer 1971, Heiland 1972, Laurs 1987].

The charge carrier dynamics are very important for the function of highly complex photo-electrochemical systems like DSSC and is nowadays typically measured with electrochemical impedance spectroscopy. An extensive paper on EIS for DSSC can be found in the work of Wang et al. [Wang 2006a].

Today, TiO_2 is a popular semiconductor for this purpose. A detailed characterization study on DSSC can be found in the PhD theses of [Lee 2012] and [Hafeez 2013], for example. An early work on TiO_2-based DSSC is presented by Anderson et al. [Anderson 1979], who used a ruthenium bipyridine complex as a dye molecule. Hagfeldt and Grätzel [Hagfeldt 2000] showed an XPS valence band spectrum of TiO_2 functionalized with a

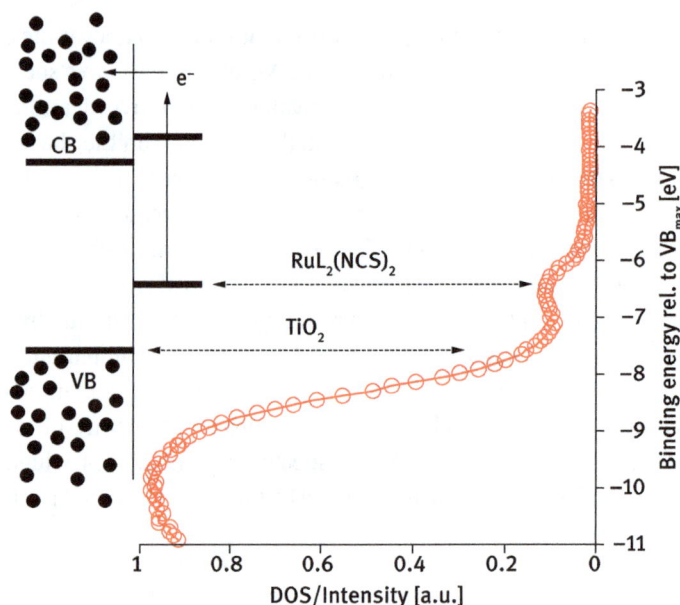

Figure 7.5: XPS valence band spectrum of TiO$_2$ photoelectrode with Ru sensitizer and inscribed energy levels, reproduced from Hagfeldt and Gratzel [Hagfeldt 2000]. Adapted with permission from Hagfeldt, A., Grätzel, M., Molecular photovoltaics. Accounts of Chemical Research 2000, 33 (5), 269–277. Copyright (2000) American Chemical Society.

ruthenium complex dye (reproduced in Figure 7.5). One problem in core level spectroscopy on ruthenium complexes is that the Ru 4d peak coincides with the C 1s peak at around 285 eV (see, e. g., [Chakroune 2005]). For systems with a low concentration of Ru such as in large complexes, it may be practically impossible to identify the Ru 4d signature when there is adsorbed carbonaceous species such as CO or CO$_2$. A potential way out of this dilemma may be the use of resonant excitation.

Extensive studies on DSSC photoelectrodes were conducted by the Jägermann group at TU Darmstadt, particularly with core level and valence band XPS. They operate the SoLiAs end station at BESSY, which is equipped with an *in situ* sample preparation chamber for electrochemical studies. For example, they looked into gap states and their interaction with dye molecules (Ru dye N3 [RuII(2,2′-bipyridyl-4,4′-dicarboxylate)2(NCS)2]) and acetonitrile solvent molecules [Schwanitz 2007a, Schwanitz 2007c]. They performed actually Ti^{3+} resonant photoemission and found a gap state, which was 2.2 eV from the leading edge of the VB, plus additional states up to the Fermi energy. These gap states can be quenched by dye and solvent. Rensmo et al. [Rensmo 1997] performed early resonant photoemission studies on the dye complex cis-bis(4,4′-dicarboxy-2,2′-bipyridine)-bis(isothiocyanato)ruthenium(II) and of its ligand 2,2′-bipyridine-4,4′-dicarboxylic acid. In parallel, they performed semiempirical calculations, and thus could map the orbital parentage of the dye and the ligands. Moreover, they related the HOMO-LUMO levels of the molecules with the VB-CB of the semiconductor.

Dopamine is one dye of interest. Its adsorption on TiO_2 (101) was recently studied with Density Functional Theory (DFT) by Stashans et al. [Stashans 2015], and on TiO_2 with DFT and Surface Enhanced Raman Scattering (SERS) by Urdaneta et al. [Urdaneta 2014]. Syres et al. [Syres 2010] have adsorbed the neurotransmitter molecule dopamine onto TiO_2 single crystal substrates and investigated it with PES and NEXAFS. The PES data suggest that the neurotransmitter molecule adsorbs on the surface in didentate geometry, whereas gap states are removed from the TiO_2 valence band, along with formation of new unoccupied states, which have a role in the charge transfer between molecular and semiconductor. Rangan et al. [Rangan 2012] performed an exhaustive NEXAFS and XPS study on chromophores (Zn(II) tetraphenylporphyrin derivatives) adsorbed on metal oxides (TiO_2(110) and ZnO(1120)), supported with DFT calculations. They sketched the position of the energy levels by UV PES and inverse PES for the chromophore and the semiconductor. Astuti et al. [Astuti 2005] studied the triplet state photosensitization of photoelectrodes by zinc-substituted cytochrome c for hydrogen production.

Kang et al. [Kang 2012] used Cu nanoparticles for enhancement of charge transfer in TiO_2 in DSSC and studied the size of the Cu nanoparticles upon steam reforming with SAXS. Cu nanoparticles are certainly not organic materials, and thus would not fit into this chapter. However, plastocyanin is a photosynthesis protein with a copper center, which plays a role in the charge transfer in many plants and cyanobacteria (see, e. g., Haehnel 1980, Haehnel 1994, Hippler 1995, Hippler 1998, Nordling 1991, Delacerda 1997, Hervas 1995, Hervas 2011, Molina-Heredia 2001, Ramesh 2002, De La Rosa 2003, Setif 2006, Rush 1988, Luis Olloqui-Sariego 2012, Luis Olloqui-Sariego 2014, Pletneva 2002). Rasmussen and Minteer investigated its use in photo-electrochemical energy generation [Rasmussen 2014]. Xiong et al. [Xiong 2013, Xiong 2015] made a systematic synthesis and processing study on the microstructure of DSSC devices. He used XRD and SAXS for the microstructure determination, but this cannot be considered a study on the organic features of the system. Wu et al. [Wu 2013] conducted a complete SAXS, Pt XANES, and XPS analysis on DSSC functionalized with Co_3O_4-Pt catalysts and combined this with electrochemical performance assessment. However, also here the analyses did not target the organic dyes in any way. A genuine application of SAXS on organic DSSC components, however, is provided by Manfredi et al. [Manfredi 2014], who used block co-polymers as solid hosts for DSSC. Neppl et al. [Neppl 2014, Neppl 2015a, Neppl 2015b] have quite recently showcased a series of laser-pump/ X-ray-probe PES experiments suited for synchrotron and free electron laser beamlines. The experiments were done on semiconductor surfaces, molecule-semiconductor interfaces, and films of semiconductor nanoparticles. Their experiments demonstrate the high sensitivity of time-resolved photoemission to photo-induced charge carrier generation, diffusion, and recombination within the space charge regions. Ivnitski et al. [Ivnitski 2008] have done a CV and angle resolved core level XPS study on bilirubin oxidase (which contains a copper center) deposited on carbon electrodes.

7.3 Proteins, enzymes, biocatalysts, living cells, and biofilms

It is possible to combine the chemistry of dye-sensitized photo-electrochemical cells with the enzyme-catalyzed biofuel cell chemistry, and thus prepare hybrid cells that can produce electric power upon illumination. De la Garza et al. [De La Garza 2003] have functionalized a SnO_2 photoanode with a porphyrin sensitizer. The photoreactions at this anode are combined with the oxidation of biological fuels like glucose or alcohols, for which a NAD(P)H/NAD(P)+ redox carrier is used. An ion-conducting membrane allows coupling these processes with the electrochemical reactions at the cathode. Such integrated biophoto electrochemical cells could have advantages over their individual independent components, i. e., dye-sensitized photo-electrochemical cells, or biofuel cells alone. Certainly, such systems are even more complex than the aforementioned DSSC or the electrocatalytic systems alone. The situation becomes even more complex when living systems, such as bacteria or microbes, are part of the system. We may then deal with complete functional architectures where it may become desirable to use spatially resolved methods (microscopy). A complete review on the development of biofuel cells can be found at [Bullen 2006].

ESCA has been shown to provide results that correlate with biological response [Ratner 1983]. The analysis of biomaterials and biosystems with XPS and ESCA is described in two books by Ratner and by Genet et al. [Genet 2008, Ratner 1983]. Miot et al. [Miot 2014] investigated the chemical interactions of microbes with mineral surfaces with STXM, micro-XRF, and conventional XANES and EXAFS. We have referred to this already in Chapter 4.5.3, where the microscopy methods were introduced. They show how spatially resolved carbon K-edges and Fe L-edges are obtained from iron bearing minerals and organic matter, for example. We have shown already in Part 2 how $K\beta$ spectra were obtained from the oxygen-evolving complex of Photosystem II. There, the Mn was the element under investigation, particularly its oxidation state and spin state. Electron spin resonance (ESR) spectroscopy has been used and is still being used for the investigation of proteins with metal centers. Lohmann [Lohmann 1965] conducted light on/off studies with ESR on *Phormidium luridum* cyanobacteria and could make conclusions on the sulfur chemistry and functionality and energy levels (compare also [Calvin 1957]).

7.3.1 Protein crystallography

Hendrickson et al. [Hendrickson 1968] crystallized and cataloged nine basic forms of the blood molecule hemoglobin from the sea lamprey *Petromyzon marinus* in 1968 and investigated them with X-ray diffraction. At about that time, Hendrickson and Lattman addressed the problem of detection of the distribution of multiple phases and how it can be quantified mathematically based on X-ray data [Hendrickson 1970]. They looked into crystalline transitions and how ligand states were affected by this. They thought that changes of the ligand states affect in a subtle way the change of polymerization sites on the monomer without much changing of the protein conformation. They later correlated their X-ray structure with physiological function and biological evolution [Love 1971].

> "Not only is X-ray diffraction analysis providing structural insight about protein function, but in addition it is helping to provide unexpected information about the biological evolution of protein molecules." [Love 1971]

One important application of XRD is protein crystallography. There, MAD phasing is used for the determination of the very complex crystal structures of proteins. The samples are typically single crystals from purified proteins, so that Laue diffraction [Ewald 1962, Von Laue 1915] is possible. MAD phasing is one of the resonant methods that I have introduced in Part 5 of this book. In principle, it should be possible to apply a Fourier transformation on the diffraction pattern with intensities F(hkl), and thus transform the reciprocal space information into real space information with atom positions $\rho(xyz)$:

$$\rho(xyz) = \frac{1}{V} \sum_{-\infty hkl}^{+\infty} |F(hkl)| \cdot e^{-2\pi i[hx+ky+lz-\varphi(hkl)]}$$

This can indeed be done on the diffraction pattern with $F(hkl)$ when also the phase factors $\varphi(hkl)$ are known [Bragg 1915a, Bragg 1915c, Bragg 1915e, Bragg 1915f, Bragg 1915g, Bragg 1922]. This is possible when the diffraction pattern is recorded with more than one X-ray wavelength, in particular, when the contrast variation is made near the X-ray absorption edges of the various elements in proteins [Karle 1980].

The bilirubin oxidase is a blue copper multicomplex that catalyzes the oxidation of bilirubin to biliverdin, and in parallel it reduces oxygen to water. This process can be utilized in biofuel cells. Cracknell et al. [Cracknell 2011] have determined the crystal structure of this complex with X-ray diffraction at the Diamond Light Source in the United Kingdom. They performed electrocatalytic and electrochemical studies on this complex. They found a hydrophilic pocket with a long narrow geometry near the T1 Cu ion, suggesting that the structure of the substrate-binding site is dynamically determined *in vivo*. The interaction between the binding pocket of Mv BOx and its highly conjugated substrate bilirubin is used to stabilize the enzyme on an electrode, more than doubling its electrocatalytic activity relative to the current obtained by simple adsorption of the protein to the carbon surface. Typical data relevant for such study are shown in Table 7.1 below. A cartoon of the crystal structure of the Mv BOx is shown in Figure 7.6.

Table 7.1: Collection and refinement statistics for Mv BOx. A Data are 95.6% complete to a resolution of 2.45 Å. The overall completeness is low due to a substantial lack of completeness in the data from 2.6–2.3 Å resolution. This arose due to the low symmetry of the crystal only being established once data collection had already been conducted, but the benefit of the additional observations in this range was such that they were still included in refinement. Reprinted from Cracknell et al. [Cracknell 2011] [blue multicopper oxidase bilirubin oxidase (BOx) from the ascomycete plant pathogen *Myrothecium verrucaria* (Mv)].

Diffraction data	
X-ray source	Diamond Light Source (I.04)
Wavelength/Å	0.97930
Detector	ADSC Q315 CCD
Diffraction conditions	100 K under N_2 cryostream
Space group	P1
Unit cell parameters/Å	$a = 52.79, b = 83.60, c = 143.15$
Angles/°	$\alpha = 89.98, \beta = 89.89, \gamma = 89.90$
Resolution limits (highest resolution shell)/Å	29.8–2.3 (2.43–2.3)
Measured reflections	188 365 (17 183)
Unique reflections	96 851 (9 203)
Completeness/%	89.7 (58.6)
R_{merge}/%	5.7 (0.097)
$I/\sigma(I)$	12.3 (5.5)
Refinement	
Program used	Phenix.refine (v. 1.6.1_357)
Average B factor/Å^2	18.63
R_{work}/%	16.97
R_{free}/%	21.63
Rms deviations from ideality:	
Bond lengths/Å	0.008
Bond angles/°	1.128
PDB code	2XLL[1]
Structure validation statistics (produced using Molprobity[25])	
Ramachandran statistics/% of residues (outliers/favored/allowed)	0/95.4/100
Clashscore (>0.4 Å overlap/1000 atoms)	13
Statistically improbable ("poor") rotamers/%	1.11
Molprobity score/percentile (among comparable structures)	1.98/91st percentile

7.3.2 Reflectometry on proteins and cells

Although protein crystallography is used for the exact structure determination, reflectometry is used for the study of the conformation, orientation, and morphology of proteins on surfaces. This is one step further to the function of proteins in components or devices. Still, reflectometry in this context is a structure determination method. Gidalevitz et al. [Gidalevitz 1999, Gidalevitz 2000] reported the protein folding at the air water interface, which they had studied with X-ray reflectometry. Specifically, they looked at alcohol dehydrogenase, glucose oxidase, and urease, which are stable in aqueous solu-

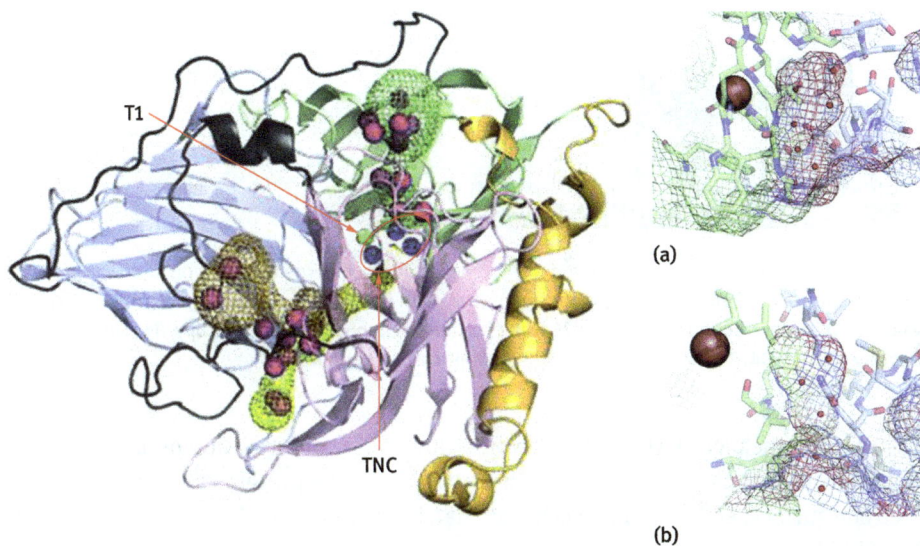

(a)

(b)

Figure 7.6: (Left) A cartoon representation of the X-ray determined crystal structure of Mv BOx. The three domains of the protein are shown in blue, pink, and green; loops linking the domains are shown in gray. The portion of the protein structure not determined in [23] in [Cracknell 2011] is shown in orange. The T1 Cu atom is shown in cyan, the TNC is shown in royal blue. Channels from the TNC to the protein surface as shown as mesh surfaces; water molecules in the vicinity of the channels are shown as red spheres. (Right) The substrate binding sites in Mv BOx (panel A) and Ec CueO (panel B), shown as a Connolly surface in mesh representation. In both cases, the T1 Cu atom is shown in orange, and atoms are shown in green or blue depending on the domain from which the originate (Figure 1 in Cracknell et al.) with heteroatoms shown in elemental colors. Water molecules are shown as red spheres. Reproduced from [Cracknell 2011] with permission of The Royal Society of Chemistry.

tion but denaturate at the water surface in contact with air, and form peptide layers of around 1 nm thickness. The reflectograms suggest that a lipid monolayer on the water surface would not prevent the proteins from unfolding. They observed a decrease in the electron density in the lipid monolayer when the protein was present, which was rationalized by presence of holes, caused by protein adsorption.

In Figure 7.7, we see a simple model of a biological membrane that was formed by detergent dialysis on an alkylsilanated electrode surface. This is basically a metal current collector coated with a conducting metal oxide and an alkylsilane layer. On top of this, Li et al. [Li 1991, Li 1994] have coated the visual receptor protein rhodopsin and a lipid layer. Membrane structures that contain the visual receptor protein rhodopsin were formed by detergent dialysis on platinum, silicon oxide, titanium oxide, and indium-tin oxide electrodes. How would you measure such assembly? You can think about doing X-ray or neutron reflectometry on it because it appears to be a regular bi-layer system. This would give account of the order, the ordering of the layer components. You certainly can study metal centers of the protein with conventional protein X-ray

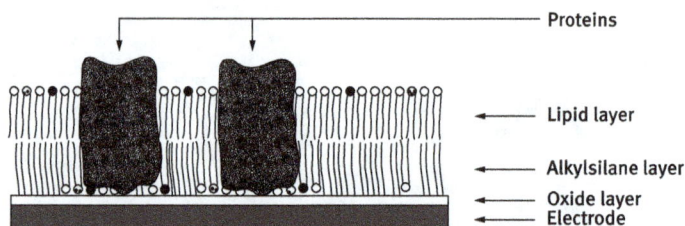

Proteins

Lipid layer

Alkylsilane layer

Oxide layer

Electrode

Figure 7.7: Model of a single surface-bound membrane formed by detergent dialysis on an alkylsilanated electrode surface [Li 1991, Li 1994]. Reprinted with permission from (Li, J. G., Downer, N. W., Smith, H. G., Evaluation of surface-bound membranes with electrochemical impedance spectroscopy. *Biomembrane Electrochemistry* **1994**, *235*, 491–510.). Copyright (1994) American Chemical Society.

spectroscopy such as EXAFS or Kβ spectroscopy and get information on bond lengths, oxidation states, and spin states. Would they differ as the proteins are attached on an electrode surface, embedded in a cell wall matrix or lipid layer? Would the structure of the lipid layer be affected by insertion of the proteins? What about the charge transfer between proteins and lipid layers and electrode and electrolyte on such an assembly?

Electrochemical impedance spectroscopy was used to evaluate the biomembrane structures and their electrical properties. A model equivalent circuit is proposed to describe the membrane-electrode interface (Figure 7.8). The data suggest that the surface structure is a complete single-membrane bilayer with coverage of 0.97 and with long-term stability [Li 1994].

The authors separate the problem in three components in a series circuit (Figure 7.8); this is a parallel circuit of resistivity (actually a conductivity R_{ox}) and capacitor in a parallel representing the metal oxide electrode support, then the same for the electrochemical double layer and wiring R_u. In between, we have the more complicated membrane structure that also includes a double layer segment within the membrane assembly, a similar component for the lipid layer and the same for the alkysilane layer. We have hence a very complex arrangement of components for the electric circuit. For the very high frequencies, the authors simplified the circuit by only using capacitive elements when the high frequency range is considered (Figure 7.7).

Is it possible to match the data from electroanalytical measurements with data from valence band PES?

It depends on the creativity of the researcher and the experimenter to come up with an experiment design to measure the proposed bioelectrode assembly. Some researchers may have a clear-cut question they want to find an answer for and then optimize the design of the experiment for this question. Other researchers have only their fundamental curiosity and investigate the system for no particular reason [Feynman 2005]. Then they observe something interesting and then the question comes up, maybe. Moreover, then they dig deeper in order to find some sense in their system.

Junghans et al. built a flow cell in which they can subject living tissue cells to neutron reflectometry under a physiological environment. In Figure 7.9, their flow cell is shown

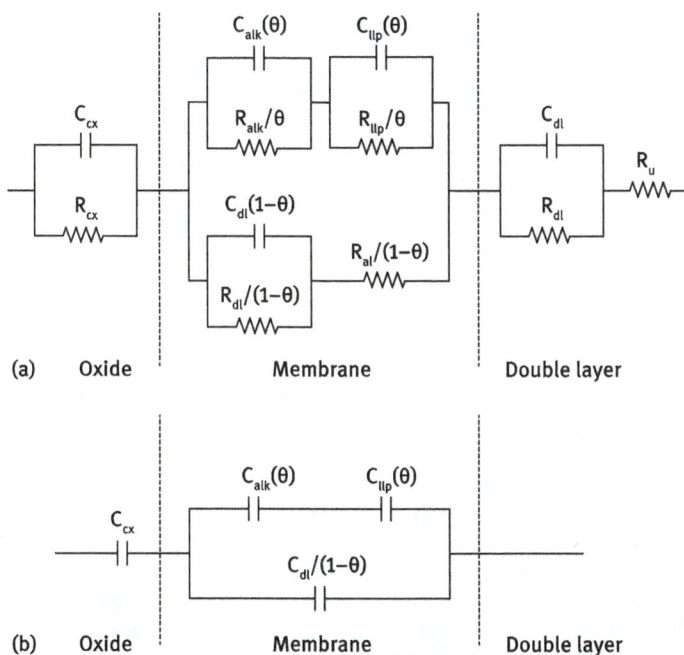

(a)　Oxide　　　　　　Membrane　　　　　　Double layer

(b)　Oxide　　　　　　Membrane　　　　　　Double layer

Figure 7.8: Proposed equivalent circuit for surface-bound membrane electrode interface. (b) Simplified equivalent circuit valid at higher frequency region. [Li 1991, Li 1994]. Reprinted with permission from (Li, J. G., Downer, N. W., Smith, H. G., Evaluation of surface-bound membranes with electrochemical impedance spectroscopy. *Biomembrane Electrochemistry* **1994**, *235*, 491–510.). Copyright (1994) American Chemical Society.

and also neutron scattering data of a film of mouse fibroblasts that are deposited on a quartz substrate. Note that when using the term "cell," we have to distinguish here the designed flow cell that is part of the experimental apparatus, and the living fibroblast cell that is the biological study object. Depending on the density of the fibroblast cell stacking, we see differences in the neutron scattering curves. There are many options we can think of in this experiment. In addition to different stacking density, we can think about using liquids with different pH. We can try to cool or heat the cell, and thus the sample. We can deposit the fibroblasts on a conducting substrate, introduce a counter electrode, and apply an electrochemical bias. We can flush the electrolyte fast, and thus impose a shear force on the fibroblast while measuring them with neutron scattering. These are just quick suggestions. Not all experiments that we do at large-scale facilities are realistic representations of situations that happen in nature. Often we have to cut down on parameters and simplify situations to models, which we can handle easier. Moreover, there is nothing wrong with this. Experience shows that these simple cases will be solved, and eventually the next steps to more realistic situations are taken, and so happens progress in science [Feynman 2005].

In Figure 7.10, we see how Junghans and associates have looked into the fibroblast cells at different temperatures (25 °C and 37 °C, which is a significant temperature differ-

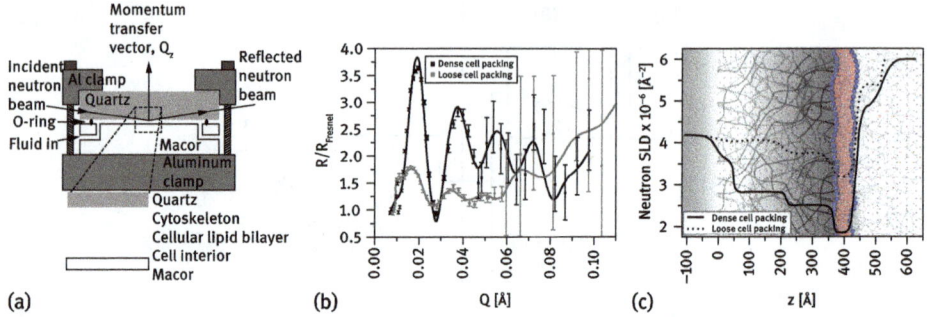

(a) (b) Q [Å] (c) z [Å]

Figure 7.9: Schematic of solid–liquid interface flow cell used for all live cell experiments and neutron reflectometry profiles and corresponding neutron scattering length distributions (SLD) of mouse fibroblasts (a). Schematic of the neutron flow cell. Confluent layer of cells is grown on the quartz substrate (see the inset), which is inserted into the flow cell. The neutron beam enters through the quartz at a glancing angle and scatters from the adhesion layer and cell membrane interfaces. Fresnel divided NR profiles (b) and corresponding SLD profiles (c) for high (black) and low (gray) cell surface densities. NR data are shown by closed squares, and error bars indicate 1 SD. The lower surface cell density is evident from the decreased scattering intensity (b) and the increased SLD in the membrane region (360–440 Å) and interior of the cell (440–600 Å) (b). [Junghans 2015] Reprinted with permission from [Junghans, A., Watkins, E. B., Barker, R. D., Singh, S., Waltman, M. J., Smith, H. L., Pocivavsek, L., Majewski, J., Analysis of biosurfaces by neutron reflectometry: From simple to complex interfaces. **2015**]. Copyright [2015], American Vacuum Society.

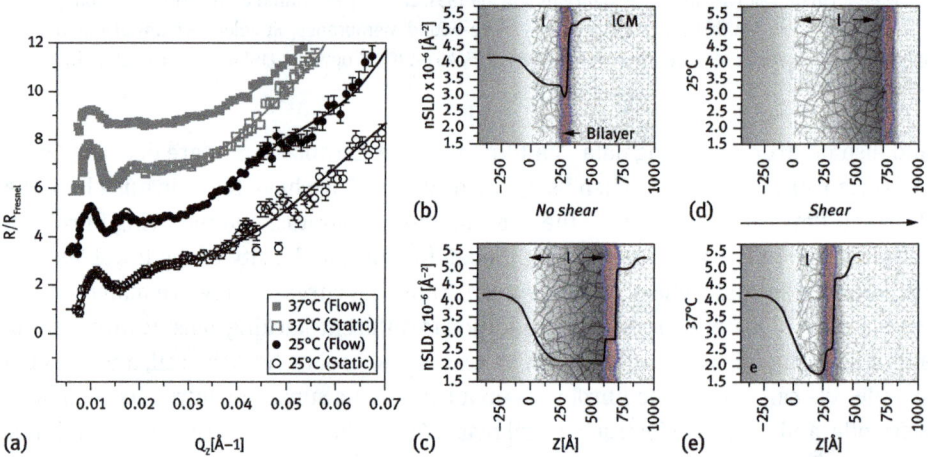

(a) Q_z[Å–1] (c) z[Å] (e) z[Å]

Figure 7.10: Differences in response to shear flow of healthy endothelial cells at ambient and physiological temperatures. Left (a) Fresnel-divided NR measurements (circles/squares with error bars) and corresponding best-fit models (solid lines) at the conditions studied. Black (open circles): 25° (static); black (closed circles): 25° (shear); gray (open squares): 37° (static); gray (closed squares): 37° (shear). The NR spectra are offset along y-axis for clarity. Right [(b)–(e)]: SLD profiles obtained by fitting the data sets using a three-box model (extracellular matrix—cell membrane—partial cell interior). [Junghans 2015] Reprinted with permission from [Junghans, A., Watkins, E. B., Barker, R. D., Singh, S., Waltman, M. J., Smith, H. L., Pocivavsek, L., Majewski, J., Analysis of biosurfaces by neutron reflectometry: from simple to complex interfaces. **2015**]. Copyright [2015], American Vacuum Society.

ence in physiological systems) and different flow rates (static vs. flowing). The conformational changes that occur due to changes in flow and temperature are qualitatively reflected in the neutron scattering curves; see the left panel of Figure 7.10. The scattering length density is plotted versus the "thickness" axis of the bilayer assembly on the right panels in Figure 7.10. We thus obtain a microscopic image, but not by imaging but by scattering and by interpreting the scattering data on the basis of a geometrical model.

The oxygen evolving complex (OEC) in photosystem II (PSII) has for a long time been subject of study in photosynthesis research. Central to the OEC is two manganese ions, which are connected by two oxygen ligands (compare, e. g., Figure 2.6). The electronic and magnetic transport along an arrangement of ions depends on their orbital overlap. The distance between atoms is therefore very important. The distance of the Mn ions in the OEC of PSII has been a matter of debate for a long time. Exposure of molecules like PSII to X-rays may cause radiation damage. This holds particularly where multiple scans need to be taken over extended periods of time because of very dilute specimen like metalloproteins. The sample may become hot and gradually deteriorate, and the atomic distances may change slightly over time [Askerka 2017].

7.3.3 Protein spectroscopy

In Part 2 of this book, I showed examples on X-ray absorption and emission on photosynthesis proteins such as hydrogenase and oxygenase. EXAFS is used as a geometric structure determination method [Gu 2003]. From this, we can learn about bonding length and coordination shells, as was explained in Chapter 3.4. $K\beta$ spectroscopy is used for the determination of the oxidation state of metal centers, and also for the determination of their spin states [Bergmann 2009, Glatzel 2001, Messinger 2001, Visser 2001]. Soft X-ray spectroscopy such as NEXAFS has widely been used for learning about the Ni electronic structure (oxidation state and spin state) in Ni hydrogenases and other enzymes [Grush 1996, Gu 2002, Ralston 2000, Wang 2001a, Wang 2000]. In Figure 7.11, we see the Fe NEXAFS spectrum of a structural model of a hydrogenase with stoichiometry $C_{12}H_{15}FeI_2O_3P$ and structure formula as shown in Figure 7.12. We basically have a coordination compound with Fe in the center of the molecule. The sample was in the form of powder and measured at BL 9.3.2 at the ALS. The NEXAFS spectrum shows clearly the Fe 2p multiplet and also the t2g and eg orbital structures. The red solid line in Figure 7.11 is the simulated multiplet spectrum with a mixture of 35 % Fe^{2+} and 65 % Fe^{3+}. The *differential of quanta* Dq for the crystal field splitting energy was taken 1.8 eV for Fe^{2+} and 1.7 eV for Fe^{3+}. The Slater–Condon parameters (Coulomb integral and exchange integral [Eder 2012]) were taken 10 % and 70 %, respectively. Finally, a Gaussian broadening of states with a width of 0.4 and 0.2 was conducted, which yielded the spectrum shown in Figure 7.11.

Vyalikh et al. [Vyalikh 2004, Vyalikh 2005, Vyalikh 2006] have investigated surface layers from *Bacillus sphaericus*, a protein template for metal nanostructures. The pro-

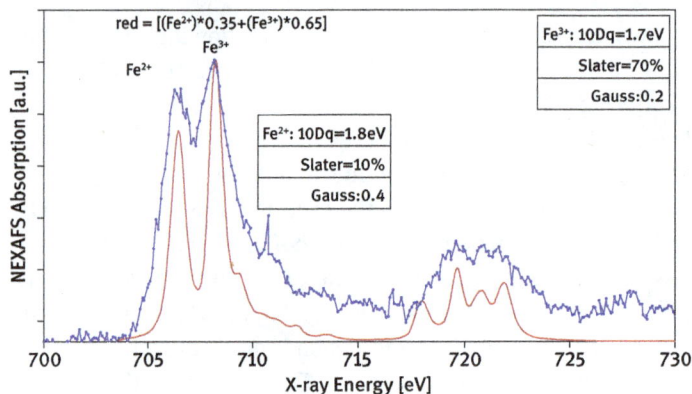

Figure 7.11: Fe 2p NEXAFS spectrum of sample B1 ($C_{21}H_{15}FeI_2O_3P$) plus the multiplet simulation. The simulation (conducted by Selma Erat, Empa/ETHZ) roughly reflects the actual Fe 2p NEXAFS spectrum. The sample B1 was received from Professor Xile Hu, EPFL Lausanne.

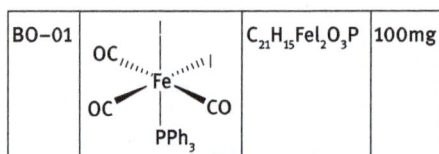

Figure 7.12: Excerpt of data sheet from three structural models of hydrogenase, showing the structure formula and stoichiometry of $C_{21}H_{15}FeI_2O_3P$ (Professor Xile Hu, EPFL). The samples were measured with NEXAFS spectroscopy in UHV at the Fe 2p energy range.

tein is two-dimensional and has semiconductor characteristics with 3 eV band gap energy. They used PES and NEXAFS for the determination of the occupied and unoccupied valence electronic states of the bacterial film. In a follow-up study [Vyalikh 2009a], they did the PES in resonant mode and found the electrons from the π clouds of aromatic rings make significant contributions to the HOMO level. They investigated with resonant PES also the charge transport [Vyalikh 2009c]. They figured a charge transport hopping mechanism based on torsional motions of the peptide backbone, and determined the lifetime of electrons excited into the LUMO to be around 100 fs. These experiments on proteins were conducted at the High Resolution Russian-German Beamline at BESSY in Berlin [Molodtsov 2009].

We will see later in the part about radiation damage how the NEXAFS of biological samples may change from scan to scan. Such changes are not so obvious when spectra are recorded with hard X-rays, such as with XANES or Kβ spectroscopy. One reason for this is that the soft X-rays are more surface sensitive, and structural changes occur rather at the surface when the samples are exposed to radiation and residual humidity from exposure to the ambient. So to speak, the radiation damages are more apparent in soft

X-ray spectroscopy. This is also because of the larger absorption cross-sections in the soft X-ray energy range where more thermal energy is dissipated in the sample.

7.3.4 Bioelectrodes, biofilms, and bioelectricity

One current field of interest is biofuel cells and microbial fuel cells where electrode supports are functionalized with microbes. Standard electrochemical testing and diagnostics is conducted on such systems such as CV, EIS, cell cycling, and degradation studies. The field of bioelectrochemistry is more advanced for biosensors (signal generation and transformation). Therefore, it serves as a very good reference for those who apply bioelectrochemistry for generation and conversion. Dye sensitization is a well-established field in solar energy conversion, but it is not considered bioelectrochemistry. Dyes are organic or bioorganic molecules that collect light and generate hole pairs upon illumination and inject electrons into semiconductors.

> "The interest and dedication towards improving the photovoltaic hydrogen technology and preparing it for commercialization unfortunately decreased again as the first modern energy crisis passed and had to wait for the next energy crisis which seems to have arrived now." [Tributsch 2008].

In very early papers of the late 1950s and early 1960s, Melvin Calvin [Calvin 1957, Calvin 1960, Calvin 1962] demonstrated how shredded plants, spinach leaves, are used to isolate chlorophyll and then distribute this over an interdigitated electrode assembly. Calvin used Aquadaq [Acheson 1910] graphite paint to make interdigitated electrodes for his experiments; see Figure 7.14. This forms a bioelectronic interface where functionalities of proteins can be directly addressed with electric signals under light exposure or in the dark. Calvin explains how the components of the cell such as chromophores, thylakoid membrane, cofactors, chlorophyll, etc., work together as functional units for photosynthesis. In particular, he states that the function of the cell originates from the connectivity of these units where electron transport and ion transport take place. Although his explanations sound nowadays somewhat naive, he appears to be the first one who explicitly rationalizes the photosynthetic apparatus as a molecular machine, even a molecular factory and industrial complex. Addressing a biological apparatus with electric stimulation was first reported by Luigi Galvani, who connected a dead frog with a galvanic cell [Galvani 1791, Piccolino 1998].

Biological systems constitute *per se* electrochemical systems because some of their vital processes are of chemical redox nature where electrons and ions are involved in charge and mass balance. It appears that Luigi Galvani was the first one who reported an electrophysiological experiment on a frog leg with a galvanic cell [Galvani 1791, Piccolino 1998]. The author of this book is not aware of any particular electrochemical processes that might have taken place at the interface of contact wires and frog leg tissue. Possibly it was what is called today "tissue injury current" [Becker 1962, Becker 1969, Becker 1982, Becker 1985, Pitkänen 2013, Seyfarth 2006, Szent-Gyorgyi 1941, Szent-Györgyi 1941]. However, from the conceptual point of view Galvani's experiment was pioneering and opened up the field of bioelectrochemistry. Inspired by Benjamin Franklin's studies on electricity, researchers investigated in the second-half of the 18th century the influence of electricity, i. e., electric fields and currents on the growth of plants. The manipulation of growth of plants by electric potentials and currents has been known over 100 years ago as electrocultures [Oswald 1933]. The so-called Becker currents are believed to promote the regeneration of damaged bones or limbs [Bassett 1964, Lavine 1968, Lavine 1969a, Lavine 1969b, Lavine 1971, Lavine 1972].

Mentovich et al. [Mentovich 2012] investigated the so-called nanojunctions from C60 and protein macromolecules. Their aim was to take control over electrical properties of protein assemblies at the molecular scale. They demonstrated this in a self-assembled protein based transistor device. They conducted the electrical measurements on the device with an XPS-based element specific method as already mentioned in [Cohen 2004].

In his experiments, Calvin uses the electric terminals merely for electroanalytical purposes. Nowadays, assemblies of proteins and inorganic electrodes are actively researched for as electrochemical sensors. In sensors, the generation and forwarding of electric signals is important for the function of the sensor device. We want to elaborate on the transition from electric signal transfer to electric power transfer, which could amount to a more difficult task than the two aforementioned ones, particularly because power transport might involve substantial dissipation of Joule (caloric, thermal) energy into the proteins that might cause denaturation and degradation.

This points to the case that electric transport, i. e., electronic and ionic charge transport, and charge transfer are necessary details to be addressed for power conversion. Hence, immobilization of proteins alone should not do the job. Rather, we have to provide that the proteins are covalently attached, or electrically wired so to speak, to the solid electrode.

> As Rabinovich has pointed out [Rabinowicz 1961], "There is no reason to assume that man will never discover inexpensive photochemical systems permitting the conversion of, say, 10 or 20 % of incident solar energy either into potential chemical energy (such as that of an explosive gas mixture) or into electrical energy."

One example where biofilms were studied with XPS in a different context than relevant for our purpose is found at Neat et al. [Neal 2004]. Nuclear waste management wants to make sure that radioactive species in nuclear waste remains confined and does not spread around in the environment. Like other metals, uranium can participate in complexation, precipitation, colloid formation, and dissolution. Microbes and bacteria such as the sulfate reducing *Desulfovibrio* spp. affect the immobilization of uranium by enzymatic reduction of U(IV) species such as $UO_2{}^{2-}$. Neal et al. attached the bacterium *Desulfovibrio desulfuricans* onto $\alpha\text{-}Fe_2O_3$ and investigated the molecular structure of uranium containing bulk aqueous phases flowing over and interacting with the biofilm with XPS.

The U4f$_{7/2}$ core region and U5f3 valence band spectra suggest that the biofilm promotes accumulation of U(VI) and U(IV) species on the substrate, whereas without biofilm only the U(VI) species are accumulated.

Charge transfer across interfaces depends on the electronic structure of these interfaces. A method for the assessment of the electronic structure of surfaces and interfaces is XPS and PES. In Brizzolara et al. [Brizzolara 1994, Brizzolara 1997], the conclusion about covalent attachment of the proton pump protein bacteriorhodopsin via genetic substitution of cysteine for serine (S35C) was based partially on the XPS core level spectra of sulfur, i. e., via detection of the chemical shift of the sulfur core level spectrum. Bacteriorhodopsin is a protein that acts as a light-driven proton pump [Balasubramanian 2013, Braun 2015a, Wang 2014]. An early valence band (VB) XPS study on a protein (D-luciferin) is presented in Wada et al. [Wada 1989], where XPS helped sketch a model for the luminescence. Prior to this, Wada et al. [Wada 1977] measured the valence band of all-trans retinal 1 and β-carotene with XPS. The results of their experiments were in agreement with the mathematically determined energy levels of the molecular orbital. Beck et al. investigated the HOMO-LUMO of beta carotene with a laboratory plasma-based NEXAFS apparatus and PES [Beck 2001, Beck 2002]. Brizzoloara investigated the chemical shift of the various components (S-N-O-C) with XPS in the methionine sulfur containing amino acid [Brizzolara 1996].

Maeda et al. [Maeda 1978] used valence band XPS for the study of a two retinal molecules. Along with molecular orbital calculations, they found that spectral differences corresponded to the differences in the molecular energy levels of the retinals. Retinal molecules are carotinoids which exist in a *trans* and in a *cis* conformation. They are part of the rhodopsin molecule and are important for the vision apparatus. Zhang et al. [Zhang 2008] verified with a patch clamp method glycine-induced electric currents in retinal slices from bull frog.

Takahashi et al. [Takahashi 1982] measured VB XPS of 13-*cis* and all-*trans* fluorophenyl-retinals (see Figure 7.13 for the structure formulae) and observed differences in the O 2s-derived parts of the VB between the two molecule forms. They interpreted their experimental results with the help of the CNDO/S molecular orbital method [Takahashi 1984].

The two external point-charge (TEPC) model for *bacteriorhodopsin* (bR) was investigated with XPS and assisted by and CNDO/S molecular orbital calculations. One of their objectives was the investigation of a point charge near the β-ionone ring. They conducted XPS measurements on fluorophenyl retinal (F-ret) and their derivatives (Schiff base, protonated Schiff base, and bacteriorhodopsin analogs (F-bR)), and paid attention to the chemical shift of the F 1s in the core level spectrum. They observed no meaningful differences among these species. This was despite numerical calculations based on an assumption of the TEPC model, which had predicted a chemical shift by around 3 eV between the F-ret and the F-bR form. This outcome raised a question on the validity of the TEPC model. The same conclusion has been reached by the present study of absorption maxima of F-ret and their derivatives. It is important to note here that calculations,

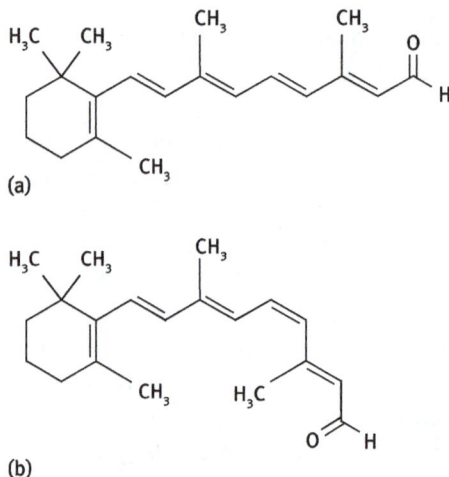

(a)

(b)

Figure 7.13: all-*trans*-Retinal (top) 11-*cis*-Retinal (bottom). Reprinted from Wikipedia.

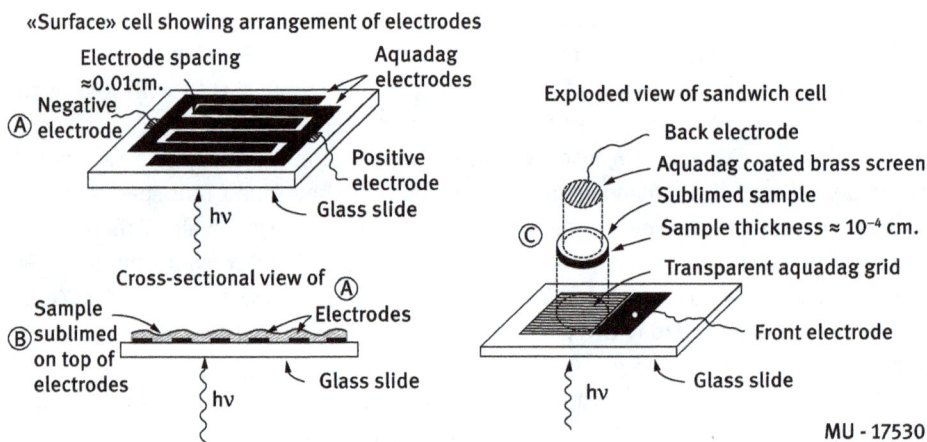

Figure 7.14: Melvin Calvin's procedure for coating a transparent glass slide substrate with interdigital electrodes from Aquadag colloidal graphite solution. On this electrode, a glass slide was then coated the bioorganic film by sublimation, with around 1-micrometer thickness. On top of this film, he put a counter electrode from a brass grid, which was also coated with Aquadag. (Reproduced from [Calvin 1960] with permission from Ernest Orlando Lawrence Berkeley National Laboratory).

not necessarily only calculations of spectra, can be very helpful in verifying or falsifying concepts and theories. By doing the calculations, we discipline ourselves to more rigor in the interpretation of our experimental results. Mathematical formalisms in any respect can be helpful here. This does not mean, however, that experimental spectra are of little value unless they are accompanied by calculated spectra. Every experimentalist is expected to understand the theory of the experimental work. Moreover, a real theoretician will have an experiment in mind that would describe the theory. In the spirit of

"Theorie ist, wenn man die Praxis versteht," it is therefore a valid statement that there is no sharp separation line between experiment and theory.

Although literature on the electronic structure of the interfaces of biological macromolecules and inorganic materials is scarce, some pioneering studies have been published. A PES and an NEXAFS study on the electronic structure of an ultrathin film of the surface layer of *Bacillus sphaericus* deposited on Si wafers showed how the highest occupied and lowest unoccupied molecular orbitals (HOMO and LUMO) constitute the band gap of the assembly [Vyalikh 2004]. In an extension of that study, these authors showed with resonant PES how charge transport evolved from torsion effects in the protein [Vyalikh 2009c].

7.3.5 Photo-electrochemical studies on cyanobacteria during γ-irradiation

We learned in the previous part about the interest on the influence of cosmic radiation on vanadium oxide thermochromic windows [Braun 2015h]. Such thermochromic window materials are of interest for the passive management of light influence. At high temperatures, they may be absorbing more light, and thus block transmission. For satellites, this can be very important. Exposure of such windows to cosmic radiation can alter their electronic structure, as we have seen by PES measurements. Consequently, their transport properties could be changed and eventually suffer.

The current study is relevant for the establishment of sustainable life in outer space. The astronaut suit has the purpose to protect astronauts from cosmic radiation. Still, the human body in this suit receives a considerable radiation dose. Is it possible to bring life into space? What effect has such cosmic radiation, e. g., on plants and algae which a future human colony in space would feed on [Baque 2013, Cockell 2011, Cwikla 2014, Lehto 2006]? Consider our Earth with its biosphere as a spaceship cruising in space self-sustainable with the necessary energy coming from the sun. This is known as Biosphere 1. Consider an actual spaceship with a biosphere, or any other isolated and self-sustaining biosphere. Such system, Biosphere 2, has been simulated near the city of Oracle in the Sonoran desert in Arizona [Allen 1991, Allen 2003, Hohler 2010], as shown in Figure 7.15. When such system is in outer space, we need to consider its exposure to cosmic radiation. With support from a European COST Action [Rea 2011], I have joined Paul Janssen from SCK·CEN, the Belgian Nuclear Research Center (Molecular and Cellular Biology, Research Unit of Microbiology), who has been working on this topic for the European Space Agency (ESA) and the International Space Station (ISS) to study this question. This question is not only relevant for life in outer space, but everywhere with increased radiation dose. This section of the book has been written mostly by Paul Janssen. I am grateful I may use the text for this book. The magnetic field orientation in the Arctica and Antarctica makes that there is less protection from cosmic radiation than in the lower numbered latitudes. We have tried to investigate biosolar cells for hydrogen pro-

Figure 7.15: Panorama view of the Biosphere 2 complex, a "greenhouse" so-to-speak, near the Santa Catalina Mountain range near Oracle, Arizona.

duction under antarctic conditions, where the summer lasts 3 months and can provide for 24 hours per day light [Aguey-Zinsou 2015].

Solar radiation has played an essential role in the origin and the early evolution of life by providing the energy required for chemical reactions [Miller 1959] and by creating genetic variation through the formation of DNA mutations. Today, only a fraction of solar radiation and cosmic radiation arrives on the Earth's surface because the radiation is shielded by the current atmosphere and by the magnetic field of the Earth. However, about 4 to 3.5 billion years (Ga) ago, while the sun was less bright than today (i. e., 74–77 % of today's bolometric luminosity [Bahcall 2001]), early life must have been subjected to much higher doses of radiation including UV, X-rays, and γ-radiation. Around that time, about 3.4 Ga ago, the first nonoxygenic photosynthetic bacteria emerged, absorbing low-energy solar radiation (i. e., near-infrared) for the oxidation of ferrous iron (Fe^{2+}), carbon (formate, oxolate, bicarbonate), and sulfur (H_2S, S_x), substrates that were abundantly present in the early Earth's crust and atmosphere. The biggest electron-donor pool, H_2O, however, remained biologically inaccessible because the liberation of electrons from H_2O requires a relatively large investment in energy. Within a time span of 700 million years, somewhere between 3.4 and 2.7 Ga, an organism evolved that was capable of oxidizing water by extracting four electrons from two molecules of water freeing O_2 as a by-product. The only viable source for this reaction is higher-energy solar radiation, i. e., visible light. Such organisms, now known as cyanobacteria, triggered the onset of an oxygenated atmosphere and are directly responsible for the evolution of aerobic metabolism, resulting in an increasingly complex life with a manifold of novel metabolic reactions previously unknown to anaerobes. In this Archean period (before 2.5 Ga), the young Sun's spectral energy distribution (SED) and Earth's atmospheric conditions were much different from today's, with different solar activity cycles and flares and very low atmospheric concentrations of oxygen, and hence ozone, as well as CO_2 [Cnossen 2007]. Although ozone is an important shielding component of UV radiation in the present atmosphere, it was postulated that atmospheric changes in CO_2 levels could have caused large differences in solar flux at the surface of the early Earth [Cnossen 2007].

Although the calculations of [Cnossen 2007] indicate that most of the X-ray and UV radiation below 200 nm were probably for a large part absorbed by the Archean Earth's atmosphere, the amount of radiation in the 200–300 nm wavelength range reaching the surface of the Archean Earth could have been several orders of magnitude larger than what is received at present. Early life, including the first photosynthetic organisms, and later on also the oxygenic cyanobacteria, must thus have been exposed to much higher levels of damaging radiation than is the case at present. Hence, Archean organisms may have learned to cope with the higher amounts of UV radiation by molecular evolution, i. e., in terms of DNA repair and the avoidance and removal of reactive oxygen species (ROS). Likewise, in response to UV abundance and a 25 % less luminous sun, these microbes may have extended their spectral range of photosynthetically active radiation (PAR) toward long-wave UV (UV-A; 315–390 nm). Such a spectral expansion has been noted on the other side of the narrow range of visible light (see the Appendix, Electromagnetic Spectrum) for various microbes, i. e., for both oxygenic and anoxygenic photosynthesis at wavelengths of up to 750 and up to 900 nm, respectively [Chen 2011a], but at the UV side, such an expansion is very rare because for photosynthesis to occur at wavelengths of 360 nm or below, a trade-off between the damaging effects of the radiation and the gain of available energy for photosynthesis must be in place. Some red macroalga, such as *Gracilaria lemaneiformis*, seem to have accomplished this feat and can utilize the energy of UV-A efficiently to drive photosynthesis under reduced levels of solar radiation [Xu 2010].

Whether the high photonic energy of electromagnetic radiation of shorter wavelengths, i. e., UV-B, UV-C, X-rays, and γ-rays, could be photosynthetically used is subject to debate. It has been shown that continuous irradiation with Co-60 gamma rays, without light, slightly increased photosynthesis in two photosynthetic organisms, *Rhodopseudomonas capsulata* and *Anacystis nidulans* [Luckey 2008], but it remains unclear whether this effect was the result of pigment activation followed by energy transfer or just the haphazard action of free radicals inadvertently inducing photosynthesis. A third option may be that electrons are not released from the pigment in the classical way, i. e., as is the case with visible light, but rather through direct interaction of the gamma ray with the pigment—or with matter surrounding the pigment—as it is well known that low-energy gamma rays can also transfer a photon to an atomic electron either by the photoelectric or the Compton effect. In this sense, ionizing radiation could be a strong source of energy, particularly in places were light is very limiting or nonexistent such as deep sea vents and deep geological deposits. Interestingly, it has been reported by Dadachova et al. [Dadachova 2007, Dadachova 2008] that exposure of the pigment melanin to high levels of Cs-137 γ-rays increases its electron transfer properties and that melanized microbes subjected to such gamma rays actually display an enhanced growth rate. Perhaps not coincidently, melanin [Greta Faccio 2016, Ihssen 2014] has, next to a possible role in electron transfer and electron energy moderation, also radioprotective properties, and melanized organisms are often dominantly present in high-radiation environments such as space stations, antarctic mountains, and in used nuclear fuel storage basins.

Early life evolved under so-called "extreme" conditions, including high levels of radiation, i. e., mainly solar UV but likely also substantial levels of γ-radiation particularly from radionuclides in the Earth's crust (radionuclides that now mostly have decayed into stable forms). Of course, the term "extreme" is somewhat misplaced as these were the normal conditions 4-3.5 Ga ago: the utilization of photonic energy from UV, X-ray and γ-radiation may very well be an old invention of nature, with low-energy radiation (light) only becoming a source for electron activation at a much later stage. From this perspective, present-time organisms may have adapted to present-time "extreme" conditions (a brighter sun, less ionizing radiation, and moderate temperatures), and many of them may have lost, in part or in full, the ability to utilize high-energy types of radiation.

The edible cyanobacterium *Arthrospira* sp. PCC 8005 was selected by the European Space Agency (ESA) as an oxygen producer as well as a nutritional end product for the life support system MELiSSA [Hendrickx 2006]. To enable whole-cell gene expression studies, its genome was fully sequenced [Janssen 2010] and gradually annotated to a finally assembled version v5 in 2014. Early experiments in 2007–2009 indicated that the PSII quantum yield and oxygen generation were significantly higher when cells were irradiated with simulated high and low LET spectra (chronic doses in the range of 1.6 to 31.4 mGy; total absorbed dose, equivalent to 10 to 200 days at the International Space Station) using a combination of ^{252}Cf (n) and ^{137}Cs (γ) sources [Morin 2009]. Using ^{60}Co alone in later experiments using acute doses of 200–3200 Gy in the RITA facility (782 Gy.h^{-1} dose rate) at SKC·CEN also showed this increase in a triple-sample executed experiment but not in subsequent irradiation experiments with 800–6000 Gy using the BRIGITTE facility (20,000 Gy.h^{-1} dose rate) at SKC·CEN [Badri 2015, Morin 2009]. The known adaptation of strain PCC 8005 to continuous intense light and its high tolerance to acute doses of ionizing radiation (with the ability to survive up to 6400 Gy of ^{60}CoCγ-rays, delivered as 527 Gy.h^{-1} in the RITA facility of SKC·CEN [Badri 2015]), makes it an ideal organism to function in bioreactors such as MELiSSA, i. e., in spacecraft or space stations where continuous operation and long-term exposure to cosmic radiation is anticipated (as noted above, radiation levels at geostationary spacecraft are at least five orders of magnitude lower). Because photosynthetic efficiency, as measured by chlorophyll fluorescence in a Dual-PAM 100 System and expressed as quantum yield, and oxygen generation both appeared to be consistently higher at low absorbed doses γ-radiation [Badri 2015], we decided to further investigate the conversion of photonic energy into electron currents for strain PCC 8005.

Because my group at Empa had already been working on bioelectrodes where metal oxide semiconductor photoelectrodes are functionalized with light harvesting proteins from cyanobacteria (phycocyanin), and all-inclusive cyanobacterial biofilms deposited on these electrodes [Bora 2012, Bora 2013a, Braun 2015b], we set out to investigate the photo-electrochemical properties of such a bioelectrode using *Arthrospira* sp. PCC 8005 cells while under irradiation with low doses of γ-rays. In particular, we have assembled a bioelectrode by growing a cyanobacterial biofilm (*Arthrospira* sp. PCC 8005) on a α-Fe_2O_3 photoelectrode. This electrode was used in a three-electrode photo-electrochemical cell with phosphate-buffered saline as pH~7 electrolyte. We have assessed the photo-electrochemical properties of the bioelectrode with wavelength-dependent chronoamperometry at 1200 mV bias potential vs. the Ag/AgCl reference electrode compared to dark current measurements and exposure to ^{60}Co γ-radiation.

The substrates were glass slides coated with a thin film of fluorine doped tin oxide. The hematite film was coated at Empa in Dübendorf by using a soft chemistry organic precursor as dip coating solution. The dip-coated glass slide was then subjected to annealing in an air-vented muffle furnace at 500 °C for 2 hours and then sent to SKC·CEN in Belgium. The *Arthrospira* sp. cultures were then coated on iron oxide slides to grow as bioelectrodes at SKC·CEN.

The bioelectrodes were mounted in a specifically designed spectro-electrochemical cell made from Teflon® equipped with three electrode terminals for counter electrode (platinum metal), reference electrode (Ag^+/AgCl), and working electrode (biofilm deposited on hematite coated on fluorine tin oxide (FTO) glass). Figure 7.16 shows how the cyanobacteria are irregularly distributed over the reddish brownish iron oxide electrode. On the left, you see such electrode clamped to a sample holder for the electric contacting. On the right, you see several dishes with electrodes surrounded by the green cyanobacteria cultures.

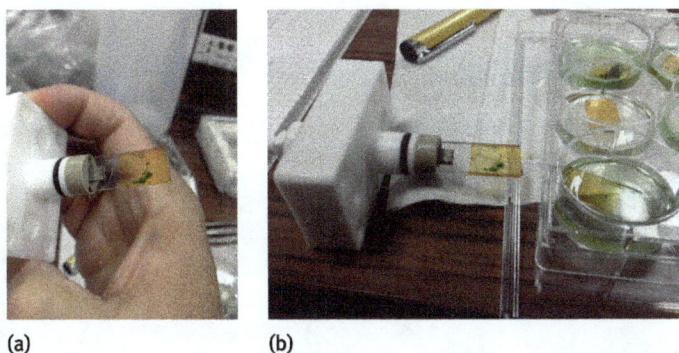

(a)   (b)

Figure 7.16: (a) Teflon sample holder with one clamped transparent FTO glass slide, coated with brownish hematite film and greenish biofilm (*Arthrospira* sp. PCC 8005, sample no. 3). (b) Biofilm was selected from 6 aliquotes.

The electrolyte was 0.05 molar phosphate-buffered saline. A potentiostat model Voltalab80 from Radiometer Analytical (Lange AG, Hegnau, CH) was used to run cyclic voltammetry and photocurrent spectra, the latter with monochromatized light from a LOT Oriel Solar Simulator (LOT Quantum Design GmbH, Darmstadt, Germany). For

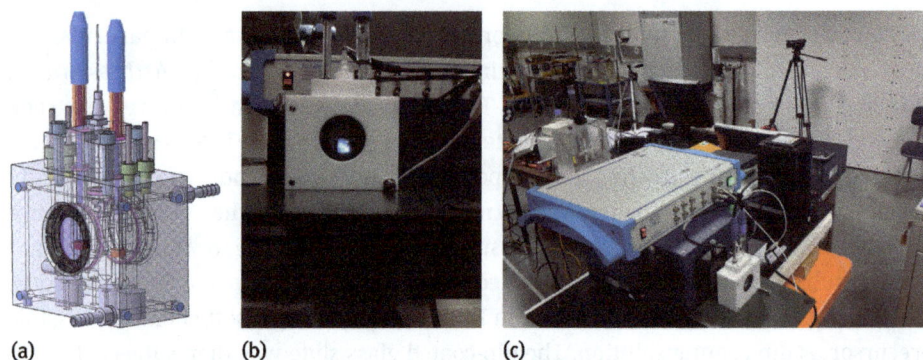

(a) (b) (c)

Figure 7.17: (a) Schematic sketch of the Teflon spectro-electrochemical cell. (b) Cell plus biofilm electrode no. 3 inside, illuminated with 540 nm light during chronoamperometry. (c) Cell connected to Voltalab80 Potentiostat and mounted in front of monochromator, inside SKC·CEN Building KAL with γ-radiation beamline.

the experiment with γ-radiation, chronoamperometry scans were recorded for specifically selected optical wavelengths, i. e., 420, 540, 600, 620, 640, 660, 680, and 700 nm for 600 seconds each at 1200 mV DC bias. Between the wavelength changes, dark current measurements were done for the same period of 600 seconds. Part of the experimental setup, which we transported from Empa to SKC·CEN, is shown in Figure 7.17.

Table 7.2 shows the parameters under which the electrodes were exposed to γ-radiation.

Table 7.2: The irradiation experiment was done by Liviu-Cristian Mihailescu of SKC·CEN. The geometry was at the horizontal installation at a distance of 96.4 cm with the following sources.

Source	Isotope	Photon energy	*K_{air} rate mGyh^{-1}
P4	Co-60	1250 keV	0.82 ± 0.04
P12	Cs-137	661 keV	9.31 ± 0.42

*K_{air} = air kerma = kinetic energy released per unit mass.

We recorded cyclic voltammograms, impedance spectra, and chronoamperograms. For the bioelectrode sample no. 3, we made a systematic study in three parts, which is outlined in detail in Table 7.3.

Our experiment is based on the photoabsorption characteristics of the *Arthrospira* sp. (Figure 7.18, left panel), with respect to which we have selected the monochromator wavelengths.

In Experiment 1, we find that the photocurrents are generally lower than the dark currents. This is a surprising finding because typically, in pristine semiconductors and in semiconductors functionalized with light harvesting proteins, the photocurrents are

Table 7.3: Experiment protocol for bioelectrode no. 3.

Experiment no.	Description
Experiment 1	Dark current for 15 seconds at 200 mV and for 3 minutes at 1200 mV, then a photocurrent at one selected wavelength for 15 seconds at 200 mV and for 3 minutes at 1200 mV, then again the same dark current protocol, then again the next photocurrent protocol. Wavelengths were 600–700 nm in steps of 20 nm.
Experiment 2	Irradiation of the same sample with Co^{60} γ-radiation for 150 minutes under open circuit potential. This dose (ca. 2 mGy) is about equal to up to 50,000 × lower (depending on the experiment) than the relevant doses mentioned in the background story. Dark current for 15 seconds at 200 mV and for 3 minutes at 1200 mV, then a photocurrent at one selected wavelength for 15 seconds at 200 mV and for 3 minutes at 1200 mV, then again the same dark current protocol, then again the next photocurrent protocol. Wavelengths were 600–700 nm in steps of 20 nm and 540 nm.
Experiment 3	Irradiation of the same sample with Co^{60} γ-radiation for 150 minutes under open circuit potential. This dose is about equal to up to 50,000 × lower (depending on the experiment) than the doses mentioned in the background story. Dark current for 15 seconds at 200 mV and for 3 minutes at 1200 mV, under irradiation of the sample with Co^{60} γ-radiation for the same time. Then— without γ-radiation—a photocurrent at one selected wavelength for 15 seconds at 200 mV and for 3 minutes at 1200 mV, then again the same dark current protocol with γ-radiation, then again the next photocurrent protocol without γ-radiation. Wavelengths were 600–700 nm in steps of 20 nm, 540 nm, and 420 nm.

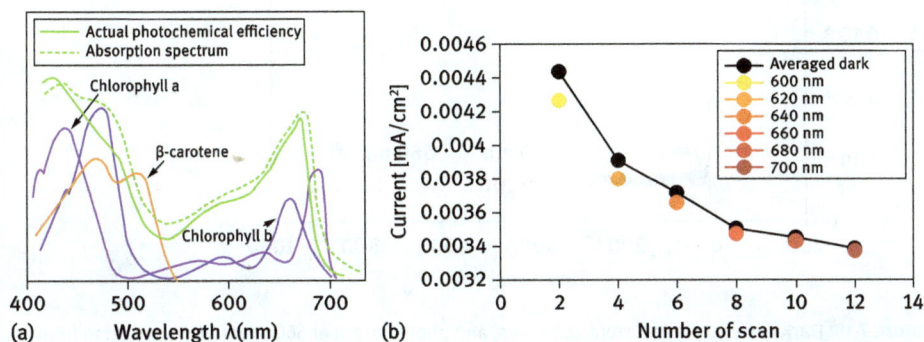

Figure 7.18: (a) Quantum efficiency of *Arthrospira* sp. (green spectra) and chlorophyll *a*, *b*, and β-carotene as reference. (b) Comparison of dark current (black top symbols) and photocurrents in the course of scanning in Experiment Part 1. The dark currents at even scan numbers 2–12 are actually the averaged dark currents from adjacent odd scan numbers $j_{2n} = (j_n + j_{n+1})/2; n = 2 \ldots 6$. The dark currents are in fact currents integrated over measurement time of round 260 seconds.

larger than the dark currents. Here, with the *Arthrospira.* sp PCC 8005 biofilm, this is not the case. A first speculation is that the stimulation of the photosynthetic apparatus in the cyanobacteria with the visible light wavelength that we have selected could cause an electrophysical activity in the cells that have a dielectric effect on the cell walls, or on the cell wall-semiconductor interface. However, this would need further investigation.

We have determined the dark currents and photocurrents by the following protocol. The current first flows for 15 seconds at 200 mV in order to equilibrate the system. This is a very short period, but because of the limited γ-radiation beam time available for our explorative experiment, we had to trim down on some experimental parameters so as to save time and use the limited resources as efficient as possible. Then the potential is switched to 1200 mV. We believe this potential is safe for the cyanobacteria where they will not denaturate. This potential was maintained for 300 more seconds. We have integrated the current from 50 to 315 seconds and considered this as the range of charge, which would be representative to the response of the bioelectrode toward the external stimulus by DC voltage and dark or light condition.

In the course of the analysis, we noticed that the current has a negative slope, which actually varied from sample to sample. This is demonstrated for three independent examples in Figure 7.19. The global decay of the current during the entire experiment Part 1–Part 3 was −4.25 μA/min.

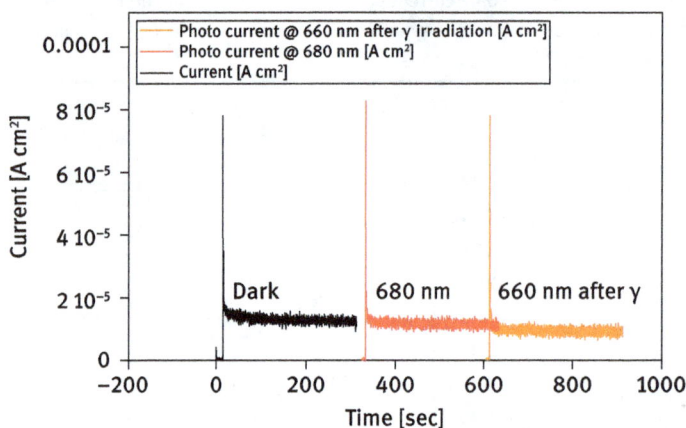

Figure 7.19: Dark current, photocurrent at 680 nm, and photocurrent at 660 nm after γ-irradiation from the bioelectrode. The current integration times where started each 50 seconds after switching to 1200 mV.

Figure 7.19 shows a comparison of dark current and photocurrents from the bioelectrode. The photocurrents were recorded at 680 nm, then γ-irradiation was employed and then another photocurrent was measured at 660 nm. These three curves are shown on the same time axis only for ease of comparison, but they were recorded at entirely different times in the course of the experiments.

During the experiment, we encountered a fundamental experimental problem; this is the current followed a Cottrell law $j^2 \sim 1/t$, which is nicely shown by the sharp decay of the current when changing the bias potential. This is the indication of a diffusion controlled process on the bioelectrode. In future follow-up experiments, we will have to allow for a wider time window to allow the currents to relax to a linear variation. We have taken care for this already for the individual currents by integrating from 50 to 315 seconds, and disregarding the currents decay near the switch-on time; see Figure 7.19.

We have plotted all dark current and all photocurrent data recorded in the three parts of the experiment in one single plot; see Figure 7.20. There, it becomes obvious that in the beginning phase of the experiment when their potential is switched on, the currents are particularly high. Therefore, the global transient is convoluted with anticipated changes in dark current to photocurrent to γ-radiation. For the time being, we have not corrected for this decay other than just forming the differences between photocurrents and dark currents. This should be still a legit procedure.

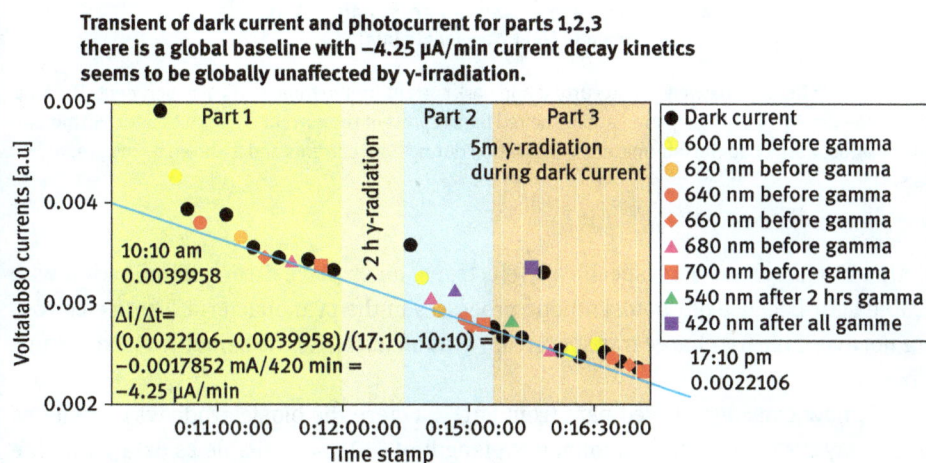

Figure 7.20: Summary of all dark currents and photocurrents in the course of the entire experiment Parts 1–3. Part 1 shows the dark and photocurrents with no γ-radiation (10:00 to 12:00 am). γ-radiation was from 12:00 to 14:30. Part 2 shows one dark current point plus a series of photocurrent points. The summary of all these points was taken as baseline for the dark current. Part 3 is the taking turns of dark current with γ-radiation and photocurrent without γ-radiation. The blue solid line shows the global baseline of current decay over time.

We have anticipated that the exposure of the biofilm to γ-rays may have an effect on the photosynthetic apparatus of the cyanobacteria—at least this was indicated by a previous extremely-high-dose γ-experiment at the RITA beamline at SKC·CEN. Therefore, the difference between photocurrent prior to γ-radiation and photocurrent after γ-radiation should be an appropriate empirical observable in this respect. We have plotted these differences in Figure 7.21.

The blue bars show how the photocurrent was lower than the dark current for the wavelengths 600 to 700 nm. One potential origin for this observation could be that a de-

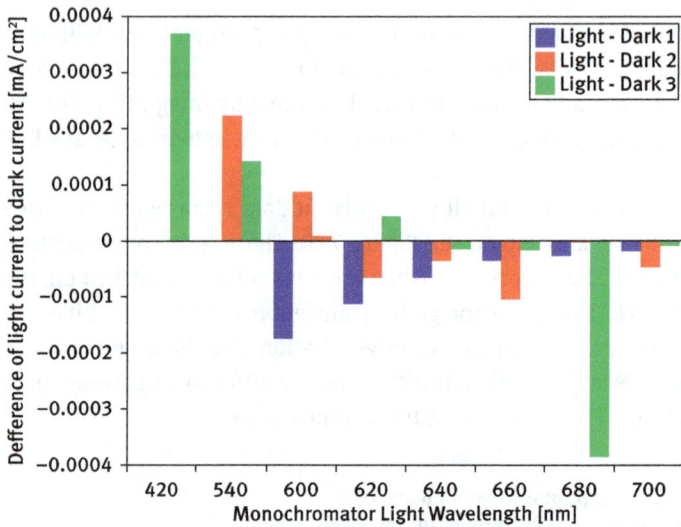

Figure 7.21: Difference between photocurrent and dark current for the three parts of experiment. The blue bars represent Part 1 prior to γ-irradiation. The red bars represent the current differences after 150 minutes γ-irradiation. Part 3 had the intermittent γ-irradiation during dark currents and is shown by the green bars.

pletion of charge carriers, specifically electrons, takes place during illumination with visible light because of photosynthetic processes in the cyanobacteria. These electrons are not available for charge transport from cyanobacteria to the hematite electrode matrix.

We now consider the red bars from Part 2, where the bioelectrode was irradiated with γ-rays. We have an additional wavelength of 540 nm available as data point, but we lack the reference point from Part 1 because this was not measured. The current difference is positive here, not negative, and thus contrasts our observations from Part 1, specifically for the data point at 600 nm.

From 600 to 620 nm in Part 2, the current difference after γ-irradiation switches from positive to negative for wavelengths 620–700 nm. Insofar, the negative current difference is maintained for the wavelengths 620–700 nm also after γ-irradiation. This negative difference is particularly strong at 660 nm.

Finally, in Part 3 where we irradiate during the dark currents, the current differences are positive for the lower wavelengths 420–620 nm and negative for wavelengths 640–700 nm. It is a noteworthy observation that the current difference at 680 nm (where, same like chlorophyll a and b and *Arthrospira* sp., have a quantum efficiency maximum) is extremely high. The current difference is extremely positive high at 420 nm, where both, too, have high quantum efficiencies. The authors of that report are concerned that at least for the wavelength of 680 nm, an artifact from the background subtraction could be causal for this very strong negative current difference.

We have not investigated the effect of γ-radiation on the pristine hematite photo-electrode, primarily because of the limited access to the resources. Therefore, emphasis was put on the bioelectrode, and not on a pristine reference electrode. It has been reported that exposure of iron oxide to γ-radiation will make chemical reduction of Fe^{3+} to Fe^{2+}. From our past experience, we would anticipate that such reduction would be favorable for the photocurrent magnitude because the Fe^{2+} acts as an electrocatalyst on the photoelectrode surface. However, likely the irradiation forms Fe^{2+} throughout the bulk of the sample, and we do not know this influence on the photocurrent.

Thus, it appears that upon exposure to γ-radiation, the negative current values become more pronounced at the higher wavelengths. This warrants further investigation with advanced planning of new experiment campaigns.

The *Arthrospira* sp. strains do not form biofilms. Therefore, we worked also on cyanobacteria that do form biofilms, in particular, the *Chlamydomonas reinhardtii* biofilms. Figure 7.22 shows MSc Niels Burzan at the synchrotron storage ring of PTB in Berlin on the preparation of Fe 3p resonant PES on these films.

Figure 7.22: MSc Niels Burzan (Empa, left) inserting the sample using the loadlock, while Dr. Hendrik Kaser (right, PTB) checks through the window flange the sample approaching the sample manipulator.

8 Complex case studies/electrochemical *in situ* studies

This chapter showcases a number of complex studies, many of which are done *operando* or *in situ* by combining electrochemical experiments with synchrotron experiments. I begin with a model study that has no relation to energy conversion. The *in situ* experiment is the combination of electrochemical copper deposition (so-called underpotential deposition) deposition on a Pt (111) single crystal from a copper chloride solution in dilute H_2SO_4 [Yee 1993]. The authors used an X-ray spectro-electrochemical flow cell, which allowed the rinsing of an electrolyte solution, and thus the cleaning of the electrolyte from the cell and electrodes. With this flow cell, the voltammetry of a pristine Pt(111) ordered electrode surface was restored after use of chloride in the system. I could not find a sketch of their cell. Given that they report *in situ* fluorescence-detected X-ray Absorption Near Edge Structure (XANES), I guess they performed their experiment not in a transmission cell but in a reflection geometry cell. They grew metallic copper films on a single crystal and monitored the electrochemical process with cyclic voltammetry.

The cell must have had an X-ray window where the X-rays with photon energy of around 9 keV could penetrate the electrolyte, which for some time contained copper ions, probed the copper layer, and an X-ray detector that would record the fluorescence signal from all materials and components in the X-ray beam. One must be cautious that the Cu fluorescence signal may originate from the deposited copper film and from the copper ions in solution. Exchanging the electrolyte during the experiment campaign, not necessarily during the actual X-ray scans, can help minimize spectral impurities from unwanted components and optimize results. Figure 8.1 shows the CV of the copper deposition in the presence of chloride in the cell (dotted spectrum) and in the absence of chloride (solid spectrum with two oxidative peaks). The panel on the right shows four Cu XANES spectra from Cu foil as reference (red spectrum on top), a Pt_3Cu alloy (blue spectrum), Cu_2O (black top spectrum), and from the Cu film deposited on Pt(111), which represented a coverage of 75 % of the substrate surface at 0.1 V. They carried out also Extended X-ray Absorption Fine Structure (EXAFS) spectroscopy and derived the Cu-Cu bond distance and Cu-O, Cu-Cl distances and concluded that Cl ions constitute a protective overlayer preventing adsorption of oxygen on the Cu either from H_2O solvent or H_2SO_4 electrolyte.

Wu et al. [Wu 1995] had followed up on this system and investigated the growth of copper on the Au(111) electrode surface with CV and XANES and EXAFS and scanning microscopy methods. They made detailed surface structure measurements for superstructures, which are typically known from ultrahigh vacuum (UHV)-based studies with Low Energy Electron Diffraction (LEED). Such experiments are relevant for model studies in heterogeneous catalysis, which are traditionally done in UHV in order to control the adsorption of gas phase reactants and impurities, and to allow for electron diffraction and electron scattering-based analytical methods, which require UHV be-

https://doi.org/10.1515/9783110794038-008

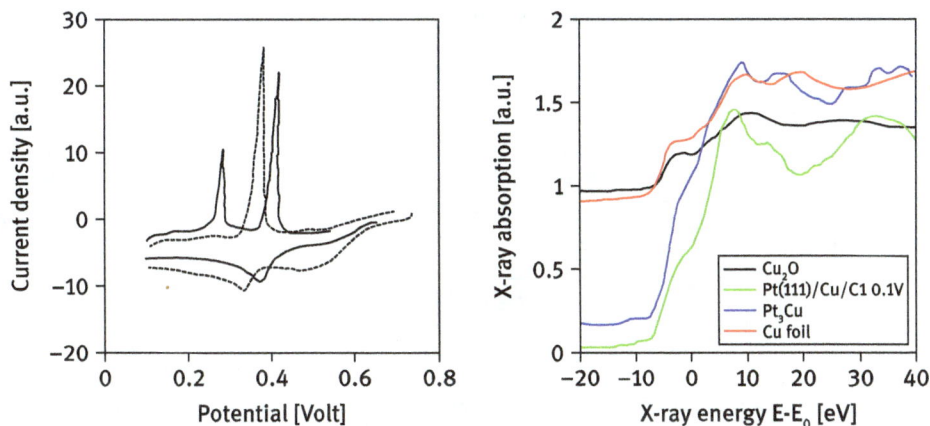

Figure 8.1: Voltammogram at Pt electrodes in contact with a 50 μM Cu^{2+} in 0.1 M H_2SO_4 solution in the presence and the absence of 1 mmol Cl-. Scan rate, $v = 2$ mV/s. Ag^+/AgCl reference electrode. Dotted line CV shows Cu deposition with chloride in the cell. Solid line CV shows Cu deposition without chloride in the cell. Reprinted (adapted) with permission from ([Yee 1993] Howell S. Yee, Hector D. Abruna, *In situ* X-ray absorption spectroscopy studies of copper underpotentially deposited in the absence and presence of chloride on platinum (111), Langmuir, 1993, 9 (9), pp. 2460–2469. DOI: 10.1021/la00033a032). Copyright (1993) American Chemical Society.

cause of the very limited mean free path of electrons (LEED, Medium Energy Electron Diffraction (MEED), Reflection High Energy Electron Diffraction (RHEED), Electron Energy Loss Spectroscopy (EELS), Auger electron spectroscopy). Therefore, catalysis experiments suffered for many decades from the fact that they were conducted in noncatalytic conditions. It was even less possible to use electron spectroscopy-based methods in liquid electrochemical environment. However, optical methods and X ray-based methods and neutron methods were certainly possible.

Be reminded that the operation and reaction conditions can be very diverse: UHV, ambient temperature gas phase, high-temperature gas phase, high-temperature gas phase electrochemistry, and liquid phase electrochemistry. These surface-based studies make sense in order to understand the atomic scale situation for catalysts in operation, for example. The platinum surface has been a model case for a very long time. Various crystal facets have a different electrochemical response in a CV [Hamann 2005], for example. Currently, more *in situ* studies for catalyst real systems and model systems are under way, which target the crystallographic structure of catalysts, the catalyst support, their triple phase boundary, the electronic structure of catalysts, and the molecular structure of reactants and reaction products.

Haubold et al. have investigated platinum as fuel cell electrocatalyst supported on carbon with Pt XANES, EXAFS, and Small Angle X-ray Scattering (SAXS). Figure 8.2 shows a schematic of the experimental setup. The spectro-electrochemical cell was made from polycarbonate Plexiglas®. The working electrode (WE) is a commercial carbon electrode

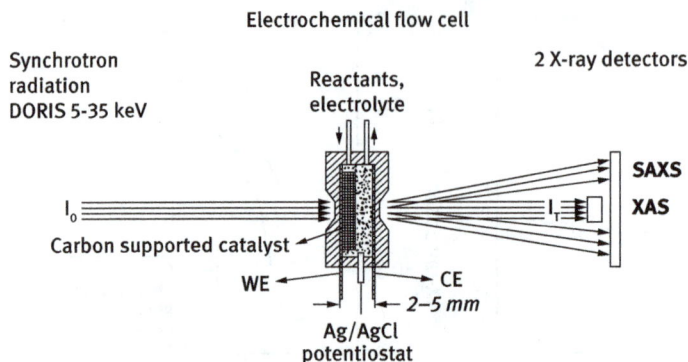

Figure 8.2: X-ray spectro-electrochemical for *in situ* experiments [Haubold 1999] at JUSIFA, Hasylab [Haubold 1989]. This is a flow cell that allows for exchange of electrolyte, and thus rinsing of the cell and cleaning of electrodes during X-ray experiments. Reproduced from [Haubold 1999] Haubold H. G., Hiller P., Jungbluth H., Vad T.: Characterization of electrocatalysts by in situ SAXS and XAS investigations. *Japanese Journal of Applied Physics Part 1-Regular Papers Brief Communications & Review Papers* 1999, 38:36–39 with permission from the Japan Society of Applied Physics. Copyright 1999.

coated with platinum catalyst (E-TEK). The counterelectrode was not disclosed but could have been a metal sheet or wire frame. Necessary like in every correct electroanalytical quantitative experiment, Haubold et al. used a reference electrode. The reference electrode was from the $Ag^+/AgCl$-type [Janz 1953, Taniguchi 1957]. The cell had terminals for exchange of electrolyte and reactants (flow cell). The cell was used at the JUSIFA Beamline at the DORIS storage ring at HASYLAB, which supplied X-ray energies ranging from 5 to 35 keV; sufficient for a transmission mode XANES, EXAFS, and anomalous small-angle X-ray scattering (ASAXS) experiment [Haubold 1999].

Figure 8.3 shows in the left panel the Pt L3 XANES spectra obtained with the catalyst in iodine containing electrolyte under electrochemical polarization. The panel in the left middle illustrates how the white line peak position of the Pt shifts depending on which potential is applied. At 400 mV oxidative potential, the peak is shifted to the right. At −400 mV in reductive direction, the peak is shifted to the left. Haubold et al. have plotted the XANES intensity as absorption cross-section $\mu \cdot D(E)$ [Haubold 1999]. The left bottom spectrum is obtained by the difference of both spectra under oxidized and reduced platinum, and thus constitutes the oxidized Pt surface layer. The electrochemical treatment leaves many signatures on the XANES spectra, such as the position and spectral weight of the white line and position of the absorption edge. They recorded spectra and also SAXS curves during the electrochemical oxidation ("combustion") of methanol and paid attention to various changes in the spectra and scattering curves, and they summarized these in the set of plots shown on the right of Figure 8.3. These changes basically reflect the changes, which we would see in the cyclic voltammogram. A more recent example for *in situ* XANES on electrochemical redox reactions in a different context, i. e., redox flow batteries, is presented by Segre et al. [Segre 2015, Timofeeva 2013].

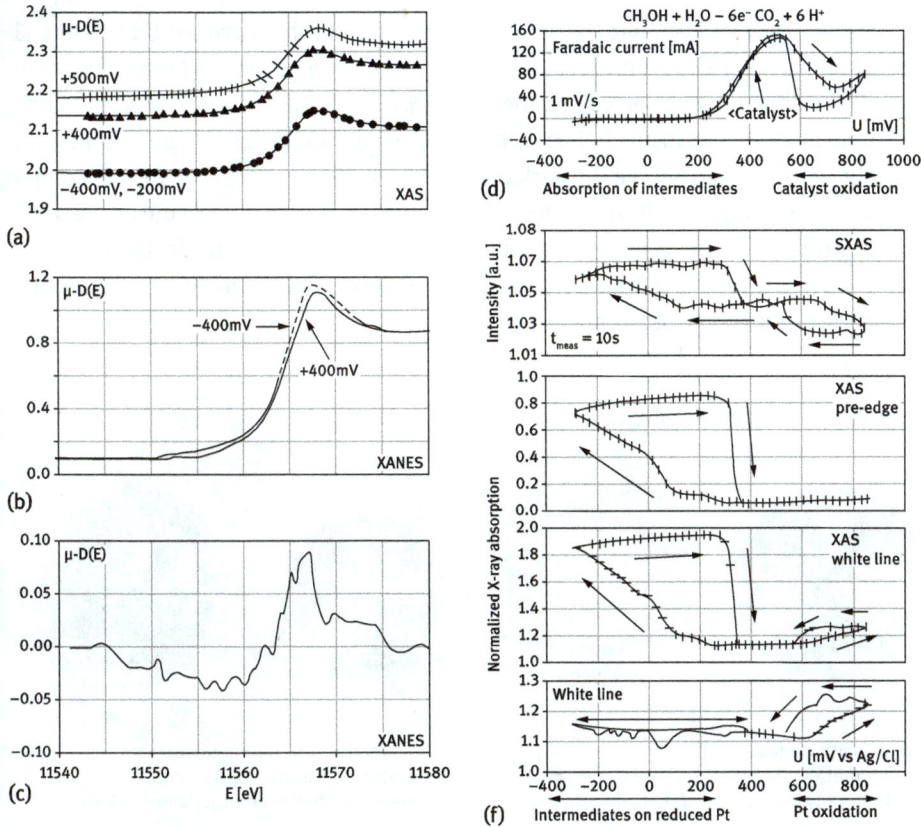

Figure 8.3: (a) Pt L3 X-ray absorption spectra from potential hold experiments at U = −400, −200, 400 and 500 m V versus Ag$^+$/AgCl. (b) Normalized absorption spectra recorded at U = −400 and 400 mV. (c) Difference spectrum. (d–f) Faradic current, integrated small-angle scattering intensity and X-ray absorption versus potential U from *in situ* studies of the electrooxidation of methanol. Reproduced from [Haubold 1999] Haubold H. G., Hiller P., Jungbluth H., Vad T.: Characterization of electrocatalysts by in situ SAXS and XAS investigations. *Japanese Journal of Applied Physics Part 1-Regular Papers Brief Communications & Review Papers* 1999, 38:36–39 with permission from the Japan Society of Applied Physics. Copyright 1999.

In the next chapters, I will show complex materials studies on batteries, supercaps, fuel cells, and photo-electrochemical systems.

8.1 Lithium ion batteries

8.1.1 Battery cell assembly and *ex situ* studies on materials and components

Figure 8.4 is a photo from a disassembled 3-V battery from the maxell® CR2016 type. It has around ¾ inch (1.9 cm) diameter. The top row shows on the left the interior of the stainless steel endcap, with white LiOH powder from the corroded lithium metal. Upon opening, the metal lithium reacts with the humidity in the air and forms the white LiOH. If the ambient had been drier, then black Li_3N from the reaction with the nitrogen would have formed instead. Next to this, we see the white separator membrane from glass fiber fabric. Then we see the dark positive electrode with active material (composition unknown, might be lithium cobalt oxide). On the top right, we see the inside of the other endcap, which black left over material from the positive electrode. In the bottom row, we seen the four components turned upside down. This is the basic macroscopic architecture of this type of battery.

Maxell CR2016 / 3V Lithium primary battery (not rechargeable)

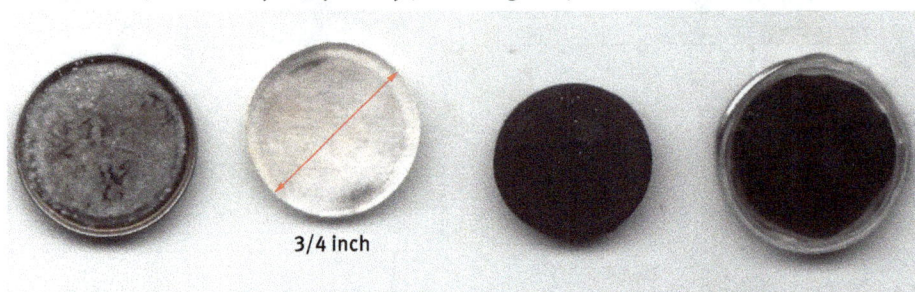

3/4 inch

End cap with negative
Lithium metal electrode   Positive electrode with (unknown)
(LiOH, already corroded)   Separator membrane   active material, carbon, binder, and
                           (glass fiber fabric) metal matrix for better conductivity   Endcap

(a)

(b)

Figure 8.4: Photographs of a disassembled Maxell CR2016 type lithium ion battery.

The electrochemical functional components are sketched in Figure 8.5 and show the cathode with $LiMn_2O_4$ active material (positive electrode), the liquid electrolyte of $LiPF_6$ crystals dissolved in dimethyl carbonate (DMC) or any other suitable solvent, and the anode, which is typically metallic lithium, or any other suitable negative electrode material.

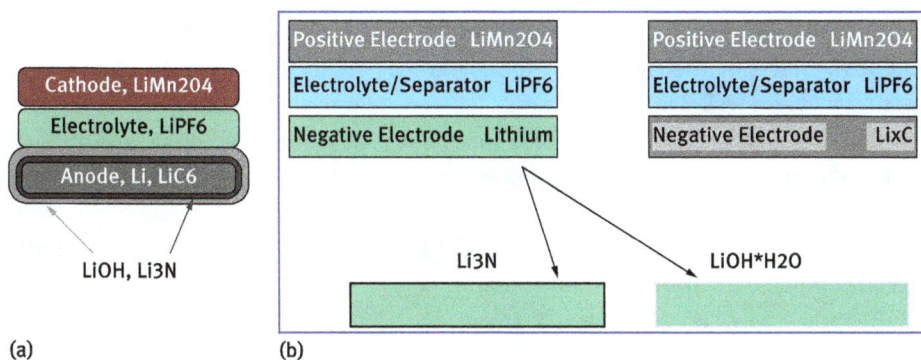

Figure 8.5: (a) Schematic of a simple battery cell electrode assembly. The anode (negative electrode) can be, e. g., metal lithium or lithium carbon. Metal lithium may form on its surface a white LiOH film in humid environment or a Li_3N film in dry air because of the high nitrogen content. The electrolyte is soaked in a porous membrane separator and may contain $LiPF_6$ dissolved in an organic solvent. The cathode (positive electrode) is a metal oxide such as $LiMn_2O_4$. Therefore, all three components may contain lithium in different molecular structure. (b) Alternative schematic for arrangement of positive and negative electrodes (cathode and anode, respectively), and electrolyte layer with separator in between electrodes. The anode can be a lithium metal foil and corrode with N_2 from air to black Li_3N, and with H_2O from air humidity to white LiOH.

Lithium battery electrodes for laboratory purposes like here can be made as follows. Lithium carbonate and MnO_2 are mixed in stoichiometric amounts to meet the cation stoichiometry and then annealed in air vented muffle furnace or oxygen supplied tube furnace at high temperature until the chemical conversion is complete to have $LiMn_2O_4$. Likely, the oxygen stoichiometry is not accurately met, and one rather obtains $LiMn_2O_{4-\delta}$. The determination of the correct phase can be done with laborious titration chemical analysis, or neutron diffractometry, Rutherford Backscattering, for example. The oxygen stoichiometry can be manipulated by tuning the temperature and the oxygen concentration during synthesis or by a subsequent treatment. The $LiMn_2O_{4-\delta}$ material is then ground in a mortar so as to obtain a fine powder. This powder is then mixed with carbon black and graphite powder, which serves as an extension of the current collector support in order to enhance the electronic conductivity of the electrode network. A binder polymer such as polyvinylidene difluoride (PVDF) is then dissolved in a solvent and mixed with the spinel carbon mixture. The way and courtesy with how these and also further processing steps are carried out determines a lot how well the battery will work. These are important steps which go beyond the often too simple posed question as to what is the "best" electrode material in terms of crystallographic structure. The theoretical "best" material, as maybe determined by density functional theory predictions, may be of little use when it is poorly prepared by an operator who is unaware of how a battery works at all scales.

The cathode powder slurry is then poured on an aluminum foil and spread homogeneously over the foil. A good help is when you have plastic stripes of several 100-

micrometer thickness, which you place and fix parallel in a 10-centimeter distance to contain the slurry. With a ruler stick gliding over the parallel stripes, you can then spread the slurry over a well-defined thickness distance. A large glass plate with flat, smooth, and clean surface is a good support for this. Figure 8.6 shows such large electrode assembly of around 10-cm × 10-cm area coated on aluminum foil, between the green gloves worn by a materials science student (Ms. Alison Fowlks, Michigan State University) during her DoE ERULF summer internship at LBNL [Fowlks 2002]. The sheet is being prepared for immediate drying on a hot plate.

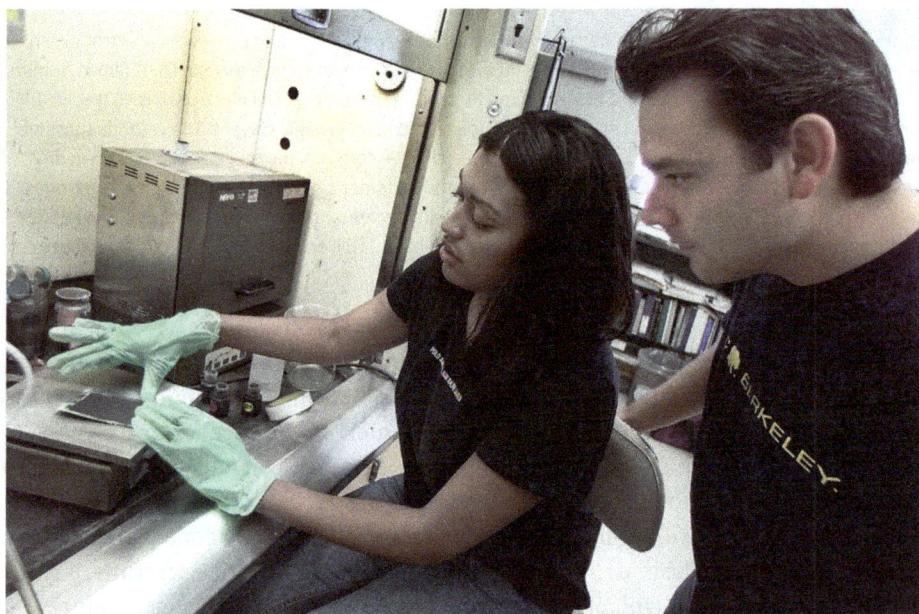

Figure 8.6: DoE ERULF Fellow Ms. Alison Fowlks (Michigan State University, PhD since 2011) and Artur Braun over lithium battery cathode fabrication at EETD, LBNL, July 2001. (Photographer Roy Kaltschmidt, Photo courtesy of Lawrence Berkeley National Laboratory, © 2010 The Regents of the University of California, Lawrence Berkeley National Laboratory). The reader is advised that protective eyewear should be worn during experiments [Young 2000].

After this initial drying, a round punch cutting tool is used to punch out electrode disks of specific diameters. Measuring the weight of either electrode disk provides then a good statistics about the weight distribution and homogeneity of the samples and allows later for the determination of the electrochemical specific capacity (compare with Figure 2.20). The electrode is then further dried in a drying furnace under controlled condition and later again measured on the fine balance. This allows for some information on the presence of humidity on the porous electrodes.

The electrodes should be transferred as quickly as possible from the drying furnace to the glove box so as to minimize exposure to moisture, the water of which would react adversely with the $LiPF_6$ electrolyte and then form deleterious HF. The glove must be filled with a chemically inert atmosphere, such as argon or helium, but not nitrogen. Nitrogen will immediately react with the lithium metal needed for battery assembly inside the glove box toward Li_3N.

You may notice from Figure 8.6 that we were not wearing laboratory safety goggles. The photo was taken in 2001. Meanwhile, both of us have become more aware of the need to use protective eyewear [Young 2000]. I want to use this instance to remind all workers to be cautious when working, particularly in the laboratory. Sources of accidents are manifold, and using safety goggles in a laboratory has meanwhile become a standard rule, sometimes limiting the discretion of the individual researcher.

Figure 8.7 shows former San Francisco State University electrical engineering student Mr. Bopamo Osaisai with his hands in the gloves of glove box at EETD Bldg. 70 at LBNL, preparing samples and utilities for lithium battery cell assembly. The glove box

Figure 8.7: Artur Braun and Mr. Bopamo Osaisai (now Software Engineer in Southern California), DoE PST Summer Fellow at LBNL, working at glove box at EETD for battery cell assembly (Photographer Roy Kaltschmidt, Photo courtesy of Lawrence Berkeley National Laboratory, © 2010 The Regents of the University of California, Lawrence Berkeley National Laboratory).

can be used for assembly and disassembly of battery cells, and also for the storage of lithium metal, battery electrolyte, electrodes, tools, microbalance, and waste container, and certainly it is possible to run in there batteries for a long time when there is an electric feed through which allows connecting batteries inside the glove with the battery cycling electronic equipment outside.

When the batteries are cycled and disassembled, their components can be visually inspected and compared with their structure and condition, which they had before cycling.

Lithium is the main player in lithium ion batteries. Interestingly, lithium has not so much been under investigation as an element in X-ray and electron spectroscopy (see Braun et al. [Braun 2007] and references therein). For a recent experimental study, see Ishii et al. [Ishii 2011]. I therefore decided to carry out a couple of NEXAFS experiments on lithium metal and compounds. Figure 8.5 illustrates that when you inspect a battery with X-ray spectroscopy, you have to be prepared to detect Li in the electrolyte and in both electrodes. The information can thus be confusing, particularly when you do an X-ray experiment in transmission where you unintentionally probe the entire battery assembly. The spectra are shown in Figure 8.8 and Figure 8.9.

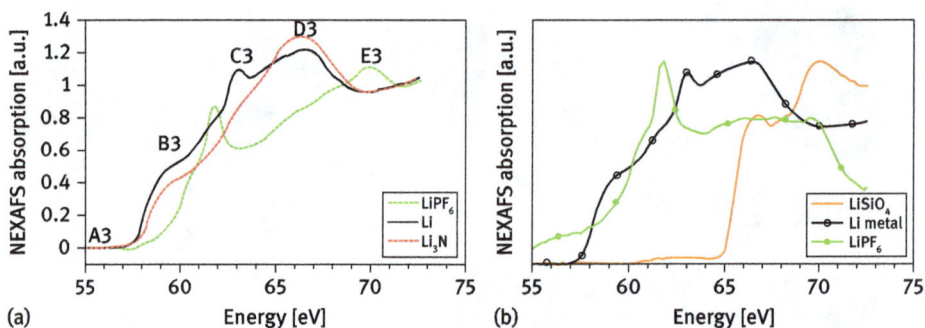

Figure 8.8: (a) Li K 1s NEXAFS spectra of Li metal, $LIPF_6$ crystals, and black Li_3N powder (Sigma Aldrich). (b) In addition to Li metal and $LIPF_6$ crystals also the NEXAFS spectrum of $LiSiO_4$.

Figure 8.9 shows the Li K-edge NEXAFS spectra of lithium metal, which one would find in the negative electrode. X-ray spectra of lithium have been measured already over 40 years ago with EELS [Alexandropoulos 1972, Cohen 1972] and with X-ray Raman spectroscopy [Cohen 1973]. When this lithium metal is corroded by exposure to nitrogen in the air, it forms black Li_3N. When there is too much humidity in the air, it forms white LiOH. I am not aware that it somehow forms a lithium oxide. Often the following question is asked by the operators of XPS instruments: "Is your sample air sensitive?" and usually it is understood that the sample can adversely react with the oxygen in the air. This here is a very good example that not the oxygen in the air, which is around 20 %, but the 78 % nitrogen is often ignored, or the traces of H_2O in the air. The lithium metal

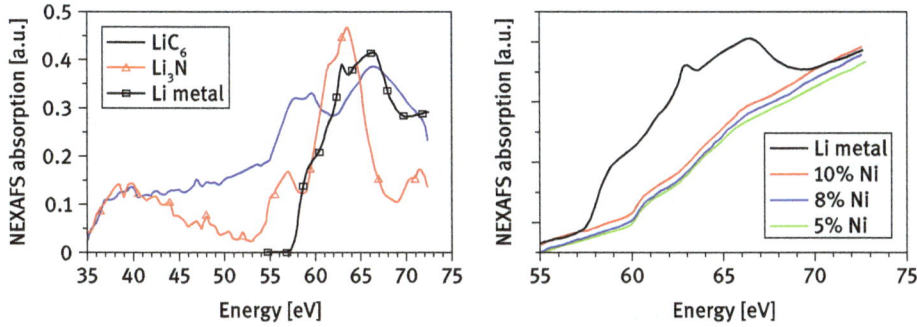

Figure 8.9: (Left) Li K 1s NEXAFS spectra of Li metal, LIPF$_6$ crystals, and Li$_3$N powder (Sigma Aldrich). (Right) Comparison of Li metal spectrum with LiMn$_{2-x}$Ni$_{0.01}$xO$_4$ (x = 5, 8, 10) spectra.

spectrum shows a relatively sharp peak at position C3, a broad hump at position D3, and a shoulder at around B3. There is a slight shoulder between B3 and C3, where the spectrum of Li in LiPF$_6$ a sharp pronounced peak [Braun 2007].

Fister et al. [Fister 2008] have calculated lithium spectra and did not reproduce one feature in my lithium spectra (from Figures 8.8 and 8.9) in their calculations. One explanation for this could be that the lithium metal, which we had measured at Beamline 4.0.2. at the ALS was already slightly corroded at the surface [Fister 2011]. We paid attention to have the lithium metal sample as pure as possible, but we cannot rule out contamination at the surface by formation of a lithium nitride film. As we notice, the X-ray absorption edge of Li is at the very low energy of around 50 eV.

The penetration depth is below 1 micrometer, and thus the spectrum contains a large portion from the structure at the surface of the sample. I have tried to illustrate this in Figure 8.5 where symbolic surface layers are drawn around the lithium sample. It is obviously possible to discriminate Li$_3$N from Li, and also the Li in the electrolyte from the Li metal. The panel on the right of Figure 8.8 shows the spectrum of LiSiO$_4$, which has no further meaning for battery applications, but it serves as a reference compound to illustrate the spectral changes depending on the coordination or ions around the Li cation.

Figure 8.9 shows on the left the NEXAFS spectra Li metal and Li$_3$N with the anode material LiC$_6$. It cannot all be ruled out that the spectra are properly aligned. We expect from the oxidized forms of elements that their spectra are shifted to higher X-ray energies. Moreover, the reduced lithium should have substantial intensity at the lower energy side. We are possibly not dealing with an entirely reduced form of Li, as mentioned already. The right panel of Figure 8.9 compares the Li K1s NEXAFS spectra of lithium metal with LiMn$_2$O$_4$, which was doped with 5 %, 8 %, and 10 % Ni. The overall intensity of the three latter spectra increases with increasing Ni content, probably because adding Ni to LiMn$_2$O$_4$ enhances the electronic conductivity. We notice here that the structures in the Li spectra become very dim and diffuse. Probably because of the strong X-ray absorption by all elements in the sample, it is difficult to make out Li structures.

When we do literature search for Li spectra, we notice that relatively early lithium X-ray spectra have been recorded. In Figure 8.10, we see a LiF EELS spectrum from Hightower et al. from Stanford University [Hightower 2000a]. With their TEM EELS, they could observe that the Li metal was melting in the electron beam. One therefore has to pay attention, as I mentioned already, when doing EELS recording with an electron microscope. The LiF EELS spectrum is aligned with the NEXAFS spectra of LiF and $LiPF_6$ from other researchers [Hightower 2000a, Hightower 2000f, Krisch 1997, Petersen 1975, Sonntag 1974, Tsuji 2001, Tsuji 2002, Tsuji 2005].

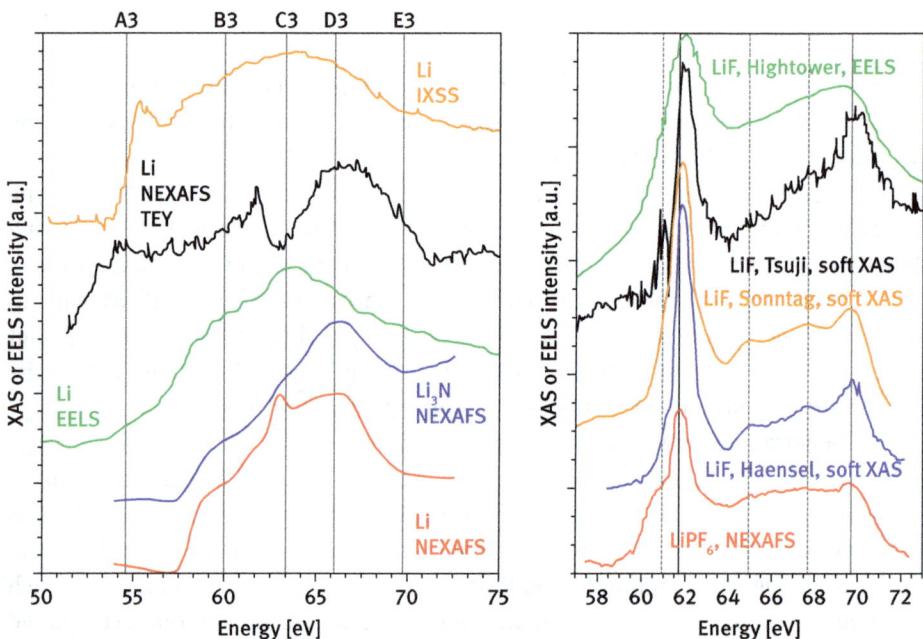

Figure 8.10: (Left) Bottom to top: our NEXAFS Li(1s) spectrum of $LiPF_6$ electrolyte, and EELS and XAS spectra from LiF [Sonntag 1974, Tsuji 2001, Hightower 2000a, Haensel 1968]. (Right) Li(1s) spectra of LiC_6 [Schulke 1988, Grunes 1983, Hightower 2000a]. Schülke's spectrum is given for perpendicular and parallel direction of the q-vector versus the graphite c-axis (dashed curves), and for the computed convolution of both (solid, top). The second spectrum from the bottom is from our own NEXAFS studies. Grunes' spectrum [Grunes 1983] was obtained by subtraction of a power law background. Reprinted from [Braun 2007] with permission from Elsevier. Copyright 2007.

Fister et al. [Fister 2008] have calculated lithium spectra and were not able to fully reproduce all features in our lithium spectra with their calculations. One explanation for this could be that the lithium metal, which we had measured at Beamline 4.0.2. at the ALS, was already slightly corroded at the surface [Fister 2011]. We paid attention to have the lithium metal sample as pure as possible but cannot rule out contamination at the surface by formation of a lithium nitride film. As we notice, the X-ray absorption edge of Li is at the very low energy of around 50 eV. The penetration depth is below 1 micrometer, and thus the spectrum contains a large portion from the structure at the surface of the sample. I have tried to illustrate this in Figure 8.5 where symbolic surface layers are drawn around the lithium sample. It is obviously possible to discriminate Li_3N from Li, and also the Li in the electrolyte from the Li metal. The panel on the right of Figure 8.9 shows the spectrum of $LiSiO_4$, which has no further meaning for battery applications, but it serves as a reference compound to illustrate the spectral changes depending on the coordination or ions around the Li cation.

8.1.2 Manufacturing and assembly of *in situ* battery cells at the synchrotron

The development of cell housing for *operando* and *in situ* spectro-electrochemical studies is important for the understanding of physicochemical mechanisms in electrochemical energy converters and storage devices such as fuel cells, batteries, electrochemical capacitors, and photo-electrochemical cells. Consequently, numerous such cells have been built and used, including also cells for soft X-ray and hard X-ray *operando* and *in situ* and studies on electrochemical systems. There are cell designs based on rigid beryllium X-ray windows, e. g., by Braun et al. [Braun 2003a], which is a versatile electrochemical *in situ* reaction cell for long-term hard X-ray experiments on battery electrodes, or Morcrette et al. [Morcrette 2002]. Selecting the appropriate materials components to optimize their experiments is work that prospective researchers should do, where necessary. This includes the choice of materials with chemical compatibility for the cell bodies, current collectors and X-ray windows, and also gaskets in order to match the chemistry of the electrode assemblies and the electrolyte in batteries, for example. When the battery electrode of interest contains, e. g., aluminum in an Al substituted $LiMn_2O_4$-based cathode, only the XAS spectrum of the Al edge should be affected by a potential aluminum current collector. Moreover, Al just happens to be a prominent current collector material in lithium batteries [Whitehead 2005]. Then one should record a XAS spectrum of the current collector to include it for the background subtraction, or chose a different current collector. When contaminations from trace elements play a role, then one can apply contrast variation by exploiting the anomalous scattering of the prospective chemical elements of interest in the study. This was done over 10 years ago with ASAXS [Braun 2001a, Braun 2001f] and for XRD at the manganese K-edge in addition to XANES and EXAFS [Braun 2003a] with one and the same spectro-electrochemical cell *operando* at three Beamlines at the APS and at SSRL. In any way, when studying chemically and structurally highly heterogeneous systems such as battery assemblies, using anomalous diffraction and scattering should be considered as option. Although it is correct that utmost caution and proper planning in the design of the *in situ* cell is important for getting high-quality X-ray data, the same holds for the electrochemistry part. The X-ray data of the batteries measured in the cell without reference electrode are insufficient in as far as they were not acquired under accurate control of the electrochemical potential. The aim of the X-ray studies is to understand the electrochemical events taking place at one particular electrode, be it the cathode or the anode, and the one of interest is then called the working electrode (WE). A three-electrode potentiostat with separated reference electrode and counterelectrode allows the potential and the current at the WE to be quantified with little interference from the other electrodes. There are many sophisticated spectro-electrochemical cells around, not only for X-ray but also for other analytical methods. Many of them contain no reference electrode, although reference electrodes are typically required when data are to be recorded under control of the electrochemical potential. This aspect is often overlooked when no electrochemist is involved in the design of the spectro-electrochemical cell. The reference electrode itself is subject to utmost design objectives and optimizations [Zhou 2007, Kramer 2004, Wu 2000]. As far as the electrochemical accuracy is concerned, even a "simple cell for *in situ* X-ray absorption spectro-electrochemistry" reported in [Farley 1999] or in [Morcrette 2002] can be superior to new ones when only optimized for X-ray aspects.

Figure 8.11 shows an X-ray spectro-electrochemical cell mounted at the diffractometer of Beamline 2-1 at SSRL [Braun 2003a]. The X-ray beam comes in from the black tube

Figure 8.11: X-ray spectro-electrochemical cell mounted at the X-ray diffractometer at BL 2-1 at Stanford Synchrotron Radiation Laboratory. Three cables lead to positive electrode, negative electrode, and reference electrode.

on the left, hits on the positive electrode plate which we see from the left side, passes the entire cell with all its components, and is diffracted to the right side where we see the detector tube. The red cable connects to the positive electrode which rests on the stainless steel plate. The thin black cable at the bottom (near the gauge microscope which helps diffractometer alignment) connects the reference electrode. The other cable is not visible, but its black plug can be seen connecting to the back plate, which is connected with the negative electrode lithium metal in the cell.

The images in Figure 8.12 show the disassembled same stainless steel/polypropylene-based *in situ* electrochemical battery cell mounted at the diffractometer of Beamline 2-1 at SSRL in Stanford for X-ray diffraction (XRD) and hard X-ray absorption spectroscopy (XANES and EXAFS), in the years 2000 and 2001. Prior to this, I used this cell for *in situ* ASAXS at the BESSRC-CAT Beamline at the Advanced Photon Source in Argonne National Laboratory [Braun 2001a, Braun 2001f]. This was the first ever reported such *in situ* ASAXS study.

The cell has two structural components, a steel disk that supports the positive electrode and a polypropylene disk that supports the lithium metal. Both disks have drilled holes that allow the X-rays to pass through, so as to have no X-ray absorption by steel and polypropylene. These holes, however, allow for ambient atmosphere to reach the cell interior. We thus need some X-ray windows over the holes, which prevent diffusion of gases into the cell but allow for high X-ray transmission. This problem is solved by using beryllium plates as X-ray windows. Beryllium as a low Z element (atomic number $Z = 4$, it has only 4 electrons and thus a high X-ray transmission) has a high X-ray transmission and is a rigid material. To prevent corrosion of the Be from contact with corrosive electrolyte, the Be plate was coated with a polyimide sticky tape (Kapton®). An aluminum disk with an aperture is fixed with bolts over the Be disk. Sealing is provided between all four disks by using in between each three concentric BUNA® O-rings.

The battery was assembled at the chemistry lab at SSRL. In 2000 and 2001, they had a glove box for their users but it was run with nitrogen, which was not compatible with lithium metal work. I therefore brought a glove bag that I had purchased from the I2R company (these bags are now sold under the Sigma Aldrich brand). This is a sealable plastic bag with built-in gloves. The bag can be filled with inert gases, such as argon or helium, and one can assemble, e. g., batteries inside this bag with ease. They were of great service for synchrotron work.

▶ Figure 8.12: (a) Polypropylene plate showing the inside face. (b) The Beryllium disks for X-ray windows were purchased from Brush Wellman Inc. (c) Polypropylene plate showing the outside face. Three O-rings in the center provide the sealing of the Be disks. The Be disk covered with Kapton foil is laying on the boundary of the PP disk and will be put concentric on the O-rings. The aluminum disk will be tightened with 8 bolts on the PP disk in eight threads. The aperture in the aluminum disk defines the maximum X-ray beam range for probing the sample. (d) The inside of the PP plate with a crumbled lithium foil, wet translucent separator, one of the three inner O-rings. The stainless steel disk has the $LiMn_2O_4$ cathode disk in the center with black $LiMn_2O_4$ active layer visible, and the two other large O-rings.

(a)

(b)

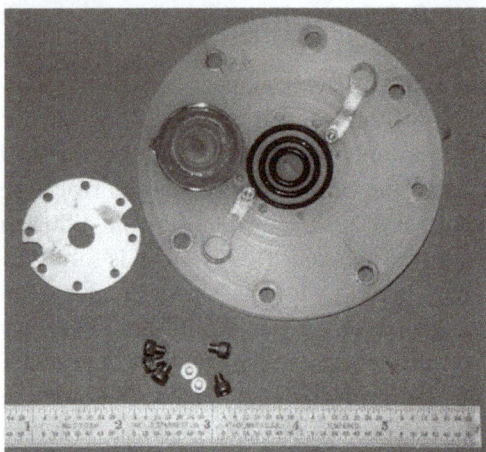
(c)

(d)

Figure 8.13 (left panel) shows how the glove bag is placed on a laboratory desk and connected via a green hose with an argon high-pressure gas cylinder. The glove bag is basically an oversized zip-lock bag through which you can insert your chemicals, your samples, your tools such as screwdrivers, tweezers, wrenches, multimeter, vials, containers, tissues for cleaning up, and so on, as shown on the right panel of Figure 8.13. It is a good idea to place all these items in a large laboratory tray in the glove bag, as is shown in the Figure 8.13. Then you float the bag from the backside with the argon gas for a while and lock the zip to close the bag and keep filling argon until the bag is filled gently like a balloon. With your hands in the glove, you can assemble your battery cell like in a conventional glove box. When the cell is assembled and still in the glove bag, you should use the multimeter and check in the bag the open circuit potential of the cell. If it is not good, you can disassemble the cell still in the glove bag and try to fix it before you remove it from the bag. When the cell works ok, simply open the bag and bring the cell to the beamline.

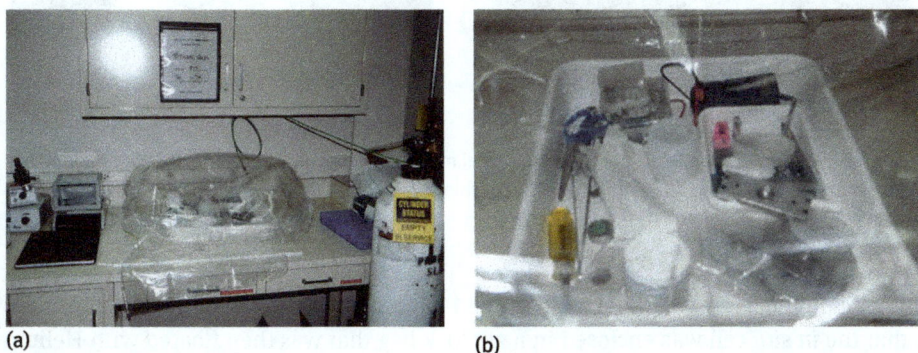

(a)   (b)

Figure 8.13: (a) Transparent plastic glove bag at SSRL chemistry lab for preparation of battery *operando* experiments at BL 2-1 and BL 2-3. Glove bag connected with Argon gas bottle. (b) The glove bag contains a plastic tray with *in situ* cell, battery electrodes, electrolyte, and assembly tools, multimeter.

On the left of Figure 8.14 you see a mobile laboratory desk with portable potentiostat (blue silver color) connected with our *in situ* spectro-electrochemical (battery) cell, controlled by a Macintosh laptop, and connected to a voltmeter showing an open circuit potential (OCV) for the cell of 2.97 V. The theoretical or ideal OCV for $LiMn_2O_4$ versus lithium would be 3.05 V. We consider this a good agreement given the circumstances under which the cell was assembled.

The Keithley instrument is used for checking the battery current, all wired to the battery *operando/in situ* cell inside the experiment hutch at SSRL Beamline 2-3. The spare *in situ* cell next to the Keithley instrument is not connected.

The desk with all instruments and the cell is placed in front of the experimental hutch at Beamline 2-1 of SSRL. The cell itself is meanwhile mounted at the diffractometer

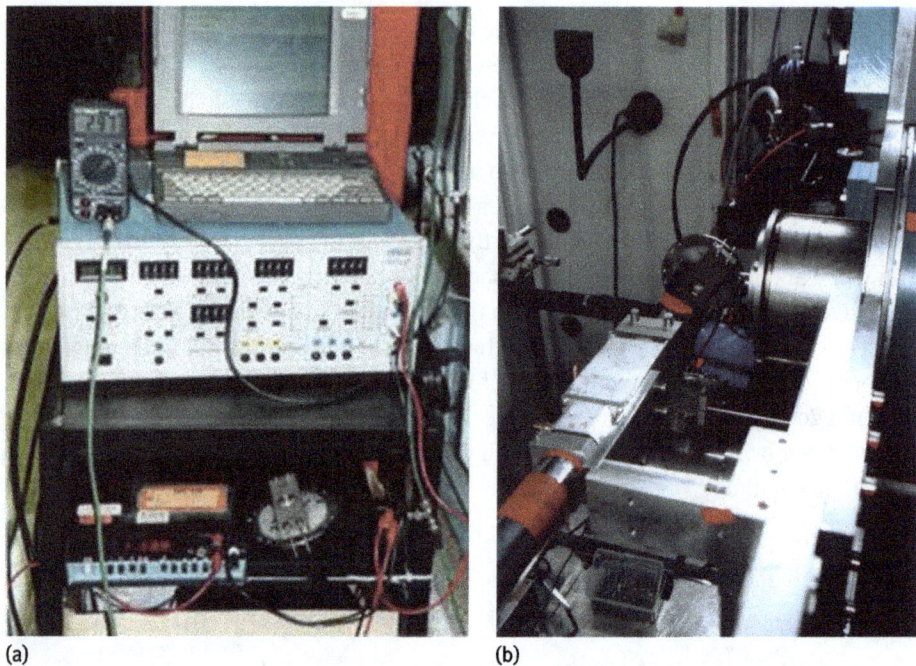

Figure 8.14: (a) Potentiostat, Macintosh laptop, digital multimeter for voltage reading, Keithley multimeter for current reading and one spare *in situ* cell on portable tray in front of (b) experimental hutch at SSRL Beamline 2-1.

of the Beamline; see right panel of Figures 8.14 and 8.11. Not shown in these photos here is that the *in situ* cell was enclosed in a zip lock bag that was then floated with Helium so as to provide protection against potentially incoming ambient air, which turned out to be not necessary. In the disassembled cell in Figure 8.12, you will see three O-rings for sealing; it turned out that those three O-rings provide enough protection and sealing against the ambient environment.

Figure 8.15 shows the experimental hutch at Beamline 2-3 while the experiment is going on inside. The X-ray detector was a gas cell proportional counter. SFSU BSc student Shawn Shrout sits at the VAX DEC work station and programs the X-ray diffraction and spectroscopy instrument control and data acquisition macros of the end-station.

Lithium manganese oxide ($LiMn_2O_4$) is a well-studied lithium battery cathode material. The manganese in it is another key player because it is subject to redox cycling between Mn^{3+} and Mn^{4+} during charging and discharging. The Mn^{3+} has some tendency to disproportionate into Mn^{4+} and Mn^{2+}, latter of which can be dissolved in liquid environment, such as in contact with a liquid battery electrolyte. This will drive structural disintegration and ultimately battery failure.

Figure 8.15: Beamline 2-3 at SSRL. The blue experimental hutch contains the X-ray diffractometer and the beamline tune. Student Shawn Shrout from San Francisco State University preparing X-ray scan macro commands on the DEC VAX computer work station on the right. Left side from the blue experiment hutch is the potentiostat controlled by a laptop, connected by feed throughs with the *operando* battery cell in the hutch mounted at the diffractometer.

8.1.3 *Operando* XANES, anomalous XRD, EXAFS, and ASAXS on a lithium ion battery

The first experiment that we did was the XANES of the battery under open circuit condition, which was 2.97 V, which can be read from the multimeter reading in Figure 8.14. We have discharged the battery and recorded XANES spectra and diffractograms. During discharging the *in situ* cell, we first recorded XANES spectra at the Mn K-edge. This is possible because the X-ray beam with energy ranging from 6500 to 6600 eV can penetrate the two beryllium X-ray windows of 380-micrometer thickness each, plus the 25 mil thick aluminum support on which the 50 micrometers thick cathode assembly is coated, plus two layers of Celgard polymer battery separator and a thin liquid layer of $LiPF_6$ dissolved in dimethyl carbonate. The left panel of Figure 8.16 shows the X-ray absorption spectra of the manganese in the *in situ* cell. Any other material in the cell as well as the cell itself leaves no characteristic signature in the spectrum.

The black spectrum was recorded at the beginning of the experiment at OCV. Discharging causes the chemical reduction of the manganese and manifests in a chemical shift of the spectrum toward lower X-ray energies, as is shown for the green spectrum

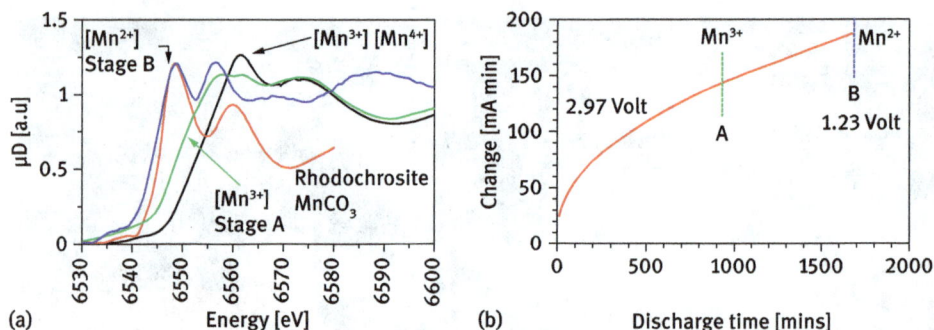

(a)

(b)

Figure 8.16: (a) Mn K-edge X-ray absorption spectra recorded from the *in situ* cell in Figure 8.11 during battery discharging. The black spectrum was recorded at OCV and signifies the $Mn^{3.5+}$ average oxidation state ($t = 0$). Upon discharging, the Mn experiences a chemical shift toward lower X-ray energies; the green spectrum signifies Mn^{2+}. Further discharging to the deleterious Mn^{2+} brings about the blue spectrum, which is validated by the white line position of Mn^{2+} mineral rhodochrosite $MnCO_3$. (b) Evolution of electric charge during discharging of the battery over 1600 minutes. OCV at $t = 0$ and position A with Mn^{3+} and position B with Mn^{2+}.

which is indicative to Mn^{3+} (position "A"). Further discharging moves the Mn spectrum further down in energy toward Mn^{2+}, shown by the blue spectrum. This was position "B" where the potential was 1.23 V. Rhodochrosite is a mineral with Mn^{2+}. Its spectrum is shown here for comparison; its white line overlaps with the white line with the blue spectrum from position "B."

The right panel of Figure 8.16 shows the evolution of the electric charge versus the discharge time. The electric charge is obtained by integrating the current passed through the cell, which we have measured with the Keithley instrument. As can be seen, the cell was discharged from 2.97 to 1.23 V over a time of 1600 minutes. At OCV, the oxidation state of the manganese is an average $Mn^{3.5+}$. We can assume that the Mn is present to 50 % as Mn^{3+} and 50 % as Mn^{4+}. With discharging, the oxidation state changes to Mn^{3+} and eventually Mn^{2+}, which is validated by the chemical shift in the XANES spectra. This information is necessary in order to validate the XRD, which we did at the same beamline during the same discharging campaign.

We have recorded current and voltage over time. The integrated current is shown in the right panel of Figure 8.16. The integration is a very simple routine, which can be done with a data plot program such as Excel, Microcal Origin, and KaleidaGraph. The evolution of the current itself can be very scattered and spiky, whereas the integration over time smooths out irregularities and yields a very uniform behavior, as noticeable in Figure 8.16 right panel. The integrated current yields the accumulated charge in the battery and can be related with the stoichiometry change [Braun 2001f]. We began at 2.97 V and discharged to 1.23 V (stage B) over 1600 minutes. About halfway at stage A, we recorded a XANES that resembles to a large extent Mn^{3+}. The spectrum at stage B is indicative to Mn^{2+}. We have taken a reference spectrum from the mineral Rhodochrosite

$MnCO_3$, which has a strong Mn^{2+} signature, at least when we compare the strong white line.

During the same campaign, we have switched from the XANES mode to the XRD mode and recorded diffractograms at various X-ray energies from 6500 to 6548 eV. For this, it was not necessary to change the mechanical diffractometer settings. Figure 8.17 shows in three panels the diffractograms. Note that we have recorded here diffractograms while we had to maintain the state of charge in the battery and while we were changing the X-ray energies for the anomalous XRD.

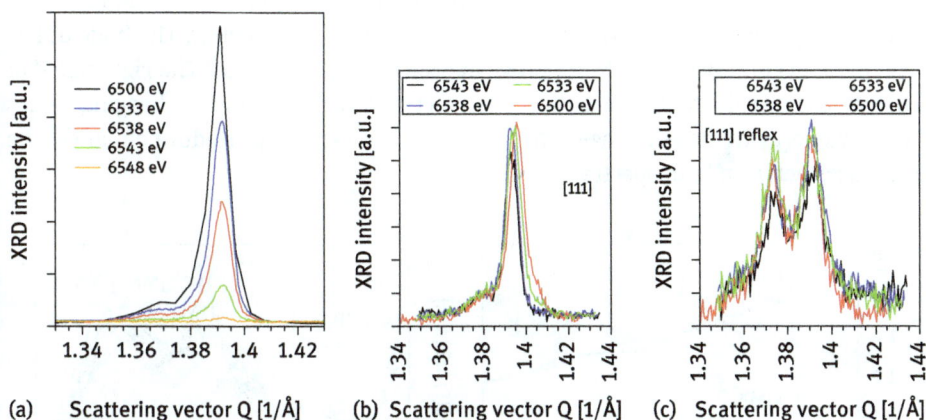

Figure 8.17: (a) [111] Bragg reflex of cubic $LiMn_2O_4$ spinel phase (Fd3m) with maximum at $Q = 1.39\,1/Å$. The X-ray energies were 6500, 6533, 6538, 6543, and 6548 eV. Small satellite peak at $Q = 1.37$ assigned to tetragonal phase of $LiMn_2O_4$ with I41/amd symmetry. Strong absorption of Mn in this energy range-peak heights shrink with increasing photon energy. (b) Diffractograms, recorded at four different photon energies, show evolution of [111] reflex during slight charging (delithiation). Peak-shift toward higher Q values is due to phase transformation, i. e., a shrinking of the unit cell by the removal of lithium from $LiMn_2O_4$ during charging, but not a result of anomalous scattering. (c) Bragg reflex after deep discharging of electrode (lithiation). Split [111] reflex indicates phase transformation from cubic to tetragonal (Verwey transition). Left satellite of twins believed to be [101] reflex from space group I41/amd.

The right panel of Figure 8.17 shows the [111] reflex measured at OCV with X-ray energies 6500, 6533, 6538, 6543, and 6548 eV. The intensity of the peak decreases as we move with the X-ray energy toward the absorption edge. The middle panel of Figure 8.17 shows the same peak for four different X-ray energies during slight charging of the battery, i. e., delithiation where the Mn should become further oxidized. The [111] peak moves to larger Q values, which indicates that the unit cell is decreasing. We would expect this from a unit cell where lithium is extracted. The right panel of Figure 8.17 shows the same Q-range after deep discharging of the battery. The small peak at $Q = 1.37\,1/Å$ that was visible at OCV has turned into a strong peak, which is as large as the principal [111] peak at $Q = 1.39\,1/Å$. We realize the extensive crystallographic change upon delithiation, i. e., a phase transformation from cubic to tetragonal. However, we cannot unambiguously see a resonant diffraction effect other than the one we observe at OCV. As I have already outlined in Part 5, it is difficult to assign the changes of the peak position and peak height the proper origin. Changes may originate from the change of the X-ray energy, hence an anomalous scattering effect, or by actual changes of the chemistry in the cathode during operation.

At a later campaign, we have made XANES and EXAFS studies on a similar cathode material at SSRL. Figure 8.18 shows the EXAFS spectra for $Li_2Mn_2O_4$ and $LiMn_2O_4$. Figure 8.18 shows in the left panel a Mn K-edge XANES spectrum with extending into the EXAFS range from 6500 to 6800 eV. The chemical shift between $LiMn_2O_4$ and $Li_2Mn_2O_4$ —the two extreme phases during the *in situ* experiment— is obvious. The huge difference in the beginning of the EXAFS range at 6580 eV is also obvious. The right panel of Figure 8.18 shows the first differential of 16 spectra recorded during discharging versus the X-ray energy in the edge region from 6530 to 6560 eV. This procedure enhances the differences between both spectra.

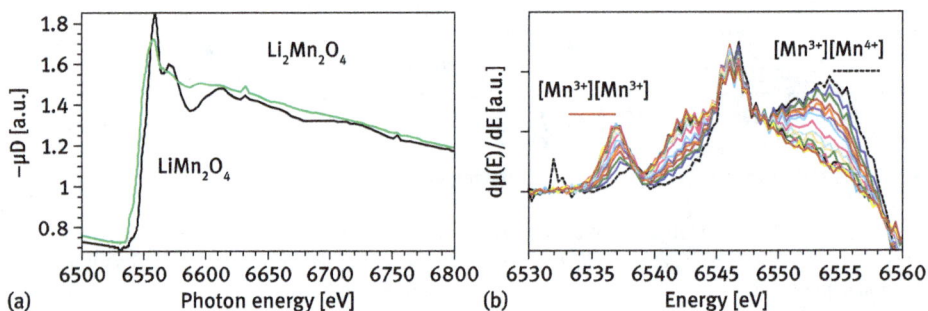

Figure 8.18: (a) Mn K-edge XANES and EXAFS range spectrum of $LiMn_2O_4$ and $Li_2Mn_2O_4$ recorded *in situ* at SSRL. (b) First derivative of the spectra recorded during the charging of the battery *in situ* cell. Sixteen spectra are shown which were recorded during charging, showing gradually changing of the spectra.

We have carried out the transformation from X-ray energy to k-vector (see Figure 8.5 [Braun 2003a], which shows EXAFS $k·\chi(k)$ of the sample while Mn was in the oxidation state of 3.39) and then performed the Fourier transformation in order to obtain the radial distribution function (shown in Figure 8.6 [Braun 2003a]) during discharging from average $Mn^{3.5+}$ to $Mn^{3.09+}$. Figure 8.19 shows the radial distribution function with color code during discharging.

The peak at $R = 2$ denotes the bond length between Mn and O, and the peak at 2.8 denotes the Mn-Mn distance. During discharging from the marker positions 1 to 25, the Mn-O distance is virtually not changing. However, the Mn-Mn distance changes considerable during this battery operation. We have not carried out a complete EXAFS analysis. A full EXAFS analysis would include computational modeling with the FEFF code [Cross 1998, Jorissen 2013], for example.

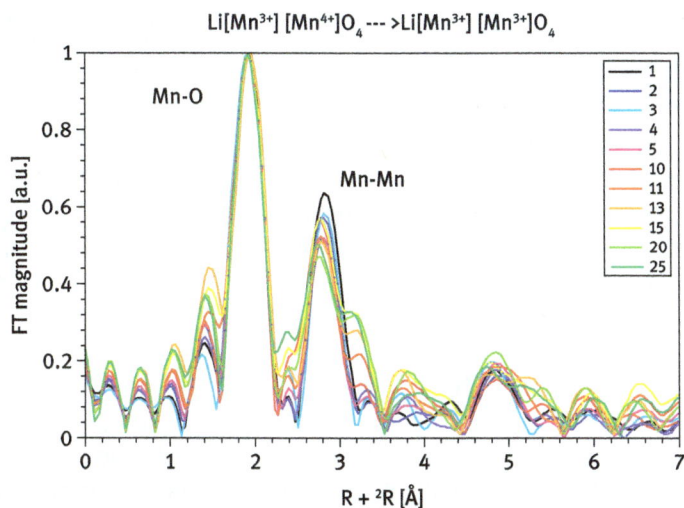

Figure 8.19: Fourier transform magnitude of EXAFS spectra of $Li_{1-x}Mn_2O_4$ obtained from a discharging battery at SSRL during 25 time stamps of operation. Changes in the Mn-Mn distances are considerable, whereas changes in the Mn-O distance are less pronounced.

Charging and discharging of a battery causes not only changes in the electronic structure and crystallographic structure of the electrodes but also changes in the microstructure. Small-angle scattering with X-rays is probably the only method capable of monitoring microstructural changes in a battery cell.

Given that our *in situ* cell is designed for X-ray transmission, the transmitted signal, i. e., the X-ray scattering curve, will contain microstructure information from all materials and components in the beam path. Although we have tried to minimize such contributions, which we can call parasitic scattering, there is still some minimum material necessary for the function of the battery cell, i. e., the aforementioned beryllium windows, aluminum foil, polymer binders, and lithium metal, in addition to the cathode material of interest.

I performed this electrochemical *in situ* lithium battery ASAXS experiment together with Soenke Seifert at BESSRC-CAT on June 21–24, 2000, Advanced Photon Source. I have presented this study already in Chapter 5.5.1 [Braun 2001a, Braun 2001f, Braun 2003a].

In order to get around the signal of these none-of-interest components, we apply X-ray contrast variation by tuning the X-ray energy to the manganese K-shell absorption edge. Then we record a number of small-angle scattering curves for a series of X-ray energies, spanning the pre-edge and absorption edge of manganese, as is shown in Chapter 5.5.1 and Figure 5.20. The ASAXS patterns were recorded with a two-dimensional X-ray detector, a CCD camera (see schematic in Figure 5.19 and Chapter 5.5.1). The recording of one two-dimensional SAXS curve took 0.1 seconds.

It was only the shutters that slowed data acquisition down to 1 second per shot. Adjustment of the X-ray energy to the next step took also some time, but altogether, the experiment for 20 X-ray energies curves took around 1 minute per charging point of the *in situ* cell. The evolution of the charge was deliberately chosen slowly. We took approximately 2000 minutes to charge and partially discharge the *in situ* cell.

For the quantitative analysis of the scattering curves, it was necessary to perform a weighted subtraction as shown in Figure 5.21 and Chapter 5.4.1. After that, the Porod background was subtracted and a global Guinier fit was applied. We applied two Guinier ranges because we could identify two significant Guinier shoulders, not withstanding that a more sophisticated analysis could yield better structural data.

8.1.4 *Operando* X-ray Raman study on a lithium ion battery

Soft X-rays provide valuable electronic structure information but are distinct surface sensitive. Electrochemical reactions take usually place at the surface of an electrode. However, in the case of lithium intercalation batteries, the redox front propagates from the electrode surface into the electrode interior. Hence, we are dealing with solid-state electrochemistry. We have learned that we can study this with hard X-ray methods that can fully penetrate and probe the electrode. However, the soft X-ray bears sometimes more valuable information than the hard X-rays. There is a way out of this dilemma by using the X-ray Raman effect, as was already outlined earlier in this book in Chapter 6.1.

Figure 8.20 illustrates how the energy resolution of soft X-rays, e. g., for a Mn L-edge can outperform the resolution of the corresponding K-edge, which is obtained with hard X-rays of 6500 eV. The black spectrum is a K-edge that was rescaled on the energy axis to fit the energy range of the Mn multiplett 2p 3/2. The latter has a significantly better

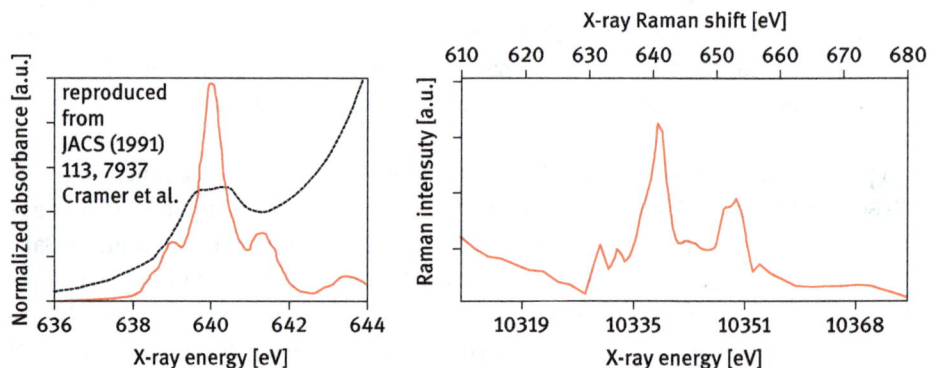

Figure 8.20: Left: Comparison of K-edge (black diffuse spectrum) and L3-edge (red peaked spectrum) of $MnCl_2$, as digitally reproduced from Cramer et al. [Cramer 1991]. Spectra are arbitrarily aligned according to Cramer et al. [S. P. Cramer, et al., Ligand field strengths and oxidation states from manganese L-edge spectroscopy, J. Am. Chem. Soc. 113 (21) (1991) 7937–7940.]. Right: X-ray Raman spectrum of Mn in the lithium battery studied in this paper, recorded around 10,300 eV, which includes the Mn L2,3 spectrum. The top axis shows the corresponding energy range for a typical soft X-ray NEXAFS spectrum.

energy resolution and shows more distinct peaks. Note, however, that a broad spectrum can be also the result of averaging different and multiple spectral features over a large enough, thick enough sample, whereas the NEXAFS method probes only a surface range of 1-micrometer thickness. The right panel shows part of the emission spectrum.

Figure 8.21 illustrates this situation by a 25-micrometer-thick active battery layer made from active spinel grains, carbon particles, and an aluminum current collector. Soft X-rays of 650 eV will not make it beyond the first micrometer of material. A conventional NEXAFS spectrum of the Mn L-edge will look like the one at the left bottom

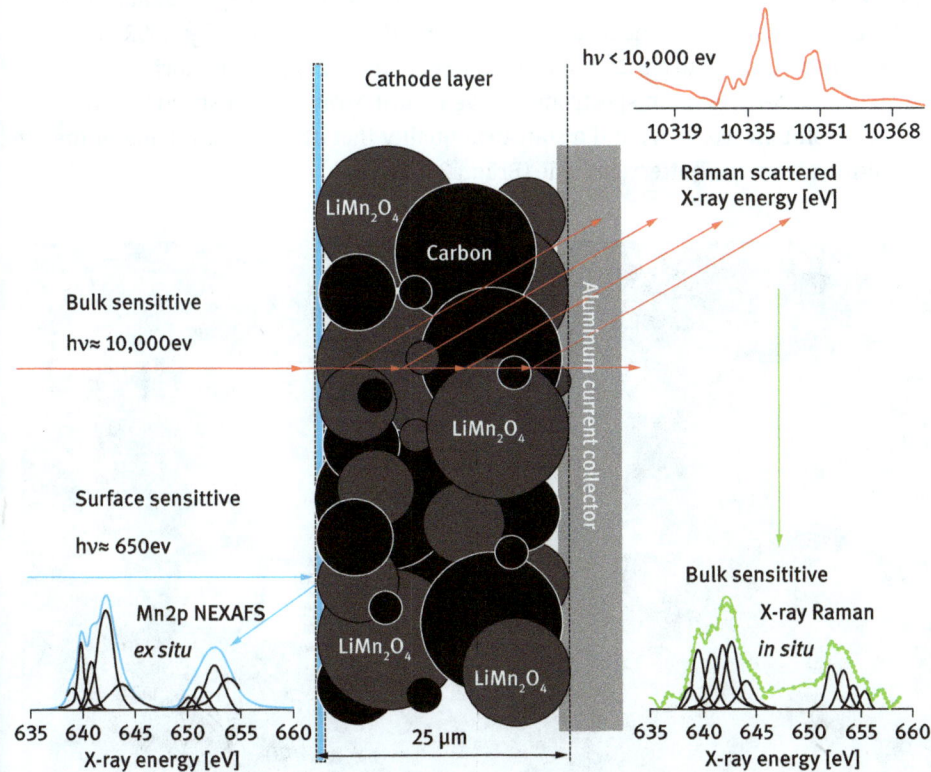

Figure 8.21: Battery assembly with LiMn$_2$O$_4$ spinel particles and carbon particles of different size, coated on a thin aluminum current collector. The soft X-rays necessary for NEXAFS spectroscopy of 650 eV photon energy penetrate and probe only the top 1 μm layer of the particles, as indicated by the blue arrows, hence not the entire electrode depth. The corresponding NEXAFS spectrum is shown on the lower left side. The hard X-rays of 10,000 eV photon energy have an attenuation length of 30 μm and penetrate and probe the entire electrode assembly, as indicated by the red arrows. The bulk sensitive X-ray emission spectrum is shown on the upper right; the extracted X-ray Raman spectrum is shown on the lower right. Reprinted from [Braun 2015j] Braun A., Nordlund D., Song S-W., Huang T-W., Sokaras D., Liu X., Yang W., Weng T-C., Liu Z.: Hard X-rays in-soft X-rays out: An operando piggyback view deep into a charging lithium ion battery with X-ray Raman spectroscopy. *Journal of Electron Spectroscopy and Related Phenomena 2015*, 200:257–263. Copyright (2015), with permission from Elsevier.

of Figure 8.21 (blue spectrum), whereas an X-ray beam of 10 keV will pass through the entire electrode assembly and carry all element specific spectroscopic information, as shown partially in the top right red spectrum, which can be utilized via the X-ray Raman effect and then "translated" into a spectroscopically equivalent L-edge spectrum, when recorded in a geometry that corresponds to the dipole approximation.

We have considered doing this in the year 2001 at the APS, but back then the spectrometer was not equipped with the necessary number of analyzer crystals so as to warrant a sufficient counting statistics for such delicate experiment.

Approximately 10 years later, the spectrometer, meanwhile permanent based at SSRL, was equipped with the double number of crystals. The experimental hutch at SSRL Beamline 6-2 during the X-ray Raman *operando* battery campaign (SSRL proposal 3573A) is shown in Figure 8.22. Using SSRL's Beamline 6-2 Si(311) monochromator and the 40-crystal X-ray Raman spectrometer, we record X-ray Raman spectra with a high energy resolution (~0.55 eV) and a statistical quality that allowed extraction of the Mn 2p multiplet from the battery cathode [Braun 2015j].

Figure 8.22: Experimental hutch at Beamline 6-2 at SSRL. The experiment takes place in the hutch and is computer controlled outside the hutch. Access to the experimental hutch is only provided to users who hold a key which they receive from the floor coordinator. Radiation safety protocols warrant that nobody is exposed to X-rays by accident.

Such synchrotron experimental hutches have interlocks which warrant that no dangerous X-rays arrive from the beamline to outside of the hutch as a measure of radiation protection. Before you can activate the beamline and open the shutters so that X-ray pass through the beam line tubes into your experiment, you have to go inside the hutch and search to make sure no person is inside. You confirm this search by pushing one or several buttons inside the hutchb which act as keys for being able later to open the Beamline shutters and have X-rays arriving at your experiment. Often there is a time limit like 30 or 60 seconds during which you have to perform this person search, leave the hutch, and lock the door. At SSRL, you will receive a key from the floor operator for which you have to sign that you received it. This key is necessary to perform the aforementioned person search and beam activation procedure.

You may not give away this key to any other experimenter unless you register this handover in a book at the beamline so that the holder of the key can be identified and made out every time.

The radiation safety measures vary from synchrotron-to-synchrotron and depend often on the technical details and probability of exposure to radiation. For synchrotrons that provide typically soft X-rays, radiation shielding is already provided by the metal beamline tubes and metal UHV chamber, and then no extra hutch is necessary.

Figure 8.23 shows a photo of the end station inside the hutch at Beamline 6-2. I have labeled the analyzer crystals array, the helium plastic bag that expels the air and thus allows for recording more counts from the sample, and the position where the *in situ* battery cell is mounted. The cell is shown magnified in the green and blue insets in Figure 8.23.

We carried out an *operando* experiment on a new designed *in situ* cell, the photos and blueprint of which are shown in Figure 8.24. The cell body is made from stainless steel for the cathode carrying side and from PEEK® plastic for the anode carrying side. Both plates have apertures for the X-rays to pass through. The apertures are closed with Be foil X-ray windows, which are coated with polyimide tape on the side exposed to the cell interior where aggressive electrolyte could do harm to uncoated beryllium. Two aluminum disks at the outer faces of the stainless steel plate and PEEK® plate press the BE foils tightly on the cell body parts with small bolts. The aluminum plate on the cathode side has an aperture with a very wide opening angle (solid angle) so that the X-ray Raman experiment can be done in reflection geometry.

The selection of the part of the spectrum for a particular element, in this case Mn, depends on the experimental geometry and solid angle. The new cell in Figure 8.24 is smaller than the previous one (which we tested at APS and SSRL in 2000 and 2001 [Braun 2003a]) with around 5 cm diameter, and only one sealing O-ring between the cell body parts and one O-ring each between aluminum plates and Be disks.

We have assembled several such cells during the X-ray Raman campaign in January 2012. At that time, battery experiments were already established at SSRL, and an Argon filled glove box was available for the increasing number of users who required

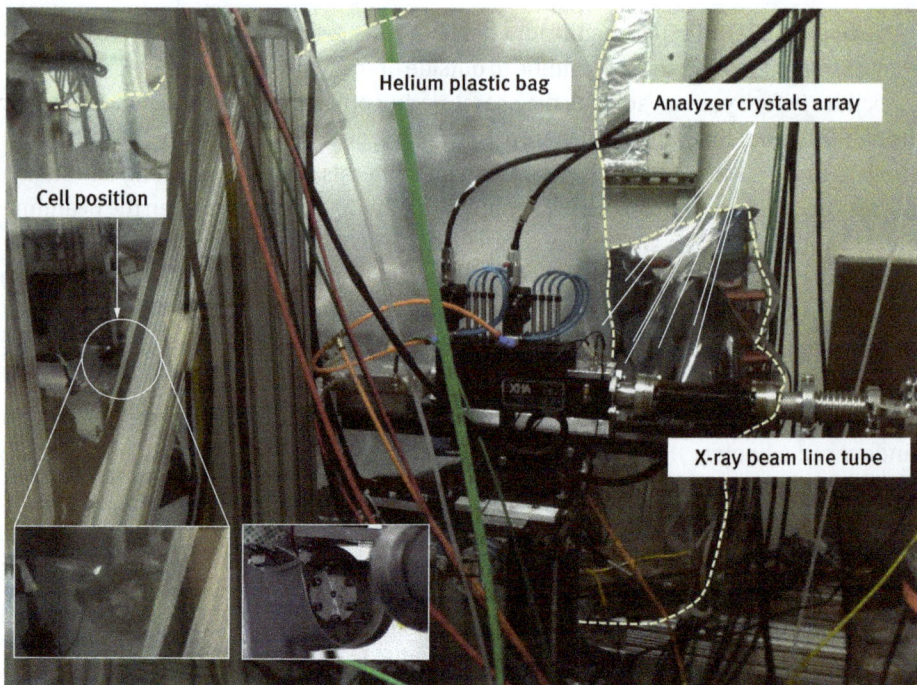

Figure 8.23: End station at Beamline 6-2 at SSRL, showing the experimental setup for X-ray Raman spectroscopy. The X-ray beam line tube is on the right side. The sample position is on the left side. The X-ray optical path between sample, spectrometer, and analyzer crystals is "bridged" by helium, which is trapped in a plastic bag, as indicated by the yellow dashed line. The green circle shows the position of the *in situ* cell in the X-ray beam. The green and blue rectangles show magnifications thereof.

(a)

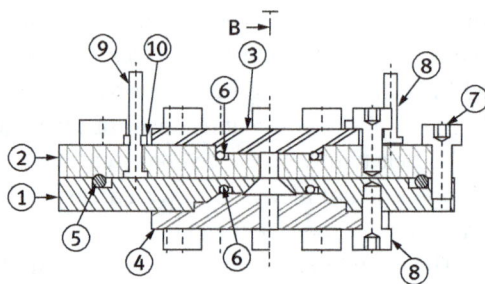

(b)

Figure 8.24: (a) Image of the front and back of two cells for X-ray Raman battery studies. (b) Design of the utilized *in situ/operando* lithium cell. 1 – stainless steel plate to receive and to contact cathode; 2 – PEEK plate; 3 – aluminum plate with X-ray window; 4 – aluminum plate with X-ray window; 5 – O-ring; 6 – O-ring; 7 – steel bolts; 8 – steel bolts; 9 – steel pin to contact lithium electrode; 10 – steel pin to contact reference electrode. Construction and manufacture of the cell was done by Empa's Urs Hintermüller and Erwin Pieper.

(a)　　　　　　　　　　　　　　　　(b)

(c)　　　　　　　　　　　　　　　　(d)

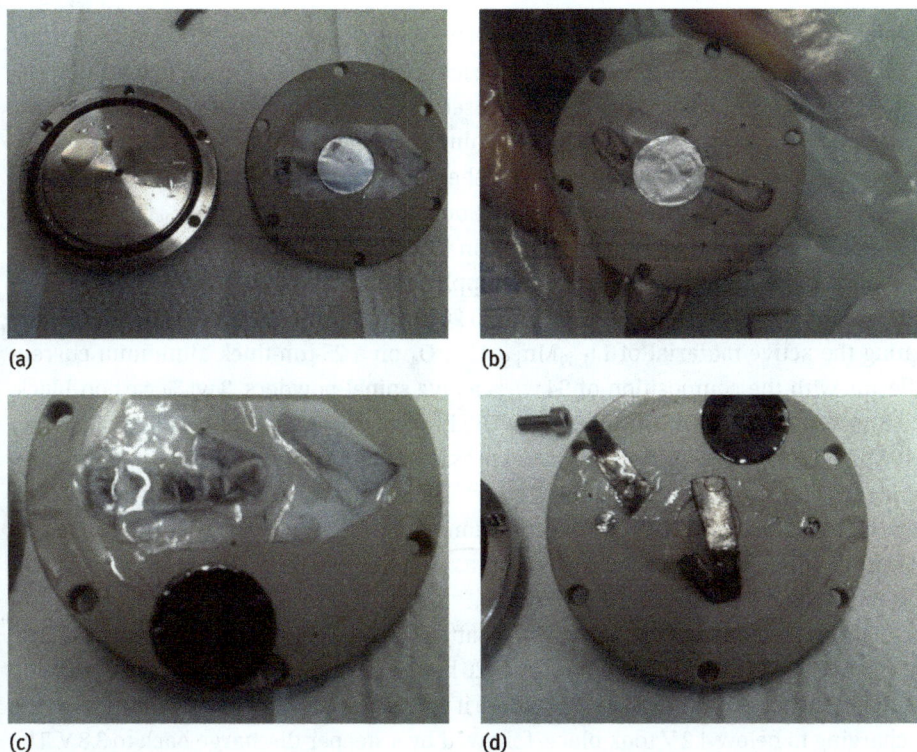

Figure 8.25: Battery *in situ* cell after disassembly. (a) Stainless steel plate for cathode side droplets from electrolyte and BUNA O-ring on the left, and PEEK® anode bearing side with separator, aluminum current collector with cathode layer underneath, and lithium stripes. (b) PEEK® disk magnification. (c) Cathode disk turned upside down, showing black spinel-carbon active layer, and wet translucent separator covering several lithium metal stripes. (d) All separators are removed, lithium counterelectrode stripe, and lithium reference electrode stripe begin turning black from exposure to nitrogen in air toward Li_3N. Note the two 1 mm wide metal bolts in the PEEK® plate, which serve as electric contacting for counter-Li electrode and reference Li electrode.

a real inert gas like Argon for their sample operation and storage. Figure 8.25 shows a disassembled cell after operation. On the upper left, we see the still electrolyte wet stainless steel cathode side plate with BUNA O-ring. Next to it, the PEEK® disk with the aluminum foil current collector, on the back of which the spinel active material. We see also the wet translucent separator foil. Upon closer view, we see how the upside down cathode (positive electrode $LiMn_2O_4$ with carbon on aluminum foil) rests on the still wet separator, and underneath we see a long stripe of lithium metal foil, the anode, shining through the separator. The lithium starts already to become black from exposure to ambient conditions. On the lower left photo, we have turned the aluminum foil around and now see the black active layer (spinel and carbon). We also see below the separator that the lithium foil is comprised of two pieces, the shorter one as the reference electrode, and the longer one as the anode (negative electrode). Observe also the 1 mm wide inner bolts on the PEEK disk which serve as electric contact terminals for the two lithium reference and counterelectrodes. Upon increased exposure to ambient air, which consists of around 80 % nitrogen, the lithium is turning into black Li_3N.

The cathode material that we used for the *operando* battery experiment was a commercial one from POSCO ES Materials in South Korea ($Li_{1.09}Mn_{1.83}Al_{0.09}O_4$) of POSCO ES Materials described as sample type P1 in [Song 2011]. Size distribution of the cathode spinel particles as disclosed by the supplier company was D10 = 9.35 μm, D50 = 19.24 μm, D90 = 34.88 μm, and Dmax = 60.26 μm. The cathode was prepared by laminating the active material of $Li_{1.09}Mn_{1.83}Al_{0.09}O_4$ on a 25-μm-thick aluminum current collector with the composition of 94 wt.% active spinel powders, 3 wt.% carbon black, and 3 wt.% polyvinylidene difluoride (PVDF) binder. The cathode laminates were dried at 100 °C overnight in a vacuum oven and pressed using a press at room temperature, resulting in a density of approximately 2 g/cm³. The electrodes were stored in the Ar-filled glove box (MOTek) with H_2O/O_2 concentration of 1 ppm [Song 2011]. The electrolyte was LP30 ($LiPF_6$ dissolved in ethylencarbonate) as supplied by Merck. The battery operation was done with a potentiostat from Biologic.

Figure 8.26 shows the evolution of the battery voltage and current. The entire charging and discharging experiment took over 20 hours. The immediate voltage was around 4 V and then increased linear to 4.3 V, where it was at rest for over 1 hour, after which the discharging to below 4.2 V took place, followed by a steeper discharge back to 3.8 V. The charge was controlled by setting an appropriate current, as can be read from the right axis in Figure 8.26. During these >20 hours, we recorded around 200 spectra. It is necessary to synchronize the electrochemical data and the X-ray spectroscopy data for such kind of experiment. The right panel of Figure 8.26 shows a visualization of this situation. The number of the spectrum has to be associated with a particular state of the battery cell during operation. We have chosen to specifically look into the positions where the

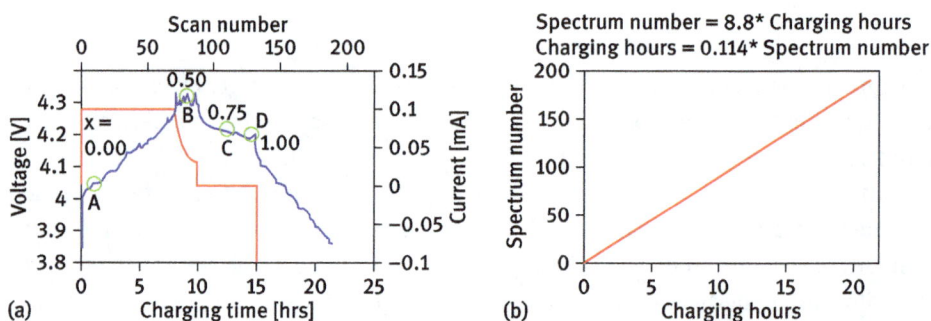

Figure 8.26: (a) Battery charging profile. Voltage (blue) and current (red) profile of the battery in the cell versus charging/discharging time (bottom axis) and X-ray Raman scan numbers (top axis). Letters A, B, C, and D denote the recording time stamp of the spectra presented here. (b) Graph for assigning the number of the spectrum to the actual charing time stamp.

battery was at start-up, at full charge, on the way to discharge, which we have labeled A–D in the left panel, in Figure 8.26. Every spectrum took around 6 minutes recording time but the counting statistics of each spectrum was yet relatively poor. It was therefore necessary to accumulate several proximate spectra and average them. Note that during this time the battery was not necessary in a static regime, as the zigzags of the blue line in Figure 8.26 left panel shows, i. e., the variation of the potential. By averaging the spectra, we therefore are also averaging over small differences in the battery charge condition. This is a trade-off which we can hardly avoid.

The spectra for the four charging state regions A, B, C, and D are shown in Figure 8.27. The X-ray Raman spectra (bulk sensitive, including the surface) are drawn in green color. They are compared with the *ex situ* obtained NEXAFS spectra (surface sensitive), which are drawn in black. From the charge state and similarity of the spec-

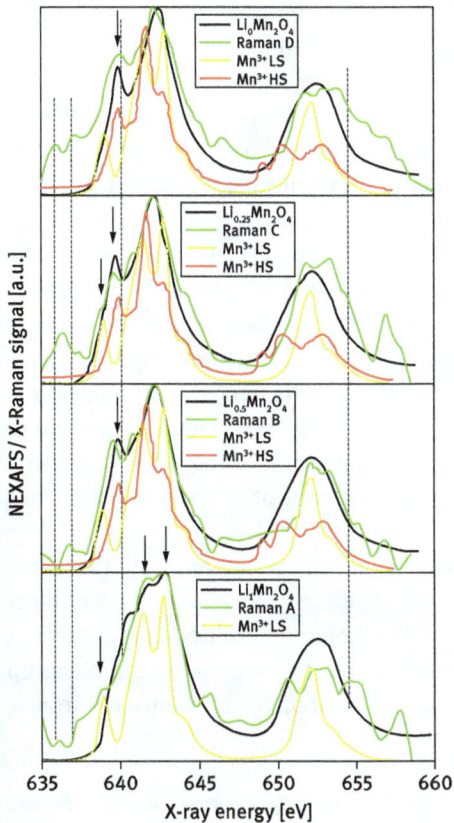

Figure 8.27: Comparison of *operando* obtained X-ray Raman spectra (green spectra) with experimental NEXAFS spectra (black spectra, [Yoon 2003]) and simulated spectra for Mn^{3+} in high spin (red spectra) and low spin (yellow spectra) configuration [Degroot 1994]. The experimental spectra are given for four different stoichiometries in $Li_x Mn_2 O_4$ with $x = 0, 0.25, 0.5,$ and 1.0.

tra, we have concluded that positions A–D correspond to the stoichiometry $LiMn_2O_4$, $Li_{0.5}Mn_2O_4$, $Li_{0.25}Mn_2O_4$, and $Li_0Mn_2O_4$, respectively. This is just a rough assignment and for sure can be improved toward higher accuracy, when more time is taken and better counting statistics and more stable battery operation is warranted. We have also drawn the computed spectra for low spin (LS) and high spin (HS) Mn^{3+} for comparison in order to verify whether a low spin, or a mixture of high spin and low spin Mn^{3+} is present [Degroot 1994]. The spectra suggest that both forms of Mn^{3+} are present at particular states during the experiment.

8.1.5 X-ray spectro-electrochemical cells for *operando* and *in situ* battery studies

You have seen by now already a number of X-ray spectro-electrochemical cells for battery studies. The suitable or perfect cell design[1] depends on the particular experimental method and also on the measurement geometry. The X-ray optical path may go straight 180° in transmission mode, but sometimes a reflection geometry is more suitable or even required. For the aforementioned X-ray Raman spectroscopy, it was necessary to collect the X-rays in a reflection mode or fluorescence mode. The sample is theoretically supposed to be concentrated in one single point with 0 radius, becomes then excited with the X-ray beam, and all scattered intensity should be fully collected in 4π direction. This is certainly hardly possible in real life. But at least the sample in the cell should scatter as much X-ray intensity as possible into the detectors. This requires a very flat design like shown in Figure 8.24, or a design with a somewhat thicker cell with an appropriate cone with a suitable solid angle. The design with the thicker plates has advantages because more material allows for better mechanical behavior and easier contacting.

My colleague Sebastian Risse at Helmholtz Zentrum Berlin (HZB) has designed such somewhat thicker cell as part of the OpMetBet project. Their HZB machine shop manufactured a number of such cells, and they were distributed among the project partners. We also received one such cell in 2023 at our project meeting at INRIM in Torino, Italy. The cell is shown in Figure 8.28.

Looking at these cells, they are all made from metal, and metals are good electronic conductors. Chances are therefore that there can be electric short circuits. You must therefore care for proper electric insulation. It also should be avoided that driplets of electrolyte can make an ionic bridge across components. Some solvents can creep up and out without you first noticing it. Think of the KCl solution for reference electrodes,

---

[1] A suitable cell design may be already a very simple one, good enough to permit the experiment in general. It maybe has no reference electrode, but allows for charging or recharging to some extent, and allows to get X-rays into the sample, and also out of the cell into the detector so that you can get an X-ray spectrum, preferably corresponding to the sample changes due to electrochemical operation.
A perfect cell has a reference electrode and allows for maximum utilization of the photon flux, and is so well sealed that you can operate it for a long time.

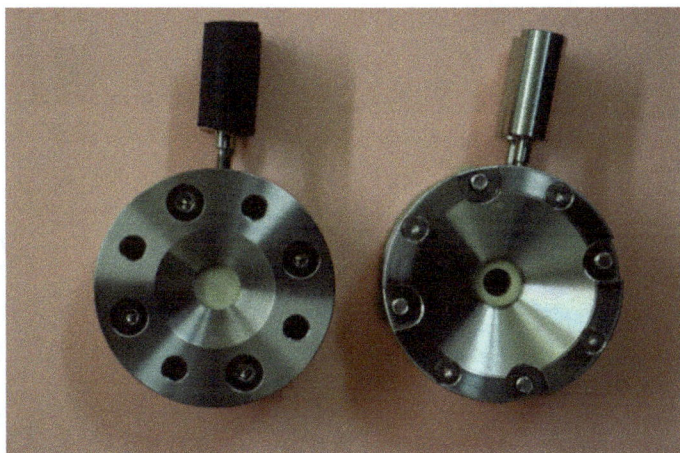

Figure 8.28: The operando cell on the left, the one with the narrow angle cone, was designed by Sebastian Risse and machined by the technicians at Helmholtz-Zentrum Berlin. Based on this design, we adapted the cell but made a wider a cone, shown on the right, to make it suited for X-ray Raman experiments by Empa's Alexey Rulev, Erich Heiniger, and Erwin Pieper.

for example. Over time, the bottle with saturated KCl solution gets empty, and crystallites accumulate around the stopper outside, although you firmly closed the bottle.

It is interesting to see how sophisticated *operando* and *in situ* cells have become over the last decades. In other fields of spectroscopy, with a larger commercial market like optical spectroscopy and electrochemistry, you can purchase well designed cells from manufacturers. Certainly, making a novel cell, or in general making scientific instruments, is part of scientific research and activity and adds to the general pleasure of finding things out.

Figure 8.29 shows a disassembled cell of our adjusted design—with the wider cone for X-ray Raman experiments. Four large disk plates from stainless steel are stacked together and fixed with bolts. In the frame of two of the disks, there are narrow screw threads, which can receive the two long thick stainless steel cylinders: one in the top and one in the bottom plate. These are then the positive and negative electrode contacts. The other ends of the two cylinders have holes, which can receive banana plugs for electric contacting. This is basically the outer architecture of the cell. The battery compartment is the PEEK ring, which receives the negative and positive electrodes, electrolyte, separators.

The two small and thin disks are commercially available button cell components. They are called volcano (cone, trumpet) spring and spacer.[2] For X-ray experiments, you have to drill a hole into the spacer to allow the X-rays to come through (e. g., by a laser

---

2 The graphical abstract in the paper from Liao and Ye [Liao 2018] shows how a standard CR2032 button cell or coin cell is built up.

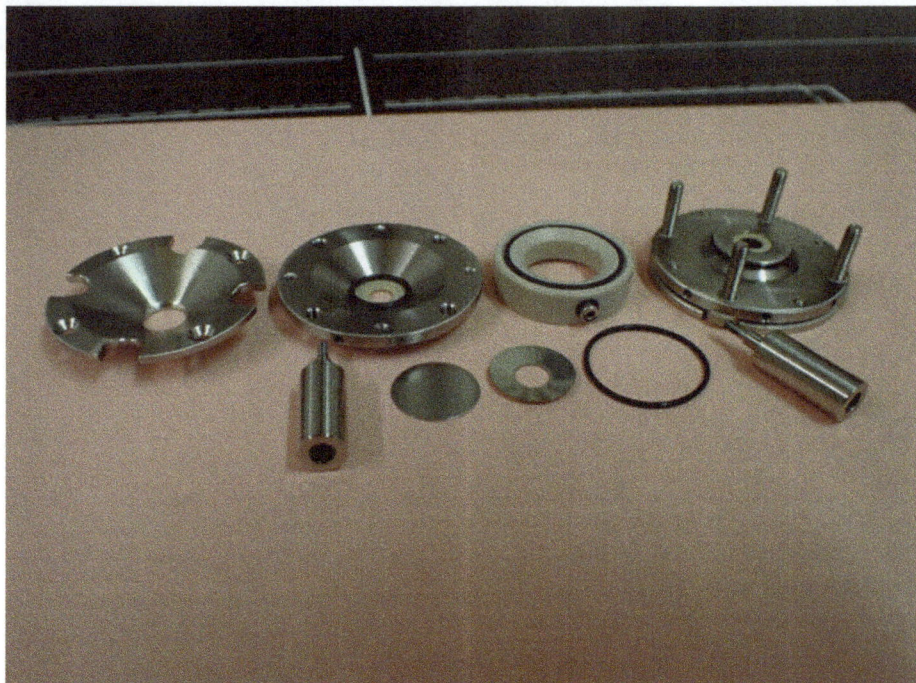

Figure 8.29: The dissassembled Empa operando cell with a wider solid angle so as to allow for X-ray Raman spectroscopy. Designed and manufactured by Empa's Alexey Rulev, Erich Heiniger, and Erwin Pieper.

cutter). The PEEK plastic is the compartment for the whole battery cell and repeats basically the coin cell geometry. The shiny metal bolt with small black O-ring with thread in that PEEK cell has multiple use. It can be used for pressure relief in soft X-ray experiments in the UHV chamber. Or you can insert electrolyte. It is not directly an electric connection, but can be used for it. Note that this sophisticated cell yet has no reference electrode; possibly, because standard coin cells do not have such. The X-ray window for this cell is by default from graphite (Optigraph GmbH, Berlin; size is 10mm x 10mm x 4 micrometer) and needs to be glued on the PEEK part of the second cone shown on the upper left, which has a small black gasket on the PEEK. It seems we can replace this small PEEK disk (it has a hole, it is basically a frame) with a beryllium disk. I prefer beryllium disks as X-ray windows.

8.1.6 Isosbestic points

When reading spectroscopy literature, you sometimes may come across the term isosbestic points. You can find such when you do a spectroscopy parameter study, such as by varying the composition of a material and then recording spectra, or changing a thermodynamic parameter of the material under investigation such as temperature, pH, pressure, and so forth. At some energy, or wavelength, or wave number or frequency, the spectra (according to IUPAC the total absorbance of a sample) may coincide at what is known as an isosbestic point [Verhoeven 1996].

They are well known in optical spectroscopy, but quite universal in all fields of spectroscopy. The existence of an isosbestic point in spectra of two different species (where species A is transformed to species B) indicates that there is a direct linear relationship in the concentration of these species. Moreover, there exists no third species C as an intermediate or a byproduct. Presence or absence of an isosbestic point can therefore be an interesting or even important information on phase transformations.

Isosbestic points are also found in X-ray spectra. But they are hardly considered and mentioned there. One of the works where the authors know about the meaning of isosbestic points is by Liu et al., who carried out *operando* soft X-ray spectroscopy on a lithium ion battery assembly at the Advanced Lightsource in Berkeley [Liu 2012]. Specifically, they investigated $Li(Co_{1/3}Ni_{1/3}Mn_{1/3})O_2$ and $LiFePO_4$ cathodes in polymer electrolytes while the electrodes were charged and discharged in the UHV chamber at the end station.

A beautiful example for an isosbestic point is shown in the graphical abstract their JACS paper on $LiFePO_4$ [Liu 2012]; see Figure 8.30. It displays the evolution of the Fe $2p_{3/2}$ absorption spectra (Fe L_3 edge) of the $LiFePO_4$ positive electrode during electrochemical delithiation, which has stoichiometry $FePO_4$. So, the lithium ions are extracted from the starting material, which has olivine structure. Let us briefly look into the stoichiometry.

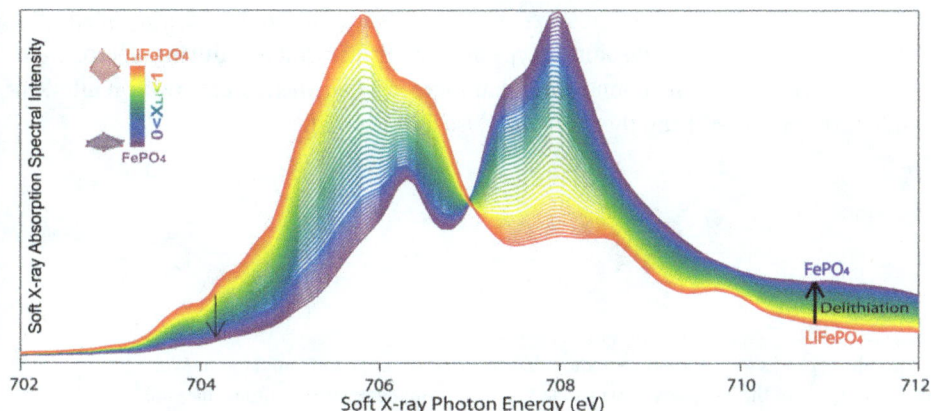

Figure 8.30: Fe 2p absorption spectra recorded from a $LiFePO_4$ (red spectrum) positive electrode during electrochemical delithiation toward $FePO_4$ (violet spectrum) positive electrode during electrochemical delithiation. Reprinted with permission from Liu et al. [Liu 2012]. Copyright 2012 American Chemical Society.

The four oxygen ions carry a total charge of −8. The phosphor ion has a charge of +5.[3] The lithium ion is +1. Therefore, the iron ion is +2. To maintain charge balance, removing of the lithium forces the iron to become oxidized from Fe^{2+} to Fe^{3+}.

The red spectrum shows the starting point before the delithiation begins. It has a high intensity t_{2g} peak at 706 eV, and a low intensity e_g peak at 708 eV. The variation of the colors in this rainbow, so to speak, denote the delithiation and Fe oxidation trend. From yellow over green to blue, the t_{2g} peak gradually disappears to the benefit of the eg peak. At 707 eV, there is a sharp point where all spectra coincide; this is the isosbestic point.

Sometimes, the spectra do not sharply coincide at one frequency; instead, the isosbestic "point" is somewhat smeared in one direction. This allows, e. g., for the determination of the temperature dependency of materials properties [Greger 2013].

Liu et al. continue later their *operando* battery work and compare two different lithium battery positive electrodes, and publish it in a Nature paper [Liu 2012]. The left panel in Figure 2 in the Nature paper shows a series of Ni L-edge spectra (labeled A–L) recorded during charge and discharge of the battery cell. The vertical dashed lines at around 851 eV and 853 eV denote t_{2g} and e_g peaks of the $2p_{3/2}$ peak (the L_3 peak) of the Ni, and thus via the relative spectral weight or peak height ratio, the oxidation state of the Ni during the course of charging and discharging.

These ratios Ni^{4+}/Ni^{2+} are plotted in panel b versus the time axis, along with the capacity of the cell as measured with their potentiostat. Both quantities show the same trend. The shape of the cycling voltage in panel c is what we generally expect from a lithium intercalation cell during cycling. Panel d shows an overlay of the L3 peaks during this charging and discharging processes. The height of the t_{2g} peak is decreasing from stages B to G and then increasing again, though not to its original height, whereas the height of the e_g peak is decreasing and then increasing again.

Liu et al. [Liu 2012] state that there is no isosbestic point in this overlay of spectra, and there is therefore a solid-solution type of phase transformation during electrochemical cycling. But close inspection shows that there is a densification of spectral intersections at around 853 eV. I call this a smeared isosbestic point.

---

3 Phosphor goes as P^{5+}. A good periodic table, such as the old one from Sargent Welch, contains the oxidation states of the elements. It is even better when you memorize all these. The American Chemical Society has published a periodic table, which I have printed in A0 format as poster. I have mounted the poster in frames in my electrochemistry laboratory and my spectroscopy laboratory. Unfortunately, they do not contain the oxidation numbers, the valencies. Maybe the ACS assumes that the chemists know them anyway. If you are interested in how a science education firm was founded over a 100 years ago, see https://www.sargentwelch.com/cms/history_of_sargent_welch

8.2 Ceramic fuel cells

As we have learned already in the previous parts, ceramic fuel cells operate at high temperatures ranging from 600 °C to 1000 °C. This is mostly because the electrolytes are solid materials, and the ion conductivity sets on only at such high temperature. The proton conducting ceramic electrolytes operate at around half this temperature, specifically around 500 °C. Here, I will present an electrochemistry *in situ* XANES study on the sulfur chemistry of a conventional SOFC anode system and an *in situ* resonant PES study on ceramic proton conductors at near ambient water vapor pressure.

8.2.1 Ceramic proton conducting electrolyte membranes

I have already introduced the ceramic proton conductors in Chapter 2.1, where I showed XANES spectra at the Y K-edges around 17 keV. We have seen very small chemical shifts depending on whether the BaZrY-oxide was dried or hydrated. This was an X-ray transmission study with distinct bulk sensitivity. We even carried out the experiment *in situ* because we heated the powder sample in a glass capillary with a fan while we supplied either dry gas or water vapor.

Same like with the previous battery studies, where the energy is stored in the bulk of the electrode and not at the surface, the function of the proton conductor is a bulk function because the protons have to diffuse through the entire proton conductor electrolyte membrane.

However, we want to investigate the properties of the proton conductor with resonant photoemission spectroscopy under operation conditions. We have done this at Beamline 9.3.2 at the Advanced Light Source using the ambient pressure XPS end station. This end station can provide gas pressures of 1000 mTorr and heat samples to 800 °C, while under electrochemical polarization.

We have carried out these experiments with a sound justification. From other studies, we know how the conductivity (ionic conductivity by protons) increases with increasing temperature. We know from temperature-dependent XRD how the unit cell volume changes. The structure and conductivity data is summarized in Figure 8.31.

Figure 8.32 shows the sample holder that was specifically designed for such high-temperature gas phase electrochemical experiments with electron spectroscopy for the ALS by Sandia National Laboratories [Whaley 2010]. It is made from high-temperature sustaining and noncorrosive steel. It has three insulated electric terminals that can be fixed with adjustable screws on particular sample positions to allow for a typical three-electrode configuration. In the center of the sample holder, we see a white ceramic pellet from $BaZr_{0.9}Y_{0.1}O_3$, which has two gold stripes coated as WE and counterelectrodes. Two tips provide the electric contacts with them.

The third tip rests on the pellet directly as reference electrode. The right panel in Figure 8.33 shows a SnO_2 pellet with gold electrodes for a gas sensor measurement. The

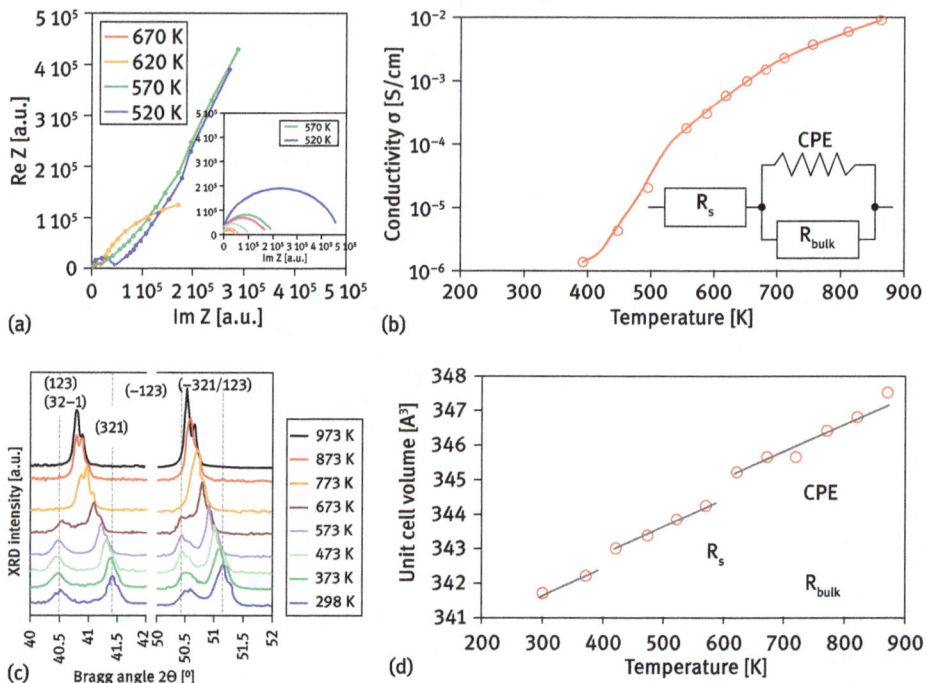

Figure 8.31: Evolution of impedance spectra during heating the sample from 298 K to 620 K (a); fraction of X-ray diffractograms for hydrated BCY20 from 298 K to 973 K (b); bulk proton conductivity of hydrated BCY20 as a function of temperature, derived from impedance spectra (c); variation of unit cell volume of hydrated BCY20 versus temperature, as derived from X-ray diffraction (d).

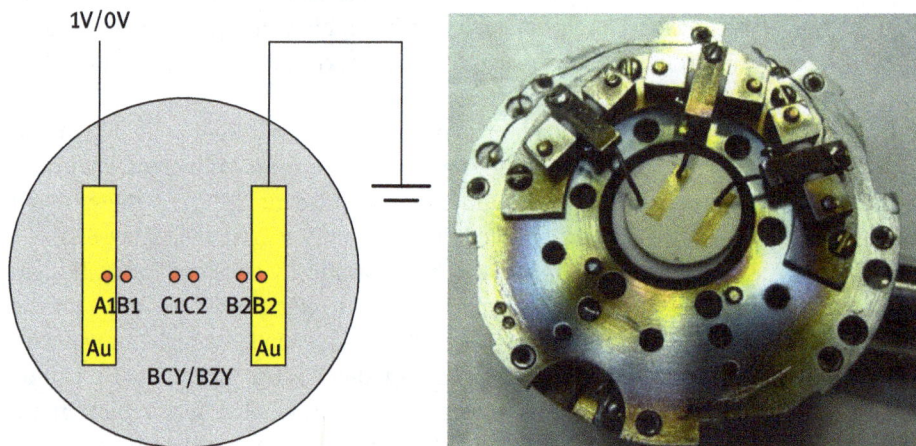

Figure 8.32: Sample holder for high-temperature gas phase electrochemical experiments [Whaley 2010] with a SnO_2 sample as gas sensor model system [Flak 2013].

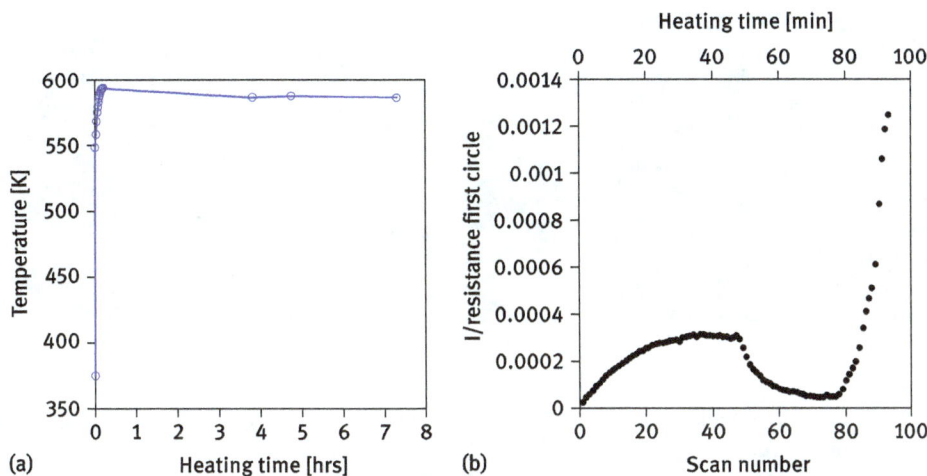

Figure 8.33: (a) Temperature profile of the proton conductor pellet sample during the ambient pressure XPS campaign. (b) The reciprocal value of the resistance (= conductivity) measured from the first semicircle in the impedance spectra for around 100 scans during ambient pressure XPS. The conductivity is steadily increasing for the first 40 scans during heating, and then decreasing upon cooling, and finally increasing again. Bookkeeping of the time stamps is necessary for the synchronization with the XPS spectra.

pellet is heated from the back electrically and exposed in the UHV chamber either to vacuum or any gas pressure.

We have inserted this sample on the sample holder via a load lock into the UHV chamber and began acquisition of data at UHV and around 300 K, recording electrochemical impedance spectroscopy (EIS) spectra and XPS spectra. Then we have increased the temperature and after some time supplied water vapor into the chamber. The spectroscopic response was noticeable for EIS and XPS. Figure 8.33 shows for two examples how we increased the temperature from ambient to just below 600 K in a very short time of a few minutes. On the right side, we see how the reciprocal of the first semicircle radius changes during the change of temperature and water vapor partial pressure.

Figure 8.34 shows a typical experimental crew setting at Beamline 9.3.2. With several computers and computer screens, not only the beamline settings are controlled but also the XPS spectrometer, the impedance analyzer, and the valves and pressure gauges for the processing and operation gases, which are filled into the UHV chamber. Once the sample is successfully inserted in the chamber and electrically connected, the experiment parameters are defined; only one researcher is necessary to guard and conduct the experiment. When one shift of 8 hours has passed, the next researcher can come to the beamline and take over the experiment. Many synchrotron and neutron sources have guesthouses and restaurants onsite so that a beam time of one week can be used efficiently.

(a)   (b)

Figure 8.34: Prof. Bongjin Simon Mun from Gwangju Institute of Science and Technology together with AGH Krakow/Empa PhD student Dorota Flak (now Scientist at Poznan University, Poland) and ETHZ/Empa PhD student Qianli Chen (now Associate Professor at University of Michigan – Shanghai Jiao Tong University Joint Institute) at end station 9.3.2 at the Advanced Light Source, Berkeley, while running ambient pressure XPS on ceramic proton conductors and gas sensors. High school student Lars Braun (Kantonsschule Glattal) operating the load lock for sample insertion.

The oxygen core level spectra carry very interesting information, as we know already by now from other studies in this book. Figure 8.35 shows the oxygen 1s spectra recorded at $t = 0$ and $t = 10$ minutes. The peak at 528 eV originates from oxygen bound in the crystal lattice of the proton conductor. The peak at 532 eV is assigned to OH. At $t = 10$ minutes, water vapor is supplied, which has a clear signature at 538.5 eV. The OH peak intensity slightly increases because more water is supplied to the system. The seemingly decreasing peak height at 528 eV is superficial because it appears diminished only because of the relative increase of the OH signature. The chemical shift toward lower binding energies could be the consequence of hole formation at the surface, or reduction of the oxygen.

On the right panel, we see three spectra recorded at scan positions 1 ($T = 373$ K), 15, and 485 ($T = 593$ K). We recorded around 500 XPS spectra during that campaign. Here, we observe that the relative spectral weight of the crystalline oxygen decreases upon heating, whereas the OH peak increases. Overall, among the probed oxygen ions, the relative portion of oxygen in an OH configuration is increasing upon heating. This is because supplied water vapor engages in filling engineered oxygen vacancies and forming OH groups. Compare Figure 8.35 with Chapter 2.3.1, where we discussed the oxygen spectra of BCY20 already.

We can summarize the findings on the oxygen spectra upon water supply and heating in Figure 8.35. In ultrahigh vacuum at 592 K, upon supplying water vapor, the temperature decreases because of the heat capacity of the water gas. This forces some cooling that we notice at the thermocouple and in the spectra. The hydroxyl peak increases. Upon further cooling, the water gas can be detected as a separate peak but the OH peak position shifts toward lower binding energies, while at the same time a new peak evolves

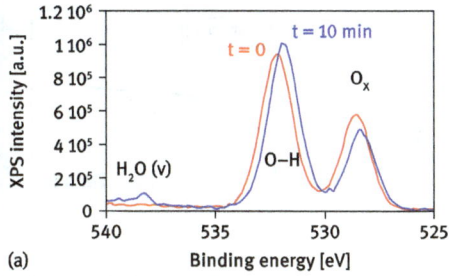

(a)

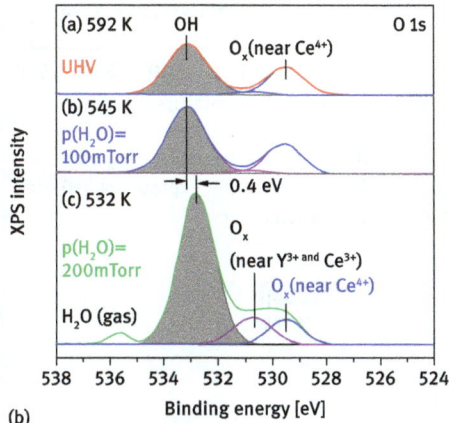

(b)

Figure 8.35: (a) O 1s core level XPS for BCY20 at 350 K (t = 0 minutes, red) and 592 K (t = 10 minutes, blue) according to the temperature profile in Figure 8.30. The blue peak at 538 eV is a response to the inlet of water vapor. (b) Comparison of three O 1s spectra recorded dry in UHV at 592 K, 100 mTorr water at 545 K, and 200 mTorr water at 532 K. Temperature changes from 592 K to 532 K due to heat capacity of injected water vapor. Photon energy = 700 eV. The spectra are normalized and aligned by the structural Ox oxygen peak (near Ce^{4+}).

near the crystal oxygen peak. We interpreted this as a signature from crystalline oxygen proximate to Y^{3+} and Ce^{3+}, and the original one proximate to Ce^{4+}.

The heating profile of the experiment is shown in Figure 8.36 as temperature versus scan number, when water vapor is being supplied. The panel on the right side shows the spectral weight, the height of the water vapor peak at around 538.5 eV. We understand from this figure that upon supply of water vapor, the temperature is decreasing from 590 K to 545 K.

We have also looked into the valence band spectra in the same campaign. We learned already in Part 5 how proper choice of excitation energy can produce resonant photoemission spectra, which have a chemical and even orbital specificity [Chen 2013d, Maiste 1995, Matolin 2009, Matsumoto 1996, Skoda 2007]. Photoemission spectra resonant to Ce (223 eV) and Y (299 eV) were obtained from dry and wet $BaCe_{0.8}Y_{0.2}O_3$ pellets in the ambient pressure mode (Figure 8.37). The Ce-resonant and Y-resonant spectra show generally the same trend. The spectra from the dry sample has a flat region up

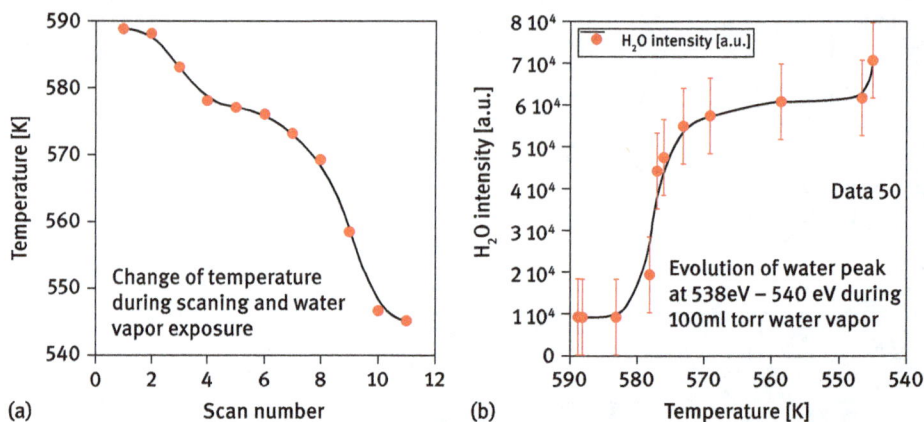

(a) Scan number (b) Temperature [K]

Figure 8.36: (a) Heating profile during 11 XPS scans during which the temperature is decreasing from 590 K to 545 K. (b) Evolution of spectral weight of the water vapor peak during water vapor exposure at 590 K to 545 K in the ambient pressure XPS chamber. Both profiles are mirroring each other.

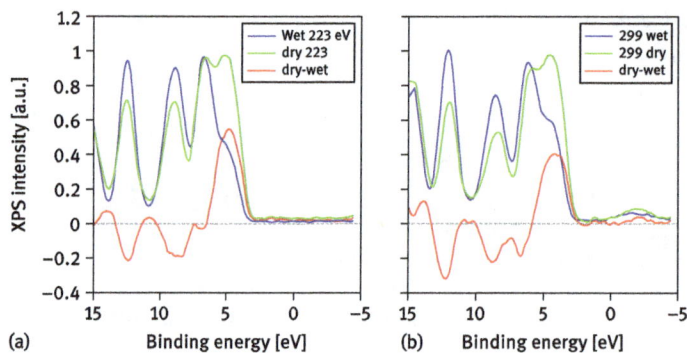

(a) Binding energy [eV] (b) Binding energy [eV]

Figure 8.37: (a) Valence band XPS spectra measured at the Ce-resonant energy for N_2 $4p_{1/2}$ at 223.2 eV (~223 eV) and (b) Y-resonant energy for M_3 $3p_{3/2}$ at 298.8 eV (~299 eV) in the dry (green spectra) and wet (= protonated, blue spectra) condition. The difference spectra are drawn in red.

to 2–3 eV binding energy, followed by a double peak structure at 4 and 6.5 eV (green spectra).

We discuss here only this double peak structure. When we expose this dry pellet to water vapor, the peak intensity at 4 eV drops dramatically to around 2/3 and ½ of its original value (blue spectra). Apparently, this is a result of the filling of engineered oxygen vacancies in the BCY20. An easy quantitative way of direct comparison of the spectra is by forming their difference spectra (red spectra = dry-wet). The red peak at 5 eV is therefore the signature of the oxygen vacancy. Let us now compare these two oxygen vacancy signatures, obtained at the Y $3p_{3/2}$ resonant energy at 299 eV and Ce $4p_{1/2}$ resonant energy at 223 eV. These two vacancy peak structures are compared in Figure 8.38. Their positions are shifted by 0.5 eV.

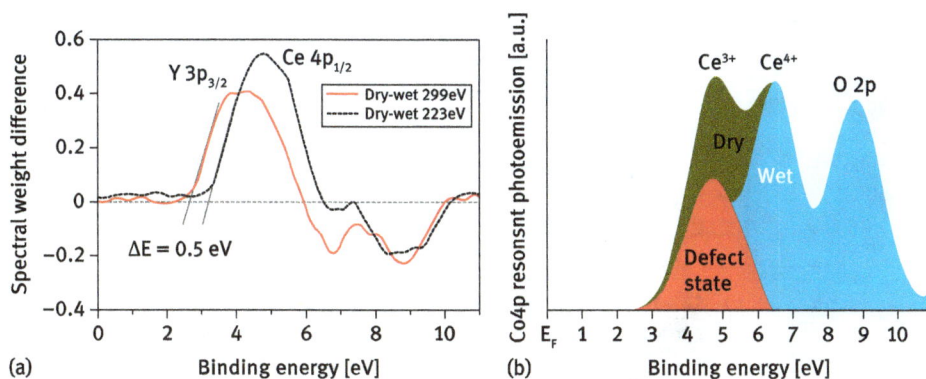

Figure 8.38: (a) Comparison of the difference spectra (dry minus wet) recorded in the Y-resonant condition at 299 eV (red) and Ce resonant condition at 223 eV (black). The spectra are shifted by around 0.5 eV. (b) Experimentally determined DOS from the resonant VB spectra. The profile is the sum of deconvoluted peaks.

Changes in the oxygen core level spectra, particularly emerging new spectral weight, suggest that oxygen ions near Y^{3+} and Ce^{3+} can be distinguished from oxygen ions near Ce^{4+}. The filling of oxygen vacancies with oxygen from water vapor is impressively reflected by the substantial decrease of the leading peak in the valence band spectra. Difference spectra of Ce $4p_{1/2}$ and Y $3p_{1/2}$ resonant VB spectra show a shift of the leading peak by 0.5 eV, which we interpret as that gap state of oxygen vacancies next to Y is 0.5 eV closer to the Fermi energy than the corresponding gap state of an oxygen vacancy next to Ce. We have illustrated this in the right panel of Figure 8.38.

We have sketched this "difference of the difference spectra" in the right panel of Figure 8.38 as a density of states (DOS), based on the deconvolution of the actual resonant photoemission spectra. The red defect state is the result of the spectral difference between dry and wet sample.

We were interested by the question whether the protons, which are ionic charge carriers in the ceramic electrolyte membranes, would face a higher resistance when the crystal lattice would be squeezed. We therefore carried out various studies on the proton conductors while we had put them under high mechanical pressure. It is then important to determine the crystallographic structure, particularly the lattice parameters, under high pressure. This was done with diamond anvil cells, through which we carried out X-ray diffraction at Beamline 12.2.2 at the Advanced Light Source [Clark 2012, Kunz 2005]. Figure 8.39 shows four diffractograms recorded while the BCY20 powder was set under pressure of 1.38 GPa to 8.9 GPa, along with the lattice parameters a, b, and c as a function of applied pressure. We noticed a linear decrease for the lattice parameters with increasing pressure.

When we compare the behavior of our wet BCY20 with results from other groups, we notice a similar trend of decreasing lattice parameter with pressure. Figure 8.40

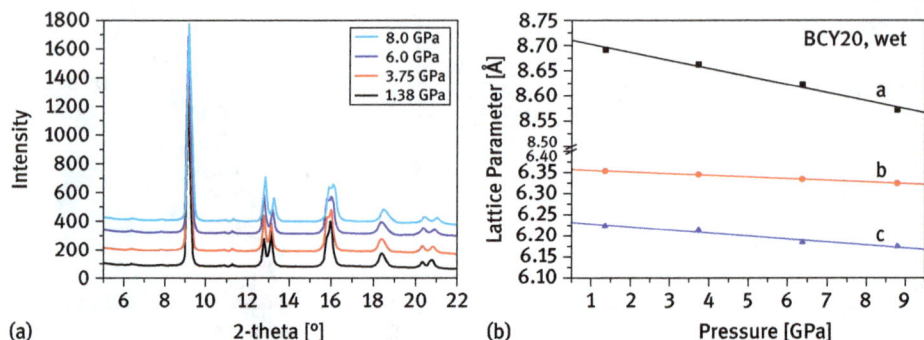

(a)

(b)

Figure 8.39: (a) X-ray diffractograms of wet (protonated) BCY20 measured in a diamond anvil cell under high pressures at 1.38, 3.75, 6.0, and 8.9 GPa at Beamline 12.2.2 at the Advanced Light Source. (b) Variation of lattice parameters versus pressure.

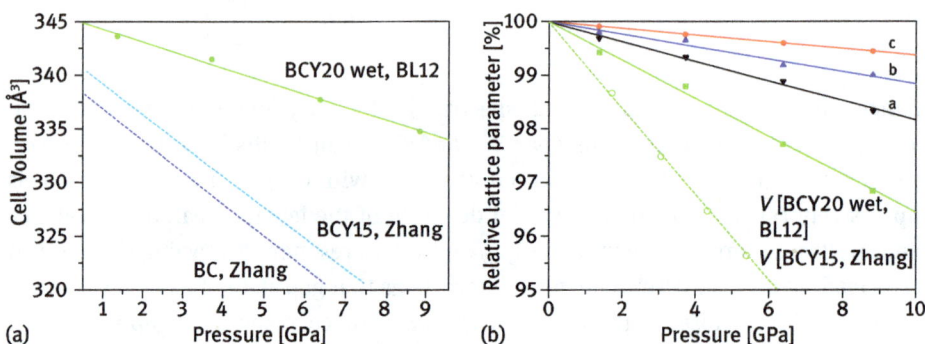

(a)

(b)

Figure 8.40: (a) Variation of the cell volume determined from the X-ray diffractograms of protonated BCY20 under high pressure at ALS on BCY20, Beamline 12 CALYPSO. The data for BC and BCY15 are from [Zhang 2007]. (b) Relative change of lattice parameters with pressure.

shows the cell volume of barium cerate (BC) and 15 % Y substituted barium cerate (BCY15) from Zhang et al. [Zhang 2007]. These two dry materials with lesser Y concentration show a more rapid decrease of lattice volume upon pressurizing. For a fair comparison of the lattice parameter variation of our BCY20 with BCY15, it was necessary to normalize the lattice parameters to 100 % and then determine the change of the relative lattice parameter versus the applied pressure, as shown in the right panel of Figure 8.40.

We conducted pressure-dependent impedance studies at high temperature and optical Raman spectroscopy measurements on the pressurized samples. It was therefore possible to relate the conductivity with the unit cell size and with the Raman modes and phonon modes. We studied the charge carrier dynamics with quasi-elastic neutron scattering at high temperature and high pressure, partially in combination with impedance spectroscopy at the neutron facility with a specifically designed *in situ* neutron cell from INCONEL metal [Chen 2010, Chen 2011c, Chen 2011d, Chen 2012b, Chen 2013c, Chen 2015b].

At Beamline 12.2.2, it is possible to combine XRD with XANES on the same instrument.[4] Figure 8.41 shows a collection of Yttrium K-edge XANES recorded from dry BCY20 at 0.08 Pa and wet BCY20 at pressures from 1.38 to 11.8 GPa. Compare these spectra with the double peaked white line of the BZY10 spectra recorded at ESRF (Chapter 2.1.1, Figure 2.9). It appears that in the BCY20 spectra, the change from dry to wet causes a considerable shift of the spectral features to higher X-ray energies. Application of pressure on the wet sample causes then a redistribution of spectral weight of peaks and shoulders.

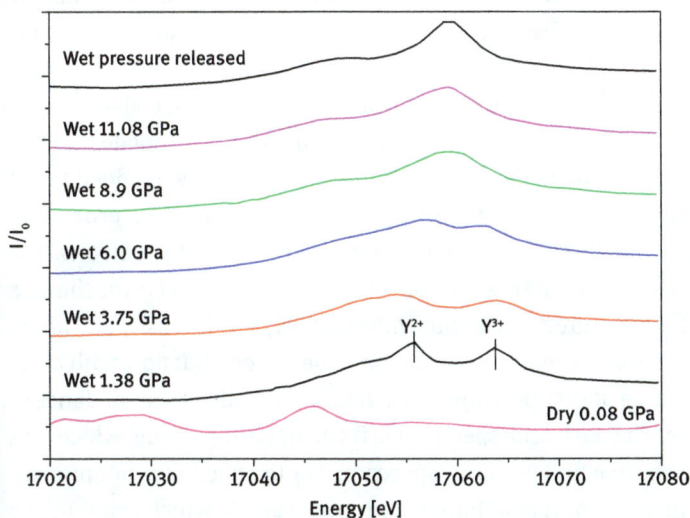

Figure 8.41: XANES spectra of dry and wet BCY20 proton conductor material obtained at Advanced Light Source Beamline 12.2.2 in a diamond anvil cell with pressures ranging from 0.08 GPa to 11.8 GPa.

8.2.2 SOFC anode poisoning and sulfur molecular structure

One advantage of SOFC is that they can run on low-cost hydrocarbon fuels in contrast to fuel cells, which require clean hydrogen or methanol. A shortcoming of these fuels is their sulfur contamination. Sulfur likes to react with the Ni catalyst in the anodes and causes deleterious structure disintegration. No high concentrations of sulfur are

---

4 Actually, this is nothing special. Rather, it is the specialization of an instrument that it is particularly well suited and tuned for one technique only, which then makes it impossible or unfeasible to carry out a different method with X-rays. My *operando* battery experiment at SSRL 25 years ago. I think it was at the easy-going and low-profile Beamline 2.3, and was done with a diffractometer, but I could easy record the XANES and also the EXAFS.

required to trigger this process because over long operation, the sulfur of course can accumulate in the anode.[5]

I wanted to carry out a simple XAS analysis on SOFC anodes with sulfur K-edge XANES. My first try was at ALS Beamline 9.3.1, which was equipped with a large format Hamamatsu silicon photodiode detector. This detector was not able to produce a sulfur XANES spectrum from our samples. With a subsequent campaign at the Swiss Light Source LUCIA Beamline [Flank 2006], we could produce sulfur XANES, probably because of the improved detector technology. These XANES were obtained by collecting the total electron yield and the fluorescence yields. A mono-element energy dispersive silicon drift diode (SDD) was used, on which a very thin window was mounted so that the fluorescence of elements down to the carbon could be detected. We found a relatively rich sulfur chemistry [Nurk 2013, Struis 2012].

Figure 8.42 shows four sulfur S 1s spectra recorded from actual SOFC electrode assemblies of a 10 cm diameter. The green spectrum was obtained from an anode that had been operated at Forschungszentrum Jülich with natural gas to which was added 1 ppm H_2S gas with the intent to add sulfur as a poisoning impurity. It shows two prominent peaks at around 2473 and 2482 eV. The corresponding intensity axis is the right one. The red spectrum is from a sample where the fuel was natural gas. The spectrum shows a wide hump indicative of many sulfur species but with relatively low intensity and abundance. The light blue spectrum was obtained from the same system, but now with a sulfur filter attached. The profile of the spectrum is flat, hence no significant abundance of sulfur was on the sample. The dark blue spectrum is from a pristine anode, which was not operated at all. Here, too, we find no sulfur spectrum. For the two latter samples, we notice upon very close inspection a small hump at around 2482 eV, which could be the trace of sulfur from exposure to ambient environment.

---

5 Long time ago in the United States, I worked for a couple of years in the fossil fuels and combustion engine community. In one of our project meetings, a professor from one of the western states in the USA said there had been a discussion between a representative from a prominent German automobile manufacturer and an American gasoline producer. The issue was that the American gasoline, or the gasoline on the American market supposedly would contain too much sulfur and then over time attack from the inside the finely-tuned German combustion engines. The removal of naturally abundant sulfur in fossil fuels is typically done by hydro-desulfurization [Katsapov 2010], one of the many large scale industrial processes in petrochemistry. So, the removal of sulfur from fossil fuels is well established. On the other hand, sulfur compounds may be added to natural gas by intention. Since natural gas is explosive and maybe you cannot smell it, a leakage in a gas pipe can be dangerous, and when the natural gas has an alarming odor, then this is an extra safety issue. This is why thiophen is added to natural gas as an odorant. There may be situations where the presence of sulfur in fuels is necessary for proper operation. For example, rubber sealing gaskets swell when in contact with sulfur containing jet fuel. I once heard that the US Air Force had problems with synthetic fuel from biomass, where such sulfur was not present. The sealing rings would therefore not swell, and thus the sealing was not proper. So, in summary, presence or absence of chemical elements in a system can be very critical.

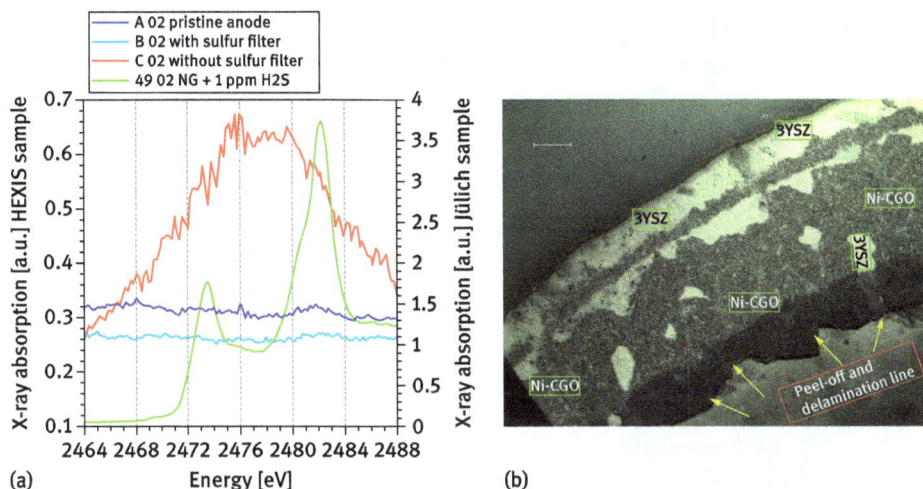

Figure 8.42: (a) Sulfur 1s XANES spectra of an anode layer from SOFC propelled with natural gas added with 1 ppm H$_2$S (green spectrum with 2 peaks), pristine anode (dark blue), anode from SOFC propelled with natural gas (red spectrum), and a SOFC anode run with a sulfur filter (light blue). (b) Micrograph of the electrode assembly after poisoning.

These experiments were done for the EU FP6 project Real-SOFC [Steinberger-Wilckens 2007]. In our consortium, we had wondered whether sulfate could be formed on a SOFC anode over SOFC operation time, but there was no analytical verification for this. Figure 8.43 shows the spectra recorded from two anode surfaces, along with a peak deconvolution. The spectroscopic assignment is based on the reference by Huffman et al. [Huffman 1991]. Our work appears to be the first evidence of sulfate on a SOFC anode after operation, as shown by the strong peak at around 2482 eV. We further identified sulfone, sulfide, and elemental sulfur, and also an organic sulfur species, specifically thiophene, which has a strong peak in both spectra. Note that we have used two arctan steps in order to account for the actual observed steps that yield the plateaus at 2475 and 2485 eV. This is typically pronounced for sulfur spectra.[6] In general, for an accurate deconvolution of spectra with many peaks, it would be appropriate to consider

---

6 Such steps are the manifestation of inelastic scattering processes. There exist routines in the software in the treatment of XPS spectra, such as in XPS Casa, for example. Also, WinXAS has this step function built in for peak deconvolution. Until recently the Multipeak Fitting Package in Wavemetrics' Igor Pro did not have the step function. Some Igor Pro users apparently expressed their need for such step function, and one user has developed a personal makro for the stepfunction that could be applied in Igor Pro. It appears that Wavemetrics followed the request and built in the function of one step as arctan and Shirley background in Igor Pro Version 9. I have used this feature in order to treat optical absorption spectra (uv vis). They had an obvious step as the background, and this was a nuisance for quantitative analysis. Taking care of the step with an arctan function made quantitative peak analysis easier.

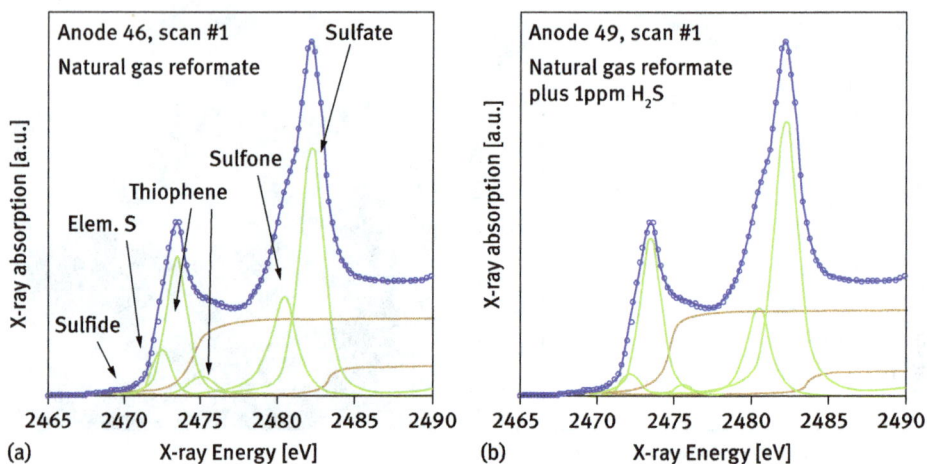

Figure 8.43: Sulfur 1s XANES spectra of anode surface from SOC propelled with natural gas (a) and natural gas plus 1 ppm H_2S added (b). Deconvolution into Voigt functions and two arctan step functions.

an absorption threshold step function for every detected peak. This, however, would make the deconvolution very laborious.

We have recorded for every sample several scans and found that the peak heights are changing from scan to scan. I interpret this as radiation damage and typical for virtually all X-ray spectroscopy experiments. It is an indication of the strong interaction of the matter with the X-ray beam and, therefore, should not be considered entirely negative. After all, this gives us some additional information of the radiation stability of the matter under investigation. We found that the relative spectral weight of the components was increasing with an increasing scan time, except for sulfate and sulfone. The relative intensity of these species was decreasing with an increasing scan number (= exposure to X-rays; see Part 10, Radiation damages) [Braun 2008c].

The environmental sciences are a field where often chemically and structurally complex samples have to be evaluated by origin (source apportionment), structure, and toxicity, for example. Principal component analysis (PCA) is a mathematical method, as the term says, to analyze a system in its particular components. This method is sometimes applied in spectroscopy. Nowadays, software programs or packages or macros are available to carry out such PCA conveniently. A flow diagram for how such PCA is done, e. g., for the decomposition of sulfur spectroscopy data is given by [Beauchemin 2002] and reproduced in Figure 8.44. A simpler procedure was already shown in Chapter 2.1 for the combination of Mn spectra [Manceau 2012].

Sulfur spectra are relevant in environmental sciences [Solomon 2003], in fossil fuel sciences [Chaturvedi 1998, Huffman 1991, Taghiei 1992], in the biosciences [He 2011, Rompel 1998], and with the advent of sulfur-based batteries also in the field of battery chemistry [Cuisinier 2013, Gorlin 2015, Lee 2007].

Data Matrix
M

2. Target transformation
- Oblique matrix rotation based on guessed vector
$$t_i = (R^T R)^{-1} R^T x_i$$
- Accept or reject the predictor vector
$$\hat{x}_i = \bar{R} t_i = (?) x_i$$
  - SPOIL value
  - F test
- Compose transformation matrix **T**

Guessed Vector
x_i

1. Principal component analysis (PCA)
- Decomposition of **M** into 2 abstract matrices
$$M = RC$$
- Determine number (n) of significant components based on statistical or empirical tests
  - Imbedded error function (IE)
  - IND function
  - F test
- Reproduction of **M** using reduced matrix space (n)
$$M \cong \bar{M} = \bar{R}\bar{C}$$

3. Estimation of Real Matrices
- Matrix of scaling coefficients
$$\hat{Y} = (T)^{-1} \bar{C}$$
- Predict the data matrix
$$\hat{D} = X_{basic} \hat{Y}$$

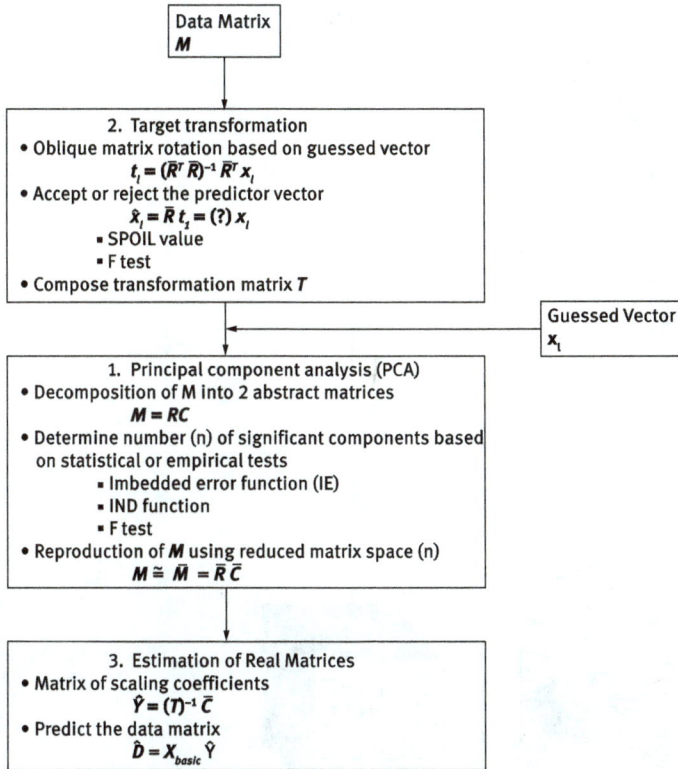

Figure 8.44: Flow diagram for the PCA of sulfur S 1s spectra [Beauchemin 2002]. Reproduced from Suzanne Beauchemin, Dean Hesterberg, and Mario Beauchemin, Principal Component Analysis Approach for Modeling Sulfur K-XANES Spectra of Humic Acids, Soil Science Society of America Journal 2001, Vol. 66 No. 1, pp. 83–91.

Among the first *in situ* sulfur XANES studies is the work from my colleagues at the University of Kentucky [Taghiei 1992]. They investigated the evolution of sulfur species in coal during pyrolysis and oxidation with *in situ* XANES spectroscopy. Figure 8.45 shows a sketch of their high-temperature gas phase reaction *in situ* cell. The cell has beryllium X-ray windows and works in fluorescent geometry. It has terminals for gas flow, water cooling, and heating coils. It is basically a furnace, which can be remotely controlled and fits in the X-ray beam of a synchrotron end station.

We have taken an *in situ* XAS cell for gas phase catalysis studies from PSI [Struis 2012] and further developed it into a high-temperature spectro-electrochemical cell. The base unit for the cell is shown in Figure 8.46. The X-ray window material was Kapton® plastic foil, which naturally set a limit on the achievable temperature of the cell. Because we wanted to use a SOFC electrode assembly, the electrolyte was the support for the anode. Current collectors were painted on the anode and cathode.

Taghiei et al.

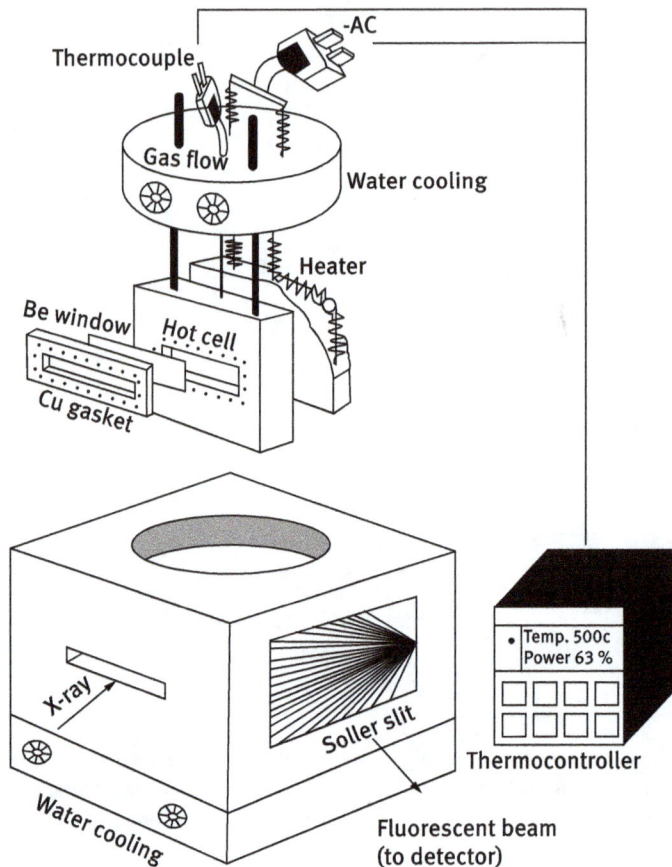

Figure 8.45: Sketch of the experimental set-up for an *in situ* EXAFS study on sulfur species during coal combustion, by [Taghiei 1992] M. Mehdi Taghiei, Frank E. Huggins, Naresh Shah, Gerald P. Huffman, *In situ* X-ray absorption fine structure spectroscopy investigation of sulfur functional groups in coal during pyrolysis and oxidation, Energy Fuels, 1992, 6 (3), 293–300. Reprinted (adapted) with permission from (Energy Fuels, 1992, 6 (3), 293–300. Copyright (1992) American Chemical Society.

Figure 8.47 shows the cell and the gas tubing and wires with feed throughs at a flange (left photo), which is inserted in a metal container that provides gas sealing and temperature control of the cell. The container is shown on the left side in the photo on the right, wired to a potentiostat (AMEL Instruments, Model 7050) and to gas supply and to a power supply for the heating element in the cell.

In the photo in Figure 8.47, you see Dr. Gunnar Nurk from Tartu University, who as a Sciex Fellow who worked at PSI Bioenergy and Catalysis Laboratory and Empa Laboratory for High Performance Ceramics for the realization of this *operando* SOFC project. In the photo, he is testing his cell under realistic non-X-ray conditions for the prepara-

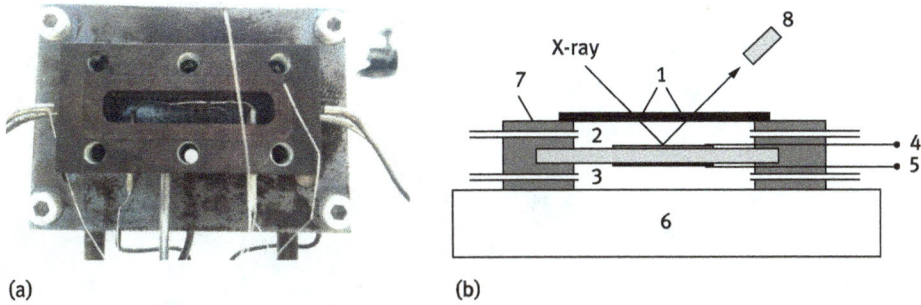

Figure 8.46: X-ray absorption spectroscopy cell (a) and sketch (b) for *operando* measurements of IT-SOFC: (1) windows for soft X-ray; (2) anode gas compartment; (3) cathode gas compartment; (4) current collector for anode; (5) current collector for cathode; (6) heating element; (7) cell body with tubing; (8) fluorescence detector.

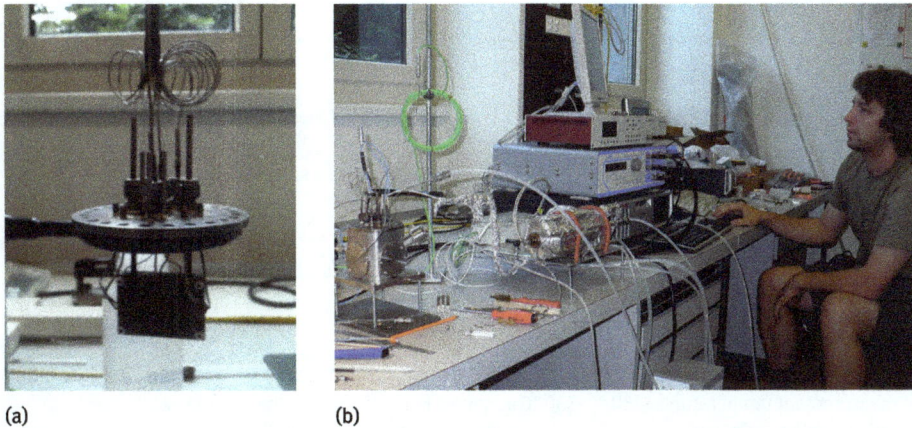

Figure 8.47: (a) *In situ* cell mounted on flange. (b) Dr. Gunnar Nurk from Tartu University and PSI, testing, and preparing at Empa the *in situ* cell.

tion (test run) of the synchrotron experiment. In Figure 8.48, we see the cell container attached to the X-ray end station Phoenix at the Swiss Light Source. The X-ray beam is coming through an X-ray tube from the source, enters the cell container, and passes the cell with the SOFC sample under operation conditions, and the X-rays are scattered by fluorescence to the detector.

We later extended the experiment and carried out sulfur XANES *in situ* at the Swiss Light Source. The large instrument panel with the six red mass flow controllers was unmounted from Empa's SOFC testing rigs and brought to the Swiss Light Source. In the front, you will see the blue-white potentiostat from VoltaLab80, which controlled the electrochemical operation of the *in situ* cell.

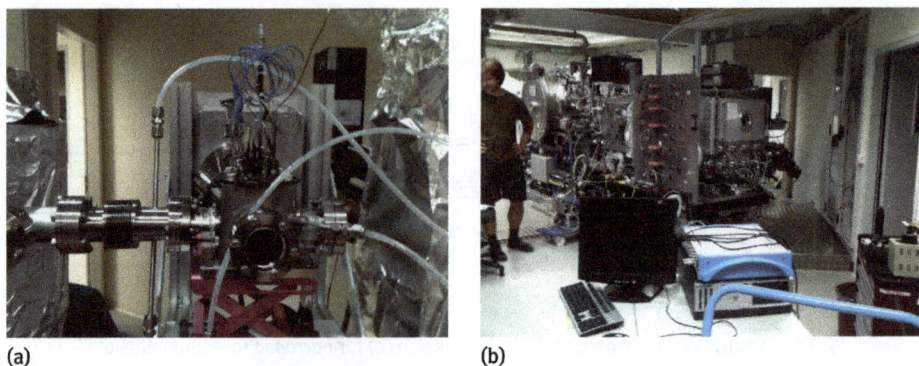

(a)   (b)

Figure 8.48: (a) Chemical gas phase reaction chamber at the Phoenix Beamline at Swiss Light Source, (b) equipped with *in situ cell* from PSI ENE Department and electrochemistry and heat and mass flow infrastructure from Empa (Dr. Gunnar Nurk).

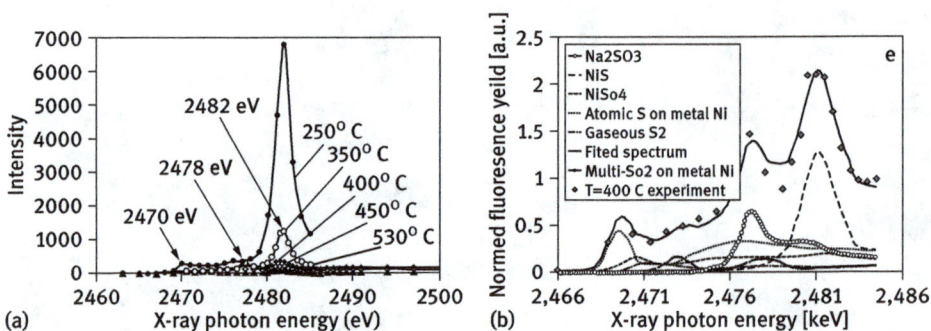

(a)   (b)

Figure 8.49: (a) Dependence of sulfur K-edge XANES spectral intensity on the X-ray photon energy at different measurement temperatures (250 °C–530 °C as shown in the figure) for the Ni-CGO anode in stream of hydrogen with 5 ppm of H_2S. (b) S1s XANES spectrum recorded *operando* at the Ni-GDC anode in presence of 5 ppm H_2S in H_2 in the working anode compartment at different temperatures (indicated in the figure). Least square fitting results and model spectra used for fitting are shown as well.

Figure 8.49 shows the S1s XANES spectra obtained *operando* when the cell was run from 250 °C to 530 °C. Note that with particular temperature, particular peaks rise in the series of spectra, such as for 2470, 2478, and 2482 eV. The right panel (b) shows the spectrum recorded at 400 °C, with the convolution into particular components and molecular structures, such as sulfide, sulfite, sulfate, elemental sulfur, and gaseous S_2. The solid top spectrum is the best fitting composition of the spectra from the aforementioned components through the actual experimental data points. Dr. Rudolf Struis used the commercial StatistiXL package for a PCA [Nurk 2013, Roberts 2009, Roberts 2016] with the number of components, eigenvalues, and cumulated variance shown in Table 8.1.

Struis used all eight *operando* XANES spectra for this PCA and found that the component PC-1 was particularly prominent in the ensemble (Table 8.1), but other compo-

Table 8.1: Cumulated variances and eigenvalues from PCA with StatistiXL in case of different number of components used in analysis by R. Struis [Nurk 2013, Struis 2012].

Number of components	Eigenvalue	Cumulated variance (%)
PC-1	7.427	92.84
PC-2	0.370	97.46
PC-3	0.106	98.78
PC-4	0.050	99.41
PC-5	0.023	99.69
PC-6	0.016	99.88
PC-7	0.008	99.98
PC-8	0.001	100.00

nents were also prominent. He carried out PCA also with the WinXAS software package [Ressler 1998], but this routine did not produce other results than with the StatistiXL package.

The fractions of the detected and evaluated molecular structures in the sulfur spectra are listed in Table 8.2. We read from there that these fractions vary depending on which temperature was in the SOFC *in situ* cell. The experiment started with 550 °C and ended with 550 °C, and the lowest temperature was 250 °C.

Table 8.2: The fitted fractions of constituting sulfur forms, the sum-of-fractions, fitted sulphate peak positions (shift relative to 2482 eV), and best-fit criterion values (%Residue).

	550 °C [e]	550 °C [s]	546 °C	530 °C	450 °C	400 °C	350 °C	250 °C
S^{6+} species	1.04	0.59	0.36	0.49	0.21	0.26	0.73	1.23
S^{4+} species	0.00	0.02	0.00	0.00	0.03	0.12	0.10	0.00
Adsorbed S on Ni^0	0.05	0.18	0.19	0.13	0.28	0.24	0.20	0.03
Molecular S_2	0.03	0.00	0.22	0.30	0.00	0.06	0.00	0.00
SO_2 on Ni^0	0.01	0.05	0.04	0.00	0.13	0.07	0.04	0.00
S^{2-} species	0.19	0.02	0.25	0.16	0.24	0.15	0.11	0.20
Related fit results:								
Sum of fractions	1.32	0.86	1.06	1.08	0.89	0.91	1.19	1.47
Residue (%)	0.250	0.113	0.0989	0.214	0.064	0.097	0.262	0.192
Shift in SO_4^{2-} (eV)[a]	−0.22	−0.45	−0.55	−0.41	−0.44	−0.35	−0.19	−0.18

[a]Relative to NiSO4 reference spectrum used in the Linear Combination Fitting (LCF). [e] = end; [s] = start

We have recorded the changes of the voltages in the SOFC *in situ* cell during the recording of the spectra and changed the gas supply to the cell, as shown in Figure 8.50. The gases were hydrogen, or various concentrations of H_2S (compare with Figure 8.2 [Sasaki 2006]).

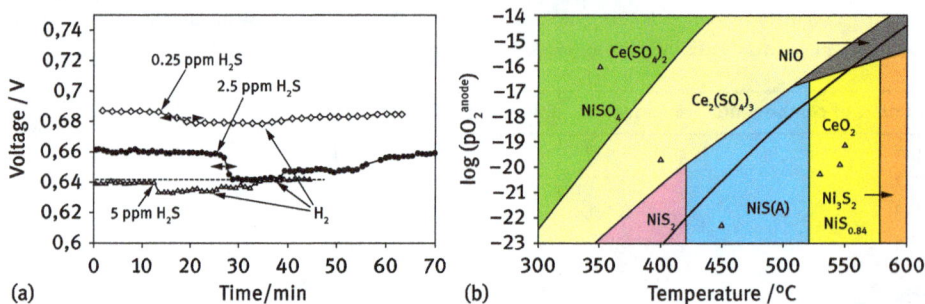

Figure 8.50: (a) Dependence of cell voltage on time during sulfur poisoning and regeneration process performed at $T = 600\,°C$ and at constant current condition (50 mA cm^{-2}) with H$_2$S concentrations of 0.25, 2.5, and 5 ppm in hydrogen (as indicated in the figure). Dotted horizontal arrows indicate durability of first poisoning step. (b) Overlay of the Ce-O-S (black) and the Ni-O-S (gray) phase diagrams as a function of temperature and oxygen partial pressure at fixed sulfur partial pressure, $P_{S2} = 2.5 \times 10^{-6}$. Triangles ($\delta$) indicate the locations/thermodynamic experimental conditions where the XAS spectra were measured. Reprinted from [Nurk 2013] with permission from Elsevier. Copyright 2013.

The thermodynamical stable species of anode materials under SOFC operation conditions in the presence of sulfur impurities have been investigated by other researchers in different contexts. Zeng et al. [Zeng 1999] and later Flytzani-Stephanopoulos [Flytzani-Stephanopoulos 2006] investigated cerium oxide as a high-temperature desulfurization sorbent. Ma et al. investigated Pt catalysts supported on CeO$_2$ with respect to their sulfur tolerance [Ma 2009]. Their results help us for the interpretation of our own data. As Ma writes [Ma 2009], "A sulfur-tolerant model of Pd/CeO$_2$ catalyst for syngas to methanol was proposed according to the reaction performance and characterization results, which might be expanded to other sulfur-containing feed gas reaction systems over metal/ceria-type oxides." We certainly expanded their knowledge beyond their own field and used it for our SOFC studies. This is one of the many examples of knowledge diffusion into related fields. We looked into the predictions set by thermodynamic calculations, as provided by Lohsoontorn et al. [Lohsoontorn 2008]. Based on their phase diagrams, we constructed a phase diagram (right panel of Figure 8.47), which shows with triangle symbols the thermodynamic (p,T)-conditions that correspond to our experiment conditions at the Swiss Light Source.

8.3 Photo-electrochemical cells (for solar hydrogen)

We have made numerous *ex situ* spectroscopy studies on materials for photoanodes, such as TiO$_2$, WO$_3$, and Fe$_2$O$_3$. I have shown in Part 5, Chapter 5 the oxygen NEXAFS spectra of iron oxide absorber layers grown on FTO glass and operated under various photo-electrochemical conditions. We found that anodization in dilute KOH forms species on the electrode surface, which produce or enhance a readily existing feature in the upper Hubbard band as revealed in the oxygen NEXAFS spectra [Bora 2011a].

8.3.1 *Operando* NEXAFS spectroscopy during water splitting

It would be interesting to analyze photoelectrodes when they are in operation, during photo-electrochemical water splitting. This means the electrode must be in contact with an electrolyte and under electrical polarization, and in a dark condition and an illuminated condition. Such an experiment is not trivial because we apply a soft X-ray method in a UHV chamber on a sample in a liquid. Our colleagues at MaxLab in Sweden and Berkeley Lab (Salmeron group and Zahid Hussain group) have succeeded to *square the circle* by designing a spectro-electrochemical cell with 100 nm thin X-ray windows from Si_3N_4. This cell is sealed against the vacuum and can sustain in a liquid electrochemical system, while soft X-rays probe the system [Jiang 2010].

Figure 8.51 shows in the top left panel a 5 × 5-mm-wide silicon wafer (Silson Ltd., United Kingdom), which has a 100-nm thin Si_3N_4 layer on its back side. On the front side, the silicon is etched away in a slit shape so that an X-ray window of 1.5 × 0.5 mm

Figure 8.51: (Upper left) 5 mm × 5 mm wide Si wafer with 100 nm thin Si_3N_4 X-ray window slit. Observe the shape of a brownish circle from iron oxide (APCVD [Kay 2006]) and one soldered wire contact as WE. (Middle) PEEK® plastic *in situ* cell housing with three electric terminals, electrolyte compartment, and liquid flow inlets. The size of the Si wafer in relation to the cell is visible in the 3rd image from the left. (Top right) Fully assembled *in situ* cell with sample film on a wafer, and blue glue on top for fixation and sealing, along with electric wires. (Center) PEEK® cell with iron oxide silicon wafer electrode assembly glued and sealed. (Bottom left) Cell mounted on load lock front view. (Bottom middle) Top view on cell with wires for control by potentiostat. (Bottom right) Cell in UHV chamber in operation as visible by the beamline video camera.

with 100-nm thickness is obtained through which a soft X-ray beam of 500 eV has 60 % transmission, for example. This Si_3N_4 window layer was coated with a 1-nm Cr layer for better adhesion of a subsequent 10 nm Au gold layer, which served as the current collector layer. On top of this, we made a 30-nm hematite layer doped with 1 % Si using the APCVD method. The round shape on the Si frame in the upper left image in Figure 8.51 indicates the circular geometry from the APCVD setup. This is the photoelectrode with the necessary soldered wire contact.

Next to this, we see one empty PEEK® cell with wires soldered to a counterelectrode and WE, ready to receive the photoelectrode shown on the left photo. The next photo shows the rear side of the PEEK® cell with two small middle holes through which liquid electrolyte is supplied via a peristaltic pump. On the right, we see the PEEK® cell with the photoelectrode glued and sealed on top. On the bottom left photo, we see the PEEK® cell applied with screws on the load lock transfer shift, and on the bottom right from top view, the insulated copper wires on the cell, over the load lock transfer shift.

Figure 1 in [Jiang 2010] shows an explosion sketch of the cell and the load lock shift. Figure 8.52 below shows a schematic representation of the cell and its function under photo-electrochemical water splitting conditions. Soft X-rays for the O1s range from 500 to 560 eV and for the Fe 2p range from 700 to 750 eV and pass after UHV through the X-ray window, metal layers, iron oxide layer, and enter the electrolyte volume.

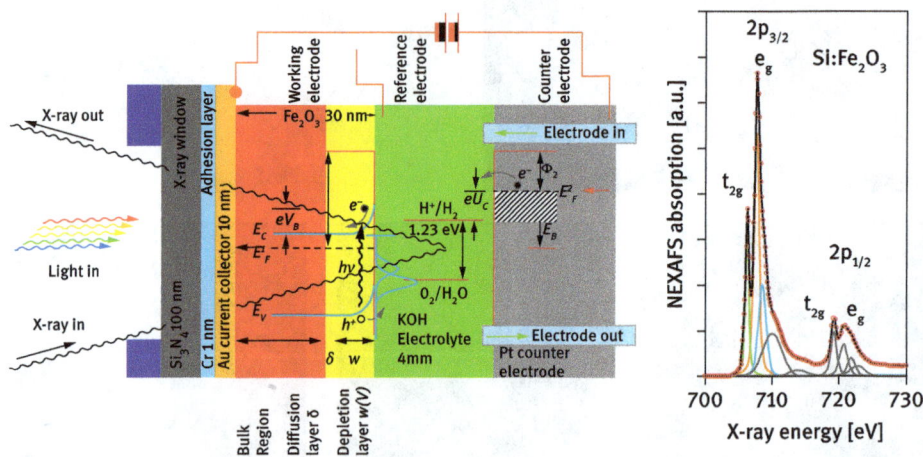

Figure 8.52: Schematic of spectro-electrochemical cell for soft X-ray *operando/in situ* experiments. X-rays enter under UHV the PEEK® cell body through 50 nm or 100 nm thin Si_3N_4 window and 1 nm Cr adhesion layer and 10 nm Au current collector and 30 nm hematite film and KOH electrolyte volume. On the left, we see an Fe 2p NEXAFS spectrum of the Si-doped iron oxide film.

When the PEEK® cell is mounted on the load shift, it can be inserted into the UHV measurement chamber, as is shown in Figure 8.53 by our colleagues Xuefei Weng and Hui Zhang from the National Synchrotron Radiation Laboratory, University of Science

(a)   (b)   (c)

(d)

Figure 8.53: (a–b) Xuefei Feng (ALS and National Synchrotron Radiation Laboratory, Hefei, 230029, China) and Hui Zhang, inserting the *in situ* cell into the UHV recipient at BL 7.2 at ALS. (c) Top right shows solar simulator light (Xenon lamp with UV filter) directed through UHV chamber window on the PEEK® cell for generation of photocurrent while NEXAFS spectra are recorded. (d) Bottom photo shows the front side of the chamber, while Xuefei Feng and Hui Zhang are inserting the sample holder on the back side.

and Technology of China, Hefei, and from the ALS. They fix the electric contacts for the potentiostat and connect the cell with the electrolyte reservoir and the peristaltic pump. The "clean" Fe 2p spectrum shown on the right of Figure 8.52 was obtained from the cell before it was filled with the electrolyte. This is a verification that the X-ray beam has hit the X-ray window properly because the photoelectrode coated on the Si_3N_4 window is the only material with iron in the vicinity of the X-ray beam.

In experiments as complex as the present one, it is necessary to check that the X-ray beam hits on the desired sample or sample position. For many *ex situ* samples where one or more samples are mounted on one sample holder, it is necessary to make sure that the samples are not mixed up. Sometimes, we use phosphorus powder that we disperse in ethanol. We then use a fine brush and paint the phosphorus-ethanol mixture around the various samples on the sample holder. Once the sample holder is in the measurement chamber, it will be moved in (x,y) direction until a bright spot from the phosphorescence can be seen when the X-ray beam hits the phosphor. This procedure allows for proper maneuvering the right sample in the X-ray beam. In the present case, we used the presence of the iron on the sample as the technical marker for the sample positioning.

The X-ray fluorescence of all matter in the X-ray beam is recorded by a detector in the UHV chamber. When we record the oxygen NEXAFS spectra, they will not contain an oxygen specific signal from the Si_3N_4 X-ray window and Cr and Au metal layers because these components do not contain oxygen. The only oxygen that the beam can probe is from the α-Fe_2O_3 hematite absorber layer and from the KOH and H_2O dilute electrolyte solution. The O1s NEXAFS spectrum recorded from an arrangement like the one shown in Figure 8.49 contains therefore the molecular structure of the photoelectrode and the electrolyte.

With the three wires on the PEEK® body, we can control the cell with a potentiostat (Biologic at ALS BL 7). Visible light necessary for photo-electrochemistry is supplied from outside the UHV chamber through a window. The top right photo in Figure 8.53 shows how a flexible lamp guides visible light from a solar simulator onto the window, and thus illuminates the entire PEEK® cell. Note that this, of course, is not a standard 1.5 AM illumination condition. However, this is not necessary for such a first time ever experiment. We do not need to have standard conditions in order to compare our material with a supposedly superior or inferior other material. It is important that we can detect and measure a physical chemical effect. Measurements with better accuracy and reproducibility under standard conditions may be made later. Note: A synchrotron X-ray beam line is normally not prepared for the standard measurement conditions in battery technology or catalysis research. This is simply not their business. Synchrotron beamlines have to be operational for the X-ray methodology standards. The equipment and instrumentation for a synchrotron end station can easily cost over 1 million dollars. Yet, more and more synchrotron and neutron facilities have started investing in the acquisition of infrastructure, e. g., for electrochemical measurements. More recently, some of these facilities have hired personnel with electrochemistry background, which makes them better prepared for user demands from the field of electrochemistry, such as Beamline 7.0.1 at

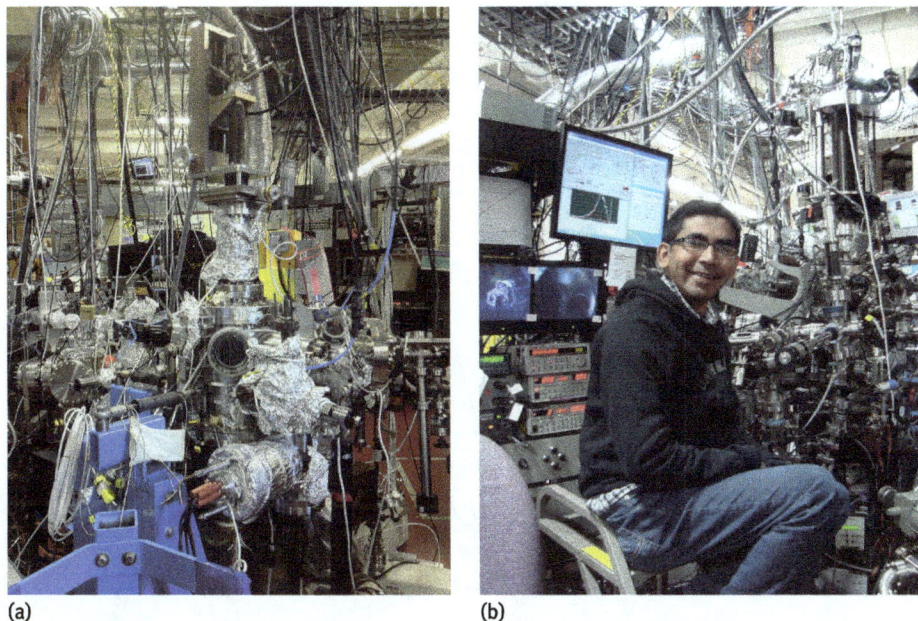

(a)   (b)

Figure 8.54: (a) End station for soft X-ray spectroscopy and scattering experiments at Beamline 7.0.1 at the Advanced Light Source, Berkeley, (b) with Empa/University of Basel PhD student Debajeet K. Bora, now Assistant Professor in Morocco [Braun 2012l].

the ALS (Figure 8.54) , where we conducted our semiconductor photo-electrochemistry experiments (Figure 8.51).

We have recorded [Braun 2012l] O1s NEXAFS spectra from the cell at potential from 100 to 900 mV versus Ag^+/AgCl reference under dark and under illumination (Figure 8.55). With the light off (dark), we see the well-known doublet of hematite with the t_{2g} orbital symmetry peak at around 529 eV and the e_g peak at around 530 eV. We notice no differences between the spectra in the left rows in Figure 8.52.

The right rows in Figure 8.55 show the spectra while a photocurrent was produced. Although the spectrum for 100 mV has a flat pre-edge region, the spectrum at 300 mV has increased and structured intensity, which can be deconvoluted into two pre-edge peaks. Upon increasing the bias to 500 mV, the intensity of the two new peaks increases further, and the deconvolution into two peaks can be done with more confidence. At 700 mV, the intensity has decreased again but the structures are still well visible. At 900 mV, only the second peak has a noticeable intensity.

It appears therefore that the two peaks evolve with a different "speed" in relation to the applied bias potential. The peak at 525 eV shows up first with higher intensity than the peak at 526 eV, but we found that with increasing potential these second peaks had a larger spectral weight.

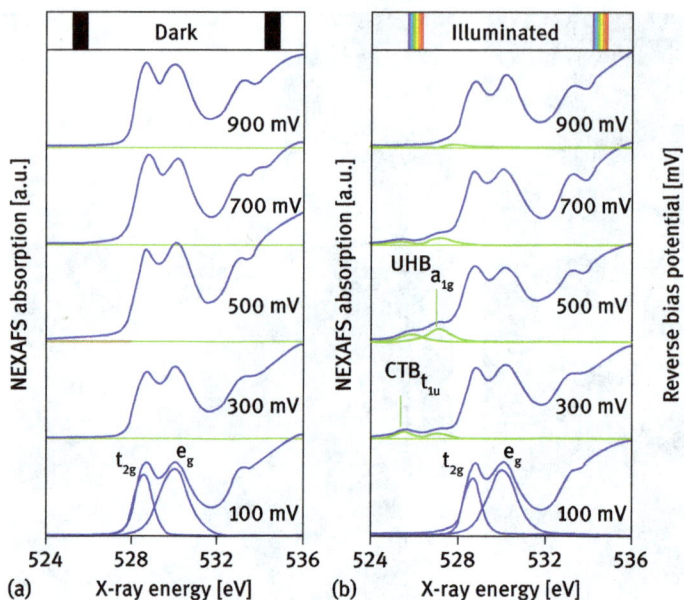

Figure 8.55: Oxygen 1s NEXAFS spectra recorded under dark (a) and (b) under solar simulator illumination with electric bias from 100 mV to 900 mV versus Ag/AgCl reference in 0.1 m KOH electrolyte, recorded *operando* at BL 7.0.1.1. Advanced Light Source [Braun 2012l].

It has turned out that the two peaks at 525 and 526 eV are hole states formed by the electron-hole pair formation upon illumination and band bending due to the electric bias. The hole state next to the Fermi energy corresponds with a transition into the charge transfer band (CTB), and the next hole state corresponds to a transition into the upper Hubbard band (UHB).

For a quantitative analysis of the variation of the peak height, the spectra were deconvoluted into Voigt functions and one step function, as demonstrated for the spectrum recorded under illumination and 300 mV in Figure 8.56, left panel. The height of hole peaks is related to the height of the t_{2g}–e_g doublet to form the relative spectral weight S; in this particular case, $S = 0.113/0.753 = 0.150$. We have done this for all spectra recorded under illumination and plotted the relative spectral weight S versus the potential, as shown in the right panel of Figure 8.56. The peak height for the CTB increases steeply and evolves like a parabolic function (square root), whereas the transition into the UHB can be modeled with a Gaussian. I have plotted this together with the photocurrent and dark current so as to be able to directly compare both quantities. The onset of photocurrent coincides with the maximum spectral weight of the CTB transition peak.

The parabolic profile of the CTB peak versus potential V reminds us of the variation of the charge carrier accumulation or depletion layer in a semiconductor, the thickness

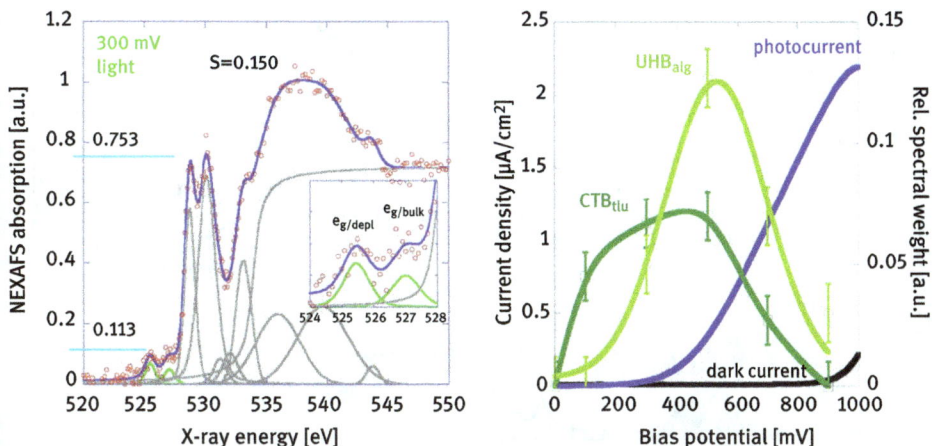

Figure 8.56: (a) Oxygen 1s NEXAFS spectrum (red open circles) recorded *operando* under illumination and 300 mV bias, with deconvolution into Voigt functions and step function, and least square fit (blue solid line). Inset shows magnified hole states in the oxygen pre-edge with deconvolution into CTB t_{1u} and UHB a_{1g} orbital symmetry peaks (green Voigt functions). (b) Dark current and photocurrent of Si-doped hematite electrode, along with the relative spectral weights S for the CTB and UHB peaks.

or width w of which scales as follows:

$$w = w_0 \sqrt{V - V_{fb}}$$

V_{fb} is the flat band potential, which is typically obtained from electrochemical impedance spectroscopy (EIS) data via the Mott–Schottky plot [Barsoukov 2005], as we will expand on further in Part 9. The Gaussian shape of the UHB peak reminds us of a peak in a typical electronic DOS. This peak originates from surface states that one can also determine with EIS [Shen 1986, Tomkiewicz 1980a, Tomkiewicz 1980d]. It is therefore quite interesting to see the parallelism between valence band X-ray spectroscopy and electro-analytical methods [Bora 2013b].

When we turn again to the original NEXAFS spectrum, we memorize that it represents actually a convolution of the spectroscopic oxygen signatures from the solid hematite electrode and the liquid H_2O-KOH electrolyte, plus potential surface intermediates. Figure 8.57 shows the O1s spectrum recorded *operando* under illumination at 500 mV bias (dark green data points and least square fit from Voigt function and arctan step deconvolution), along with an O1s spectrum of α-Fe_2O_3 nanoparticle powder recorded in UHV in a dark condition (red solid line, with the $t_{2g}e_g$ doublet deconvoluted as red dotted peaks). The spectral difference between the pre-edge (525–527.5 eV) of this powder sample and the running water splitting device at 500 mV under illumination stands out by the two electron hole peaks that perform the water splitting.

We have also plotted the spectra of liquid water and gas phase water that we have traced from data from Myneni et al. [Hetenyi 2004, Myneni 2002]. We see now that a

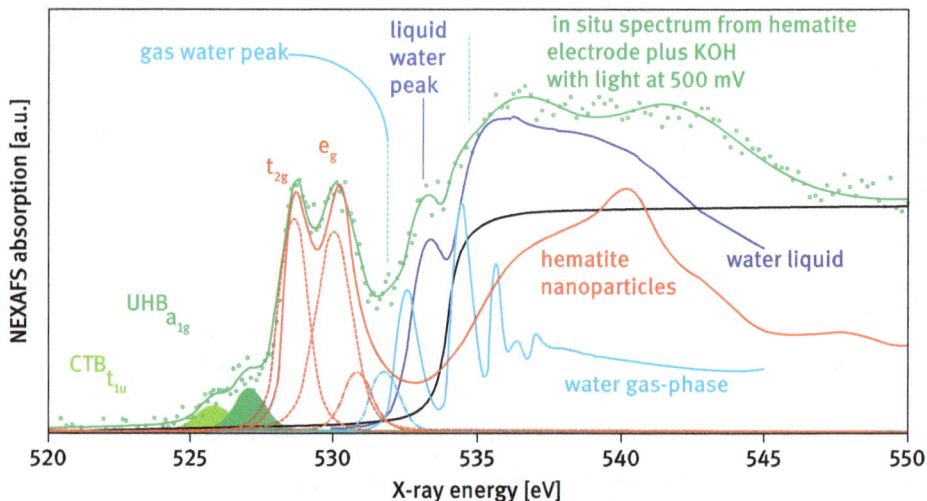

Figure 8.57: O1s spectra of *operando* experiment under light at 500 mV (dark green spectrum) [Braun 2012I], compared with three spectra from literature, i. e., *ex situ* α-Fe$_2$O$_3$ nanoparticle powder (red solid spectrum), liquid water (dark blue spectrum), and water in the gas phase (light blue spectrum) (reproduced from [Hetenyi 2004, Myneni 2002]).

peak at 533 eV in our *operando* spectrum corresponds to a prominent peak in the water spectrum. I therefore believe this is a "water oxygen peak." When in the phase transformations from ice to water and from water to water vapor, more and more hydrogen bonds are broken; this peak shifts to lower X-ray energies. We observe such shift in the spectra from 100 to 500 mV bias potential, and from 500 to 900 mV shifting back to higher X-ray energies.

Comparing this with the water gas phase spectrum, we notice that the intensity minimum at approximately 532 eV has a shoulder at the higher energy flank. We remember this intensity minimum from the first *ex situ* NEXAFS study where we identified a new feature that was grown on iron oxide after PEC anode operation at 600 mV. We speculated this was α FeOOH-type species [Bora 2011a]. I have deconvoluted this region with two states, as is shown by the two gray dotted Voigt functions at around 530.5 and 532 eV. We have seen these two peaks already in Figures 8.55 and 8.56 in the 300-mV spectrum. The peaks there had lower intensity than here at 500 mV. We remember that the height of this intensity minimum scales with the bias potential.

We can therefore with one *operando* study cover a wide range of processes that are going on during PEC water splitting [Braun 2016c]. Figure 8.58 summarizes the spectroscopic findings so far. Panel a shows the evolution of the photocurrent with bias potential versus the reversible hydrogen electrode (RHE) reference. The linear increase of current from 1000 to 1500 mV is due to an ohmic resistivity in the electrode assembly. The water splitting onset potential is 1500 mV versus RHE. Panel b shows the CTB and UHB peaks from the *operando* NEXAFS campaign. Panel c shows a capacitive surface

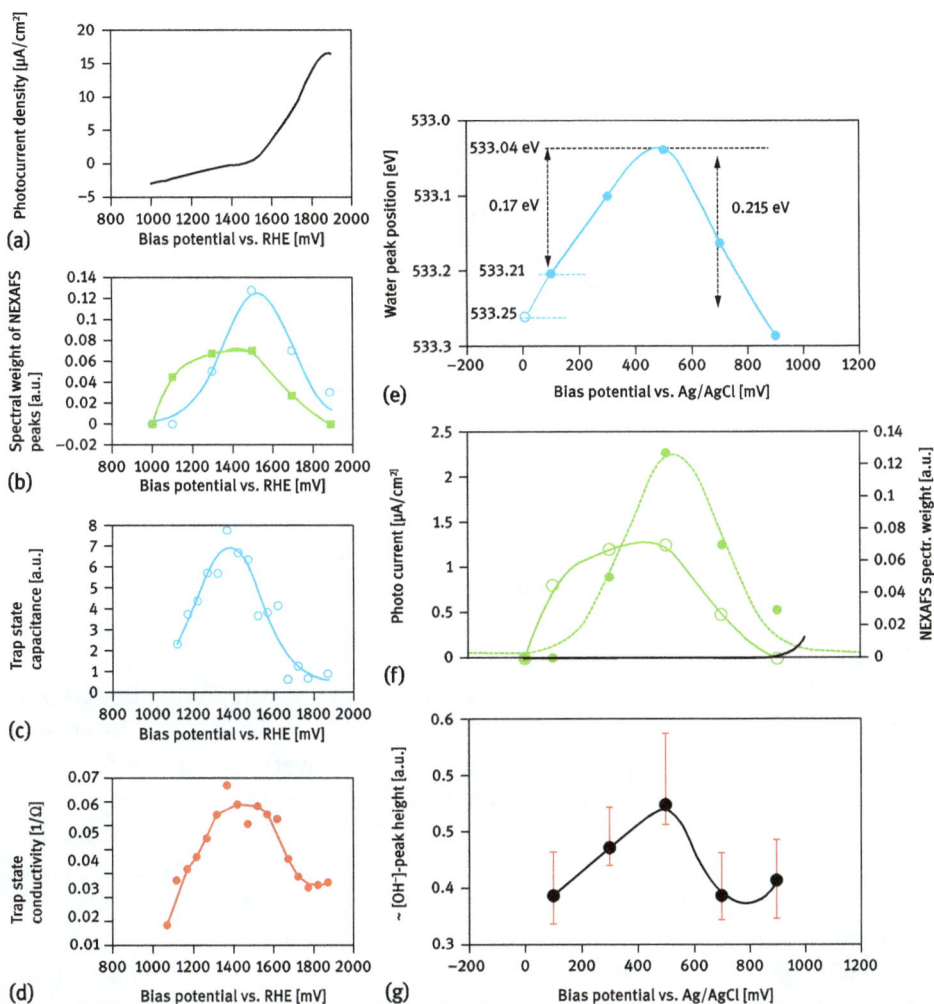

Figure 8.58: (a) Photocurrent density of the hematite film measured *ex situ* after *operando* experiment at ALS. (b) Variation of the spectral weight of hole peaks in the oxygen NEXAFS spectra versus bias potential in RHE. The CTB hole peak variation is fitted with a parabolic function up to 1500 mV; the UHB hole peak is fitted with a Gaussian. (c) Trap state capacitance of the film measured with *ex situ* EIS. (d) Trap state conductivity measured *ex situ* with EIS. (e) Shift of the "water peak" versus bias potential in Ag/AgCl reference. (f) Comparison of NEXAFS hole peak spectral weights (same like Figure 8.5) with light current and dark current densities of dip-coated Si-doped hematite film. (g) Variation of the –OH NEXAFS peak in the *operando* spectra with bias potential.

trap state, which was obtained with EIS. Beyond the maximum of this trap state, which can be considered like an electronic DOS, the PEC water splitting sets on, when compared with panels a and b. The corresponding trap state conductivity, which was same as the capacity determined with Randles circuit fitting, is shown in panel d and varies parallel with c. The aforementioned shift of the so-called water peak or hydrogen bond breaking peak is shown in panel d. We have a steep increase of the energy position for this process toward lower X-ray energies—note the reversed abscissa axis direction. The maximum for this shift to lower X-ray energies is observed at 500 mV potential, confirming that the breaking of hydrogen bonds is maximal in this potential range. For higher potentials, the position of this peak is shifting back to higher X-ray energies. The maximum for this process is at the same energy like the maximum of the UHB holes from the NEXAFS study; see panel f. Finally, in panel g, we have plotted the spectral weight of the hydroxyl peak in the intensity minimum at around 532 eV as a function of bias potential. The intensity, and thus the abundance of [OH]$^-$ increases with the potential up to the water splitting onset potential of 500 mV. Beyond this potential, the peak height, and thus the relative abundance for [OH]$^-$ decreases again.

8.3.2 *Operando* photoelectron spectroscopy during water splitting

In the last section, we have investigated the water splitting *in situ* and *operando* with NEXAFS spectroscopy, which like any X-ray absorption method probes the unoccupied electronic states.

We have looked into the water splitting also with XPS, which probes the occupied electronic states. Because we are dealing with electrons as probes for XPS, substantial experimental constraints apply because electrons have a very short mean free path. The *operando* cell from the previous section cannot be used for XPS. Ambient pressure XPS, however, is a solution to this problem. I have demonstrated this method already in Chapter 2.3.2. We need here an "open" cell so that electrons coming from the sample can travel to the detector. For this, we cannot use a conventional PEC cell where WE and counterelectrode are separated by a liquid electrolyte bridge.

Yelin Hu (Empa, EPFL LPI) and Florent Boudoire (Empa, Uni Basel Chemistry) have designed a wedge-shaped lateral photoelectrode assembly for this purpose, as shown in Figure 8.59. First, FTO glass was coated with a hematite layer. The deposition geometry was deliberately chosen "off-center" to form a film with an inhomogeneous thickness. This is a well-known trick for making films for thickness-dependent studies. As a matter of fact, making homogeneous films is usually more difficult and of no particular advantage in an experiment like this. One-half of the film is then covered with Nafion® solution to form an electrolyte layer. On top of the Nafion®, we sputter a thin Pt layer as a counterelectrode. The NAFION® layer needs some minimum humidity for operation, but this is warranted from exposure to the ambient before bringing into the UHV chamber. Under water vapor conditions during the experiment, the NAFION® will not dry out.

PhD students Boudoire and Hu have tested this cell assembly at Empa prior to the actual *operando* XPS measurement at the synchrotron. They sprayed 0.1 molar KOH solution with a manual evaporizer at the fully wired electrode assembly while it was un-

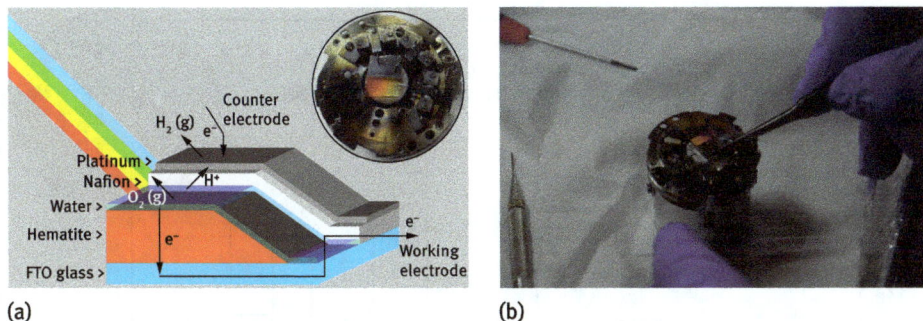

(a)   (b)

Figure 8.59: (a) Schematic of wedge-shaped photoelectrode sample for ambient pressure XPS during photo electrochemical water splitting. (b) Fixing and contacting the wedge shaped sample on the sample holder for the near ambient pressure experiment at Beamline 9.3.2 at the Advanced Light Source.

der illumination by the solar simulator. Figure 8.57 shows the photocurrent at 1500 mV between hematite WE and Pt counterelectrode versus the experiment time. They have "chopped" the light from the solar simulator and found that the transition from dark to light causes spikes in the current, confirming the formation of electron hole pairs. They also find a clear difference between cark current and photo current. We have thus a working photoelectrode assembly ready for deployment in the ambient pressure XPS measurement at BL 9.3.2 at the ALS. Figure 8.59 shows the AP-XPS Beamline 9.3.2. at the ALS and the photo of the sample holder with our electrode assembly under operation conditions.

We carried out ambient pressure XPS in the Fe 2p resonant mode because we wanted to make sure that we can record the valence band (VB) for the relevant electrochemical interface in our experiment. The value of this spectroscopic trick may not be underestimated, particularly not when one is dealing with analytically very complex systems such as operating electrochemical devices. Figure 8.62 (top panel) shows the VB spectrum of the sample referred to in Figures 8.56–8.61 recorded with excitation energies of 706 and 707 eV. By changing the photon energy from 706 to 707 eV, the intensity flank at a binding energy of 3.5 eV develops into a sharp shoulder. This is the spectral signature of Fe^{3+}. At 2 eV binding energy, we see a low-intensity hump from Fe^{2+}. The bottom panel of Figure 8.52 shows the Fe^{3+} resonant VB spectra under four different conditions: 1 and 1.5 V in dark and in light conditions. We see no difference in the spectra under these four different situations.

The spectra in Figure 8.62 were recorded when the sample was in a dry condition. The spectra undergo a drastic change when water vapor is supplied into the UHV chamber, as shown in the spectra in Figure 8.63. The signatures from Fe^{2+} at 2 eV and Fe^{3+} at 3.5 eV are still noticeable, but now with roughly the same intensity. The sample is under 1 V bias and becomes illuminated, which causes the intensity at binding energies of 8 eV and 10.5 eV to slightly decrease, whereas the intensity in the range from 2 to 7.5 eV in-

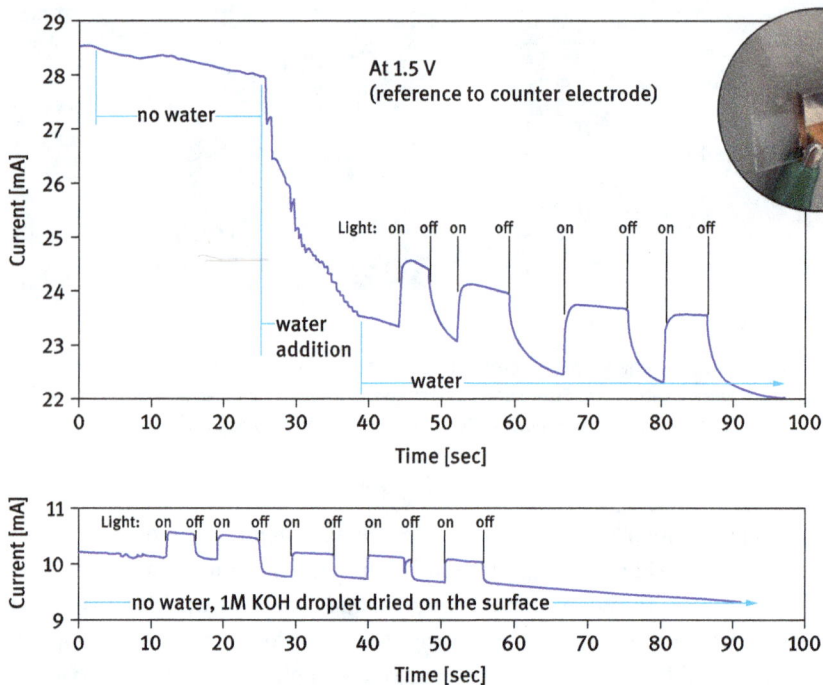

Figure 8.60: Preparation of beamtime experiment with a simple electrochemical experiment on the wedge-shaped electrode, which is wired to counterelectrode and WE contacts; see red and green crocodile clamps in the photo. The top panel shows the light and dark current over time, initially without water, then with addition of water by gentle spraying. The bottom panel shows the current versus time of the same sample when the water was dried, but after adding a drop of KOH solution, which remains hygroscopic after drying and has minimum residual humidity, allowing for photo-electrochemical light on/off experiments, as evident from the data.

creases upon illumination. This could be an effect of increased photoconductivity (see Chapter 5.6.3, Fe 3p resonant PES on hematite at BESSY). The binding energy range affected by the illumination is the one at and beyond the oxygen O 2p bonding peak at around 6–8 eV.

The right top panel of Figure 8.63 shows additionally the spectra at 1.5 V. Now the O 2p bonding peak is shifting by almost 2 eV to higher binding energies and the shape of the spectral range is considerable changing. As 1.5 V is the necessary voltage for water splitting on iron oxide, we have to attribute the pronounced redistribution of spectral weight to the electrochemical reactions taking place on the electrode surface. Although the bonding peak is shifted to higher energies, the Fe 2p signatures shift toward the Fermi energy, revealing that the water splitting involves hole doping. We use this shift of the Fe 2p signatures later for further data reduction of the spectra.

We have recorded the spectra for the voltage range from 1.0 to 1.7 V in increments of 0.1 V, under dark and light conditions. At this point, I should disclose that the mea-

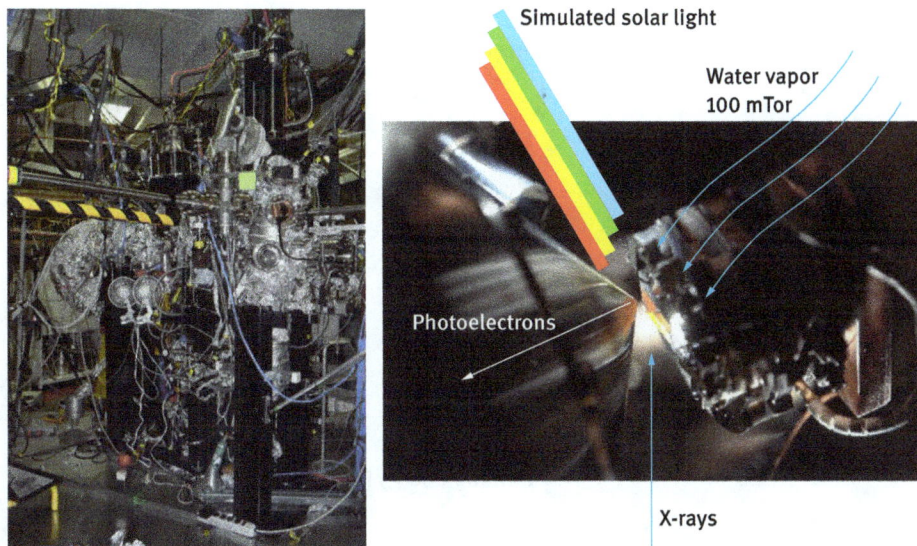

Figure 8.61: (Left) Beamline end station 9.3.2 at the Advanced Light Source Berkeley, where the photo-electrochemical XPS measurements were carried out. (Right) Magnified image from the sample and sample holder in the UHV chamber under water pressure condition of 150 mTorr and exposure to light, connected with the potentiostat and subject to resonant photoemission spectroscopy.

surement chamber at end station 9.3.2 at the ALS was not designed for experiments where the sample in the measurement position should be illuminated with an additional light source, such as for our solar cell experiments. The lamp was therefore directed from outside through a chamber window in the sample direction, where the sample then received scattered and reflected visible light. The light intensity on the sample, which we could not further quantify, was therefore low. The differences in the dark and illuminated spectra will be more pronounced when 1.5 AM intensity would hit the sample. The 1.5 air mass (1.5 AM) is a reference illumination condition that is very important for solar cell studies. When different materials and systems are compared, it is necessary to do this on a common reference. For electrochemistry, such reference is, e. g., the reversible hydrogen electrode RHE, against which all electrochemical potentials are referenced. The references for 1.5 AM are basically derived from astronomical data [Bird 1983, Gueymard 2001, Gueymard 2002, Gueymard 2003a, Gueymard 2003b, Gueymard 2004a, Gueymard 2004b, Gueymard 2004c, Hulstrom 1985].

For the further quantification of the spectral changes versus the voltage and the condition dark/light, we have plotted the shift of the Fe 2p peaks versus the voltage. This shift follows a different slope depending on whether we look at the dark or light spectra (Figure 8.64). Since the slope for the Fe 2p shift in light condition is around as half the slope of the Fe 2p shift in the dark condition, we can say the response of the Fe 2p

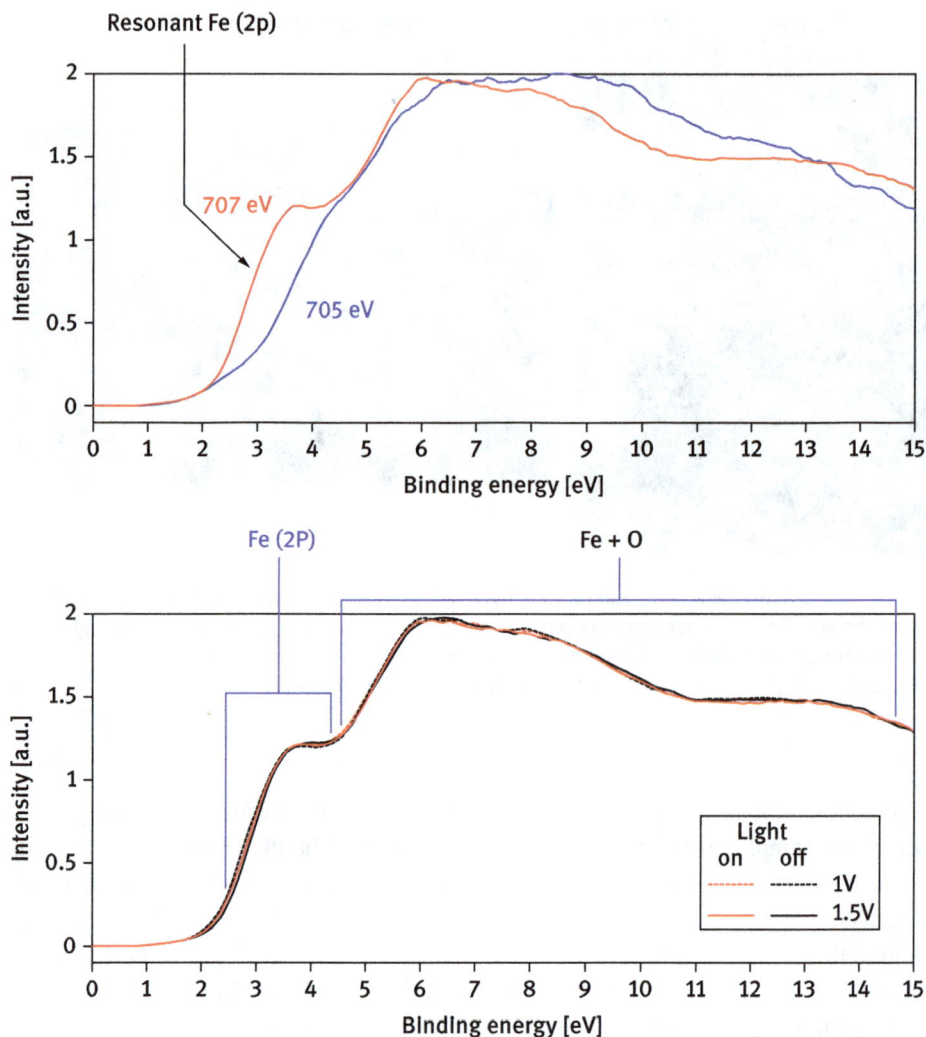

Figure 8.62: (Top) VB XPS spectra of the sample shown in Figures 8.56–8.58 recorded at photon energies of 705 and 707 eV. (Bottom) Fe 2p resonant VB XPS spectra recorded in light-on and light-off conditions at 1 and 1.5 V bias. The sample was in the dry condition during this part of the experiment.

signature to light is rather "stiff." The application of a large potential difference of 0.7 V in light is less effective in shifting the signature than application of the same potential in the dark.

The shift of the O 2p peak is investigated in Figure 8.65. We have here a different pattern for which we need to find a different analysis and data treatment approach. The data reduction and the data analysis for these near ambient pressure XPS data were done and inspired by Empa/Uni Basel PhD student Mr. Florent Boudoire [Boudoire 2015],

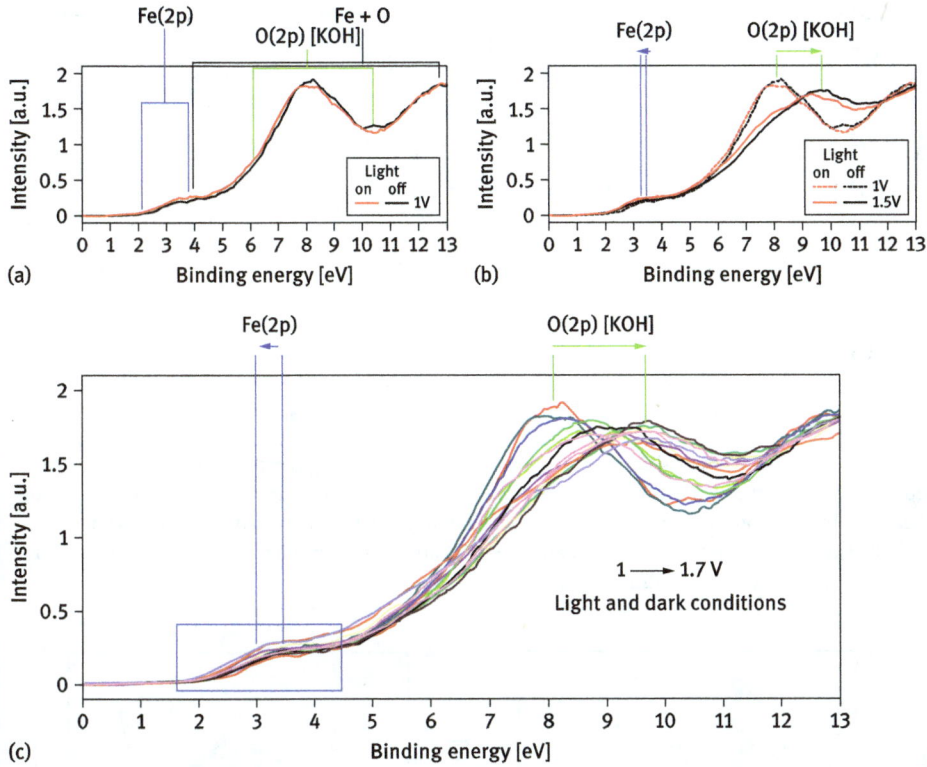

Figure 8.63: Fe^{3+} resonant VB spectra recorded with 707 eV photon energy at 150 mTorr water vapor pressure and 1 to 1.7 V polarization in dark (off) and light (on) condition at Beamline 9.3.2 at the Advanced Light Source. (a) Spectra recorded at 1 V in dark (black spectrum) and light (red spectrum) with spectral assignment of VB features. (b) Comparison of dark and light spectra at 1 and 1.5 V. (c) All spectra recorded in a light and dark condition from 1 to 1.7 V.

and the manuscript for publication was finalized by Empa/EPFL PhD student Mr. Yelin Hu. The left panel of Figure 8.65 shows the shift of the O 2p peak in dark linear increasing with an increasing potential from 1 to 1.2 V and then following a saturation plateau. The corresponding behavior under light shows fewer differences.

With a simple linear background subtraction (subtraction of the line indicated in green), we can transform the characteristics of the shift shown in the left panel into a shape that is reminiscent of the dark current and light current, as shown in the right panel of Figure 8.65. We subtract another straight line with the same slope from the profile of the O 2p peak shift under the light condition, and thus obtain the same renormalization as with the dark shift.

The outcome of this transformation is very interesting. The profiles for the O 2p peak shift versus potential for the dark and illuminated condition are at large identical with the profiles for the dark current and light current (photocurrent), as is demonstrated

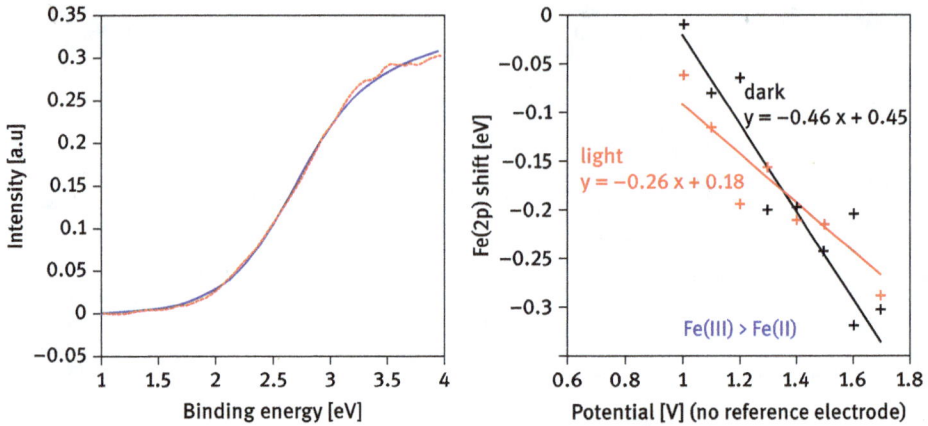

Figure 8.64: (Left) Experimental data for VB maximum region (red data points) of a spectrum with a least square fit (blue) for the data treatment. (Right) Experimentally determined shifts of the Fe 2p as a function of external bias potential from 1 to 1.7 V for dark (black data points) and light (red data points) condition. The solid lines (red for light and black for dark condition) are linear least square fits for the determination of the slope of the shift of the Fe 2p signature.

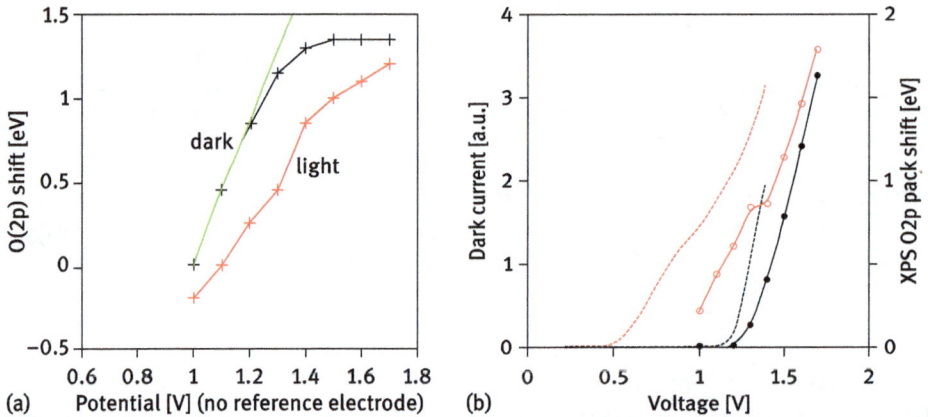

Figure 8.65: (a) Shift of the O 2p peak in the resonant VB XPS spectra of the sample in Figures 8.56–8.58 under dark (black curve) and illuminated (red curve) conditions. Green line is linear least square fit for background subtraction. (b) Comparison of dark current (dotted black curve) and light current (photocurrent, dotted red curve) with the background subtracted O 2p peak shift under dark (thick solid black curve) and illuminated (thick red curve) conditions.

in the right panel of Figure 8.65. The saddle point that is typical for the light current or photocurrent in iron oxide PEC data is fully reproduced in the O 2p shift under light conditions.

8.4 Bioelectrochemical systems

We learned in Part 7 about biological samples. Protein crystallography is probably the most developed application of an X-ray method to biological systems. Protein crystallography is typically done by multianomalous diffraction phasing [Hendrickson 1979, Karle 1980]. The next well-known method is protein spectroscopy with XANES and EX-AFS [Gu 2003], including NEXAFS spectroscopy on metaloproteins and enzymes known in photosynthesis [Grush 1996, Gu 2002, Ralston 2000, Wang 2001a, Wang 2000]. The reader is also reminded of the works with $K\beta$ spectroscopy [Messinger 2001]. STXM has recently been used for the investigation of biofilms, e. g., [Lawrence 2012], and the references in the part on imaging and tomography in this book. Finally, I mention the NRVS method, which I explained already in Part 5 of this book. Here, I present two XPS studies on proteins, which were done at SSRL and at ALS.

8.4.1 Fe 3p resonant VB XPS on phycocyanin adsorbed on iron oxide

Phycocyanin is a light harvesting protein (enzyme) in blue-green algae and cyanobacteria. We found that this enzyme enhances the photocurrent of the iron oxide photoanode when we adsorb it on the surface (Figure 8.66). It turned out that it makes a difference whether phycocyanin is just physisorbed on the photoanode, or whether a linker molecule is used for the attachment of the enzyme. To some extent, better attachment caused higher photocurrent [Bora 2012]. By intuition, most researchers would focus on the measuring of the photocurrent of such biohybrid electrodes. However, there is more to study than just the photocurrent. The oxygen gas and hydrogen gas production and also the dark current measurements provide information on the bioelectric interfaces.

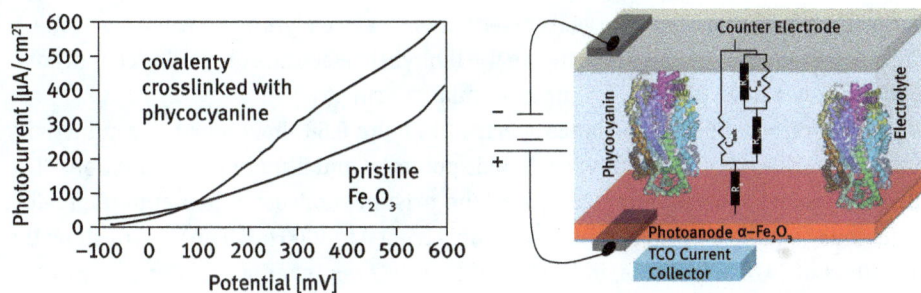

Figure 8.66: (Left) Photocurrent density of a pristine hematite electrode (blue) and such electrode coated with light harvesting C-phycocyanin (red). (Right) Schematic of an iron oxide absorber coated on a transparent conducting oxide current collector. On top of the absorber, in contact with the electrolyte are the adsorbed or chemisorbed or physisorbed phycocyanin molecules. The counterelectrode is metal platinum. Inscribed in the sketch is the electric Randles' circuit.

Figure 8.67: α-Fe$_2$O$_3$ single crystals (SurfaceNet GmbH, D-48432 Rheine, Germany) coated with CPC. The two photos on the right show the single crystal with CPC fixed and contacted with silver paste on the back to an FTO glass. Lacomite Varnish (Agar Scientific) was used to cover the silver paste on the back from exposure to electrolyte.

We used α-Fe$_2$O$_3$ single crystals and coated them with phycocyanin (Figure 8.67) in three different ways. The first way was simple physisorption by drop casting of a CPC solution on the single crystal surface. The second way was using CPC, which had a histidine tag, and the third was CPC mixed with melanin, an organic semiconductor. When organic or bioorganic components are used in photo-electrochemical studies, one has to be cautious with the choice of electrolyte. Strong caustic KOH with pH 13–14 will denature the CPC but the chromophores may still work, as we have found in our studies [Bora 2012].

What do we need to get a "high photocurrent"? Certainly, we need to get a high conversion of photons to electron-hole pairs in the absorber material, i. e., the photoelectrode. We need to have a good charge transfer between the phycocyanin and the photoelectrode, irrespective of the optical properties of the CPC. A very high optical activity of CPC would be of little use if the produced electron-hole pairs would become ineffective because of poor charge transfer between electrolyte or between electrode. It is therefore necessary to also investigate the biohybrid electrode under dark conditions. So to speak, we have to check for the molecular "wiring."

The electrochemical impedance spectra in Figure 8.68 show how the conductivity of the assemblies changes with varying bias potential and illumination conditions. The upper left panel shows the impedance of the pristine, noncoated hematite photoelectrode under a dark condition at bias potentials from 0 to 1000 mV. The spectra show the signatures of two semicircles, indicative of two different charge transfer processes in the electrode. We have not deconvoluted the spectra but indicated the range and position of the two semicircles from around 0 to 1000 Ω and from 1000 to 3300 Ω. The global "steepness" of the impedance curve is decreasing upon increasing bias. At 1000 mV, the steep global behavior has turned into a third semicircle with large radius. The upper right panel shows the corresponding impedance spectra under illumination conditions

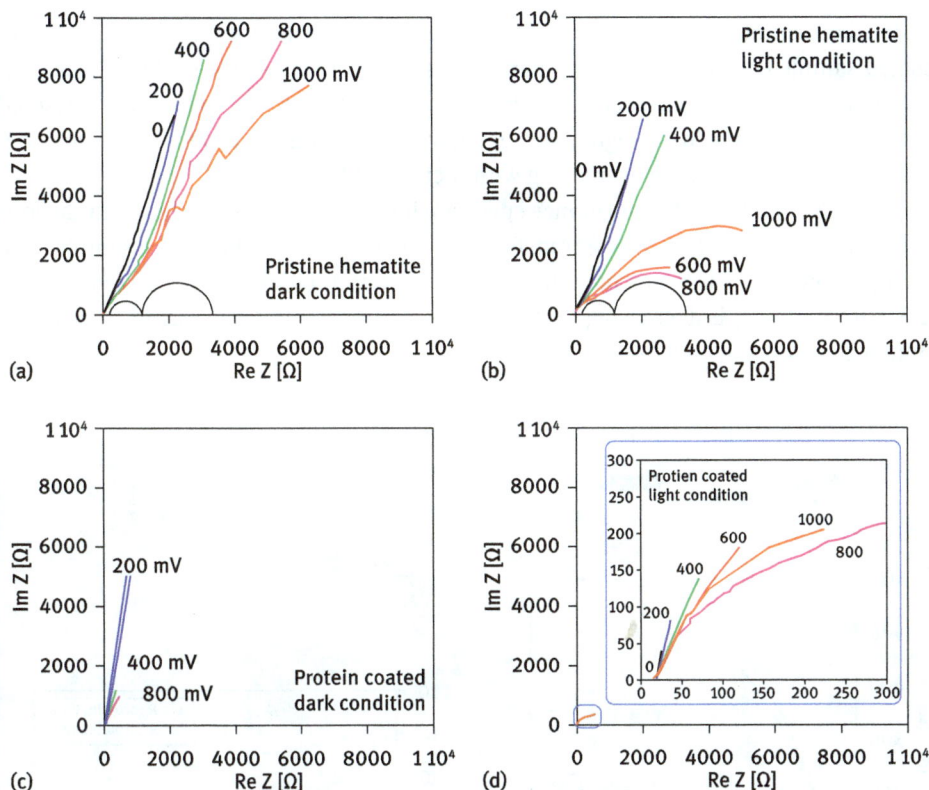

Figure 8.68: Top: Nyquist plots of impedance spectra of porous pristine hematite film under dark (a) and under illuminated (b) condition with DC bias from 0 to 1000 mV (indicated with bold numerals 0, 200, 400, 600, 800, 1000 mV). The two thin black semicircles with radii of 1000 and 2200 ω drawn on the Re Z axis indicate the electrode resistance and the charge transfer resistance, which becomes particularly obvious for 600, 800, and 1000 mV under light condition. Bottom: Nyquist plots of impedance spectra of protein (C-phycocyanin)-coated porous hematite films under dark (c) and under illuminated (d) condition, recorded with DC bias ranging from 0 to 1000 mV (bias numbers are shown in bold in the plots). For all plots, the data window corresponds to 104 ω. The inset in the right panel shows a magnification of the impedance range of the protein-coated film under illumination (data window size 300 ω). Reprinted with permission from [Braun 2015b] Copyright (2015) Wiley.

at 1.5 AM. The spectra for 0 and 200 mV have the similar magnitude of 6000 to 8000 Ω. At 400 mV, the height drops from 8000 Ω to below 6000 Ω. At 600 and 800 mV, we notice a drastic decrease of the impedance, i. e., a drastic increase in conductivity.

Let us now turn to the bottom row in Figure 8.68. On the left, we have again the dark measurement, but with phycocyanin coated on the electrode. The impedance values all range below 5000 Ω. For the higher bias values of 400 and 800 mV, we see impedance values of lower than 1000 Ω. Obviously, the CPC coating causes significantly increased charge transfer between the electrolyte and electrode. Moreover, we have learned this under dark conditions. When we look into the light-dependent impedance of the CPC coated sample on the right, we can hardly make out any impedance values. Note that impedance is a complex resistance, and low impedance means therefore high conductivity. The inset in the lower right panel of Figure 8.68 demonstrates very nicely how the radii of the semicircles are decreasing with increasing bias potential.

We have modeled the impedance spectra with Randles circles as shown in the left panel of Figure 8.69. The charge transfer resistance R_{CT} of the CPC-coated sample was then plotted versus the bias potential for the dark and light experiment (right panel in Figure 8.69). R_{CT} is decreasing for the dark experiment and for the light experiment similar to an exponential decay.

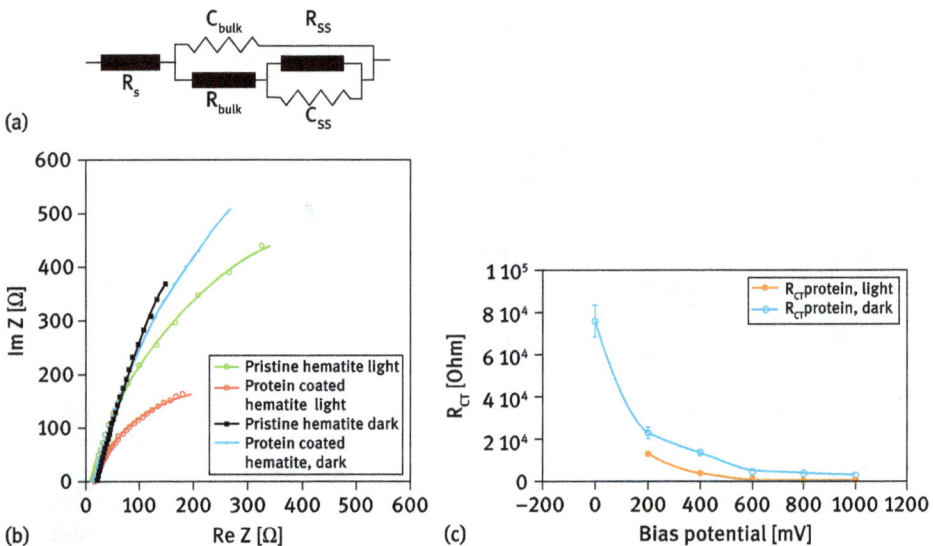

Figure 8.69: Electric impedance circuit (a) with serial resistance R_S, bulk resistance R_{bulk}, bulk capacitance C_{bulk}, and a parallel circuit from surface state capacitance C_{SS} and resistance R_{SS}, originating from surface states. The capacitances are interpreted as constant phase elements. (b) Nyquist plot of impedance spectra of protein-coated hematite film under dark and under illuminated condition with DC bias of 1 V. The frequency range for the least square fit (solid lines) extends from 63291 to 10 Hz. (c) Charge transfer resistance R_{CT} for the protein-coated electrode in a light and dark condition for bias potentials from 0 mV to 1000 mV.

We have looked into these protein-coated electrodes with photoemission spectroscopy. We chose again an iron resonant excitation energy, in this particular case Fe 3p at around 57 eV, in order to have an elemental marker, which allowed us to distinguish the iron oxide in the valence band—CPC interface from the bulk of the organic material on the electrode. The protein films were relatively thick, in the range of 1 micrometer or more. The excitation energy of 50–60 eV is therefore too low to penetrate the thick areas of the CPC layer, and also the photoelectrons cannot pass through thick regions of the CPC layer. Only at extremely thin regions or at open spots in the film can we expect an iron resonant XPS signal.

Figure 8.70 shows SSRL photoemission end station at BL8-1, with Florent Boudoire moving the samples into measurement position. The Fe 3p resonant VB PES spectra for the three different prepared CPC films are shown in Figure 8.71. The top panel shows the CPC with the histidine tag, which supposedly should provide a better covalent attachment to the photoelectrode. We have measured this with two excitation energies, 53.2 and 55.8 eV.

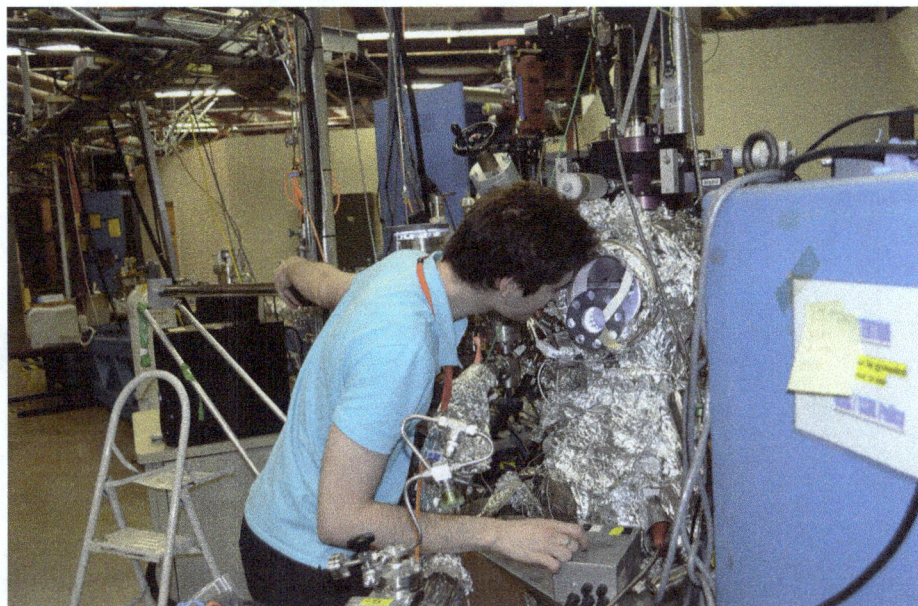

Figure 8.70: Uni Basel/Empa PhD student Florent Boudoire at Beamline 8-1 of Stanford Synchrotron Radiation Laboratory after preparation for resonant photoemission studies on lithium batteries, photoelectrochemical cells and bioelectrodes.

We have modeled the impedance spectra with Randles circles as shown in the left panel of Figure 8.69. The charge transfer resistance R_{CT} of the CPC-coated sample was then plotted versus the bias potential for the dark and light experiment (right panel of Figure 8.66). R_{CT} is decreasing for the dark experiment, and for the light experiment, it is similar to an exponential decay.

We have looked into these protein-coated electrodes with photoemission spectroscopy. We chose again an iron resonant excitation energy, in this particular case Fe 3p at around 57 eV, in order to have an elemental marker, which allowed us to distinguish in the valence band the iron oxide—CPC interface from the bulk of the organic material on the electrode. The protein films were relatively thick, in the range of 1 micrometer

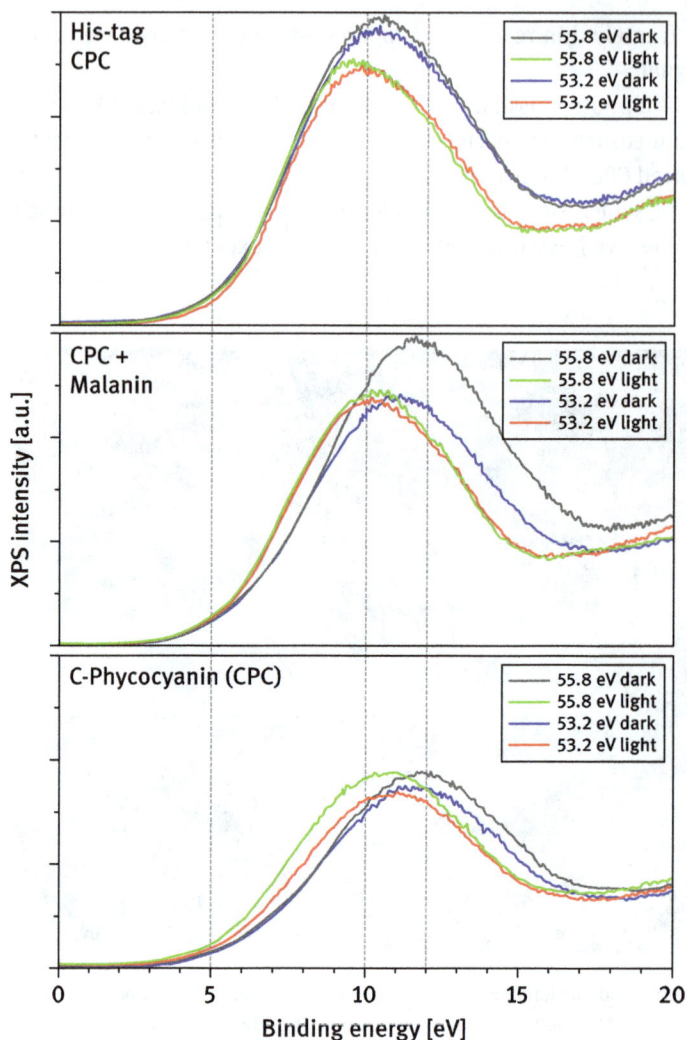

Figure 8.71: Comparison of VB PES spectra measured at 53.2 eV and 55.8 eV in dark and light conditions for the three different protein coatings. The steep straight lines intersecting near 5 eV show the VB position. The sample coated with αHisPC and melanin shows only 0.6 eV shift on VB position while PC only shows 1.0 eV under illumination. In the case of αHisPC, no shift in the VB position is observed.

or more. The excitation energy of 50–60 eV is therefore too low to penetrate the thick areas of the CPC layer, and also the photoelectrons cannot pass through thick regions of the CPC layer. Only at extremely thin regions or at open spots in the film can we expect an iron resonant XPS signal.

Figure 8.67 shows SSRL photoemission end station at BL8-1, with Florent Boudoire moving the samples into measurement position. The Fe 3p resonant VB PES spectra for the three different prepared CPC films are shown in Figure 8.68. The top panel shows the CPC with the histidine tag, which supposedly should provide better covalent attachment to the photoelectrode. We have measured this with two excitation energies, 53.2 eV and 55.8 eV.

In the dark condition, the intensity maximum of the spectrum from the PC film shifts from 12 eV to 11.5 eV toward E_F, when PC was copolymerized with melanin (Figure 8.68). With αHisPC+Mel on hematite, the intensity maximum shifts to 10.3 eV. A similar trend is shown in the case of illumination with the solar simulator. With increasing improvement of processing technology, the maxima of the VB spectra move closer to E_F, reflecting the changes in the electronic structure with increased DOS near E_F. The charge transfer between hematite and phycocyanin might be improved in αHisPC because histidine coordinates to Fe. We do not know whether the shift toward the E_F is a result of hole doping due to PC+Mel in the films. In addition, the His-tag presumably improves entrapment of the Histagged protein into the melanin network, as the formation of histidine-tyrosine bonds have been observed upon tyrosinase-catalyzed crosslinking [Hellman 2011]. The correlation between improved electric transport, increased the photocurrent, and increased DOS in the VB near E_F is obvious.

When we subject the protein-hematite assembly to 1.5 AM simulated solar light, the spectral weight of the intensity range near the O 2p bonding peak decreases and shifts by up to 0.7 eV toward E_F. We interpret this shift as the hole-doped DOS resultant from the exposure of visible light. With the light on, the electron-hole pairs can be created in the conduction band (CB) and the holes stays near the surface generating a p-type DOS. A similar trend of binding energy shift is observed when silicon is doped [Sezen 2011]. Our PES studies on proteins are in principal not different from studies on synthetic organic molecules and polymers, which are frequently used in solar cells and light emitting diodes. A comprehensive review on such VB spectroscopy studies is presented by Koch et al. [Koch 2012].

Among our protein functionalization processes, the coating with PC only is the least complex one. For the comparison of the PES spectra in Figure 8.68, we determine the position of the VB by extrapolation of the spectral flank near E_F toward zero intensity—as a practical metric. The intercept is then by our definition the VB position. The VB position of the PES spectra shifts from 6 to 5 eV when the sample is illuminated with the solar simulator. This 1 eV shift thus originates from the illumination effect, i. e., a charge carrier (electron hole pair) generation.

When we turn to the PC+Mel-coated hematite, the VB position shifts from 5.6 to 5 eV, i. e., the shift is only 0.6 eV upon illumination. In the case of the HisPC+Mel coating, there is no shift of the VB spectrum when the light is switched on. Interestingly, for all films, the position of the VB in the illumination condition is 5 eV. Hence, it appears that light has no influence on the charge transfer as far as the interface of the protein film with the hematite is concerned [Faccio 2015].

In addition to the electric charge transfer, which is the scope of this paper, fluorescence resonance energy transfer (FRET) could hypothetically occur between phycocyanin and hematite, and thus increase the photocurrent density. The efficiency of the FRET mechanism is inversely proportional to the sixth power of the distance between light donor and light acceptor, making it extremely sensitive to distances and virtually negligible for small distances. Due to the architectures of our bio-hybrid electrodes, where agarose, CDI, and His-tag virtually constitute "spacers" between PC and hematite, it appears improbable that FRET could occur across these spacers.

8.4.2 *Operando* NAP-XPS spectroscopy on an illuminated algal biofilm

I want to begin this section with the colonization of bacterial cultures. Pigments and proteins adsorbed on or linked to surfaces can degrade depending on their specific half-life under the local physicochemical conditions. If we want to use them on photoelectrodes, they have to be replaced so as to warrant further charge transport. Likewise, photosynthetic organisms constantly replace pigments and proteins in their photosystems that have been inactivated by reactive species such as reactive oxygen species.

Evolution and growth of such complex biohybrid architectures is investigated in fields, which are not related with photo-electrochemistry and artificial photosynthesis. Yet, their investigation may turn out valuable for what we propose here, in particular as the understanding of the covalent attachment and charge transfer between inorganic framework and living matter is concerned [Athreya 1999].

Making use of the natural renewal process in living organisms as coating material instead of isolated proteins or whole-cell extracts could potentially reduce maintenance costs and increase the life time of a PEC system for small scale residential or for large scale industrial use (compare [Hindersin 2013, Hindersin 2014, Leupold 2013a, Leupold 2013b]). Owing to their fast life cycle and robustness, model planktonic algae are used, e. g., in the production of biofuel. Ideally, the algae for use in a PEC would adhere easily to the photoelectrode, have a broad ecological niche regarding water chemistry and incident light intensity, and substantially minimize electrical losses to the pristine material. The cyanobacterium *Anabaena* sp. has recently been shown to be capable of several energy-generating growth modes [Stebegg 2012]. It forms filaments that produce a large surface in contact with the surrounding medium and attaches to surfaces in surface waters where it is found in naturally occurring biofilms. *Anabaena* fixes nitrogen, which may open further applications when used in a PEC system.

It appears that *Anabaena* and similar organisms could be suitable enhancers for photocurrents in PEC. We established *Anabaena* sp. at two different cell densities on hematite films on FTO and measured it—coated on iron oxide layers—with impedance spectroscopy under light and dark conditions at bias potentials from 0 to 1400 mV versus Ag/AgCl. Figure 8.69 shows how the impedance under dark ranges in the 100 kΩ range for the potentials up to 800 mV, but experiences a decrease to the 1 kΩ range in the 600 to 800 mV range and above upon illumination.

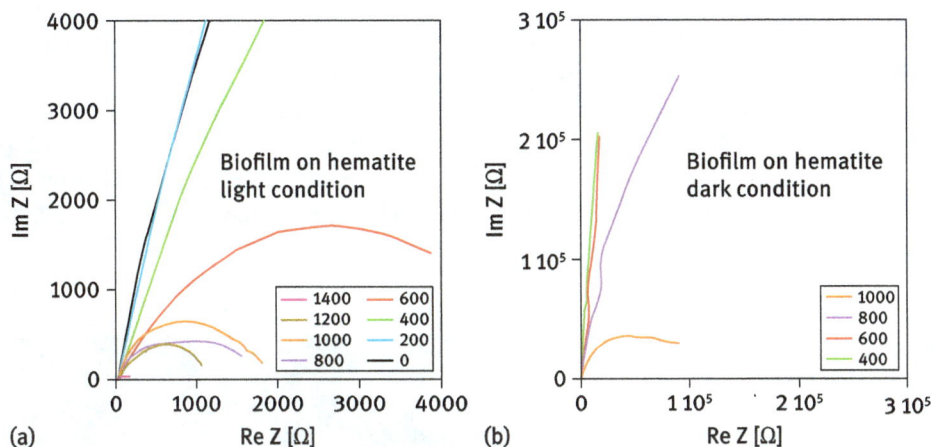

Figure 8.72: Impedance spectra of a biofilm on a hematite photoelectrode (b) in dark and (a) light conditions. Two electrodes with low and high concentration film were subject to EIS in 0.05 molar phosphate buffer saline (PBS) from 100 kHz to 10 mHz in dark and 1.5 AM light from 0 to 1200 mV DC bias versus the Ag+/AgCl reference. Reprinted with permission from ([Braun 2015b] Copyright (2015) Wiley).

Upon comparison with the impedance spectra of the pristine (this is iron oxide not coated with a biofilm) and CPC-coated film, we find that for 800 and 1000 mV the biofilm PEC assembly has a lower impedance upon illumination than the pristine hematite electrode. This impedance is still larger than the impedance for the CPC-coated electrode. The different impedance response under illumination for the pristine electrode and the electrode coated with a biofilm and a protein film is demonstrated in Figure 8.70.

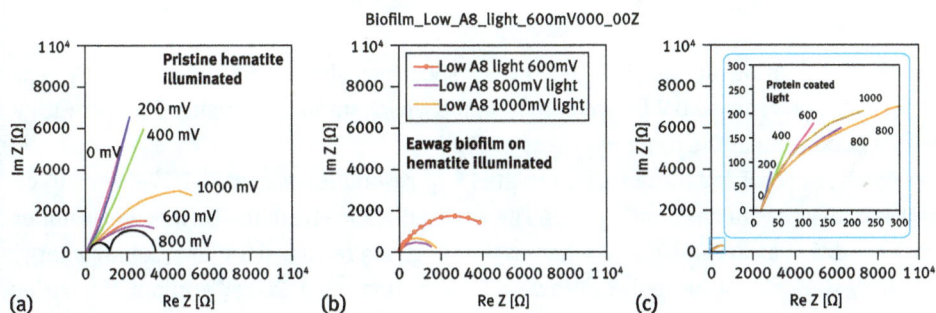

Figure 8.73: Comparison of impedance spectra from pristine iron oxide (a) and biofilm (b) and protein-coated iron oxide (c) films. The films were measured in phosphate buffer saline (PBS) electrolyte.

We have subjected this biofilm-electrode to AP-XPS spectroscopy with water vapor pressure of 150 mTorr and under electrochemical polarization from 0 C to 1.5 V in dark and under light (Ambient Pressure XPS Beamline 9.3.2 at the Advanced Light Source in

Berkeley, California [Ogletree 2002]); see Figure 8.71. As already mentioned in the previous section, the sample holder at End Station 9.3.2 is not situated that one can directly point a light source on the sample. It was therefore necessary to illuminate the sample environment as much as possible. For this purpose, we took several high flux flood lights (halogen bulbs) and shone those to the sample. This is demonstrated in the right panel of Figure 8.71. For making dark experiments, the windows in the measurement chamber are closed with plastic caps.

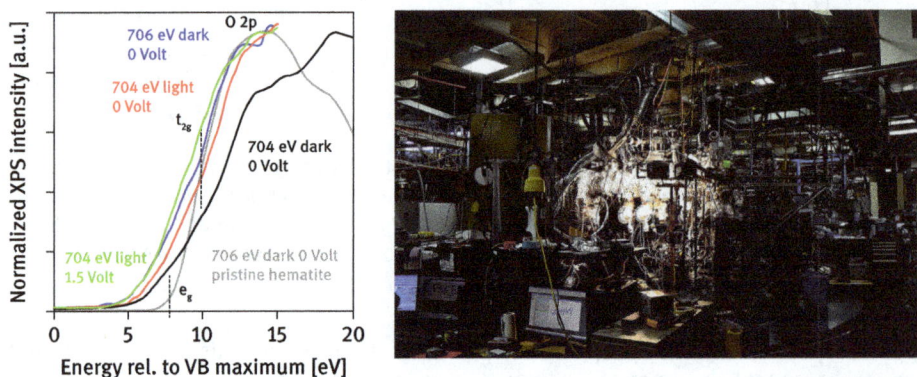

Figure 8.74: (Left) *Operando* Fe 2p resonant valence band XPS spectra of biofilm deposited on hematite, recorded with photon energies of 704, 706, and 707 eV at 100 mTorr water vapor pressure at 0 and 1500 mV in dark and light condition. (Right) Advanced Light Source Berkeley, Beamline 9.3.1 during an XPS valence band study on a biofilm grown on metal oxide semiconductor photoelectrode. The design of the end station did not allow for direct pointing an optical light source. Therefore, the end station was illuminated with a set or halogen lamps to provide as much light on the sample as possible [Braun 2015b].

Under illumination by visible light, the VB spectrum shifts by 1.8 eV toward E_F. The difference spectrum at 0 V between light on/off yields intensity ranging from 5 to 15 eV with a maximum at the binding energy of 11 eV.

We have recorded these spectra in the Fe 2p resonant range at 704, 706, and 707 eV at 0 and 1.5 V bias in dark and in light. The widest distance from the Fermi level is under dark condition at 704 and 707 eV, whereas the spectrum recorded under resonant energy of 706 eV shows a substantial shift toward E_F. Illumination at 704 eV without bias yields a noticeable shift toward E_F. Illumination at the photon energies 706 and 707 eV yields a larger shift. Applying the 1500-mV bias, e. g., at 704 eV increases the shift toward E_F substantially.

Here, it is important to note that XPS studies are generally extremely surface sensitive. Hence, the above spectra show predominantly the biofilm and only little spectral intensity comes from the hematite underneath. This is why we tune the X-ray energy to an absorption threshold of the iron, i. e., Fe 2p resonant around 706 eV.

Upon comparison with spectra from pristine hematite recorded under identical conditions (p_{H2O}, T, U, E_{photon}, light on/off), we can discriminate the contributions to some extent.

8.5 Electrochemical double layer capacitors

Electrochemical double layer capacitors store electrostatic energy in the double layer, which is formed on the surface of an electrode in an electrolyte [Bockris 1963, Braun 1999a, Conway 1999, Kötz 2002]. The energy stored in such supercapacitor is overwhelmingly larger than in any conventional dielectric capacitor. In the ideal case, the energy from such a supercap or ultracap originates only from the electrostatic energy in the double layer. High surface area materials such as porous carbon are very good electrode materials for supercaps. Some supercap electrode materials such as ruthenium oxide have a so-called pseudo-capacity, which originates from chemical storage associated with the redox potential of the electrode material. Such pseudo-capacitance is not always welcome in supercaps because the underlying chemical reactions in the electrode can cause structural disintegration of the electrode material. Glassy carbon is a high surface supercap electrode material provided its internal porosity becomes connected in an extensive pore network [Braun 1999a, Braun 2000b, Braun 2003i, Miklos 1980]. Figure 8.72 shows an electron micrograph of an electrode assembly from glassy carbon, which is at its surface oxidized with open pores and in the core (bulk) not oxidized with closed pores [Braun 2010h].

8.5.1 X-ray diffraction and wide angle scattering on glassy carbon

We have seen the X-ray diffractograms on glassy carbon monoliths and powders in Figure 3.5 and Chapter 3.2 on X-ray wide angle scattering [Braun 1999c, Braun 2002a]. The thermal gas phase oxidation causes structural changes in the glassy carbon that can be monitored with XRD and WAXS. Figure 4 in [Braun 2000a] shows XRD diffractogram displaying the (002)-peak of nonoxidized and oxidized GC around 25 K. During oxidation, the peak is shifted toward larger diffraction angles.

8.5.2 Small-angle scattering on supercapacitor electrodes with X-rays and with neutrons

Since porosity and internal surface is a key property of supercapacitor electrodes, they were investigated with small-angle scattering as already shown in Chapter 3.3.1 [Braun 1999c] and USANS in Chapter 3.3.2 [Braun 2004a]. I want to present here an alternative SAXS data evaluation method based on Rosiwal's integration principle and

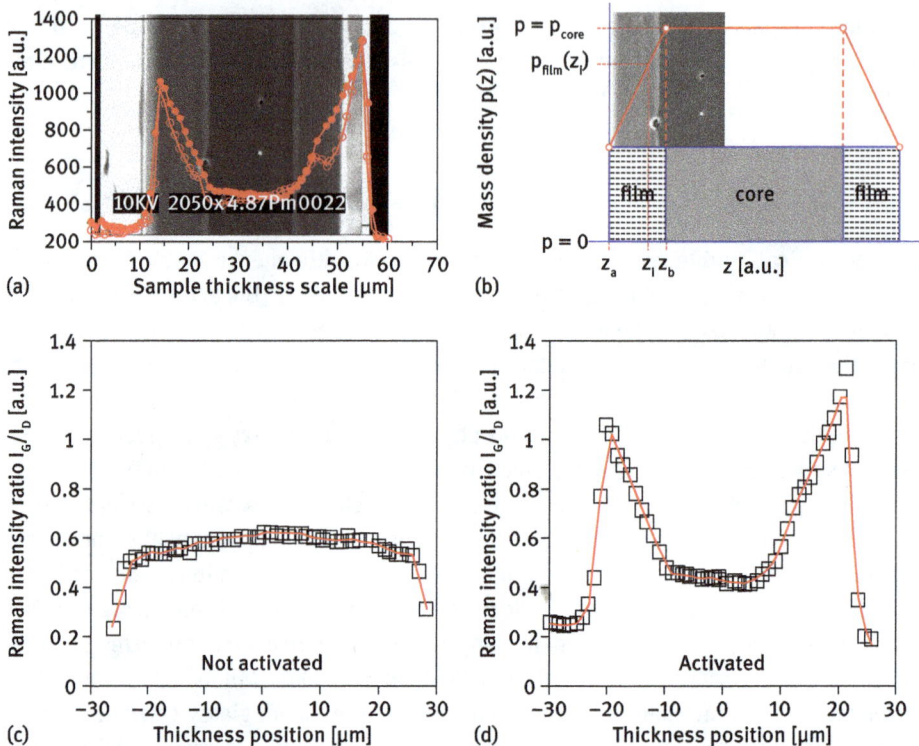

Figure 8.75: (a) Scanning electron micrograph of a cross-section of a nominally 60-micrometer glassy carbon electrode thermally gas phase activated in air. The gray color stripes on the 10–25 and 40–50 micrometers marks denote the porous active films for electrolyte filling and charge storage. The black part in the middle between 25 and 40 micrometers is the unactivated bulk material. The red data points are optical Raman intensities from two line scans across the cross-section, denoting the density gradient. (b) Schematic showing an overlay of the SEM micrograph and the calculated mass density gradient across the porous active film (open porosity) and porous (closed porosity) bulk (core). (c) Raman line scan across the cross-section of a nonactivated glassy carbon electrode. (d) Raman line scan across a partially activated glassy carbon electrode, showing the porosity gradient across the active film. Same data as shown in the left top panel.

demonstrated by Professor Wilfried Gille of University of Halle in Germany [Gille 2003]. Because of the intimate relationship between small-angle scattering and the geometric, stereological theory of chord length distribution density for isotropic uniform random chords, information on the porosity can be obtained, regardless whether the pores are open and accessible, or not.

The isotropic small-angle scattering intensity $I(h)$ is related with the correlation function $\gamma(r)$ via

$$I(h) = \frac{\int_O^L 4\pi r^2 \cdot \gamma(r) \cdot \sin(h \cdot r)/(h \cdot r)dr}{\int_O^L 4\pi r^2 \cdot \gamma(r)dr}$$

The second derivative of the correlation function $y(r)''$ contains the distribution densities $\phi(r)$ and $f(r)$. Rosiwal established a mathematical formalism [Rosiwal 1898] that allows for the "derivation" of $y(r)''$. We determined this quantity for the nonoxidized and fully oxidized glassy carbon sheets. Figure 8.73 shows the second derivatives of the correlation function, $\gamma(r)''$ versus the real space parameter and in the insets the actual correlation function $\gamma(r)$.

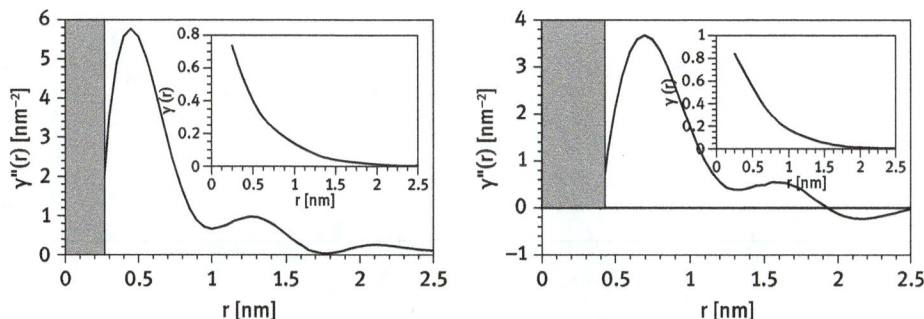

Figure 8.76: Second derivative of the correlation function, $\gamma(r)''$, and in the insets the correlation function $\gamma(r)$ for nonoxidized glassy carbon (left) and for the oxidized glassy carbon (right).

With the availability of the correlation function and the interpretation of its second derivative for the porosity, we can simulate an indicator function $i(r)$ and the fluctuation $\eta(r)$, along with a function of the occupancy $Z(r)$ and the chord distribution function $\gamma(r)$. Figure 8.74 shows the results from a linear simulation model for both glassy carbon electrodes. The model parameters that we used for the oxidized glassy carbon were an average (mean) 0.7 nm for $<l>$ and 1.5 nm for the mean $<m>$ for the chore segments and a porosity p of 0.33. The results are shown in the top panel of Figure 8.74. The bottom panel of Figure 8.74 shows the simulation result for the nonoxidized glassy carbon with $>l> = 0.3$ nm and $<m> = 1.3$ nm, with $p = 0.25$.

8.5.3 NEXAFS spectroscopy on thermally oxidized glassy carbon plates

The variation of the X-ray density as determined from XRD, and the variation of the exponent of decay in the SAXS curves suggested that during early stages of thermal activation of glassy carbon, an intermediate form of carbon grows, which may have a different molecular structure than the carbon porous film or the glassy carbon bulk. I have therefore subjected glassy carbon sheets to C 1s NEXAFS spectroscopy at Beamline 9.3.2 at the Advanced Light Source. Figure 8.75 shows the NEXAFS spectra of graphite as reference material, nonactivated glassy carbon, 1 minute activate, and 3 and 4 hours acti-

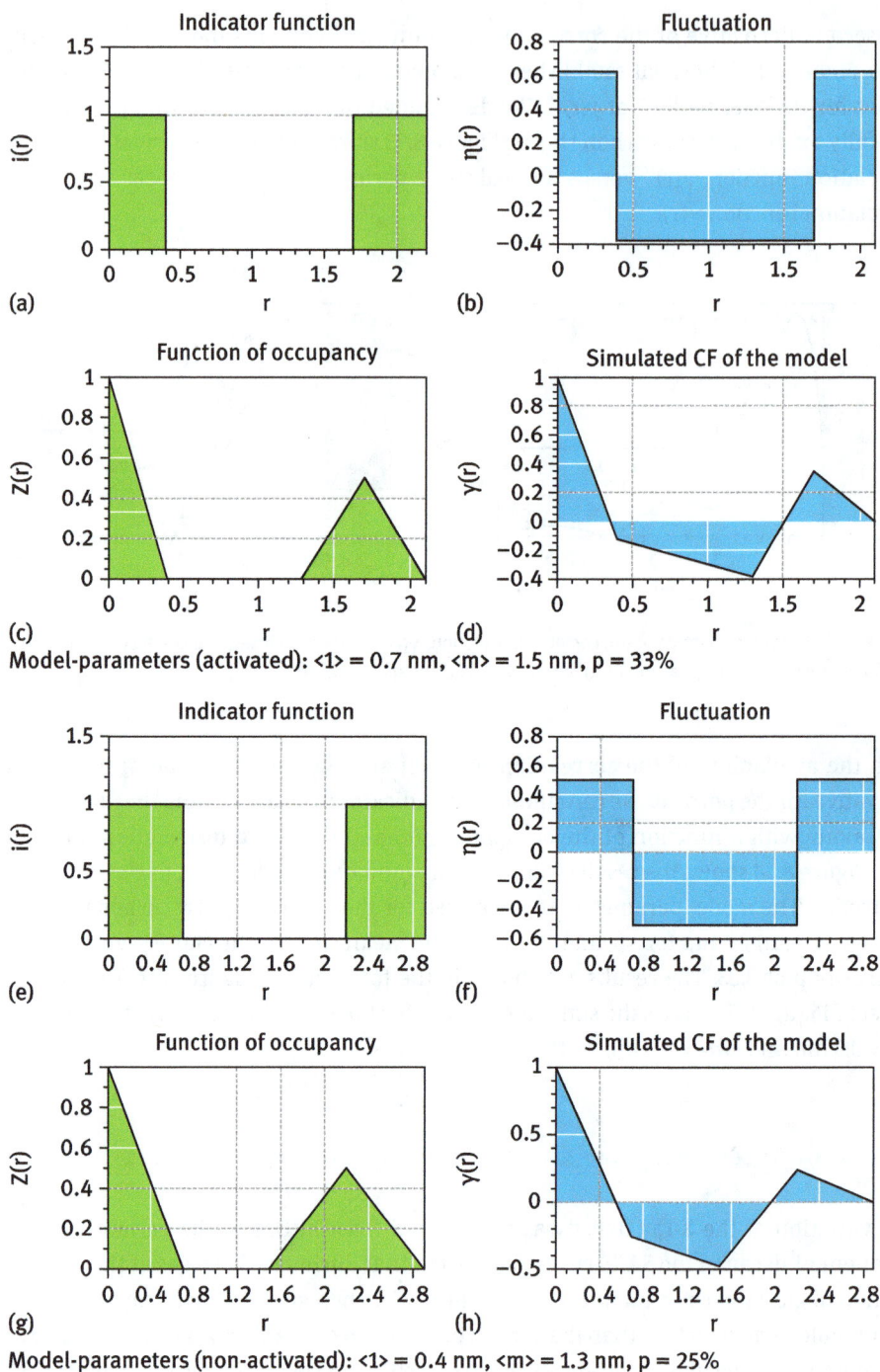

Figure 8.77: Results of the linear simulation model for the oxidized (top panel) and nonoxidized (bottom panel) glassy carbon with input parameters $<l> = 0.7$ nm, $<m> = 1.5$ nm, and $p = 0.33$; $<l> = 0.3$ nm, $<m> = 1.3$ nm, and $p = 0.25$, respectively [Gille 2003].

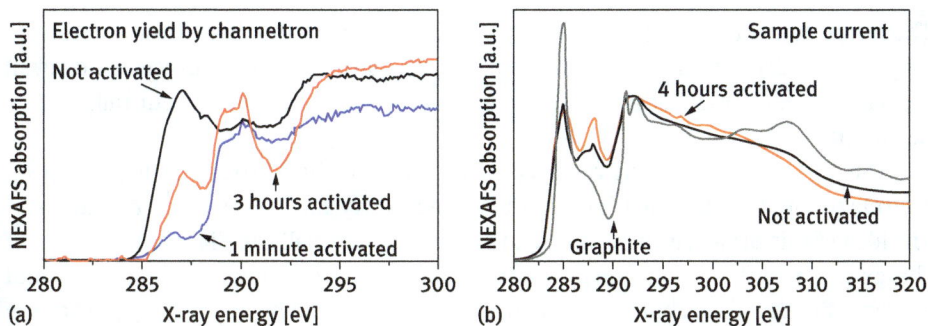

Figure 8.78: (a) NEXAFS spectra of nonactivated (black), 3-hour activated (red), and 1-minute activated (blue curve) glassy carbon detected with photoelectrons in channeltron at Beamline 9.3.2 at the Advanced Light Source. (b) Sample current NEXAFS spectra of graphite (gray), not activated (black curve) and 4-hour activated (orange curve) glassy carbon.

vated glassy carbon. The spectra in the left panel were recorded with the highly surface sensitive electron yield mode where photoelectrons are detected by the Channeltron.

The sample activated for only one minute has a very small graphite peak at around 285 eV, whereas the nonactivated sample (zero minutes activated) has a very strong such graphite peak. It is striking that after 1 minute of activation, the surface of glassy carbon develops a molecular structure, which neither reminiscent of the bulk glassy carbon nor reminiscent of the film glassy carbon. This material has the lowest X-ray density and the smallest SAXS exponent of decay. The right panel shows the less surface sensitivity sample current NEXAFS spectra. We notice substantial differences from graphite, but the graphite-like character of glassy carbon is obvious for the nonactivated and the highly activated glassy carbon. Other than the graphite, glassy carbon has a strong peak from aliphatic carbon at around 288 eV. The 4 hours activated glassy carbon did not contain any noticeable amount of bulk glassy carbon anymore, but it has a relatively large fraction of aliphatic carbon, as evidenced from the strong peak at around 288 eV.

8.6 "OpMetBat": A battery metrology project

Metrology is the science of measuring. It is an independent scientific field, but closely related with the physical sciences such as physics and chemistry. It is the science of how to measure, and how to measure correctly and with accuracy and precision. It is the science of tracing back measurements of systems to basic nature constants.

When you apply statistical methods to your data analysis, this is already a metrology procedure. When you convert technical quantities to scientific ones, like absolute units, this fits in the field of metrology. When you meet the technicians and engineers in a machine shop and show them your sketch of a device or instrument, which they should build for you. A metal plate may contain a hole, an X-ray window, or aperture.

The technician will likely ask you, what is the tolerance for that hole. How accurate do they have to drill—with millimeter precision, or with micrometer precision, or even higher precision? Often, as a scientist, you do not have such issues on your radar in the beginning.

The OpMetBat is a joint research project within the European Partnership on Metrology. OpMetBat stands for "Operando metrology for energy storage materials." The idea of this project is to be able to analyze (better, actually, to diagnose) the state of charge and "health" of a lithium ion battery with a simple impedance measurement and maybe a simple "X-ray shot." This diagnostic method should be applied, e. g., to battery manufacturing assembly lines, or as on-board diagnostics in battery-powered vehicles or stationary batteries.

The principal partners in this project are the national metrology institutes. The project coordinator is Dr. Burkhard Beckhoff from the Physikalische Technische Bundesanstalt (PTB), the German metrology institute. Another leading metrology institute in the project is the National Physics Laboratory (NPL) in the United Kingdom. Note that OpMetBat is a "European" project, funded by the European Commission under the Green Deal Programme, with project number 21GRD01. The term Green Deal stands for the transformational change of the economy and society toward zero carbon dioxide emission by global population through abandoning of the fossil fuels—by the year 2050 [Krämer 2020].

The project partners, as already indicated, are not only from countries, which are members of the European Union; see Table 8.3. NPL from the United Kingdom, Empa from Switzerland, and Mersin Universty from Turkey are contributing and participating guests, so to speak. They do not get funding from the European Union budget, unlike the partners from EU member countries. Instead, these countries may provide their own funding for the partner. The Swiss Confederation has provided a back-up plan for its Swiss research institutes and universities that allows them to receive funding from the Swiss Government so that we can participate in EU projects.[7]

The OpMetBat project is motivated by the "European Green Deal" target that Europe shall be free of CO_2 emissions by 2050. More over, the automobiles shall be emission free by the year 2035. It is therefore believed that electrification of the automotive sector away from fossil fuels is the right way to meet this goal. In this respect are batteries as the principal electric storage devices. The search for better performing batteries is an ongoing quest. Performance includes also extended lifetime and cycling stability. There is a need for what is termed "traceable" analytical methods to find the quantitative relation between structure, function and degradation. This shall make us able to make better batteries by intelligent design, and not by empirical serendipity.

According to the International Organization of Legal Metrology [Metrology 2007], metrological traceability means *property of a measurement result whereby the result can*

---

7 Swiss transitional measures for the Horizon package (2021–2027).

Table 8.3: Partners in the OpMetBat project. Türkiye, United Kingdom, and Switzerland are not EU member states. They are participating and they are funded by their national funding schemes.

Number of components	Partner	Country
PTB	Physikalisch Technische Bundesanstalt	Germany
CEA	Commissariat à l'énergie atomique et aux énergies alternatives	France
CMI	Czech Metrology Institute	Czech Republic
INRIM	Istituto Nazionale di Ricerca Metrologica	Italia
HZB	Helmholtz Zentrum Berlin	Germany
JSI	Institut "Jožef Stefan"	Slovenia
MEU	Mersin	Türkiye
UNG	University of Nova Gorica	Slovenia
UNIROMA1	Sapienza University di Roma	Italia
WWU	Westfälische Wilhelms-Universität Münster	Germany
OG	OptiGraph GmbH	Germany
ELECTRO	0.001	United Kingdom
Empa	Eidgenössische Materialprüfungs- und Forschungsanstalt	Switzerland
NPL	National Physics Laboratory	United Kingdom

be related to a reference through a documented unbroken chain of calibrations, each contributing to the measurement uncertainty.

We learn in this book how important it is for understanding the material that we know as the elemental composition, the oxidation states, and molecular coordination of elements, nature of bonds between ions, and so on. This all can be studied and quantified very well with synchrotron based X-ray methods. The aim of the OpMetBat project is to carry out these studies by the combination of electrochemical and X-ray techniques in real time during operation of the battery. In the past, degradation was investigated by monitoring the state of charge (SoC) via the current and voltage transients. But structural studies were carried out *post mortem*, which means the battery cells were disassembled and inspected after they were considered or diagnosed "dead" after operation, based on the electrochemical performance data.

However, an accurate analysis of the materials of a dead cell is difficult when disassembling the cell and exposing it to an environment. More sophisticated technology is required to do the structural studies of the cell while the cell is still fully sealed and assembled and the materials in their expected and designed environment, such as in contact with the electrolyte. This is the *in situ* situation. It is even better when one can monitor the structural changes while such cell is in true device operation, which is called *operando*.

These very advanced *operando* synchrotron methods are not necessary ready yet for traceability measurements as defined by the aforementioned International Organization of Legal Metrology. Also, there is an interest in hybrid *operando* methods, where a variety of different measurands are simultaneously probed during battery operation.

For this to become real, novel battery cell designs are necessary, which can satisfy the requirements for electroanalytical (electric feedthroughs, for example) and X-ray optical (thin windows, for example) experiments. A robust and validated *operando* metrology framework shall be developed and transferred to manufacturers in the battery industry.

While the focus is on X-ray methods, particularly those of synchrotron X-rays, we do apply also neutron methods for structure determination, specifically neutron diffraction. The reason for this is the following: lithium batteries contain lithium as the key chemical element and player, apparently. But lithium is an element with three electrons only. The scattering cross-section of lithium for X-rays is relatively poor because, while X-rays strongly interact (and thus are scattered) with electrons, lithium has only few, only three of them per atom. It is by default difficult therefore to detect and study the light elements with X-ray methods.

The scattering cross-sections of the elements with respect to interactions with neutrons is very different and highly diverse. It is possible to enhance the scattering contrast by selecting particular isotopes of the elements. For example, coherent scattering such as neutron diffraction works very well for hydrogen studies, when deuterium is used as isotope. Whereas quasi-elastic neutron scattering (QENS) is incoherent, there it is better to use hydrogen as an isotope.

For the OpMetBat project, it is necessary to carry out DFT calculations in order to theoretically determine the electronic structure of electrode materials, and then simulate X-ray spectra and compare them with experimental spectra. For the DFT calculations, the geometric positions of the atoms in the unit cell are necessary input data. Determination of these positions is the job of the X-ray crystallographers. But the positions of lithium atoms or ions is not so easy because they are hardly visible with X-rays for the aforementioned reasons. Neutron diffraction, however, gives very precise data. It is also easier with neutrons to determine the positions of the oxygen, or even their vacancies, based on the Rietveld refinement.

I had therefore suggested to carry out complementary neutron diffraction, which we soon did at ISIS in the United Kingdom, and SINQ at PSI in Switzerland, as soon we had beam time available.

Back to the original objectives of the project, one of them is establishing a good practice guide for current and emerging methods on *operando* spectroscopy, including also some assessment of radiation damage. A second objective is the development of novel dynamic electrochemical methods, which should be combined with other spectroscopy methods. The term "dynamic" here means that fast time dependent electrochemical methods shall be used, particularly impedance spectroscopy, and maybe also switched on/off during experiments. The former operates in the frequency domain, whereas the latter operates in the time domain. Both domains are linked via Fourier transformation.

Like most projects, also OpMEtBat is organized in work packages. The table (Table 8.4) lists all work packages of the project.

Work Packages (WP) in the OpMetBat project.

Work Package	Description
WP1	Traceable *ex situ* characterization of high-capacity energy storage materials
WP2	Establishing good practice guide for current operando spectroscopy and diffraction methods
WP3	Development of novel dynamic electrochemical analysis approaches for combination with operando spectroscopy and dimensional metrology
WP4	Development of novel operando instrumentation and hybrid methodologies for multiparameter characterization
WP5	Creating impact
WP6	Management and coordination

9 Correlation of electronic structure and conductivity

Comprehension and control of transport properties are essential for the functionality and integrity of electrochemical energy storage and conversion devices. Such devices, as we know, include batteries, fuel cells, electrolyzers, solar cells, supercapacitors, photo electrochemical cells, and so on. Transport properties in this context include the electric charge transport by electrons, electron holes, ions, such as protons, oxygen, but also oxygen vacancies, and polarons. It also includes mass transport such as fluids and liquids (keyword multiphase flow), but also ions. So, here we may have a coupling of mass transport with electric transport. The multiphase flow may be coupled with heat transfer and other radiative transfer, including optical (light) transport.

All these transport properties depend on the structure of the involved materials. However, structure is not a well-defined term. The various readers of this book may have a different perception of what is typically meant by structure. Structure may have a different meaning to a microscopist compared to a spectroscopist or someone who does neutron scattering. I consider structure, where a material is concerned, as something multiscale and multidimensional, which starts at the atomic scale and extends to the design of components and architecture of devices.

This part of the book deals only with the electronic structure of metal oxides as far as the electronic conductivity is concerned. We can determine this with X-ray and electron spectroscopy. To some extent, we can do this also for the ionic conductivity in solid electrolytes. The mass transport by fluids and liquids is determined, e. g., in porous electrodes, by the microstructure. We have treated the microstructure of porous media with small-angle scattering. It is possible to mathematically model the transport through porous media with software packages. However, this topic is beyond the scope of this book.

The device architecture, if sufficiently small, can be interrogated with microscopy and imaging, including damages and larger defects. The mass transfer depends on the pore topology in porous electrodes. This can be probed by small-angle X-ray and neutron scattering, as we have learned. With so-called multiscale physics, modeling engineers can calculate heat and mass transfer of a given component or material. The experimental data from small angle scattering are valuable input for such calculations.

The conductivity of materials is the most important property when it is used as an electrode or as an electrolyte, or as an insulator. For electrodes, the electronic conductivity is the relevant one. The conductivity of metal electrodes is well established and typically part of the curriculum in undergraduate physics courses. The conductivity is provided by the structure of the materials, and the charge carriers are electrons, holes, ions, and vacancies.

Let us consider the conductivity of compounds such as metal oxides, as in batteries and fuel cell electrodes. We recall from Chapter 1 that their conductivities may be of po-

https://doi.org/10.1515/9783110794038-009

laronic nature [Jung 2000], with small polaron hopping relation $\sigma T \sim \sigma_0 \exp(-W_H/k_B T)$ with hopping energy W_H.

For simplicity, I will call this the polaron activation energy. Figure 9.1 shows a density contour plot of the polaron activation energies of the conductivity of $La_{(1-x)}Sr_x$ $Fe_{(1-y)}Ni_yO_{3-\delta}$ pellets as determined from temperature-dependent four-point DC measurements. We have not investigated the entire stoichiometry range of this material because it is not a single phase for all (x,y) combinations. In the probed range, the highest activation energies are centered in $x = 0.25$ for Sr and $y = 0.1$ for y.

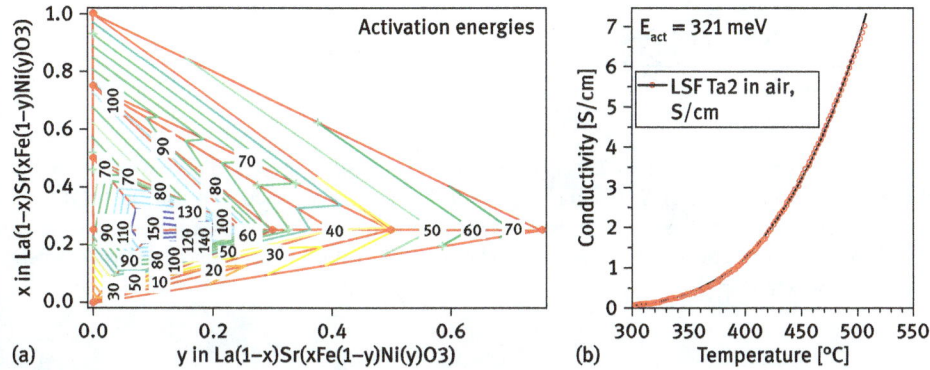

Figure 9.1: (a) Density contour plot for the polaron activation energies [meV] of LaSrFeNi oxide as a function of the Sr-content x and Ni-content y. (b) Four-point DC conductivity of $LaSrFeTa_{0.2}O_3$ in air atmosphere. The solid line is a fit to an exponential.

The electric conductivity of materials is primarily based on their electronic structure. In particular, the density of states (DOS) at the Fermi level (Fermi energy E_F) determines not only the electric properties but also the magnetic and optical properties. The Zaanen–Sawatzky–Allen (ZSA) theory has addressed the electronic band gaps and electronic structure of transition metal compounds [Zaanen 1985]. It describes the dependence of the conductivity (band) gap and the nature of electron and hole states on the Coulomb energy U and for transition-metal compounds. Their concept is summarized in a phase diagram (Figure 9.2), where conductivity regions are plotted along the coordinates of potential energy U (d-d Coulomb and exchange interaction), charge transfer T, and the hybridization interaction δ.

On the basis of experimental data and band structure modeling, it is possible to measure or estimate the necessary parameters for the ZSA scheme. This can be done with X-ray spectroscopy and DFT calculations at an almost arbitrary accuracy. Although the ZSA scheme is a standard conceptual tool for condensed matter physicists, it is not being used by electrochemists and device engineers. This is mostly because the three parameters δ,U,T cannot be easy determined and because they are not standard quantities in energy conversion and storage technology. A simpler alternative, at least for the

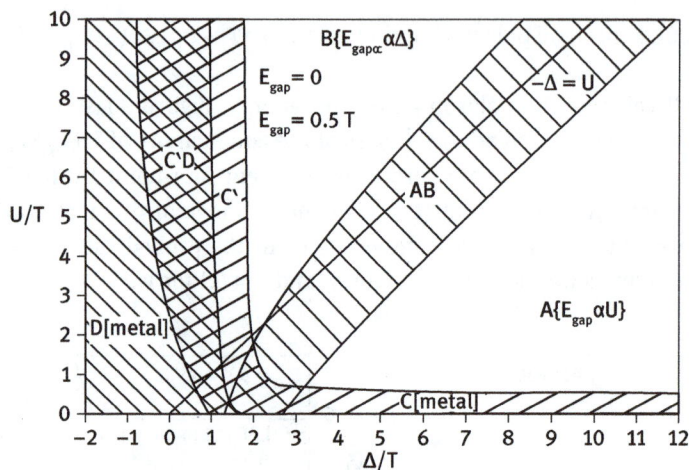

Figure 9.2: Zaanen–Sawatzky–Allen (ZSA) diagram exhibiting the regions of conducting and insulating phases. The thick line separates the semiconductors from the metals type conductivity [Zaanen 1985]. Reprinted figure with permission from Zaanen J., Sawatzky G. A., Allen J. W.: Band gaps and electronic structure of transition-metal compounds. *Physical Review Letters* 1985, 55:418–421. Copyright (1985) by the American Physical Society.

ABO_3-type perovskite materials, is the Nakamura map, which requires only the valence states and the ionic radii of the A and B site elements. We have exercised this concept for the system PrSrMnIn oxide, where we saw that Mn had varying oxidation state and Pr had virtually constant oxidation state, as was determined with XANES (see Chapter 2.1, Figure 2.1).

In the last chapter of this book, I present several studies where the variation and fine structure of X-ray spectra shows peculiarities that are reflected in the electric conductivity data. This is very important and suitable because this means that one can, in principle, determine the electric properties of a material, a component, and an electrode of a device without electrically probing the sample material. There may be occasions and instances where this is quite suitable when direct electric measurements are not possible. Moreover, these X-ray spectroscopy experiments can be considered noninvasive to a large extent.[1] We have seen how even very complex electrochemical processes can be probed *operando* and *in situ*.

When XPS experiments are carried out with the sample under a DC electric bias, it is possible to observe spectral shifts, which allow information on the Fermi level pinning. This so-called surface charge spectroscopy allows the derivation of conductivity

---

1 One early idea in the OpMetBat project was that the battery could be diagnosed without electric contact, when it was possible to use X-rays as probe. Note that X-rays are not necessary noninvasive. They can cause radiation damage.

information to some extent. The charging potential associated with, e. g., a local impurity in the sample can be used to estimate the (spatial?) depth location of the impurity. The degree of charging provides also information on the insulating properties of a dielectric layer in the electrode assembly [Lau 1989]. One practical reference can be found by Hagai Cohen [Cohen 2004].

9.1 Hole conductivity in SOFC cathodes

We recall in Chapter 5.4 where we investigated the hole doping of iron perovskites with oxygen ligand spectroscopy via the NEXAFS data. Figure 9.3 shows such spectrum for $La_{0.5}Sr_{0.5}FeO_3$. The pre-edge of the oxygen spectra shows a nice "hole" peak, which will become bigger with increasing substitution of the La^{3+} by Sr^{2+} in the parent compound $LaFeO_3$. This is an insulator, and thus certainly not suited as an electrode material [Braun 2009a, Braun 2011a, Braun 2012d]. But by doping or substitution, one can make it electronically conducting. The increase in Sr^{2+} on the La^{3+} sites forces the Fe^{3+} into a nominal Fe^{4+} valence, which is spectroscopically interpreted as $Fe3d^5L$, with the conducting ligand hole L. The very sharp peak at 532 eV in Figure 9.3 is not from the lanthanum strontium ferrite (LSF). Rather, it is the oxygen signature from the conducting sticky carbon tape underneath, which was used to fix the powder particles for the NEXAFS experiment.

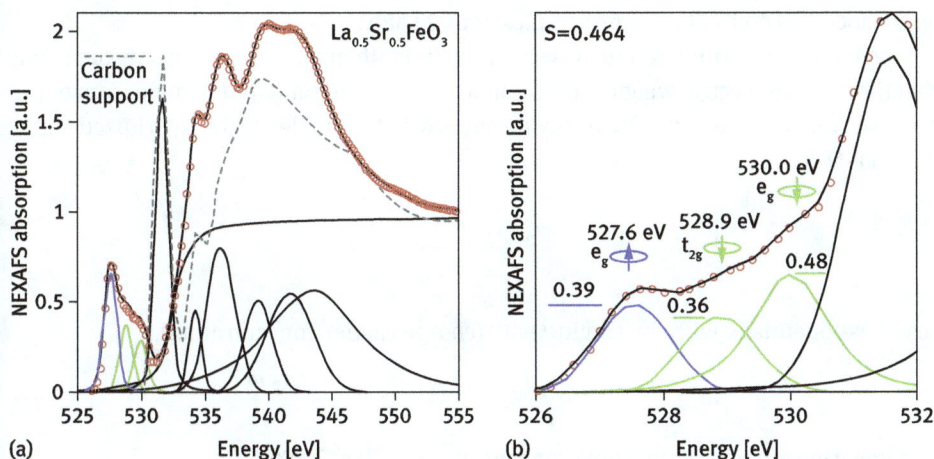

Figure 9.3: (a) Oxygen NEXAFS spectrum (red open symbols) of $La_{0.5}Sr_{0.5}FeO_3$ measured in UHV at 300 K with sample current detection mode. The spectrum is deconvoluted into Voigt functions. The blue peak is the hole doping peak. The two green peaks is the e_g-t_{2g} doublet from Fe 3d–O 2p hybridized states. The dotted spectrum is the oxygen signature of the sticky conducting carbon tape substrate underneath the powder sample. (b) Part of spectrum from 20% Ti substituted LSF with peak positions and indicated peak height.

This may sound counterintuitive, as carbon has no oxygen, and thus no oxygen signature in its spectrum. However, the carbon tape contains oxygen groups. This is a very illustrative example for why a researcher should pay attention to all routines in the experiments and also perform experiments and tests, which for a normal mind would make no sense. The right panel of Figure 9.3 shows the original spectrum with such carbon tape peak at around 532 eV highlighted by deconvolution. Once the peak is deconvoluted and included in the least square fit of the entire spectrum, the peak position, height, and width can be used to draw the peak in a separate file and subtract it from the original experimental spectrum. This subtraction was done for the left panel of Figure 9.3, and the peak was plotted into the panel for demonstration purposes.[2]

In particular, the estimation of the spectral weight of the green peak at 530 eV is influenced by the carbon tape peak, which is the reason why it was necessary to remove this contribution by least square fitting. The actual range of interest is the blue and green indicates peaks.

We have determined the relative spectral weight of the hold doping peak and the doublet for a number of iron perovskites for which the necessary oxygen NEXAFS spectra were available [Braun 2009a]. When we plot their four-point DC conductivity versus this relative spectral weight—let us call it S-ratio—then we find a linear relationship in a semilog plot. The conductivity of the samples therefore scales exponentially with the S-ratio. When we compare the ratio with the formal electron hole concentration, this is the ratio of Fe^{4+} ions over Fe^{3+} ions, and we find a linear relation. Both relations are shown in the left panels of Figure 9.4 [Braun 2009a]. Tables 9.1 and 9.2 summarize the stoichiometry, gas processing conditions, conductivity data, and structure data as well as the spectroscopic data for the samples of that study.

As we try to formulate our observations in mathematical equations, it turns out that the relative spectral weight S of this hole peak, when compared with the sum of the spectral weight of the typically nicely developed e_g-t_{2g} doublet of the hybridized states of Fe 3d–O 2p,

$$S = \frac{e_g^\uparrow}{e_g^\downarrow + t_{2g}^\downarrow}$$

scales exponentially with the conductivity (charge carrier concentration),

$$\sigma \sim \sigma_0 \cdot \exp(U \cdot \tau)$$

and linearly with the relative hole concentration $\tau = [Fe^{4+}]/[Fe^{3+}]$,

$$S(\tau) = S_0 + a \cdot \tau.$$

---

2 I use the WinXAS program for peak deconvolution [Ressler 1998].

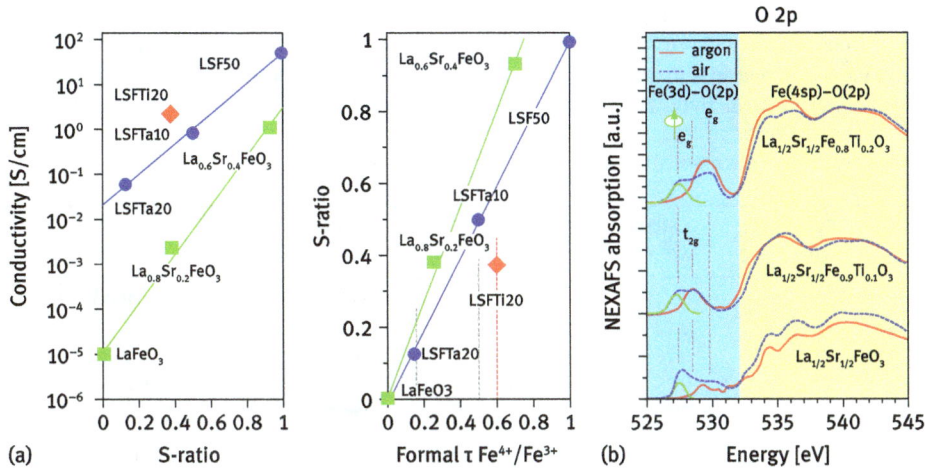

Figure 9.4: (Left) The conductivity or iron perovskite samples (LaSrFe oxide doped with Ti, Ta) plotted versus the relative spectral weight (S-ratio) from oxygen NEXAFS spectra. Note the semilogarithmic abscissa, which yields an exponential relationship between conductivity and relative hole weight. (Middle) The S-ratio scales linear with the formal hole concentration [Fe4+]/[Fe3+]. Reprinted from [Braun 2009a] [Braun A., Bayraktar D., Erat S., Harvey A. S., Beckel D., Purton J. A., Holtappels P., Gauckler L. J., Graule T.: Pre-edges in oxygen (1s) X-ray absorption spectra: a spectral indicator for electron hole depletion and transport blocking in iron perovskites. *Applied Physics Letters* 2009, 94.], with the permission of AIP Publishing. (Right) Oxygen NEXAFS spectra of LaSrFe oxide heat treated in argon (red spectra) and in air (blue spectra). The green peak is obtained from least square fitting and denotes the oxygen hole peak. Reprinted with permission from [Braun 2012d] (Braun A., Erat S., Bayraktar D., Harvey A., Graule T.: Electronic origin of conductivity changes and isothermal expansion of Ta- and Ti-substituted $La_{1/2}Sr_{1/2}$Fe-oxide in oxidative and reducing atmosphere. *Chemistry of Materials* 2012, 24:1529–1535). Copyright (2012) American Chemical Society.

Table 9.1: Oxygen deficiency in air at 1173 K, formal hole concentration $\tau = Fe^{4+}/Fe^{3+}$, experimental hole concentration in air at 1173 K, conductivity at 300 K, peak ratio L_2/L_3, and branching ratio $L_3/(L_2 + L_3)$.

Sample	τ_{form}	Gas	δ_{grav} @ 1173 K	τ_{grav} at 1173 K	σ [S/cm] E_a [meV]	L_2/L_3	$L_3/(L_2+L_3)$
LSF50	1.00	Argon			0.075	0.24	0.81
LSF50	1.00	Air	0.078	0.52	50 / 180	0.29	0.67
LSFTi20	0.60	Argon			–	0.23	0.81
LSFTi20	0.60	Air	0.063	0.28	2.28 / 262	0.27	0.68
LSFTa10	0.50	Argon			0.017	0.24	0.81
LSFTa10	0.50	Air	0.058	0.26	0.82 / 342	0.28	0.66

LSF is certainly not the only hole conducting material. It was actually the high-temperature superconducting cuprate materials that were of interest for hole doping and where the holes were found in oxygen NEXAFS spectra [Hong 2002, Learmonth 2007, Pellegrin 1996].

Table 9.2: Relative hole concentrations τ from atomic multiplet simulation of spectra heated in air (10Dq = 1.85 eV) and in argon (10Dq = 1.70 eV). Data were obtained from the NEXAFS spectra least square fitting of the corresponding Fe 2p spectra to the right panel of Figure 9.4. Reprinted with permission from [Braun 2012d] (Braun A., Erat S., Bayraktar D., Harvey A., Graule T.: Electronic origin of conductivity changes and isothermal expansion of Ta- and Ti-substituted $La_{1/2}Sr_{1/2}$Fe-oxide in oxidative and reducing atmosphere. *Chemistry of Materials* 2012, 24:1529–1535). Copyright (2012) American Chemical Society.

	Calculated with $\delta = 0$	Calculated δ@ 1173 K	10Dq = 1.85 eV	10DQ = 1.70 eV	
	Fe^{4+}/Fe^{3+}	Fe^{4+}/Fe^{3+} in air	Fe^{4+}/Fe^{3+} in air	Fe^{4+}/Fe^{3+} in argon	Oxidation change
LSF 50	1	0.52	1	1	stable
LSFTi20	0.56	0.28	0.67	0.43	Reduced in argon
LSFTa10	0.43	0.25	0.67	0.43	Reduced in argon

Another example is the O 2p hole doping in $PrBaCo_2O_{6-\delta}$ (PBCO). This material has shown some peculiar magnetic behavior depending on oxygen stoichiometry [Ganorkar 2011]. We received the samples from Professor K. Conder and Dr. E. Pomjakushina at PSI. Substoichiometric praseodymium barium cobalt oxide ($PrBaCo_2O_{6-\delta}$) has been subject to soft X-ray absorption spectroscopy at the oxygen K-edge and cobalt L-edge. The oxygen near edge X-ray absorption fine structure spectra show an additional hole doping peak before the doublet from O 2p-Co3d hybridized states, the relative spectral weight of which corresponds to the conductivity, lattice constant, and Curie temperature. The polycrystalline $PrBaCo_2O_{6-\delta}$ was prepared by a solid-state reaction. Starting materials of Pr_6O_{11}, Co_3O_4, and $BaCO_3$ with 99.99 % purity were mixed and ground, followed by heat treatment at 1000 °C–1150 °C in air during at least 100 h with several intermediate grindings followed by oxidation in 1.5 bar of oxygen at 800 °C. Oxygen content was measured by iodometric titration.

Then an alumina crucible with powdered sample of precisely known weight and oxygen stoichiometry was placed in a quartz ampoule together with a precisely weighed copper powder in an another crucible. The ampoule was then sealed under vacuum and heated up to 850 °C during 10 h followed by a slow cooling (5 K/h) in order to ensure a homogenous oxygen distribution. For all samples, phase purity was confirmed with an X-ray diffractometer (SIEMENS D500). NEXAFS spectra were recorded at Beamline 9.3.2 of the Advanced Light Source [Grass 2010, Hussain 1996], Lawrence Berkeley National Laboratory. The energy resolution of the beamline is $E/\delta E = 3000$. During measurements, the base pressure of the main chamber was maintained at 10^{-9} Torr. The spectra were recorded with an energy resolution of around 0.1 eV and in an ultrahigh vacuum of 10^{-9} Torr.

Figure 9.5 shows the oxygen 1s NEXAFS spectra of the three samples with different oxygen deficiency. Close inspection of the spectrum from the sample with the least oxygen deficiency, i. e., $PrBaCo_2O_{5.69}$, shows that the pre-edge needs three peaks for satisfactory deconvolution of the pre-edge structure. It is well known that the oxygen spec-

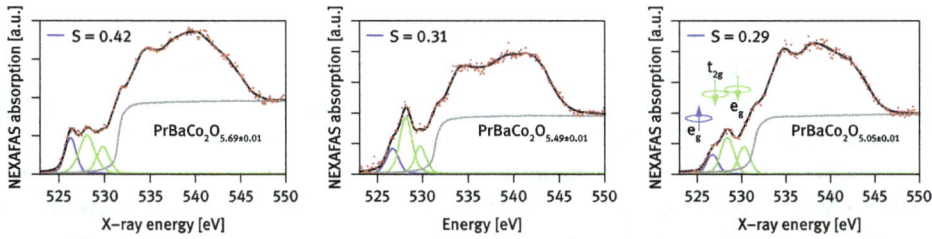

Figure 9.5: Oxygen 1s NEXAFS spectra of three different PrBaCo oxide powder samples with deconvoluted relevant peaks for the determination of the relative spectral weight S. The blue peaks denote the hole states.

tra from materials with perovskite structure often shown an additional peak before the doublet, which is assigned to doped electron holes. This hole peak is located at 526.3 eV in Figure 9.5. The doublet, which often is formed in 3d metal oxides from hybridized O 2p–Me 3d states near the Fermi energy, has its peaks at 528.15 eV and 529.7 eV.

In analogy to previous quantitative oxygen NEXAFS study on iron perovskites, we determine the relative spectral weight of the electron hole doping peak at 526 eV with respect to the e_g-t_{2g} doublet at 528 and 530 eV, respectively (S-ratio). Figure 9.6 shows the S-ratio for the three samples and the lattice cell parameters from a wider range of samples. The data were taken from Podlesnyak et al. [Podlesnyak 2005, Podlesnyak 2006]. Along the change of oxygen deficiency, we observe the change of the valence of the central cation from Co^{3+} to Co^{4+}. We see that the higher oxidation state cation Co^{4+} has also the higher oxygen hole peak and, therefore, the higher S-ratio. In the right panel of Figure 9.6, we compare the S-ratio with the conductivity of the samples. Both quantities show roughly the same trend.

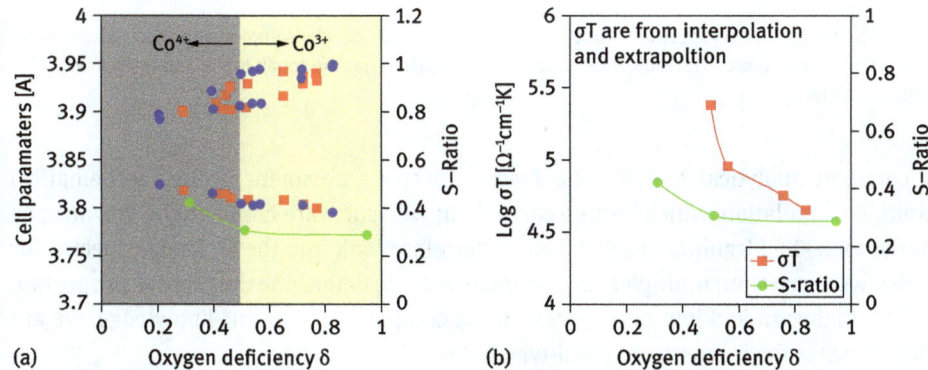

Figure 9.6: (a) Comparison of lattice parameters and relative spectral weight S for hole doping. Lattice parameters of $PrBaCo_2O_{5+\delta}$ as a function of oxygen content for the slowly cooled SQ (bullets) and rapidly quenched RQ (squares) compounds. The lines are guides to eyes. (b) Comparison of the S-ratio and the electric conductivity of PBCO at 300 K, latter obtained from extrapolation of other samples. The data are taken from [Podlesnyak 2006].

We see here again how the conductivity and, e. g., the lattice constants vary with the relative spectral weight of the holes. The oxygen deficiency here is the process parameter that controls these properties. The influence of the oxygen inhomogeneity on the magnetic properties of $RBaCo_2O_{5+\delta}$ was pointed out as well [Ganorkar 2011].

We have also measured the cobalt L-edges of the PBCO samples. Figure 9.7 shows the well-resolved Co 2p multiplet, which is split in the conjugated t_{2g} and e_g orbital symmetry peaks. The red spectrum is from the sample with the lowest oxidation state, and the green spectrum from the sample with the highest oxidation state. At least nominally, this should be the case. We have shifted the green spectrum on the intensity axis just to avoid overlapping of the curves. The higher oxidation state species should have a relatively higher Co $2p_{3/2}$ peak (high branching ratio), but we do not see this here. It is not so easy to make out relevant differences in the three spectra in Figure 9.7, whereas the hole structures in the oxygen pre-edges are very easy to make out.

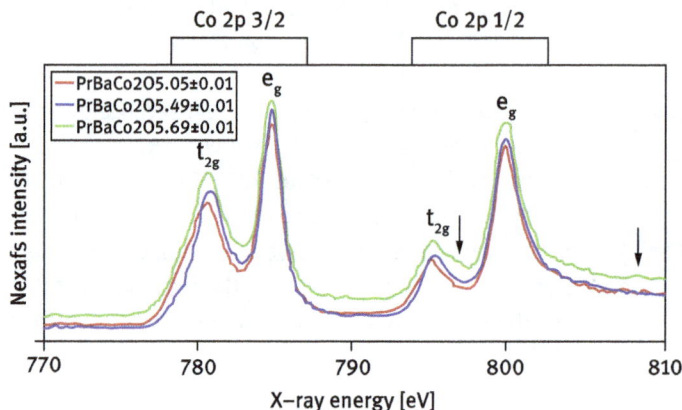

Figure 9.7: Co 2p (L-edge) spectra of three different PrBaCo oxide powder samples with spectral assignment of prominent peaks. The red spectrum with δ = 5.05±0.01 has the cobalt in the nominally lowest oxidation state.

A common analytical tool for the L-edge spectra assessment is the mathematical multiplet simulation, which is often carried out with software based on the Cowan code [De Groot 2008, Mcguinness 2007]. Let us therefore look into the Fe L-edge spectra for which we carried out multiplet simulation in order to determine the relative proportion of Fe^{3+} high spin and low spin species in the compound. With this knowledge, we are able to assess the influence on the d-type holes.

This approach was tested for the related system where Ni substitution was done on the B-site of the iron perovskite [Erat 2009]. Figure 9.8 shows the Fe L-edge spectra of $LaFeO_3$ and 25 % Ni substituted $LaFeO_3$ (spectra with blue symbols). We also calculated the spectra (black curve). The calculated spectra are formed by broadening of the discrete transitions with a Gaussian or Lorentzian, which account for the life-time broad-

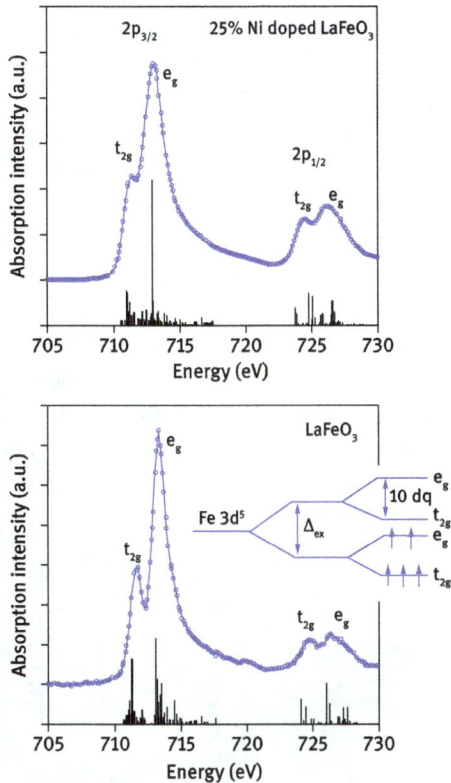

Figure 9.8: Comparison between experimental (blue line with open symbols) and simulated (black line, bottom) Fe2,3 absorption spectra for $LaFeO_3$ and $LaFe_{0.75}Ni_{0.25}O_3$. Inset shows the electronic configurations of HS Fe 3d5 in the ground state and modified after Abbate et al. [Abbate 1992]. Fe L edge split into L3 ($2p_{3/2}$) due to spin orbit coupling and additionally split into t_{2g} and e_g levels due to crystal field effect. Reprinted from [Erat 2010c] [Erat S., Braun A., Piamonteze C., Liu Z., Ovalle A., Schindler H., Graule T., Gauckler L. J.: Entanglement of charge transfer, hole doping, exchange interaction, and octahedron tilting angle and their influence on the conductivity of $La_{1-x}Sr_xFe_{0.75}Ni_{0.25}O_{3-\delta}$: A combination of X-ray spectroscopy and diffraction. *Journal of Applied Physics* 2010, 108], with the permission of AIP Publishing.

ening and instrument broadening. The similarity of the experimental spectra and the calculated spectra is in general overwhelming, given that the Cowan code is based on very simple quantum mechanical principles.

Analysis of the spectra by mathematically modeling them with multiplet simulation [De Groot 2008] suggests that the holes may be of p-type or d-type, depending on the extent of substitution. Here, we observe that the conductivity increase with increasing Sr substitution is paralleled until for Sr content from 0 % to 50 %. The formal hole concentration and S-ratio are quite parallel. Note that we do not derive quantities from the spectra when we calculate them. Rather, we assume a particular molecular environment, which includes the relevant chemical elements, the geometric coordination, bond lengths, and based on the corresponding atomic orbital the spectra are calculated, and then compared with the experimental ones. Step-by-step, the models then are improved until a very good visual fit is present between theory and experiment.

We can then use the L-edge spectra including simulation in order to check for the presence of 3d-type holes, which reside on the Fe, and compare those with the O 2p-type hole from the oxygen that we determined from the S-ratio in the oxygen NEXAFS spectra.

Figure 9.9 shows such comparison. The left panel shows the formal hole concentration Fe^{4+}/Fe^{3+} and the corresponding conductivity of the samples, plotted versus the relative Sr content in $La_{1-x}Sr_xFe_{0.75}Ni_{0.25}O_{3-\delta}$. The right panel shows the formal hole concentration for the O 2p holes, compared with the conductivity. The latter show a better quantitative correlation with the conductivity than the former ones. It appears therefore that the conductivity in this lanthanum strontium iron nickel oxide (LSFN) sample is governed by the O 2p holes.

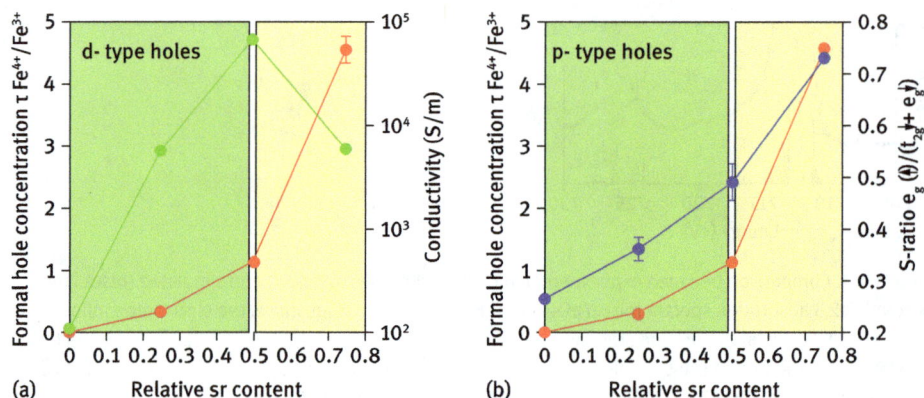

Figure 9.9: (a) Correlation of d-type holes derived from Fe(2p) and (b) p-type holes derived from O(1s) NEXAFS spectra with electrical conductivity of $La_{1-x}Sr_xFe_{0.75}Ni_{0.25}O_{3-\delta}$. These data were obtained by atomic multiplet simulation of experimental NEXAFS spectra. Reprinted from [Erat 2009] [Erat S., Braun A., Ovalle A., Piamonteze C., Liu Z., Graule T., Gauckler L. J.: Correlation of O (1s) and Fe (2p) near edge X-ray absorption fine structure spectra and electrical conductivity of $La_{1-x}Sr_xFe_{0.75}Ni_{0.25}O_{3-\delta}$. *Applied Physics Letters* 2009, 95.], with the permission of AIP Publishing.

A principally different approach is by linking the conductivity data with X-ray crystallography data. Structure refinement of X-ray or neutron diffraction data allows for very accurate determination of the atomic positions and sites, e. g., by using Rietveld refinement [Rietveld 1966, Rietveld 1967, Rietveld 1969, Rietveld 2010, Rietveld 2014]. Charge transfer in metal oxides is governed by exchange interactions, which in turn, are regulated by the overlap of metal orbitals and ligand orbitals.

One metric for perovskite type materials is the Goldschmidt tolerance factor [Goldschmidt 1926], which is determined by geometrical principles based on the ionic radii. r_A is the radius of the A-cation. r_B is the radius of the B-cation. r_0 is the radius of the anion, e. g., oxygen:

$$t = \frac{r_A + r_B}{\sqrt{2}(r_B + r_0)}$$

For LSFN, the tolerance factor t scales linear over the entire Sr substitution range, and it is slightly offset by a constant difference between high-spin and low-spin nickel Ni^{3+}. When we determine then the distance between Ni and O in the lattice, we notice a deviation from linearity with increasing Sr content. For $x > 0.5$, this distance remains almost linear. For a full assessment and understanding of the conductivity variation with stoichiometry, e. g., one needs to take such observations into account.

As I am showing in Figure 9.10, the A-site Coulomb potential decreases linear with increasing Sr content. This is in line with our expectation because La is 3+ and doped with lower charged Sr^{2+} having larger ionic radius. The distance between Fe/Ni–O decreases linear up to 50 % Sr doping, and for $0.50 < x \leq 1.0$, it approaches a constant value of around 1.925 Å. Therefore, the variation of the Fe/Ni–O distance deviates from the linear behavior of Z_A/r_A for $x > 0.50$, possibly because the system is not completely ionic but becomes more covalent [Erat 2010a, Erat 2010c].

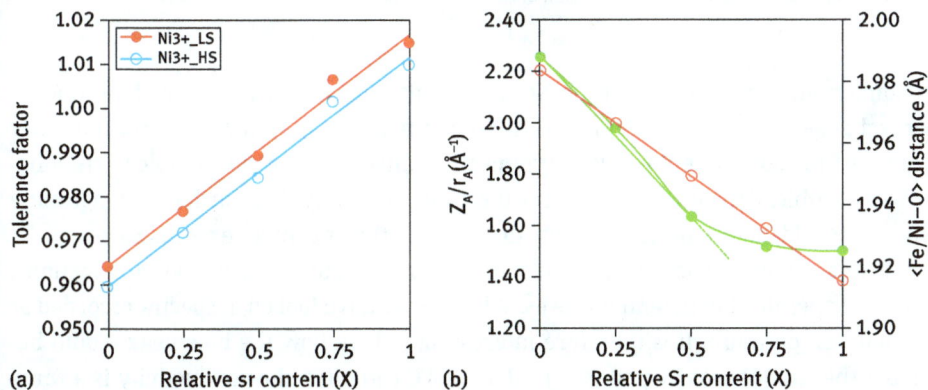

Figure 9.10: (a) Goldschmidt tolerance factor, and (b) A-site Coulomb potential and Fe/Ni–O distance of $La_{1-x}Sr_xFe_{0.75}Ni_{0.25}O_{3-\delta}$. Reprinted from [Erat 2010c] [Erat S., Braun A., Piamonteze C., Liu Z., Ovalle A., Schindler H., Graule T., Gauckler L. J.: Entanglement of charge transfer, hole doping, exchange interaction, and octahedron tilting angle and their influence on the conductivity of $La_{1-x}Sr_xFe_{0.75}Ni_{0.25}O_{3-\delta}$: A combination of X-ray spectroscopy and diffraction. *Journal of Applied Physics* 2010, 108], with the permission of AIP Publishing.

The perovskite materials are very good model systems for showing how the relative proportions of elements A, B, and oxygen have influence on the incline of octahedral and the effect on the superexchange angle, with concomitant orbital overlap and changes in the electronic transport properties.

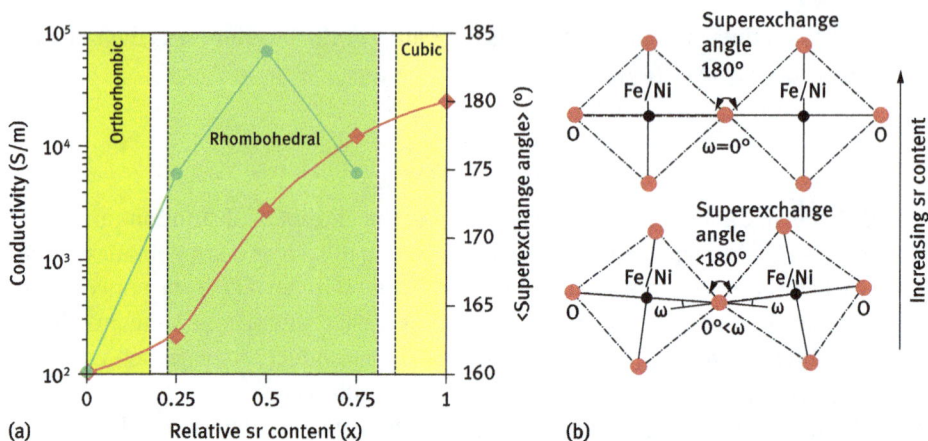

Figure 9.11: Electrical conductivity at 300 K and superexchange angle depending on Sr content. A fragment of the rhombohedral/orthorhombic symmetry after [24] with a schematic representation of the average superexchange angle, <Ni/Fe–O–Ni/Fe>, and the average tilting angle, <ω> which changes in to cubic symmetry with 180° superexchange angle and 0° tilting angle at high Sr doping. Reprinted from [Erat 2010c] [Erat S., Braun A., Piamonteze C., Liu Z., Ovalle A., Schindler H., Graule T., Gauckler L. J.: Entanglement of charge transfer, hole doping, exchange interaction, and octahedron tilting angle and their influence on the conductivity of $La_{1-x}Sr_xFe_{0.75}Ni_{0.25}O_{3-\delta}$: A combination of X-ray spectroscopy and diffraction. *Journal of Applied Physics* 2010, 108], with the permission of AIP Publishing.

At some point, this causes also phase transitions. Figure 9.11 illustrates how the superexchange angle, as determined from X-ray diffraction, e. g., increases continuously from 160 ° to 180 ° while the crystal symmetry changes from orthorhombic to rhombohedral to cubic. Concomitant increases the conductivity by three orders of magnitude when Sr is added, but the conductivity drops when the Sr content exceeds 50 %.

In this section, we have dealt with materials relevant for SOFC cathode applications, but these operate at high temperatures. Whereas we have looked at spectra recorded at ambient temperature. It is therefore interesting to look how the hole peak would behave if the sample was heated. After all, in SOFC cathodes, the conductivity is a function of temperature. Figure 9.12 shows the oxygen NEXAFS spectrum of $LaFe_{3/4}Ni_{1/4}O_3$ recorded at 500 °C (773 K), deconvoluted in prominent peaks. We have also recorded the spectra for lower temperatures down to ambient conditions, in UHV, at BESSY II. We recorded the spectra in the electron yield (sample current) and in the photon yield (with a photodiode). At high temperatures, the photodiode sensed the temperature of the sample, and the spectroscopic signals became overshadowed from this effect. We found that the spectral weight of the hole peak increases with increasing temperature. The middle panel of Figure 9.12 shows the relative spectral weight S of the hole peak and the four-point DC conductivity as a function of temperature in Arrhenius presentation (1000/T). For the conductivity, as determined from the slope of the Arrhenius plot, the activation energy is 1760 meV, whereas the activation energy for the spectral weight S evo-

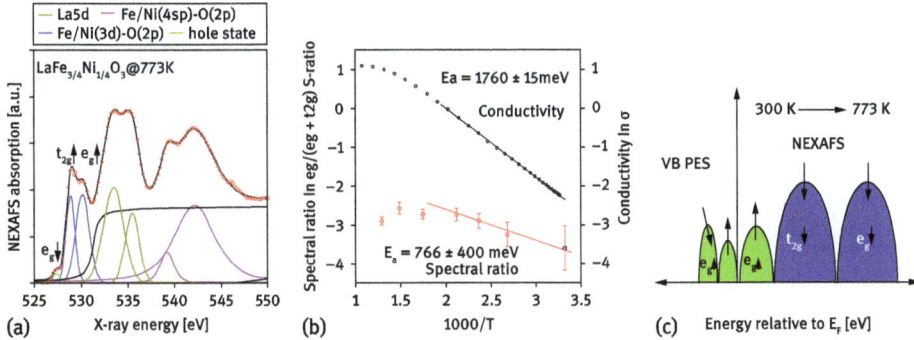

Figure 9.12: (a) O 1s NEXAFs spectrum of $LaFe_{3/4}Ni_{1/4}O_3$ recorded at BESSY II in UHV at 773 K. (b) Conductivity and spectral weight S of hole peak in Arrhenius presentation. (c). Schematic for electronic DOS as a function of annealing from 300 K to 773 K. Reproduced and reprinted from [Braun 2011d] [Braun A., Erat S., Ariffin A. K., Manzke R., Wadati H., Graule T., Gauckler L. J.: High temperature oxygen near edge X-ray absorption fine structure valence band spectra and conductivity of $LaFe_{3/4}Ni_{1/4}O_3$ from 300 to 773 K. *Applied Physics Letters* 2011c, 99.] with the permission of AIP Publishing.

lution is 766±400 meV. We would naively expect the activation energies to be the same, but this is obviously not the case. Maybe there are parasitic processes in the transport properties—a macroscopic quantity—which are not reflected at the molecular scale by the hole peaks [Erat 2010c].

We can use this temperature-dependent information for the sketch of the electronic DOS and how it is influenced by temperature. I have adopted this from Hiroki Wadati, who to the best of my knowledge is the first to sketch such temperature-dependent DOS [Wadati 2006]. The spectral weight of the t_{2g}–e_g doublet in the NEXAFS spectra is decreasing upon annealing, whereas the hole peak—closer to the Fermi level—increases. From Wadati, we know that for a comparable LSF material, the e_g orbital symmetry peak decreases with increasing temperature, whereas a small peak near E_F increases. Hence, occupied and unoccupied states near E_F are increasing with different extent with increasing T, at least for the system we have studied here.

This outcome is encouraging for further spectroscopy studies at higher temperatures. In the course of various conductivity studies on SOFC cathode materials, it was found that the conductivity characteristic had inhomogeneous behavior, which was particularly clear when the temperature-dependent conductivity was expressed in an Arrhenius plot, where steps and kinks and changes in the slope were observed. In the daily lab life, these were attributed typically to malfunctioning of the meter, with which the conductivity was measured. It turned out this was not true. There was another reason that resided in the material that causes these kinks and steps in the conductivity profile.

Figure 9.13 shows the conductivity of a single crystal slab with nominal stoichiometry $La_{0.9}Sr_{0.1}FeO_{3-\delta}$. In the temperature range from 300 to 1250 K, temperatures 357, 573, and 700 K show deviations from an "orderly" behavior in a step, a change in slope, and a change in sign of slope, respectively. We have recorded photoemission spectra in

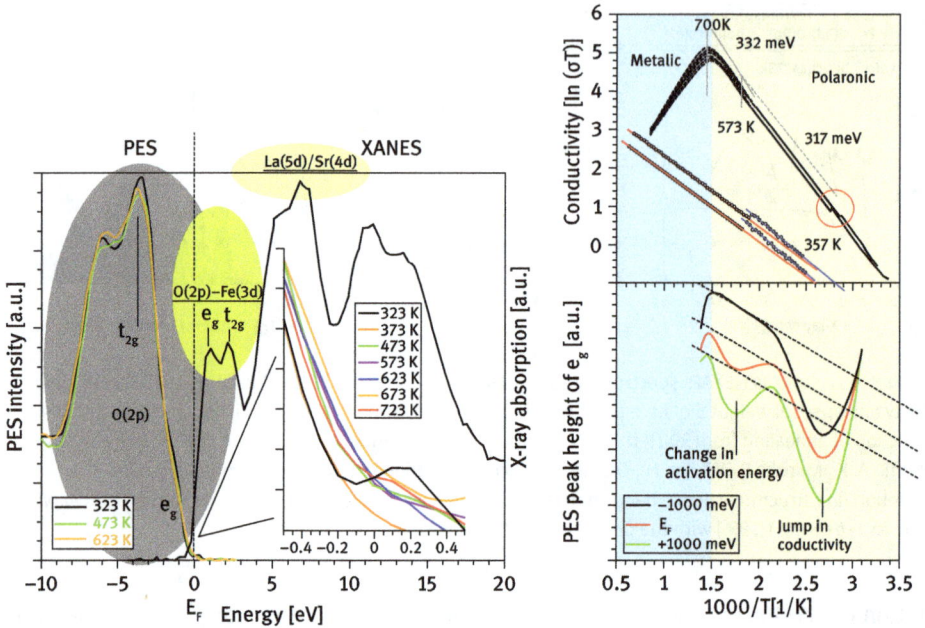

Figure 9.13: (Left) VB PES spectra of $La_{0.9}Sr_{0.1}FeO_{3-\delta}$ single crystal for 323, 473, and 673 K. The inset shows further spectra in smaller energy range for T up to 723 K. The middle panel shows clearly the redistribution of spectral weight from the region around −1 to −2 eV toward the Fermi level. (Right, top) The four-point DC conductivity in Arrhenius presentation. The data in the red circle are magnified and plotted at the lower left portion of the top panel for better discrimination (axes are not applicable for this portion). (Right, bottom) Variation of the spectral weight (PES height) at three prominent energy positions in the PES spectra, plotted versus reciprocal temperature. Stark changes are reflected in both spectral weight and conductivity plot. Reprinted from [Braun 2008d] [Braun A., Richter J., Harvey A. S., Erat S., Infortuna A., Frei A., Pomjakushina E., Mun B. S., Holtappels P., Vogt U., Conder K., Gauckler L. J., Graule T.: Electron hole-phonon interaction, correlation of structure, and conductivity in single crystal $La_{0.9}Sr_{0.1}FeO_{3-\delta}$. *Applied Physics Letters* 2008, 93.], with the permission of AIP Publishing.

the valence band region of this single crystal while it was heated in the UHV chamber. We identified minute changes that we could quantify after normalization of the spectra. Comparison of the spectra versus temperatures shows then that the characteristic changes in the conductivity data were paralleled by characteristic changes in the photoemission spectroscopy (PES) data.

If one carries out such high temperature studies, it can make a difference whether a bulk material like a thick slab or whether a thin film is investigated. In the next study, we investigated a thin PLD film and measured its conductivity and its PES valence band (VB) characteristic. In the case of very thin films, high temperature can initiate processes, which affect the conductivity considerably. This is demonstrated in the following study where a strained LaSrFeNi oxide film was heated and cooled, as shown in Figure 9.14. In this study, too, the conductivity and PES measurements were done in separate experiments [Braun 2009m].

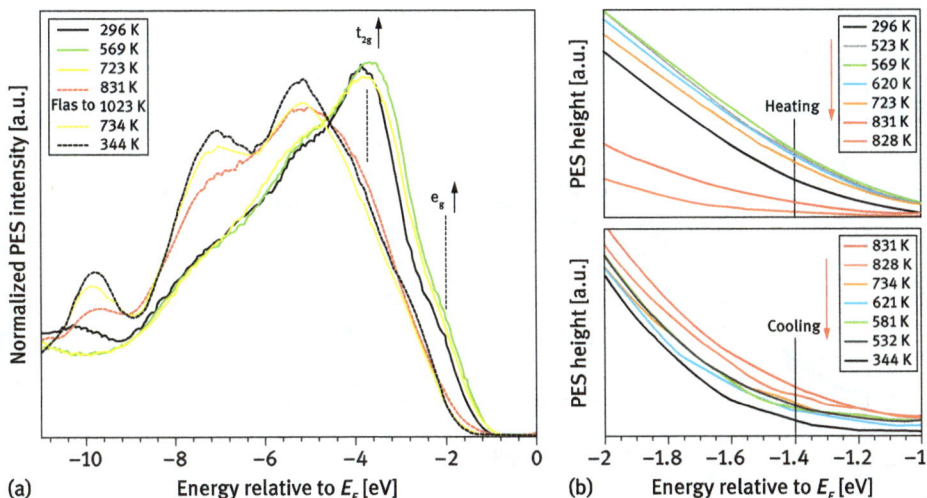

Figure 9.14: (a) VB PES spectra of a strained LSFN film deposited on SrTiO$_3$ and heated to 1023 K and cooled down again to ambient temperature. Between 723 and 831 K, a drastic change in spectral shape and a shift toward higher binding energies occur, revealing chemical changes in the film. (b) Magnified range near E_F showing the change of height during heating and cooling in UHV. Spectra recorded at SSRL. Reprinted from [Braun 2009m] [Braun A., Zhang X., Sun Y., Muller U., Liu Z., Erat S., Ari M., Grimmer H., Mao S., Graule T.: Correlation of high temperature X-ray photoemission valence band spectra and conductivity in strained LaSrFeNi oxide on SrTiO3(110). *Applied Physics Letters* 2009d, 95.], with the permission of AIP Publishing.

The shift of the spectrum toward higher binding energy indicates that the electronic conductivity will decrease (Figure 9.14). This is possibly because of interdiffusion of Ti from the SrTiO$_3$ substrate into the LSFN film. Possibly, the Ti^{4+} "eats up" the electron holes that otherwise provide the high conductivity in LSF-based electrodes. We remember from previous studies that substitution by Ti^{4+} and Ta^{5+} the hole concentration in LSF will be negatively affected [Braun 2009a, Braun 2011a, Braun 2012d]. However, we have not further tested this, so this remains a speculation.

The four-point DC conductivity (van der Pauw method) of the film during annealing and cooling is shown in Figure 9.15. The first conductivity scan is the curve with the green symbols. The conductivity increases with a relatively low slope and changes to sharper increase at around 750 K. The second scan (curve with red symbols) on the same film shows the same general profile, but now an oscillation at around 550 to 600 K, which is only shallow visible in the first scan. At 830 K, the conductivity shows a "jump" of about ½ unit in the plot and then continues the same profile like before. In the third scan, the previous mentioned oscillation turns into a sharp decrease, and the conductivity recovers in a plateau, which does not, however, recover its previous maximum value of around 3 units, and rather stays at 1 unit. For comparison, I have included the conductivity of a sintered slab, which had the nominally same stoichiometry. The max-

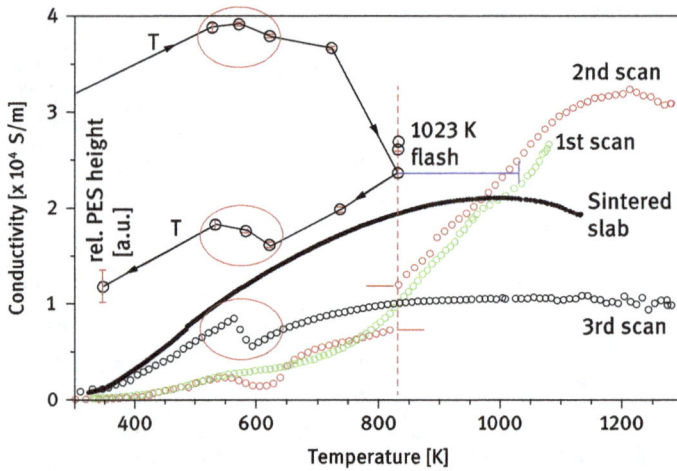

Figure 9.15: Electronic conductivity of thin strained LSFN film between 300 K and approximately 1200 K for three scans as obtained from the van der Pauw method (open red, green, and black symbols). The solid dark line is the conductivity of a single crystal slab for reference. The arrows connect the PES height measured from the VB spectra. Reprinted from [Braun 2009m] [Braun A., Zhang X., Sun Y., Muller U., Liu Z., Erat S., Ari M., Grimmer H., Mao S., Graule T.: Correlation of high temperature X-ray photoemission valence band spectra and conductivity in strained LaSrFeNi oxide on SrTiO3(110). *Applied Physics Letters* 2009d, 95.], with the permission of AIP Publishing.

imum conductivity of the slab at high temperatures is higher than that of the film after the third scan, but the profile versus T is similar.

On the left upper portion of Figure 9.15, the relative heights of the PES intensity at −1.4 eV are plotted. This energy was deliberately chosen as a reference because here the spectral changes were close enough to E_F and yet large enough for comparison.

Note that we have looked into one heating and cooling cycle of the strained LSFN film. The relative PES height (no axis provided) increases steadily from 300 K to around 600 K and then decreases again. At 830 K, we have "flash" annealed the film up to 1023 K for a short time. Upon cooling, the PES height steadily decreases. However, note the red circles drawn around the oscillation in the conductivity at 550 K and the similar behavior of the PES height in the heating and cooling half cycle. We have therefore a clear qualitative correlation between conductivity and spectral weight in the PES spectra in the course of annealing and cooling. This accounts also for the estimated chemical changes which go on in the film by annealing.

The aforementioned samples were related with the electronic structure and conductivity of iron perovskite materials for solid oxide fuel cell applications. We have shown the dependency of the conductivity from the composition and from the operation temperature, along with the corresponding spectroscopic response. We have done a similar spectroscopy study with Fe 2p-resonant PES [Erat 2010h].

9.2 SOFC chromium poisoning

The metal interconnect plates in SOFC stacks contain chromium that may form volatile chromium oxide species at high temperature, which then may diffuse and react with the cathode material, causing decreased conductivity with overall decrease of performance. This is known as chromium poisoning and typically attributed to microstructure changes or formation of new phases in the cathode material, particularly spinels with decreased conductivity. We investigated this effect with traditional solid-state ionics methods [Tsekouras 2014] but also with soft X-ray spectroscopy [Tsekouras 2015]. This was a collaboration with Professor D. D. Sarma at Indian Institute of Science in Bangalore and Dr. Kevin Prince at ELETTRA (Proposal No. 20130288), funded by the Indo Swiss Joint Research Program (ISJRP) [Braun 2011h]. Dr. George Tsekouras (Empa) planned and carried out the experiments and did the analyses shown here.

George investigated $(La_{0.8}Sr_{0.2})_{0.98}Mn_{1-x}Cr_xO_3$ with $x = 0, 0.05, 0.1$ with soft X-ray synchrotron radiation at room and elevated temperature. The O 1s NEXAFS spectra show (Figure 9.16) that low-level Cr substitution results in lowered hybridization between O 2p orbitals and the metal M 3d and M 4sp orbitals. The Mn L3-edge resonant photoemission spectra show a decrease of the Mn 3d–O 2p hybridization upon Cr substitution.

Further data analysis is based on the deconvolution of O K-edge NEXAFS spectra, but with an approach that has not been published elsewhere to the best of our knowledge. We accounted for the effects of exchange and crystal field splitting (see Figure 9.17)

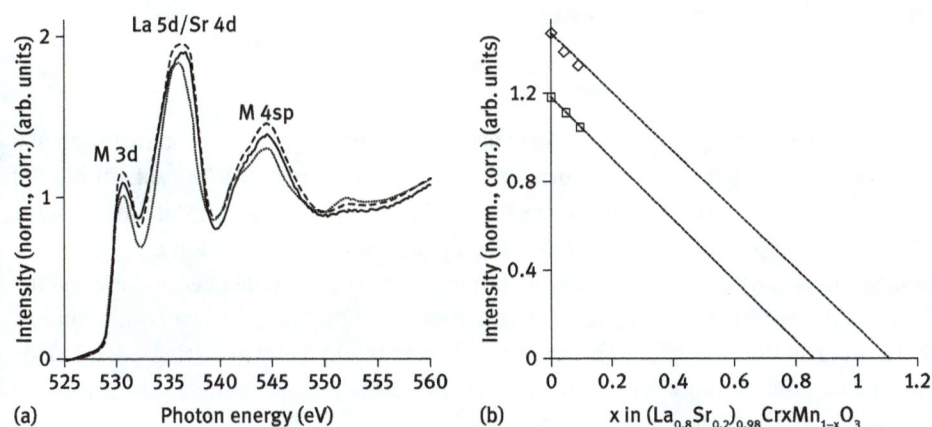

(a)   Photon energy (eV)

(b)   x in $(La_{0.8}Sr_{0.2})_{0.98}CrxMn_{1-x}O_3$

Figure 9.16: (a) O 1s near-edge X-ray absorption fine structure spectra of $(La_{0.8}Sr_{0.2})_{0.98}Mn_{1-x}Cr_xO_3$ compositions (—) $x = 0$ (–) $x = 0.05$ (•••) $x = 0.1$. (b) Room temperature, normalized and corrected O 1s near-edge X-ray absorption fine structure O 2p–M 3d (□) and O 2p–M 4sp (◊) hybrid orbital peak intensities vs. extent of chromium substitution in the $(La_{0.8}Sr_{0.2})_{0.98}Mn_{1-x}Cr_xO_3$ model series, as obtained by peak deconvolution. M denotes Mn and Cr. Dashed lines are least square fits extrapolated to zero intensity. Reprinted from [Tsekouras 2015] [Tsekouras G., Boudoire F., Pal B., Vondracek M., Prince K. C., Sarma D. D., Braun A.: Electronic structure origin of conductivity and oxygen reduction activity changes in low-level Cr-substituted (La, Sr)MnO3. *Journal of Chemical Physics* 2015, 143.], with the permission of AIP Publishing.

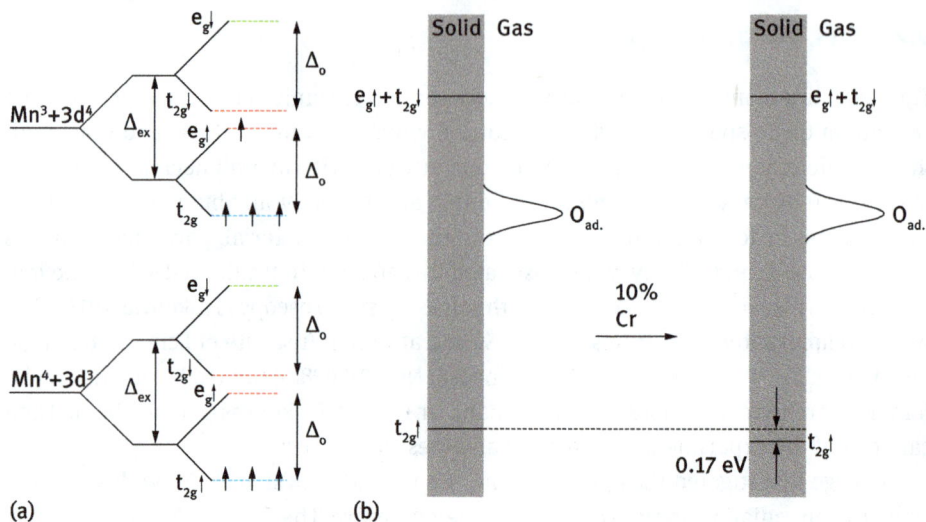

Figure 9.17: (a) Schematic representation of effects of exchange splitting (δ_{ex}) and octahedral field splitting (δ_O) on the 3d orbitals in Mn^{3+} ($3d^4$) and Mn^{4+} ($3d^3$) in the $(La,Sr)MnO_3$ SOFC cathode. Because δ_{ex} is only slightly larger versus δ_O, superposition of $e_g\uparrow$ and $t_{2g}\downarrow$ orbitals occurs. Effects of Jahn–Teller distortion in Mn^{3+} omitted for clarity. (b) Schematic of $t_{2g}\uparrow$ and ($e_g\uparrow + t_{2g}\downarrow$) hybrid O 2p–Mn 3d orbital energy levels at room temperature from deconvoluted O 1s near-edge X-ray absorption fine structure spectra of $(La_{0.8}Sr_{0.2})_{0.98}MnO_3$ and $(La_{0.8}Sr_{0.2})_{0.98}Mn_{0.9}Cr_{0.1}O_3$. One electronic state from the adsorbed oxygen on the cathode surface is also shown. Reprinted from [Tsekouras 2015] [Tsekouras G., Boudoire F., Pal B., Vondracek M., Prince K. C., Sarma D. D., Braun A.: Electronic structure origin of conductivity and oxygen reduction activity changes in low-level Cr-substituted (La, Sr)MnO3. *Journal of Chemical Physics* 2015, 143.], with the permission of AIP Publishing.

where the pre-edge region was described using the nominally filled $t_{2g}\uparrow$ state (not shown in this book; see Figure 9.3 in the original paper [Tsekouras 2015]). The substitution of LSM with 10 % Cr substitution produces a 0.17-eV lowering in the energy of the $t_{2g}\uparrow$ state. This energy corresponds well with the rise by 0.15 eV in the activation energy for the oxygen reduction reaction. The aforementioned decreased orbital overlap between the O 2p and the Mn 3d states is in qualitative agreement with the lower electronic conductivity. An orbital-level understanding of the thermodynamically predicted a solid oxide fuel cell cathode poisoning mechanism involving low-level chromium substitution on the B-site of $(La,Sr)MnO_3$ is presented.

It appears therefore that the well-known effect of chromium poisoning in SOFC cathodes has its molecular origin in the changes of the hybridization of states in the cathode material when Cr accumulates. We are looking here at a 0.15-eV difference, which I believe is a very big change. Changes of the electronic structure happen upon substitution of manganese by chromium, the 3d orbitals of which engages with the oxygen ligands not in substantial bond hybridization, preventing substantially the charge transport. Microstructural changes in cathodes upon Cr exposure, which are reported in literature, may therefore be only observations becoming obvious after the actual Cr poisoning has set on already at a level that cannot be observed with microstructural studies.

This study provides an orbital-level understanding of the thermodynamically predicted SOFC cathode poisoning mechanism upon the low level Cr substitution on the B-site of (La,Sr)MnO3.

9.3 Proton conductivity in IT-SOFC electrolytes

I began the first part of this book with hard X-ray XANES studies on proton conductors. It is not so trivial to correlate electronic structure changes with changes in the ionic-protonic conductivity, as opposed to the electronic conductivity in the cathode or anode materials. Let me therefore start out with the synthesis of the ceramic proton conductor. The parent material may be barium cerate with stoichiometry $BaCeO_{3-\delta}$ and perovskite crystal structure. Substitution of the Ce^{4+} by Y^{3+} forces the material to form oxygen vacancies in excess to the readily existent δ, which we can reasonable assume to be $\delta \neq 0$ already for the parent compound. When we expose, e. g., $BaCe_{0.8}Y_{0.2}O_{3-\delta}$ to water vapor, the H_2O molecules can enter the material, and we can write the stoichiometry $BaCe_{0.8}Y_{0.2}O_{3-\delta} \cdot H_2O$, which reads like a hydrate. The oxygen from the water molecule can fill an oxygen vacancy. The two protons from the water molecule will then settle in the vicinity of the just filled oxygen vacancy and form two OH hydroxyl bonds. We now can write the stoichiometry like a hydrate: $BaCe_{0.8}Y_{0.2}O_{3-\delta}H_{0.02}$, for example. The protons have become part of the crystal lattice. This means the protons are structural protons. We have witnessed this process by the oxygen core level spectra and the resonant VB spectra recorded by the ambient pressure XPS studies. When we now heat the sample, we notice a decrease of the hydroxyl group concentration and also a shift of the OH peak in the oxygen spectra [Chen 2013d]. The hydrogen bonds in the hydroxyl groups melt away, which we can interpret as the release of protons from the crystal lattice, whereas they become protonic charge carriers. These spectroscopic observations are accompanied by an onset of electric conductivity, which depends on the presence of protons, which are not bound anymore to the crystal lattice.

Figure 9.18 shows how we have identified the defect state associated with oxygen vacancies in the resonant VB PES spectra, which in turn however accounts for conductivity only when it is filled by oxygen ions, which bring along protons by water molecules. This is not so surprising because we are dealing here not with electronic conductivity but with ionic conductivity, specifically proton conductivity. This is verified by conductivity experiments based on impedance spectroscopy or quasi-elastic neutron scattering [Chen-Wiegart 2012b, Qianli 2013]. The temperature at which this happens is around 650 to 700 K for the BZY and BCY materials in this book. In this temperature range, we observe changes in the thermal expansion coefficients [Braun 2009j] and the onset of proton conductivity [Braun 2009g]. We have thus a thermal activated proton transport. The charge carrier dynamics of this proton conductor system bears some similarities

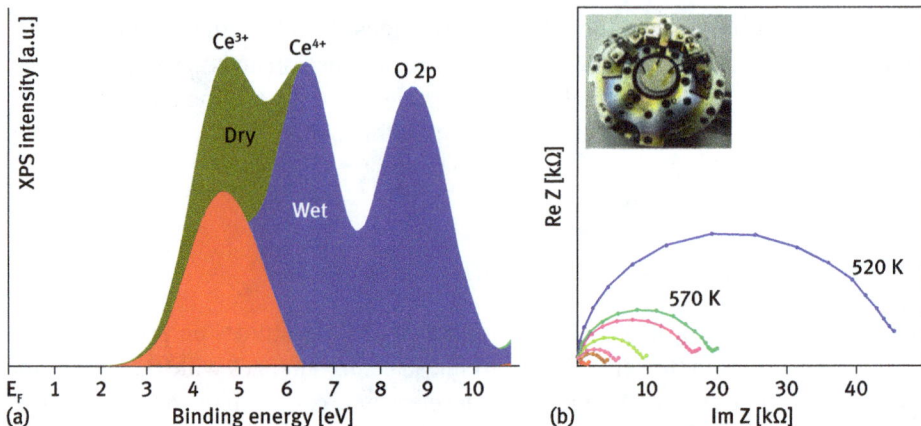

Figure 9.18: (a) Electronic DOS of BCY20 obtained from resonant VB PES spectra at high water pressure (blue shaded region) and dry condition (brown shaded region). The red shaded region is obtained by the subtraction "dry-wet" spectra, constituting the "Ce^{3+} defect state" from oxygen vacancies. The spectra show the smooth least square fit modeling output. (b) The right panel shows the impedance spectra and their growing semicircle radius with decreasing temperature, indicative of the decreasing proton conductivity [Chen 2013d].

with the polaron conductivity in perovskites, which are used in the SOFC cathodes. This is an interesting observation and analogy because it points us to the polaronic nature of transport in this case. We have determined the jump rates and jump times for the proton conductors at elevated temperatures with quasi-elastic neutron scattering (QENS). The temperature dependence of these jump rates varies very well with the Holstein polaron model for electrons [Chen 2013a]. It appears therefore, at least phenomenological, that the protons in ceramic proton conductors BCY20 form proton polarons with the crystal lattice.

9.4 Lithium ion batteries

We have used the hole peak methodology with the O 1s NEXAFS spectra also for battery cathodes (positive electrodes) for the Polish Swiss Research Program project LiBEV [Braun 2012i]. The method was used for analyzing the nature of the nonmetal–metal transition in the Li_xCoO_2 oxide ($0.53 \leq x \leq 1$). Stoichiometric $LiCoO_2$ is a semiconductor, and extraction of the lithium ions causes the system to become metallic. The distortion of the oxygen octahedron significantly modifies top valence states. We postulate that the nonmetal–metal transition can be interpreted on the basis on the Anderson-type transition [Milewska 2014].

Figure 9.19 shows the O 1s spectrum of Li_xCoO_2 with $x = 1.00$, 0.99, 0.98, 0.97, 0.94, and 0.60. It shows the valence band from hybridized O 2p and Co 3d states, which amounts in the well-known doublet from states with e_g and t_{2g} orbital symmetry at

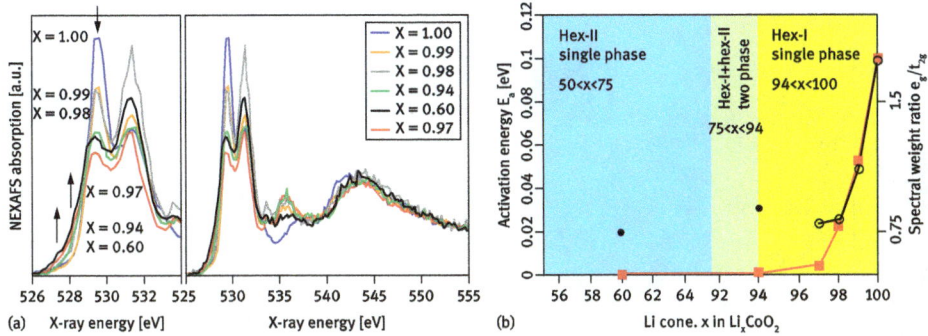

Figure 9.19: (a) Comparison of the NEXAFS spectra for Li_xCoO_2 (x = 1.00, 0.99, 0.98, 0.97, 0.94, and 0.60). Note the change of the peak heights at 529 and 531 eV in the left magnified portion of the spectrum. (b) Variation of the spectral weight e_g/t_{2g} determined from the NEXAFS spectra as a function of x in Li_xCoO_2, and on the left axis the activation energy for the electronic conductivity. Samples and data are the same like in [Molenda 2015].

around 530 eV. Depending on the Li concentration x, the respective peak heights and the resulting relative spectral weight between e_g and t_{2g} undergo changes. In this case, we formed the relative spectral weight S as e_g/t_{2g} and plotted it in the right panel of Figure 9.19.

Here, S deviates for x = 0.94 from the trend given by the spectra of the samples with higher lithium concentration. Maybe this is so because of the coexistence of the hex-II phase in this concentration region. There are somewhat more subtle yet systematic changes in the pre-edge region. The spectral weight just before the e_g peak becomes slightly but systematically enhanced with decreasing lithium concentration x. A similar observation has been made in the delithiated single crystals of Li_xCoO_2 [Mizokawa 2013]. The two spectral features (marked by the two arrows at 527 and 528 eV pointing up) originate from two absorption processes of Co^{4+} [Galakhov 2009]. With x decreasing from 1.00 to 0.97, the spectral weight is increasing homogeneously. Beyond x = 0.97, the hole intensity is decreasing at even lower x increasing again. This change in variation at around 0.97 fits the variation of the activation energy of electrical conductivity.

The insertion of Li in the parent compound causes two phase transitions in the Li concentration range x from 56 to 100 %. The phase transition extends from a region with a single hexagonal phase to a region with two different hexagonal phases to another single hexagonal phase. In the same right panel is the activation energy plotted. There is a steep increase of activation energy and spectral weight in the narrow concentration range from 97 % to 100 %. The lower concentrations have a flat region.

We made a similar study for the related system Na_xCoO_2. This material has been of interest for solid-state physicists and solid-state chemists from a fundamental point of view because of metal-insulator transitions. Also here, we find a global correlation between the spectral weight ratio for holes as observed in the pre-edge of the oxygen spectra and the conductivity. This is shown in Figure 9.20. Note that we have plotted

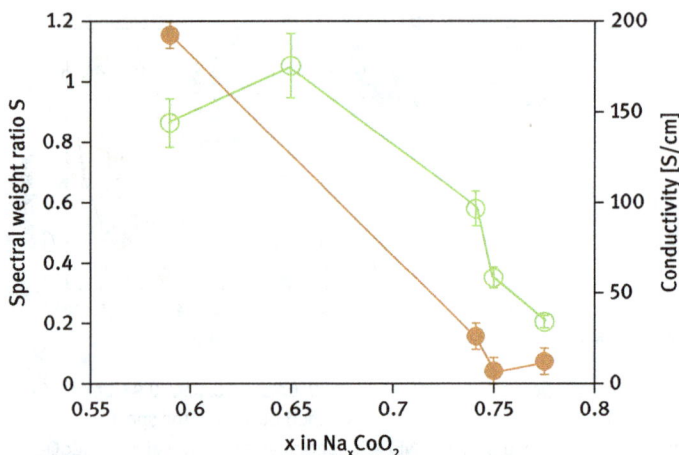

Figure 9.20: Correlation of spectral weight ratio S of holes from O 1s NEXAFS spectra and electric conductivity versus the sodium concentration in NaxCoO2 from samples and data published in [Molenda 2015].

there the conductivity and not the activation energy. Note that the electric conductivity is basically the product of charge carrier concentration and their mobility. Moreover, both terms may be temperature dependent. The charge carrier concentration should be detectable via the S-ratio. In theory, it should be possible to make well-calibrated (metrological) experiments with a perfect match of X-ray spectra information and conductivity information based also on a sound theoretical or mathematical quantitative description. Maybe in the future somebody will present such treatment.

9.5 Water splitting photoelectrodes

From the perspective of device applications, semiconductor photo-electrochemistry represents one of the most difficult fields of physical chemistry. The photoelectrode is a light absorber that by virtue of its semiconductor band gap absorbs light and converts the light energy into electric energy as excitonic electron-hole pairs. The energy bands of the absorber material must be so that the charge transport can take place with ease to the electric current collectors. This primarily settles all issues about photovoltaic (PV) applications.

The situation becomes more complex when the photoelectrode is supposed to perform chemical work. A prominent example is the water oxidation at photoelectrode surfaces for the so-called solar water splitting (PEC). Here, the principles of electrochemistry apply. Specifically, electron holes produced in the absorber need to be driven to the electrode surface, where they chemically react with water in the electrochemical double layer according to the chemical reaction equation

$$2h^+ + H_2O \Leftrightarrow O_2 + 2H^+$$

Although the thermodynamic energy required for water dissociation is known to 1.23 V, kinetic barriers at the electrode surface amount into overpotentials, which shift the voltage required for water oxidation by electrolysis (EL) to typically over 1.8 V. Here, then the tasks of electrocatalysis come into play for the decrease of these overpotentials. Because the photoelectrode is inserted into a water containing electrolyte, its Fermi level will adjust to the Redox potential of the electrolyte, thus bending the energy bands of the photoelectrode at the surface.

The electronegativity of the aqueous electrolyte will cause polarization at the photoelectrode surface, which in turns causes formation of a charge carrier depletion layer in the subsurface region of the photoelectrode. In the case of the α-Fe_2O_3 photoelectrode, the conduction band is below the water oxidation potential. This is the reason why an external bias is needed to drive the water oxidation reaction on the iron oxide photoelectrode. This bias is yet smaller than the aforementioned voltage for electrolysis, which is one of the reasons that PEC is considered economically more viable than PV+EL.

It is the photogenerated electron holes that have the power to oxidize the water molecule. Kennedy and Frese [Frese 1978, Kennedy 1976, Kennedy 1977, Kennedy 1978a, Kennedy 1978b] have investigated the iron oxide photoanode (and also $BaTiO_3$ photoanode, [Kennedy 1976]) with optical and electrochemical methods and concluded that there should be a Fe 3d-3d hole with lesser energy in addition to an O 2p type hole. We have carried out an *operando* NEXAFS experiment for water splitting (Chapter 8.3) and found additional spectral weight in the same oxygen pre-edge region, where we found hole states in the iron perovskites (Chapter 9.1).

We found two hole states. At about the same time, Bisquert et al. published an impedance study [Klahr 2012], which was based on the early concepts by [Tomkiewicz 1980a, Tomkiewicz 1980d]. Basically, you run impedance spectra at various DC bias potentials as if you want to reproduce the CV of the system. With subsequent modeling of Randles' circuits, you find out capacitive states that can be interpreted to some extent as the electronic DOS. The energy axes from the Mott–Schottky type impedance study and the corresponding X-ray energies from XPS or XAS have to be aligned correspondingly.

Figure 9.21 (top) shows the evolution of the spectral weight S of the two O 1s NEXAFS hole states with applied bias potential, along with the photocurrent density. The bottom panel shows the surface trap state (orange) and the corresponding trap resistance (conductivity, red), along with the photocurrent. It is remarkable how our UHB hole state (top panel, orange) matches position, shape, and width of the surface trap state (bottom panel, orange). We witness here a clear quantitative correlation of information from electroanalytical methods and X-ray spectroscopy methods.

Upon close investigation of the NEXAFS spectra, we find further correlations. Figure 9.22 shows on the left top panel the photocurrent evolution. Underneath, we remember the CTB and UHB hole peaks from the NEXAFS *operando* experiment [Braun 2012l]. Below that, we have the trap state capacitance (blue) and trap state conductivity (red), as was determined by Empa/EPFL PhD student Yelin Hu [Braun 2010e] with

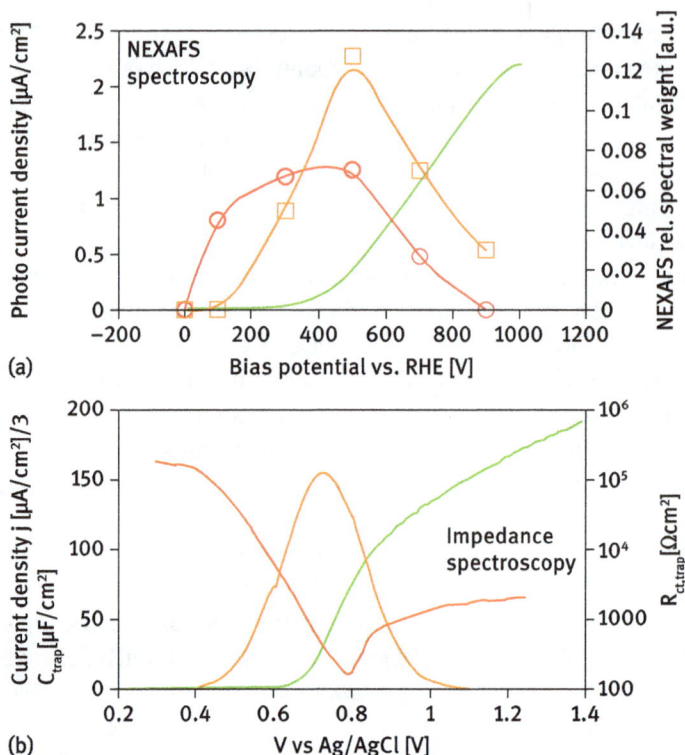

Figure 9.21: (a) Relative spectral weights S of hole transitions into CTB (red circles) and UHB (orange squares) [Braun 2012l] obtained from *operando* NEXAFS spectra during PEC water splitting on a-Fe$_2$O$_3$. Solid lines are guides to the eye. The green line is the photocurrent. (b) Defect concentration obtained from fitting of EIS spectra from Klahr et al. [Klahr 2012]. The parallelism of the hole state and defect state data extracted from X-ray spectroscopy and electroanalytical methods, as evidenced by both figures, is compelling [Bora 2013b].

impedance spectroscopy following the method presented by [Shen 1986, Tomkiewicz 1980a, Tomkiewicz 1980d, Tomkiewicz 1980g, Ullman 1980]. On the top right, we have plotted the shift of the "water peak" in the O 1S NEXAFS spectra (the hydrogen bond breaking peak).

This peak shifts upon increasing bias potential to lower X-ray energies and at 500 mV shift back to higher energies. The bottom right panel shows the evolution of the hydroxyl peak, which actually resides in the intensity minimum of the O 1s NEXAFS spectra. However, its intensity scales with applied bias. The height of this peak increases while the hydrogen bond breaking peak shifts to lower energies. Obviously, the chemical changes that go on with the water molecule and the OH groups at the photoanode surface are clearly corresponding with one another.

Therefore, the correlation that we find are not only of electronic nature but also of chemical nature [Braun 2016c].

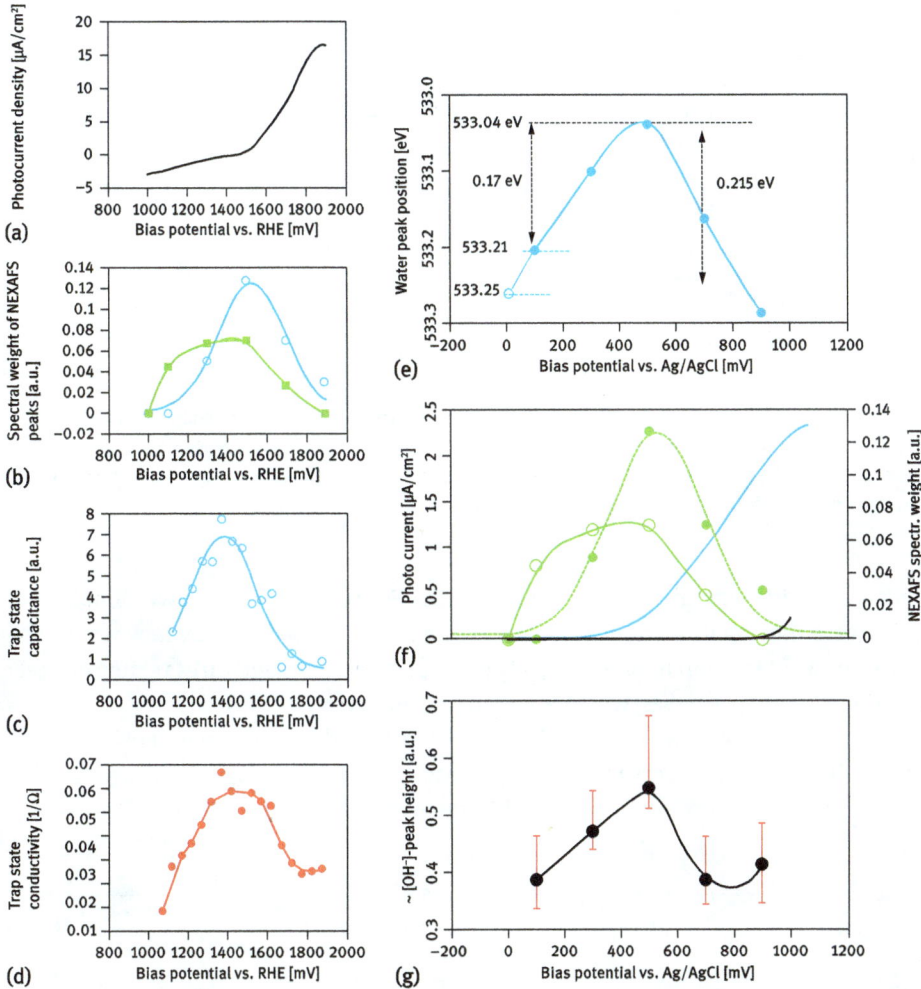

Figure 9.22: (a) Photocurrent density of the hematite film measured *ex situ* after the *operando* experiment at ALS. (b) Variation of the spectral weight of hole peaks in the oxygen NEXAFS spectra versus bias potential in RHE reference. The CTB hole peak variation is fitted with a parabolic function (square root vs. potential) up to 1500 mV; the UHB hole peak is fitted with a Gaussian profile. (c) Trap state capacitance of the film measured with *ex situ* EIS. (d) Trap state conductivity measured *ex situ* with EIS. (e) Shift of the "water peak" versus bias potential in Ag$^+$/AgCl reference. (f) Variation of NEXAFS hole peak spectral weights (same like Figure 9.22 b) along with photocurrent variation. (g) Variation of the –OH NEXAFS peak in the *operando* spectra with bias potential.

Let me finally turn to the *operando* XPS experiments that we did at BL9.3.2 at the ALS. Here, we used Fe-resonant photoemission in order to make the spectroscopic changes more visible by contrast variation (Figure 9.23). The O 2p peak was shifting upon changing bias toward the water oxidation potential and illuminating significantly

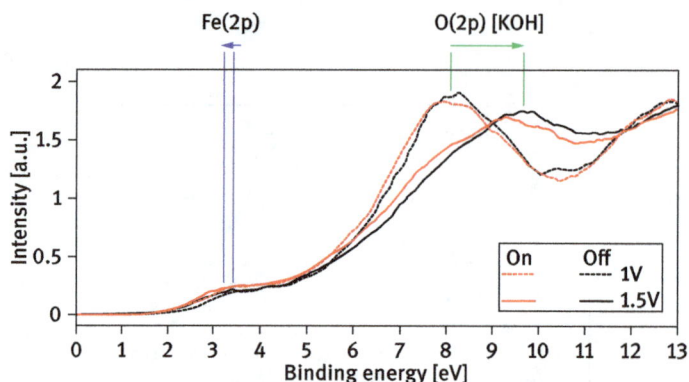

Figure 9.23: Fe 2p resonant photoemission valence band spectra of a photo electrochemical cell with α-Fe$_2$O$_3$ photoanode under 1.0 V (the dotted spectra) and 1.5 V (solid spectra) DC bias under an illuminated condition (the two red curves) and under a dark condition (the two black curves). The spectra were recorded under 100 mTorr water vapor *operando* conditions at the BL 9.3.2. at the ALS.

to higher binding energies. At the same time, the Fe 2p signatures at lower binding energies were moving toward the Fermi level. We demonstrated already how we did the background subtraction in order to get "pure" spectroscopic signatures related entirely with the relevant processes, which we want to study.

Figure 9.24 shows the well-known dark current and light current densities of the iron oxide photoanode versus the bias potential. On the right panel, we see the shift of the O 2p peak that we have observed in the *operando* XPS experiment. The general

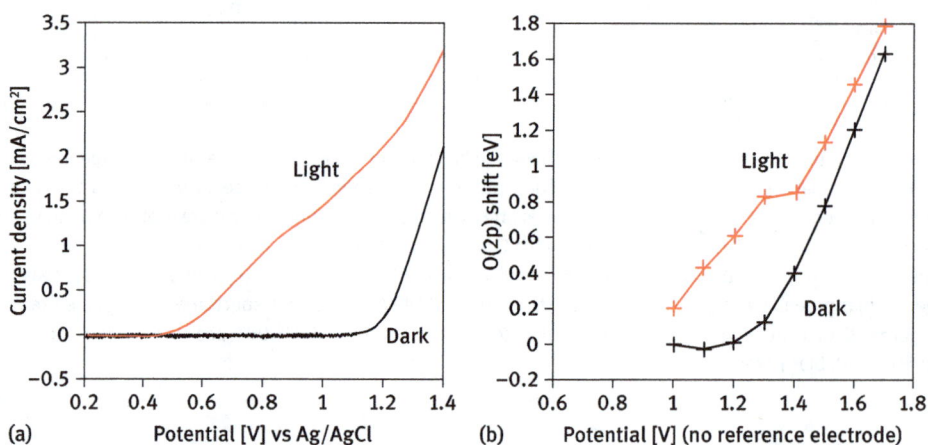

Figure 9.24: (a) Light current (red curves) and dark current (black curve) densities of α-Fe$_2$O$_3$ photoelectrodes in KOH. (b) Evolution of the shift of the O 2p bonding peak in Fe 2p resonant VB XPS spectra recorded *operando* under 150 mTorr water vapor pressure and corresponding illumination and DC biasing conditions.

curvatures of this spectral shift and the currents is gratifyingly similar. We have thus another example of how well electroanalytical methods and X-ray methods arrive at similar data.

Note, however, that the mere current density evolution provides no answer about the structural or electronic origin of the observed signals, whereas the X-ray core level spectra or resonant VB spectra allow insights from which element and from which orbitals the device effect originates.

10 Radiation damages

For a fair assessment of the radiation-based methods presented here in this book, it is necessary to have a look at their shortcomings. One frequently heard question when it comes to X-ray-based methods is that of "radiation damages." We know from our visits to physicians for medical exams that human exposure to X-rays is subject to legal monitoring, controlling, and regulation. This is because the X-rays used for medical imaging are harmful to our tissue and DNA when critical X-ray dose limits are exceeded.

Considerate researchers therefore acknowledge and address radiation damage issues. The *Journal of Electron Spectroscopy and Related Phenomena*, published in 2009 in its volume 170, a special issue on radiation damage. The March 2015 issue of the *Journal of Synchrotron Radiation* includes papers, which were presented at the 8th International Workshop on X-ray Radiation Damage to Biological Crystalline Samples Hamburg, Germany, April 10–12, 2014. In March 2016, the 9th International Workshop on X-ray Radiation Damage to Biological Crystalline Samples took place at Maxlab in Lund, Sweden. It is therefore clear that scientific communities pay attention to this issue.

X-rays do not only interact adverse with human tissue. X-rays interact with all materials, and this makes them excellent probes for the analyses we have seen in this book. The photoelectric effect creates electron hole pairs, which can cause chemical changes in the irradiated material. Depending on the X-ray wavelength and dose, however, the irradiated material can become permanently modified or even damaged. Experienced X-ray and electron spectroscopists become aware of this, e. g., when they run several scans for spectra and notice gradual and systematic changes in peak positions or peak intensities.

Some materials are very prone to structural changes upon irradiation with X-rays. A metastable compound can decay during X-ray exposure. A cation in the compound with an unusual oxidation state can relax into a more stable oxidation state upon X-ray exposure. This manifests in a chemical shift, which is observed as a shift of the absorption edge, for example. Another possibility is that the sample under investigation has absorbed humidity and ambient gases like CO_2, O_2, and N_2 prior to the measurement. The sample may be put into the UHV chamber but it still has the aforementioned residual gases. Upon exposure to X-rays during the X-ray spectroscopy experiment, the sample may become hot, and this causes chemical reactions of the sample with adsorbed gases at the sample surface. And this can be detected in the measured spectra. For example, the oxygen of adsorbed water molecules can fill oxygen vacancies, and with the protons hydroxyl species can be formed on the sample surface, which is obvious from the XPS or NEXAFS core level spectra. The necessary activation energy for this reaction can come from the power of the X-ray beam or electron beam.

Figure 10.1 lists three synthetic hydrogenase models, which I received from Professor Xile Hu at EPFL for inspection at the synchrotron. They has to be synthesized at EPFL in the glove box and were not supposed to be exposed to ambient conditions—first, because they are sensitive to the ambient conditions and will degrade, and second, because

https://doi.org/10.1515/9783110794038-010

Sample structure	Chemical formula	Quantity
BO-01	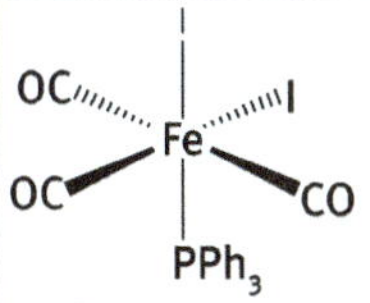 $C_{21}H_{15}FeI_2O_3P$	100mg
BO-02	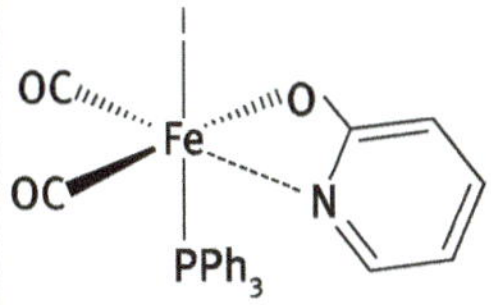 $C_{25}H_{21}FeINO_3P$	100mg
BO-03	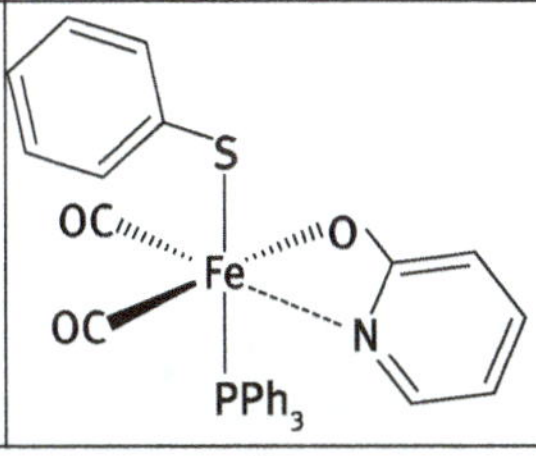 $C_{31}H_{26}FeNO_3PS$	100mg

Caution: strong smell

Figure 10.1: Synthetic hydrogenase structural models with their structural formula and stoichiometry measured at the Advanced Light Source with Fe 2p NEXAFS spectroscopy. Samples were obtained from Professor Xile Hu, EPF Lausanne.

their smell is very strong. Figure 10.2 shows a series of Fe 2p NEXAFS scans, which were run at the Advanced Light Source. For each of the three specimens, we notice substantial variations in the relative peak heights of the $2p_{1/2}$ and $2p_{3/2}$ doublets. In the middle panel in Figure 10.2, I have labeled the first, second, and third scans. The e_g peak intensity at 706 eV increases from the first to second scan. The spectrum from the third scan shows even a decrease of the t_{2g} peak. This is typical radiation damage. When we are averaging the three scans in order to get a better statistics, we are averaging out the differences in the spectra. The first spectrum is probably one that we should be interested in. There are ways to minimize this effect. For example, the first scan is made, and then the sample is moved versus the beam spot position and the second scan is taken at a sample position, which was not yet exposed to the beam, and so on. Then we have a series of virgin-like spectra from fresh sample positions.

We often measured environmental specimen from ambient air sampling campaigns, which were collected on specifically designed filters. The intense X-ray beam from a third generation synchrotron source can easily burn this sample spot. Somebody had the idea to mount the round filter paper on a motor, which would rotate the filter versus the fixed X-ray beam, so that within some minutes or so, several dozens of fresh spots on the filter were probed and measured. This was a slick method. But the beam perforated the filter azimuthal and eventually the filter fell off from its frame that held it so far. The beam at the APS was so harsh that we sometimes

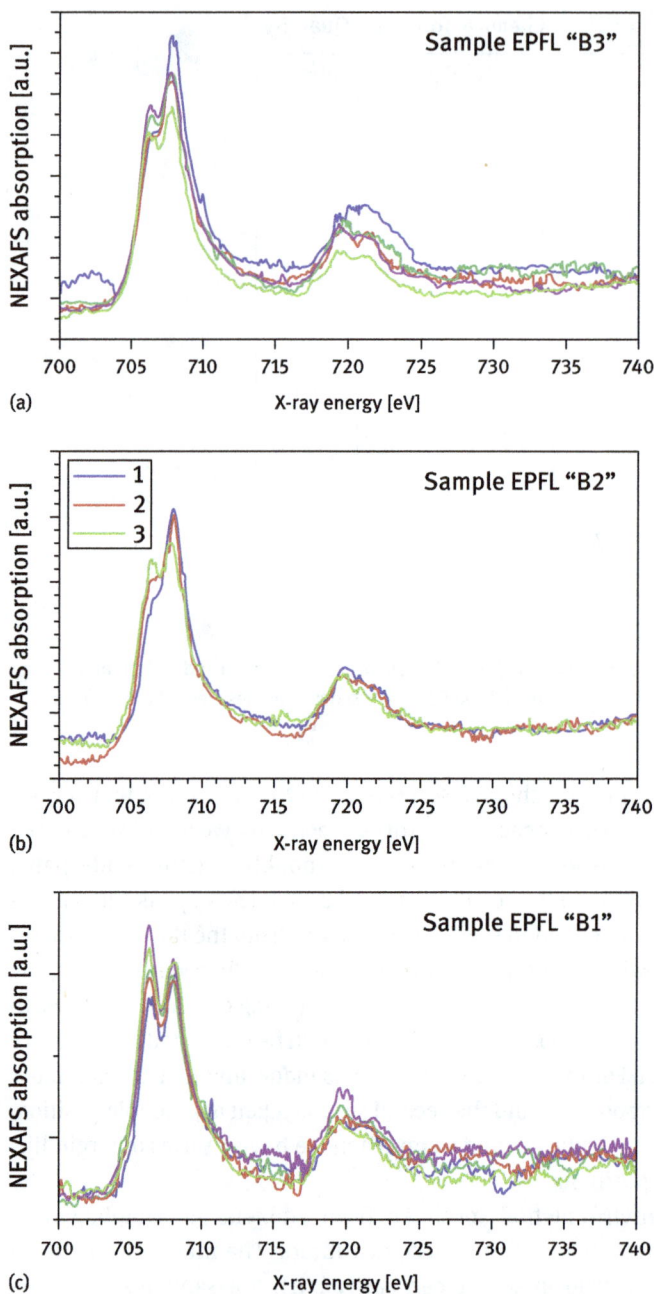

(a)

(b)

(c)

Figure 10.2: Fe 2p NEXAFS spectra of the structural hydrogenase models described in Figure 10.1. Spectra were recorded at the Advanced Light Source. Sample B3 (a) was measured with 5 consecutive scans, which show a general similar shape, but fine structures such as the e_g-t_{2g} doublet between 705 and 710 eV show alterations. Same holds for the 2p1/2 doublet at around 720 eV. Hydrogenase B2 (b) shows systematic increase of the e_g peak at 706 eV with explicit double peak formation at the 3rd scan. Hydrogenase B1 (c) shows substantial changes during scans at the e_g peak at 706 eV.

needed to put absorber plates in the beam to attenuate it so as to protect the samples. The beam can be so intense that it melts the organic electrolyte and separator in an *in situ* battery cell. In protein spectroscopy, the biological samples are often cooled down to liquid nitrogen or liquid helium temperature. This has two purposes. One is to cool down to minimize radiation damage. The other purpose is to minimize molecular vibrations, which typically cause broadening of the spectral features. These are described by temperature dependent Debye–Waller factor, which smears out peaks [Debye 1913a, Debye 1913c, Waller 1923a, Waller 1923c].

The power from electron beams is often underestimated, sometimes neglected, and even completely ignored. The damages to the samples can be particularly severe when transmission electron spectroscopy with EDX or EELS is applied. The results of elemental mapping in EDX images is not necessary affected by the heating from the electron beam. But TEM-EELS spectra may sometimes be considered as "high-temperature" spectra because the electron beam (>10 keV, several mA current) heats up the sample. Lithium has been reported melting in the TEM beam [Hightower 2000a, Hightower 2000f], for example. Carbon soot typically looks like graphite in EELS, because the organic and aliphatic volatiles are evaporated by the electron beam, whereas STXM spectra show a wider variety of molecular species in soot [Braun 2005b]. Therefore, if one wants to make high-temperature EELS studies with a TEM setup on SOFC materials, the experimenter has to know that the sample is at an elevated temperature anyway just because of the electron beam heating the sample. Further heating will therefore not show the anticipated changes in the spectra.

Materials are more forgiving in this respect when they are exposed to neutron radiation. Neutrons cause basically no radiation damage.[1] However, neutrons may react with some specific nuclei in the probed material and then γ-rays are produced. The γ-rays may then cause radiation damage as a secondary effect, depending on the chemical composition of the probed material. Attention must be paid to the possibility that materials may become neutron activated. The probed sample becomes then radioactive and needs to stay in quarantine until radiation levels are safe again.

Whenever energy dissipation plays a role, we should be concerned about radiation damages, no matter which probes are used. This is definitely not limited to X-rays.

Below I show a study where radiation damages were discovered during C1s NEXAFS studies (STXM) on diesel soot extracts. Repeated scans produced systematic changes in the spectra. Since we used STXM, the intensity of the beam was very strong and the changes occurred very fast. Dr. Sue Wirick at NSLS was familiar with this kind of change

---

1 With the exception of nuclear reactions, such as by neutron capture. One example is the neutron capture of the natural isotope of gold, ^{197}Au. The incoming neutron, if captured, will form the heavier isotope ^{198}Au in a nuclear excited state. This will rapidly decay to the ground state of ^{198}Au, and a γ quantum will be emitted. The isotope ^{198}Au is known as a β-emitter, which means it will releases an electron. This means that the isotope ^{198}Hg mercury is formed. So, when your valuable gold is turned into less worth "quicksilver" mercury, this is a real damage of value, literally. And certainly a radiation damage.

and was reminded of a previous study, which she had done on alginic acid. With continuing exposure to the STXM beam, she had observed the same changes in alginic acid like we observed in our diesel soot extracts.

The experiments were conducted at the National Synchrotron Light Source in Brookhaven National Laboratory, Upton, NY, USA. The instrument was a scanning transmission X-ray microscope (STXM), available at Beamline X1A [Kirz 1994a, Kirz 1994b, Zhang 1994]. The beam was defocused to 1 micrometer. Spectra from 270 to 310 eV were recorded in 0.1-eV steps with 0.12-second dwell time. An image of a sample region is obtained by scanning the region with the detector at constant X-ray energy, while the X-rays are absorbed by the sample and the sample holder, which is only around 1 micrometer thick in total. The spectrum of an empty sample holder is subtracted for reference. This technique is bulk specific to thin samples. Images recorded at other X-ray energies can be combined to a stack.

This data set can then be used to assign every pixel in the image a NEXAFS spectrum, provided sufficient energy range was covered. With 400 data points, the overall exposure time per spectrum was 480 seconds (8 minutes). The detector was a photo diode. For X-ray energy calibration, the instrument was flushed with CO_2 gas, the absorption edge of which is well known.

To measure the spectra of extracts, one drop (ca. 1 μL) of the extract was placed on an ultrathin silicon nitride sample holder. The thickness of the stain from the soot extract after evaporation of the water under ambient conditions was 1 micrometer. Thin films of sodium alginate were made by dissolving alginic acid sodium salt in water and drying the solution on the sample holder in ambient atmosphere.

Figure 10.3 (left) shows the STXM image of a stain of diesel soot extract, covering about a 3 μm × 3 μm area. The sample plus silicon nitride substrate is thin enough to permit significant transmission of photons with 280 eV energy. The yellow color in the lower left corner is indicative of the substrate, free of sample. Blue regions are from the soot extract. The red color indicates a saturation of the absorption signal due to too thick sample, and hence too strong absorption.

The right panel in Figure 10.3 shows five carbon NEXAFS spectra obtained this way, plus the peak assignment supplemented by Table 10.1. The carboxyl peak is decreasing while the carbonate peak is increasing with continuing exposure to radiation. We can call this radiation damage. However, we can express this and turn this also into something more positive, because we measure how the diesel soot extract film is chemically converted upon irradiation. After all, diesel soot with this volatile material is brought by the exhaust pipe into the atmosphere and then subject to irradiation by visible light and UV light.

Since the structure of alginic acid is well known (Figure 10.4, left panel, shows the variation of spectra of alginic acid during irradiaiotn by soft X-rays, and right panel variation of peak heights from diesel soot extract experiment), we can propose a scenario

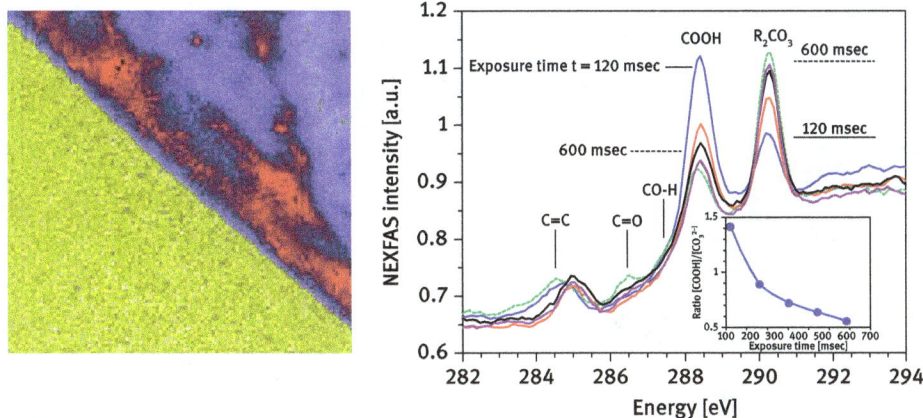

Figure 10.3: (left) STXM image of 3×3 μm^2 area. Upper right half indicates the diesel soot extract stain. Bottom left is the empty sample holder area for reference. (right) Changes observed in the NEXAFS spectra of soot particles as a function of exposure in a scanning transmission X-ray microscope. As the radiation dose on the sample is increased, the COOH peak at 288.5 eV decreases in height while the R_2CO_3 peak at 290.2 eV increases in height, suggesting a breaking of carboxyl groups and formation of carbonate within the sample. The radiation dose for the acquisition of one NEXAFS spectrum was approximately 10^6 gray for 8 minutes of irradiation. The beam size defocused 10 micrometers is 1 micrometer, and the 120-msec dwell time is per pixel or in the case of point spectra per 0.1-eV steps [Braun 2009i].

Table 10.1: Assignment of NEXAFS peak positions and molecular species; Reprinted from the Journal of Electron Spectroscopy and Related Phenomena 170, A. Braun, A. Kubatova, S. Wirick, S. B. Mun, Radiation damage from EELS and NEXAFS in diesel soot and diesel soot extracts, 42–48, Copyright (2009), with permission from Elsevier.

eV	Transition	Functionality
283.7	1s-π*	quinone
	1s-π*	protonated/alkylated
284.9–285.5		Aromatic and PNA
	1s-π*	Carbonyl substituted
285.8–286.4		Aromatic, phenolic
	1s-π*	Aromatic C-OH
		Ketone-C=O
287.1–287.4		aliphatic
	1s-π*	Aromatic carbonyl
287.7–288.3		C=O
287.6–288.2	1s – 3p/σ*	CH_3, CH_2, CH
288.2–288.6	1s-π*	COOH
289.3–289.5	1s – 3p/σ*	C-OH, alcohol

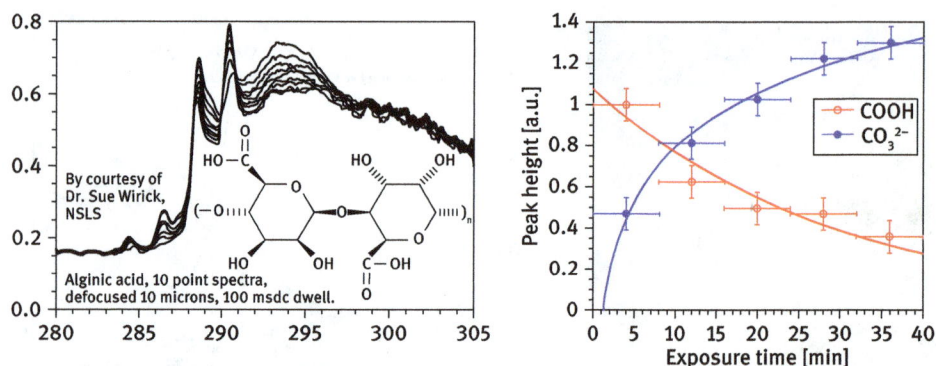

Figure 10.4: (Left) C(1s) NEXAFS spectra of alginic acid films, with changing peak heights upon irradiation. The inset shows the structure formula of alginic acid. (right) Peak heights for consumed carboxyl (open symbols) peak and evolving carbonate (filled symbols) resonances for determination of the reaction rate constants. Solid lines are least square fits.

where the COOH groups react with oxygen radicals to form an organo-carbonate and water:

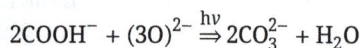

$$2COOH^- + (3O)^{2-} \overset{h\nu}{\Rightarrow} 2CO_3^{2-} + H_2O$$

We believe that a similar chemical reaction scenario occurs in the soot extracts when exposed to the beam. Since the STXM is not a sealed system, presence of some air is considered inevitable unless the STXM instrument becomes sealed. This effect is not observed in NEXAFS spectra recorded when the sample was in an ultrahigh-vacuum chamber. This opens a new perspective to environmental and atmospheric scientists. For instance, the interaction of oxygen, radiation, and soot basically could be studied *in situ* at a synchrotron beamline.

11 Background subtraction

Most of our experimental scientific work is based on careful data analysis. Such analysis requires treatment of the experimental raw data, usually referred to as data reduction. Important information may be hidden in your data. You may have to extract that information in many, sometimes painstaking steps, and sometimes extraction of data is easy. I noticed over the years that not necessarily all science majors are acquainted with or have a feeling for such data reduction.

11.1 Optical absorption spectra of phycocyanin on a photoelectrode

Some of our work on bioelectrodes included optical spectroscopy. The question was whether the absorbance of the iron oxide photoelectrode would be different from the absorbance of such electrode when functionalized with the light harvesting antenna protein phycocyanin. The next question was whether the adsorbed phycocyanin would be desorbed or degraded during operation in a PEC cell. For example, we used an aggressive caustic electrolyte of pH 13.85 in many of our studies on iron oxide, for which this is a fine condition. However, proteins may denaturate under such alkaline conditions. Hence, one of the concerns of some of our colleagues is that this is a study made by PhD student Debajeet K. Bora. Figure 11.1 shows the absorbance spectra of the pristine hematite electrode (black spectrum A) and hematite electrode adsorbed with phycocyanin (blue spectrum B), and the spectra of both samples after they had been operated in a PEC cell (red spectrum A after PEC study and pink spectrum B after PEC study). We therefore want to compare the red and the pink spectrum.

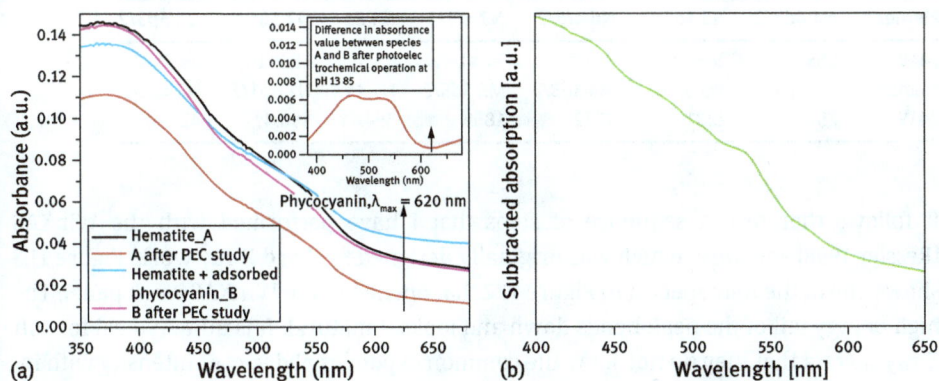

Figure 11.1: (a) Optical absorption spectra of hematite (black, A), hematite plus phycocyanin (blue, B), hematite after operation in a PEC cell (red spectrum), and protein-adsorbed hematite after operation in a PEC cell (pink spectrum). The spectrum in the top right corner is from pure phycocyanin. (b) Difference spectrum of hematite plus phycocyanin minus the hematite spectrum. It resembles the curvature of phycocyanin.

https://doi.org/10.1515/9783110794038-011

11.2 NEXAFS data on CuWO₄ from BESSY-II

In the course of the analysis of NEXAFS spectra from CuWO$_4$ samples, there was some inconsistency with the peak assignment.

We inspected tungsten copper oxide samples from the University of Hawaii at BESSY-II with the SurICat end station. The beamline at which we used this end station had a limited X-ray energy range. The core level peak positions of the elements are listed in the X-ray Data Booklet, which is a very handy tool for the X-ray spectroscopist [Thompson 2009]. The X-ray energy range at the beamline was limited so that we could not detect the tungsten 5p peaks at 490 and 423 eV. We saw peaks in the spectra at around 265 eV, and there was confusion that elements in the sample could produce such peaks. First, you would think of carbon 1s at 285, plus a misalignment of the end station X-ray optics, which frequently happens. In this case here, we would have an X-ray energy offset by 20 eV. The next scenario would be that the peak originates from tungsten N4 4d3/2 at 255.9 eV and N5 4d5/2 at 243.5 eV. This would only be possible if there was an energy scale offset by −10 eV. We have to trust the beamline technicians at BESSY-II who are known for doing their utmost professional job, including the X-ray energy axis calibration. Table 11.1 shows the core levels of tungsten, and none of their energies match. The same holds with copper and with oxygen. Thus, where does the peak at 265 eV come from? Because the curvature of the spectra was somewhat unnatural, I felt it would be necessary to "massage" the spectra somewhat—a term which I learned during an X-ray Raman beam time at Argonne Lab from a well-known spectroscopist.

Table 11.1: Low-energy core level position for element 74 tungsten according to Thompson et al. [Thompson 2009].

Element	N4 4d3/2	N5 4d5/2	N6 4f5/2	N7 4f7/2	O1 5s	O2 5p1/2	O3 5p3/2
74 W	255.9†	243.5†	33.6*	31.4†	75.6†	45.3*b†	36.8†
Element	M2 3p1/2	M3 3p3/2	M4 3d3/2	M5 3d5/2	N1 4s	N2 4p1/2	N3 4p3/2
74 W	2575	2281	1872	1809	594.1†	490.4†	423.6†

It follows therefore a sequence of steps that I have performed with the WinXAS [Ressler 1998] software, which was originally designed for hard X-ray work. Figure 11.3 shows one of the four spectra in Figure 11.2 that opened in the WinXAS work panel. The high-energy tail of the peak bends down and looks unnatural. It is quite typical for soft X-ray spectra that they do not show the common exponential decay of intensity with increasing X-ray energy. It needs therefore some experience and maybe also some courage when dealing with NEXAFS spectra in the energy range below say 400 eV. We previously saw in Part 8 some strange trend in the Li 1s spectra of lithium manganite. Using WinXAS' routines, we guess an energy position in the spectrum where we want to apply a Victoreen fit. This is demonstrated in Figure 11.4 where a blue curve segment is indicated

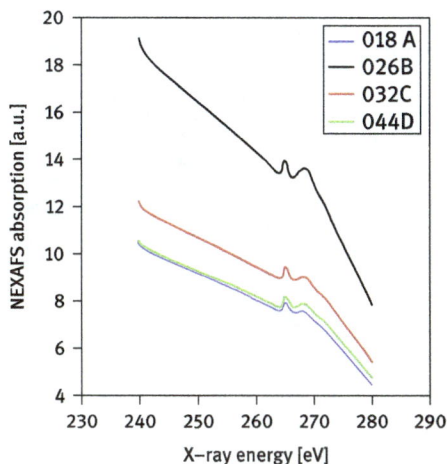

Figure 11.2: Preparatory analysis of NEXAFS spectra of four different samples with nominal CuWO$_4$ stoichiometry from BESSY-II. The peaks at 265 eV were unassigned because they did not match copper, tungsten, and also not oxygen [Thompson 2009].

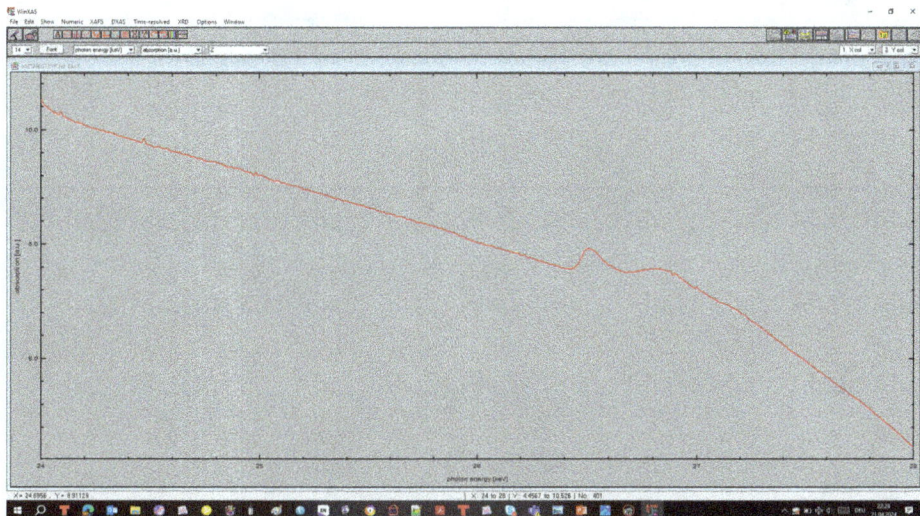

Figure 11.3: NEXAFS spectrum of CuWO$_4$ sample opened in WinXAS window.

in the red spectrum. Figure 11.5 shows how the Victoreen function looks like, which we will apply. We will use this as a subtraction operation, which will straighten out the high-energy tail of the peak we are investigating. Certainly, the remaining low-energy part of the blue Victoreen function will have an effect on the low-energy part of the red spectrum once it is subtracted. However, at present we ignore this for the sake of making progress with the center of the spectrum that bears the unassigned peak. Figure 11.6 was taken when the Victoreen function was subtracted from the entire spectrum. We are looking now at a peak that has a flat and high high-energy tail, which is reminiscent of an ionization threshold by absorption. This looks pretty good now. The low-energy tail of the red spectrum looks now—after subtraction of the Victoreen function— steep up

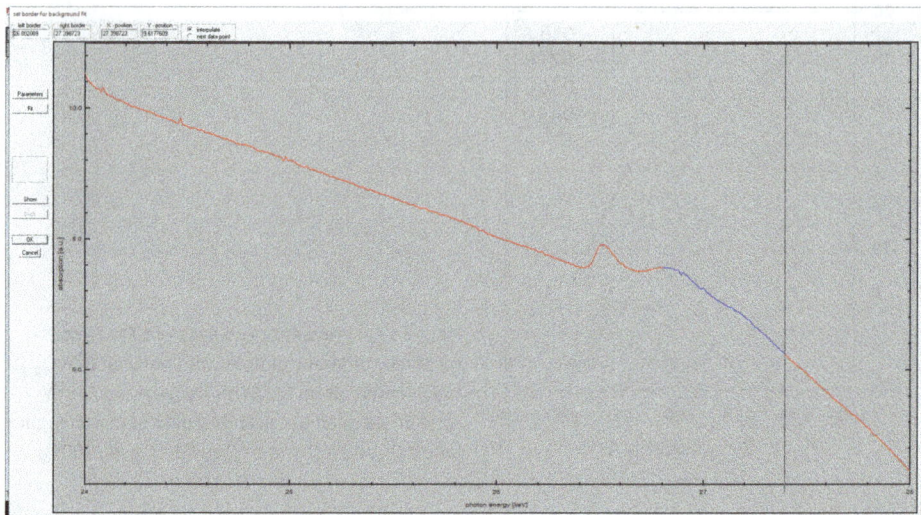

Figure 11.4: NEXAFS spectrum of $CuWO_4$. The blue line right from the vertical separation line indicates where to the spectrum a Victoreen spline curve for later subtraction is applied.

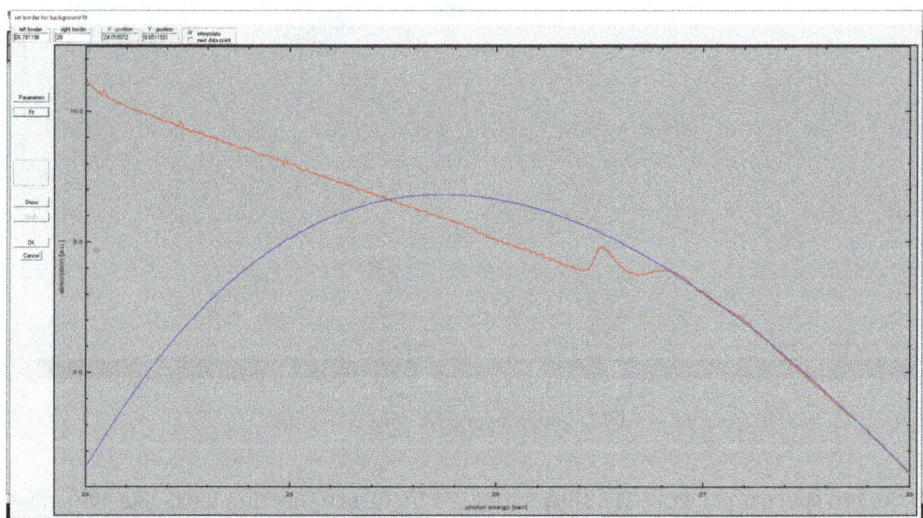

Figure 11.5: NEXAFS spectrum of $CuWO_4$ (red curve) with Victoreen spline curve applied to the high energy tail of the peak structure.

and bothers us more than it did before. We guess now that there is some X-ray absorption edge before the peak structure. We therefore mark what we consider as a sound spectrum with blue and simple cutoff of the remaining and ill-shaped red part; see Figure 11.7. Upon completion of this cutoff of, Fig. 11.8 shows—is this magic?—a "healthy" spectrum. To the experienced spectroscopist, the spectrum in Figure 11.8 looks like an

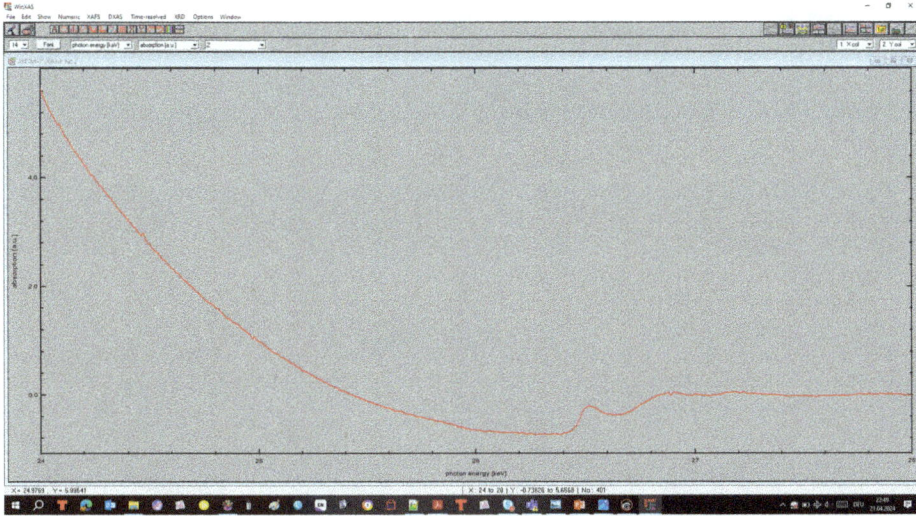

Figure 11.6: NEXAFS spectrum of CuWO$_4$ (red) after the Victoreen spline curve from Figures 11.4 and 11.5 has been subtracted from the spectrum. The high-energy tail of the spectrum is not rectified and horizontal.

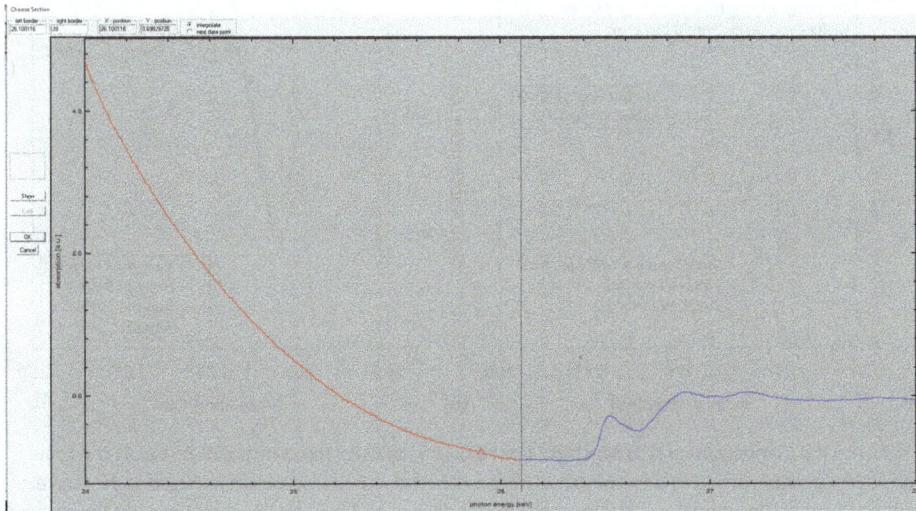

Figure 11.7: NEXAFS spectrum from CuWO$_4$ sample with the blue range/red range separation for energy cut-off of spectrum.

oxygen O1s NEXAFS spectrum. It just happens to be at the wrong X-ray energy: 265 eV instead of 528 eV. Note, we have found this oxygen-like structure at half of the ordinary energy position. We are witnessing here an X-ray optical artifact that is known from acoustics. Figure 11.9 compares this ½ tone O 1s spectrum of CuWO$_4$ (red spectrum on

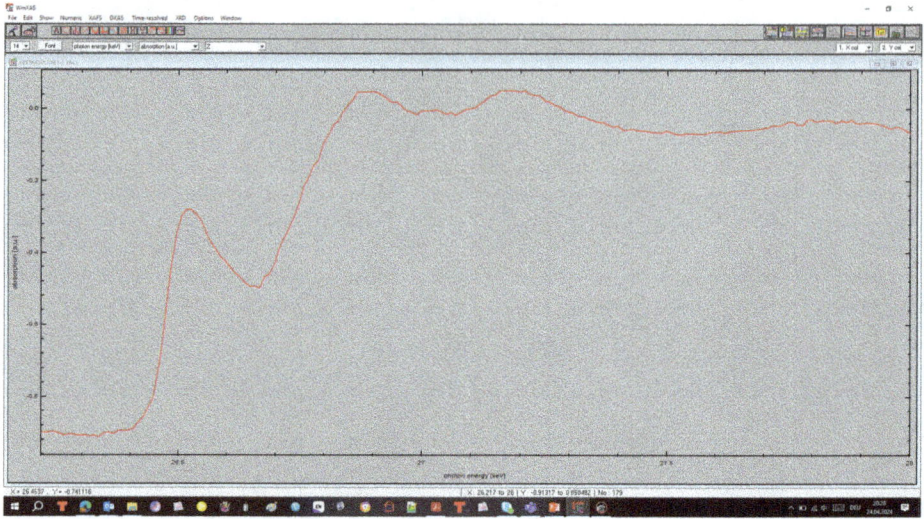

Figure 11.8: NEXAFS spectrum of $CuWO_4$ after rectification and subtraction of high-energy tail and cut-off of low-energy tail. The experienced NEXAFS spectroscopist sees now an oxygen spectrum in the "wrong" energy range.

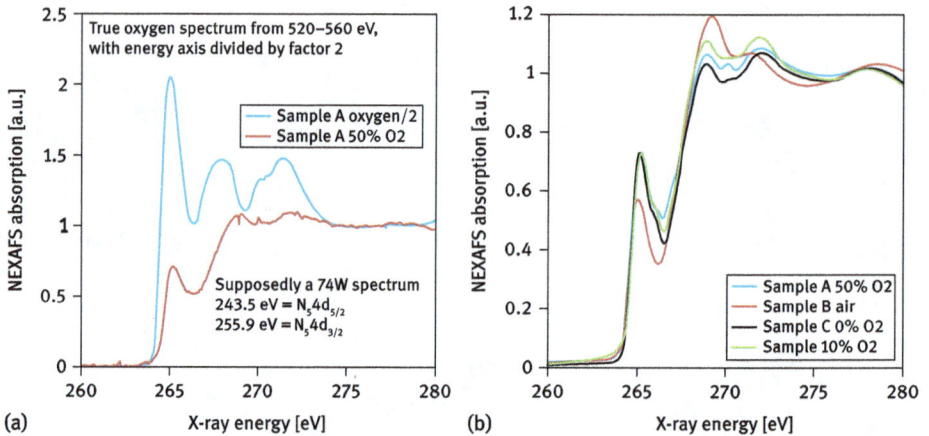

(a)

(b)

Figure 11.9: (a) Comparison of O 1s spectrum from a $CuWO_4$ sample detected at 265 eV, i. e., ½ the common energy position of 528 eV (red spectrum), with a conventional O 1s spectrum detected at the "correct" energy of 528 eV. The energy axis of this spectrum was divided by 2 for ease of comparison. (b) Comparison of the O 1s spectra from four samples subject to different gas phase treatment, after background subtraction. These spectra were detected at ½ the correct energy for O 1s spectra. Spectra were recorded at BESSY-II.

the bottom) with a conventional O 1s spectrum (blue spectrum on top), the energy scale of which we have divided by 2. The overall shape is identical. Differences may be originating from different signal detection, for example. Figure 11.10 shows the four spectra from Figure 11.2 after the background subtraction explained in Figures 11.3 to 11.8. The samples, which had been under different gas phase treatment, show an overall similarity but systematic differences in spectral fine structures.

11.3 Treatment of optical Raman spectra

We had a project on thin ion conducting yttrium stabilized zirconia films (YSZ) films that were pulsed laser deposited on silicon wafers [Braun 2010g]. The film thicknesses were 10, 20, and 50 nm. We were interested in whether there could be a shift of the optical Raman spectra. The experiments were carried out by Edvinas Navickas. The rationale for this was that the films were maybe strained like epitaxial strained films. Figure 11.10 shows in three plots the Raman spectra of the three films deposited on the silicon wafer, plus the spectrum from the empty silicon wafer, plus the difference spectra where the Raman spectrum of the silicon wafer was subtracted each time. The three difference spectra are summarized in Figure 11.11 for visual comparison and further analysis. We do not provide here a spectral assignment of the Raman resonances, but note that the spectra of YSZ-film plus silicon wafer look virtually identical to the Raman spectrum of the silicon wafer. This is because we have a strong Raman scattering contribution from the thick wafer and a correspondingly small contribution from the very thin film. However, the subtraction provides features that we can further analyze. It was necessary for this purpose to study a bulk Raman spectrum of YSZ; see Fig. 3 of Naumenko et al. [Naumenko 2008]. The peak at above 610 wavenumbers is identified as a vibration of the type $E^+_{1\alpha}$.

When comparing your own data with literature data, sometimes it helps to directly overlay spectra or scattering curves or diffractograms on a common scale. The fastest way to do this is pasting your own spectra, e. g., in Microsoft Powerpoint as a picture. Then you copy the spectrum from the other work that you have found in literature and paste it over the first (your own) spectrum in the same PowerPoint window. It is necessary that the spectra have the same energy axis or wavenumber axis. When this is indeed the case, you may have to stretch or compress one of both figures in the energy direction so as to match two arbitrary energy positions for both spectra. These are your anchor points. Usually, you then see already a good match. Maybe you have to adjust the signal height in the vertical axis as well. This all can be done in a couple of minutes, and is a first practical help for judging whether you are on the same page with your colleagues somewhere who worked on the same or similar system.

There may be a situation where you want to work in other researchers' spectra or diffractograms in your own collection of data. I have done this in the complex case studies in this book where I compared historic Li NEXAFS and EELS spectra with my own spectra. For this purpose, I used a digitalization software [Tummers 2006]. This is particularly helpful for those cases where no spectra database exist. The authors of the underlying publications will likely be happy when you properly cite their papers and, depending on the circumstances, obtain the copyright license from the publishers [Garfield 1975].

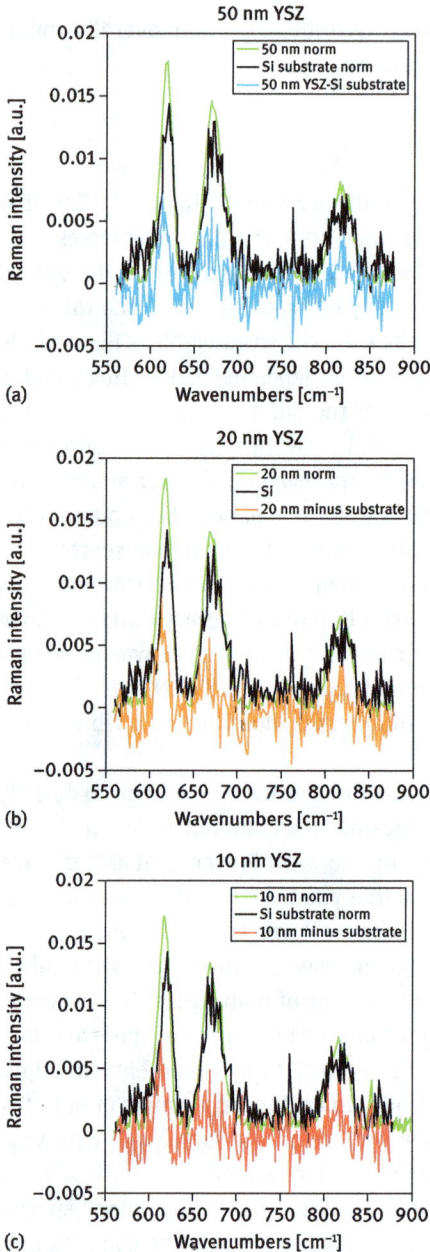

(a)

(b)

(c)

Figure 11.10: Optical Raman spectra of Si wafer, YSZ films deposited on them with 10-, 20-, and 50-nm thickness, and the difference spectra. (a) Spectrum of 50 nm YSZ (green), Si wafer (black), and difference spectrum (blue). (b) Spectrum of 20 nm YSZ (green), Si wafer (black), and difference spectrum (orange). (c) Spectrum of 10 nm YSZ (green), Si wafer (black), and difference spectrum (red).

What we learn from this study is that the Raman mode for the YSZ film varies with the film thickness, as we can see in Figure 11.12. It appears that the thick film of 50 nm has the "highest pitch" because it has the highest Raman frequency. If we can trust our extrapolation to 0, then the Raman shift is more than six wavenumbers. We

Raman spectra of thin YSZ films on Si single crystal substrates

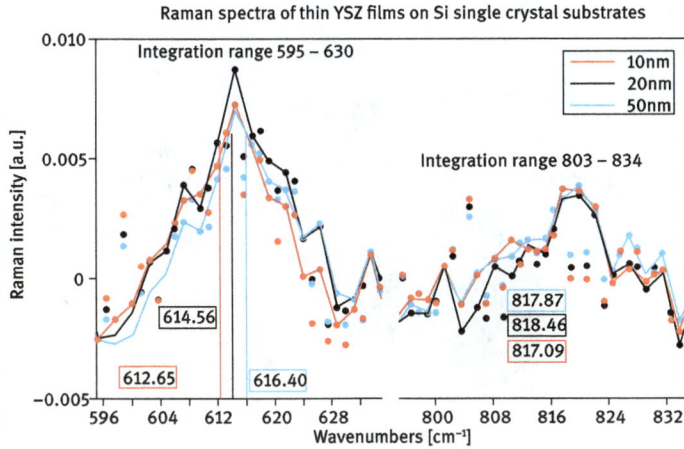

Figure 11.11: Raman intensity for the three films after Si wafer subtraction. The peak areas were integrated from 595 to 630 cm^{-1}, and from 803 to 834 cm^{-1}. Then the center of gravity was determined (first statistical moment). The peak positions were 612.65 cm^{-1} for the 10-nm film, and 614.56 and 616.40 cm^{-1} for the 20- and 50-nm films, respectively.

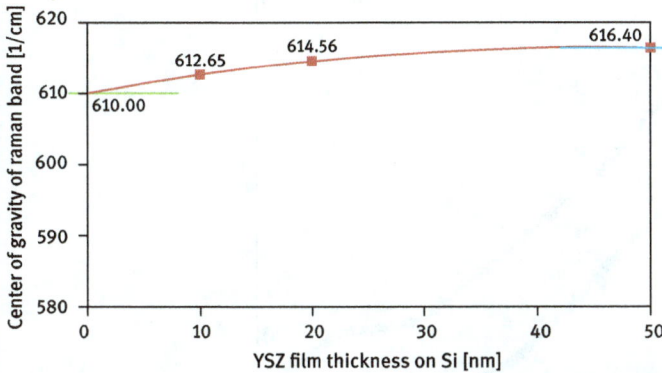

Figure 11.12: Variation of the center of gravity determined for the Raman peak position for YSZ film thicknesses of 10, 20, and 50 nm. The red line is a spline extrapolation, which at 0 nm thickness, gives 610 cm^{-1}.

had made a similar study on proton conducting electrolytes, which we had set under mechanical pressure up to several GPa. With increasing pressure, the wavenumbers were increasing. Upon hydration, there was also a slight increase in the wavenumber [Chen 2011d, Chen 2013a].

11.4 Background subtraction for an empty cell

I am showing here an X-ray spectro-electrochemical cell, which I received from Rüdiger Kötz (PSI) for testing at the Advanced Photon Source for USAXS. The cell is shown in Figure 11.13. It has a PEEK® base plate and a metal Ti body, another PEEK® body and

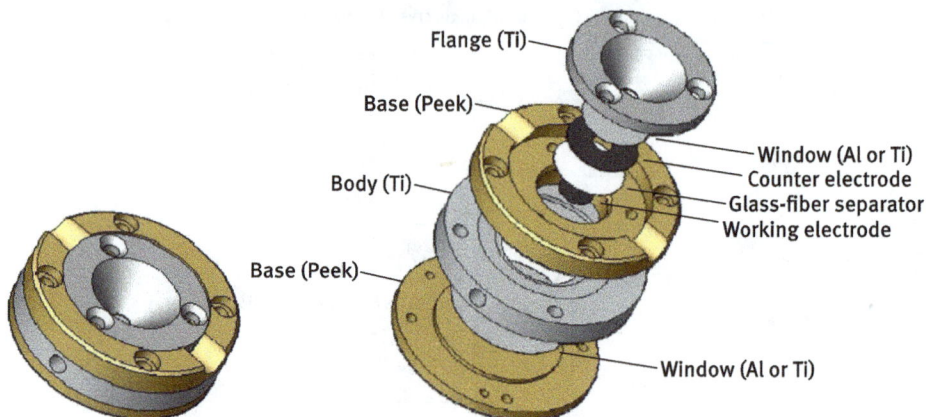

Figure 11.13: X-ray spectro-electrochemical cell for *operando* and *in situ* studies for supercapacitor investigations. Image by courtesy of Rüdiger Kötz, PSI.

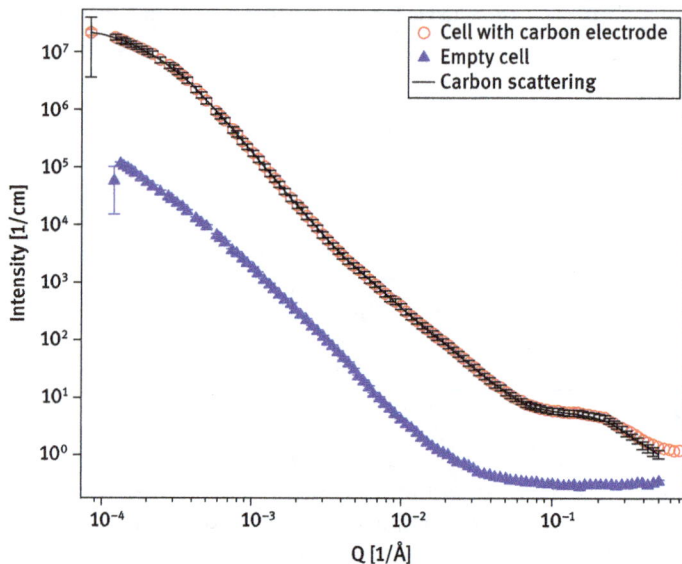

Figure 11.14: USAXS curves from the empty *in situ* cell (blue scattering curve) and the cell filled with a high surface area carbon electrode (red circles). The black solid line is the difference between both scattering curves. Cell and carbon electrode were obtained from Rüdiger Kötz (PSI). Data were recorded at UNICAT APS with Jan Ilavsky (APS) and Andrew Allen (NIST).

a conus flange from Ti metal. The cell was designed for supercapacitor studies. The figure shows a small working electrode disk, a glass fiber separator, and a counterelectrode ring disk plus an X-ray window from Al or Ti. The cell contained a carbon electrode only. I measured this cell plus carbon at UNICAT with Jan Ilavsky and Andrew Allen. The USAXS curves are shown in Figure 11.14. The blue scattering curve is from the "empty cell."

The beam sees only the Al or Ti X-ray window. The other cell components are not in the X-ray beam. The cell plus the carbon electrode produces the scattering curve denoted by the red data circles. The high surface area and porosity of the carbon causes a significant small-angle scattering, around 1 to 2 orders of magnitude larger than the blue scattering curve from the empty cell. The solid black line is the difference of both curves. Only in the high Q range do we see a noticeable difference from both scattering curves.

A Appendix

A.1 X-ray physics Nobel prizes

The first Nobel Prize in Physics was awarded to physicist, Dr. Wilhelm Conrad Röntgen, in 1901, for the discovery of the X-rays in 1895 [Röntgen 1901]. Röntgen's X-ray source was a Hittorf cathode tube where he could image the skeleton on a film.

Philipp Eduard Anton von Lenard received the Nobel Prize in Physics 1905 for his work on cathode rays and claimed therefore a share for the Röntgen's Nobel Prize.

Max von Laue received the prize 1914 "for the discovery of X-ray diffraction in crystals." The Nobel Prize in Physics 1915 went to Sir William Henry Bragg and William Lawrence Bragg "for their services in the analysis of crystal structure by means of X-rays." Today, X-ray diffraction is undoubted the most important X-ray based characterization method in the laboratory—for phase determination and crystal structure analysis.

The Nobel Prize in Physics 1917 went to Charles Glover Barkla "for his discovery of the characteristic Röntgen radiation of the elements." This property allows for element specific chemical analyses like EDAX, which even nowadays is used in electron microscopes. Closely related to this property is the Nobel Prize in Physics 1918 to Max Karl Ernst Ludwig Planck "in recognition of the services he rendered to the advancement of Physics by his discovery of energy quanta." This provides the important relation between X-ray energy and X-ray wavelength $E = h\nu = hc/\lambda$.

The Nobel Prize in Physics 1921 went to Albert Einstein "for his services to Theoretical Physics, and especially for his discovery of the law of the photoelectric effect." In this effect, an electron is emitted from an electrode material when it is hit by electromagnetic waves of sufficient high energy. This is an important detection method in X-ray spectroscopy experiments. Moreover, the reversal of this effect is that when electrons hit a target, it may emit electromagnetic waves, such as in an X-ray tube. It is therefore closely related to the prizes from 1914, 1915, and 1917.

The trend continues with the Nobel Prize in Physics 1922 to Niels Henrik David Bohr "for his services in the investigation of the structure of atoms and of the radiation emanating from them." This radiation includes the X-rays, and the structure of the atoms to some extent is important for discussing most of the X-ray analytical methods.

The Nobel Prize in 1924 in Physics went to Karl Manne Georg Siegbahn "for his discoveries and research in the field of X-ray spectroscopy."

The Nobel Prize in 1925 in Physics went to James Franck and Gustav Ludwig Hertz "for their discovery of the laws governing the impact of an electron upon an atom." This shows that the energy levels of atoms are discrete, and important for X-ray spectroscopy.

The Nobel Prize in Physics 1927 went to Arthur Holly Compton "for his discovery of the effect named after him." Compton scattering is the inelastic scattering of a photon by a quasi-free charged particle, usually an electron. Compton scattering is often

https://doi.org/10.1515/9783110794038-012

considered a "side effect" in X-ray spectroscopy and sometimes needs to be identified, quantified, and subtracted for the correct quantitative interpretation of X-ray spectra.

The Nobel Prize in Physics 1929 went to Prince Louis-Victor Pierre Raymond de Broglie "for his discovery of the wave nature of electrons." This one is particularly important in relation to matter waves such as from neutrons, which is another very important probe for materials analysis such as neutron scattering and spectroscopy. I mention here directly the Nobel Prize in Physics 1935 to James Chadwick "for the discovery of the neutron" as well. The Nobel Prize in Physics 1937 went to Clinton Joseph Davisson and George Paget Thomson "for their experimental discovery of the diffraction of electrons by crystals." The universality of interference and the wave particle dualism is a very fundamental insight in Physics from the theoretical and practical perspective.

The Nobel Prize in Physics 1930 went to Sir Chandrasekhara Venkata Raman "for his work on the scattering of light and for the discovery of the effect named after him" does not really belong here, but the Raman effect is not only limited to optical wavelengths but also observed in X-ray emission spectroscopy, along with the Compton scattering.

The Nobel Prize in Physics 1932 went to Werner Karl Heisenberg "for the creation of quantum mechanics, the application of which has, among other things, led to the discovery of the allotropic forms of hydrogen." Heisenberg's basic theory is so general that it entered all X-ray scattering theory.

The Nobel Prize in Physics 1933 went to Erwin Schrödinger and Paul Adrien Maurice Dirac "for the discovery of new productive forms of atomic theory." The Schrödinger equation is the start for all calculations of the energy levels in atoms, molecules, and solid matter. Closely related is the Nobel Prize in Physics 1954 to Max Born "for his fundamental research in quantum mechanics, especially for his statistical interpretation of the wavefunction." Born's approximation of the time dependent Schrödinger equation is the entry for the scattering theory, which is important for X-ray spectroscopy.

The Nobel Prize in Physics 1981 went to Kai M. Siegbahn (son of Karl M. G. Siegbahn) "for his contribution to the development of high-resolution electron spectroscopy," because it is a synchrotron-based spectroscopy method, i. e., X-ray photoelectron spectroscopy.

The Nobel Prize in Physics 1992 went to Georges Charpak "for his invention and development of particle detectors, in particular, the multiwire proportional chamber" for its general utility as X-ray detectors at synchrotrons, although not widely used.

I want to mention here the Nobel Prize in Physics 1994 "for pioneering contributions to the development of neutron scattering techniques for studies of condensed matter" to Bertram N. Brockhouse and "for the development of neutron spectroscopy" and to Clifford G. Shull "for the development of the neutron diffraction technique." This is a direct application correspondence to the X-ray methods.

Although the field of chemistry has tremendously benefited from X-ray methods, the field itself was awarded with Nobel Prizes where X-rays were critical for the success.

The Nobel Prize in Physiology or Medicine 1946 went to Hermann Joseph Muller "for the discovery of the production of mutations by means of X-ray irradiation," noteworthy in general with respect to the wide field of occupational health and safety when working with X-rays.

The Nobel Prize in Chemistry 1936 went to Petrus (Peter) Josephus Wilhelmus Debye "for his contributions to our knowledge of molecular structure through his investigations on dipole moments and on the diffraction of X-rays and electrons in gases."

The Nobel Prize in Physiology or Medicine 1962 went to Francis Harry Compton Crick, James Dewey Watson, and Maurice Hugh Frederick Wilkins "for their discoveries concerning the molecular structure of nucleic acids and its significance for information transfer in living material." Here, I want to mention that Watson and Wilkins based their conclusions on the X-ray crystallography results that they had taken from Mrs. Rosalind E. Franklin, who had made pioneering contributions to the field of X-ray crystallography on crystals and on disordered materials, specifically carbon. She was not awarded with the Nobel Prize, although she published her independent results on the structure of the DNA jointly with her PhD student Raymond Gosling in April 1953.

The Nobel Prize in Chemistry 1964 went to Dorothy Crowfoot Hodgkin "for her determinations by X-ray techniques of the structures of important biochemical substances."

The Nobel Prize in Chemistry 1985 went to Herbert A. Hauptman and Jerome Karle "for their outstanding achievements in the development of direct methods for the determination of crystal structures."

The Nobel Prize in Chemistry 1988 went to Johann Deisenhofer, Robert Huber, and Hartmut Michel "for the determination of the three-dimensional structure of a photosynthetic reaction center."

Finally, is the Nobel Prize in Chemistry 1959 to Jaroslav Heyrovský "for his discovery and development of the polarographic methods of analysis," basically an extension of cyclic voltammetry. He developed this method in 1922. During WW1, Heyrovsky worked in a hospital as a chemist and a radiologist. At about the same time, Karl Siegbahn worked with electricity and magnetism and later turned to his new field of X-ray spectroscopy. Thus, it is interesting to observe how intertwined the careers of researchers in the early 20th century would eventually evolve, as shown by this particular example.

In 1920, Walther Nenrst received the Nobel Prize in Chemistry in recognition of his work in thermochemistry.

The Nobel Prize in Chemistry 1918 was awarded to Fritz Haber "for the synthesis of ammonia from its elements," but Haber authored a textbook in 1898 about electrochemistry. Apart from the Nobel Prize, Haber received many honors during his life. At Max von Laue's instigation, the Institute for Physical and Electrochemistry at Berlin-Dahlem was renamed the Fritz Haber Institute after his death.

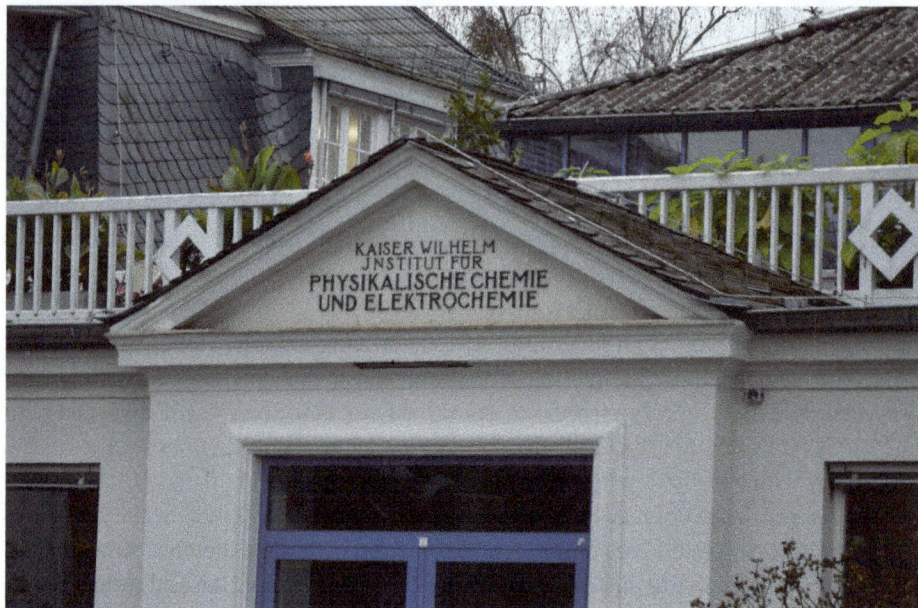

Figure A.1: Front side of the original building of the Kaiser Wilhelm Institut für Physikalische Chemie und Elektrochemie, now Fritz Haber Institut der Max Planck Gesellschaft in Berlin-Dahlem, Germany.

A.2 Synchrotron centers worldwide

Table A.1: List of synchrotron centers worldwide in alphabetical order by country.

Name	Country	Website
The African Lightsource (project phase only)	Africa	https://www.africanlightsource.org/
CANDELA	Armenia	http://www.candle.am/index.html
Australian Synchrotron	Australia	http://www.synchrotron.org.au
Laboratorio Nacional de Luz Sincrotron	Brazil	http://www.lnls.br/
Canadian Light Source	Canada	http://www.lightsource.ca
Beijing Synchrotron Radiation Facility	China	http://bsrf.ihep.cas.cn/
National Synchrotron Radiation Laboratory	China	http://www.nsrl.ustc.edu.cn/
SSRF – Shanghai Synchrotron Radiation Facility	China	http://e-ssrf.sari.ac.cn/
Institute for Storage Ring Facilities	Denmark	http://www.isa.au.dk/
European Synchrotron Radiation Facility	France	http://www.esrf.eu
SOLEIL	France	http://www.synchrotron-soleil.fr/
Angstromquelle Karlsruhe - ANKA	Germany	https://www.ibpt.kit.edu/KIT_Light_Source.php
BESSY II – Helmholtz-Zentrum Berlin	Germany	http://www.helmholtz-berlin.de/
Dortmund Electron Storage Ring Facility	Germany	http://www.delta.tu-dortmund.de/
ELSA – Electron Stretcher Accelerator	Germany	http://www-elsa.physik.uni-bonn.de/elsa-facility_en.html

Table A.1 (continued)

Name	Country	Website
Metrology Light Source	Germany	http://www.ptb.de/mls/
PETRA III at DESY	Germany	http://photon-science.desy.de
INDUS	India	https://www.rrcat.gov.in/technology/accel/indus/ioas/index.html
Iranian Light Source Facility	Iran	http://ilsf.ipm.ac.ir/
DAFNE	Italy	http://web.infn.it/Dafne_Light/
Elettra Synchrotron Light Laboratory	Italy	http://www.elettra.eu
Aichi Synchrotron Radiation Center	Japan	http://www.astf-kha.jp/synchrotron/en/
Hiroshima Synchrotron Radiation Center	Japan	http://www.hsrc.hiroshima-u.ac.jp/index.html
Photon Factory	Japan	http://pfwww.kek.jp/
Ritsumeikan University SR Center	Japan	http://www.ritsumei.ac.jp/acd/re/src/index.htm
Saga Light Source	Japan	http://www.saga-ls.jp/?page=206
SPring-8	Japan	http://www.spring8.or.jp/en/
Ultraviolet Synchrotron Orbital Radiation Facility	Japan	http://www.uvsor.ims.ac.jp/defaultE.html
SESAME	Jordan	https://www.sesame.org.jo/
Pohang Light Source II	Korea	http://paleng.postech.ac.kr
SOLARIS	Poland	https://synchrotron.uj.edu.pl/en_GB/start
Dubna Electron Synchrotron	Russia	http://wwwinfo.jinr.ru/delsy/
Kurchatov Synchrotron Radiation Source	Russia	http://www.nrcki.ru/e/engl.html
Siberian Synchrotron Research Centre	Russia	https://www.inp.nsk.su/budker-institute-of-nuclear-physics
TNK	Russia	http://www.niifp.ru/page/sinhrotron
Singapore Synchrotron Light Source	Singapore	http://ssls.nus.edu.sg/index.html
ALBA	Spain	http://www.cells.es/
MAX IV Laboratory	Sweden	https://www.maxiv.lu.se/
Swiss Light Source	Switzerland	http://www.psi.ch/sls/
National Synchrotron Radiation Research Center	Taiwan	https://www.nsrrc.org.tw/english/index.aspx
Synchrotron Light Research Institute	Thailand	http://www.slri.or.th
Diamond Light Source	United Kingdom	http://www.diamond.ac.uk/
Advanced Light Source	USA	http://www-als.lbl.gov/
Advanced Photon Source	USA	http://www.aps.anl.gov
Center for Advanced Microstructures and Devices	USA	http://www.camd.lsu.edu/
Cornell High Energy Synchrotron Source	USA	http://www.chess.cornell.edu/
National Synchrotron Light Source II	USA	http://www.bnl.gov/ps/
Stanford Synchrotron Radiation Lightsource	USA	http://www-ssrl.slac.stanford.edu
Synchrotron Ultraviolet Radiation Facility	USA	http://physics.nist.gov/MajResFac/SURF/SURF/index.html

A.3 Electromagnetic spectrum

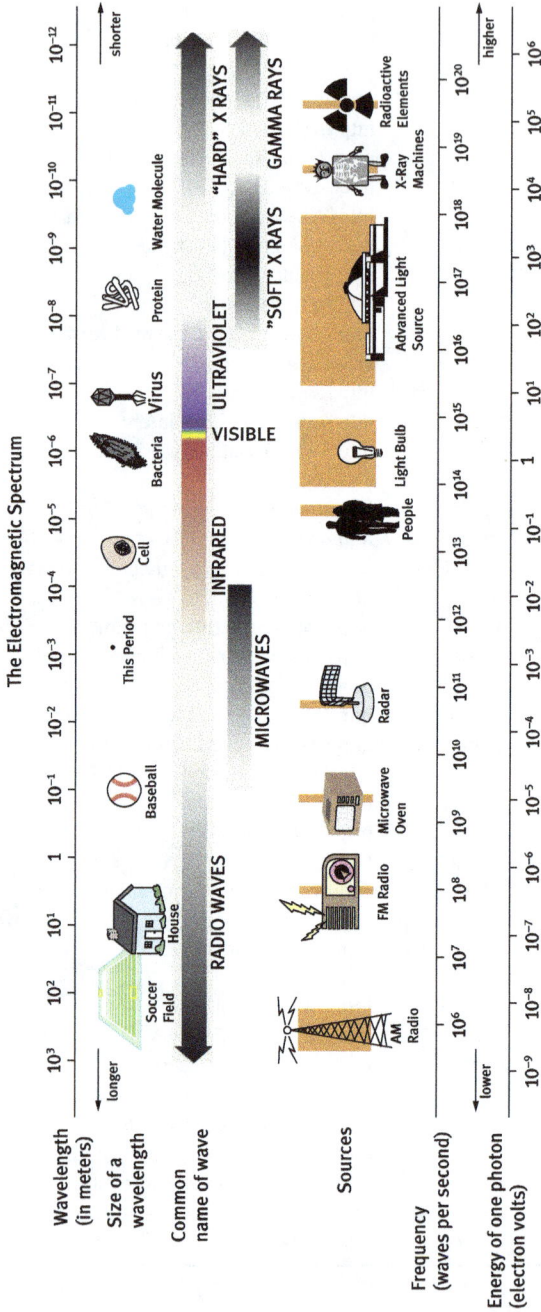

Figure A.2: Courtesy of the Advanced Light Source, Lawrence Berkeley National Laboratory.

A.4 Kα, β X-ray energies

Table A.2: Experimental Kα X-ray energies adopted from [Nave 2016] and http://physics.nist.gov/ PhysRefData. *Interpolated from nearby elements. #means large deviation between theory and experiment.

Z	Element	$K\alpha_2$ / eV	$K\alpha_1$ / eV	Z	Element	$K\alpha_2$ / eV	$K\alpha_1$ / eV
10	Ne	848.61	848.61	53	I	28317.52	28612.32
11	Na	1040.98	1040.98	54	Xe	29458.25	29778.78
12	Mg	1253.437	1253.69	55	Cs	30625.40	30973.13
13	Al	1486.295	1486.71	56	Ba	31816.62	32193.26
14	Si	1739.394	1739.985	57	La	33034.38	33442.12
15	P	2012.70*	2013.68*	58	Ce	34279.28	34720.00
16	S	2306.700	2307.89	59	Pr	35550.59	36026.71
17	Cl	2620.846	2622.440	60	Nd	37360.74	37360.74
18	Ar	2955.566	2957.682	61	Pm	38171.55	38725.11
19	K	3311.196	3313.95	62	Sm	39523.39	40118.48
20	Ca	3688.128	3691.72	63	Eu	40902.33	41542.63
21	Sc	4085.95	4090.74	64	Gd	42309.30	42996.72
22	Ti	4504.92	4510.90	65	Tb	43744.62	44482.75
23	V	4944.671	4952.22	66	Dy	45208.27	45998.94
24	Cr	5405.54	5414.81	67	Ho	46699.98	47547.10
25	Mn	5887.69	5898.80	68	Er	48221.61	49127.24
26	Fe	6391.03	6404.01	69	Tm	49772.67	50741.48
27	Co	6915.54	6930.38	70	Yb	51354.60	52389.48
28	Ni	7461.03	7478.25	71	Lu	52965.57	54070.39
29	Cu	8027.84	8047.82	72	Hf	54612.0#	55790.8#
30	Zn	8615.82	8638.91	73	Ta	56277.6	57533.2
31	Ga	9224.84	9251.67	74	W	57981.77	59318.85
32	Ge	9855.42	9886.52	75	Re	59718.57	61141.00
33	As	10507.50	10543.27	76	Os	61487.27	63001.07
34	Se	11181.53	11222.52	77	Ir	63287.29	64896.2
35	Br	11877.75	11924.36	78	Pt	65123.3	66832.9
36	Kr	12595.42	12648.00	79	Au	66990.73	68804.50
37	Rb	13335.88	13395.49	80	Hg	68895.1	70819.5
38	Sr	14098.03	14165.20	81	Tl	70832.5	72872.5
39	Y	14882.94	14958.54	82	Pb	72805.42	74970.11
40	Zr	15690.65	15774.91	83	Bi-209	74816.21	77109.2
41	Nb	16521.28	16615.16	84	Po-209	76864.4*	79292.9*
42	Mo	17374.29	17479.37	85	At-210	78944*	81514*
43	Tc	18250.9	18367.2*	86	Rn-222	81066*	83783*
44	Ru	19150.49	19279.16	87	Fr-223	83232*	86105*
45	Rh	20073.67	20216.12	88	Ra-226	85436	88476
46	Pd	21020.15	21177.08	89	Ac-227	87676*	90884.8*
47	Ag	21990.30	22162.91	90	Th-232	89957.04	93347.38
48	Cd	22984.05	23173.98	91	Pa-231	92283.4	95866.4

Table A.2 (continued)

Z	Element	$K\alpha_2$ / eV	$K\alpha_1$ / eV	Z	Element	$K\alpha_2$ / eV	$K\alpha_1$ / eV
49	In	24002.03	24209.75	92	U-238	94650.84	98431.58
50	Sn	25044.04	25271.36	93	Np-237	97068.4	*
51	Sb	26110.78	26358.86	94	Pu-239	99523.2	*
52	Te	27201.99	27472.57	94	Pu-244	99529.4	*

A.5 Periodic table of elements

The Periodic Table of the Elements

1																		2
H Hydrogen 1.00794																		**He** Helium 4.003
3	4											5	6	7	8	9	10	
Li Lithium 6.941	**Be** Beryllium 9.012182											**B** Boron 10.811	**C** Carbon 12.0107	**N** Nitrogen 14.00674	**O** Oxygen 15.9994	**F** Fluorine 18.9984032	**Ne** Neon 20.1797	
11	12											13	14	15	16	17	18	
Na Sodium 22.989770	**Mg** Magnesium 24.3050											**Al** Aluminum 26.981538	**Si** Silicon 28.0855	**P** Phosphorus 30.973761	**S** Sulfur 32.066	**Cl** Chlorine 35.4527	**Ar** Argon 39.948	
19	20	21	22	23	24	25	26	27	28	29	30	31	32	33	34	35	36	
K Potassium 39.0983	**Ca** Calcium 40.078	**Sc** Scandium 44.955910	**Ti** Titanium 47.867	**V** Vanadium 50.9415	**Cr** Chromium 51.9961	**Mn** Manganese 54.938049	**Fe** Iron 55.845	**Co** Cobalt 58.933200	**Ni** Nickel 58.6934	**Cu** Copper 63.546	**Zn** Zinc 65.39	**Ga** Gallium 69.723	**Ge** Germanium 72.61	**As** Arsenic 74.92160	**Se** Selenium 78.96	**Br** Bromine 79.904	**Kr** Krypton 83.80	
37	38	39	40	41	42	43	44	45	46	47	48	49	50	51	52	53	54	
Rb Rubidium 85.4678	**Sr** Strontium 87.62	**Y** Yttrium 88.90585	**Zr** Zirconium 91.224	**Nb** Niobium 92.90638	**Mo** Molybdenum 95.94	**Tc** Technetium (98)	**Ru** Ruthenium 101.07	**Rh** Rhodium 102.90550	**Pd** Palladium 106.42	**Ag** Silver 107.8682	**Cd** Cadmium 112.411	**In** Indium 114.818	**Sn** Tin 118.710	**Sb** Antimony 121.760	**Te** Tellurium 127.60	**I** Iodine 126.90447	**Xe** Xenon 131.29	
55	56	57	72	73	74	75	76	77	78	79	80	81	82	83	84	85	86	
Cs Cesium 132.90545	**Ba** Barium 137.327	**La** Lanthanum 138.9055	**Hf** Hafnium 178.49	**Ta** Tantalum 180.9479	**W** Tungsten 183.84	**Re** Rhenium 186.207	**Os** Osmium 190.23	**Ir** Iridium 192.217	**Pt** Platinum 195.078	**Au** Gold 196.96655	**Hg** Mercury 200.59	**Tl** Thallium 204.3833	**Pb** Lead 207.2	**Bi** Bismuth 208.98038	**Po** Polonium (209)	**At** Astatine (210)	**Rn** Radon (222)	
87	88	89	104	105	106	107	108	109	110	111	112	113	114					
Fr Francium (223)	**Ra** Radium (226)	**Ac** Actinium (227)	**Rf** Rutherfordium (261)	**Db** Dubnium (262)	**Sg** Seaborgium (263)	**Bh** Bohrium (262)	**Hs** Hassium (265)	**Mt** Meitnerium (266)	(269)	(272)	(277)							

58	59	60	61	62	63	64	65	66	67	68	69	70	71
Ce Cerium 140.116	**Pr** Praseodymium 140.90765	**Nd** Neodymium 144.24	**Pm** Promethium (145)	**Sm** Samarium 150.36	**Eu** Europium 151.964	**Gd** Gadolinium 157.25	**Tb** Terbium 158.92534	**Dy** Dysprosium 162.50	**Ho** Holmium 164.93032	**Er** Erbium 167.26	**Tm** Thulium 168.93421	**Yb** Ytterbium 173.04	**Lu** Lutetium 174.967
90	91	92	93	94	95	96	97	98	99	100	101	102	103
Th Thorium 232.0381	**Pa** Protactinium 231.03588	**U** Uranium 238.0289	**Np** Neptunium (237)	**Pu** Plutonium (244)	**Am** Americium (243)	**Cm** Curium (247)	**Bk** Berkelium (247)	**Cf** Californium (251)	**Es** Einsteinium (252)	**Fm** Fermium (257)	**Md** Mendelevium (258)	**No** Nobelium (259)	**Lr** Lawrencium (262)

Courtesy: TRIUMF—Canada's national laboratory for particle and nuclear physics.

A.6 Electronic configuration of the chemical elements

Table A.3: Electron configurations of the elements. The electron configurations of elements indicated in red are exceptions because of the added stability associated with half-filled and filled subshells. The electron configurations of the elements indicated in blue are also anomalous, but the reasons for the observed configurations are more complex. Adopted from UC Davis ChemWiki.

Atomic symbol electron number configuration	Atomic symbol electron number configuration	Atomic symbol electron number configuration
1 H $1s^1$	37 Rb [Kr]$5s^1$	73 Ta [Xe]$6s^2 4fl^4 5cf3$
2 He $1s^2$		74 W [Xe]$6s24fl^4 5d^4$
3 Li [He]$2s^1$	38 Sr [Kr]$5s^2$	75 Re [Xe]$6s^2 4f1^4 5d^5$
4 Be [He]$2s^2$	39 Y [Kr]$Ss24d^1$	76 Os [Xe]$6s^2 4fl^4 5di$
5 B [He]$2s^2 2p^1$	40 Zr [Kr]$5s^2 4d2$	77 Ir [Xe]$6s^2 4fl^4 5cf$
6 C [He]$2s^2 2p^2$	41 Nb [Kr]$5s^1 4d^4$	78 Pt [Xe]$6s^1 4fl^4 5cJ9$
7 N [He]$2s^2 2p^3$	42 Mo [Kr]$5s^1 4d^5$	79 Au [Xe]$6s^1 4fl^4 5d^{10}$
8 0 [He]$2s^2 2p^4$	43 Tc [Kr]$5s^2 4d^5$	80 Hg [Xe]$6s^2 4fl^4 5d^{10}$
9 F [He]$2s^2 2p^5$	44 Ru [Kr]$5s^1 4cf$	81 Tl [Xe]$6s^2 4fl^4 5d^{10} 6p^1$
10 Ne [He]$2s^2 2p^6$		82 Pb [Xe]$6s^2 4fl^4 5d^{10} 6p^2$
11 Na [Ne]$3s^1$	45 Rh [Kr]$5s^1 4<fl$	83 Bi [Xe]$6s^2 4fl^4 5d^{10} 6p^3$
12 Mg [Ne]$3s^2$	46 Pd [Kr]$4d^{10}$	84 Po [Xe]$6s^2 4fl^4 5d^{10} 6p^4$
13 Al [Ne]$3s^2 3p^1$	47 Ag [Kr]$5s^1 4d^{10}$	85 At [Xe]$6s^2 4fl^4 5d^{10} 6p^5$
14 Si [Ne]$3s^2 3p^2$	48 Cd [Kr]$5s^2 4d^{10}$	86 Rn [Xe]$6s^2 4fl^4 5d^{10} 6p^6$
15 P [Ne]$3s^2 3p^3$	49 In [Kr]$5s^2 4d^{10} 5p^1$	
16 S [Ne]$3s^2 3p^4$	50 Sn [Kr]$5s^2 4d^{10} 5p^2$	87 Fr [Rn]$7s^1$
17 Cl [Ne]$3s^2 3p^5$		88 Ra [Rn]$7s^2$
18 Ar [Ne]$3s^2 3p^6$	51 Sb [Kr]$5s^2 4d^{10} 5p^3$	89 Ac [Rn]$7s^2 6d^1$
19 K [Ar]$4s^1$	52 Te [Kr]$5s^2 4d^{10} 5p^4$	90 Th [Rn]$7s^2 6d2$
20 Ca [Ar]$4s^2$	53 I [Kr]$5s^2 4d^{10} 5p^5$	91 Pa [Rn]$7s^2 5f26d^1$
21 Sc [Ar]$4s^2 3d^1$	54 Xe [Kr]$5s^2 4d^{10} 5p^6$	92 u [Rn]$7s^2 5f36d^1$
22 n [Ar]$4s^2 3d2$	55 Cs [Xe]$6s^1$	93 Np [Rn]$7s^2 5f'6d^1$
23 V [Ar]$4s^2 3dl$	56 Ba [Xe]$6s^2$	94 Pu [Rn]$7s^2 5f0$
24 Cr [Ar]$4s^1 3d^5$	57 La [Xe]$6s^2 5d^1$	95 Am [Rn]$7s^2 5f7$
25 Mn [Ar]$4s^2 3d5$	58 Ce [Xe]$6s^2 4fl5d^1$	96 Cm [Rn]$7s^2 5f76d^1$
26 Fe [Ar]$4s^2 3cf$	59 Pr [Xe]$6s^2 4f3$	97 Bk [Rn]$7s^2 5f9$
27 Co [Ar]$4s^2 3cf$	60 Nd [Xe]$6s^2 4f'$	98 Cf [Rn]$7s^2 5fl^0$
28 Ni [Ar]$4s^2 3d8$	61 Pm [Xe]$6s^2 4f5$	99 Es [Rn]$7s^2 5f1^1$
29 Cu [Ar]$4s^1 3d^{10}$	62 Sm [Xe]$6s^2 4.f0$	100 Fm [Rn]$7s^2 5f1^2$
30 Zn [Ar]$4s^2 3d^{10}$	63 Eu [Xe]$6s^2 4f$	101 Md [Rn]$7s^2 5f1^3$
31 Ga [Ar]$4s^2 3d^{10} 4p^1$	64 Gd [Xe]$6s^2 4f75d^1$	102 No [Rn]$7s^2 5f1^4$
32 Ge [Ar]$4s^2 3d^{10} 4p^2$	65 Tb [Xe]$6s^2 4f9$	103 Lr [Rn]$7s^2 5f1^4 6d^1$
33 As [Ar]$4s^2 3d^{10} 4p^3$	66 Dy [Xe]$6s^2 4f1°$	104 Rf [Rn]$7s^2 5f1^4 6d'$
34 Se [Ar]$4s^2 3d^{10} 4p^4$	67 Ho [Xe]$6s^2 4f1^1$	105 Db [Rn]$7s^2 5f1^4 6cf3$

Table A.3 (continued)

Atomic symbol electron number configuration	Atomic symbol electron number configuration	Atomic symbol electron number configuration
35 Br [Ar]$4s^2 3d^{10} 4p^5$	68 Er [Xe]$6s^2 4f1^2$	106 Sg [Rn]$7s^2 5f1^4 6d4$
36 Kr [Ar]$4s^2 3d^{10} 4p^6$	69 Tm [Xe]$6s^2 4f1^3$	107 Bh [Rn]$7s^2 5f1^4 6d5$
	70 Yb [Xe]$6s^2 4f1^4$	108 Hs [Rn]$?s^2 5f1^4 6di$
	71 Lu [Xe]$6s^2 4f1^4 5d^1$	109 Mt [Rn]$7s^2 5f1^4 6cf$
	72 Hf [Xe]$6s^2 4f1^4 5d2$	110 Ds [Rn]$7s^1 : 5f1^4 6cf$
		111 Rg [Rn]$7s^1 5f1^4 6d^{10}$

A.7 X-ray data analysis software

UNIFIT
WinXAS
EXAFSPAK
https://www.esrf.fr/computing/scientific/exafs/intro.html

Bibliography

[Abbate 1992] Abbate M, Degroot FMF, Fuggle JC, Fujimori A, Strebel O, Lopez F, Domke M, Kaindl G, Sawatzky GA, Takano M, Takeda Y, Eisaki H, Uchida S: Controlled-valence properties of La1-Xsrxfeo3 and La1-Xsrxmno3 studied by soft-X-ray absorption-spectroscopy. *Physical Review B* 1992, 46:4511–4519.

[Acheson 1910] Acheson EG: *A Pathfinder: Discovery, Invention and Industry; How the World Came to Have Aquadag and Oildag; Also Carborundum, Artificial Graphite.* New York: The Press Scrap Book; 1910.

[Ade 1990] Ade H, Kirz J, Hulbert SL, Johnson ED, Anderson E, Kern D: X-ray spectromicroscopy with a zone plate generated microprobe. *Applied Physics Letters* 1990, 56:1841–1844.

[Ade 1992] Ade H, Zhang X, Cameron S, Costello C, Kirz J, Williams S: Chemical contrast in X-ray microscopy and spatially resolved XANES spectroscopy of organic specimens. *Science* 1992, 258:972–975.

[Aguey-Zinsou 2015] Aguey-Zinsou K-F, Braun A, Cowan DA, Diale MM, Grossmann G, Janssen P, Makhalanyane T, Ponomarev A, Пономарев A, Reutemann J: *Algal Solar Hydrogen Fuel Reactor for Renewable Logistics in the Antarctic.* Research Proposal. Antarctica: Antarctic Circumnavigation Expedition ("ACE"); 2015.

[Agui 2005] Agui A, Butorin SM, Kaambre T, Sathe C, Saitoh T, Moritomo Y, Nordgren J: Resonant Mn L emission spectra of layered manganite $La_{1.2}Sr_{1.8}Mn_2O_7$. *Journal of the Physical Society of Japan* 2005, 74:1772–1776.

[Ahsan 2009] Ahsan M, Tesfamichael T, Bell J, Blackford MG: Microstructural characterization of electron beam evaporated tungsten oxide films for gas sensing applications. In *16th AINSE Conference on Nuclear and Complementary Techniques of Analysis; 2009; Lucas Heights, Sydney.*

[Alexandropoulos 1972] Alexandropoulos NG, Kuriyama M, Cohen GG: Investigation of energy-loss spectrum of li for K/KF = 2.52. *Bulletin of the American Physical Society* 1972, 17:271.

[Allen 2004] Allen AJ, Dobbins TA, Zhao F, Virkar A, Ilavsky J, Almer J, Decarlo F: Characterization of solid oxide fuel cell layers by computed X-ray microtomography and small-angle scattering. *Ceramic Engineering and Science Proceedings* 2004, 25:275–280.

[Allen 2005] Allen AJ: Characterization of ceramics by X-ray and neutron small-angle scattering. *Journal of the American Ceramic Society* 2005, 88:1367–1381.

[Allen 2009] Allen AJ, Ilavsky J, Braun A: Multi-scale microstructure characterization of solid oxide fuel cell assemblies with ultra small-angle X-ray scattering. *Advanced Engineering Materials* 2009, 11:495–501.

[Allen 2014] Allen AJ, Ilavsky J, Jemian PR, Braun A: Evolution of electrochemical interfaces in solid oxide fuel cells (SOFC): a Ni and Zr resonant anomalous ultra-small-angle X-ray scattering study with elemental and spatial resolution across the cell assembly. *RSC Advances* 2014, 4:4676–4690.

[Allen 1991] Allen JL: *Biosphere 2: The Human Experiment.* Penguin Books; 1991.

[Allen 2003] Allen JP, Nelson M, Alling A: The legacy of Biosphere 2 for the study of biospherics and closed ecological systems. *Space Life Sciences: Closed Artificial Ecosystems and Life Support Systems* 2003, 31:1629–1639.

[Alleno 1999] Alleno E, Godart C, Fisher B, Genossar J, Patlagan L, Reisner GM: Valence of Pr in $Y_{0.8}Pr_{0.2}Ba_2Cu_3O_7$ and $PrBa_2Cu_3$-$xCoxO_7$ (0 <= x <= 1). *Physica. B, Condensed Matter* 1999, 259–61:530–532.

[Allisson 1930a] Allisson F: Oxydo-Reduktionen mit Chlorophyll und anderen Sensibilatoren. *Helvetica Chimica Acta* 1930, 13:18.

[Allisson 1930b] Allisson F: Oxydo-Reduktionen mit Chlorophyll und anderen Sensibilatoren – ETH E-Collection. *Thesis.* ETH Zürich, Chemie; 1930.

[Amano 2012] Amano T, Muramatsu Y, Sano N, Denlinger JD, Gullikson EM: Chemical state analysis of entrapped nitrogen in carbon nanohorns using soft X-ray emission and absorption spectroscopy. *Journal of Physical Chemistry C* 2012, 116:6793–6799.

[Amsalem 2015] Amsalem P, Heimel G, Oehzelt M, Koch N: The interface electronic properties of organic photovoltaic cells. *Journal of Electron Spectroscopy and Related Phenomena* 2015, 204:177–185.

https://doi.org/10.1515/9783110794038-013

[Anastasi 2008]   Anastasi PaF, Howard AM: Sample holder for use in e. g. X-ray microscopy fields, has functional structure that provides additional functionality which allows selected parameters of sample on support surface to be measured and/or controlled with electrodes. Great Britain: Silson Ltd; 2008.

[Anderson 1972]   Anderson PW: More is different – broken symmetry and nature of hierarchical structure of science. *Science* 1972, 177:393–396.

[Anderson 1979]   Anderson S, Constable EC, Dareedwards MP, Goodenough JB, Hamnett A, Seddon KR, Wright RD: Chemical modification of a titanium(iv) oxide electrode to give stable dye sensitization without a supersensitizer. *Nature* 1979, 280:571–573.

[Andrews 2013]   Andrews JC, Weckhuysen BM: Hard X-ray spectroscopic nano-imaging of hierarchical functional materials at work. *ChemPhysChem* 2013, 14:3655–3666.

[Anselmo 2013]   Anselmo AS, Dzwilewski A, Svensson K, Moons E: Molecular orientation and composition at the surface of spin-coated polyfluorene: fullerene blend films. *Journal of Polymer Science. Part B, Polymer Physics* 2013, 51:176–182.

[Arima 1993]   Arima T, Tokura Y, Torrance JB: Variation of optical gaps in perovskite-type 3d transition-metal oxides. *Physical Review B* 1993, 48:17006–17009.

[Askerka 2017]   Askerka M, Brudvig GW, Batista VS: The O2-Evolving Complex of Photosystem II: Recent Insights from Quantum Mechanics/Molecular Mechanics (QM/MM), Extended X-ray Absorption Fine Structure (EXAFS), and Femtosecond X-ray Crystallography Data. *Accounts of Chemical Research* 2017, 50:41–48.

[Assmus 1995]   Assmus A: Early History of X Rays. *Beam Line – A Periodical of Particle Physics* 1995, 25:10–24.

[Astuti 2005]   Astuti Y, Palomares E, Haque SA, Durrant JR: Triplet state photosensitization of nanocrystalline metal oxide electrodes by zinc-substituted cytochrome c: application to hydrogen evolution. *Journal of the American Chemical Society* 2005, 127:15120–15126.

[Athreya 1999]   Athreya SA, Martin DC: Impedance spectroscopy of protein polymer modified silicon micromachined probes. *Sensors and Actuators. A, Physical* 1999, 72:203–216.

[Atkinson]   Atkinson A, Braun A, Mai A, Bronin D, Montinaro D, De Haart LGJ, Lefebvre-Joud F, Steinberger-Wilckens R, Hjelm J, Holtappels P, Kiviaho J, Van Herle J, Vels Hansen K, Bucheli O, Baranek P, Hocker T: *(SOFC-Life) Solid Oxide Fuel Cells – Integrating Degradation Effects into Lifetime Prediction Models*. Europe: European Commission.

[Atuchin 2011a]   Atuchin VV, Galashov EN, Khyzhun OY, Kozhukhov AS, Pokrovsky LD, Shlegel VN: Structural and electronic properties of $ZnWO_4(010)$ cleaved surface. *Crystal Growth & Design* 2011, 11:2479–2484.

[Atuchin 2011b]   Atuchin VV, Troitskaia IB, Khyzhun OY, Bekenev VL, Solonin YM: Electronic properties of $h\text{-}WO_3$ and $CuWO_4$ nanocrystalsas determined from X-ray spectroscopy and first-principles band-structure calculations. *International Journal of Applied Physics and Mathematics* 2011, 1:5.

[Augustsson 2004]   Augustsson A: Soft X-ray Emission Spectroscopy of Liquids and Lithium Battery Materials. Uppsala, Faculty of Science and Technology; 2004.

[Augustsson 2005]   Augustsson A, Zhuang GV, Butorin SM, Osorio-Guillen JM, Dong CL, Ahuja R, Chang CL, Ross PN, Nordgren J, Guo JH: Electronic structure of phospho-olivines $Li_x FePO_4$ (x = 0,1) from soft-X-ray-absorption and – emission spectroscopies. *Journal of Chemical Physics* 2005, 123.

[Augustynski 2008]   Augustynski J, Braun A, Grätzel M, Kuznetsov A, Mendes A, Meda L, Rothschild A, Van De Krol R, Weidenkaff A: *NANOPEC (Nanostructured Photoelectrodes for Energy Conversion)*. Europe: European Commission; 2008.

[Azad 2008]   Azad AK, Savaniu C, Tao S, Duval S, Holtappels P, Ibberson RM, Irvine JTS: Structural origins of the differing grain conductivity values in BaZr(0.9)Y(0.1)O(2.95) and indication of novel approach to counter defect association. *Journal of Materials Chemistry* 2008, 18:3414–3418.

[Badri 2015]   Badri H, Monsieurs P, Coninx I, Wattiez R, Leys N: Molecular investigation of the radiation resistance of edible cyanobacterium Arthrospira sp PCC 8005. *MicrobiologyOpen* 2015, 4:187–207.

[Baer 2010]   Baer M, Weinhardt L, Marsen B, Cole B, Gaillard N, Miller E, Heske C: Mo incorporation in WO_3 thin film photoanodes: tailoring the electronic structure for photoelectrochemical hydrogen production. *Applied Physics Letters* 2010, 96.

[Bahadur 2010] Bahadur R, Russell LM, Prather K: Composition and morphology of individual combustion, biomass burning, and secondary organic particle types obtained using urban and coastal ATOFMS and STXM-NEXAFS measurements. *Aerosol Science and Technology* 2010, 44:551–562.

[Bahcall 2001] Bahcall JN, Pinsonneault MH, Basu S: Solar models: Current epoch and time dependences, neutrinos, and helioseismological properties. *The Astrophysical Journal* 2001, 555:990–1012.

[Balasubramanian 2013] Balasubramanian S, Wang P, Schaller RD, Rajh T, Rozhkova EA: High-performance bioassisted nanophotocatalyst for hydrogen production. *Nano Letters* 2013, 13:3365–3371.

[Baque 2013] Baque M, De Vera JP, Rettberg P, Billi D: The BOSS and BIOMEX space experiments on the EXPOSE-R2 mission: endurance of the desert cyanobacterium Chroococcidiopsis under simulated space vacuum, Martian atmosphere, UVC radiation and temperature extremes. *Acta Astronautica* 2013, 91:180–186.

[Barbir 2012] Barbir F: *PEM Fuel Cells: Theory and Practice*. Academic Press; 2012.

[Barsoukov 2005] Barsoukov E, Macdonald JR: *Impedance Spectroscopy: Theory, Experiment, and Applications*, 2nd Edition. Hoboken NJ: John Wiley & Sons, Inc.; 2005.

[Bartlett 2011] Bartlett BM, Yourey JE: Electrodeposition and photoelectrochemistry of $CuWO_4$ thin-film photoanodes: toward an inexpensive, Earth-abundant water oxidation photocatalyst. *Abstracts of Papers—American Chemical Society* 2011, 242.

[Bassett 1964] Bassett CA, Pawluk RJ, Becker RO: Effects of electric currents on bone in vivo. *Nature* 1964, 204:652–654.

[Batterman 1964] Batterman BW: Effect of dynamical diffraction in X-ray fluorescence scattering. *Physical Review. A, General Physics* 1964, 133:A759–A764.

[Baturina 2011] Baturina OA, Epshteyn A, Northrup PA, Swider-Lyons KE: Insights into PEMFC performance degradation from HCl in air. *Journal of the Electrochemical Society* 2011, 158.

[Bauer 1971] Bauer W, Heiland G: Spectral sensitization of photoconductivity on surface of zinc oxide crystals under clean conditions. *Journal of Physics and Chemistry of Solids* 1971, 32:2605–2611.

[Bearden 1967] Bearden JA, Burr AF: Reevaluation of X-ray atomic energy levels. *Reviews of Modern Physics* 1967, 39:125–142.

[Beaucage 1995] Beaucage G: Approximations leading to a unified exponential power-law approach to small-angle scattering. *Journal of Applied Crystallography* 1995, 28:717–728.

[Beauchemin 2002] Beauchemin S, Hesterberg D, Beauchemin M: Principal component analysis approach for modeling sulfur K-XANES spectra of humic acids. *Soil Science Society of America Journal* 2002, 66:83–91.

[Beck 2001] Beck M, Stiel H, Leupold D, Winter B, Pop D, Vogt U, Spitz C: Evaluation of the energetic position of the lowest excited singlet state of beta-carotene by NEXAFS and photoemission spectroscopy. *Biochimica Et Biophysica Acta. Bioenergetics* 2001, 1506:260–267.

[Beck 2002] Beck M: Charakterisierung einer XUV-Laserplasmaquelle und ihre Anwendung in der NEXAFS-Spektroskopie an organischen Molekülen. Technische Universität Berlin, Fakultät II (Mathematik und Naturwissenschaften); 2002.

[Becker 1962] Becker RO, Bachman CH, Slaughter WH: Longitudinal direct-current gradients of spinal nerves. *Nature* 1962, 196:675–676.

[Becker 1969] Becker RO: Phylogenetic and functional significance of dc potentials in living organisms. *Psychophysiology* 1969, 5:573.

[Becker 1982] Becker RO, Marino AA: *Electromagnetism and Life*. State University of New York Press; 1982.

[Becker 1985] Becker RO: A theory of the interaction between dc and elf electromagnetic-fields and living organisms. *Journal of Bioelectricity* 1985, 4:133–140.

[Beckwith 2015] Beckwith MA, Ames W, Vila FD, Krewald V, Pantazis DA, Mantel C, Pecaut J, Gennari M, Duboc C, Collomb M-N, Yano J, Rehr JJ, Neese F, Debeer S: How accurately can extended X-ray absorption spectra be predicted from first principles? Implications for modeling the oxygen-evolving complex in photosystem II. *Journal of the American Chemical Society* 2015, 137:12815–12834.

[Ben-Zvi 1988] Ben-Zvi I, Gover A, Jerby E, Sokolowski JS, Wachtel J: Design of a tandem accelerator free electron laser. *Nuclear Instruments & Methods in Physics Research. Section A, Accelerators, Spectrometers, Detectors and Associated Equipment* 1988, 268:561–566.

[Ben-Zvi 1990] Ben-Zvi I, Elkonin BV, Fruchtman A, Sokolowski JS, Gover A, Jerby E, Kleinman H, Mandelbaum B, Rosenberg A, Shiloh J, Hazak G, Shahal O: Status of the Rehovot en tandem accelerator free electron laser. *Nuclear Instruments & Methods in Physics Research. Section A, Accelerators, Spectrometers, Detectors and Associated Equipment* 1990, 287:93–98.

[Benko 1982] Benko FA, Maclaurin CL, Koffyberg FP: $CUWO_4$ and CU_3WO_6 as anodes for the photoelectrolysis of water. *Materials Research Bulletin* 1982, 17:133–136.

[Berejnov 2012] Berejnov V, Martin Z, West M, Kundu S, Bessarabov D, Stumper J, Susac D, Hitchcock AP: Probing platinum degradation in polymer electrolyte membrane fuel cells by synchrotron X-ray microscopy. *Physical Chemistry Chemical Physics* 2012, 14:4835–4843.

[Bergmann 2002] Bergmann U, Glatzel P, Cramer SP: Bulk-sensitive XAS characterization of light elements: from X-ray Raman scattering to X-ray Raman spectroscopy. *Microchemical Journal* 2002, 71:221–230.

[Bergmann 2009] Bergmann U, Glatzel P: X-ray emission spectroscopy. *Photosynthesis Research* 2009, 102:255–266.

[Bianconi 1987] Bianconi A, Marcelli A, Dexpert H, Karnatak R, Kotani A, Jo T, Petiau J: Specific intermediate-valence state of insulating 4f compounds detected by I3 X-ray absorption. *Physical Review B* 1987, 35:806–812.

[Biesinger 2015] X-ray Photoelectron Spectroscopy (XPS) Reference Pages [http://www.xpsfitting.com/].

[Bird 1983] Bird RE, Hulstrom RL, Lewis LJ: Terrestrial solar spectral data sets. *Solar Energy* 1983, 30:563–573.

[Bittencourt 2005] Bittencourt C, Felten A, Mirabella F, Ivanov P, Llobet E, Silva MaP, Nunes LaO, Pireaux JJ: High-resolution photoelectron spectroscopy studies on WO_3 films modified by Ag addition. *Journal of Physics. Condensed Matter* 2005, 17:6813–6822.

[Blanton 2015] Blanton TN, Koestner R: Characterization of NAFION proton exchange membrane films using wide-angle X-ray diffraction. *International Centre for Diffraction Data* 2015:9.

[Bluhm 2006] Bluhm H, Andersson K, Araki T, Benzerara K, Brown GE, Dynes JJ, Ghosal S, Gilles MK, Hansen HC, Hemminger JC, Hitchcock AP, Ketteler G, Kilcoyne ALD, Kneedler E, Lawrence JR, Leppard GG, Majzlan J, Mun BS, Myneni SCB, Nilsson A, Ogasawara H, Ogletree DF, Pecher K, Salmeron M, Shuh DK, Tonner B, Tyliszczak T, Warwick T, Yoon TH: Soft X-ray microscopy and spectroscopy at the molecular environmental science beamline at the Advanced Light Source. *Journal of Electron Spectroscopy and Related Phenomena* 2006, 150:86–104.

[Bockris 1963] Bockris JO, Devanathan MaV, Muller K: On structure of charged interfaces. *Proceedings of the Royal Society of London. Series A, Mathematical and Physical Sciences* 1963, 274:55–79.

[Bolling 2012] Bolling AK, Totlandsdal AI, Sallsten G, Braun A, Westerholm R, Bergvall C, Boman J, Dahlman HJ, Sehlstedt M, Cassee F, Sandstrom T, Schwarze PE, Herseth JI: Wood smoke particles from different combustion phases induce similar pro-inflammatory effects in a co-culture of monocyte and pneumocyte cell lines. *Particle and Fibre Toxicology* 2012, 9.

[Bonnemann 2002] Bonnemann H, Waldofner N, Haubold HG, Vad T: Preparation and characterization of three-dimensional Pt nanoparticle networks. *Chemistry of Materials* 2002, 14:1115–1120.

[Bora 2011a] Bora DK, Braun A, Erat S, Ariffin AK, Loehnert R, Sivula K, Toepfer J, Graetzel M, Manzke R, Graule T, Constable EC: Evolution of an oxygen near-edge X-ray absorption fine structure transition in the upper hubbard band in alpha-Fe_2O_3 upon electrochemical oxidation. *Journal of Physical Chemistry C* 2011, 115:5619–5625.

[Bora 2011f] Bora DK, Braun A, Erni R, Fortunato G, Graule T, Constable EC: Hydrothermal treatment of a hematite film leads to highly oriented faceted nanostructures with enhanced photocurrents. *Chemistry of Materials* 2011, 23:2051–2061.

[Bora 2012] Bora DK, Rozhkova EA, Schrantz K, Wyss PP, Braun A, Graule T, Constable EC: Functionalization of nanostructured hematite thin-film electrodes with the light-harvesting membrane protein C-phycocyanin yields an enhanced photocurrent. *Advanced Functional Materials* 2012, 22:490–502.

[Bora 2013a] Bora DK, Braun A, Constable EC: "In rust we trust". Hematite – the prospective inorganic backbone for artificial photosynthesis. *Energy & Environmental Science* 2013, 6:407–425.

[Bora 2013b] Bora DK, Hu Y, Thiess S, Erat S, Feng X, Mukherjee S, Fortunato G, Gaillard N, Toth R, Gajda-Schrantz K, Drube W, Graetzel M, Guo J, Zhu J, Constable EC, Sarma DD, Wang H, Braun A: Between photocatalysis and photosynthesis: synchrotron spectroscopy methods on molecules and materials for solar hydrogen generation. *Journal of Electron Spectroscopy and Related Phenomena* 2013, 190:93–105.

[Bota 2002] Bota A, Goerigk G, Drucker T, Haubold HG, Petro J: Anomalous small-angle X-ray scattering on a new, nonpyrophoric Raney-type Ni catalyst. *Journal of Catalysis* 2002, 205:354–357.

[Boudoire 2015] Boudoire F: Self-assembled photonic mesostructures for water splitting photoanodes. *Doctoral thesis*. Universität Basel, Chemistry; 2015.

[Bouguer 1729] Bouguer P: *Essai d'optique sur la gradation de la lumière*. Paris: Claude Jombert; 1729.

[Bozzini 2011] Bozzini B, Guerrieri M, Capotondi F, Sgura I, Tondo E: Electrochemical preparation of particles for X-ray free electron laser based diffractive imaging. *International Journal of Electrochemical Science* 2011, 6:2609–2631.

[Bozzini 2012a] Bozzini B, Abyaneh MK, Amati M, Gianoncelli A, Gregoratti L, Kaulich B, Kiskinova M: Soft X-ray imaging and spectromicroscopy: new insights in chemical state and morphology of the key components in operating fuel-cells. *Chemistry* 2012, 18:10196–10210.

[Bozzini 2012b] Bozzini B, Amati M, Gianoncelli A, Gregoratti L, Kaulich B, Kiskinova M: New energy sources: in-situ characterisation of fuel cell and supercapacitor components. Complementary studies using transmission, fluorescence and photoelectron microscopy and imaging. In *11th International Conference on X-ray Microscopy (XRM); 2013 Aug 05–10; Shanghai Synchrotron Radiat Facil, Shanghai, People's Republic of China*. 2012.

[Bozzini 2013] Bozzini B, Gianoncelli A, Kaulich B, Mele C, Prasciolu M, Kiskinova M: In situ soft X-ray microscopy study of Fe interconnect corrosion in ionic liquid-based nano-PEMFC half-cells. *Fuel Cells* 2013, 13:196–202.

[Bragg 1915a] Bragg WH: Bakerian Lecture: X-rays and Crystal Structure. *Philosophical Transactions of the Royal Society of London A* 1915, 215:22.

[Bragg 1915c] Bragg WH: The distribution of the electrons in atoms. *Nature* 1915, 95:344.

[Bragg 1915e] Bragg WH: The structure of magnetite and the spinels. *Nature* 1915, 95:561.

[Bragg 1915f] Bragg WH: The relation between certain X-rag wave-lengths and their absorption coefficients. *Philosophical Magazine* 1915, 29:407–412.

[Bragg 1915g] Bragg WH: The structure of the spinel group of crystals. *Philosophical Magazine* 1915, 30:305–315.

[Bragg 1922] Bragg WL: *Nobel Lecture: The Diffraction of X-rays by Crystals*. Stockholm; 1922.

[Bragg 1929] Bragg WL: The determination of parameters in crystal structures by means of Fourier series. *Proceedings of the Royal Society of London. Series A, Containing Papers of a Mathematical and Physical Character* 1929, 123:537–559.

[Braun 1999a] Braun A: Development and characterization of glassy carbon electrodes for a bipolar electrochemical double layer capacitor. Dissertation ETH Zürich no° 13292, 1999.

[Braun 1999c] Braun A, Bartsch M, Schnyder B, Kotz R, Haas O, Haubold HG, Goerigk G: X-ray scattering and adsorption studies of thermally oxidized glassy carbon. *Journal of Non-Crystalline Solids* 1999, 260:1–14.

[Braun 2000a] Braun A, Bartsch M, Geiger F, Schnyder B, Kotz R, Haas O, Carlen M, Christen T, Ohler C, Unternahrer P, Krause E: A study on oxidized glassy carbon sheets for bipolar supercapacitor electrodes. In *New Materials for Batteries and Fuel Cells. Volume 575*. Edited by Doughty DH, Nazar LF, Arakawa M, Brack HP, Naoi K; 2000: 369–380: *Materials Research Society Symposium Proceedings*.

[Braun 2000b] Braun A, Bartsch M, Schnyder B, Kotz R: A model for the film growth in samples with two moving reaction frontiers – an application and extension of the unreacted-core model. *Chemical Engineering Science* 2000, 55:5273–5282.

[Braun 2001a] Braun A, Seifert S, Cairns EJ: In situ anomalous small angle X-ray scattering of LiMn$_2$O$_4$ during electrochemical delithiation. *Abstracts of Papers—American Chemical Society* 2001, 221:U689–U689.

[Braun 2001f] Braun A, Seifert S, Thiyagarajan P, Cramer SP, Cairns EJ: In situ anomalous small angle X-ray scattering and absorption on an operating rechargeable lithium ion battery cell. *Electrochemistry Communications* 2001, 3:136–141.

[Braun 2002a] Braun A, Bartsch M, Schnyder B, Kotz R, Haas O, Wokaun A: Evolution of BET internal surface area in glassy carbon powder during thermal oxidation. *Carbon* 2002, 40:375–382.

[Braun 2002c] Braun A, Shah N, Huggins FE, Huffman GP, Kelly K, Sarofim AF, Wirick S, Jacobsen CJ: Investigation of fine particulate matter from a diesel engine using scanning transmission X-ray microspectroscopy. *Abstracts of Papers—American Chemical Society* 2002, 224:U569–U569.

[Braun 2002d] Braun A, Wang HX, Bergmann U, Tucker MC, Gu WW, Cramer SP, Cairns EJ: Origin of chemical shift of manganese in lithium battery electrode materials – a comparison of hard and soft X-ray techniques. *Journal of Power Sources* 2002, 112:231–235.

[Braun 2003a] Braun A, Shrout S, Fowlks A, Al. E: Electrochemical in situ reaction cell for X-ray scattering, diffraction and spectroscopy. *Journal of Synchrotron Radiation* 2003, 10:320–325.

[Braun 2003i] Braun A, Bärtsch M, Merlo O, Schnyder B, Schaffner B, Kötz R, Haas O, Wokaun A: Exponential growth of electrochemical double layer capacitance in glassy carbon during thermal oxidation. *Carbon* 2003, 41:759–765.

[Braun 2003j] Braun A, Wokaun A, Hermanns HG: Analytical solution to a growth problem with two moving boundaries. *Applied Mathematical Modelling* 2003, 27:47–52.

[Braun 2004a] Braun A, Kohlbrecher J, Bartsch M, Al. E: Small-angle neutron scattering and cyclic voltammetry study on electrochemically oxidized and reduced pyrolytic carbon. *Electrochimica Acta* 2004, 49:1105–1112.

[Braun 2004b] Braun A, Shah N, Huggins F, Al. E: A study of diesel PM with X-ray microspectroscopy. *Fuel* 2004, 83:997–1000.

[Braun 2005a] Braun A: Carbon speciation in airborne particulate matter with C (1s) NEXAFS spectroscopy. *Journal of Environmental Monitoring* 2005, 7:1059–1065.

[Braun 2005b] Braun A, Huggins FE, Shah N, Chen Y, Wirick S, Mun SB, Jacobsen C, Huffman GP: Advantages of soft X-ray absorption over TEM-EELS for solid carbon studies – a comparative study on diesel soot with EELS and NEXAFS. *Carbon* 2005, 43:117–124.

[Braun 2005d] Braun A, Ilavsky J, Dunn BC, Jemian PR, Huggins FE, Eyring EM, Huffman GP, Iucr: Ostwald ripening of cobalt precipitates in silica aerogels? An ultra-small-angle X-ray scattering study. *Journal of Applied Crystallography* 2005, 38:132–138.

[Braun 2005g] Braun A, Ilavsky J, Seifert S, Jemian PR: Deformation of diesel soot aggregates as a function of pellet pressure: A study with ultra-small-angle X-ray scattering. *Journal of Applied Physics* 2005, 98.

[Braun 2005h] Braun A, Shah N, Huggins F, Al. E: X-ray scattering and spectroscopy studies on diesel soot from oxygenated fuel under various engine load conditions. *Carbon* 2005, 43:2588–2599.

[Braun 2006a] Braun A: Some comments on "Soot surface area evolution during air oxidation as evaluated by small angle X-ray scattering and CO$_2$ adsorption". *Carbon* 2006, 44:1313–1315.

[Braun 2006b] Braun A, Holtappels P: *X-ray and Electrochemical Studies on Solid Oxide Fuel Cells and Related Materials (HiTempEchem)*. Switzerland: European Commission; 2006.

[Braun 2006c] Braun A, Huggins FE, Kelly KE, Mun BS, Ehrlich SN, Huffman GP: Impact of ferrocene on the structure of diesel exhaust soot as probed with wide-angle X-ray scattering and C(1s) NEXAFS spectroscopy. *Carbon* 2006, 44:2904–2911.

[Braun 2006d] Braun A, Wirick A, Kubatova A, Mun BS, Huggins FE: Photochemically induced decarboxylation in diesel soot extracts. *Atmospheric Environment* 2006, 40:5837–5844.

[Braun 2007] Braun A, Wang H, Shim J, Lee SS, Cairns EJ: Lithium K(1s) synchrotron NEXAFS spectra of lithium-ion battery cathode, anode and electrolyte materials. *Journal of Power Sources* 2007, 170:173–178.

[Braun 2008a]   Braun A, Huggins FE, Kubatova A, Wirick S, Maricq MM, Mun BS, Mcdonald JD, Kelly KE, Shah N, Huffman GP: Toward distinguishing woodsmoke and diesel exhaust in ambient particulate matter. *Environmental Science & Technology* 2008, 42:374–380.

[Braun 2008c]   Braun A, Janousch M, Sfeir J, Kiviaho J, Noponen M, Huggins FE, Smith MJ, Steinberger-Wilckens R, Holtappels P, Graule T: Molecular speciation of sulfur in solid oxide fuel cell anodes with X-ray absorption spectroscopy. *Journal of Power Sources* 2008, 183:564–570.

[Braun 2008d]   Braun A, Richter J, Harvey AS, Erat S, Infortuna A, Frei A, Pomjakushina E, Mun BS, Holtappels P, Vogt U, Conder K, Gauckler LJ, Graule T: Electron hole-phonon interaction, correlation of structure, and conductivity in single crystal $La_{0.9}Sr_{0.1}FeO_3$-delta. *Applied Physics Letters* 2008, 93.

[Braun 2009a]   Braun A, Bayraktar D, Erat S, Harvey AS, Beckel D, Purton JA, Holtappels P, Gauckler LJ, Graule T: Pre-edges in oxygen (1s) X-ray absorption spectra: a spectral indicator for electron hole depletion and transport blocking in iron perovskites. *Applied Physics Letters* 2009, 94.

[Braun 2009g]   Braun A, Duval S, Ried P, Embs J, Juranyi F, Straessle T, Stimming U, Hempelmann R, Holtappels P, Graule T: Proton diffusivity in the $BaZr_{0.9}Y_{0.1}O_3$-delta proton conductor. *Journal of Applied Electrochemistry* 2009, 39:471–475.

[Braun 2009i]   Braun A, Kubatova A, Wirick S, Mun SB: Radiation damage from EELS and NEXAFS in diesel soot and diesel soot extracts. *Journal of Electron Spectroscopy and Related Phenomena* 2009, 170:42–48.

[Braun 2009j]   Braun A, Ovalle A, Pomjakushin V, Cervellino A, Erat S, Stolte WC, Graule T: Yttrium and hydrogen superstructure and correlation of lattice expansion and proton conductivity in the $BaZr_{0.9}Y_{0.1}O_{2.95}$ proton conductor. *Applied Physics Letters* 2009, 95.

[Braun 2009l]   Braun A, Ovalle A, Pomjakushin V, Cervellino A, Erat S, Stolte WC, Graule T: Yttrium and hydrogen superstructure and correlation of lattice expansion and proton conductivity in the $BaZr_{0.9}Y_{0.1}O_{2.95}$ proton conductor. 2009.

[Braun 2009m]   Braun A, Zhang X, Sun Y, Muller U, Liu Z, Erat S, Ari M, Grimmer H, Mao S, Graule T: Correlation of high temperature X-ray photoemission valence band spectra and conductivity in strained LaSrFeNi oxide on $SrTiO_3$(110). *Applied Physics Letters* 2009, 95.

[Braun 2010a]   Braun A: Depth profile analysis of a cycled lithium ion manganese oxide battery electrode via the valence state of manganese, with soft X-ray emission spectroscopy. *Journal of Power Sources*, 2010, 195:7644–7648.

[Braun 2010b]   Braun A, Akurati KK, Fortunato G, Reifler FA, Ritter A, Harvey AS, Vital A, Graule T: Nitrogen doping of TiO_2 photocatalyst forms a second e(g) state in the oxygen 1s NEXAFS pre-edge. *Journal of Physical Chemistry C* 2010, 114:516–519.

[Braun 2010d]   Braun A, Gajda-Schrantz K, Dombi A: *Nanobio-Interfaces for Photocatalytic Solar Hydrogen (NIPSH)*. Switzerland and Hungary: Swiss State Secretariat for Education, Research and Innovation (SERI); 2010.

[Braun 2010e]   Braun A, Grätzel M: *Defects in the Bulk and on Surfaces and interfaces of Metal Oxides with Photoelectrochemical Properties: In-Situ Photoelectrochemical and Resonant X-Ray and Electron Spectroscopy Studies*. Empa Dübendorf, Switzerland EPFL Lausanne, Switzerland: Swiss National Science Foundation; Research Grant http://p3.snf.ch/Project-132126.

[Braun 2010f]   Braun A, Mun BS, Sun Y, Liu Z, Groening O, Maeder R, Erat S, Zhang X, Mao SS, Pomjakushina E, Conder K, Graule T: Correlation of conductivity and angle integrated valence band photoemission characteristics in single crystal iron perovskites for 300 K < T < 800 K: Comparison of surface and bulk sensitive methods. *Journal of Electron Spectroscopy and Related Phenomena* 2010, 181:56–62.

[Braun 2010g]   Braun A, Navickas E, Tamulevicius S: *Proton Conductivity in Thin Ceramic Films (PROFIL)*. Switzerland: Sciex Research Grant with Lithuania.

[Braun 2010h]   Braun A, Seifert S, Ilavsky J: Highly porous activated glassy carbon film sandwich structure for electrochemical energy storage in ultracapacitor applications: study of the porous film structure and gradient. *Journal of Materials Research* 2010, 25:1532–1540.

[Braun 2011a]   Braun A, Erat S, Ariffin AK, Manzke R, Wadati H, Graule T, Gauckler LJ: High temperature oxygen near edge x-ray absorption fine structure valence band spectra and conductivity of $LaFe_3/4Ni_1/4O_3$ from 300 to 773 K. *Applied Physics Letters* 2011, 99:202112.

[Braun 2011c] Braun A, Constable EC, Heier J, Toth R: Reaction-diffusion processes for the growth of patterned structures and architectures: a bottom-up approach for photoelectrochemical electrodes. Switzerland: Swiss National Science Foundation; Research Grant http://p3.snf.ch/Project-137868.

[Braun 2011d] Braun A, Erat S, Ariffin AK, Manzke R, Wadati H, Graule T, Gauckler LJ: High temperature oxygen near edge X-ray absorption fine structure valence band spectra and conductivity of $LaFe_3/4Ni_1/4O_3$ from 300 to 773 K. *Applied Physics Letters* 2011, 99.

[Braun 2011e] Braun A, Erat S, Zhang X, Chen Q, Huang T-W, Aksoy F, Loehnert R, Liu Z, Mao SS, Graule T: Surface and bulk oxygen vacancy defect states near the Fermi level in 125 nm WO_3-delta/TiO_2 (110) films: a resonant valence band photoemission spectroscopy study. *Journal of Physical Chemistry C* 2011, 115:16411–16417.

[Braun 2011h] Braun A, Sarma DD: Indo-Swiss Joint Research Programme. Empa Dübendorf, Switzerland. IIS Bengaluru, India. Indo-Swiss Joint Research Programme.

[Braun 2012a] Braun A, Akgul FA, Chen Q, Erat S, Huang T-W, Jabeen N, Liu Z, Mun BS, Mao SS, Zhang X: Observation of substrate orientation-dependent oxygen defect filling in thin WO_3-delta/TiO_2 pulsed laser-deposited films with in situ xps at high oxygen pressure and temperature. *Chemistry of Materials* 2012, 24:3473–3480.

[Braun 2012c] Braun A, Chen Q, Flak D, Fortunato G, Gajda-Schrantz K, Graetzel M, Graule T, Guo J, Huang T-W, Liu Z, Popelo AV, Sivula K, Wadati H, Wyss PP, Zhang L, Zhu J: Iron resonant photoemission spectroscopy on anodized hematite points to electron hole doping during anodization. *ChemPhysChem* 2012, 13:2937–2944.

[Braun 2012d] Braun A, Erat S, Bayraktar D, Harvey A, Graule T: Electronic origin of conductivity changes and isothermal expansion of Ta- and Ti-Substituted $La_1/2Sr_1/2Fe$-oxide in oxidative and reducing atmosphere. *Chemistry of Materials* 2012, 24:1529–1535.

[Braun 2012h] Braun A, Ihssen J: *Biomimetic Photoelectrochemical Cells for Solar Hydrogen Generation (BioPEC)*. Switzerland: VELUX Foundation; Research Grant no° 790.

[Braun 2012i] Braun A, Molenda J: Positive electrode materials for Li-Ion batteries for electric vehicles applications – LiBEV. Empa Dübendorf, Switzerland. AGH Krakow, Poland: Polish Swiss Research Program. Swiss Contribution; Research Grant.

[Braun 2012l] Braun A, Sivula K, Bora DK, Zhu J, Zhang L, Graetzel M, Guo J, Constable EC: Direct observation of two electron holes in a hematite photoanode during photoelectrochemical water splitting. *Journal of Physical Chemistry C* 2012, 116:16870–16875.

[Braun 2013a] Braun A: Between photocatalysis and photosynthesis: synchrotron spectroscopy methods on molecules and materials for solar hydrogen generation. *Journal of Electron Spectroscopy and Related Phenomena* 2013, 190:93–105.

[Braun 2013c] Braun A, Constable EC, Diale MM, Rohwer E, Ozoemena KI, Roduner E: *Production of Liquid Solar Fuels from CO_2 and Water: Using Renewable Energy Resources*. Switzerland and South Africa: Swiss South Africa Joint Research Programme; Research Grant http://p3.snf.ch/Project-149031.

[Braun 2015a] Braun A: X-ray and neutron spectroscopy methods for photosynthesis and photoelectrochemistry. In *Solar Energy for Fuels. Volume 371*. Edited by Chan CK, Tüysüz H. Springer Cham Heidelberg New York Dordrecht London: Springer International Publishing; 2015: 261–274: *Solar Energy for Fuels*.

[Braun 2015b] Braun A, Boudoire F, Bora DK, Faccio G, Hu Y, Kroll A, Mun BS, Wilson ST: Biological components and bioelectronic interfaces of water splitting photoelectrodes for solar hydrogen production. *Chemistry* 2015, 21:4188–4199.

[Braun 2015h] Braun A, Madiba GI, Maaza M: *Vanadium Oxide for Thermochromic Windows (Collaboration Empa/ETHZ and iThemba LABS/UNISA)*. Switzerland: Eidgenössische Stipendienkommission für ausländische Studierende ESKAS.

[Braun 2015j] Braun A, Nordlund D, Song S-W, Huang T-W, Sokaras D, Liu X, Yang W, Weng T-C, Liu Z: Hard X-rays in-soft X-rays out: an operando piggyback view deep into a charging lithium ion battery with X-ray Raman spectroscopy. *Journal of Electron Spectroscopy and Related Phenomena* 2015, 200:257–263.

[Braun 2016a] Braun A: Structure and transport properties in ceramic fuel cells (SOFC), components, and materials. In *Structural Characterization Techniques: Advances and Applications in Clean Energy*. Edited by Malavasi L. Pan Stanford Publishing; 2016.

[Braun 2016c] Braun A, Hu Y, Boudoire F, Bora DK, Sarma DD, Graetzel M, Eggleston CM: The electronic, chemical and electrocatalytic processes and intermediates on iron oxide surfaces during photoelectrochemical water splitting. *Catalysis Today* 2016, 260:72–81.

[Braun 2019] Braun A: *Electrochemical Energy Systems - Foundations, Energy Storage and Conversion*. Boston, Berlin: Walter de Gruyter GmbH; 2019.

[Brillson 2010] Brillson LJ: *Surfaces and Interfaces of Electronic Materials*. Wiley-VCH Verlag GmbH & Co. KGaA; 2010.

[Brizzolara 1994] Brizzolara RA, Boyd JL, Thorne RE, Tate AE: Purple membrane by X-ray photoelectron spectroscopy. *Surface Science Spectra* 1994, 3:169–174.

[Brizzolara 1996] Brizzolara RA: Methionine by X-ray photoelectron spectroscopy. *Surface Science Spectra* 1996, 4:96–101.

[Brizzolara 1997] Brizzolara RA, Boyd JL, Tate AE: Evidence for covalent attachment of purple membrane to a gold surface via genetic modification of bacteriorhodopsin. *Journal of Vacuum Science & Technology. A. Vacuum, Surfaces, and Films* 1997, 15:773–778.

[Bullen 2006] Bullen RA, Arnot TC, Lakeman JB, Walsh FC: Biofuel cells and their development. *Biosensors & Bioelectronics* 2006, 21:2015–2045.

[Bullett 1983] Bullett DW: Bulk and surface electron-states in WO_3 and tungsten bronzes. *Journal of Physics. C. Solid State Physics* 1983, 16:2197–2207.

[Busbey 1999] Busbey A: IGOR Pro 3.1. *Geotimes* 1999, 44:43–43.

[Butler 2011] Butler LG, Schillinger B, Ham K, Dobbins TA, Liu P, Vajo JJ: Neutron imaging of a commercial Li-ion battery during discharge: application of monochromatic imaging and polychromatic dynamic tomography. *Nuclear Instruments & Methods in Physics Research. Section A, Accelerators, Spectrometers, Detectors and Associated Equipment* 2011, 651:320–328.

[Cairns 2016] Cairns EJ: Everything Matters - Sulfur (YouTube); 2016. Available from: https://www.youtube.com/watch?v=UB8epvEIdH8.

[Caliebe 1998] Caliebe WA, Kao CC, Hastings JB, Taguchi M, Uozumi T, De Groot FMF: 1s2p resonant inelastic X-ray scattering in alpha-Fe_2O_3. *Physical Review B* 1998, 58:13452–13458.

[Calvin 1957] Calvin M, Sogo PB: Primary quantum conversion process in photosynthesis – electron spin resonance. *Science* 1957, 125:499–500.

[Calvin 1960] Calvin M: Some photochemical and photophysical reactions of chlorophyll ad its relatives. In *McCollum-Pratt Symposium on Light and Life*. pp. 67. Johns Hopkins University, Baltimore; 1960:67.

[Calvin 1962] Calvin M, Androes GM: Primary quantum conversion in photosynthesis. *Science* 1962, 138:867–873.

[Cardona 1978] Cardona M, Ley L: *Photoemission in Solids. I. General Principles*. Berlin: Springer-Verlag; 1978.

[Cavalleri 2002] Cavalleri M, Ogasawara H, Pettersson LGM, Nilsson A: The interpretation of X-ray absorption spectra of water and ice. *Chemical Physics Letters* 2002, 364:363–370.

[Ceppi 2014] Ceppi S, Mesquita A, Pomiro F, Miner EVP, Tirao G: Study of K beta X-ray emission spectroscopy applied to $Mn((2 - x))V((1 + x))O_4$ (x = 0 and 1/3) oxyspinel and comparison with XANES. *Journal of Physics and Chemistry of Solids* 2014, 75:366–373.

[Chakroune 2005] Chakroune N, Viau G, Ammar S, Poul L, Veautier D, Chehimi MM, Mangeney C, Villain F, Fievet F: Acetate- and thiol-capped monodisperse ruthenium nanoparticles: XPS, XAS, and HRTEM studies. *Langmuir* 2005, 21:6788–6796.

[Chameides 2002] Chameides WL, Bergin M: Climate change – soot takes center stage. *Science* 2002, 297:2214–2215.

[Chang 2011] Chang Y, Braun A, Deangelis A, Kaneshiro J, Gaillard N: Effect of thermal treatment on the crystallographic, surface energetics, and photoelectrochemical properties of reactively cosputtered copper tungstate for water splitting. *Journal of Physical Chemistry C* 2011, 115:25490–25495.

[Chaturvedi 1998] Chaturvedi S, Rodriguez JA, Brito JL: Characterization of pure and sulfided $NiMoO_4$ catalysts using synchrotron-based X-ray absorption spectroscopy (XAS) and temperature-programmed reduction (TPR). *Catalysis Letters* 1998, 51:85–93.

[Che 2015] Che QT, Zhu ZF, Chen N, Zhai X: Methylimidazolium group – modified polyvinyl chloride (PVC) doped with phosphoric acid for high temperature proton exchange membranes. *Materials & Design* 2015, 87:1047–1055.

[Chen-Wiegart 2012a] Chen-Wiegart KY-C, Harris WM, Lombardo JJ, Chiu WKS, Wang J: Oxidation states study of nickel in solid oxide fuel cell anode using X-ray full-field spectroscopic nano-tomography. *Applied Physics Letters* 2012, 101:4.

[Chen-Wiegart 2012b] Chen-Wiegart Y-CK, Cronin JS, Yuan Q, Yakal-Kremski KJ, Barnett SA, Wang J: 3D Non-destructive morphological analysis of a solid oxide fuel cell anode using full-field X-ray nano-tomography. *Journal of Power Sources* 2012, 218:348–351.

[Chen-Wiegart 2013] Chen-Wiegart Y-CK, Liu Z, Faber KT, Barnett SA, Wang J: 3D analysis of a $LiCoO_2$-$Li(Ni_1/3Mn_1/3Co_1/3)O$-2 Li-ion battery positive electrode using X-ray nano-tomography. *Electrochemistry Communications* 2013, 28:127–130.

[Chen-Wiegart 2014] Chen-Wiegart Y-CK, Demike R, Erdonmez C, Thornton K, Barnett SA, Wang J: Tortuosity characterization of 3D microstructure at nano-scale for energy storage and conversion materials. *Journal of Power Sources* 2014, 249:349–356.

[Chen-Wiegart 2012c] Chen-Wiegart YCK, Shearing P, Yuan QX, Tkachuk A, Wang J: 3D morphological evolution of Li-ion battery negative electrode $LiVO_2$ during oxidation using X-ray nano-tomography. *Electrochemistry Communications* 2012, 21:58–61.

[Chen 2015a] Chen H, Wang Z, Gao K, Hou Q, Wang D, Wu Z: Quantitative phase retrieval in X-ray Zernike phase contrast microscopy. *Journal of Synchrotron Radiation* 2015a, 22:1056–1061.

[Chen 2011a] Chen M, Blankenship RE: Expanding the solar spectrum used by photosynthesis. *Trends in Plant Science* 2011, 16:427–431.

[Chen 2010] Chen Q, Braun A, Ovalle A, Savaniu C-D, Graule T, Bagdassarov N: Hydrostatic pressure decreases the proton mobility in the hydrated $BaZr_{0.9}Y_{0.1}O_3$ proton conductor. *Applied Physics Letters* 2010, 97:041902 (041903 pp.).

[Chen 2011c] Chen Q, Braun A, Yoon S, Bagdassarov N, Graule T: Effect of lattice volume and compressive strain on the conductivity of BaCeY-oxide ceramic proton conductors. *Journal of the European Ceramic Society* 2011, 31:2657–2661.

[Chen 2011d] Chen Q, Huang T-W, Baldini M, Hushur A, Pomjakushin V, Clark S, Mao WL, Manghnani MH, Braun A, Graule T: Effect of compressive strain on the Raman modes of the dry and hydrated $BaCe_{0.8}Y_{0.2}O_3$ proton conductor. *Journal of Physical Chemistry C* 2011, 115:24021–24027.

[Chen 2012b] Chen Q, Holdsworth S, Embs J, Pomjakushin V, Frick B, Braun A: High-temperature high pressure cell for neutron-scattering studies. *High Pressure Research* 2012, 32:471–481.

[Chen 2013a] Chen Q: *Effects of Pressure on the Proton – Phonon Coupling in Metal Oxides with Perovskite Structure*. ETH Zürich, Physics; 2013.

[Chen 2013c] Chen Q, Banyte J, Zhang X, Embs JP, Braun A: Proton diffusivity in spark plasma sintered $BaCe_{0.8}Y_{0.2}O_3$ (-) (delta): In-situ combination of quasi-elastic neutron scattering and impedance spectroscopy. *Solid State Ionics* 2013, 252:2–6.

[Chen 2013d] Chen Q, El Gabaly F, Akgul FA, Liu Z, Mun BS, Yamaguchi S, Braun A: Observation of oxygen vacancy filling under water vapor in ceramic proton conductors in situ with ambient pressure XPS. *Chemistry of Materials* 2013, 25:4690–4696.

[Chen 2015b] Chen Q, Braun A: Elucidating the biography of a proton in a proton conductor with neutrons and X-rays. In *Swiss Neutron News*, vol. 45. pp. 9. Schweizerische Gesellschaft für Neutronenstreuung; 2015:9.

[Cheng 2014] Cheng Y, Suhonen H, Helfen L, Li J, Xu F, Grunze M, Levkin PA, Baumbach T: Direct three-dimensional imaging of polymer-water interfaces by nanoscale hard X-ray phase tomography. *Soft Matter* 2014, 10:2982–2990.

[Chiou 2011] Chiou JW, Chang SY, Huang WH, Chen YT, Hsu CW, Hu YM, Chen JM, Chen CH, Kumar K, Guo JH: The characterization of Cr secondary oxide phases in ZnO films studied by X-ray spectroscopy and photoemission spectroscopy. *Applied Surface Science* 2011, 257:4863–4866.

[Chung 2013] Chung D-W, Ebner M, Ely DR, Wood V, Garcia RE: Validity of the Bruggeman relation for porous electrodes. *Modelling and Simulation in Materials Science and Engineering* 2013, 21.

[Claessen 2009] Claessen R, Sing M, Paul M, Berner G, Wetscherek A, Mueller A, Drube W: Hard X-ray photoelectron spectroscopy of oxide hybrid and heterostructures: a new method for the study of buried interfaces. *New Journal of Physics* 2009, 11.

[Clark 2012] Clark SM, Macdowell AA, Knight J, Kalkan B, Yan J, Chen B, Williams Q: Beamline 12.2.2: An Extreme Conditions Beamline at the Advanced Light Source. http://dxdoiorg/101080/089408862012736832 2012, 25:2.

[Cnossen 2007] Cnossen I, Sanz-Forcada J, Favata F, Witasse O, Zegers T, Arnold NF: Habitat of early life: solar X-ray and UV radiation at Earth's surface 4–3.5 billion years ago. *Journal of Geophysical Research: Planets* 2007, 112.

[Cockcroft 1997] Choice of X-ray Target [http://pd.chem.ucl.ac.uk/pdnn/inst1/anode.htm].

[Cockell 2011] Cockell CS, Rettberg P, Rabbow E, Olsson-Francis K: Exposure of phototrophs to 548 days in low Earth orbit: microbial selection pressures in outer space and on early earth. *The ISME Journal* 2011, 5:1671–1682.

[Cody 1995a] Cody GD, Botto RE, Ade H, Behal S, Disko M, Wirick S: Inner-shell spectroscopy and imaging of a subbituminous coal - in-situ analysis of organic and inorganic microstructure using C(1s)-NEXAFS, Ca(2p)-NEXAFS, and C1(2s)-NEXAFS. *Energy & Fuels* 1995, 9:525–533.

[Cody 1995b] Cody GD, Botto RE, Ade H, Behal S, Disko M, Wirick S: c-NEXAFS microanalysis and scanning-X-ray microscopy of microheterogeneities in a high-volatile a bituminous coal. *Energy & Fuels* 1995, 9:75–83.

[Cody 1995c] Cody GD, Botto RE, Ade H, Wirick S: Soft-x-ray microanalysis and microscopy – a unique probe of the organic-chemistry of heterogeneous solids. *Abstracts of Papers—American Chemical Society* 1995, 210:1-Fuel.

[Cody 2003] Cody GD, Alexander CMOD, Wirick S: Approaches to establishing the chemical structure of extraterrestrial organic solids. In *Workshop on Cometary Dust in Astrophysics. August 10–15, 2003. Crystal Mountain, Washington*; 2003:60571–60572.

[Cohen 1972] Cohen GG, Kuriyama M, Alexandropoulos NG: Energy-loss spectrum of lithium metal in region of intermediate momentum-transfer. *Solid State Communications* 1972, 10:95-+.

[Cohen 1973] Cohen GG, Alexandropoulos NG, Kuriyama M: Relation between X-ray-raman and soft-X-ray-absorption spectra. *Physical Review B* 1973, 8:5427–5431.

[Cohen 2004] Cohen H: Chemically resolved electrical measurements using X-ray photoelectron spectroscopy. *Applied Physics Letters* 2004, 85:1271–1273.

[Colella 1996] Colella R: X-ray and neutron interferometry: Basic principles and applications. *X-ray and Neutron Dynamical Diffraction: Theory and Applications* 1996, 357:369–380.

[Conway 1999] Conway BE: *Electrochemical Supercapacitors.* Springer; 1999.

[Cosslett 1982] Cosslett VE: Walter Hoppe 65. *Ultramicroscopy* 1982, 9:1–2.

[Coster 1924] Coster D: Über die Absorptionsspektren im Röntgengebiet. *Zeitschrift für Physik* 1924, 25:16.

[Cracknell 2011] Cracknell JA, Mcnamara TP, Lowe ED, Blanford CF: Bilirubin oxidase from *Myrothecium verrucaria*: X-ray determination of the complete crystal structure and a rational surface modification for enhanced electrocatalytic O-2 reduction. *Dalton Transactions* 2011, 40:6668–6675.

[Cramer 1991] Cramer SP, Degroot FMF, Ma Y, Chen CT, Sette F, Kipke CA, Eichhorn DM, Chan MK, Armstrong WH, Libby E, Christou G, Brooker S, Mckee V, Mullins OC, Fuggle JC: Ligand-field strengths and oxidation-states from manganese l-edge spectroscopy. *Journal of the American Chemical Society* 1991, 113:7937–7940.

[Cramer 1992] Cramer SP, Chen J, George SJ, Vanelp J, Moore J, Tensch O, Colaresi J, Yocum M, Mullins OC, Chen CT: Soft X-ray spectroscopy of metalloproteins using fluorescence detection. *Nuclear Instruments & Methods in Physics Research. Section A, Accelerators, Spectrometers, Detectors and Associated Equipment* 1992, 319:285–289.

[Cramer 1997] Cramer SP, Ralston CY, Wang HX, Bryant C: Bioinorganic applications of X-ray multiplets – the impact of Theo Thole's work. *Journal of Electron Spectroscopy and Related Phenomena* 1997, 86:175–183.

[Cromer 1965] Cromer DT, Waber JT: Scattering factors computed from relativistic Dirac-Slater wave functions. *Acta Crystallographica* 1965, 18:104–109.

[Cromer 1967] Cromer DT, Mann JB: Compton scattering factors for spherically symmetric free atoms. *Journal of Chemical Physics* 1967, 47:1892–1893.

[Cromer 1970] Cromer DT, Liberman D: Relativistic calculation of anomalous scattering factors for X-rays. *Journal of Chemical Physics* 1970, 53:1891–1898.

[Cromer 1981] Cromer DT, Liberman DA: Anomalous dispersion calculations near to and on the long-wavelength side of an absorption edge. *Acta Crystallographica. Section A, Crystal Physics, Diffraction, Theoretical and General Crystallography* 1981, A37:267–268.

[Cross 1998] Cross JO, Newville M, Rehr JJ, Sorensen LB, Bouldin CE, Watson G, Gouder T, Lander GH, Bell MI: Inclusion of local structure effects in theoretical X-ray resonant scattering amplitudes using ab initio X-ray-absorption spectra calculations. *Physical Review B* 1998, 58:11215–11225.

[Cuisinier 2013] Cuisinier M, Cabelguen P-E, Evers S, He G, Kolbeck M, Garsuch A, Bolin T, Balasubramanian M, Nazar LF: Sulfur speciation in li-s batteries determined by operando X-ray absorption spectroscopy. *Journal of Physical Chemistry Letters* 2013, 4:3227–3232.

[Cwikla 2014] Cwikla J, Milroy S, Reider D, Skelton T: Pioneering Mars: turning the red planet green with earth's smallest settlers. *The American Biology Teacher* 2014, 76:300–305.

[Dadachova 2007] Dadachova E, Bryan RA, Huang XC, Moadel T, Schweitzer AD, Aisen P, Nosanchuk JD, Casadevall A: Ionizing radiation changes the electronic properties of melanin and enhances the growth of melanized fungi. *PLoS ONE* 2007, 2.

[Dadachova 2008] Dadachova E, Casadevall A: Ionizing radiation: how fungi cope, adapt, and exploit with the help of melanin. *Current Opinion in Microbiology* 2008, 11:525–531.

[Daillant 1999] Daillant J, Gibaud A: *X-ray and Neutron: Principles and Applications*. Springer; 1999.

[Dare-Edwards 1983] Dare-Edwards MP, Goodenough JB, Hamnett A, Trevellick PR: Electrochemistry and photoelectrochemistry of iron(iii) oxide. *Journal of the Chemical Society. Faraday Transactions I* 1983, 79:2027–2041.

[Dau 2001] Dau H, Iuzzolino L, Dittmer J: The tetra-manganese complex of photosystem II during its redox cycle – X-ray absorption results and mechanistic implications. *Biochimica Et Biophysica Acta. Bioenergetics* 2001, 1503:24–39.

[Day 2009] Day DA, Takahama S, Gilardoni S, Russell LM: Organic composition of single and submicron particles in different regions of western North America and the eastern pacific during INTEX-B 2006. *Atmospheric Chemistry and Physics* 2009, 9:5433–5446.

[De Groot 2008] De Groot F, Kotani A: *Core Level Spectroscopy of Solids*. CRC Press; 2008.

[De Groot 1990] De Groot FMF, Fuggle JC, Thole BT, Sawatzky GA: 2p X-ray absorption of 3d transition-metal compounds – an atomic multiplet description including the crystal field. *Physical Review. B, Condensed Matter* 1990, 42:5459–5468.

[De La Garza 2003] De La Garza L, Jeong G, Liddell PA, Sotomura T, Moore TA, Moore AL, Gust D: Enzyme-based photoelectrochemical biofuel cell. *Journal of Physical Chemistry. B* 2003, 107:10252–10260.

[De La Rosa 2003] De La Rosa MA: Electron transfer from cytochrome bf to photosystem i as mediated by cytochrome c6 and plastocyanin. *European Biophysics Journal* 2003, 32:169–169.

[De Souza 2012] De Souza RA, Metlenko V, Park D, Weirich TE: Behavior of oxygen vacancies in single-crystal $SrTiO_3$: equilibrium distribution and diffusion kinetics. *Physical Review B* 2012, 85.

[Debye 1913a] Debye P: Interference of X rays and heat movement. *Annalen der Physik* 1913, 43:49–95.

[Debye 1913c] Debye P: Interferenz von Röntgenstrahlen und Wärmebewegung. *Annalen der Physik* 1913, 348.

[Degroot 1994] Degroot FMF: X-ray-absorption and dichroism of transition-metals and their compounds. *Journal of Electron Spectroscopy and Related Phenomena* 1994, 67:529–622.

[Delacerda 1997] Delacerda B, Navarro JA, Hervas M, Delarosa MA: Changes in the reaction mechanism of electron transfer from plastocyanin to photosystem I in the cyanobacterium *Synechocystis* sp. PCC 6803 as induced by site-directed mutagenesis of the copper protein. *Biochemistry* 1997, 36:10125–10130.

[Diaz 2007] Diaz J, Monteiro OR, Hussain Z: Structure of amorphous carbon from near-edge and extended X-ray absorption spectroscopy. *Physical Review B* 2007, 76.

[Dirac 1927a] Dirac PaM: On quantum algebra. *Proceedings of the Cambridge Philosophical Society* 1927, 23:412–418.

[Dirac 1927b] Dirac PaM: The quantum theory of the emission and absorption of radiation. *Proceedings of the Royal Society of London. Series A, Containing Papers of a Mathematical and Physical Character* 1927, 114:243–265.

[Dockery 1993] Dockery DW, Pope CA, Xu XP, Spengler JD, Ware JH, Fay ME, Ferris BG, Speizer FE: An association between air-pollution and mortality in 6 united-states cities. *The New England Journal of Medicine* 1993, 329:1753–1759.

[Dommann 2003] Dommann M: Durchsicht, Einsicht, Vorsicht: eine Geschichte der Röntgenstrahlen 1896–1963. University of Zürich, Faculty of Arts Institute of History; 2003.

[Doumerc 1981] Doumerc JP, Dance JM, Chaminade JP, Pouchard M, Hagenmuller P, Krussanova M: An example of alternating anti-Ferromagnetic chains with spin 1/2 – $CUWO_4$. *Materials Research Bulletin* 1981, 16:985–990.

[Drits 1991] Drits VA, Tchoubar A: *X-ray Diffraction by Disordered Lamellar Structures*. New York: Springer Verlag; 1991.

[Duke 2003] Duke CB: The birth and evolution of surface science: child of the union of science and technology. *Proceedings of the National Academy of Sciences of the United States of America* 2003, 100:3858–3864.

[Dumschat 1995] Dumschat J, Wartmann G, Felner I: L(II,III) Near-edge study of tetravalent PR-OXIDES – $PRBAO_3$ and PRO_2. *Physica B* 1995, 208:313–315.

[Dunitz 1957a] Dunitz JD, Orgel LE: Electronic properties of transition-metal oxides .1. Distortions from cubic symmetry. *Journal of Physics and Chemistry of Solids* 1957, 3:20–29.

[Dunitz 1957b] Dunitz JD, Orgel LE: Electronic properties of transition-metal oxides .2. Cation distribution amongst octahedral and tetrahedral sites. *Journal of Physics and Chemistry of Solids* 1957, 3:318–323.

[Ebel 1975] Ebel H, Gurker N: Deconvolution of Xps-valence band spectrum of gold. *Physics Letters A* 1975, A 50:449–450.

[Eberhardt 2014] Eberhardt SH, Marone F, Stampanoni M, Buechi FN, Schmidt TJ: Quantifying phosphoric acid in high-temperature polymer electrolyte fuel cell components by X-ray tomographic microscopy. *Journal of Synchrotron Radiation* 2014, 21:1319–1326.

[Ebner 2013] Ebner M, Geldmacher F, Marone F, Stampanoni M, Wood V: X-ray tomography of porous, transition metal oxide based lithium ion battery electrodes. *Advanced Energy Materials* 2013, 3:845–850.

[Eder 2012] Eder R: Multiplets in transition metal ions. In *Correlated Electrons: From Models to Materials*. Edited by Pavarini E, Koch E, Anders F, Jarrell M. Jülich: Forschungszentrum Jülich GmbH, Institute for Advanced Simulation; 2012.

[Eller 2013] Eller J: *X-ray Tomographic Microscopy of Polymer Electrolyte Fuel Cells*. ETH Zürich, 2013.

[Ensling 2006] Ensling D: Photoelektronenspektroskopische Untersuchung der elektronischen Struktur dünner Lithiumkobaltoxidschichten. *Doctoral thesis*. Technische Universität Darmstadt, Material- und Geowissenschaften; 2006.

[Ensling 2010] Ensling D, Thissen A, Laubach S, Schmidt PC, Jaegermann W: Electronic structure of $LiCoO_2$ thin films: a combined photoemission spectroscopy and density functional theory study. *Physical Review B* 2010, 82.

[Erat 2009] Erat S, Braun A, Ovalle A, Piamonteze C, Liu Z, Graule T, Gauckler LJ: Correlation of O (1s) and Fe (2p) near edge X-ray absorption fine structure spectra and electrical conductivity of $La_{1-x}SrxFe_{0.75}Ni_{0.25}O_3$-delta. *Applied Physics Letters* 2009, 95.

[Erat 2010a] Erat S: On the origin of enhanced conductivity in LaSrFe-oxide upon Ni-doping. ETH Zürich, D-MAT; 2010.

[Erat 2010c] Erat S, Braun A, Piamonteze C, Liu Z, Ovalle A, Schindler H, Graule T, Gauckler LJ: Entanglement of charge transfer, hole doping, exchange interaction, and octahedron tilting angle and their influence on the conductivity of $La_{1-x}SrxFe_{0.75}Ni_{0.25}O_3$-delta: a combination of X-ray spectroscopy and diffraction. *Journal of Applied Physics* 2010, 108.

[Erat 2010h] Erat S, Wadati H, Aksoy F, Liu Z, Graule T, Gauckler LJ, Braun A: Iron-resonant valence band photoemission and oxygen near edge X-ray absorption fine structure study on $La_{1-x}SrxFe_{0.75}Ni_{0.25}O_3$-delta. *Applied Physics Letters* 2010, 97.

[Eriksson 2014] Eriksson M, van der Veen JF, Quitmann C: Diffraction-limited storage rings - a window to the science of tomorrow. *Journal of Synchrotron Radiation* 2014, 21:837–842.

[Ewald 1921] Ewald PP: Die Berechnung optischer und elektrostatischer Gitterpotentiale. *Annalen der Physik* 1921, 369:253–287.

[Ewald 1962] Ewald PP: Laue's discovery of X-ray diffraction by crystals. In *Fifty Years of X-ray Diffraction*. Edited by Ewald PP. Boston, MA: Springer US; 1962: 733. [Ewald PP (series editor)].

[Faccio 2015] Faccio G, Gajda-Schrantz K, Ihssen J, Boudoire F, Hu Y, Mun BS, Bora DK, Thöny-Meyer L, Braun A: Charge transfer between photosynthetic proteins and hematite in bio-hybrid photoelectrodes for solar water splitting cells. *Nano Convergence* 2015, 2:11.

[Fadley 2013] Fadley CS: Hard X-ray photoemission with angular resolution and standing-wave excitation. *Journal of Electron Spectroscopy and Related Phenomena* 2013, 190:165–179.

[Farley 1999] Farley NRS, Gurman SJ, Hillman AR: Simple cell for in situ X-ray absorption spectroelectrochemistry. *Electrochemistry Communications* 1999, 1:449–452.

[Fedorovskaya 2014] Fedorovskaya EO, Bulusheva LG, Kurenya AG, Asanov IP, Rudina NA, Funtov KO, Lyubutin IS, Okotrub AV: Supercapacitor performance of vertically aligned multiwall carbon nanotubes produced by aerosol-assisted CCVD method. *Electrochimica Acta* 2014, 139:165–172.

[Feigin 1987] Feigin LA, Svergun DI: *Structure Analysis by Small-Angle X-ray and Neutron Scattering*. New York: Plenum Press; 1987.

[Fenter 2000] Fenter P, Cheng L, Rihs S, Machesky M, Bedzyk MJ, Sturchio NC: Electrical double-layer structure at the rutile-water interface as observed in situ with small-period X-ray standing waves. *Journal of Colloid and Interface Science* 2000, 225:154–165.

[Ferreira 2004] Ferreira KN, Iverson TM, Maghlaoui K, Barber J, Iwata S: Architecture of the photosynthetic oxygen-evolving center. *Science* 2004, 303:1831–1838.

[Fey 1994] Fey GTK, Li W, Dahn JR: $LINIVO_4$ – A 4.8 Volt electrode material for lithium cells. *Journal of the Electrochemical Society* 1994, 141:2279–2282.

[Feynman 2005] Feynman RP: *The Pleasure of Finding Things Out: The Best Short Works of Richard P. Feynman (Helix Books)*. Basic Books; 2005.

[Figueroa 2005a] Figueroa SJA, Requejo FG, Lede EJ, Lamaita L, Peluso MA, Sambeth JE: XANES study of electronic and structural nature of Mn-sites in manganese oxides with catalytic properties. *Catalysis Today* 2005, 107–108:849–855.

[Figueroa 2005e] Figueroa SJA, Requejo FG, Lede EJ, Lamaita L, Peluso MA, Sambeth JE: XANES study of electronic and structural nature of Mn-sites in manganese oxides with catalytic properties. *Catalysis Today* 2005, 107–08:849–855.

[Fink 1989] Fink J: Recent developments in energy-loss spectroscopy. *Advances in Electronics and Electron Physics* 1989, 75:121–232.

[Fister 2008] Fister TT, Seidler GT, Shirley EL, Vila FD, Rehr JJ, Nagle KP, Linehan JC, Cross JO: The local electronic structure of alpha-Li3N. *Journal of Chemical Physics* 2008, 129.

[Fister 2011] Fister TT, Schmidt M, Fenter P, Johnson CS, Slater MD, Chan MKY, Shirley EL: Electronic structure of lithium battery interphase compounds: comparison between inelastic X-ray scattering measurements and theory. *Journal of Chemical Physics* 2011, 135.

[Flak 2013] Flak D, Braun A, Mun BS, Park JB, Parlinska-Wojtan M, Graule T, Rekas M: Spectroscopic assessment of the role of hydrogen in surface defects, in the electronic structure and transport properties of TiO_2, ZnO and SnO_2 nanoparticles. *Physical Chemistry Chemical Physics* 2013, 15:1417–1430.

[Flanders 1911] Flanders T: Speakers give sound advice. In *Syracuse Post Standard*. pp. 18; 1911:18.

[Flank 2006] Flank AM, Cauchon G, Lagarde P, Bac S, Janousch M, Wetter R, Dubuisson JM, Idir M, Langlois F, Moreno T, Vantelon D: LUCIA, a microfocus soft XAS beamline. *Nuclear Instruments & Methods in Physics Research. Section B, Beam Interactions With Materials and Atoms* 2006, 246:269–274.

[Flytzani-Stephanopoulos 2006] Flytzani-Stephanopoulos M, Sakbodin M, Wang Z: Regenerative adsorption and removal of H_2S from hot fuel gas streams by rare earth oxides. *Science* 2006, 312:1508–1510.

[Fowlks 2002] Fowlks A, Braun A: In situ X-ray absorption spectroscopy on Mn-oxide based lithium battery electrodes. In *Journal of Undergraduate Research, Volume II, 2002. Volume In situ X-ray Absorption Spectroscopy on Mn-Oxide Based Lithium Battery Electrodes*. Edited by Faletra P, Franz K, Clark T, Manning K, Musick C, Sokolski K, Walbridge SE. United States: DOESC (USDOE office of Science (SC) (United States)); 2002: 117.

[Franklin 1951a] Franklin RE: The structure of graphitic carbons. *Acta Crystallographica* 1951, 4:253–261.

[Franklin 1951c] Franklin RE: Crystallite growth in graphitizing and non-graphitizing carbons. *Proceedings of the Royal Society of London. Series A, Mathematical and Physical Sciences* 1951, 209:196–218.

[Franklin 1951e] Franklin RE: Les Carbones graphitisables et non-graphitisables. *Comptes Rendus Hebdomadaires Des Séances De L'Académie Des Sciences* 1951, 232:232–234.

[Frese 1978] Frese KW, Kennedy JH: Extrinsic hole lifetimes for N-type alpha-Fe_2O_3 from photocurrent transients. *Journal of the Electrochemical Society* 1978, 125:C160–C160.

[Frisch 2014] Frisch J: *Electronic Properties of Interfaces in Polymer Based Organic Photovoltaic Cells*. Humboldt-Universität zu Berlin, Mathematisch-Naturwissenschaftliche Fakultät; 2014.

[Fuggle 1980] Fuggle JC, Martensson N: Core-level binding-energies in metals. *Journal of Electron Spectroscopy and Related Phenomena* 1980, 21:275–281.

[Fujimori 1986] Fujimori A, Saeki M, Kimizuka N, Taniguchi M, Suga S: Photoemission satellites and electronic-structure of FE_2O_3. *Physical Review B* 1986, 34:7318–7333.

[Fujimori 1987] Fujimori A, Kimizuka N, Taniguchi M, Suga S: Electronic-structure of Fexo. *Physical Review B* 1987, 36:6691–6694.

[Fukunaga 2005] Fukunaga H, Kishimi M, Ozaki T, Sakai T: Non-foam nickel electrode with quasi-three-dimensional substrate for Ni-MH battery. *Journal of the Electrochemical Society* 2005, 152:A126–A131.

[Furrer 1998] Furrer A: *Complementarity Between Neutron and Synchrotron X-ray Scattering*. Summer school edition. Singapore, New Jersey, London, Hong Kong: World Scientific Publishing Co. Pte. Ltd.; 1998.

[Furth 1933] Furth R: About some relationships between classical and quantum statistical mechanics. *Zeitschrift für Physik* 1933, 81:143–162.

[Gaillard 2010] Gaillard N, Cole B, Kaneshiro J, Miller EL, Marsen B, Weinhardt L, Baer M, Heske C, Ahn K-S, Yan Y, Al-Jassim MM: Improved current collection in WO_3:Mo/WO_3 bilayer photoelectrodes. *Journal of Materials Research* 2010, 25:45–51.

[Gaillard 2012] Gaillard N, Chang Y, Braun A, DeAngelis A: Copper tungstate ($CuWO_4$)-based materials for photoelectrochemical hydrogen production. In *Materials Research Society Symposium Proceedings*. vol. 1446; 2012: 19–24.

[Gaillard 2013] Gaillard N, Chang Y, DeAngelis A, Higgins S, Braun A: A nanocomposite photoelectrode made of 2.2 eV band gap copper tungstate (CuWO$_4$) and multi-wall carbon nanotubes for solar-assisted water splitting. *International Journal of Hydrogen Energy* 2013, 38:3166–3176.

[Gajda-Schrantz 2013] Gajda-Schrantz K, Tymen S, Boudoire F, Toth R, Bora DK, Calvet W, Graetzel M, Constable EC, Braun A: Formation of an electron hole doped film in the alpha-Fe$_2$O$_3$ photoanode upon electrochemical oxidation. *Physical Chemistry Chemical Physics* 2013, 15:1443–1451.

[Galakhov 2009] Galakhov VR, Neumann M, Kellerman DG: Electronic structure of defective lithium cobaltites Li (x) CoO$_2$. *Applied Physics. A, Materials Science & Processing* 2009, 94:497–500.

[Galakhov 2013] Galakhov VR, Shamin SN, Mironova EM, Uimin MA, Yermakov AY, Boukhvalov DW: Electronic structure and resonant X-ray emission spectra of carbon shells of iron nanoparticles. *JETP Letters* 2013, 96:710–713.

[Galvani 1791] Galvani A: *De viribus electricitatis in motu musculari commentarius*. pp. 72. Bononiae: Ex Typographia Instituti Scientiarium; 1791:72.

[Ganduglia-Pirovano 2007] Ganduglia-Pirovano MV, Hofmann A, Sauer J: Oxygen vacancies in transition metal and rare earth oxides: current state of understanding and remaining challenges. *Surface Science Reports* 2007, 62:219–270.

[Ganorkar 2011] Ganorkar S, Priolkar KR, Sarode PR, Banerjee A: Effect of oxygen content on magnetic properties of layered cobaltites PrBaCo(2)O(5+delta). *Journal of Applied Physics* 2011, 110.

[Garcia 2004] Garcia J, Subias G: The Verwey transition – a new perspective. *Journal of Physics. Condensed Matter* 2004, 16:R145–R178.

[Garfield 1975] Garfield E: Libraries need a copyright clearinghouse – ISI has one they can use. *Current Contents* 1975:5–7.

[Garzon 2007] Garzon FH, Lau SH, Davey JR, Borup R: Micro and nano X-ray tomography of PEM fuel cell membranes after transient operation. *ECS Transactions* 2007, 11:10.

[Gelb 2012] Gelb J: Functionality to failure: materials engineering in the 4th dimension. *Advanced Materials & Processes* 2012, 170:14–18.

[Gelmukhanov 1994] Gelmukhanov F, Agren H: Resonant inelastic X-ray-scattering with symmetry-selective excitation. *Physical Review A* 1994, 49:4378–4389.

[Genc 2013] Genc A, Kovarik L, Gu M, Cheng H, Plachinda P, Pullan L, Freitag B, Wang C: Xeds stem tomography for 3D chemical characterization of nanoscale particles. *Ultramicroscopy* 2013, 131:24–32.

[Genet 2008] Genet MJ, Dupont-Gillain CC, Rouxhet PG: *XPS analysis of biosystems and Biomaterials*. 2008.

[Gengenbach 1996] Gengenbach TR, Chatelier RC, Griesser HJ: Correlation of the nitrogen 1s and oxygen 1s XPS binding energies with compositional changes during oxidation of ethylene diamine plasma polymers. *Surface and Interface Analysis* 1996, 24:611–619.

[George 2001] George SD, Metz M, Szilagyi RK, Wang HX, Cramer SP, Lu Y, Tolman WB, Hedman B, Hodgson KO, Solomon EI: A quantitative description of the ground-state wave function of Cu-A by X-ray absorption spectroscopy: comparison to plastocyanin and relevance to electron transfer. *Journal of the American Chemical Society* 2001, 123:5757–5767.

[Gidalevitz 1999] Gidalevitz D, Huang ZQ, Rice SA: Protein folding at the air-water interface studied with X-ray reflectivity. *Proceedings of the National Academy of Sciences of the United States of America* 1999, 96:2608–2611.

[Gidalevitz 2000] Gidalevitz D, Rice SA: Folding of water-soluble proteins at the air-water interface studied with X-ray reflectivity. *Abstracts of Papers—American Chemical Society* 2000, 219:U527–U528.

[Gille 2003] Gille W, Braun A: SAXS chord length distribution analysis and porosity estimation of activated and non-activated glassy carbon. *Journal of Non-Crystalline Solids* 2003, 321:89–95.

[Glatter 1982] Glatter O, Kratky O: *Small Angle X-ray Scattering*. Academic Press; 1982.

[Glatzel 7 May 2002] Glatzel P: New developments in hard X-ray spectroscopy. In *Berkeley Spectroscopy Club*. pp. 32. University of California, Davis: Glatzel, Pieter; 7 May 2002:32.

[Glatzel 2001] Glatzel P, Bergmann U, De Groot FMF, Cramer SP: Influence of the core hole on K beta emission following photoionization or orbital electron capture: a comparison using MnO and (Fe$_2$O$_3$)-Fe-55. *Physical Review B* 2001, 64.

[Glatzel 2004] Glatzel P, Bergmann U, Yano J, Visser H, Robblee JH, Gu WW, De Groot FMF, Christou G, Pecoraro VL, Cramer SP, Yachandra VK: The electronic structure of Mn in oxides, coordination complexes, and the oxygen-evolving complex of photosystem II studied by resonant inelastic X-ray scattering. *Journal of the American Chemical Society* 2004, 126:9946–9959.

[Glatzel 2005] Glatzel P, Yano J, Bergmann U, Visser H, Robblee JH, Gu WW, De Groot FMF, Cramer SP, Yachandra VK: Resonant inelastic X-ray scattering (RIXS) spectroscopy at the MnK absorption pre-edge – a direct probe of the 3d orbitals. *Journal of Physics and Chemistry of Solids* 2005, 66:2163–2167.

[Glatzel 2009] Glatzel P, Sikora M, Smolentsev G, Fernandez-Garcia M: Hard X-ray photon-in photon-out spectroscopy. *Catalysis Today* 2009, 145:294–299.

[Glatzel 2013] Glatzel P, Weng TC, Kvashnina K, Swarbrick J, Sikora M, Gallo E, Smolentsev N, Mori RA: Reflections on hard X-ray photon-in/photon-out spectroscopy for electronic structure studies. *Journal of Electron Spectroscopy and Related Phenomena* 2013, 188:17–25.

[Gloskovskii 2012] Gloskovskii A, Stryganyuk G, Fecher GH, Felser C, Thiess S, Schulz-Ritter H, Drube W, Berner G, Sing M, Claessen R, Yamamoto M: Magnetometry of buried layers-Linear magnetic dichroism and spin detection in angular resolved hard X-ray photoelectron spectroscopy. *Journal of Electron Spectroscopy and Related Phenomena* 2012, 185:47–52.

[Goerigk 1997] Goerigk G, Haubold HG, Schilling W: Kinetics of decomposition in copper-cobalt: a time-resolved ASAXS study. *Journal of Applied Crystallography* 1997, 30:1041–1047.

[Goerigk 2003] Goerigk G, Haubold HG, Lyon O, Simon JP: Anomalous small-angle X-ray scattering in materials science. *Journal of Applied Crystallography* 2003, 36:425–429.

[Goldschmidt 1926] Goldschmidt VM: Die Gesetze der Krystallochemie. vol. 14, Originalaufsätze Und Berichte Reine Und Technisch Angewandte Chemie Und Physikalische Chemie edition. pp. 477–485; 1926:477–485.

[Goldstein 2003] Goldstein J, Newbury DE, Joy DC, Lyman CE, Echlin P, Lifshin E, Sawyer L, Michael JR: *Scanning Electron Microscopy and X-ray Microanalysis*. New York: Springer Science + Business Media; 2003.

[Golks 2012] Golks F, Stettner J, Gruender Y, Krug K, Zegenhagen J, Magnussen OM: Anomalous potential dependence in homoepitaxial Cu(001) electrodeposition: an in situ surface X-ray diffraction study. *Physical Review Letters* 2012, 108.

[Golks 2015] Golks F, Gruender Y, Stettner J, Krug K, Zegenhagen J, Magnussen OM: In situ surface X-ray diffraction studies of homoepitaxial growth on Cu(001) from aqueous acidic electrolyte. *Surface Science* 2015, 631:112–122.

[Goodenough 1995] Goodenough JB: Solid electrolytes. *Pure and Applied Chemistry* 1995, 67:931–938.

[Goodenough 1998] Goodenough JB: Jahn-Teller phenomena in solids. *Annual Review of Materials Science* 1998, 28:1–27.

[Goodenough 2004] Goodenough JB: Electronic and ionic transport properties and other physical aspects of perovskites. *Reports on Progress in Physics* 2004, 67:1915–1993.

[Goodenough 2008] Goodenough JB: Goodenough-Kanamori rule. *Scholarpedia* 2008, 3.

[Gorlin 2015] Gorlin Y, Siebel A, Piana M, Huthwelker T, Jha H, Monsch G, Kraus F, Gasteiger HA, Tromp M: Operando characterization of intermediates produced in a lithium-sulfur battery. *Journal of the Electrochemical Society* 2015, 162:A1146–A1155.

[Goroff 1997] Goroff NS: IGOR Pro, Version 3 (Mac). *Journal of the American Chemical Society* 1997, 119:10567–10567.

[Granroth 2011] Granroth S, Olovsson W, Holmstroem E, Knut R, Gorgoi M, Svensson S, Karis O: Understanding interface properties from high kinetic energy photoelectron spectroscopy and first principles theory. *Journal of Electron Spectroscopy and Related Phenomena* 2011, 183:80–93.

[Grass 2010] Grass ME, Karlsson PG, Aksoy F, Lundqvist M, Wannberg B, Mun BS, Hussain Z, Liu Z: New ambient pressure photoemission endstation at Advanced Light Source beamline 9.3.2. *Review of Scientific Instruments* 2010, 81.

[Gray 1998] Gray E: Structure determination by neutron powder diffraction. Proceedings of the Sixth Summer School on Neutron Scattering, Complementarity Between Neutron and Synchrotron X-ray Scattering: World Scientific; 1998:61–82.

[Greger 2013] Greger M, Kollar M, Vollhardt D: Isosbestic points: How a narrow crossing region of curves determines their leading parameter dependence. *Physical Review B* 2013, 87.

[Greta Faccio 2016] Charge transfer between photosynthetic proteins and hematite in bio-hybrid photoelectrodes for solar water splitting cells. SpringerLink [http://link.springer.com/article/10.1186/s40580-014-0040-4].

[Grolimund 1999] Grolimund D, Trainor TP, Fitts JP, Kendelewicz T, Liu P, Chambers SA, Brown GE: Identification of Cr species at the aqueous solution-hematite interface after Cr(VI)-Cr(III) reduction using GI-XAFS and Cr L-edge NEXAFS. *Journal of Synchrotron Radiation* 1999, 6:612–614.

[Grunes 1983] Grunes LA, Gates IP, Ritsko JJ, Mele EJ, Divincenzo DP, Preil ME, Fischer JE: Valence and core electronic excitations in lic6. *Physical Review B* 1983, 28:6681–6686.

[Grünzweig 2012] Grünzweig C: Dynamic neutron radiography measurement of a two-stroke engine running fired at 10000 rpm., 19 Sep 2012 edition. YouTube: YouTube; 2012.

[Grush 1995] Grush MM, Christou G, Hamalainen K, Cramer SP: Site-selective XANES and EXAFS – a demonstration with manganese mixtures and mixed-valence complexes. *Journal of the American Chemical Society* 1995, 117:5895–5896.

[Grush 1996] Grush MM, Chen J, Stemmler TL, George SJ, Ralston CY, Stibrany RT, Gelasco A, Christou G, Gorun SM, Pennerhahn JE, Cramer SP: Manganese L-edge X-ray absorption spectroscopy of manganese catalase from *Lactobacillus plantarum* and mixed valence manganese complexes. *Journal of the American Chemical Society* 1996, 118:65–69.

[Gu 2002] Gu W, Wang H, Cramer SP, Friedrich S, Funk T, Grahame D, Ragsdale S: Ni L-edge X-ray absorption spectroscopy of acetyl-CoA synthases and model compounds. *Abstracts of Papers—American Chemical Society* 2002, 223:468-INOR 468.

[Gu 2003] Gu WW, Jacquamet L, Patil DS, Wang HX, Evans DJ, Smith MC, Millar M, Koch S, Eichhorn DM, Latimer M, Cramer SP: Refinement of the nickel site structure in *Desulfovibrio gigas* hydrogenase using range-extended EXAFS spectroscopy. *Journal of Inorganic Biochemistry* 2003, 93:41–51.

[Guay 2005] Guay D, Stewart-Ornstein J, Zhang XR, Hitchcock AP: In situ spatial and time-resolved studies of electrochemical reactions by scanning transmission X-ray microscopy. *Analytical Chemistry* 2005, 77:3479–3487.

[Gueymard 2001] Gueymard CA: Parameterized transmittance model for direct beam and circumsolar spectral irradiance. *Solar Energy* 2001, 71:325–346.

[Gueymard 2002] Gueymard CA, Myers D, Emery K: Proposed reference irradiance spectra for solar energy systems testing. *Solar Energy* 2002, 73:443–467.

[Gueymard 2003a] Gueymard CA: Direct solar transmittance and irradiance predictions with broadband models. Part I: detailed theoretical performance assessment. *Solar Energy* 2003, 74:355–379.

[Gueymard 2003b] Gueymard CA: Direct solar transmittance and irradiance predictions with broadband models. Part II: validation with high-quality measurements. *Solar Energy* 2003, 74:381–395.

[Gueymard 2004a] Gueymard CA: The sun's total and spectral irradiance for solar energy applications and solar radiation models. *Solar Energy* 2004, 76:423–453.

[Gueymard 2004b] Gueymard CA: Direct solar transmittance and irradiance predictions with broadband models. Part I: detailed theoretical performance assessment (vol. 74, p. 355, 2003). *Solar Energy* 2004, 76:513–513.

[Gueymard 2004c] Gueymard CA: Direct solar transmittance and irradiance predictions with broadband models: Part II: validation with high-quality measurements (vol. 74, p. 381, 2003). *Solar Energy* 2004, 76:515–515.

[Guinier 1939] Guinier A: La diffraction des rayons X aux tres petits angles; application a l'etude de phenomenes ultramicroscopiques. *Doctoral thesis*. Universite de Paris, Faculté des sciences de Paris; 1939.

[Guinier 1955] Guinier A, Fournet G: *Small Angle Scattering of X-rays*. John Wiley; 1955.

[Guinier 1964] Guinier A: *Theorie et technique de la radiocristallographie*. 3rd edition. Paris: Dunod; 1964.

[Gullikson 1995] X-ray Interactions With Matter [http://henke.lbl.gov/optical_constants/].

[Guo 2006] Guo JH: Understanding the electronic properties of hydrogen storage materials with photon-in/photon-out soft-X-ray spectroscopy. In *Conference on Solar Hydrogen and Nanotechnology; Aug 14–17; San Diego, CA*. Spie-Int Soc Optical Engineering; 2006: U28–U38.

[Guo 2010] Guo JH, Kastanov S, Soderstom J, Glans PA, West M, Learmonth T, Chiou JW, Luo Y, Nordgren J, Smith K, Pong WF, Cheng H, Griffiss JM: Electronic structure study of the bases in DNA duplexes by in situ photon-in/photon-out soft X-ray spectroscopy. *Journal of Electron Spectroscopy and Related Phenomena* 2010, 181:197–201.

[Guo 2008] Guo Y, Wang H, Xiao Y, Vogt S, Thauer RK, Shima S, Volkers PI, Rauchfuss TB, Pelmenschikov V, Case DA, Alp EE, Sturhahn W, Yoda Y, Cramer SP: Characterization of the Fe site in iron-sulfur cluster-free hydrogenase (Hmd) and of a model compound via nuclear resonance vibrational spectroscopy (NRVS). *Inorganic Chemistry* 2008, 47:3969–3977.

[Haas 2009] Haas O, Vogt UF, Soltmann C, Braun A, Yoon WS, Yang XQ, Graule T: The Fe K-edge X-ray absorption characteristics of $La_{1-x}Sr_xFeO_3$-delta prepared by solid state reaction. *Materials Research Bulletin* 2009, 44:1397–1404.

[Haehnel 1980] Haehnel W, Hesse V, Propper A: Electron-transfer from plastocyanin to p700 – function of a subunit of photosystem-i reaction center. *FEBS Letters* 1980, 111:79–82.

[Haehnel 1994] Haehnel W, Jansen T, Gause K, Klosgen RB, Stahl B, Michl D, Huvermann B, Karas M, Herrmann RG: Electron-transfer from plastocyanin to photosystem-i. *EMBO Journal* 1994, 13:1028–1038.

[Haensel 1968] Haensel R, Kunz C, Sonntag B: Measurement of photoabsorption of lithium halides near lithium k edge. *Physical Review Letters* 1968, 20:262–264.

[Hafeez 2013] Hafeez A: *Precious Metal-Free Dye-Sensitized Solar Cells*. Dalhousie University, 2013.

[Hagfeldt 2000] Hagfeldt A, Gratzel M: Molecular photovoltaics. *Accounts of Chemical Research* 2000, 33:269–277.

[Hamann 2005] Hamann CF, Vielstich W: *Elektrochemie*. 4th edition. Weinheim: Wiley-VCH Verlag GmbH & Co. KGaA; 2005.

[Harris 2001] Harris PJF: Rosalind Franklin's work on coal, carbon, and graphite. *Interdisciplinary Science Reviews* 2001, 26:204–210.

[Harris 2013] Harris SJ, Lu P: Effects of inhomogeneities—nanoscale to mesoscale—on the durability of Li-Ion batteries. *Journal of Physical Chemistry C* 2013, 117:6481–6492.

[Haubold 1994] Haubold H-G, Gebhardt R, Buth, Goerigk G: Structural characterization of compositional and density inhomogeneities by ASAXS. In *Resonant Anomalous X-Ray Scattering: Theory and Applications*. Edited by Materlik G, Sparks CJ, Fischer K. Amsterdam: Elsevier; 1994: 295–304.

[Haubold 1975] Haubold HG: Measurement of diffuse x-ray-scattering between reciprocal-lattice points as a new experimental-method in determining interstitial structures. *Journal of Applied Crystallography* 1975, 8:175–183.

[Haubold 1976] Haubold HG: Application of x-ray diffuse-scattering to structure determinations of point-defects. *Revue de Physique Appliquée* 1976, 11:73–81.

[Haubold 1989] Haubold HG, Gruenhagen K, Wagener M, Jungbluth H, Heer H, Pfeil A, Rongen H, Brandenberg G, Moeller R, Matzerath J, Hiller P, Halling H: Jusifa—a new user-dedicated ASAXS beamline for materials science. *Review of Scientific Instruments* 1989, 60:1943–1946.

[Haubold 1995] Haubold HG, Wang XH: ASAXS studies of carbon-supported electrocatalysts. *Nuclear Instruments & Methods in Physics Research. Section B, Beam Interactions With Materials and Atoms* 1995, 97:50–54.

[Haubold 1996] Haubold HG, Jungbluth H, Hiller P, Wang XH, Goerigk G, Schilling W: *Investigation of electrocatalysts by in situ energy dispersive X-ray absorption and small angle scattering*. 1996.

[Haubold 1997] Haubold HG, Wang XH, Goerigk G, Schilling W: In situ anomalous small-angle X-ray scattering investigation of carbon-supported electrocatalysts. *Journal of Applied Crystallography* 1997, 30:653–658.

[Haubold 1999] Haubold HG, Hiller P, Jungbluth H, Vad T: Characterization of electrocatalysts by in situ SAXS and XAS investigations. *Japanese Journal of Applied Physics Part 1: Regular Papers Brief Communications and Review Papers* 1999, 38:36–39.

[Haubold 2001] Haubold HG, Vad T, Jungbluth H, Hiller P: Nano structure of NAFION: A SAXS study. *Electrochimica Acta* 2001, 46:1559–1563.

[He 2011] He H, Xia J-L, Huang G-H, Jiang H-C, Tao X-X, Zhao Y-D, He W: Analysis of the elemental sulfur bio-oxidation by *Acidithiobacillus ferrooxidans* with sulfur K-edge XANES. *World Journal of Microbiology and Biotechnology* 2011, 27:1927–1931.

[Heiland 1972] Heiland G, Bauer W, Neuhaus M: Spectrally sensitized photoconductivity of zinc oxide crystals. *Photochemistry and Photobiology* 1972, 16:315–324.

[Hellman 2011] Hellman M, Mattinen ML, Fu BA, Buchert J, Permi P: Effect of protein structural integrity on cross-linking by tyrosinase evidenced by multidimensional heteronuclear magnetic resonance spectroscopy. *Journal of Biotechnology* 2011, 151:143–150.

[Hellwig 2011] Hellwig C, Fronczek DN, Sörgel S, Bessler WG: Physically based modeling and simulation of a LiFePO$_4$-based lithium-ion battery. In *ModVal8 - 8th Symposium on Fuel Cell Modeling and Experimental Validation*. Edited by Kulikovsky A. Bonn: Forschungszentrum Jülich GmbH; 2011.

[Hellwig 2013] Hellwig CA: Modeling, simulation and experimental investigation of the thermal and electrochemical behavior of a LiFePO$_4$-based lithium-ion battery. *Dissertation*. Universität Stuttgart, Chemie; 2013.

[Hempelmann 1995] Hempelmann R, Karmonik C, Matzke T, Cappadonia M, Stimming U, Springer T, Adams MA: Quasi-elastic neutron-scattering study of proton diffusion in Srce$_{0.95}$YB$_{0.05}$H$_{0.02}$O$_{2.985}$. *Solid State Ionics* 1995, 77:152–156.

[Hendrickson 1968] Hendrickson WA, Love WE, Murray GC: Crystal forms of lamprey hemoglobin and crystalline transitions between ligand states. *Journal of Molecular Biology* 1968, 33:829–842.

[Hendrickson 1970] Hendrickson WA, Lattman EE, Iucr: Representation of phase probability distributions for simplified combination of independent phase information. *Acta Crystallographica. Section B, Structural Crystallography and Crystal Chemistry* 1970, 26:136–143.

[Hendrickson 1979] Hendrickson WA: Phase information from anomalous-scattering measurements. *Acta Crystallographica. Section A* 1979, 35:245–247.

[Hendrickson 1980] Hendrickson WA, Konnert JH: Diffraction analysis of motion in proteins. *Biophysical Journal* 1980, 32:645–647.

[Hendrickson 1981] Hendrickson WA, Teeter MM: Structure of the hydrophobic protein crambin determined directly from the anomalous scattering of sulfur. *Nature* 1981, 290:107–113.

[Hendrickx 2006] Hendrickx L, De Wever H, Hermans V, Mastroleo F, Morin N, Wilmotte A, Janssen P, Mergeay M: Microbial ecology of the closed artificial ecosystem MELiSSA (Micro-Ecological Life Support System Alternative): reinventing and compartmentalizing the Earth's food and oxygen regeneration system for long-haul space exploration missions. *Research in Microbiology* 2006, 157:77–86.

[Henke 1993a] Henke BL, Gullikson EM, Davis JC: X-ray interactions – photoabsorption, scattering, transmission, and reflection at E = 50–30,000 EV, Z = 1–92. *Atomic Data and Nuclear Data Tables* 1993, 54:181–342.

[Henke 1993b] Henke BL, Gullikson EM, Davis JC: X-ray interactions – photoabsorption, scattering, transmission and reflection AT E = 50–30,000 EV, Z = 1–92 (vol. 54, p. 181, 1993). *Atomic Data and Nuclear Data Tables* 1993, 55:349–349.

[Herman 2006] Herman GT: *Fundamentals of Computerized Tomography: Image Reconstruction from Projections*. Springer; 2006.

[Hermann 1935] Hermann G: The natural-philosophical fundaments of quantum mechanics. *Naturwissenschaften* 1935, 23:718–721.

[Hervas 1995] Hervas M, Navarro MA, Diaz A, Bottin H, Delarossa MA: *Mechanism of Electron Transfer from Plastocyanin and Cytochrome c6 to Photosystem I in a Number of Evolutionarily Differentiated Organisms.* 1995. Kluwer Academic Publ, Dordrecht, Netherlands.

[Hervas 2011] Hervas M, Navarro JA: Effect of crowding on the electron transfer process from plastocyanin and cytochrome c(6) to photosystem I: a comparative study from cyanobacteria to green algae. *Photosynthesis Research* 2011, 107:279–286.

[Hetenyi 2004] Hetenyi B, De Angelis F, Giannozzi P, Car R: Calculation of near-edge X-ray-absorption fine structure at finite temperatures: spectral signatures of hydrogen bond breaking in liquid water. *Journal of Chemical Physics* 2004, 120:8632–8637.

[Heymann 2014] Heymann K, Lehmann J, Solomon D, Liang B, Neves E, Wirick S: Can functional group composition of alkaline isolates from black carbon-rich soils be identified on a sub-100 nm scale? *Geoderma* 2014, 235:163–169.

[Hightower 2000a] Hightower A, Ahn CC, Fultz B: Electron energy loss spectrometry on lithiated graphite. In *Microbeam Analysis 2000, Proceedings.* Edited by Williams DB, Shimizu R. *Institute of Physics Conference Series;* 2000: 226–226.

[Hightower 2000f] Hightower A, Ahn CC, Fultz B, Rez P: Electron energy-loss spectrometry on lithiated graphite. *Applied Physics Letters* 2000, 77:238–240.

[Himpsel 1984] Himpsel FJ, Morar JF, Mcfeely FR, Pollak RA, Hollinger G: Core-level shifts and oxidation-states of Ta and W – electron-spectroscopy for chemical-analysis applied to surfaces. *Physical Review B* 1984, 30:7236–7241.

[Hindersin 2013] Hindersin S, Leupold M, Kerner M, Hanelt D: Irradiance optimization of outdoor microalgal cultures using solar tracked photobioreactors. *Bioprocess and Biosystems Engineering* 2013, 36:345–355.

[Hindersin 2014] Hindersin S, Leupold M, Kerner M, Hanelt D: Key parameters for outdoor biomass production of *Scenedesmus obliquus* in solar tracked photobioreactors. *Journal of Applied Phycology* 2014, 26:2315–2325.

[Hippler 1995] Hippler M, Drepper F, Haehnel W: *The Oxidizing Site of Photosystem I Modulates the Electron Transfer from Plastocyanin to P700.* 1995. Kluwer Academic Publ, Dordrecht, Netherlands.

[Hippler 1998] Hippler M, Drepper F, Haehnel W, Rochaix JD: The N-terminal domain of PsaF: precise recognition site for binding and fast electron transfer from cytochrome c(6) and plastocyanin to photosystem I of *Chlamydomonas reinhardtii. Proceedings of the National Academy of Sciences of the United States of America* 1998, 95:7339–7344.

[Hiszpanski 2014] Hiszpanski AM, Loo Y-L: Directing the film structure of organic semiconductors via post-deposition processing for transistor and solar cell applications. *Energy & Environmental Science* 2014, 7:592–608.

[Hitchcock 1994] Hitchcock AP, Mancini DC, Ennis L, Tulumello D, Gill J, Stewart-Ornstein J: Gas phase core excitation database. In *J Electron Spectroscopy and Related Phenomena 67 (1994) 1–123.* Uppsala University and McMaster University; 1994.

[Hitchcock 2005] Hitchcock AP, Zhang X, Guay D: In situ STXM: Studies of wet electrochemical systems under potential control. In *The 8th International Conference on X-ray Microscopy – XRM2005.* Egret Himeji, Japan; 2005.

[Hitchcock 2014] Hitchcock AP, Toney MF: Spectromicroscopy and coherent diffraction imaging: focus on energy materials applications. *Journal of Synchrotron Radiation* 2014, 21:1019–1030.

[Hochst 1976] Hochst H, Hufner S, Goldmann A: XPS-valence bands of iron, cobalt, palladium and platinum. *Physics Letters A* 1976, 57:265–266.

[Hohler 2010] Hohler S: The environment as a life support system: the case of Biosphere 2. *History and Technology* 2010, 26:39–58.

[Hollmark 2010] Hollmark HM, Duda LC, Dahbi M, Saadoune I, Gustafsson T, Edstrom K: Resonant soft X-ray emission spectroscopy and X-ray absorption spectroscopy on the cathode material $LiNi_{0.65}Co_{0.25}Mn_{0.1}O_2$. *Journal of the Electrochemical Society* 2010, 157:A962–A966.

[Holzner 2010] Holzner C, Feser M, Vogt S, Hornberger B, Baines SB, Jacobsen C: Zernike phase contrast in scanning microscopy with X-rays. *Nature Physics* 2010, 6:883–887.

[Hong 2002] Hong IP, Lin JY, Chen JM, Chatterjee S, Liu SJ, Gou YS, Yang HD: Possible evidence for the existence of the Fehrenbacher-Rice band: O K-edge XANES study on $Pr_{1-x}Ca_xBa_2Cu_3O_7$. *Europhysics Letters* 2002, 58:126–132.

[Horne 2000] Horne CR, Bergmann U, Grush MM, Perera RCC, Ederer DL, Callcott TA, Cairns EJ, Cramer SP: Electronic structure of chemically-prepared $Li_xMn_2O_4$ determined by Mn X-ray absorption and emission spectroscopies. *Journal of Physical Chemistry B* 2000, 104:9587–9596.

[Hosemann 1952] Hosemann R, Bagchi SN: Existenzbeweis für eine eindeutige Röntgenstrukturanalyse durch Entfaltung. I. Entfaltung zentrosymmetrischer endlicher Massenverteilungen. *Acta Crystallographica* 1952, 5:749–762.

[Hosemann 1962] Hosemann R, Bagchi SN: *Direct Analysis of Diffraction by Matter*. Amsterdam: North Holland; 1962.

[Huang 2019] Huang X, Wang ZL, Knibbe R, Luo B, Ahad SA, Sun D, Wang LZ: Cyclic voltammetry in lithium-sulfur batteries-challenges and opportunities. *Energy Technology* 2019, 7.

[Huang 2021] Huang N, Deng H, Liu B, Wang D, Zhao Z: Features and futures of X-ray free-electron lasers. *Innovation (NY)* 2021, 2:100097.

[Huffman 1991] Huffman GP, Mitra S, Huggins FE, Shah N, Vaidya S, Lu FL: Quantitative-analysis of all major forms of sulfur in coal by X-ray absorption fine-structure spectroscopy. *Energy & Fuels* 1991, 5:574–581.

[Hüfner 1995] Hüfner S: *Photoelectron Spectroscopy. Principles and Applications*. Springer; 1995.

[Hüfner 2003] Hüfner S: *Photoelectron Spectroscopy – Principles and Applications | Stephan Hüfner | Springer*. Berlin, Heidelberg: Springer Verlag; 2003.

[Hulstrom 1985] Hulstrom R, Bird R, Riordan C: Spectral solar irradiance data sets for selected terrestrial conditions. *Solar Cells* 1985, 15:365–391.

[Hussain 2016] Hussain H, Torrelles X, Cabailh G, Rajput P, Lindsay R, Bikondoa O, Tillotson M, Grau-Crespo R, Zegenhagen J, Thornton G: Quantitative structure of an acetate dye molecule analogue at the TiO_2-acetic acid interface. *Journal of Physical Chemistry C* 2016, 120:7586–7590.

[Hussain 1996] Hussain Z, Huff WRA, Kellar SA, Moler EJ, Heimann PA, Mckinney W, Padmore HA, Fadley CS, Shirley DA: High resolution soft X-ray bending magnet beamline 9.3.2 with circularly polarized radiation capability at the advanced light source. *Journal of Electron Spectroscopy and Related Phenomena* 1996, 80:401–404.

[Ibach 1982] Ibach H, Mills DL: *Electron Energy Loss Spectroscopy and Surface Vibrations*. Academic Press, New York, London, Paris, San Diego, San Francisco, São Paulo, Sydney, Tokyo, Toronto 1982.

[Ihssen 2014] Ihssen J, Braun A, Faccio G, Gajda-Schrantz K, Thöny-Meyer L: Light harvesting proteins for solar fuel generation in bioengineered photoelectrochemical cells. *Current Protein and Peptide Science* 2014, 15: 374–384.

[Ilavsky 2006] Atomic scattering factors and attenuation using Cromer-Liberman code for Igor Pro ver. 4.0 [http://usaxs.xray.aps.anl.gov/staff/ilavsky/AtomicFormFactors.html].

[Ilavsky 2009] Ilavsky J, Jemian PR: Irena: Tool suite for modeling and analysis of small-angle scattering. *Journal of Applied Crystallography* 2009, 42:347–353.

[Ilavsky 2012] Ilavsky J: Nika: Software for two-dimensional data reduction. *Journal of Applied Crystallography* 2012, 45:324–328.

[Ilkiv 2013] Ilkiv B, Petrovska S, Sergiienko R, Shibata E, Nakamura T, Zaulychnyy Y, Ilkiv B, Petrovska S, Sergiienko R, Shibata E, Nakamura T, Zaulychnyy Y: X-ray spectral investigation of carbon nanocapsule and graphite nanosheet electronic structures. *International Scholarly Research Notices* 2013.

[Ingall 2011] Ingall ED, Brandes JA, Diaz JM, de Jonge MD, Paterson D, McNulty I, Elliott WC, Northrup P: Phosphorus K-edge XANES spectroscopy of mineral standards. *Journal of Synchrotron Radiation* 2011, 18:189–197.

[Ishii 2011] Ishii H, Nakanishi K, Watanabe I, Ohta T, Kojima K: Improvement of ultra soft X-ray absorption spectroscopy and photoelectron spectroscopy beamline for studies on related materials and cathodes of lithium ion batteries. *E-Journal of Surface Science and Nanotechnology* 2011, 9:416–421.

[Ivnitski 2008] Ivnitski D, Artyushkova K, Atanassov P: Surface characterization and direct electrochemistry of redox copper centers of bilirubin oxidase from fungi *Myrothecium verrucaria*. *Bioelectrochemistry* 2008, 74:101–110.

[Iwanenko 1944] Iwanenko D, Pomeranchuk I: On the maximal energy attainable in a betatron. *Physical Review* 1944, 65:343–343.

[Janssen 2010] Janssen PJ, Morin N, Mergeay M, Leroy B, Wattiez R, Vallaeys T, Waleron K, Waleron M, Wilmotte A, Quillardet P, De Marsac NT, Talla E, Zhang CC, Leys N: Genome sequence of the edible cyanobacterium *Arthrospira* sp PCC 8005. *Journal of Bacteriology* 2010, 192:2465–2466.

[Janz 1953] Janz GJ, Taniguchi H: The silver-silver halide electrodes - preparation, stability, reproducibility, and standard potentials in aqueous and non-aqueous media. *Chemical Reviews* 1953, 53:397–437.

[Jenkins 1976] Jenkins GM, Kawamura K: *Polymeric Carbons – Carbon Fibre, Glass and Char*. 1st edition. Cambridge UK: Cambridge University Press; 1976.

[Jerliu 2013] Jerliu B, Doerrer L, Hueger E, Borchardt G, Steitz R, Geckle U, Oberst V, Bruns M, Schneider O, Schmidt H: Neutron reflectometry studies on the lithiation of amorphous silicon electrodes in lithium-ion batteries. *Physical Chemistry Chemical Physics* 2013, 15:7777–7784.

[Jiang 2010] Jiang P, Chen J-L, Borondics F, Glans P-A, West MW, Chang C-L, Salmeron M, Guo J: In situ soft X-ray absorption spectroscopy investigation of electrochemical corrosion of copper in aqueous $NaHCO_3$ solution. *Electrochemistry Communications* 2010, 12:820–822.

[Jimenez-Mier 2004] Jimenez-Mier J, Ederer DL, Schuler T: Chemical effects in the manganese 3s -> 2p X-ray emission that follows resonant and nonresonant photon production of a 2p hole. *Physical Review B* 2004, 70.

[Johansson 2007] Johansson GA, Tyliszczak T, Mitchell GE, Keefe MH, Hitchcock AP: Three-dimensional chemical mapping by scanning transmission X-ray spectromicroscopy. *Journal of Synchrotron Radiation* 2007, 14:395–402.

[Jorissen 2013] Jorissen K, Rehr JJ: New developments in FEFF: FEFF9 and JFEFF. In *15th International Conference on X-ray Absorption Fine Structure. Volume 430*. Edited by Wu ZY; 2013: *Journal of Physics Conference Series*.

[Jung 2000] Jung WH, Wakai H, Nakatsugawa H, Iguchi E: Small polarons in $La_{2/3}TiO_3$-delta. *Journal of Applied Physics* 2000, 88:2560–2563.

[Junghans 2015] Junghans A, Watkins EB, Barker RD, Singh S, Waltman MJ, Smith HL, Pocivavsek L, Majewski J: Analysis of biosurfaces by neutron reflectometry: from simple to complex interfaces. 2015.

[Kak 2001] Kak AC, Slaney M: *Principles of Computerized Tomographic Imaging*. Society of Industrial and Applied Mathematics; 2001.

[Kallfass 2012] Kallfass C, Hoch C, Hilger Andre, Manke I: Short-circuit and overcharge behaviour of some lithium ion batteries. In *9th International Multi-Conference on Systems, Signals and Devices (SSD)*. pp. 1–5. Chemnitz, Germany: IEEE; 2012:1–5.

[Kamata 1973] Kamata K, Nakamura T: Graphical discrimination of localized or itinerant electrons in perovskite compounds ABO_3. *Journal of the Physical Society of Japan* 1973, 35:1558–1558.

[Kamata 1974] Kamata K, Nakamura T, Sata T: On the state of d-electrons in perovskite-type compounds ABO_3. *Bulletin of Tokyo Institute of Technology* 1974, 120:73.

[Kaminski 2004] Kaminski P: Stanford linear accelerator, shown in an aerial digital orthoimage. The two roads seen near the accelerator are California Interstate 280 (to the East) and Sand Hill Road (along the Northwest). In *Photo* (File:Stanford-Linear-Accelerator-Usgs-Ortho-Kaminski-5900.Jpg ed., vol. (5,900 × 1,480 pixels, file size: 1.88 MB, MIME type: image/jpeg). https://en.wikipedia.org/wiki/SLAC_National_Accelerator_Laboratory: United States Geological Survey; 2004).

[Kanamura 1992] Kanamura K, Tamura H, Takehara Z: XPS analysis of a lithium surface immersed in propylene carbonate solution containing various salts. *Journal of Electroanalytical Chemistry* 1992, 333:127–142.

[Kang 2012] Kang HY, Wang HP: Cu@C dispersed TiO_2 for dye-sensitized solar cell photoanodes. *Applied Energy* 2012, 100:144–147.

[Kardjilov 2011] Kardjilov N, Manke I, Hilger A, Strobl M, Banhart J: Neutron imaging in materials science. *Materials Today* 2011, 14:248–256.

[Karle 1980] Karle J: Some developments in anomalous dispersion for the structural investigation of macromolecular systems in biology. *International Journal of Quantum Chemistry* 1980, 18:357–367.

[Karmonik 1995] Karmonik C, Hempelmann R, Matzke T, Springer T: Proton diffusion in strontium cerate ceramics studied by quasi-elastic neutron-scattering and impedance spectroscopy. *Zeitschrift Fur Naturforschung Section A: a Journal of Physical Sciences* 1995, 50:539–548.

[Katsapov 2010] Katsapov GY, Braun A: Deuterium tracer experiments prove the thiophenic hydrogen involvement during the initial step of thiophene hydrodesulfurization. *Catalysis Letters* 2010, 138:224–230.

[Kay 2006] Kay A, Cesar I, Graetzel M: New benchmark for water photooxidation by nanostructured alpha-Fe_2O_3 films. *Journal of the American Chemical Society* 2006, 128:15714–15721.

[Kennedy 1976] Kennedy JH, Frese KW: Photooxidation of water at barium-titanate electrodes. *Journal of the Electrochemical Society* 1976, 123:1683–1686.

[Kennedy 1977] Kennedy JH, Frese KW: Photooxidation of water at alpha-Fe_2O_3 electrodes. *Journal of the Electrochemical Society* 1977, 124:C130–C130.

[Kennedy 1978a] Kennedy JH, Frese KW: Photo-oxidation of water at alpha-Fe_2O_3 electrodes. *Journal of the Electrochemical Society* 1978, 125:709–714.

[Kennedy 1978b] Kennedy JH, Frese KW: Flatband potentials and donor densities of polycrystalline alpha-Fe_2O_3 determined from Mott-Schottky plots. *Journal of the Electrochemical Society* 1978, 125:723–726.

[Khyzhun 2000] Khyzhun OY: XPS, XES and XAS studies of the electronic structure of tungsten oxides. *Journal of Alloys and Compounds* 2000, 305:1–6.

[Kinyangi 2006] Kinyangi J, Solomon D, Liang B, Lerotic M, Wirick S, Lehmann J: Nanoscale biogeocomplexity of the organomineral assemblage in soil: application of STXM microscopy and C 1s-NEXAFS spectroscopy. *Soil Science Society of America Journal* 2006, 70:1708–1718.

[Kirz 1974] Kirz J: Phase zone plates for X-rays and extreme UV. *Journal of the Optical Society of America* 1974, 64:301–309.

[Kirz 1994a] Kirz J, Ade H, Anderson E, Buckley C, Chapman H, Howells M, Jacobsen C, Ko CH, Lindaas S, Sayre D, Williams S, Wirick S, Zhang X: New results in soft-X-ray microscopy. *Nuclear Instruments & Methods in Physics Research. Section B, Beam Interactions With Materials and Atoms* 1994, 87:92–97.

[Kirz 1994b] Kirz J, Jacobsen C, Lindaas S, Williams S, Zhang XD, Anderson E, Howells M: Soft X-ray Microscopy at the National Synchrotron Light-Source. In *Synchrotron Radiation in the Biosciences.* Edited by Chance B, Deisenhofer J, Ebashi S, Goodhead DT, Helliwell JR, Huxley HE, Iizuka T, Kirz J, Mitsui T, Rubenstein E, et al. New York: Oxford University Press; 1994: 563–571.

[Klahr 2012] Klahr B, Gimenez S, Fabregat-Santiago F, Hamann T, Bisquert J: Water oxidation at hematite photoelectrodes: the role of surface states. *Journal of the American Chemical Society* 2012, 134:4294–4302.

[Kline 2006] Kline SR: Reduction and analysis of SANS and USANS data using IGOR Pro. *Journal of Applied Crystallography* 2006, 39:895–900.

[Koch 2012] Koch N: Electronic structure of interfaces with conjugated organic materials. *Physica Status Solidi (RRL)—Rapid Research Letters* 2012, 6:277–293.

[Koehler 1957] Koehler WC, Wollan EO: Neutron-diffraction study of the magnetic properties of perovskite-like compounds $LABO_3$. *Journal of Physics and Chemistry of Solids* 1957, 2:100–106.

[Kojima 1999] Kojima I, Li B: Structural characterization of thin films by X-ray reflectivity. *The Rigaku Journal* 1999, 16:11.

[Koningsberger 1988] Koningsberger DC, Prins R: *X-ray Absorption: Principles, Applications, Techniques of EXAFS, SEXAFS and XANES*. 1st edition: Wiley-Interscience; 1988.

[Kotschau 1998] Kotschau IM, Dahn JR: In situ X-ray study of $LiMnO_2$. *Journal of the Electrochemical Society* 1998, 145:2672–2677.

[Kötz 2002] Kötz R: Doppelschichtkondensatoren – Technik, Kosten, Perspektiven (PDF Download Available). In *Siebentes Kasseler Symposium Energie – Systemtechnik, Erneuerbare Energien und Rationelle Energieverwendung, Energiespeicher und Energietransport*, vol. Tagungsband 2002. Kassel, Germany; 2002.

[Kowalczy 1972] Kowalczy S, Pollak R, Shirley DA, Ley L: High-resolution xps spectra of ir, PT and AU valence bands. *Physics Letters A* 1972, A 41:455–456.

[Kramer 2004] Kramer D, Mcevoy AJ, Schneider I, Kuhn H, Wokaun A, Scherer GG: Test and measurement methods for fuel cell technology. *CHIMIA* 2004, 58:851–856.

[Krämer 2020] Krämer L: Planning for Climate and the Environment: the EU Green Deal. *Journal for European Environmental & Planning Law* 2020, 17:267–306.

[Krisch 1997] Krisch MH, Sette F, Masciovecchio C, Verbeni R: Momentum transfer dependence of inelastic X-ray scattering from the Li K edge. *Physical Review Letters* 1997, 78:2843–2846.

[Kristiansen 2013] Kristiansen PT, Rocha TCR, Knop-Gericke A, Guo JH, Duda LC: Reaction cell for in situ soft X-ray absorption spectroscopy and resonant inelastic X-ray scattering measurements of heterogeneous catalysis up to 1 atm and 250 degrees C. *Review of Scientific Instruments* 2013, 84:12.

[Kuiper 1998] Kuiper P, Guo JH, Sathe C, Duda LC, Nordgren J, Pothuizen JJM, De Groot FMF, Sawatzky GA: Resonant X-ray raman spectra of Cu dd excitations in $Sr_2CUO_2Cl_2$. *Physical Review Letters* 1998, 80:5204–5207.

[Kunz 2005] Kunz M, Macdowell AA, Caldwell WA, Cambie D, Celestre RS, Domning EE, Duarte RM, Gleason AE, Glossinger JM, Kelez N, Plate DW, Yu T, Zaug JM, Padmore HA, Jeanloz R, Alivisatos AP, Clark SM: A beamline for high-pressure studies at the advanced light source with a superconducting bending magnet as the source. *Journal of Synchrotron Radiation* 2005, 12:650–658.

[Kupper 2015] Kupper C, Wussler S, Buqa H, Bessler W: Combined modelling and experimental characterization of titanate-based lithium-ion batteries. In *Kraftwerk Batterie 2015*. Aachen: EnergieAgentur.NRW; 2015.

[Kurmaev 1986] Kurmaev EZ, Shamin SN, Kolobova KM, Shulepov SV: X-ray-emission spectra of carbon materials. *Carbon* 1986, 24:249–253.

[Kurmaev 1999] Kurmaev EZ, Korotin MA, Galakhov VR, Finkelstein LD, Zabolotzky EI, Efremova NN, Lobachevskaya NI, Stadler S, Ederer DL, Callcott TA, Zhou L, Moewes A, Bartkowski S, Neumann M, Matsuno J, Mizokawa T, Fujimori A, Mitchell J: X-ray emission and photoelectron spectra of $Pr_{0.5}Sr_{0.5}MnO_3$. *Physical Review B* 1999, 59:12799–12806.

[Laing 1981] An introduction to the scope, potential and applications of X-ray analysis [http://www.mx.iucr.org/iucr-top/comm/cteach/pamphlets/2/2.html].

[Lalic 2011] Lalic MV, Popovic ZS, Vukajlovic FR: Ab initio study of electronic, magnetic and optical properties of $CuWO_4$ tungstate. *Computational Materials Science* 2011, 50:1179–1186.

[Lamanna 2015] Lamanna J, Hussey DS, Jacobson DL, Baöltic E: Simultaneous Neutron and X-Ray Tomography for Advanced Battery Research. vol. Abstract MA2015–02, 91: Electrochemical Society (ECS); 2015.

[Lau 1989] Lau WM: Use of surface charging in X-ray photoelectron spectroscopic studies of ultrathin dielectric films on semiconductors. *Applied Physics Letters* 1989, 54:338–340.

[Laurs 1987] Laurs H, Heiland G: Electrical and optical-properties of phthalocyanine films. *Thin Solid Films* 1987, 149:129–142.

[Lauw 2010] Lauw Y, Rodopoulos T, Gross M, Nelson A, Gardner R, Horne MD: Electrochemical cell for neutron reflectometry studies of the structure of ionic liquids at electrified interface. *Review of Scientific Instruments* 2010, 81.

[Lavine 1968] Lavine LS, Lustrin I, Shamos MH: The influence of direct electric current in vivo. *Calcified Tissue Research* 1968:Suppl:9-Suppl:9.

[Lavine 1969a] Lavine LS, Lustrin I, Shamos MH: Experimental model for studying effect of electric current on bone in vivo. *Nature* 1969, 224:1112–1113.

[Lavine 1969b] Lavine LS, Shamos MH, Lustrin I, Moss M: Influence of direct electric current on bone formation. *Journal of the American Medical Association* 1969, 208:1482.

[Lavine 1971] Lavine LS, Lustrin I, Shamos MH, Moss ML: Influence of electric current on bone regeneration in-vivo. *Acta Orthopaedica Scandinavica* 1971, 42:305–314.

[Lavine 1972] Lavine LS, Shamos MH, Liboff AR, Lustrin I, Rinaldi RA: Electric en-hancement of bone healing. *Science* 1972, 175:1118–1121.

[Lawrence 2012] Lawrence JR, Dynes JJ, Korber DR, Swerhone GDW, Leppardd GG, Hitchcock AP: Monitoring the fate of copper nanoparticles in river biofilms using scanning transmission X-ray microscopy (STXM). *Chemical Geology* 2012, 329:18–25.

[Leach 1988] Leach S: Chapter 6. In *The Free Electron Laser, Offspring of Synchrotron Radiation*. 1988: 89–152.

[Learmonth 2007] Learmonth T, Mcguinness C, Glans PA, Downes JE, Schmitt T, Duda LC, Guo JH, Chou FC, Smith KE: Observation of multiple Zhang-Rice excitations in a correlated solid: resonant inelastic X-ray scattering study of Li_2CuO_2. *EPL* 2007, 79.

[Lee 2012] Lee KE: *Investigation of the Dye-Anatase Interface in Dye Sensitized Solar Cells*. McGill University, Department of Mining and Materials Engineering; 2012.

[Lee 2008] Lee SJ, Lim N-Y, Kim S, Park G-G, Kim C-S: X-ray imaging of water distribution in a polymer electrolyte fuel cell. *Journal of Power Sources* 2008, 185:867–870.

[Lee 2014] Lee V, Berejnov V, West M, Kundu S, Susac D, Stumper J, Atanasoski RT, Debe M, Hitchcock AP: Scanning transmission X-ray microscopy of nano structured thin film catalysts for proton-exchange-membrane fuel cells. *Journal of Power Sources* 2014, 263:163–174.

[Lee 2007] Lee Y-J, Prange A, Lichtenberg H, Rohde M, Dashti M, Wiegel J: In situ analysis of sulfur species in sulfur globules produced from thiosulfate by thermoanaerobacter sulfurigignens and thermoanaerobacterium thermosulfurigenes. *Journal of Bacteriology* 2007, 189:7525–7529.

[Lehmann 2001] Lehmann EH, Vontobel P, Wiezel L: Properties of the radiography facility NEUTRA at SINQ and its potential for use as European reference facility. *Nondestructive Testing and Evaluation* 2001, 16:12.

[Lehmann 2005] Lehmann J, Liang BQ, Solomon D, Lerotic M, Luizao F, Kinyangi J, Schafer T, Wirick S, Jacobsen C: Near-edge X-ray absorption fine structure (NEXAFS) spectroscopy for mapping nano-scale distribution of organic carbon forms in soil: application to black carbon particles. *Global Biogeochemical Cycles* 2005, 19.

[Lehto 2006] Lehto KA, Lehto HJ, Kanervo EA: Suitability of different photosynthetic organisms for an extraterrestrial biological life support system. *Research in Microbiology* 2006, 157:69–76.

[Leung 2010] Leung BO, Brash JL, Hitchcock AP: Characterization of biomaterials by soft X-ray spectromicroscopy. *Materials* 2010, 3:3911–3938.

[Leung 2012] Leung BO, Hitchcock AP, Cornelius RM, Brash JL, Scholl A, Doran A: Using X-PEEM to study biomaterials: protein and peptide adsorption to a polystyrene-poly(methyl methacrylate)-b-polyacrylic acid blend. *Journal of Electron Spectroscopy and Related Phenomena* 2012, 185:406–416.

[Leupold 2013a] Leupold M, Hindersin S, Gust G, Kerner M, Hanelt D: Influence of mixing and shear stress on *Chlorella vulgaris*, *Scenedesmus obliquus*, and *Chlamydomonas reinhardtii*. *Journal of Applied Phycology* 2013, 25:485–495.

[Leupold 2013b] Leupold M, Hindersin S, Kerner M, Hanelt D: The effect of discontinuous airlift mixing in outdoor flat panel photobioreactors on growth of *Scenedesmus obliquus*. *Bioprocess and Biosystems Engineering* 2013, 36:1653–1663.

[Lewens 2007] Lewens T: Launch the Icebergs! Perutz and the Secret of Life by Georgina Ferry. *London Review of Books* 2007, 29:2.

[Li 1991] Li J, Downer NW, Smith GH: Evaluation of surface-bound membranes with electrochemical impedance spectroscopy. Contract N00014-88-C-0193 edition. pp. 41. Worcester, MA 01460: TSI Mason Research Institute Biochemistry Department; 1991:41.

[Li 1994] Li JG, Downer NW, Smith HG: Evaluation of surface-bound membranes with electrochemical impedance spectroscopy. *Biomembrane Electrochemistry* 1994, 235:491–510.

[Li 2010] Li T, Zhong G, Fu R, Yang Y: Synthesis and characterization of Nafion/cross-linked PVP semi-interpenetrating polymer network membrane for direct methanol fuel cell. *Journal of Membrane Science* 2010, 354:189–197.

[Li 2012] Li W, Herve M, Edouard P: Use of different rapid mixing devices for controlling the properties of magnetite nanoparticles produced by precipitation. *Journal of Crystal Growth* 2012, 342:21–27.

[Li 2014] Li Y, El Gabaly F, Ferguson TR, Smith RB, Bartelt NC, Sugar JD, Fenton KR, Cogswell DA, Kilcoyne ALD, Tyliszczak T, Bazant MZ, Chueh WC: Current-induced transition from particle-by-particle to concurrent intercalation in phase-separating battery electrodes. *Nature Materials* 2014, 13:1149–1156.

[Liao 2018] Liao J, Ye Z: Nontrivial effects of "trivial" parameters on the performance of lithium–sulfur batteries. *Batteries* 2018, 4.

[Liang 2006] Liang B, Lehmann J, Solomon D, Kinyangi J, Grossman J, O'neill B, Skjemstad JO, Thies J, Luizao FJ, Petersen J, Neves EG: Black carbon increases cation exchange capacity in soils. *Soil Science Society of America Journal* 2006, 70:1719–1730.

[Liang 2008] Liang B, Lehmann J, Solomon D, Sohi S, Thies JE, Skjemstad JO, Luizao FJ, Engelhard MH, Neves EG, Wirick S: Stability of biomass-derived black carbon in soils. *Geochimica Et Cosmochimica Acta* 2008, 72:6069–6078.

[Liesegang 1906] Liesegang RE: An apparent chemical remote action. *Annalen der Physik* 1906, 19:395–406.

[Liesegang 1914] Liesegang RE: Silver chromate rings spirals. I. The necessity of H-ions for the banding of silver chromats. *Zeitschrift Fur Physikalische Chemie–Stochiometrie Und Verwandtschaftslehre* 1914, 88:1–12.

[Liesegang 1915] Liesegang RE: Rhythmic crystallisation. *Naturwissenschaften* 1915, 3:500–502.

[Liesegang 1939] Liesegang RE: Spiral formation in precipitates of jellies. *Kolloid-Zeitschrift* 1939, 87:57–58.

[Lifshitz 1961] Lifshitz IM, Slyozov VV: The kinetics of precipitation from supersaturated solid solutions. *Journal of Physics and Chemistry of Solids* 1961, 19:35–50.

[Liu 2009] Liu S, Takahama S, Russell LM, Gilardoni S, Baumgardner D: Oxygenated organic functional groups and their sources in single and submicron organic particles in MILAGRO 2006 campaign. *Atmospheric Chemistry and Physics* 2009, 9:6849–6863.

[Liu 2012] Liu X, Liu J, Qiao R, Yu Y, Li H, Suo L, Hu YS, Chuang YD, Shu G, Chou F, Weng TC, Nordlund D, Sokaras D, Wang YJ, Lin H, Barbiellini B, Bansil A, Song X, Liu Z, Yan S, Liu G, Qiao S, Richardson TJ, Prendergast D, Hussain Z, de Groot FM, Yang W: Phase transformation and lithiation effect on electronic structure of Li(x)FePO4: an in-depth study by soft X-ray and simulations. *Journal of the American Chemical Society* 2012, 134:13708–13715.

[Liu 2013] Liu X, Wang D, Liu G, Srinivasan V, Liu Z, Hussain Z, Yang W: Distinct charge dynamics in battery electrodes revealed by in situ and operando soft X-ray spectroscopy. *Nature Communications* 2013, 4.

[Liu 2015] Liu YS, Glans PA, Chuang CH, Kapilashrami M, Guo JH: Perspectives of in situ/operando resonant inelastic X-ray scattering in catalytic energy materials science. *Journal of Electron Spectroscopy and Related Phenomena* 2015, 200:282–292.

[Lohmann 1965] Lohmann W: Primary quantum conversion process in photosynthesis - electron spin resonance studies. *Naturwissenschaften* 1965, 52:260–261.

[Lohsoontorn 2008] Lohsoontorn P, Brett DJL, Brandon NP: Thermodynamic predictions of the impact of fuel composition on the propensity of sulphur to interact with Ni and ceria-based anodes for solid oxide fuel cells. *Journal of Power Sources* 2008, 175:60–67.

[Love 1971] Love WE, Klock PA, Lattman EE, Padlan EA, Ward KBJ, Hendrickson WA: The structures of lamprey and bloodworm hemo globins in relation to their evolution and function. Edited by Watson JD. Cold Spring Harbor Symposia on Quantitative Biology, Vol Xxxvi. Structure and Function of Proteins at the Three-Dimensional Level, Xxv+644p. Long Island, NY: Illus Cold Spring Harbor Laboratory; 1971:349–357.

[Lu 2001] Lu L, Sahajwalla V, Kong C, Harris D: Quantitative X-ray diffraction analysis and its application to various coals. *Carbon* 2001, 39:1821–1833.

[Luckey 2008] Luckey TD: The evidence for gamma ray photosynthesis. *21st Century Science and Technology* 2008, Fall-Winter 2008:3.

[Luis Olloqui-Sariego 2012] Luis Olloqui-Sariego J, Frutos-Beltran E, Roldan E, De La Rosa MA, Jose Calvente J, Diaz-Quintana A, Andreu R: Voltammetric study of the adsorbed thermophilic plastocyanin from *Phormidium laminosum* up to 90 degrees C. *Electrochemistry Communications* 2012, 19:105–107.

[Luis Olloqui-Sariego 2014] Luis Olloqui-Sariego J, Moreno-Beltran B, Diaz-Quintana A, De La Rosa MA, Jose Calvente J, Andreu R: Temperature-driven changeover in the electron-transfer mechanism of a thermophilic plastocyanin. *Journal of Physical Chemistry Letters* 2014, 5:910–914.

[Lye 1980] Lye RC, Phillips JC, Kaplan D, Doniach S, Hodgson KO: White lines in l-edge X-ray absorption-spectra and their implications for anomalous diffraction studies of biological-materials. *Proceedings of the National Academy of Sciences of the United States of America: Biological Sciences* 1980, 77:5884–5888.

[Ma 2009] Ma YC, Ge QJ, Li WZ, Xu HY: Methanol synthesis from sulfur-containing syngas over Pd/CeO_2 catalyst. *Applied Catalysis. B, Environmental* 2009, 90:99–104.

[Maaza 1995] Maaza M, Gorenstein A, Sella C, Bridou F, Pardo B, Corno J, Roger G, Bohnke O, Julien C: Roughness effect on the lithium diffusivity in WO_3 thin films. In *Solid State Ionics Iv. Volume 369*. Edited by Nazri GA, Tarascon JM, Schreiber M; 1995: 125–130: *Materials Research Society Symposium Proceedings*.

[Maeda 1978] Maeda K, Ihara H, Chujo R, Ando I, Inoue Y, Toda M: X-ray photoelectron-spectra of retinals. *Japanese Journal of Applied Physics* 1978, 17:298–300.

[Mai 2009] Mai A, Iwanschitz B, Weissen U, Denzler R, Haberstock D, Nerlich V, Sfeir J, Schuler A: Status of Hexis SOFC stack development and the Galileo 1000 N Micro-CHP system. In *Solid Oxide Fuel Cells 11*. Edited by Singhal SC, Yokokawa H; ECS Transactions, vol. 25. 2009: 149–158.

[Mai 2011] Mai A, Iwanschitz B, Weissen U, Denzler R, Haberstock D, Nerlich V, Schuler A: Status of Hexis' SOFC Stack Development and the Galileo 1000 N Micro-CHP System. In *Solid Oxide Fuel Cells 12 (Sofc Xii)*. ECS Transactions, vol. 35. 2011: 87–95.

[Mai 2013] Mai A, Iwanschitz B, Schuler JA, Denzler R, Nerlich V, Schuler A: Hexis' SOFC System Galileo 1000 N - lab and field test experiences. In *Solid Oxide Fuel Cells 13*. Edited by Kawada T, Singhal SC; ECS Transactions, vol. 57. 2013: 73–80.

[Maiste 1995] Maiste A, Kikas A, Ruus R, Saar A, Ausmees A, Elango M: 3d-resonant photo- and Auger emission of Ce in CeO_2. *Journal of Electron Spectroscopy and Related Phenomena* 1995, 76:583–587.

[Manceau 2012] Manceau A, Marcus MA, Grangeon S: Determination of Mn valence states in mixed-valent manganates by XANES spectroscopy. *The American Mineralogist* 2012, 97:816–827.

[Manfredi 2014] Manfredi N, Bianchi A, Causin V, Ruffo R, Simonutti R, Abbotto A: Electrolytes for quasi solid-state dye-sensitized solar cells based on block copolymers. *Journal of Polymer Science. Part A, Polymer Chemistry* 2014, 52:719–727.

[Manke 2007] Manke I, Banhart J, Haibel A, Rack A, Zabler S, Kardjilov N, Hilger A, Melzer A, Riesemeier H: In situ investigation of the discharge of alkaline $Zn-MnO_2$ batteries with synchrotron X-ray and neutron tomographies. *Applied Physics Letters* 2007, 90.

[Manke 2011] Manke I, Markoetter H, Toetzke C, Kardjilov N, Grothausmann R, Dawson M, Hartnig C, Haas S, Thomas D, Hoell A, Genzel C, Banhart J: Investigation of energy-relevant materials with synchrotron X-rays and neutrons. *Advanced Engineering Materials* 2011, 13:712–729.

[Maria 2004] Maria SF, Russell LM, Gilles MK, Myneni SCB: Organic aerosol growth mechanisms and their climate-forcing implications. *Science* 2004, 306:1921–1924.

[Marpet 1996] Marpet MI: Igor Pro 3.0 graph drawing package. *SciTech Journal* 1996, 6:25–29.

[Marsen 2007] Marsen B, Cole B, Miller EL: Influence of sputter oxygen partial pressure on photoelectrochemical performance of tungsten oxide films. *Solar Energy Materials and Solar Cells* 2007, 91:1954–1958.

[Martin 1991]   Martin JM, Mansot JL: EXELFS analysis of amorphous and crystalline silicon carbide. *Journal of Microscopy* 1991, 162:171–178.

[Martin 2010]   Martin JJ, Wu J, Yuan XZ, Wang H: *Species, Temperature, and Current Distribution Mapping in Polymer Electrolyte Membrane Fuel Cells*. New York, Dordrecht, Heidelberg, London: Springer; 2010.

[Materlik 1984]   Materlik G, Zegenhagen J: X-ray standing wave analysis with synchrotron radiation applied for surface and bulk systems. *Physics Letters A* 1984, 104:47–50.

[Materlik 1987]   Materlik G, Schmah M, Zegenhagen J, Uelhoff W: Structure determination of adsorbates on single-crystal electrodes with X-ray standing waves. *Berichte Der Bunsen-Gesellschaft-Physical Chemistry Chemical Physics* 1987, 91:292–296.

[Matolin 2009]   Matolin V, Matolinova I, Sedlacek L, Prince KC, Skala T: A resonant photoemission applied to cerium oxide based nanocrystals. *Nanotechnology* 2009, 20.

[Matsumoto 1996]   Matsumoto M, Soda K, Ichikawa K, Taguchi Y, Jouda K, Kageyama M, Tanaka S, Sata N, Tezuka Y, Shin S, Kimura S, Aita O: Photoemission study of CeO_2 and $SrCeO_3$. *Journal of Electron Spectroscopy and Related Phenomena* 1996, 78:179–182.

[Matzke 1996]   Matzke T, Stimming U, Karmonik C, Soetratmo M, Hempelmann R, Guthoff F: Quasielastic thermal neutron scattering experiment on the proton conductor $SrCe_{0.95}Yb_{0.05}H_{0.02}O_{2.985}$. *Solid State Ionics* 1996, 86–8:621–628.

[Maurellis 2000]   Maurellis AN, Cravens TE, Gladstone GR, Waite JH, Acton LW: Jovian X-ray emission from solar X-ray scattering. *Geophysical Research Letters* 2000, 27:1339–1342.

[Mauritz 2004]   Mauritz KA, Moore RB: State of understanding of nafion. *Chemical Reviews* 2004, 104:4535–4585.

[Mayer 2005]   Mayer T, Lebedev M, Hunger R, Jaegermann W: Elementary processes at semiconductor/electrolyte interfaces: perspectives and limits of electron spectroscopy. *Applied Surface Science* 2005, 252:31–42.

[Mcguinness 2007]   Robert D. Cowan's Atomic Structure Code [https://www.tcd.ie/Physics/people/Cormac. McGuinness/Cowan/].

[Mcintosh 2004]   Mcintosh S, Gorte RJ: Direct hydrocarbon solid oxide fuel cells. *Chemical Reviews* 2004, 104:4845–4865.

[Mclachlan 1989]   Mclachlan DS: Measurement and analysis of a model dual conductivity medium using a generalized effective medium theory. *Physica A* 1989, 157:188–191.

[Mcneill 2013]   Mcneill CR, Ade H: Soft X-ray characterisation of organic semiconductor films. *Journal of Materials Chemistry C* 2013, 1:187–201.

[Meirer 2015]   Meirer F, Morris DT, Kalirai S, Liu Y, Andrews JC, Weckhuysen BM: Mapping metals incorporation of a whole single catalyst particle using element specific X-ray nanotomography. *Journal of the American Chemical Society* 2015, 137:102–105.

[Memioglu 2014]   Memioglu F, Bayrakceken A, Oznuluer T, Ak M: Conducting car-bon/polymer composites as a catalyst support for proton exchange membrane fuel cells. *International Journal of Energy Research* 2014, 38:1278–1287.

[Mentovich 2012]   Mentovich E, Belgorodsky B, Gozin M, Richter S, Cohen H: Doped biomolecules in miniaturized electric junctions. *Journal of the American Chemical Society* 2012, 134:8468–8473.

[Messinger 2001]   Messinger J, Robblee JH, Bergmann U, Fernandez C, Glatzel P, Visser H, Cinco RM, Mcfarlane KL, Bellacchio E, Pizarro SA, Cramer SP, Sauer K, Klein MP, Yachandra VK: Absence of Mn-centered oxidation in the S-2 -> S-3 transition: Implications for the mechanism of photosynthetic water oxidation. *Journal of the American Chemical Society* 2001, 123:7804–7820.

[Metin 2010]   Metin H, Ari M, Erat S, Durmus S, Bozoklu M, Braun A: The effect of annealing temperature on the structural, optical, and electrical properties of CdS films. *Journal of Materials Research* 2010, 25:189–196.

[Metrology 2007]   Metrology IOOL: *International Vocabulary of Metrology – Basic and General Concepts and Associated Terms (VIM)*. Paris: Bureau International de Métrologie Légale, 11, rue Turgot - 75009 Paris - France 2007.

[Miklos 1980]   Miklos J, Mund K, Naschwitz W: Doppelschichtkondensator. Office EP ed., vol. DE 30 11 701. Germany: Siemens aktiengesellschaft; 1980.

[Milewska 2014]   Milewska A, Swierczek K, Tobola J, Boudoire F, Hu Y, Bora DK, Mun BS, Braun A, Molenda J: The nature of the nonmetal-metal transition in Li_xCoO_2 oxide. *Solid State Ionics* 2014, 263:110–118.

[Miller 1959]   Miller SL, Urey HC: Organic compound synthesis on the primitive earth. *Science* 1959, 130:245–251.

[Miot 2014]   Miot J, Benzerara K, Kappler A: Investigating microbe-mineral interactions: recent advances in X-ray and electron microscopy and redox-sensitive methods. *Annual Review of Earth and Planetary Sciences* 2014, 42:271–289.

[Mishra 1999]   Mishra SK, Ceder G: Structural stability of lithium manganese oxides. *Physical Review B* 1999, 59:6120–6130.

[Mizokawa 2013]   Mizokawa T, Wakisaka Y, Sudayama T, Iwai C, Miyoshi K, Takeuchi J, Wadati H, Hawthorn DG, Regier TZ, Sawatzky GA: Role of oxygen holes in Li_xCoO_2 revealed by soft X-ray spectroscopy. *Physical Review Letters* 2013, 111.

[Molenda 2015]   Molenda J, Baster D, Milewska A, Swierczek K, Bora DK, Braun A, Tobola J: Electronic origin of difference in discharge curve between Li_xCoO_2 and Na_xCoO_2 cathodes. *Solid State Ionics* 2015, 271:15–27.

[Molina-Heredia 2001]   Molina-Heredia FP, Hervas M, Navarro JA, De La Rosa MA: A single arginyl residue in plastocyanin and in cytochrome c(6) from the cyanobacterium *Anabaena* sp PCC 7119 is required for efficient reduction of photosystem I. *Journal of Biological Chemistry* 2001, 276:601–605.

[Molodtsov 2009]   Molodtsov SL, Fedoseenko SI, Vyalikh DV, Iossifov IE, Follath R, Gorovikov SA, Brzhezinskaya MM, Dedkov YS, Puttner R, Schmidt JS, Adamchuk VK, Gudat W, Kaindl G: High-resolution Russian-German beamline at BESSY. *Applied Physics. A, Materials Science & Processing* 2009, 94:501–505.

[Morcrette 2002]   Morcrette M, Chabre Y, Vaughan G, Amatucci G, Leriche JB, Patoux S, Masquelier C, Tarascon JM: In situ X-ray diffraction techniques as a powerful tool to study battery electrode materials. *Electrochimica Acta* 2002, 47:3137–3149.

[Morin 2009]   Morin N: Studies on the response to spaceflight related conditions in the cyanobacterium Arthrospira sp. PCC8005 using a genomic approach. Université de Liège, 2009.

[Morita 1987]   Morita S, Otsuka I, Okada T, Yokoyama H, Iwasaki T, Mikoshiba N: Construction of a scanning tunneling microscope for electrochemical studies. *Japanese Journal of Applied Physics, Part 1 (Regular Papers and Short Notes)* 1987, 26:L1853–1855.

[Moseley 1913a]   Moseley H, Darwin CG: The reflection of the X rays. *Nature* 1913, 90:594–594.

[Moseley 1913b]   Moseley HGJ: The high-frequency spectra of the elements. *Philosophical Magazine* 1913, 26:1024–1034.

[Moseley 1913c]   Moseley HGJ, Darwin CG: The reflexion of the X-rays. *Philosophical Magazine* 1913, 26:210–232.

[Muhlbauer 2009]   Muhlbauer S, Binz B, Jonietz F, Pfleiderer C, Rosch A, Neubauer A, Georgii R, Boni P: Skyrmion lattice in a chiral magnet. *Science* 2009, 323:915–919.

[Müller 2014]   Müller M, Choudhury S, Gruber K, Cruz VB, Fuchsbichler B, Jacob T, Koller S, Stamm M, Ionov L, Beckhoff B: Sulfur X-ray absorption fine structure in porous Li–S cathode films measured under argon atmospheric conditions. *Spectrochimica Acta, Part B: Atomic Spectroscopy* 2014, 94-95:22–26.

[Musk 2014]   All our patent are belong to you [Web Page]; 2014. 12 June 2014. Available from: https://www.teslamotors.com/de_CH/blog/all-our-patent-are-belong-you.

[Myneni 2002]   Myneni S, Luo Y, Naslund LA, Cavalleri M, Ojamae L, Ogasawara H, Pelmenschikov A, Wernet P, Vaterlein P, Heske C, Hussain Z, Pettersson LGM, Nilsson A: Spectroscopic probing of local hydrogen-bonding structures in liquid water. *Journal of Physics. Condensed Matter* 2002, 14:L213–L219.

[Nakamura 2006]   Nakamura T: Potential map of ABO3. Richter J ed., [Richter 2006] Personal fax letter communication between Jörg Richter and Tetsuro Nakamura in 30 June 2006. pp. 16; 2006:16.

[Nakata 2015] Nakata A, Fukuda K, Murayama H, Tanida H, Yamane T, Arai H, Uchimoto Y, Sakurai K, Ogumi Z: Operando X-ray fluorescence imaging for zinc-based secondary batteries. *Electrochemistry* 2015, 83:849–851.

[Nakayama 1988] Nakayama N, Moritani I, Shinjo T, Fujii Y, Sasaki S: Anomalous X-ray-scattering study of composition profile in Fe/Mn superlattice films. *Journal of Physics F. Metal Physics* 1988, 18:429–442.

[Nanda 1999] Nanda J, Kuruvilla BA, Sarma DD: Photoelectron spectroscopic study of CdS nanocrystallites. *Physical Review B* 1999, 59:7473–7479.

[Nanda 2001] Nanda J, Sarma DD: Photoemission spectroscopy of size selected zinc sulfide nanocrystallites. *Journal of Applied Physics* 2001, 90:2504–2510.

[Naumenko 2008] Naumenko AP, Berezovska NI, Biliy MM, Shevchenko OV: Vibrational analysis and Raman spectra of tetragonal zirconia. *Physics and Chemistry of Solid State* 2008, 9:5.

[Nave 2016] Experimental K-alpha X ray energies [http://hyperphysics.phy-astr.gsu.edu/hbase/tables/kxray.html].

[Nazri 1985] Nazri G, Muller RH: In situ X-ray-diffraction of surface-layers on lithium in nonaqueous electrolyte. *Journal of the Electrochemical Society* 1985, 132:1385–1387.

[Neal 2004] Neal AL, Amonette JE, Peyton BM, Geesey GG: Uranium complexes formed at hematite surfaces colonized by sulfate-reducing bacteria. *Environmental Science & Technology* 2004, 38:3019–3027.

[Necas 2009] Necas D, Zajickova L, Franta D, Stahel P, Mikulik P, Meduna M, Valtr M: Optical characterization of ultra-thin iron and iron oxide films. *E-Journal of Surface Science and Nanotechnology* 2009, 7:486–490.

[Nelms 1955] Nelms AT, Oppenheim I: Data on the atomic form factor - computation and survey. *Journal of Research of the National Bureau of Standards* 1955, 55:53–62.

[Nelson 2011] Nelson GJ, Harris WM, Izzo JR, Jr., Grew KN, Chiu WKS, Chu YS, Yi J, Andrews JC, Liu Y, Pianetta P: Three-dimensional mapping of nickel oxidation states using full field X-ray absorption near edge structure nanotomography. *Applied Physics Letters* 2011, 98.

[Nelson Weker 2015] Nelson Weker J, Toney MF: Emerging in situ and operando nanoscale X-ray imaging techniques for energy storage materials. *Advanced Functional Materials* 2015, 25:1622–1637.

[Nemsak 2014] Nemsak S, Shavorskiy A, Karslioglu O, Zegkinoglou I, Rattanachata A, Conlon CS, Keqi A, Greene PK, Burks EC, Salmassi F, Gullikson EM, Yang S-H, Liu K, Bluhm H, Fadley CS: Concentration and chemical-state profiles at heterogeneous interfaces with sub-nm accuracy from standing-wave ambient-pressure photoemission. *Nature Communications* 2014, 5:7.

[Neppl 2014] Neppl S, Shavorskiy A, Zegkinoglou I, Fraund M, Slaughter DS, Troy T, Ziemkiewicz MP, Ahmed M, Gul S, Rude B, Zhang JZ, Tremsin AS, Glans P-A, Liu Y-S, Wu CH, Guo J, Salmeron M, Bluhm H, Gessner O: Capturing interfacial photoelectrochemical dynamics with picosecond time-resolved X-ray photoelectron spectroscopy. *Faraday Discussions* 2014, 171:219–241.

[Neppl 2015a] Neppl S, Gessner O: Time-resolved X-ray photoelectron spectroscopy techniques for the study of interfacial charge dynamics. *Journal of Electron Spectroscopy and Related Phenomena* 2015, 200:64–77.

[Neppl 2015b] Neppl S, Liu YS, Wu CH, Shavorskiy A, Zegkinoglou I, Troy T, Slaughter DS, Ahmed M, Tremsin AS, Guo JH, Glans PA, Salmeron M, Bluhm H, Gessner O: Toward ultrafast in situ X-ray studies of interfacial photoelectrochemistry. In *Ultrafast Phenomena Xix. Volume 162*. Edited by Yamanouchi I, Cundiff S, Devivieriedle R, Kuwatagonokami M, Dimauro L; 2015: 325–328: *Springer Proceedings in Physics*.

[Newville 2023] Diffraction Anomalous Fine Structure [Web Page]; 2023. 29 December 2023. Available from: http://cars9.uchicago.edu/dafs/diffkk/.

[Nitschke 2013] Nitschke J, Tham J, Ropulos B: ZEISS to acquire Xradia to complement its microscopy business. 13 June 2013 edition. finance.yahoo.com: @YahooFinance; 2013.

[Nordling 1991] Nordling M, Sigfridsson K, Young S, Lundberg LG, Hansson O: Flash-photolysis studies of the electron-transfer from genetically modified spinach plastocyanin to photosystem-I. *FEBS Letters* 1991, 291:327–330.

[Nurk 2011] Nurk G, Holtappels P, Figi R, Wochele J, Wellinger M, Braun A, Graule T: A versatile salt evaporation reactor system for SOFC operando studies on anode contamination and degradation with impedance spectroscopy. *Journal of Power Sources* 2011, 196:3134–3140.

[Nurk 2013] Nurk G, Huthwelker T, Braun A, Ludwig C, Lust E, Struis RPWJ: Redox dynamics of sulphur with Ni/GDC anode during SOFC operation at mid- and low-range temperatures: An operando S K-edge XANES study. *Journal of Power Sources* 2013, 240:448–457.

[Ocana 1998] Ocana M, Caballero A, Gonzalez-Elipe AR, Tartaj P, Serna CJ: Valence and localization of praseodymium in Pr-doped zircon. *Journal of Solid State Chemistry* 1998, 139:412–415.

[Ogletree 2002] Ogletree DF, Bluhm H, Lebedev G, Fadley CS, Hussain Z, Salmeron M: A differentially pumped electrostatic lens system for photoemission studies in the millibar range. *Review of Scientific Instruments* 2002, 73:3872–3877.

[Oswald 1933] Oswald W: Beiträge zur Theorie der Elektrokultur. ETH zürich, 1933.

[Ou 2015] Ou Q-D, Li C, Li Y-Q, Tang J-X: Photoemission spectroscopy study on interfacial energy level alignments in tandem organic light-emitting diodes. *Journal of Electron Spectroscopy and Related Phenomena* 2015, 204:186–195.

[Pai 2022] Pai R, Singh A, Tang MH, Kalra V: Stabilization of gamma sulfur at room temperature to enable the use of carbonate electrolyte in Li-S batteries. *Communications Chemistry* 2022, 5:17.

[Pan 1983] Pan HK, Yarusso DJ, Knapp GS, Cooper SL: EXAFS studies of nickel nafion ionomers. *Journal of Polymer Science. Part B, Polymer Physics* 1983, 21:1389–1401.

[Pandey 2005] Pandey PK, Bhave NS, Kharat RB: Spray deposition process of polycrystalline thin films of $CuWO_4$ and study on its photovoltaic electrochemical properties. *Materials Letters* 2005, 59:3149–3155.

[Paolucci 1991] Paolucci G, Santoni A, Comelli G, Prince KC, Agostino RG: M-4,5 absorption-edge of Ag, Pd, and Rh by reflection electron-energy-loss spectroscopy – role of nondipole transitions. *Physical Review B* 1991, 44:10888–10891.

[Parent 2002] Parent P, Laffon C, Mangeney C, Bournel F, Tronc M: Structure of the water ice surface studied by X-ray absorption spectroscopy at the OK-edge. *Journal of Chemical Physics* 2002, 117:10842–10851.

[Parratt 1954a] Parratt LG: Surface studies of solids by total reflection of X-rays. *Physical Review* 1954, 95:359–369.

[Parratt 1954b] Parratt LG: Solid surface studies by total reflection of X-rays. *Physical Review* 1954, 95:617–617.

[Patterson 1935] Patterson AL: A direct method for the determination of the components of interatomic distances in crystals. *Zeitschrift für Kristallographie - Crystalline Materials* 1935, 90:517–542.

[Pecher 2005] Pecher K, Baer DR, Mccready D, Engelhard M, Lopatin S, Browning N: Spectroscopic Characterization of nano-magnetite: facts and mystery about an illusive mineral phase. In *Goldschmidt Conference on Transition Metal Precipitates*; 2005.

[Peisert 2015] Peisert H, Uihlein J, Petraki F, Chasse T: Charge transfer between transition metal phthalocyanines and metal substrates: the role of the transition metal. *Journal of Electron Spectroscopy and Related Phenomena* 2015, 204:49–60.

[Pellegrin 1996] Pellegrin E, Zaanen J, Lin HJ, Meigs G, Chen CT, Ho GH, Eisaki H, Uchida S: O 1s near-edge X-ray absorption of $La_{2-x}Sr_xNiO_4$+delta: holes, polarons, and excitons. *Physical Review B* 1996, 53:10667–10679.

[Perret 1971] Perret R, Ruland W: The evaluation of multiple X-ray small-angle scattering. *Journal of Applied Crystallography* 1971, 4:444–451.

[Perutz 1962] Perutz MF, Kendrew JC: *Nobel Lecture: X-Ray Analysis of Haemoglobin*. Amsterdam: Elsevier Publishing Company; 1962.

[Perutz 1990] Perutz MF: How W. L. Bragg invented X-ray analysis. *Acta Crystallographica. Section A, Foundations of Crystallography* 1990, 46:633–643.

[Peter 1990] Peter LM: Dynamic aspects of semiconductor photoelectrochemistry. *Chemical Reviews* 1990, 90:753–769.

[Petersen 1975] Petersen H: Observation of real Li K absorption-edge. *Physical Review Letters* 1975, 35:1363–1366.

[Pfeiffer 2017] Pfeiffer F: X-ray ptychography. *Nature Photonics* 2017, 12:9–17.

[Philip Ewels 2016] Philip Ewels, Thierry Sikora, Virginie Serin, Chris P. Ewels, Lajaunie L: A complete overhaul of the electron energy-loss spectroscopy and X-ray absorption spectroscopy database: eelsdb.eu. *Microscopy and Microanalysis* 2016.

[Piccolino 1998] Piccolino M: Animal electricity and the birth of electrophysiology: the legacy of Luigi Galvani. *Brain Research Bulletin* 1998, 46:381–407.

[Piper 2011] Piper LFJ, Preston ARH, Cho SW, Demasi A, Chen B, Laverock J, Smith KE, Miara LJ, Davis JN, Basu SN, Pal U, Gopalan S, Saraf L, Kaspar T, Matsuura AY, Glans PA, Guo JH: Soft X-ray spectroscopic study of dense strontium-doped lanthanum manganite cathodes for solid oxide fuel cell applications. *Journal of the Electrochemical Society* 2011, 158:B99–B105.

[Pitkänen 2013] Pitkänen M: Quantum model for the direct currents of Becker. *Journal of Non-Locality* 2013, 2:29.

[Pletneva 2002] Pletneva EV, Crnogorac MM, Kostic NM: Mimicking biological electron transport in sol-gel glass: photoinduced electron transfer from zinc cytochrome c to plastocyanin or cytochrome c mediated by mobile inorganic complexes. *Journal of the American Chemical Society* 2002, 124:14342–14354.

[Podlesnyak 2005] Podlesnyak A, Streule S, Medarde M, Conder K, Pomjakushina E, Mesot J: Effect of oxygen nonstoichiometry on structural and magnetic properties of $PrBa-CO_2O_5$+delta. *Physica. B, Condensed Matter* 2005, 359:1348–1350.

[Podlesnyak 2006] Podlesnyak A, Conder K, Pomjakushina E, Mirmelstein A: Layered cobalt perovskites: current topics and future promises. In *Frontal Semiconductor Research. Edited by Chang OT.* Huntington, New York: Nova Science Publishers, Inc.; 2006: 171–209.

[Porod 1951] Porod G: Die rontgenkleinwinkelstreuung von dichtgepackten kolloiden systemen .1. *Kolloid-Zeitschrift and Zeitschrift Fur Polymere* 1951, 124:83–114.

[Porod 1952a] Porod G: Die rontgenkleinwinkelstreuung von dichtgepakten kolloiden systemen .2. *Kolloid-Zeitschrift and Zeitschrift Fur Polymere* 1952, 125:51–57.

[Porod 1952b] Porod G: Die rontgenkleinwinkelstreuung von dichtgepackten kolloiden systemen .2. *Kolloid-Zeitschrift and Zeitschrift Fur Polymere* 1952, 125:108–122.

[Prize 2014] Nobel Prize [https://www.nobelprize.org/nobel_prizes/facts/].

[Puxley 1994] Puxley DC, Squire GD, Bates DR: A new cell for in-situ X-ray-diffraction studies of catalysts and other materials under reactive gas atmospheres. *Journal of Applied Crystallography* 1994, 27:585–594.

[Pynn 1990] Pynn R: Neutron scattering: a primer. *Los Alamos Science* 1990, 19.

[Qian 2014] Qian F, Wang H, Ling Y, Wang G, Thelen MP, Li Y: Photoenhanced electrochemical interaction between shewanella and a hematite nanowire photoanode. *Nano Letters* 2014, 14:3688–3693.

[Qianli 2013] Qianli C, Banyte J, Xinyu Z, Embs JP, Braun A: Proton diffusivity in spark plasma sintered $BaCe_{0.8}Y_{0.2}O_3$-delta: in-situ combination of quasi-elastic neutron scattering and impedance spectroscopy. *Solid State Ionics, Diffusion and Reactions* 2013, 252:2–6.

[Qiu 2012] Qiu G, Joshi AS, Dennison CR, Knehr KW, Kumbur EC, Sun Y: 3-D pore-scale resolved model for coupled species/charge/fluid transport in a vanadium redox flow battery. *Electrochimica Acta* 2012, 64:46–64.

[Rabinowicz 1961] Rabinowicz E: Photochemical utilization of light energy. *Proceedings of the National Academy of Sciences of the United States of America* 1961, 47:1296–1303.

[Rajagopal 2010] Rajagopal S, Bekenev VL, Nataraj D, Mangalaraj D, Khyzhun OY: Electronic structure of $FeWO_4$ and $CoWO_4$ tungstates: first-principles FP-LAPW calculations and X-ray spectroscopy studies. *Journal of Alloys and Compounds* 2010, 496:61–68.

[Ralston 2000] Ralston CY, Wang HX, Ragsdale SW, Kumar M, Spangler NJ, Ludden PW, Gu W, Jones RM, Patil DS, Cramer SP: Characterization of heterogeneous nickel sites in CO dehydrogenases from *Clostridium thermoaceticum* and *Rhodospirillum rubrum* by nickel L-edge X-ray spectroscopy. *Journal of the American Chemical Society* 2000, 122:10553–10560.

[Ramesh 2002] Ramesh VM, Guergova-Kuras M, Joliot P, Webber AN: Electron transfer from plastocyanin to the photosystem I reaction center in mutants with increased potential of the primary donor in *Chlamydomonas reinhardtii*. *Biochemistry* 2002, 41:14652–14658.

[Rangan 2012] Rangan S, Coh S, Bartynski RA, Chitre KP, Galoppini E, Jaye C, Fischer D: Energy alignment, molecular packing, and electronic pathways: zinc(II) tetraphenylporphyrin derivatives adsorbed on TiO_2(110) and ZnO(11–20) surfaces. *Journal of Physical Chemistry C* 2012, 116:23921–23930.

[Rasmussen 2014] Rasmussen M, Minteer SD: Investigating the mechanism of thylakoid direct electron transfer for photocurrent generation. *Electrochimica Acta* 2014, 126:68–73.

[Ratner 1983] Ratner BD: Surface characterization of biomaterials by electron-spectroscopy for chemical-analysis. *Annals of Biomedical Engineering* 1983, 11:313–336.

[Rayleigh 1910] Rayleigh L: The incidence of light upon a transparent sphere of dimensions comparable with the wave-length. *Proceedings of the Royal Society of London. Series A, Containing Papers of a Mathematical and Physical Character* 1910, 84:25–46.

[Rea 2011] Rea G, Braun A, C. CE, Janssen P, Su B-L: *Photosynthetic Proteins for Technological Applications: Biosensors and Biochips (PHOTOTECH)*. Europe: COST. European Cooperation in Science and Technology; 2011.

[Rehr 1991] Rehr JJ, Deleon JM, Zabinsky SI, Albers RC: Theoretical X-ray absorption fine-structure standards. *Journal of the American Chemical Society* 1991, 113:5135–5140.

[Rehr 1998] Rehr JJ, Ankudinov A, Zabinsky SI: New developments in NEXAFS/EXAFS theory. *Catalysis Today* 1998, 39:263–269.

[Rehr 2001] Rehr JJ, Ankudinov AL: Progress and challenges in the theory and interpretation of X-ray spectra. *Journal of Synchrotron Radiation* 2001, 8:61–65.

[Rehr 2003] Rehr JJ, Ankudinov AL: New developments in the theory and interpretation of X-ray spectra based on fast parallel calculations. *Journal of Synchrotron Radiation* 2003, 10:43–45.

[Rensmo 1997] Rensmo H, Sodergren S, Patthey L, Westermark K, Vayssieres L, Kohle O, Bruhwiler PA, Hagfeldt A, Siegbahn H: The electronic structure of the cis-bis(4,4'-dicarboxy-2,2'-bipyridine)-bis(isothiocyanato) ruthenium(II) complex and its ligand 2,2'-bipyridyl-4,4'-dicarboxylic acid studied with electron spectroscopy. *Chemical Physics Letters* 1997, 274:51–57.

[Ressler 1998] Ressler T: WinXAS: a program for X-ray absorption spectroscopy data analysis under MS-Windows. *Journal of Synchrotron Radiation* 1998, 5:118–122.

[Richter 2008a] Richter J: Mixed Conducting Ceramics for High Temperature Electrochemical Devices. *Dissertation*. ETH Zürich, D-MAT; 2008.

[Richter 2008e] Richter J, Braun A, Harvey AS, Holtappels P, Graule T, Gauckler L: Valence changes of manganese and praseodymium in $Pr_{1-x}Sr_xMn_{1-y}In_yO_3$-delta perovskites upon cation substitution as determined with XANES and ELNES. *Physica. B, Condensed Matter* 2008, 403:87–94.

[Rietveld 1966] Rietveld HM: A method for including line profiles of neutron powder diffraction peaks in determination of crystal structures. *Acta Crystallographica* 1966, S 21:A228.

[Rietveld 1967] Rietveld HM: Line profiles of neutron powder-diffraction peaks for structure refinement. *Acta Crystallographica* 1967, 22:151–152.

[Rietveld 1969] Rietveld HM: A profile refinement method for nuclear and magnetic structures. *Journal of Applied Crystallography* 1969, 2:65–71.

[Rietveld 2010] Rietveld HM: The Rietveld method: a retrospection. *Zeitschrift für Kristallographie* 2010, 225:545–547.

[Rietveld 2014] Rietveld HM: The Rietveld method. *Physica Scripta* 2014, 89.

[Robblee 2001] Robblee JH, Cinco RM, Yachandra VK: X-ray spectroscopy-based structure of the Mn cluster and mechanism of photosynthetic oxygen evolution. *Biochimica Et Biophysica Acta. Bioenergetics* 2001, 1503:7–23.

[Roberts 2009] statistiXL: Statistical Power for Microsoft Excel [http://www.statistixl.com/features/features.aspx].

[Roberts 2016] statistiXL: Statistical Power for Microsoft Excel [http://www.statistixl.com/features/features.aspx].

[Robinson 1992a] Robinson IK, Tweet DJ: Surface X-ray-diffraction. *Reports on Progress in Physics* 1992, 55:599–651.

[Robinson 1992b] Robinson KM, Robinson IK, Ogrady WE: Electrochemically induced surface-roughness on Au(100) studied by surface X-ray-diffraction. *Electrochimica Acta* 1992, 37:2169–2172.

[Rodriguez 1998] Rodriguez JA, Chaturvedi S, Hanson JC, Albornoz A, Brito JL: Electronic properties and phase transformations in $CoMoO_4$ and $NiMoO_4$: XANES and time-resolved synchrotron XRD studies. *Journal of Physical Chemistry B* 1998, 102:1347–1355.

[Rodriguez 2000] Rodriguez MA, Ingersoll D, Daoughty DH: An electrochemical cell for in-situ X-ray characterization. *Advances in X-ray Analysis* 2000, 42.

[Rodriguez 2013] Rodriguez JA, Hanson JC, Stacchiola D, Senanayake SD: In situ/operando studies for the production of hydrogen through the water-gas shift on metal oxide catalysts. *Physical Chemistry Chemical Physics* 2013, 15:12004–12025.

[Rodriguez 2013a] Rodriguez JA, Hanson JC, Chupas PJ: Pair distribution function analysis of high-energy X-ray scattering data. In *In-situ Characterization of Heterogeneous Catalysts*. Edited by Chapman KW, Chupas PJ. John Wiley & Sons, Inc.; 2013.

[Roland 1952] Strukturformel von Nafion [Von Roland 1952 – Roland 1952, Gemeinfrei, https://commons.wikimedia.org/w/index.php?curid=8704957].

[Rompel 1998] Rompel A, Cinco RM, Latimer MJ, Mcdermott AE, Guiles RD, Quintanilha A, Krauss RM, Sauer K, Yachandra VK, Klein MP: Sulfur K-edge X-ray absorption spectroscopy: a spectroscopic tool to examine the redox state of S-containing metabolites in vivo. *Proceedings of the National Academy of Sciences of the United States of America* 1998, 95:6122–6127.

[Röntgen 1901] Wilhelm Conrad Röntgen - Nobel Lecture [https://www.nobelprize.org/nobel_prizes/physics/laureates/1901/rontgen-lecture.html].

[Rosiwal 1898] Rosiwal A: Über geometrische Gesteinsanalysen. *KK Verh Geol Reichsanst* 1898, 5 and 6.

[Roth 2015] Roth F, Knupfer M: Electronic excitation spectrum of doped organic thin films investigated using electron energy-loss spectroscopy. *Journal of Electron Spectroscopy and Related Phenomena* 2015, 204:23–28.

[Rueff 2012] GALAXIES beamline [http://www.synchrotron-so-leil.fr/images/Image/Recherche/LignesLumiere/GALAXIES/AuVB_HR(1).png].

[Ruiz-Fuertes 2008] Ruiz-Fuertes J, Errandonea D, Segura A, Manjon FJ, Zhu Z, Tu CY: Growth, characterization, and high-pressure optical studies of $CuWO_4$. *High Pressure Research* 2008, 28:565–570.

[Ruland 1971] Ruland W: Small-angle scattering of two-phase systems: determination and significance of systematic deviations from Porod's law. *Journal of Applied Crystallography* 1971, 4:70–73.

[Ruland 1974] Ruland W: The effect of finite slit heights on the determination of systematic deviations from Porod's law. *Journal of Applied Crystallography* 1974, 7:383–386.

[Rush 1988] Rush JD, Levine F, Koppenol WH: The electron-transfer site of spinach plastocyanin. *Biochemistry* 1988, 27:5876–5884.

[Russell 2002] Russell LM, Maria SF, Myneni SCB: Mapping organic coatings on atmospheric particles. *Geophysical Research Letters* 2002, 29.

[Saikubo 2005] Saikubo A, Igaki J, Kato Y, Kometani R, Kanda K, Matsui S: Soft X-ray emission and absorption spectra of DLC film formed by FIB-CVD method. *Digest of Papers Microprocesses and Nanotechnology 2005 2005 International Microprocesses and Nanotechnology Conference (IEEE Cat No 05EX1180)* 2005:98–99.

[Saitoh 1993a] Saitoh T, Bocquet AE, Mizokawa T, Namatame H, Fujimori A, Abbate M, Takeda Y, Takano M: Photoemission and X-ray-absorption study of la1-xsrxmno3. *Japanese Journal of Applied Physics Part 1-Regular Papers Short Notes and Review Papers* 1993, 32:258–260.

[Saitoh 1993b] Saitoh T, Bocquet AE, Mizokawa T, Namatame H, Fujimori A, Abbate M, Takeda Y, Takano M: Photoemission and X-ray absorption study of La 1-xSr xMnO 3. *Japanese Journal of Applied Physics, Supplement* 1993, 32:258–260.

[Saitoh 1995] Saitoh T, Bocquet AE, Mizokawa T, Namatame H, Fujimori A, Abbate M, Takeda Y, Takano M: Electronic-structure of la1-xsrxmno3 studied by photoemission and X-ray-absorption spectroscopy. *Physical Review B* 1995, 51:13942–13951.

[Saitoh 1999] Saitoh T, Bocquet AE, Mizokawa T, Namatame H, Fujimori A, Takeda Y, Takano M: Strontium-doped lanthanum manganese oxides studied by XPS. *Surface Science Spectra* 1999, 6:292–301.

[Sakdinawat 2010] Sakdinawat A, Attwood D: Nanoscale X-ray imaging. *Nature Photonics* 2010, 4:840–848.

[Samant 1988] Samant MG, Toney MF, Borges GL, Blum L, Melroy OR: Grazing-incidence X-ray-diffraction of lead monolayers at a silver (111) and gold (111) electrode-electrolyte interface. *Journal of Physical Chemistry* 1988, 92:220–225.

[Sandford 2006] Sandford SA, Aleon J, Alexander CMOD, Araki T, Bajt S, Baratta GA, Borg J, Bradley JP, Brownlee DE, Brucato JR, Burchell MJ, Busemann H, Butterworth A, Clemett SJ, Cody G, Colangeli L, Cooper G, D'hendecourt L, Djouadi Z, Dworkin JP, Ferrini G, Fleckenstein H, Flynn GJ, Franchi IA, Fries M, Gilles MK, Glavin DP, Gounelle M, Grossemy F, Jacobsen C, Keller LP, Kilcoyne ALD, Leitner J, Matrajt G, Meibom A, Mennella V, Mostefaoui S, Nittler LR, Palumbo ME, Papanastassiou DA, Robert F, Rotundi A, Snead CJ, Spencer MK, Stadermann FJ, Steele A, Stephan T, Tsou P, Tyliszczak T, Westphal AJ, Wirick S, Wopenka B, Yabuta H, Zare RN, Zolensky ME: Organics captured from comet 81P/wild 2 by the stardust spacecraft. *Science* 2006, 314:1720–1724.

[Sandford 2007] Sandford SA, Brownlee DE: Response to comment on "Organics captured from comet 81p/wild 2 by the stardust spacecraft". *Science* 2007, 317.

[Santra 2009] Santra PK, Viswanatha R, Daniels SM, Pickett NL, Smith JM, O'brien P, Sarma DD: Investigation of the internal heterostructure of highly luminescent quantum dot-quantum well nanocrystals. *Journal of the American Chemical Society* 2009, 131:470–477.

[Sapra 2006] Sapra S, Nanda J, Pietryga JM, Hollingsworth JA, Sarma DD: Unraveling internal structures of highly luminescent PbSe nanocrystallites using variable-energy synchrotron radiation photoelectron spectroscopy. *Journal of Physical Chemistry B* 2006, 110:15244–15250.

[Sarma 2010] Sarma DD, Nag A, Santra PK, Kumar A, Sapra S, Mahadevan P: Origin of the enhanced photoluminescence from semiconductor CdSeS nanocrystals. *Journal of Physical Chemistry Letters* 2010, 1:2149–2153.

[Sasaki 2006] Sasaki K, Susuki K, Iyoshi A, Uchimura M, Imamura N, Kusaba H, Teraoka Y, Fuchino H, Tsujimoto K, Uchida Y, Jingo N: H2S poisoning of solid oxide fuel cells. *Journal of the Electrochemical Society* 2006, 153:A2023–A2029.

[Schiavello 1977] Schiavello M, Pepe F, Cannizzaro M, Derossi S, Tilley RJD: Adsorption of oxygen and propene over a series of tungsten-oxides. *Zeitschrift für Physikalische Chemie-Frankfurt* 1977, 106:45–56.

[Schlenter 1996] Schlenter M: Röntgenkleinwinkelstreuung und XANES-Messungen an kohlenstoffgeträgerten platinkatalysatoren. RWTH Aachen, Physik; 1996.

[Schmidt 2018] Schmidt S, Sallard S, Borca C, Huthwelker T, Novak P, Villevieille C: Phosphorus anionic redox activity revealed by operando P K-edge X-ray absorption spectroscopy on diphosphonate-based conversion materials in Li-ion batteries. *Chemical Communications (Camb)* 2018, 54:4939–4942.

[Schönborn 1998] Schönborn H-B: *Lok 2000. Re 460/465 - modernste Elektrolok der Schweiz*. München: GeraMond Verlag; 1998.

[Schulke 1988] Schulke W, Berthold A, Kaprolat A, Guntherodt HJ: Evidence for interlayer band shifts upon lithium intercalation in graphite from inelastic x-ray-scattering. *Physical Review Letters* 1988, 60:2217–2220.

[Schwanitz 2007a] Schwanitz K, Mankel E, Hunger R, Mayer T, Jaegermann W: Photoelectron spectroscopy at the solid-liquid interface of dye-sensitized solar cells: Unique experiments with the solid-liquid interface analysis system SoLiAS at BESSY. *Chimia* 2007, 61:796–800.

[Schwanitz 2007c] Schwanitz K, Weiler U, Hunger R, Mayer T, Jaegermann W: Synchrotron-induced photoelectron spectroscopy of the dye-sensitized nanocrystalline TiO_2/electrolyte interface: band gap states and their interaction with dye and solvent molecules. *Journal of Physical Chemistry C* 2007, 111:849–854.

[Schwartz 2010] Schwartz RE, Russell LM, Sjostedt SJ, Vlasenko A, Slowik JG, Abbatt JPD, Macdonald AM, Li SM, Liggio J, Toom-Sauntry D, Leaitch WR: Biogenic oxidized organic functional groups in aerosol particles from a mountain forest site and their similarities to laboratory chamber products. *Atmospheric Chemistry and Physics* 2010, 10:5075–5088.

[Schwenzer 2011] Schwenzer B, Zhang JL, Kim S, Li LY, Liu J, Yang ZG: Membrane development for vanadium redox flow batteries. *ChemSusChem* 2011, 4:1388–1406.

[Seah 1979] Seah MP, Dench WA: Quantitative electron spectroscopy of surfaces: a standard data base for electron inelastic mean free paths in solids. *Surface and Interface Analysis* 1979, 1:2–11.

[Segre 2015] Segre CU, Katsoudas JP, Timofeeva EV, Ramani VK, Singh D, Pelliccione CJ, Ding Y, Aryal S, Beaver NM, Li Y, Sen S: In situ X-ray absorption spectroscopy for batteries: discovery of new mechanisms and materials. In *ECS Meeting*, vol. Abstract MA2015-01 1932: The Electrochemical society; 2015.

[Setif 2006] Setif P: *Electron Transfer From the Bound Iron-Sulfur Clusters to Ferredoxin/Flavodoxin: Kinetic and Structural Properties of Ferredoxin/Flavodoxin Reduction by Photosystem I.* Doordrecht, The Netherlands: Springer; 2006.

[Seyfarth 2006] Seyfarth EA: Julius Bernstein (1839–1917): Pioneer neurobiologist and biophysicist. *Biological Cybernetics* 2006, 94:2–8.

[Sezen 2011] Sezen H, Suzer S: Communication: Enhancement of dopant dependent X-ray photoelectron spectroscopy peak shifts of Si by surface photovoltage. *Journal of Chemical Physics* 2011, 135.

[Shakya 2013] Shakya KM, Liu S, Takahama S, Russell LM, Keutsch FN, Galloway MM, Shilling JE, Hiranuma N, Song C, Kim H, Paulson SE, Pfaffenberger L, Barmet P, Slowik J, Prevot ASH, Dommen J, Baltensperger U: Similarities in STXM-NEXAFS spectra of atmospheric particles and secondary organic aerosol generated from glyoxal, alpha-pinene, isoprene, 1,2,4-trimethylbenzene, and d-limonene. *Aerosol Science and Technology* 2013, 47:543–555.

[Shapiro 2014] Shapiro DA, Yu YS, Tyliszczak T, Cabana J, Celestre R, Chao WL, Kaznatcheev K, Kilcoyne ALD, Maia F, Marchesini S, Meng YS, Warwick T, Yang LL, Padmore HA: Chemical composition mapping with nanometre resolution by soft X-ray microscopy. *Nature Photonics* 2014, 8:765–769.

[Shearing 2011] Shearing P, Yan W, Harris SJ, Brandon N: In situ X-ray spectroscopy and imaging of battery materials. *The Electrochemical Society Interface* 2011, 20:43–47.

[Shearing 2010] Shearing PR, Howard LE, Jorgensen PS, Brandon NP, Harris SJ: Characterization of the 3-dimensional microstructure of a graphite negative electrode from a Li-ion battery. *Electrochemistry Communications* 2010, 12:374–377.

[Shearing 2013] Shearing PR, Eastwood DS, Bradley RS, Gelb J, Cooper SJ, Tariq F, Brett DJL, Brandon NP, Withers PJ, Lee PD: Exploring electrochemical devices using X-ray microscopy: 3D microstructure of batteries and fuel cells. In *Microscopy and Analysis*, vol. 27; 2013.

[Shen 1986] Shen WM, Tomkiewicz M, Cahen D: Impedance study of surface optimization of n-culnse2 in photoelectrochemical solar-cells. *Journal of the Electrochemical Society* 1986, 133:112–116.

[Sherwood 1997] Sherwood PMA: Extracting more chemical information from X-ray photoelectron spectroscopy by using monochromatic X rays. *Journal of Vacuum Science & Technology. A. Vacuum, Surfaces, and Films* 1997, 15:520–525.

[Shimakawa 1997] Shimakawa Y, Numata T, Tabuchi J: Verwey-type transition and magnetic properties of the $LiMn_2O_4$ spinels. *Journal of Solid State Chemistry* 1997, 131:138–143.

[Shiraishi 1997] Shiraishi Y, Nakai I, Tsubata T, Himeda T, Nishikawa F: In situ transmission X-ray absorption fine structure analysis of the charge-discharge process in LiMn$_2$O$_4$, a rechargeable lithium battery material. *Journal of Solid State Chemistry* 1997, 133:587–590.

[Shui 2013] Shui JL, Okasinski JS, Kenesei P, Dobbs HA, Zhao D, Almer JD, Liu DJ: Reversibility of anodic lithium in rechargeable lithium-oxygen batteries. *Nature Communications* 2013, 4:7.

[Simunek 1988] Simunek A, Wiech G: Angle-resolved X-ray-emission spectra (ARXES) of carbon in potassium graphite c8k. *Physica Status Solidi. B, Basic Research* 1988, 149:765–774.

[Sines 1987] Sines G, Sakellarakis YA: Lenses in antiquity. *American Journal of Archaeology* 1987, 91:191–196.

[Sinha 2006] Sinha PK, Halleck P, Wang CY: Quantification of liquid water saturation in a PEM fuel cell diffusion medium using X-ray microtomography. *Electrochemical and Solid-State Letters* 2006, 9:A344–A348.

[Skoda 2007] Skoda M, Libra J, Sutara F, Tsud N, Skala Y, Sedlacek L, Chab V, Prince KC, Matolin V: A resonant photoemission study of the Ce and Ce-oxide/Pd(111) interfaces. *Surface Science* 2007, 601:4958–4965.

[Skorupska 2005] Skorupska K, Lublow M, Kanis M, Jungblut H, Lewerenz HJ: Electrochemical preparation of a stable accumulation layer on Si: a synchrotron radiation photoelectron spectroscopy study. *Applied Physics Letters* 2005, 87.

[Solarska 2010a] Solarska R, Alexander BD, Braun A, Jurczakowski R, Fortunato G, Stiefel M, Graule T, Augustynski J: Tailoring the morphology of WO(3) films with substitutional cation doping: effect on the photoelectrochemical properties. *Electrochimica Acta* 2010, 55:7780–7787.

[Solarska 2010d] Solarska R, Heel A, Ropka J, Braun A, Holzer L, Ye J, Graule T: Nanoscale calcium bismuth mixed oxide with enhanced photocatalytic performance under visible light. *Applied Catalysis. A, General* 2010, 382:190–196.

[Solomon 2003] Solomon D, Lehmann J, Martinez CE: Sulfur K-edge XANES spectroscopy as a tool for understanding sulfur dynamics in soil organic matter. *Soil Science Society of America Journal* 2003, 67:1721–1731.

[Solomon 2012] Solomon D, Lehmann J, Harden J, Wang J, Kinyangi J, Heymann K, Karunakaran C, Lu Y, Wirick S, Jacobsen C: Micro- and nano-environments of carbon sequestration: multi-element STXM-NEXAFS spectromicroscopy assessment of microbial carbon and mineral associations. *Chemical Geology* 2012, 329:53–73.

[Solonin 2001] Solonin YM, Khyzhun OY, Graivoronskaya EA: Nonstoichiometric tungsten oxide based on hexagonal WO3. *Crystal Growth & Design* 2001, 1:473–477.

[Somogyi 2015] Somogyi A, Mocuta C: Possibilities and challenges of scanning hard X-ray spectro-microscopy techniques in material sciences. *AIMS Materials Science* 2015, 2:40.

[Song 2010] Song J, Hong BL, Zheng J, Lin P, Zheng MS, Wu QH, Dong QF, Sun SG: Electronic properties of LiMn$_2$-x Ti (x) O-4. *Applied Physics. A, Materials Science & Processing* 2010, 98:455–460.

[Song 2011] Song JW, Nguyen CC, Choi H, Lee KH, Han KH, Kim YJ, Choy S, Song SW: Impacts of surface Mn valence on cycling performance and surface chemistry of Li- and Al-substituted spinel battery cathodes. *Journal of the Electrochemical Society* 2011, 158:A458–A464.

[Song 2013] Song MK, Cairns EJ, Zhang Y: Lithium/sulfur batteries with high specific energy: old challenges and new opportunities. *Nanoscale* 2013, 5:2186–2204.

[Sonntag 1974] Sonntag BF: Observations of forbidden soft-X-ray transitions: Li K absorption in LiF. *Physical Review B* 1974, 9:3601–3602.

[Stadtmueller 2015] Stadtmueller B, Schroeder S, Kumpf C: Heteromolecular metal-organic interfaces: electronic and structural fingerprints of chemical bonding. *Journal of Electron Spectroscopy and Related Phenomena* 2015, 204:80–91.

[Stashans 2015] Stashans A, Marcillo F, Castillo D: Dopamine adsorption configurations on anatase (101) surface. *Surface Review and Letters* 2015, 22.

[Stebegg 2012] Stebegg R, Wurzinger B, Mikulic M, Schmetterer G: Chemoheterotrophic growth of the cyanobacterium *Anabaena* sp strain PCC 7120 dependent on a functional cytochrome c oxidase. *Journal of Bacteriology* 2012, 194:4601–4607.

[Steinberger-Wilckens 2007] Steinberger-Wilckens R, Tietz F, Smith MJ, Mougin J, Rietveld B, Bucheli O, Van Herle J, Rosenberg R, Zahid M, Holtappels P: Real-SOFC – a joint European effort in understanding SOFC degradation. *ECS Transactions* 2007, 7:67–76.

[Stevens 2001] Stevens DA, Dahn JR: The mechanisms of lithium and sodium insertion in carbon materials. *Journal of the Electrochemical Society* 2001, 148:A803–A811.

[Stewart-Ornstein 2007] Stewart-Ornstein J, Hitchcock AP, Cruz DH, Henklein P, Overhage J, Hilpert K, Hale JD, Hancock REW: Using intrinsic X-ray absorption spectral differences to identify and map peptides and proteins. *Journal of Physical Chemistry B* 2007, 111:7691–7699.

[Stöhr 1992] Stöhr J: *NEXAFS Spectroscopy*. Berlin Heidelberg: Springer-Verlag; 1992.

[Struis 2012] Struis R, Nurk G, Braun A, Huthwelker T, Janousch M, Lust E, Ludwig C: Probing the fate of sulphur in a working solid-oxide fuel cell anode using S K-edge XAS. In *PSI Scientific Highlights 2011*. pp. 60–61. Paul Scherrer Institut; 2012:60–61.

[Suzuki 2006] Suzuki S, Kwon SK, Saito M, Kamimura T, Miyuki H, Waseda Y: Atomic-scale structure of beta-FeOOH containing chromium by anomalous X-ray scattering coupled with reverse Monte Carlo simulation. *Corrosion Science* 2006, 48:1571–1584.

[Swanson 1951] Swanson HE, Tatge E: Standard X-ray diffraction patterns. *Journal of Research of the National Bureau of Standards* 1951, 46:318–327.

[Swanson 1966] Swanson HE, Morris MC, Evans EH: Standard X-ray diffraction powder patterns. Commerce USDO ed., vol. 25. Washington, D.C.: National bureau of standards; 1966.

[Syres 2010] Syres K, Thomas A, Bondino F, Malvestuto M, Gratzel M: Dopamine adsorption on anatase TiO_2(101): a photoemission and NEXAFS spectroscopy study. *Langmuir* 2010, 26:14548–14555.

[Szent-Gyorgyi 1941] Szent-Gyorgyi A: Towards a new biochemistry?. *Science* 1941, 93:609–611.

[Szent-Györgyi 1941] Szent-Györgyi A: The study of energy-levels in biochemistry. *Nature* 1941, 148:157–159.

[Taghiei 1992] Taghiei MM, Huggins FE, Shah N, Huffman GP: In situ X-ray absorption fine-structure spectroscopy investigation of sulfur functional-groups in coal during pyrolysis and oxidation. *Energy & Fuels* 1992, 6:293–300.

[Takahama 2007] Takahama S, Gilardoni S, Russell LM, Kilcoyne ALD: Classification of multiple types of organic carbon composition in atmospheric particles by scanning transmission X-ray microscopy analysis. *Atmospheric Environment* 2007, 41:9435–9451.

[Takahama 2008] Takahama S, Gilardoni S, Russell LM: Single-particle oxidation state and morphology of atmospheric iron aerosols. *Journal of Geophysical Research. Atmospheres* 2008, 113.

[Takahama 2013] Takahama S, Johnson A, Morales JG, Russell LM, Duran R, Rodriguez G, Zheng J, Zhang R, Toom-Sauntry D, Leaitch WR: Submicron organic aerosol in Tijuana, Mexico, from local and Southern California sources during the CalMex campaign. *Atmospheric Environment* 2013, 70:500–512.

[Takahashi 1984] Takahashi M, Takahashi T, Tokunaga F, Murano K, Tsujimoto K, Sagawa T: X-ray photoelectron spectroscopy (XPS) of bacteriorhodopsin analogues synthesized from fluorophenyl retinals. *Journal of the Physical Society of Japan* https://doi.org/10.1143/JPSJ.53.1557 1984, 53:8.

[Takahashi 1982] Takahashi T, Sato M, Kono S, Tokunaga F, Murano K, Tsujimoto K, Sagawa T: X-ray photoemission-study of 13-cis and all-trans fluorophenyl-retinal. *Journal of the Physical Society of Japan* 1982, 51:3332–3336.

[Takano 1999] Takano M, Kanno R, Takeda T: A chemical contribution to the search for novel electronic properties in transition metal oxides: $LiNiO_2$. *Materials Science & Engineering. B, Solid-State Materials for Advanced Technology* 1999, 63:6–10.

[Tamm 1932] Tamm I: On the Possible Bound States of Electrons on a Crystal Surface. *Phys Z Soviet Union* 1932, 1:3.

[Taniguchi 1957] Taniguchi H, Janz GJ: Preparation and reproducibility of the thermal-electrolytic silver-silver chloride electrode. *Journal of the Electrochemical Society* 1957, 104:123–127.

[Tarascon 2004] Tarascon JM, Delacourt C, Prakash AS, Morcrette M, Hegde MS, Wurm C, Masquelier C: Various strategies to tune the ionic/electronic properties of electrode materials. *Dalton Transactions* 2004:2988–2994.

[Taylor 2003] Taylor G: The phase problem. *Acta Crystallographica. Section D, Biological Crystallography* 2003, 59:1881–1890.

[Thackeray 1998] Thackeray MM, Shao-Horn Y, Kahaian AJ, Kepler KD, Vaughey JT, Hackney SA: Structural fatigue in spinel electrodes in high voltage (4V) Li/LixMn$_2$O$_4$ cells. *Electrochemical and Solid-State Letters* 1998, 1:7–9.

[Thibault 2008] Thibault P, Dierolf M, Menzel A, Bunk O, David C, Pfeiffer F: High-resolution scanning x-ray diffraction microscopy. *Science* 2008, 321:379–382.

[Thomas 2012] Thomas JM: Centenary: The birth of X-ray crystallography. *Nature* 2012, 491:186–187.

[Thompson 2009] Thompson A, Attwood D, Gullikson E, Howells M, Kim K-J, Kirz J, Kortright J, Lindau I, Pianetta P, Robinson A, Scofield J, Underwood J, Vaughan D, Williams G, Winick H: *X-ray Data Booklet.* Berkeley, California: Lawrence Berkeley National Laboratory; 2009.

[Timofeeva 2013] Timofeeva EV, Katsoudas JP, Segre CU, Singh D: Rechargeable nanofluid electrodes for high energy density flow battery. In *techConnect 2013, NSTI Nanotechnology Conference and Expo,* vol. Cleantech 2013. pp. 363–366. Washington D.C.; 2013:363–366.

[Tohji 1989a] Tohji K, Udagawa Y: Observation of X-ray raman-scattering. *Physica B* 1989, 158:550–552.

[Tohji 1989b] Tohji K, Udagawa Y: X-ray raman-scattering as a substitute for soft-X-ray extended X-ray-absorption fine-structure. *Physical Review B* 1989, 39:7590–7594.

[Tomkiewicz 1980a] Tomkiewicz M: The nature of surface-states on chemically modified TiO$_2$ electrodes. *Journal of the Electrochemical Society* 1980, 127:1518–1525.

[Tomkiewicz 1980d] Tomkiewicz M: Surface-states on chemically modified TiO$_2$ electrodes. *Surface Science* 1980, 101:286–294.

[Tomkiewicz 1980g] Tomkiewicz M, Silberstein RP, Pollak FH: Determination of the potential distribution at the CdSe-electrolyte interface by electrolyte electroreflectance spectroscopy. *Bulletin of the American Physical Society* 1980, 25:265–265.

[Toner 2005] Toner B, Fakra S, Villalobos M, Warwick T, Sposito G: Spatially resolved characterization of biogenic manganese oxide production within a bacterial biofilm. *Applied and Environmental Microbiology* 2005, 71:1300–1310.

[Toney 1991] Toney MF, Gordon JG, Samant MG, Borges GL, Wiesler DG, Yee D, Sorensen LB: In situ surface X-ray-scattering measurements of electrochemically deposited bi on silver(111) - structure, compressibility, and comparison with ex situ low-energy electron-diffraction measurements. *Langmuir* 1991, 7:796–802.

[Toth 2011] Toth R: Self-organization processes to pattern thin films: A bottom-up approach for photoelectrodes. Switzerland: Swiss National Science Foundation; 2011. Research Grant http://p3.snf.ch/project-139698.

[Toth 2016] Toth R, Walliser RM, Lagzi I, Boudoire F, Duggelin M, Braun A, Housecroft CE, Constable EC: Probing the mystery of liesegang band formation: revealing the origin of self-organized dual-frequency micro and nanoparticle arrays. *Soft Matter* 2016, 12:8367–8374.

[Totlandsdal 2012] Totlandsdal AI, Herseth JI, Bolling AK, Kubatova A, Braun A, Cochran RE, Refsnes M, Ovrevik J, Lag M: Differential effects of the particle core and organic extract of diesel exhaust particles. *Toxicology Letters* 2012, 208:262–268.

[Tremsin 2011] Tremsin AS, Mcphate JB, Vallerga JV, Siegmund OHW, Feller WB, Lehmann E, Dawson M: Improved efficiency of high resolution thermal and cold neutron imaging. *Nuclear Instruments & Methods in Physics Research. Section A, Accelerators, Spectrometers, Detectors and Associated Equipment* 2011, 628:415–418.

[Tributsch 1968] Tributsch H: Eine elektrochemische methode zum studium der spektralen sensibilisierung und heterogener photochemischer reaktionen an ZnO-Elektroden. Technische Universität München, 1968.

[Tributsch 1969a] Tributsch H: Electrochemistry of sensitization and photochemical properties of pseudoisocyanine dye aggregates on ZnO electrodes. *Berichte Der Bunsen-Gesellschaft Fur Physikalische Chemie* 1969, 73:9.

[Tributsch 1969b] Tributsch H, Gerischer H: Use of semiconductor electrodes in study of photochemical reactions. *Berichte Der Bunsen-Gesellschaft Fur Physikalische Chemie* 1969, 73:850–854.

[Tributsch 1969c] Tributsch H, Gerischer H: Electrochemical investigations on mechanism of sensitization and supersensitization of ZnO monocrystals. *Berichte Der Bunsen-Gesellschaft Fur Physikalische Chemie* 1969, 73:10.

[Tributsch 1971] Tributsch H, Calvin M: Electrochemistry of excited molecules - photo-electrochemical reactions of chlorophylls. *Photochemistry and Photobiology* 1971, 14:95–112.

[Tributsch 2008] Tributsch H: Photovoltaic hydrogen generation. *International Journal of Hydrogen Energy* 2008, 33:5911–5930.

[Tributsch 2015] Tributsch H: Personal communication of Helmut Tributsch with Artur Braun. 17–19 August 2015: international exploratory workshop on photoelectrochemistry, catalysis and X-ray spectroscopy SNF International workshop on edition. Empa Dübendorf; 2015.

[Tsekouras 2014] Tsekouras G, Braun A: Conductivity and oxygen reduction activity changes in lanthanum strontium manganite upon low-level chromium substitution. *Solid State Ionics* 2014, 266:19–24.

[Tsekouras 2015] Tsekouras G, Boudoire F, Pal B, Vondracek M, Prince KC, Sarma DD, Braun A: Electronic structure origin of conductivity and oxygen reduction activity changes in low-level Cr-substituted (La, Sr)MnO$_3$. *Journal of Chemical Physics* 2015, 143:114701–114705.

[Tsuji 2001] Tsuji J, Kojima K, Ikeda S, Nakamatsu H, Mukoyama T, Taniguchi K: Li K-edge spectra of lithium halides. *Journal of Synchrotron Radiation* 2001, 8:554–556.

[Tsuji 2002] Tsuji J, Nakamatsu H, Mukoyama T, Kojima K, Ikeda S, Taniguchi K: Lithium K-edge XANES spectra for lithium compounds. *X-Ray Spectrometry* 2002, 31:319–326.

[Tsuji 2005] Tsuji J, Fujita M, Haruyama Y, Kanda K, Matsui S, Ozawa N, Yao T, Taniguchi K: Excitation energy dependence for the Li 1s X-ray photoelectron spectra of LiMn$_2$O$_4$. *Analytical Sciences* 2005, 21:779–781.

[Tucker 2000] Tucker MC, Braun A, Bergmann U, Wang H, Glatzel P, Reimer JA, Cramer SP: 7Li MAS-NMR, X-ray spectroscopy and electrochemical studies of LiMn$_2$O$_4$-basedspinels for lithium rechargeable batteries. In *Interfaces, Phenomena, and Nanostructures in Lithium Batteries; 11-13 Dec 2000; Argonne IL.* Edited by Landgrebe AR, Klingler RJ. United States, Pennington, NJ: Electrochemical Society; 2000: 68–79.

[Tucker 2002a] Tucker MC, Doeff MM, Richardson TJ, Finones R, Cairns EJ, Reimer JA: Hyperfine fields at the Li site in LiFePO$_4$-type olivine materials for lithium rechargeable batteries: a Li-7 MAS NMR and SQUID study. *Journal of the American Chemical Society* 2002, 124:3832–3833.

[Tucker 2002b] Tucker MC, Doeff MM, Richardson TJ, Finones R, Reimer JA, Cairns EJ: Li-7 and P-31 magic angle spinning nuclear magnetic resonance of LiFePO$_4$-type materials. *Electrochemical and Solid-State Letters* 2002, 5:A95–A98.

[Tucker 2002c] Tucker MC, Kroeck L, Reimer JA, Cairns EJ: The influence of covalence on capacity retention in metal-substituted spinels - Li-7 NMR, SQUID, and electrochemical studies. *Journal of the Electrochemical Society* 2002, 149:A1409–A1413.

[Tucker 2002d] Tucker MC, Reimer JA, Cairns EJ: A Li-7 NMR study of capacity fade in metal-substituted lithium manganese oxide spinels. *Journal of the Electrochemical Society* 2002, 149:A574–A585.

[Tummers 2006] Tummers B, Van Der Laan J, Huyser K: DataThief III., vol. III. A data digitalization program; 2006.

[Ullman 1980] Ullman DL, Tomkiewicz M: The potential distribution at the TiO$_2$ aqueous-electrolyte interface - discussion. *Journal of the Electrochemical Society* 1980, 127:1321–1322.

[Urdaneta 2014] Urdaneta I, Keller A, Atabek O, Palma JL, Finkelstein-Shapiro D, Tarakeshwar P, Mujica V, Calatayud M: Dopamine adsorption on TiO$_2$ anatase surfaces. *Journal of Physical Chemistry C* 2014, 118:20688–20693.

[Vad 2002] Vad T, Haubold HG, Waldofner N, Bonnemann H: Three-dimensional Pt-nanoparticle networks studied by anomalous small-angle X-ray scattering and X-ray absorption spectroscopy. *Journal of Applied Crystallography* 2002, 35:459–470.

[Vartanyants 2001] Vartanyants IA, Kovalchuk MV: Theory and applications of X-ray standing waves in real crystals. *Reports on Progress in Physics* 2001, 64:1009–1084.

[Verhoeven 1996] Verhoeven JW: Glossary of terms used in photochemistry (IUPAC Recommendations 1996). *Pure and Applied Chemistry* 1996, 68:2223–2286.

[Vijayakumar 2015] Vijayakumar M, Govind N, Li B, Wei X, Nie Z, Thevuthasan S, Sprenkle V, Wang W: Aqua-vanadyl ion interaction with Nafion® membranes. *Frontiers in Energy Research* 2015, 3:1.5.

[Visser 2001] Visser H, Anxolabehere-Mallart E, Bergmann U, Glatzel P, Robblee JH, Cramer SP, Girerd JJ, Sauer K, Klein MP, Yachandra VK: MnK-edge XANES and K beta XES studies of two Mn-oxo binuclear complexes: investigation of three different oxidation states relevant to the oxygen-evolving complex of photosystem II. *Journal of the American Chemical Society* 2001, 123:7031–7039.

[Von Laue 1915] Von Laue M: Nobel Lecture: *Concerning the Detection of X-Ray Interferences. Nobel Lectures, Physics 1901–1921 Edition.* Stockholm: Elsevier Publishing Company; 1915.

[Vong 1988] Vong MSW, Sermon PA, Self VA, Grant K, Blackburn AJ: An x-ray cell for in situ study of solid-state transformations. *Journal of Physics. E, Scientific Instruments* 1988, 21:495–496.

[Vyalikh 2004] Vyalikh DV, Danzenbacher S, Mertig M, Kirchner A, Pompe W, Dedkov YS, Molodtsov SL: Electronic structure of regular bacterial surface layers. *Physical Review Letters* 2004, 93.

[Vyalikh 2005] Vyalikh DV, Kirchner A, Danzenbacher S, Dedkov YS, Kade A, Mertig M, Molodtsov SL: Photoemission and near-edge X-ray absorption fine structure studies of the bacterial surface protein layer of bacillus sphaericus NCTC 9602. *Journal of Physical Chemistry B* 2005, 109:18620–18627.

[Vyalikh 2006] Vyalikh DV, Kirchner A, Kade A, Danzenbaecher S, Dedkov YS, Mertig M, Molodtsov SL: Spectroscopic studies of the electronic properties of regularly arrayed two-dimensional protein layers. *Journal of Physics. Condensed Matter* 2006, 18:S131–S144.

[Vyalikh 2009a] Vyalikh DV, Kummer K, Kade A, Blueher A, Katzschner B, Mertig M, Molodtsov SL: Site-specific electronic structure of bacterial surface protein layers. *Applied Physics. A, Materials Science & Processing* 2009, 94:455–459.

[Vyalikh 2009c] Vyalikh DV, Maslyuk VV, Bluher A, Kade A, Kummer K, Dedkov YS, Bredow T, Mertig I, Mertig M, Molodtsov SL: Charge transport in proteins probed by resonant photoemission. *Physical Review Letters* 2009, 102.

[Wada 1977] Wada N, Sagawa T: Xps measurements of valence band in all-trans retinal-1 and beta-carotene. *Journal of the Physical Society of Japan* 1977, 43:2107–2108.

[Wada 1989] Wada N, Shibata R, Shibazaki M, Suzuki N: X-ray photoelectron valence band studies on firefly d-(-)-luciferin. *Photochemistry and Photobiology* 1989, 49:513–518.

[Wadati 2006] Wadati H: *Photoemission Studies of Perovskite-Type Transition-Metal Oxides in Epitaxial Thin Films.* University of Tokyo, Physics; 2006.

[Wagner 1961] Wagner C: Theorie der Alterung von Niederschlagen durch Umlosen (Ostwald-Reifung). *Zeitschrift für Elektrochemie* 1961, 65:581–591.

[Waller 1923a] Waller I: On the question of the influence of thermal motion on the interference of X-rays. *Zeitschrift für Physik* 1923, 17:398–408.

[Waller 1923c] Waller I: Zur Frage der Einwirkung der Wärmebewegung auf die Interferenz von Rön. *Zeitschrift für Physik* 1923, 17.

[Walliser 2015a] Walliser RM: *Towards Cheap and Sustainable Energy Sources by Exploiting Self-Organized Catalyst Micro and Nanostructures.* Universität Basel, Philosophisch- Naturwissenschaftliche Fakultät; 2015.

[Walliser 2015b] Walliser RM, Boudoire F, Orosz E, Toth R, Braun A, Constable EC, Racz Z, Lagzi I: Growth of nanoparticles and microparticles by controlled reaction-diffusion processes. *Langmuir* 2015, 31:1828–1834.

[Wang 2001a] Wang H, Patil DS, Ralston CY, Bryant C, Cramer SP: L-edge X-ray magnetic circular dichroism of Ni enzymes: direct probe of Ni spin states. *Journal of Electron Spectroscopy and Related Phenomena* 2001, 114:865–871.

[Wang 2013a] Wang H, Young AT, Guo J, Cramer SP, Friedrich S, Braun A, Gu W: Soft X-ray absorption spectroscopy and resonant inelastic X-ray scattering spectroscopy below 100 eV: probing first-row transition-metal M-edges in chemical complexes. *Journal of Synchrotron Radiation* 2013, 20:614–619.

[Wang 2013b] Wang H, Young AT, Guo J, Cramer SP, Friedrich S, Braun A, Gu W, Iucr: Soft X-ray absorption spectroscopy and resonant inelastic X-ray scattering spectroscopy below 100 eV: probing first-row transition-metal M-edges in chemical complexes. *Journal of Synchrotron Radiation* 2013, 20:614–619.

[Wang 2013d] Wang H, Young AT, Jinghua G, Cramer SP, Friedrich S, Braun A, Weiwei G: Soft X-ray absorption spectroscopy and resonant inelastic X-ray scattering spectroscopy below 100 eV: probing first-row transition-metal M-edges in chemical complexes. *Journal of Synchrotron Radiation* 2013, 20:614–619.

[Wang 1997] Wang HX, Peng G, Miller LM, Scheuring EM, George SJ, Chance MR, Cramer SP: Iron L-edge X-ray absorption spectroscopy of myoglobin complexes and photolysis products. *Journal of the American Chemical Society* 1997, 119:4921–4928.

[Wang 1998] Wang HX, Ge PH, Riordan CG, Brooker S, Woomer CG, Collins T, Melendres CA, Graudejus O, Bartlett N, Cramer SP: Integrated X-ray L absorption spectra. Counting holes in Ni complexes. *Journal of Physical Chemistry B* 1998, 102:8343–8346.

[Wang 2000] Wang HX, Ralston CY, Patil DS, Jones RM, Gu W, Verhagen M, Adams M, Ge P, Riordan C, Marganian CA, Mascharak P, Kovacs J, Miller CG, Collins TJ, Brooker S, Croucher PD, Wang K, Stiefel EI, Cramer SP: Nickel L-edge soft X-ray spectroscopy of nickel-iron hydrogenases and model compounds - evidence for high-spin nickel(II) in the active enzyme. *Journal of the American Chemical Society* 2000, 122:10544–10552.

[Wang 2001c] Wang HX, Patil DS, Gu WW, Jacquamet L, Friedrich S, Funk T, Cramer SP: L-edge X-ray absorption spectroscopy of some Ni enzymes: probe of Ni electronic structure. *Journal of Electron Spectroscopy and Related Phenomena* 2001, 114:855–863.

[Wang 1992] Wang J, Ocko BM, Davenport AJ, Isaacs HS: In situ X-ray-diffraction and X-ray-reflectivity studies of the au(111) electrolyte interface - reconstruction and anion adsorption. *Physical Review B* 1992, 46:10321–10338.

[Wang 2013e] Wang J, Chen-Wiegart Y-CK, Wang J: In situ chemical mapping of a lithium-ion battery using full-field hard X-ray spectroscopic imaging. *Chemical Communications* 2013, 49:6480–6482.

[Wang 2014] Wang P, Dimitrijevic NM, Chang AY, Schaller RD, Liu YZ, Rajh T, Rozhkova EA: Photoinduced electron transfer pathways in hydrogen-evolving reduced graphene oxide-boosted hybrid nano-bio catalyst. *ACS Nano* 2014, 8:7995–8002.

[Wang 2006a] Wang Q, Ito S, Graetzel M, Fabregat-Santiago F, Mora-Sero I, Bisquert J, Bessho T, Imai H: Characteristics of high efficiency dye-sensitized solar cells. *Journal of Physical Chemistry B* 2006, 110:25210–25221.

[Wang 2006b] Wang XL, Zhang HM, Zhang JL, Xu HF, Zhu XB, Chen J, Yi BL: A bi-functional micro-porous layer with composite carbon black for PEM fuel cells. *Journal of Power Sources* 2006, 162:474–479.

[Wang 2021] Wang H, Braun A, Cramer SP, Gee LB, Yoda Y: Nuclear Resonance Vibrational Spectroscopy: A Modern Tool to Pinpoint Site-Specific Cooperative Processes. *Catalysts* 2021, 11.

[Warschkow 2002] Warschkow O, Ellis DE, Hwang JH, Mansourian-Hadavi N, Mason TO: Defects and charge transport near the hematite (0001) surface: an atomistic study of oxygen vacancies. *Journal of the American Ceramic Society* 2002, 85:213–220.

[Wei 2008] Wei Z, Dunbar Z, Masel R: Microct X-ray imaging of water movement in a PEM fuel cell. *Fuel Cells and Alternative Fuel Systems* 2008:80–80.

[Weihe 1997] Weihe H, Gudel HU: Quantitative interpretation of the Goodenough-Kanamori rules: a critical analysis. *Inorganic Chemistry* 1997, 36:3632–3639.

[Weinhardt 2008] Weinhardt L, Blum M, Baer M, Heske C, Cole B, Marsen B, Miller EL: Electronic surface level positions of WO$_3$ thin films for photoelectrochemical hydrogen production. *Journal of Physical Chemistry C* 2008, 112:3078–3082.

[Weinhardt 2013] Weinhardt L, Blum M, Fuchs O, Pookpanratana S, George K, Cole B, Marsen B, Gaillard N, Miller E, Ahn KS, Shet S, Yan Y, Al-Jassim MM, Denlinger JD, Yang W, Baer M, Heske C: Soft X-ray and electron spectroscopy to determine the electronic structure of materials for photoelectrochemical hydrogen production. *Journal of Electron Spectroscopy and Related Phenomena* 2013, 190:106–112.

[Whaley 2010] Whaley JA, Mcdaniel AH, El Gabaly F, Farrow RL, Grass ME, Hussain Z, Liu Z, Linne MA, Bluhm H, Mccarty KF: Note: Fixture for characterizing electrochemical devices in-operando in traditional vacuum systems. *Review of Scientific Instruments* 2010, 81.

[Whitehead 1996] Whitehead AH, Edstrom K, Rao N, Owen JR: In situ X-ray diffraction studies of a graphite-based Li-ion battery negative electrode. *Journal of Power Sources* 1996, 63:41–45.

[Whitehead 2005] Whitehead AH, Schreiber M: Current collectors for positive electrodes of lithium-based batteries. *Journal of the Electrochemical Society* 2005, 152:A2105–A2113.

[Whitfield 2005a] Whitfield PS, Davidson IJ, Cranswick LMD, Swainson IP, Stephens PW: Investigation of possible superstructure and cation disorder in the lithium battery cathode material $LiMn_1/3N_1/3Co_1/3O_2$ using neutron and anomalous dispersion powder diffraction. *Solid State Ionics* 2005, 176:463–471.

[Whitfield 2005c] Whitfield PS, Davidson IJ, Cranswick LMD, Swainson IP, Stephens PW: Untangling cation ordering in complex lithium battery cathode materials - simultaneous refinement of X-ray, neutron and resonant scattering data. *Advances in X-ray Analysis* 2005, 49:6.

[Wilson 2001] Wilson KR, Rude BS, Catalano T, Schaller RD, Tobin JG, Co DT, Saykally RJ: X-ray spectroscopy of liquid water microjets. *Journal of Physical Chemistry B* 2001, 105:3346–3349.

[Wise 2015] Wise AM, Ohldag H, Chueh WC, Turner J, Toney MF, Weker JN: Development of a soft X-ray ptychography beamline at SSRL and its application in the study of energy storage materials. In *X-Ray Nanoimaging: Instruments And Methods II*. Proceedings of SPIE, vol. 9592. SPIE; 2015.

[Wong-Ng 2005] Wong-Ng W, Allen AJ, Fischer DA, Cook LP: Advanced materials for energy applications. In *Materials Science and Engineering Laboratory - FY 2005 Programs and Accomplishments*. Gaithersburg: NIST; 2005.

[Wong 1984a] Wong J, Lytle FW, Messmer RP, Maylotte DH: K-edge absorption spectra of selected vanadium compounds. *Physical Review B* 1984, 30:5596.

[Wong 1984c] Wong J, Lytle FW, Messmer RP, Maylotte DH: K-edge absorption-spectra of selected vanadium compounds. *Physical Review B* 1984, 30:5596–5610.

[Wu 2013] Wu C-Y, Liu Y-T, Huang P-C, Luo T-JM, Lee C-H, Yang Y-W, Wen T-C, Chen T-Y, Lin T-L: Heterogeneous junction engineering on core-shell nanocatalysts boosts the dye-sensitized solar cell. *Nanoscale* 2013, 5:9181–9192.

[Wu 2000] Wu QW, Lu WQ, Prakash J: Characterization of a commercial size cylindrical Li-ion cell with a reference electrode. *Journal of Power Sources* 2000, 88:237–242.

[Wu 1995] Wu S, Lipkowski J, Tyliszczak T, Hitchcock AP: Effect of anion adsorption on early stages of copper electrocrystallization at Au(111) surface. *Progress in Surface Science* 1995, 50:227–236.

[Xiao 2006] Xiao YM, Fisher K, Smith MC, Newton WE, Case DA, George SJ, Wang HX, Sturhahn W, Alp EE, Zhao JY, Yoda Y, Cramer SP: How nitrogenase shakes - Initial information about P-cluster and FeMo-cofactor normal modes from nuclear resonance vibrational Spectroscopy (NRVS). *Journal of the American Chemical Society* 2006, 128:7608–7612.

[Xiao 2014] Xiao Y, Cho C: Experimental investigation and discussion on the mechanical endurance limit of nafion membrane used in proton exchange membrane fuel cell. *Energies* 2014, 7:6401–6411.

[Xiao 2004] Xiao YN, Wittmer DE, Izumi F, Mini S, Graber T, Vaccaro PJ: Determination of cations distribution in Mn_3O_4 by anomalous X-ray powder diffraction. *Applied Physics Letters* 2004, 85:736–738.

[Xiong 2013] Xiong Y: Mesoporous metal-oxides for dye sensitized solar cells and photocatalysts. University of Bath, Department of Chemistry; 2013.

[Xiong 2015] Xiong Y, He D, Jin Y, Cameron PJ, Edler KJ: Ordered mesoporous particles in titania films with hierarchical structure as scattering layers in dye-sensitized solar cells. *Journal of Physical Chemistry C* 2015, 119:22552–22559.

[Xu 2010] Xu JT, Gao KS: Use of UV-A energy for photosynthesis in the red macroalga gracilaria *lemaneiformis*. *Photochemistry and Photobiology* 2010, 86:580–585.

[Yamada 1999] Yamada A, Tanaka M, Tanaka K, Sekai K: Jahn-Teller instability in spinel Li-Mn-O. *Journal of Power Sources* 1999, 81:73–78.

[Yamaoka 2004] Yamaoka H, Oura M, Taguchi M, Morikawa T, Takahiro K, Terai A, Kawatsura K, Vlaicu AM, Ito Y, Mukoyama T: $K\beta$ Resonant X-ray emission spectroscopy for Fe, Fe_2O_3 and Fe_3O_4. *Journal of the Physical Society of Japan* 2004, 73:3182–3191.

[Yan 2012] Yan B, Lim C, Yin L, Zhu L: Three dimensional simulation of galvanostatic discharge of $LiCoO_2$ cathode based on X-ray nano-CT images. *Journal of the Electrochemical Society* 2012, 159:A1604–A1614.

[Yan 2013] Yan B, Lim C, Yin L, Zhu L: Simulation of heat generation in a reconstructed $LiCoO_2$ cathode during galvanostatic discharge. *Electrochimica Acta* 2013, 100:171–179.

[Yee 1993] Yee HS, Abruna HD: In-situ x-ray-absorption spectroscopy studies of copper underpotentially deposited in the absence and presence of chloride on platinum(111). *Langmuir* 1993, 9:2460–2469.

[Yoon 2003] Yoon WS, Chung KY, Oh KH, Kim KB: Changes in electronic structure of the electrochemically Li-ion deintercalated $LiMn_2O_4$ system investigated by soft X-ray absorption spectroscopy. *Journal of Power Sources* 2003, 119:706–709.

[Yoshizawa 2006] Yoshizawa N, Tanaike O, Hatori H, Yoshikawa K, Kondo A, Abe T: TEM and electron tomography studies of carbon nanospheres for lithium secondary batteries. *Carbon* 2006, 44:2558–2564.

[Young 2000] Young JA: Getting students to wear safety goggles. *Journal of Chemical Education* 2000, 77:1214–1214.

[Yourey 2011] Yourey JE, Bartlett BM: Electrochemical deposition and photoelectrochemistry of $CuWO_4$, a promising photoanode for water oxidation. *Journal of Materials Chemistry* 2011, 21:7651–7660.

[Yufit 2011] Yufit V, Shearing P, Hamilton RW, Lee PD, Wu M, Brandon NP: Investigation of lithium-ion polymer battery cell failure using X-ray computed tomography. *Electrochemistry Communications* 2011, 13:608–610.

[Zaanen 1985] Zaanen J, Sawatzky GA, Allen JW: Band-gaps and electronic-structure of transition-metal compounds. *Physical Review Letters* 1985, 55:418–421.

[Zech 2021] Zech C, Hönicke P, Kayser Y, Risse S, Grätz O, Stamm M, Beckhoff B: Polysulfide driven degradation in lithium–sulfur batteries during cycling – quantitative and high time-resolution operando X-ray absorption study for dissolved polysulfides probed at both electrode sides. *Journal of Materials Chemistry A* 2021, 9:10231–10239.

[Zegenhagen 1991] Zegenhagen J: Surface-structure analysis with X-ray standing waves. *Physica Scripta* 1991, T39:328–332.

[Zegenhagen 2010] Zegenhagen J, Detlefs B, Lee T-L, Thiess S, Isern H, Petit L, Andre L, Roy J, Mi Y, Joumard I: X-ray standing waves and hard X-ray photoelectron spectroscopy at the insertion device beamline ID32. *Journal of Electron Spectroscopy and Related Phenomena* 2010, 178:258–267.

[Zegenhagen 2013] Zegenhagen J, Kazimirov A (Eds.): *The X ray Standing Wave Technique: Principles and Applications*. Series on Synchrotron Radiation Techniques and Applications, vol. 7. World Scientific; 2013.

[Zeng 1999] Zeng Y, Zhang S, Groves FR, Harrison DP: High temperature gas desulfurization with elemental sulfur production. *Chemical Engineering Science* 1999, 54:3007–3017.

[Zerulla 2009] Zerulla D, Chassé T: Structure and self-assembly of alkanethiols on III–V semiconductor (110) surfaces. *Journal of Electron Spectroscopy and Related Phenomena* 2009, 172:78–87.

[Zhang 2013] Zhang J, Wu J, Zhang H, Zhang J: *PEM Fuel Cell Testing and Diagnosis*. Elsevier; 2013.

[Zhang 2007] Zhang JZ, Zhao YS, Xu HW, Li BS, Weidner DJ, Navrotsky A: Elastic properties of yttrium-doped $BaCeO_3$ perovskite. *Applied Physics Letters* 2007, 90.

[Zhang 1994] Zhang X, Ade H, Jacobsen C, Kirz J, Lindaas S, Williams S, Wirick S: Micro-XANES - chemical contrast in the scanning-transmission X-ray microscope. *Nuclear Instruments & Methods in Physics Research. Section A, Accelerators, Spectrometers, Detectors and Associated Equipment* 1994, 347:431–435.

[Zhang 2008]   Zhang XJ, Ge LH, Liu J, Yang XL: Functional expression of the glycine transporter 1 on bullfrog retinal cones. *NeuroReport* 2008, 19:1667–1671.

[Zhang 2004]   Zhang Z: *Atomic Scale X-Ray Studies of the Electrical Double Layer Structure at the Rutile Tio$_2$ (110) – Aqueous Interface*. Northwestern University, Department of Materials Science and Engineering; 2004.

[Zhang 2017]   Zhang W, Kjaer KS, Alonso-Mori R, Bergmann U, Chollet M, Fredin LA, Hadt RG, Hartsock RW, Harlang T, Kroll T, Kubicek K, Lemke HT, Liang HW, Liu Y, Nielsen MM, Persson P, Robinson JS, Solomon EI, Sun Z, Sokaras D, van Driel TB, Weng TC, Zhu D, Warnmark K, Sundstrom V, Gaffney KJ: Manipulating charge transfer excited state relaxation and spin crossover in iron coordination complexes with ligand substitution. *Chemical Science* 2017, 8:515–523.

[Zhong 2015]   Zhong S, Zhong JQ, Wee ATS, Chen W: Molecular orientation and electronic structure at organic heterojunction interfaces. *Journal of Electron Spectroscopy and Related Phenomena* 2015, 204:12–22.

[Zhou 2007]   Zhou J: *Lithium Metal Microreference Electrodes and their Applications to Li-ion Batteries*. Technische Universiteit Eindhoven, Department of Chemical Engineering; 2007.

[Zhu 2008]   Zhu W, Dunbar ZW, Masel RI: MicroCT X-ray imaging of water movement in a PEM fuel cell. In *Proton Exchange Membrane Fuel Cells 8, Pts 1 and 2. Volume 16*. Edited by Fuller T, Shinohara K, Ramani V, Shirvanian P, Uchida H, Cleghorn S, Inaba M, Mitsushima S, Strasser P, Nakagawa H, et al.; ECS Transactions; 2008: 995–1000.

Index

https://doi.org/10.1515/9783110794038-014

www.ingramcontent.com/pod-product-compliance
Lightning Source LLC
Chambersburg PA
CBHW080118220326
41598CB00032B/4879